Ichthyology in Context (1500–1880)

Intersections

INTERDISCIPLINARY STUDIES IN EARLY MODERN CULTURE

VOLUME 87 – 2024

The titles published in this series are listed at *brill.com/inte*

Ichthyology in Context (1500–1880)

Edited by

Paul J. Smith
Florike Egmond

BRILL

LEIDEN | BOSTON

Cover illustrations: (central image) A pufferfish. Drawing (anonymous) made in Frankfurt, before 1565. Gessner-Platter Animal Albums, Album III C 22, fol. 130. © Amsterdam, University Library; (background image) Sadeler (I) Johan after Maarten de Vos, *The Fifth Day of the Creation*, detail. Print, Rijksmuseum Amsterdam. RP-P-OB-5394.

Library of Congress Cataloging-in-Publication Data

Names: Smith, P. J. (Paul J.), editor. | Egmond, Florike, illustrator.
Title: Ichthyology in context (1500–1880) / edited by Paul J. Smith, Florike Egmond.
Description: Leiden ; Boston : Brill, 2024 | Series: Intersections, 1568–1181 ; volume 87 | Includes bibliographical references and indexes.
Identifiers: LCCN 2023045781 (print) | LCCN 2023045782 (ebook) | ISBN 9789004681170 (hardback) | ISBN 9789004681187 (ebook)
Subjects: LCSH: Ichthyology—Early works to 1800. | Ichthyology—History—19th century. | Aquatic animals—Social aspects—Europe—History. | Fishes—Social aspects—Europe—History. | Aquatic animals in art. | Fishes in art.
Classification: LCC QL614.8 .I24 2024 (print) | LCC QL614.8 (ebook) | DDC 597—dc23/eng/20231102
LC record available at https://lccn.loc.gov/2023045781
LC ebook record available at https://lccn.loc.gov/2023045782

Typeface for the Latin, Greek, and Cyrillic scripts: "Brill". See and download: brill.com/brill-typeface.

ISSN 1568-1181
ISBN 978-90-04-68117-0 (hardback)
ISBN 978-90-04-68118-7 (e-book)
DOI 10.1163/9789004681187

This book is printed on acid-free paper and produced in a sustainable manner.

Contents

Acknowledgments

The present volume stems from the research project *A New History of Fishes. A long-term approach to fishes in science and culture, 1550–1880*. This project was funded by NWO (Dutch Research Council) and carried out at Leiden University Center for the Arts in Society (LUCAS) and at Naturalis Biodiversity Center. Thirteen contributions to this volume stem from the closing conference *Towards a Cultural History of Early Modern Ichthyology* (1500–1800) (Leiden University, 14–16 October 2021), and have been supplemented with eleven new contributions. We would like to thank Gera van Bedaf and Ivo Romein, both affiliated with Brill, for their advice and help with the material processing of the manuscript, in particular with regard to the many illustrations of this volume. We also thank Meredith McGroarty for her critical reading of several chapters of this volume. Finally, our thanks also go to the authors of this volume, who (although coming from very different disciplines, and accustomed to different methods of publishing) have stepped out of the comfort zone of their own discipline, and have managed to work together from a genuine interdisciplinary attitude.

Figures

Notes on the Editors

Paul J. Smith

is Emeritus Professor of French literature at Leiden University. His research focuses on 16th-, 17th-, and 20th-century French literature, its reception in the Netherlands, French and Dutch fable and emblem books, literary rhetoric and intermediality. He has also published on animal symbolism and early modern zoology, and its presence in art and literature. His main book publications include *Voyage et écriture. Etude sur le* Quart Livre *de Rabelais* (1987), *Het schouwtoneel der dieren. Embleemfabels in de Nederlanden (1567–ca. 1670)* (2006), *Dispositio. Problematic Ordering in French Renaissance Literature* (2007), and *Réécrire la Renaissance, de Marcel Proust à Michel Tournier. Exercices de lecture rapprochée* (2009). He is co-author of *Francis Ponge: lectures et méthodes* (2004), editor of *Editer et traduire Rabelais à travers les âges* (1997) and *Translating Montaigne* (2011), and co-editor of *Lectures de René Char* (1990), *Fabuleux La Fontaine* (1996), *Le paradoxe en linguistique et en littérature* (1996), *Montaigne and the Low Countries (1580–1700)* (2007), *Early Modern Zoology. The Construction of Animals in Science, Literature and the Visual Arts* (2007), *Zoology in Early Modern Culture* (2014), *Emblems and the Natural World* (2017), *Natural History in Early Modern France* (2018), *Langues hybrides: expérimentations linguistiques et littéraires (XVe–début XVIIe siècle)* (2019), *Early Modern Catalogues of Imaginary Books. A Scholarly Anthology* (2020), and *Ronsard and Du Bartas in Early Modern Europe* (2021). Currently he is the director of two NWO research projects: *A New History of Fishes. A long-term approach to fishes in science and culture, 1550–1880*, and *Aesopian Fables 1500–2010: Word, Image, Education*. He is chief editor of *Neophilologus*, and member of the editorial board of *Intersections*, *Montaigne Studies*, and *Deshima*.

Florike Egmond

is a retired historian affiliated with Leiden University and living in Rome. She has worked for successive NWO-funded research projects in the field of natural history that were based at Leiden University. Her main research interests are the history of natural history, collecting and depicting naturalia in early modern Europe, gardens as botanical collections, the visual history of science, and methodological aspects of the early modern natural sciences studied in the context of the history of medicine, communication and exchange. She is currently involved in research on botanical collections, images of fossils, marine history, and networks of naturalists centred on Rome. Her main publications include the monographs *Eye for Detail. Images of Plants and Animals in Art and Science* (2017), *The World of Carolus Clusius. Natural History in the Making, 1550–1610* (2010), and *Het Visboek. De wereld volgens Adriaen Coenen* (2005). She is the co-author of *The Mammoth and the Mouse. Microhistory and Morphology* (1997). She edited and introduced *Conrad Gessners Thierbuch. Die Originalzeichnungen* (2018), and co-edited *Das Naturalien-Kabinett des Ulisse Adrovandi. 1000 Aquarelltafeln aus dem 16. Jahrhundert* (2024); *Carolus Clusius. Towards a cultural history of a Renaissance naturalist* (Amsterdam, 2007); *Correspondence and Cultural Exchange in Early Modern Europe 1400–1700* (2007); *The Whale Book. Whales and other marine animals as described by Adriaen Coenen in 1585* (2003); and *Bodily Extremities. Preoccupations with the Human Body in Early Modern European Culture* (2003).

Notes on the Contributors

Cristina Brito
is Associate Professor at the History Department at NOVA FCSH, Lisbon, and a senior researcher at CHAM – Center for the Humanities. She is a CO-PI of the ERC Synergy Grant *4-OCEANS: Human History of Marine Life* (GA: 951649). Cristina Brito has an interdisciplinary, comparative, and cross-cultural approach to her research. Her scientific interests include early modern marine environmental history, local and global perceptions about and uses of the seas, Atlantic and oceanic histories, humans and nonhumans entanglements, the Anthropocene, and the Blue Humanities. She coordinates several research projects, is a member of the Board of OPI – Oceans Past Initiative, and is active in several networks and research projects, such as the UNESCO Chair on Oceans' Cultural Heritage and the H2020 RISE project CONCHA.

Tobias Bulang
studied German and philosophy at the TU Dresden and The Ohio State University in Columbus, Ohio (USA). In 2002 he obtained his PhD in Dresden with a work about literature and historicism. Between 2003 and 2012 he had different engagements at the Universities of Göttingen, Dresden, Zurich, and Munich. In 2009 he obtained his *Habilitation* in Zurich with a study on mediaeval and early modern encyclopaedic poetry. Since 2012 he has been a Professor for Medieval German Literature at the University of Heidelberg.

João Paulo S. Cabral
earned his degree in History at Open University and a PhD in Biology at Oporto University. Between 1991 and 2022, he was university professor at Oporto University, teaching biology and history of biology. His research interests have focused on archaeomalacology and history of botany, resulting in the publication of articles on molluscs found in Portuguese archaeological sites, and several books and articles on the life and work of Garcia de Orta (ca. 1501–1568), Father Jerónimo Lobo SJ (1593–1678), Gabriel Grisley (–after 1669), Domingos Vandelli (1735–1816), Gonçalo Sampaio (1865–1937), Rocha Peixoto (1866–1909) and Américo Pires de Lima (1886–1963), and a book, *Natural History in the Enlightenment*, published in 2022.

Dorothee Fischer
is a university lecturer and research assistant at the Department of Art History at the Universität Trier, Germany. She has studied at the Universität Konstanz

(Germany), the University College Cork (Ireland) and the Universität Wien (Austria). Her research focuses on human-animal relations in art (history). With a particular interest in aquatic animals, she has published on animals' agency in contemporary art as well as on collecting animals in the early modern period. She is currently working on her doctoral thesis regarding the representation of fish in 18th-century natural collections.

Holger Funk
is an independent scholar. He has published numerous articles and a few books on zoology and botany from antiquity to the 18th century. Currently he is preparing a nomenclaturally updated electronical version of D.W. Thompson's *Glossary of Greek Fishes* (1947). In addition, he writes contributions to Christopher Scharpf's ETYFish Project (https://etyfish.org/).

Dirk Geirnaert
is an independent researcher in the domain of Dutch historical literature and linguistics. He was one of the editors of the *Woordenboek der Nederlandsche Taal*, the counterpart of the *Oxford English Dictionary*, describing the Dutch vocabulary from 1500 up to and including 1975. He is co-author of *Het Antwerps liedboek*, edition by D.E. van der Poel a.o., with a reconstruction of the tunes by L.P. Grijp (2004) and, with prof. R. Lievens, of *Van zondeval tot hemel, staties uit de heilsgeschiedenis*, a study and edition of an unknown fifteenth century work by the Brugean rhetorician Anthonis de Roovere on the life and passion of Christ (2018). Furthermore he has published many articles on Middle Dutch fragments, on Dutch literature of the Middle Ages and the sixteenth century (especially in Bruges) and on allied products of medieval and renaissance figurative arts.

Philippe Glardon
has a doctorate in history from the University of Lausanne. His fields of research include the history of pre-scientific and scientific discourses and vocabularies, the history of modern botany and zoology (Renaissance–19th century) and the history of scientific illustration. His thesis is on the natural history of the 16th century. He teaches history of science at the École Polytechnique de Lausanne (EPFL) and is a lecturer at the University of Lausanne (Faculty of Biology and Medicine) since 2009.

Justin R. Hanisch
is a biologist for the Government of Alberta and also a collector and independent scholar of antiquarian books on fishes. He obtained a PhD. in ecology in 2016

from the University of Alberta under the supervision of Drs. William M. Tonn. His professional research interest includes monitoring impacts of anthropogenic activities on freshwater ecosystems. As a bibliophile and independent scholar, Hanisch's primary interest is pre-20th century books, manuscripts, and ephemera on the natural and social history of fishes. He has previously authored an exhibition catalogue, *A Contemplative Angler, Selections from the Bruce P. Dancik Collection of Angling Books* (2018).

Bernardo Jerosch Herold
graduated in Chemical Engineering 1956 at Instituto Superior Técnico (IST), Lisbon University, and achieved his PhD in Chemistry at Heidelberg University in 1961. He was Professor of Industrial Organic Chemistry IST 1962–2003. He belongs to the Centro de Química Estrutural, Instituto Superior Técnico, Lisbon University and is a Member of the Lisbon Academy of Sciences. He is interested in the history of science and natural history of Portugal and in early modern travel accounts.

Rob Lenders
is an assistant professor at the Institute for Science in Society (ISiS) and at the Radboud Institute of Biological and Environmental Sciences (RIBES) (Nijmegen, the Netherlands). He is an animal ecologist by training, specialised mainly in fish, amphibians and reptiles. His teaching and research focusses on changing human – nature relationships, especially but not exclusively, from a historical perspective, spanning the entire Holocene. To this end, he works intensively with historians, archaeologists, linguists, sociologists and philosophers. He is one of the initiators of the ATHENA project (Access Tool for Historical ecology and ENvironmental Archaeology).

Alan Moss
received his PhD degree at Radboud University Nijmegen in 2022. His thesis focuses on seventeenth-century travelogues, more specific on the cultural and religious encounters and confrontations of Dutch travellers on the Grand Tour. He works as a researcher for the Dutch National Archives (Nationaal Archief), where he focusses on the War Archives of the Dutch Red Cross. His main research interests lie in the field of early modern history of travel, nationalism and national thought, cultural exchange, peace studies, pamphlet culture, and humour and satire.

Doreen Mueller
is assistant professor of Japanese art and material culture at Leiden University. She earned a PhD in the History of Art at SOAS, University of London. Her research explores the intersections of visual culture, social and environmental history with a focus on how representations of famine and natural disasters were used for moral admonition and social criticism in early modern Japan. She has published on the production of knowledge about animals and plants in early modern East Asia, and the nature of political caricature in nineteenth-century Japan.

Johannes Müller
is assistant professor of German and Literary Studies. His research interests include environmental history, migration and the circulation of knowledge and science in the early modern world.

Martien J.P. van Oijen
was trained as a biologist at Leiden University. He was curator of Fishes at the Rijks Museum van Natuurlijke History RMNH, Head of the Department of Vertebrates at the Nationaal Natuurhistorisch Museum and senior researcher at Naturalis Biodiversity Center, Leiden. Before he was employed at the RMNH he spent three years in Tanzania investigating haplochromine cichlids of Lake Victoria. On the bass this research he earned his Ph.D. As curator of Fishes, he specialized in the work of the Dutch ichthyologist Pieter Bleeker, the Japanese fishes from the Siebold collection, and the history of the RMNH collection. He was involved in a number of exhibitions on the Siebold collection and published a Dutch translation of von Siebold's account of the court journey of 1826.

Pietro Daniel Omodeo
is a cultural historian of science and professor of historical epistemology at Ca' Foscari University of Venice. Since 2023, he is the holder of the UNESCO Chair for Water, Heritage, and Sustainable Development. He is the PI of the ERC project EarlyModernCosmology (2017–2023, GA 725883), the FARE project EarlyGeoPraxis (funded by the Italian Ministry of University and Research) and the Max Planck Partner Group in Venice The Water City. Among other publications, he is the author of *Defending Descartes in Brandenburg-Prussia* (2022); *Political Epistemology* (2019), and (with Jürgen Renn) *Science in Court Society* (2019).

Anne Overduin-de Vries

is a biologist who earned her PhD at Utrecht University. Her thesis focusses on the cognitive capacities in the visual domain of non-human primates. She is currently working as the project coordinator on the "Malesian butterfly project" at Naturalis Biodiversity Center. Previously she coordinated the citizen science project "Fishing in the past" hosted by Naturalis.

Theodore W. Pietsch

is professor emeritus and curator emeritus of fishes in the School of Aquatic and Fishery Sciences, and Burke Museum of Natural History and Culture, at the University of Washington in Seattle. He was educated at the University of Michigan (B.A., 1967), University of Southern California (M.S. and Ph.D., 1969 and 1972), and Harvard University (Postdoctoral Fellowship, 1973–1975). He has conducted research in marine ichthyology, especially the biosystematics, zoogeography, and reproductive biology of coral reef and deep-sea fishes. He has published extensively on the biology of fishes as well as the history of science.

Cynthia M. Pyle

was trained as a biologist, and worked at MIT and Harvard prior to becoming a scholar of the 15th and 16th Century European Renaissance (Ph.D. Columbia University). She was an Attachée de Recherche in the Centre National de la Recherche Scientifique, Paris; a Research Associate in the Department of the History of Science, Harvard University, and she founded the Renaissance Studies Certificate Program at the Graduate Center of the City University of New York. She is currently a Research Affiliate at New York University, and the Co-Chair of the Columbia University Seminar in the Renaissance.

Marlise Rijks

is a historian specialized in art, science, and technology of the early modern period. She is assistant professor at the VUB (Vrije Universiteit Brussels) and postdoc researcher at Ghent University, working on the project *Printing Images in the Early Modern Low Countries. Patents, Copyrights, and the Separation of Art and Technology* (funded by the Research Foundation Flanders, FWO). In the years 2016–2020 she worked at Leiden University on the NWO-funded project *A New History of Fishes. A Long-term Approach to Fishes in Science and Culture, 1550–1800*. She has published widely on the history of collections, natural history, and workshop knowledge. Her recently published monograph is titled *Artists' and Artisans' Collections in Early Modern Antwerp. Catalysts of Innovation* (2022).

Ronny Spaans
is Associate Professor in Nordic Literature at the University of Oslo. In 2015 he earned his PhD at the same university with his thesis on the poetry of the Dutch writer Joannes Six van Chandelier (1620–1695). The thesis appeared in 2020 as monography at Amsterdam University Press: *Dangerous Drugs. The Self-Presentation of the Merchant-Poet Joannes Six van Chandelier (1620–1695)*.

Robbert Striekwold
is a PhD student at Naturalis Biodiversity Center and Leiden University, where he studies the history of Dutch 19th-century ichthyology. He studied biology and history of science in Utrecht and Stockholm. His research interests include the history of modern biology and palaeontology in relation to collections.

Melinda Susanto
is a PhD candidate in history at Leiden University. Her current research investigates how botanical and medical knowledge played a role in interactions between Asian courts and the Dutch East India Company in the early modern period. Her research interests include the intersections between natural history and visual and material culture, and the role of South and Southeast Asia in global circulations of knowledge.

Didi van Trijp
is historian of science and curator. She earned her PhD at Leiden University in 2021 as part of the NWO-funded project *A New History of Fishes*. Her doctoral dissertation examined the cultural history of eighteenth-century natural historical studies into fish, and is under contract to appear with Brill (in the series *Emergence of Natural History*). Her research is concerned with the history of collecting and the material and visual culture of science, particularly of natural history in the early modern period. A recent fellowship at the Science History Institute in Philadelphia focused on techniques to convey iridescence in natural historical plates and drawings.

Sabina Tsapaeva
is a scientific researcher and freelance lecturer at the Institute of German studies of the University of Hamburg, where she teaches history of the German and Low German languages, Medieval studies and Contact Linguistics. She previously was a research assistant and freelance lecturer of Low German Philology at the University of Rostock. Her research interests are, among others, historical linguistics, historical semantics, historical lexicology and lexicography,

Middle Low German language and intralingual translation of historical texts. Her publications include *Das Rostocker Tierepos 'Reynke Vosz de olde' (Ludwig Dietz, 1539) im Kontext der niederdeutschen 'Reynke de vos' – Überlieferung des 15. und 16. Jahrhunderts* [...] (2018). She is the author (2017, 2021) and the editor (2019, 2020, 2023) of *Mittelniederdeutsches Wörterbuch*.

Ching-Ling Wang
is curator of Chinese art at the Rijksmuseum, Amsterdam. He received his doctoral degree from Freie Universität, Berlin. He worked as research fellow at the Kunsthistorisches Institut of the Max-Planck Institut in Florence, and was curator of Chinese art at the Museum für Asiatische Kunst, Staatliche Museen zu Berlin. His research focuses on Chinese painting, Qing court art, and the exchanges between China and Europe in the field of visual culture.

CHAPTER 1

Introduction: Towards a Cultural History of Early Modern Ichthyology (1500–1880)

Paul J. Smith

1 A New History of Fish

The present volume stems from the research project *A New History of Fishes. A long-term approach to fishes in science and culture, 1550–1880*. This project, which was funded by the NWO (Dutch Research Council) and carried out at Leiden University and at Naturalis Biodiversity Center, started in 2017 and was concluded in 2022. Here are a few quotes from the project description, published by NWO,[1] which are important for the theme of this volume:

> This project proposes a new history of European ichthyological knowledge over three centuries (1550–1880). It is a new history because no long-term history of expert fish knowledge has been written since the early 19th century (Cuvier). Moreover, we argue for long-term continuities rather than Foucauldian epistemological breaks. It is new, furthermore, because our project is aligned with the New History of Science: we look at "science" in context – and, therefore, for the early modern period at expert fish knowledge (manifested in collecting practices and information exchange via texts, objects and images) […]. Following this approach we hope to answer our central questions: How and where did ichthyology develop as a scientific discipline; how did it take shape as a field of expert knowledge in the cultural context of early modern and modern Europe.

It went without saying that these perspectives could not be realized within a single research project. A scholarly volume was therefore provided, in which not only would several findings arising from the project be presented (these are in this volume the contributions of Florike Egmond, Anne Overduin-de Vries, Marlise Rijks, Paul Smith, Robbert Striekwold, and Didi van Trijp), but a number of researchers from various countries, both starting and established scholars, were also asked to contribute. The expertise present in this volume

1 https://www.nwo.nl/en/projects/360-55-090.

is highly interdisciplinary and transdisciplinary: it includes not only biologists and historians of science but also historians, art historians, literary scholars, and linguists, and thus covers many cultural and social domains in which ichthyology[2] is embedded. The two editors of this volume, Florike Egmond and I, have divided the twenty-four contributions (which vary greatly in subject matter and methodology) into four more or less thematically coherent sections ("Beginnings", "Depicting", "Fish and Society in Europe", and "Ichthyological Knowledge from Afar"), which also do justice to the changing socio-cultural context from a long-term perspective. This "longue durée" spans more than three centuries: it begins with Paolo Giovio's *De Romanis piscibus* (1524), the first printed monograph devoted to fish, and continues until the mid-19th century, a period in which ichthyology diversified, as happens in all branches of natural history, under the impetus of, among others, Linnaean classification, the comparative anatomy of Cuvier, emerging Darwinism, the narrative natural history of Buffon and Brehm, and, most importantly, the increasing knowledge of exotic, non-European plants and animals. Knowledge about the exotic natural world was gathered with the help of indigenous plant and animal knowledge generated outside Europe, and its transfer was often bilateral, as the present volume will demonstrate. Not only did natural knowledge come to Europe from the East and the West, but it also travelled the other way around, from Europe to far abroad.

In order to give an impression of the scope, the diversity, and the changeability of this context, and the intricate mutual interaction between ichthyology and visual and literal culture, this introduction briefly zooms in on five striking examples from early modern painting, literature, and ichthyology itself: (1) a painting by Jan Brueghel the Elder, *Diana and the Nymphs after the Hunt* (ca. 1621), in which the interaction between ichthyological knowledge and painting is made the main theme; (2) two chapters from the work of François Rabelais: one (1532) seriously celebrating natural history, and within it the knowledge of aquatic life, and the other (1564) satirically expressing the author's scepticism about natural history, and particularly ichthyology; (3) two passages from the "Foreword to the Reader" of the authoritative *Historia animalium*, Book IV (1558), in which Conrad Gessner positions ichthyology and

2 In the present introduction (and in several articles of this volume), the term "ichthyology" is used for facility. In fact, this term is anachronistic: if its first occurrence (in Latin) was already encountered occasionally in the 16th century, it came into common use only in the 18th century. The term used more frequently in the 16th and 17th centuries is "(natural) history of fish", including the study not only of fish, but of all forms of aquatic animal life, such as marine mammals, sea turtles, crustaceans, shellfish, and (nonexistent) sea monsters.

makes a case for the use of illustrations in natural history; (4) a fish market scene by Frans Snyders and Anthonie Van Dyck (ca. 1521) in which realism, exoticism, and religious symbolism come together; and, finally, (5) the frontispiece that the illustrator Paul van Somer (II) engraved for the *Historia piscium* (1686), by Francis Willughby and John Ray. This frontispiece illustrates, through the image, the innovative word-image content of the *Historia piscium*.

2 Jan Brueghel and the Fascination for the World of Fish

The painting *Diana and the Nymphs after the Hunt* (ca. 1621) [Fig. 1.1] by Jan Brueghel the Elder (1568–1625) depicts a traditional subject: the goddess Diana and her nymphs resting after their hunt. However, contrary to the pictorial tradition, the women do not rest by sleeping or by bathing in nearby water, but by fishing. In the background are nymphs enthusiastically swimming and fishing, and in the right foreground Diana and two nymphs are excitedly inspecting the freshly caught fish. In doing so, they ignore the piled-up hunting booty of birds and land animals, leaving them literally to the left; all their attention is on the floundering fish being pulled from the net. Even the hunting dogs focus exclusively on the caught fish.

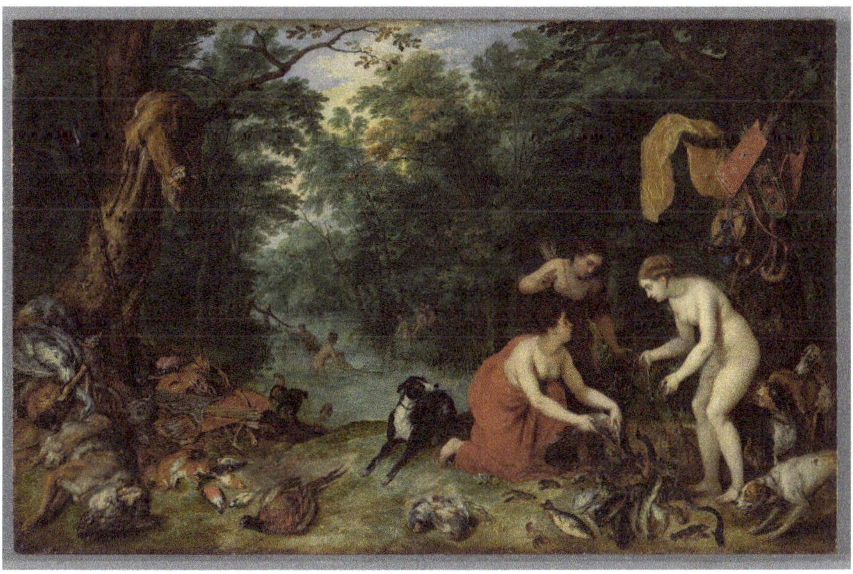

FIGURE 1.1 Jan Brueghel the Elder, *Diana and the Nymphs after the Hunt* (ca. 1621). Neuburg an der Donau, Staatsgalerie. bpk / Bayerische Staatsgemäldesammlungen

This fixation on fish and fishing reflects man's natural, deep-seated fascination with the world of fish and other *aquatilia* – a world with which man is familiar in his daily life, from fishing and the fish trade to the kitchen, but which at the same time remains unknown because it is hidden under water. The spectator watches along with the women. The fish (as well as the pile of hunted animals and birds) are realistically depicted – a number of fish species, especially the larger specimens, are recognizable: the spectator can easily distinguish a pike, eel, carp, and a crayfish. However, the large fish held by the left nymph cannot be identified, as her hand makes identification impossible. The discussion between the three women seems to be about this particular fish. As a spectator we are thus invited to interpret a painting about interpreting characters. Let us accept this invitation, because this multi-layered painting in many ways exemplifies the subject of the present volume.

3 Rabelais: Problematizing Ichthyology

Before discussing Brueghel's painting in more detail, it is necessary to take a step back in time. The preference given by Brueghel's nymphs to fish over the other animals reflects a broader general tendency in natural history that was already visible at the beginning of the 16th century. The most striking example of this tendency is the above-mentioned *De Romanis piscibus* (1524; several editions) by Paolo Giovio (1483–1552), which, as the first printed monograph devoted to fish, precedes those devoted to the other animal fields of natural history.

The growing interest in ichthyology at the expense of the other areas of natural history can also be traced from an unexpected corner, namely literature: the French writer François Rabelais (?–1553), physician, Erasmian humanist, and author of a series of comic narratives about the giants Gargantua and Pantagruel. Rabelais can be regarded as a critical witness to the scientific developments of his time. A passage important for our argument is found in his *Pantagruel* (1532), chapter 8, which contains the letter of education that father Gargantua writes to his son Pantagruel, who is going to study in Paris. The idealized educational programme of young Pantagruel mentions ichthyology as the first part of natural history – before the fields that we now call ornithology, botany, and mineralogy:

> Et quand à la congnoissance des faictz de nature, je veulx que te y adonne curieusement, qu'il n'y ayt mer, riviere, ny fontaine, dont tu ne congnoisse les poissons, tous les oyseaulx de l'air, tous les arbres, arbustes et

> fructices des foretz, toutes les herbes de la terre, tous les metaulx cachez au ventre des abysmes, les pierreries de tout l'Orient et midy, rien ne te soit incongneu.[3]

> And as for the knowledge of nature's works, I want you to devote yourself to that with care: let there be no sea, stream, or spring, whose fish you do not know; all the birds of the air, all the trees, shrubs, and bushes of the forests, all the herbs of the earth, all the metal hidden in the bowels of the depths, the precious stones of the entire Orient and Southern Hemisphere: let nothing be unknown to you.[4]

Rabelais' interest in the field of ichthyology is also expressed in his other works, albeit less enthusiastically. For example, his posthumously published *Fifth Book* (1564) contains a satire on the uncertainties of knowledge, especially in the field of natural history. Driven by curiosity, the giant Pantagruel and his travelling companions arrive in the Land of Satin, where they find a great number of extraordinary animals of all kinds. All these animals turn out to be fictional because they are not real, but rather have lifelike depictions in tapestries. These images were created at the instigation of the allegorical character Ouy-Dire (Hearsay). After first seeing land animals and birds (the description of which often goes back to Pliny, whom Rabelais elsewhere calls a 'liar'),[5] the tour group goes to the centre of the land – the core of natural history – which is Aristotelian ichthyology.[6] As Alcofribas, Rabelais's narrator, recounts:

> Passans quelque peu avant en ce pays de tapisserie, vismes la mer mediterranee, ouverte et descouverte jusques aux abismes, tout ainsi comme au gouffre Arabic se descouvrit la mer Erithrée, pour faire chemin aux Juifs issans d'Egypte. Là je recongnu Triton sonnant de sa grosse conche, Glaucus, Proteus, Nereus, et mille autres dieux et monstres marins. Vismes aussi nombre infiny de poissons en especes diverses, dansans, volans, voltigeans, combatans, mangeans, respirans, belutans, chassans, dressans escarmouches, faisans embuscades, composans tresves, marchandans, jurans, s'esbatans. En un coing là prés vismes Aristoteles tenant une lanterne, en semblable contenance que l'on peint l'hermite

3 Rabelais François, *Œuvres complètes*, ed. M. Huchon (Paris: 1994) 244–245.
4 Rabelais François, *The Complete Works*, transl. D.M. Frame (Berkeley – Los Angeles: 1991) 161.
5 Rabelais, *Œuvres Complètes* 22.
6 For the importance of ichthyology for Aristotle, see Leroi A.M., *The Lagoon: How Aristotle Invented Science* (New York: 2014).

> prés sainct Christofle, espiant, considerant, le tout redigeant par escrit. Derriere luy estoient, comme records de sergents, plusieurs autres Philosophes, Appianus, Heliodorus, Atheneus, Porphirius, Pancrates, Archadian, Numenius, Possidonius, Ovidius, Oppianus, Olympius, Seleucus, Leonides, Agathocles, Theophraste, Damostrate, Mutianus, Nymphodorus, Elianus, cinq cens autres gens, aussi de loisir comme fut Chrysippus, ou Aristarchus de Sole, lequel demeura cinquante huit ans à contempler l'estat des abeilles, sans autre chose faire. Entre iceux j'y advisay Pierre Gylles lequel tenoit un urinal en main, considerant en profonde contemplation l'urine de ces beaux poissons.[7]

> Pushing on a little farther into the land of tapestry, we saw the Mediterranean Sea opened up and uncovered down to its deepest abysses, even as in the Persian Gulf the Red Sea opened up to make a roadway for the Jews coming out of Egypt. There I recognized Triton sounding his great shell horn, Glaucus, Proteus, Nereus, and myriad other gods and monsters of the sea. We also saw an infinite number of fish of various kinds, dancing, flying, curveting, [fighting], eating, breathing, screwing, hunting, skirmishing, laying ambushes, arranging truces, bargaining, swearing, disporting.
>
> In a nook nearby I saw Aristotle holding a lantern in a posture like that in which they paint the hermit next to Saint Christopher, closely watching, considering, putting it all down in writing. Behind him, like sergeants' witnesses, were many other philosophers: Appian, Heliodoros, Athenaeus, Porphyrius, Pancrates[,] Arcadi[us], Numenius[,] Posidonius, Ovid, Oppian, Olympius, Seleucus, Leonides, Agathocles, Theophrastus, Damostratus, Mutianus, Nymphodorus, Aelian, also five hundred idle folk, as was Chrysippus, or Aristarchus of Sola, who stayed fifty-eight years contemplating the state of the bees, without doing anything else. Among these I noticed Pierre Gilles, who, holding a urinal in his hand, was deeply contemplating the urine of these fine fish.[8]

By including Aristotle and his disciples in the centre of the Land of Satin, Rabelais gives a critical reflection on ichthyology. The question is who and what exactly he is criticizing in this particularly dense episode. Aristotle's position as the figurehead of knowledge does not seem to be up for debate. However,

7 Rabelais, *Œuvres Complètes* 803. For references to existing fish images from antiquity and the Renaissance and to the medieval iconology of Saint Christopher, see Smith P.J., "Rabelais ichtyologue", in Garnier I. et al. (eds.), *Narrations fabuleuses. Mélanges en l'honneur de Mireille Huchon* (Paris: 2022) 439–452.

8 Rabelais, *Complete Works* 681.

this does not apply to his disciples, some of whom are known only by name and others being non-existent, invented by Rabelais. These disciples are reprehensible, for they blindly follow Aristotle without scrutiny and without any attempt at personal observation. Among the ichthyologists gathered around Aristotle, only one contemporary writer is mentioned, namely Pierre Gilles (1490–ca. 1555), author of a French-Latin lexicon of fish names (*De Gallicis et Latinis nominibus piscium*) and translator into Latin of Aelian's *De natura animalium*. His lexicon and translation were printed together in 1535 by Sébastien Gryphe (Gryphius), humanist printer and publisher in Lyon. Gilles' lexicon and translation may have been for correction in the hands of Rabelais, who at the time was working as a proof reader at the printing house of Gryphe. At first glance, Gilles seems to belong to the group of followers of Aristotle in the Land of Satin. But at the same time he distances himself from them because he is the only ichthyologist who does not servilely copy Aristotle, but also works independently through autopsy and experimental observation. The figure of Pierre Gilles is indeed ambiguous: on the one hand, as a publisher and translator of Aelian (whom Rabelais elsewhere calls a 'tiercelet de menterie'[9] ('a tiercel [expert] in lying')),[10] he is a transmitter of ancient ichthyology; on the other hand, he is an innovator because of his lexicon, in which he pays particular attention to the fish of Marseille and their Provençal names. The latter will certainly not be criticized by Rabelais, fond as he is of dialectical lexicography. What is ridiculed by Rabelais through the figure of Gilles is both the practical difficulty, or even impossibility, of certain kinds of experimental research (how do you get fish urine?) and their usefulness.

Essentially, Rabelais' episode is about scientific reliability. The lying, unreliable narrator Alcofribas makes no distinction between really existing animals (elephant, rhinoceros, chameleon), imaginary or doubtful but well-known animals (unicorn, griffin), and impossible, invented animals (Half Lent, and the animal with two backs, also known from Shakespeare's *Othello*). The eyewitness testimony does not concern live animals but images of animals. These images are based solely on textual testimonials, not made *de visu*. In turn, those testimonies are also unreliable because they are not drawn from personal observation but from hearsay and the authority of others. In this dizzying game of truth and fiction, Rabelais does not seem to want to take a stand. Instead, he makes the reader reflect on the usefulness and reliability of natural history writings, whether illustrated or not. Stimulating critical reflection, not through self-positioning but through humour that puts things in perspective, is one of the constants in Rabelais' work.

9 Rabelais, *Œuvres Complètes* 800.

10 Rabelais, *Complete Works* 679.

4 Conrad Gessner and the Visual Turn

Rabelais' latent scepticism about science, which can also be heard elsewhere in his work, would certainly not have been shared by his contemporary, the Swiss physician and naturalist Conrad Gessner (1516–1565), the world's most influential naturalist of that time. For Gessner, it is precisely man's unfamiliarity with the world of fish that constitutes the greatest challenge in natural history. In his authoritative *Historia animalium*, Book IV, devoted to fish (1558), he recounts in retrospective the moment he discovered this area around 1548:

> Cæpi enim profecto ante decennium (que maxima ætates humanae pars est) de omni animalium genere multa subinde observare, et condendis de ipsorum natura voluminibus materiam omni studio praeparare. Prae caeteris autem Aquatilium historia me fatigavit, magis omninovaria et multplex, difficiliorque (mihi praesertim mediterraneo et pene ad summas alpes undique a mari remote homini) quam reliquorum animalium.[11]
>
> Indeed, a decade ago (which is a very large part of a person's life), I began to make many observations about all kinds of animals and to prepare the material for every kind of study by means of volumes about their nature. Above all other volumes, the history of aquatic animals kept me breathless, which is altogether very diverse and varied and also more difficult than the history of the other animals (especially for me, as someone who lives in the interior of the country near the highest Alps and is cut off from the sea on all sides).[12]

It is therefore not surprising that he waited a long time to publish his book about fish. Only after finishing his three other volumes (on mammals, oviparous quadrupeds (reptiles), and birds, respectively) did he decide to publish this work, which, with its thousand thickly printed folio pages, is the most voluminous and richly illustrated volume in the series. In his "Preface to the Reader" Gessner relates how he discovered the works of Pierre Belon (1517–1564), Guillaume Rondelet (1507–1566), and Ippolyto Salviani (1514–1572) – authors upon whom he would greatly rely.

With Gessner, Belon, Rondelet, and Salviani a new era begins. As Florike Egmond states in her contribution to this volume, we are dealing with a 'visual

11 Gessner Conrad, "Praefatio ad lectorem", in *Historia animalium liber IIII. qui est de piscium et aquatilium animantium natura* (Zurich, Christoph Froschauer: 1558) fol. b3r.

12 Translation by Holger Funk (personal communication).

turn': 'the interest in *aquatilia* [did not] suddenly exploded [...], but their visual representation [did].' Gessner was well aware of this visual turn, as can be read in this quotation of disarming frankness from the "Preface to the Reader":

> Primus nostris temporibus Paulus Iovius, ut Piscium historiam excolere coepit, ita et picturas eorum fieri curavit, ut ipse refert: quas tamen typis publicatas non puto. Inde post multos annos cum ego omnino negligi ab omnibus hoc argumentum putarem, (Rondeletium enim, Bellonium, et Salvianum, idem moliri nondum cognoveram), plurimas in Italia et apud nos piscium picturas mihi comparavi. Sed dum in Quadrupedum Aviumque historia, et aliis quibusdam libris aedendis haero, illi quos iam nominavi, me anteverterunt, quod mihi certe non ingratum fuit. Ab illis enim icones sum mutuatus, quibus vel ipse carebam, vel quae ab ipsis accuratius mihi expresse videbantur. Plurimas quidem e Rondeletii libris, paucas e Bellonii opere, paucissimas, nempe unam aut alteram a Salviani. Non modo quod tardius liber eius ad me pervenisset, sed quia non plures e centum illis, quas dedit deesse mihi videbantur.[13]

> The first to care about drawings of fish in our time was Paolo Giovio (Paulus Iovius), when he started elaborating a fish history, as he himself reports. However, I do not believe that these drawings have appeared in print. Because I believed for many years that the pictorial representation [of fish] was generally neglected by all (for I was not yet aware of the fact that Rondelet, Belon, and Salviani were concerned with this subject), I acquired myself a good many fish pictures in Italy and here in our home country. But while those I have already mentioned have forestalled me in publishing the natural history of the quadrupeds and birds and certain other books, this was certainly not unwelcome for me. For I borrowed from them the pictures that I myself lacked or that seemed to me to be printed more carefully in their books; most of them come from Rondelet's books, a few from the work of Belon, least of all, that is, only the stray image, from Salviani. Not only because his book came to me only with a delay, but because I did not seem to lack many of the one hundred he presented.[14]

13 Gessner, "Praefatio ad lectorem", in *Historia animalium liber IIII. qui est de piscium et aquatilium animantium natura* fol. b3r.

14 Translation by Holger Funk (personal communication).

Whereas the works of Giovio and Gilles were still unillustrated, in the 1550s the frequent and systematic use of illustrations gave a real boost to ichthyology. We see the appearance of a rapid succession of eight significant, richly illustrated publications which epitomize the avant-garde role ichthyology played within early modern zoology: Belon publishes two illustrated ichthyological works in French (1551 and 1555) and one in Latin (1553); Rondelet two in Latin (1554–1555) and one in French (1558); Salviani one in Latin (1558); and Gessner (1516–1565) two in Latin (1558 and 1560).[15] The differences from publications in the other segments of illustrated zoology are also striking. Illustrated ichthyological works not only appeared earlier, but they also exceeded contemporary publications on other animals in number and volume: the 1550s saw the appearance of only one work on mammals (Gessner 1551); two on birds (Belon 1555 and Gessner 1555); and one on reptiles (Gessner 1554). Their many updated editions, reissues of their illustrations, and translations into vernacular languages make it clear that these illustrated books on fish were read widely by a readership far beyond only those who knew Latin. As the title page of Gessner's *Fischbuoch*,[16] the widely read German translation, indicates, the illustrations were intended not only for naturalists, but also 'zum nutz und guten allen Arzeten, Maleren, Weyleuten und Kochen' (for the benefit and good of all doctors, painters, farmers, and cooks). This is how fish illustrations entered en masse into the world of painting.[17]

5 Jan Brueghel and Science

Now, back to Brueghel's painting. The painting not only presents a passive reflection of the scientific, specifically ichthyological, interest in the 16th and early 17th centuries, but actively thematizes this interest as a subject. This becomes especially clear when we look at Brueghel's other paintings with

15 Similar observations in Zucker A., "Fonctions des classes dans les traités ichtyologiques de P. Belon et G. Rondelet: empreinte ou alibi antique?" in Gontier T. (ed.), *Animal et animalité dans la philosophie de la Renaissance et de l'Âge classique* (Louvain – Paris: 2005) 7–32 (here: 7–8). See also Glardon P., *L'histoire naturelle au XVIe siècle. Introduction, étude et édition critique de* La nature et diversité des poissons *de Pierre Belon (1555)* (Geneva: 2011) 6–7.

16 Gessner Conrad, *Fischbuoch*, transl. Conrad Forrer (Zurich, Christoph Froschauer: 1563), title page.

17 For a general historical overview of the entry of natural history into South-Netherlandish painting and printmaking, see Rikken M., *Dieren verbeeld. Diervoorstellingen in tekeningen, prenten en schilderijen door kunstenaars uit de Zuidelijke Nederlanden tussen 1550 en 1630* (PhD dissertation Leiden University: 2016).

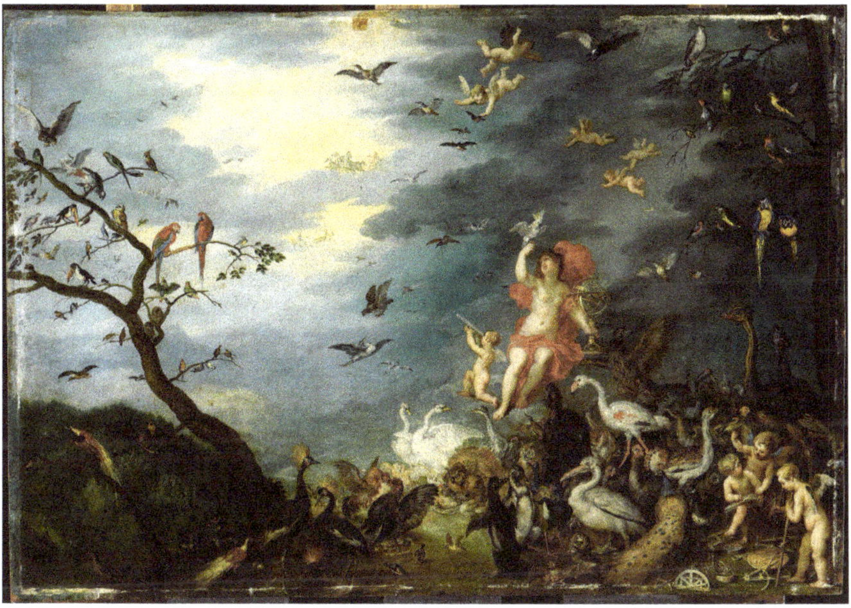

FIGURE 1.2 Jan Brueghel the Elder, *The Allegory of Air* (1621). Louvre
PHOTO © RMN-GRAND PALAIS (MUSÉE DU LOUVRE) / FRANCK RAUX

natural history themes from the same period. One can think of his painting *Allegory of the Element Air* (1621) [Fig. 1.2], commissioned by the Milanese Cardinal Federico Borromeo (1564–1631), who was himself fascinated by the advancement of science. As in *Diana and the Nymphs*, scientific knowledge is personified in this painting by characters from classical mythology: The goddess Aurora is assisted by many putti, who are equipped with measuring instruments (including the newly invented telescope) with which they study nature.[18] In this painting, the connection between art and science is particularly present. That is, the painting should be seen as a direct response to Carolus Clusius' *Exoticorum libri decem* (1605) – it incorporates many animals from Clusius' work, depicting them better and more faithfully (especially the cassowary and the penguin), and moreover in colour. In addition, he depicts many more bird species than authoritative books on birds (those by Belon, Gessner, and Aldrovandi, including Clusius) could ever depict or describe. This can be seen in particular in Brueghel's parrots, of which only a few species are described in contemporary ornithological literature. Moreover, he cuts

18 For a detailed analysis of the painting, see Rikken M. – Smith P.J., "Jan Brueghel's *Allegory of air* (1621) from a natural historical perspective", *Netherlands Yearbook for History of Art* 61 (2011) 86–116.

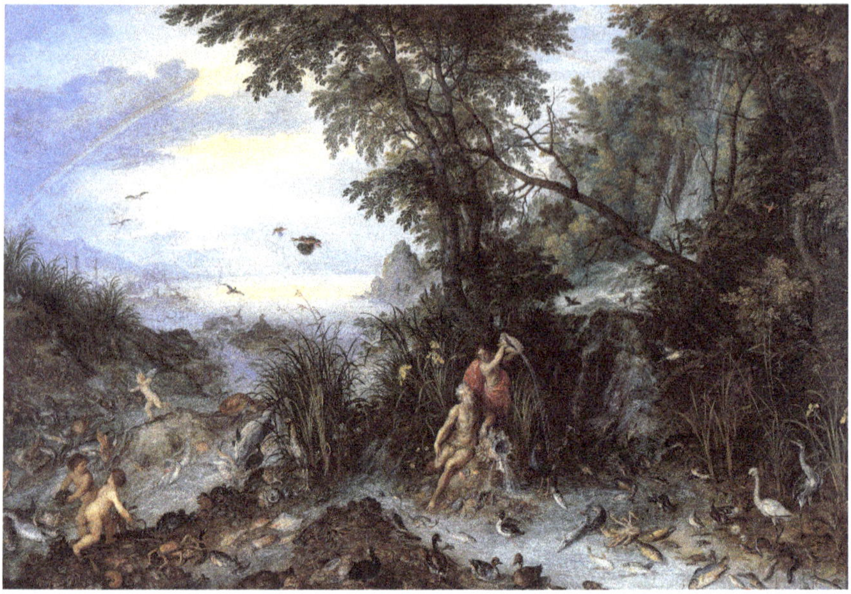

FIGURE 1.3 Jan Brueghel the Elder, *The Allegory of Water* (1614). Milan Ambrosiana

corners on points that Clusius has doubts about. For example, he shows that the crowned crane is not a 'pavo marinis' (sea peacock), as Clusius assumes, on the basis of a drawing sent to him by Jacques Plateau, one of his correspondents, and he answers the question of whether birds of paradise have or don't have legs by clearly depicting both birds with legs in the foreground. And, by pontifically depicting a large egg on the soil next to the two birds, he debunks the myth that the egg of the bird of paradise is said to be laid by the female in a hollow in the back of the male. In the painting *Allegory of Water* [Fig. 1.3], which is part of another, but thematically identical series of the four elements, Brueghel depicts a putto aiming with his bow and arrow at a monstrous sea creature [Fig. 1.4] that comes straight out of Pierre Belon's book of fishes [Fig. 1.5],[19] namely a mantis shrimp (which, by the way, with its actual length of only 20 cm, has a depiction that is much larger and more frightening than the animal actually is).

These examples, including the painting *Diana and the Nymphs*, can be seen as so many pleas for the power of painting, which appears not to be inferior to

19 Belon Pierre, "Cigale de mer", *La nature et diversité des poissons* (Paris, Charles Estienne: 1555) 353. See also Rondelet Guillaume, *L'histoire entiere des poissons. Premiere partie* (Lyon, Macé Bonhomme: 1558) 397.

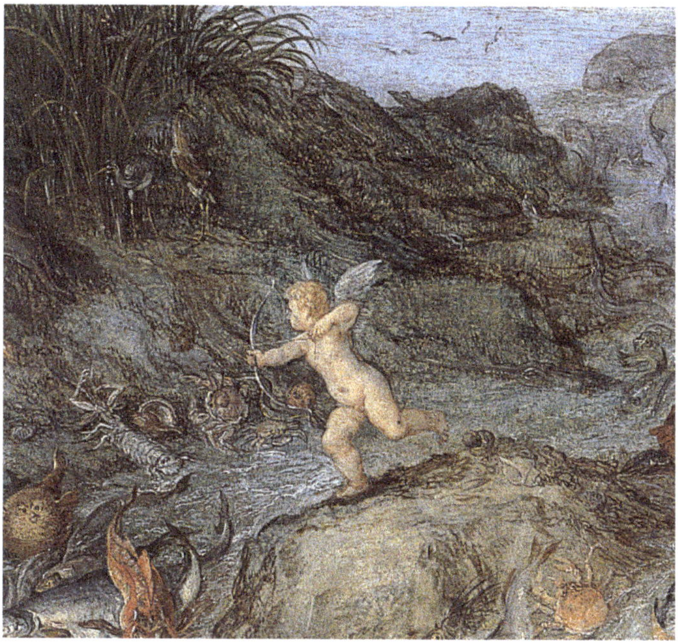

FIGURE 1.4 Putto and mantis shrimp (detail of Fig. 1.3)

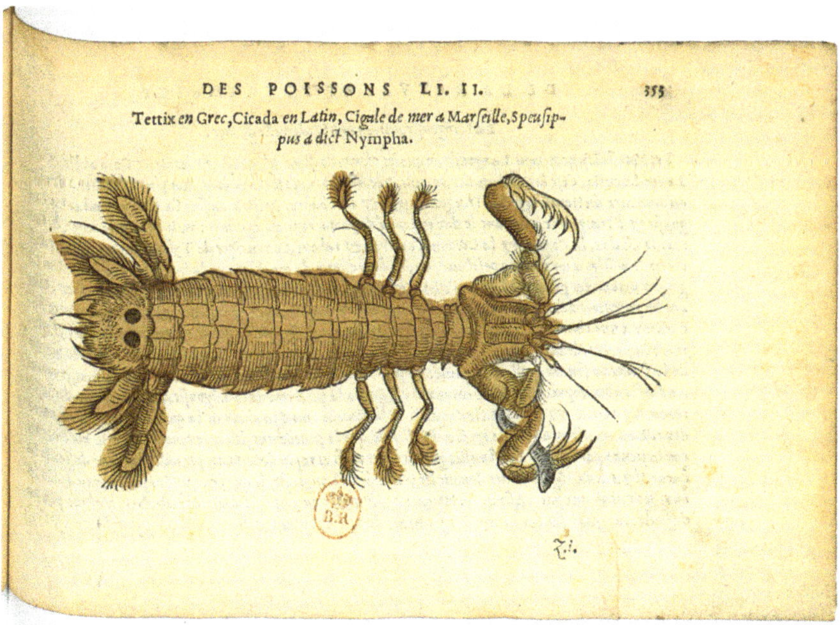

FIGURE 1.5 Belon Pierre, 'Cigale de mer', *La nature et diversité des poissons* (Paris, Charles Estienne: 1555) 353. Bibliothèque nationale de France

natural history writing. Brueghel seems to indicate that painting is capable of depicting nature more quickly and precisely than natural history can. Painting sees itself not as a slavish representation of natural history, but as a stimulating partner that advances knowledge about nature.

From this perspective, a few more striking aspects of the painting can be highlighted, which also play a role in contemporary ichthyology. It is striking, for example, that, in contrast to the aforementioned pictorial allegories on the elements air and water, which show a colourful mixture of animals from all corners of the world, there is a strong regionalism in *Diana and the Nymphs*: all animal species depicted – terrestrial animals, birds, and fish – come from a southern or central European region. It is also remarkable that all the fish depicted are freshwater fish (while the fish of *Allegory of Water* are both marine and freshwater). Here, too, a link can be made with 16th-century ichthyology, which was initially aimed at a specific region: Giovio and Salviani wrote their ichthyological works on fish in Rome and the surrounding area, Pierre Gilles focused upon the fish of Marseille and Provence, and the German historian Gregor Mangolt (1498–1578 (?)) wrote his *Fischbuch* (1557) on the freshwater fishes of Lake Constance (Switzerland). It was only later that attention turned to a more comprehensive and encyclopaedic study of fishes – of which the works of Belon, Rondelet, Gessner, and Aldrovandi are the best examples.

Finally, Brueghel's painting is also meta-pictorial. It indicates what the attention of the modern painter of his time is (or should be) focused on, namely the realistic depiction of nature, in this case of fish. In this, Brueghel joins four new subgenres, which would take off from the end of the 16th century: albums of watercolour drawings, such as the one by Joris Hoefnagel (1542–1600/1601), who still relies heavily on Gessner's illustrations, albums of fish prints, such as those produced by Adriaen Collaert (ca. 1560–1618), fish market scenes, made famous by Joachim Beuckelaer (1534–ca. 1574) among others, and, somewhat later, fish still lifes, traditionally connected to the name of Clara Peeters (fl. 1607–1621).

The realism of these paintings is so great that they provide a possible (and indirect) source of information about the occurrence of certain fish species on the fish market, and therefore in a certain region.[20] However, this realism is often problematic because it can be partial or feigned: in many cases it is demonstrable that the depicted scene could not have been depicted *ad vivum* – for instance when the painter demonstrably makes repeated use of

20 On painting as a tool for the study of historical biodiversity, see the article by Anne Overduin-de Vries and Paul Smith in the present volume.

the same model sketches – and this is the case of the carp in *Diana and the Nymphs*, which is depicted frequently in other paintings by Brueghel.

6 The *Fish Market* by Snyders and Van Dyck

With the fish paintings by Frans Snyders (1579–1657) we see a different elaboration of realism – in his paintings realism is not limited to local fish species but extends to foreign, even exotic species, in which he largely surpasses Brueghel's *Allegory of the Element Water*. From this perspective, let us take a look at *The Fish Market* (ca. 1621) by Frans Snyders and Anthony Van Dyck (1599–1641) [Fig. 1.6]. What is striking in this painting is the meticulousness with which the animals are depicted. The fish are at first glance depicted directly from nature ("naar 't leven", true to life). All species of the huge pile of fish on display can be identified: on the table we see a huge sturgeon, and all kinds of other North Sea fish, as well as a dead otter, and under the table we find a small harbour seal, and a tub with carp. And at the top left, hanging from above, are a whole salmon, a salmon steak, two halibuts, one with the typical binding head to tail, and two bunches of herring, one smoked and the other fresh. All this is very realistic at first glance. But on the round table in the foreground are many exotic shells that are not found in the North Sea and which do not belong in a regular fish market, at least not in Western Europe. There is also an animal depicted that is very special, namely a horseshoe crab. This animal was a collector's item in curiosity cabinets.[21] One can distinguish two specimens lying on their backs, one on top of the pile of fish, the other on the bottom of the pile [Fig. 1.7]. So, whereas the individual fish are realistically rendered (or suggest that they are), the larger whole in which they are depicted is problematic, to say the least.

Perhaps the horseshoe crab and the exotic shells are only intended to suggest a *couleur locale*, a scene from the Mediterranean region – but even then the combination of North Sea fauna and tropical animals remains problematic. It may also be that because of the unusual combination of the fish species, the painters want to indicate that something special is happening here, in the daily life of the fishermen. And for that we have to consider more closely the scene in the background. There is no certainty which event is depicted here, but it is probably a biblical scene, namely Matthew 19, in which Peter finds a coin in the mouth of a fish he caught, with which he can pay his taxes. Here we

21 Rijks M., "A Painter, a Collector, and a Horseshoe Crab: Connoisseurs of Art and Nature in Early Modern Antwerp", *Journal of the History of Collections* 31.2 (2019) 343–361.

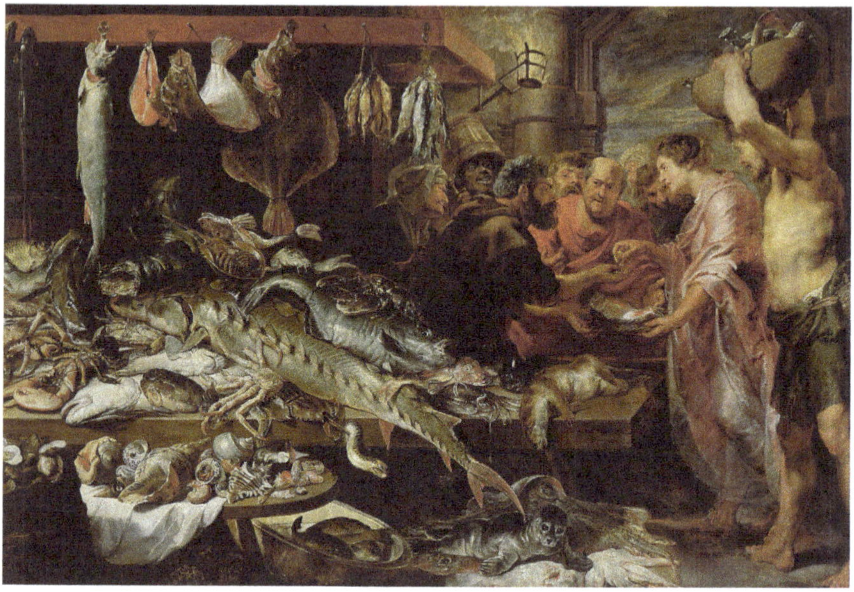

FIGURE 1.6 Frans Snyders and Anthonis van Dyck, *Fishmarket* (ca. 1621). Kunsthistorisches Museum Vienna. Wikimedia Commons

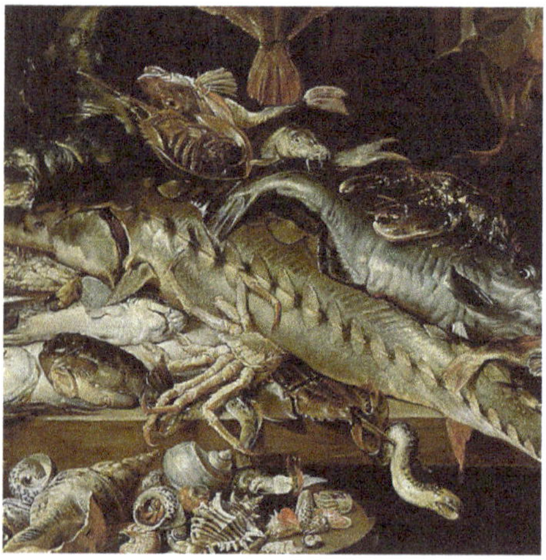

FIGURE 1.7
Two horseshoe crabs (detail of Fig. 1.6)

see the moment when Peter, traditionally poor and depicted with a wild beard and curly hair, hands the coin to the richly dressed tax collector, who looks very surprised.[22]

22 See Uchtmann D. – Haag S. (eds.), *The Pleasures of the Table in Art: Thirty-Eight Works from the Kunsthistorisches Museum in Vienna* (Vienna: 2011).

The paintings by Brueghel and Snyders/Van Dyck show how differently and also ambivalently the ichthyological knowledge can be used: to position the painter as a *pictor doctus*, capable of competing with the natural historians, or to illustrate his skill, able to depict in a laborious oil painting such a fleeting subject as a fish, in a lifelike manner. Ichthyological knowledge can also be used to depict a mythological or biblical scene, or, as with Beuckelaer and his followers, a realistically depicted fish market, which is at the same time an erotic scene containing the apparent sexual symbolism of the fishmonger holding a salmon steak to his middle finger.[23] In this type of painting it often remains unclear whether the eroticism depicted is intended merely for entertainment, or for edifying rejection of the depicted symbolism.[24]

7 The Frontispiece of Willughby and Ray's *Historia piscium*

The *Historia piscium* (1686), written by Francis Willughby and John Ray,[25] reflects the beginnings of a new area in the history of ichthyology. This work has, at the head of its illustrations section [Fig. 1.8], a frontispiece that was made by the Dutch painter and illustrator Paul van Somer (II) (1649–1714/1716). This frontispiece does what ideally every paratextual element should do:[26] It is designed to inform the potential reader of its contents, as well as to entice him to read it. Once it has pulled the reader across the threshold of opening the book, it can serve as a reader's guide, putting it, in the words of Gérard Genette, 'at the service of a better reception for the text and a more pertinent reading of it'.[27] Van Somer's engraving shows a number of fishermen coming ashore and displaying the freshly caught fish in a teeming heap. In the foreground some mythological sea gods are depicted for decoration, which are somewhat reminiscent of Rabelais' fish tapestry: we find Triton with his great conch and

23 See, for instance, the *Fish Market* (not dated) by a follower of Beuckelaer (Bonnefantenmuseum, Maastricht).

24 See for this multifold symbolism and ambivalence Helmus M.H. (ed.), *Vis. Stillevens van Hollandse en Vlaamse meester 1550–1700* (Utrecht: 2004) and Slechte H., *Vis in beeld* (Zwolle: 2019).

25 Willughby Francis – Ray John, *De historia piscium libri quatuor* (Oxford, Sheldonian Theatre: 1686).

26 The next two sentences are partially quoted from my "Title Prints and Paratexts in the Emblematic Fable Books of the Gheeraerts Filiation (1567–1617)", in Bossier P. – Scheffer R. (eds.), *Soglie testuali. Funzioni del paratesto nel secondo Cinquecento e oltre / Textual Thresholds: Function of Paratexts in the Late Sixteenth Century and Beyond* (Rome: 2010) 157–200 (here: 157).

27 Genette G., *Paratexts: Thresholds of Interpretation* (Cambridge: 1997) 1–2.

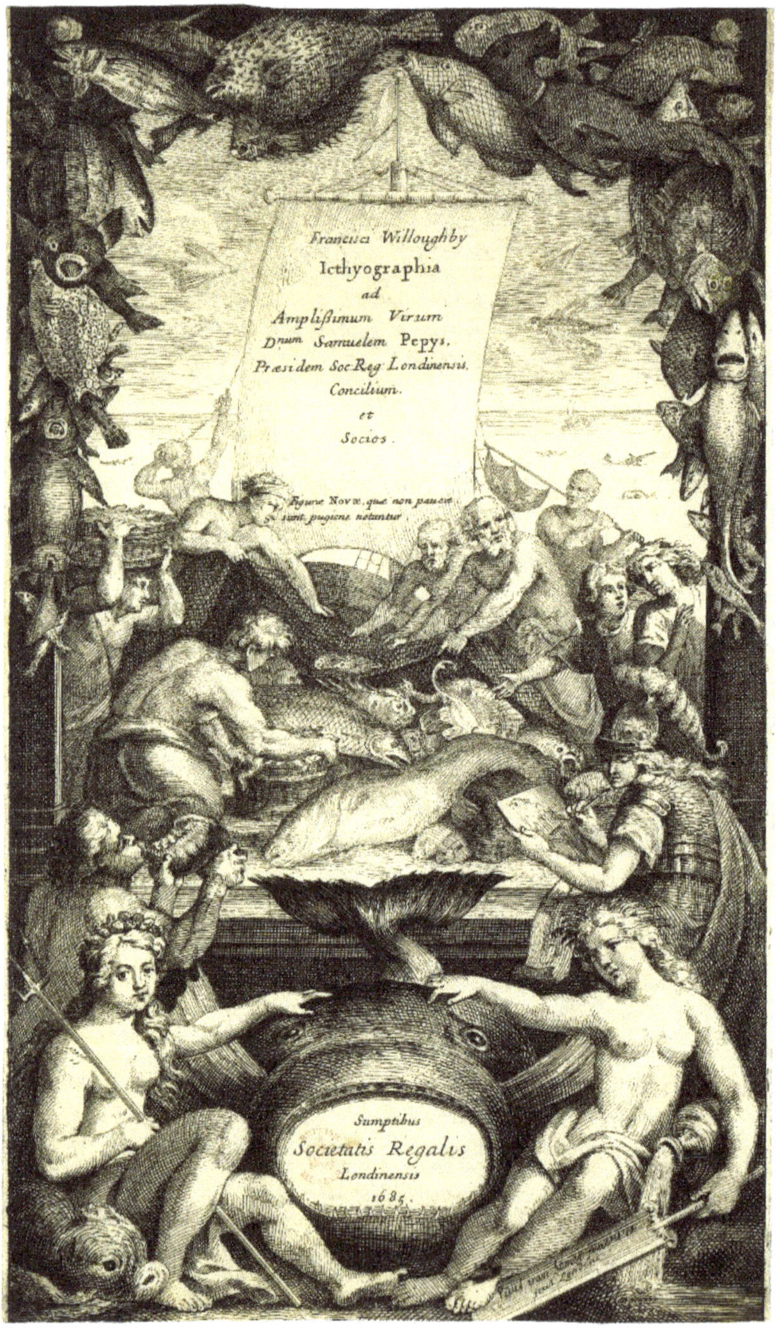

FIGURE 1.8 Paul van Somer (II), Frontispiece of Willughby Francis – Ray John, *De historia piscium libri quatuor* (Oxford, e theatro Sheldoniano: 1686)
COURTESY OF THE BRITISH MUSEUM, LONDON

the Nereids, as well as the mass of living and jumping fish of all species. And we note, just as in Rabelais' Land of Satin, the presence of people who observe. It is here too that the differences between Rabelais' fiction and Van Somer's engraving become apparent. Aristotle and his disciples have disappeared; they are, so to speak, substituted with the goddess Minerva and two observers, who are in modern dress and have individualized features. Just as Pierre Gilles was recognizable in the tapestry described by Rabelais, it is perhaps possible here to recognize the two authors of the book, Willughby and Ray, in their youth, whose portraits are known.[28]

Whoever these two observers may be, the fact is that they, along with Minerva, personify some essential aspects of the *Historia piscium*. These concern, first of all, the acquisition of information, which takes place from autopsy, direct observation, which may also include dissection where possible. The frontispiece seems to give practical advice to the reader wishing to learn about ichthyology: this direct observation, so essential to the knowledge of fish, is ideally done at fish markets, which provide the ichthyologist researcher with a daily supply of fish to study.[29] This good advice is put into practice by Willughby and Ray themselves. It is known that they found much of their information at fish markets. In their travel diaries they describe in detail their visits to the fish markets – those of Rome and Venice especially[30] – so it is perhaps not by chance that the fishermen represented look more Mediterranean than Nordic.

What is also important is that the acquisition of information should be from the common people, i.e. the fishermen themselves. This last aspect is essential for Ray and Willughby: for example, when Ray was travelling through the

28 On this hypothesis, see Smith P.J. – Trijp D. van, "Dynamiques européennes de l'humanisme érudit dans l'histoire naturelle. Le cas de l'ichtyologie, de Belon, Rondelet et Gessner à Willughby et Ray", in Crouzet D. – Crouzet-Pavan É. – Desan P. – Revest C. (eds.), *L'humanisme à l'épreuve de l'Europe (XVe–XVIe siècle). Histoire d'une transmutation culturelle* (Ceyzérieu: 2019) 167–181 (here: 177–179).

29 Willugby and Ray were not the first to encourage visits to fishmongers: Belon, Rondelet, Gessner, and Aldrovandi preceded them in this. See Trijp D. van, *Captured on Paper: Fish Books, Natural History and Questions of Demarcation in Eighteenth-Century Europe (ca. 1680–1820)* (PhD dissertation Leiden University: 2021) 80–81, and Egmond F., "On Northern Shores: Sixteenth-Century Observations of Fish and Seabirds (North Sea and North Atlantic)", in MacGregor A. (ed.), *Naturalists in the Field: Collecting, Recording and Preserving the Natural World from the Fifteenth to the Twenty-First Century* (Leiden – Boston: 2018) 129–148.

30 Greengrass M. – Hildyard D. – Preston C.D. – Smith P.J., "Science on the Move: Francis Willughby's Expeditions", in Birkhead T. (ed.), *Virtuoso by Nature: The Scientific Worlds of Francis Willughby FRS (1635–1672)* (Leiden – Boston: 2016) 142–226 (here: 183–184).

British Isles to compile a list of English words and a catalogue of local birds and fish, he consulted 'one of the ancientest and most experienced fishermen' in Cornwall.[31] This recourse to indigenous science, both within and outside Europe, is already noticeable in the 16th and early 17th centuries, for example with Gessner when it comes to the Swiss aquatic fauna, or, in the case of Brazilian fauna, with reporters such as Jean de Léry (*Histoire d'un voyage faict en la terre du Bresil* (1578)) and Georg Marcgraf (*Historia Naturalis Brasiliae* (1648)).[32]

The figure of Minerva symbolizes the importance of the visual arts in the context of science communication. The goddess is drawing, from direct observation, a fish which is draped in front of her. Images are an essential aspect of the *Historia piscium*: Although not all the images are unpublished (several having been copied from the works of Gessner, Rondelet, or Salviani), some of the engravings were certainly made especially for this book. These engravings are largely made according to drawings purchased by Willughby and Ray on their journey through Europe, as well as drawings sent to the Royal Society, or directly from specimens in the collection of the Royal Society.[33] The importance of the new illustrations is explicitly stated in the frontispiece: 'Figurae Novae, quae non paucae sunt, † notantur' (The new images, of which there are not a few, are marked with a cross).

The frontispiece shows yet another element relating to the images of the book: the garland of depicted fish adorning the sides and top of the frontispiece. These images are striking in their realism. A significant portion of these images correspond directly to the fish illustrations found in the book. Thus, from left to right, we note the following species or genera: an angular roughshark, a boxfish, a unicorn fish, two species of pufferfish, a hammerhead shark,

---

31 Ray John, *A Collection of English words, not generally used* […] (London, H. Bruges for Tho. Barell: 1674). 97. See Trijp D. van, "Fresh Fish: Observation up Close in Late Seventeenth-Century England", *Notes and Records. The Royal Society Journal of the History of Science* 75 (2021) 311–332 (here: 325).

32 See my articles "Léry et les poissons: une lecture rapprochée des stratégies descriptives", *Le Verger* 25 (December 2022). http://cornucopia16.com/blog/2023/01/06/bouquet-xxv-lhistoire-dun-voyage-faict-en-la-terre-du-bresil-de-jean-de-lery/ and "Marcgraf's Fish in the *Historia Naturalis Brasiliae* and the Rhetorics of Autoptic Testimony", in De Campos Françozo M. (ed.), *Toward an Intercultural Natural History of Brazil. The* Historia Naturalis Brasiliae *Reconsidered* (New York: 2023) 122–141.

33 See Kusukawa S., "*Historia piscium* (1686) and Its Sources", in Birkhead T. (ed.), *Virtuoso by Nature: The Scientific Worlds of Francis Willughby* FRS *(1635–1672)* (Leiden – Boston: 2016) 305–334.

a turbot, and a species of dogfish.[34] Some of these fish are exotic and belong, like the horseshoe crabs of Snyders and Van Dyck, in the *cabinets de curiosités* of the time. The frontispiece's garland of fish indeed resembles the pictorial representations of the well-known cabinets of Ole Worm or Ferrante Imperato. Van Somer seems to indicate how important collections are for the ichthyology of the 17th and 18th centuries. The frontispiece is also an expression of the growing interest in tropical fish. Willughby wanted to make another study trip to the New World. But due to his untimely death nothing came of that, and Ray had to make do with the descriptions and illustrations of Georg Marcgraf's *Historia Naturalis Brasiliae*, which were often integrated into the *Historia piscium*. This interest, already present in Willughby and Ray, would become characteristic of 18th- and 19th-century ichthyology.

8 The Present Volume

As mentioned, the contributions are divided into four more or less thematically related and chronological sections. In the section "Beginnings", a detailed overview is given of the ichthyology of the years 1520–1550. Holger Funk reviews the history of ichthyology from Aristotle through the 16th century from the point of view of true-to-life realism, zooming in on a controversy between Rondelet and Salviani. The development of ichthyology in the first half of the 16th century, outlined above, is given a new and more precise periodization by Philippe Glardon, with an emphasis on the work of Paolo Giovio. Onomastics and etymology play an essential role in this development, as Tobias Bulang shows in Gessner's work and in the fictional work of the German Rabelais translator Johann Fischart. From a lexicographical perspective, Bernardo Jerosch Herold and João Paulo S. Cabral demonstrate the same in the ichthyological notes of Leonhardt Thurneysser zum Thurn, a German traveller in Portugal.

In the section "Depicting", Florike Egmond provides an extensive overview and a cultural-historical contextualization of all European ichthyological images that had not appeared in print. Cynthia M. Pyle studies the role of 16th-century fish images drawn in the margins of a medieval manuscript. Marlise Rijks gives an overview of the above-mentioned pictorial subgenres of fish market scenes, fish still lifes, and some others which at the time of Jan

34 In the *Historia piscium*, the angular roughshark is represented in table B3 ('Centrina supina'), the boxfish on I17 ('Piscis triangularis ex toto maculosus'), the unicorn fish on O4 ('Monoceros piscis'), the pufferfish on I5 and I16, the hammerhead shark on B1 ('Zygaena Salviani'), the turbot on F2 ('Rhombus maximus') and the dogfish on B6 ('Canis galeus').

Brueghel had fish as their subject in the Low Countries. Anne M. Overduin-de Vries and Paul J. Smith report on an experimental study of fish motifs in early modern Netherlandish paintings, using citizen science (crowd sourcing).

The section "Fish and Society in Europe" draws knowledge about fish into a broader cultural-historical and social context: literature, fisheries, fish consumption, and medicine. In this context, Dirk Geirnaert analyses some 16th-century Bruges poems of praise for the fisherman's profession. The use of fish in medical prescriptions, as found in late medieval German-language pharmacopoeias, is the subject of Sabina Tsapaeva's contribution. Pietro Daniel Omodeo analyses how Venetian city administrators used the knowledge of local fishermen to regulate their city's water management. Cristina Brito provides an interpretive overview of reports of whale landings in early modern Portugal. In the three other articles of this section, the step is taken to the 18th and 19th centuries. Ronny Spaans investigates how the knowledge of fish and fishing is used as an argument in a nationalist epic from 18th-century Norway. From the perspective of historical ecology, Rob Lenders analyses and demystifies the Western European myth of the maid and the cheap salmon against the background of changing aquatic biodiversity and the concept of the 'shifting baseline syndrome'. Finally, Paul Smith conducts newspaper-based research into the changing public opinion regarding seals (commercial hunting object, fisherman's competitor, iconic cuddly animal of the nascent animal protection movement ...) in the Netherlands in the 18th and 19th centuries.

With the articles of Spaans, Lenders, and Smith we have entered a new period, which is ushered in by Willughby and Ray's *Historia piscium*. As mentioned above, this period is characterized by an increasing diversification in method and description. This is particularly evident when it comes to the description of unknown fish species, which forms the main topic of the last section of our volume, "Ichthyological Knowledge from Afar". This section emphasizes the importance of collections and knowledge of newly discovered fish species – knowledge for which people are increasingly dependent on indigenous knowledge from distant lands. Incidentally, this knowledge transfer is bilateral because the knowledge does not only flow from the Far East and the Far West to Europe, but also vice versa, from Europe to the East and West. For example, Melinda Susanto focuses on an object, the nautilus, one of the *naturalia* that is integrated from the East into the culture of the West, where it is "artificialized", made into an object of art. The ichthyological descriptions and drawings of the unknown account (1698) of a journey made by François de Meyer from La Rochelle (France) to Guadeloupe are presented and analysed by Paul Smith, Didi van Trijp, and Alan Moss. Dorothee Fisher focuses on a single fish, namely the pufferfish, and its integration into German collections and ichthyological work. Theodore W. Pietsch and Justin R. Hanisch provide a synoptic overview of the research into Louis Renard's famous book on tropical fish (1719), supplemented with recent discoveries on the book. Johannes Müller examines some descriptions of fish from India in the authoritative work of Marcus Elieser Bloch – descriptions that could only be created by systematically drawing on indigenous knowledge. Ching-Ling Wang examines how illustrated ichthyological information from the West is incorporated into the ichthyology of 18th- and 19th-century China. *Mutatis mutandis*, Martien van Oijen does the same for early modern Japan, focusing more specifically on the Dutch contributions to Japanese ichthyology. Doreen Mueller addresses early modern Japanese whale knowledge, which 'was a vibrant field of intersecting local, cross-regional, Sino-Japanese, and Western epistemic practices.' Finally, Robbert Striekwold looks at the debates about scientific fish illustrations in the 19th century, zooming in on the ichthyological work of Hermann Schlegel.

9 Perspectives

In response to the research questions outlined in the NWO project description quoted in the beginning of this introduction, the present collection shows that the "longue durée" perspective is rewarding for mapping the developments of ichthyology towards an independent discipline within zoology. It shows that developments are never abrupt and absolutely innovative, because every innovation is rooted in a tradition, rebels against it, and takes over elements of it, in a modified form or not. Innovations can never be explained exclusively from a contemporary, teleological perspective, which falsely explains the development of a discipline in terms of linear progress – a progress that would inevitably and without detours lead to the contemporary scientific knowledge of nature.[35] On the contrary: the course of development of the natural sciences does not follow a straight path, but proceeds via side roads, detours, and dead ends, and in this respect is no different from the evolution in nature itself.

The contributions to this volume also show that developments do not take place in a vacuum but are always products of non-natural history factors, such as the biographical circumstances of the individual researcher trying to

35 See Enenkel K.A.E. – Smith P.J., "Introduction", in idem (eds.), *Early Modern Zoology: The Construction of Animals in Science, Literature and the Visual Arts* (Leiden – Boston: 2007) 1–12.

position himself as a scientific *persona*,[36] or the more general cultural-historical contexts from which the developments arise. These contexts can be religious, political, economic, or art-aesthetic. And as said before, these contexts can themselves be influenced by knowledge about fish – this is especially true, as various contributions show, for the visual arts and literature.

As is clear, most of the contributions are case studies, and if not, they offer overview studies on specific areas of ichthyology. Both of the editors of this volume express the hope that the contributions, however different in subject matter and methodology they may be, will be extrapolated to wider research into the subjects of current scholarly attention, both within environmental humanities and in other fields, such as historical ecology, with its focus on water and the ocean, and the relationship between biodiversity and local knowledge, both within and outside Europe. We hope that the interdisciplinary and intradisciplinary design of this volume will ultimately result in a new cultural history of ichthyology.

Bibliography

Belon Pierre, *La nature et diversité des poissons* (Paris, Charles Estienne: 1555).

Daston L. – Sibum O., "Introduction: Scientific Personae and Their Histories", *Science in Context* 16.1–2 (2003) 1–8.

Egmond F., "On Northern Shores: Sixteenth-Century Observations of Fish and Seabirds (North Sea and North Atlantic)", in MacGregor A. (ed.), *Naturalists in the Field: Collecting, Recording and Preserving the Natural World from the Fifteenth to the Twenty-First Century* (Leiden – Boston: 2018) 129–148.

Enenkel K.A.E. – Smith P.J., "Introduction", in idem (eds.), *Early Modern Zoology: The Construction of Animals in Science, Literature and the Visual Arts* (Leiden – Boston: 2007) 1–12.

Genette G., *Paratexts: Thresholds of Interpretation* (Cambridge: 1997).

Gessner Conrad, *Historiae animalium liber IIII. qui est de piscium et aquatilium animantium natura* (Zurich, Christoph Froschauer: 1558).

Gessner Conrad, *Fischbuoch*, transl. Conrad Forrer (Zurich, Christoph Froschauer: 1563).

36 See Daston L. – Sibum O., "Introduction: Scientific Personae and Their Histories", *Science in Context* 16.1–2 (2003) 1–8, and Paul H., "Introduction: Scholarly Personae: What They Are and Why They Matter", in Paul H. (ed.), *How to Be a Historian: Scholarly Personae in Historical Studies* (Manchester: 2019) 1–14.

Glardon P., *L'histoire naturelle au XVIe siècle. Introduction, étude et édition critique de* La nature et diversité des poissons *de Pierre Belon (1555)* (Geneva: 2011).

Greengrass M. – Hildyard D. – Preston C.D. – Smith P.J., "Science on the Move: Francis Willughby's Expeditions", in Birkhead T. (ed.), *Virtuoso by Nature: The Scientific Worlds of Francis Willughby FRS (1635–1672)* (Leiden – Boston: 2016) 142–226.

Helmus M.H. (ed.), *Vis. Stillevens van Hollandse en Vlaamse meester 1550–1700* (Utrecht: 2004).

Kusukawa S., "*Historia piscium* (1686) and Its Sources", in Birkhead T. (ed.), *Virtuoso by Nature: The Scientific Worlds of Francis Willughby FRS (1635–1672)* (Leiden – Boston: 2016) 305–334.

Leoi A.M., *The Lagoon: How Aristotle Invented Science* (New York: 2014).

Paul H., "Introduction: Scholarly Personae: What They Are and Why They Matter", Paul H. (ed.), *How to Be a Historian: Scholarly Personae in Historical Studies* (Manchester: 2019) 1–14.

Rabelais François, *The Complete Works*, transl. D.M. Frame (Berkeley – Los Angeles: 1991).

Rabelais François, *Œuvres complètes*, ed. M. Huchon (Paris: 1994).

Ray John, *A Collection of English words, not generally used* [...] (London, H. Bruges for Tho. Barell: 1674).

Rikken M., *Dieren verbeeld. Diervoorstellingen in tekeningen, prenten en schilderijen door kunstenaars uit de Zuidelijke Nederlanden tussen 1550 en 1630* (PhD dissertation Leiden University: 2016).

Rikken M. – Smith P.J., "Jan Brueghel's *Allegory of air* (1621) from a Natural Historical Perspective", *Netherlands Yearbook for History of Art* 61 (2011) 86–116.

Rondelet Guillaume, *L'histoire entiere des poissons. Premiere partie* (Lyon, Macé Bonhomme: 1558).

Slechte H., *Vis in beeld* (Zwolle: 2019).

Smith P.J., "Title Prints and Paratexts in the Emblematic Fable Books of the Gheeraerts Filiation (1567–1617)", in Bossier P. – Scheffer R. (eds.), *Soglie testuali. Funzioni del paratesto nel secondo Cinquecento e oltre / Textual Thresholds: Function of Paratexts in the Late Sixteenth Century and Beyond* (Rome: 2010) 157–200.

Smith P.J., "Rabelais ichtyologue", in Garnier I. et al. (eds.), *Narrations fabuleuses. Mélanges en l'honneur de Mireille Huchon* (Paris: 2022) 439–452.

Smith P.J., "Léry et les poissons: une lecture rapprochée des stratégies descriptives", *Le Verger* 25 (December 2022). http://cornucopia16.com/blog/2023/01/06/bouquet-xxv-lhistoire-dun-voyage-faict-en-la-terre-du-bresil-de-jean-de-lery/.

Smith P.J., "Marcgraf's Fish in the *Historia Naturalis Brasiliae* and the Rhetorics of Autoptic Testimony", in De Campos Francozo M. (ed.), *Toward an Intercultural Natural History of Brazil. The* Historia Naturalis Brasiliae *Reconsidered* (New York: 2023) 122–141.

Smith P.J. – Trijp D. van, "Dynamiques européennes de l'humanisme érudit dans l'histoire naturelle. Le cas de l'ichtyologie, de Belon, Rondelet et Gessner à Willughby et Ray", in Crouzet D. – Crouzet-Pavan É. – Desan P. – Revest C. (eds.), *L'humanisme à l'épreuve de l'Europe (XVe–XVIe siècle). Histoire d'une transmutation culturelle* (Ceyzérieu: 2019) 167–181.

Trijp D. van, "Fresh Fish: Observation up Close in Late Seventeenth-Century England", *Notes and Records. The Royal Society Journal of the History of Science* 75 (2021) 311–332.

Trijp, D. van, *Captured on Paper. Fish Books, Natural History and Questions of Demarcation in Eighteenth-Century Europe (ca. 1680–1820)* (PhD dissertation Leiden University: 2021).

Uchtmann D. – Haag S. (eds.), *The Pleasures of the Table in Art: Thirty-Eight Works from the Kunsthistorisches Museum in Vienna* (Vienna: 2011).

Willughby Francis – Ray John, *De historia piscium libri quatuor* (Oxford, Shaldonian Theatre: 1686).

Zucker A., "Fonctions des classes dans les traités ichtyologiques de P. Belon et G. Rondelet: empreinte ou alibi antique?" in Gontier T. (ed.), *Animal et animalité dans la philosophie de la Renaissance et de l'Âge classique* (Louvain – Paris: 2005) 7–32.

PART 1

Beginnings

CHAPTER 2

Fish Images True to Life and a 16th-Century Controversy between Rondelet and Salviani. Essay and Documentation of the Sources

Holger Funk

1 The Aristotelian Legacy and the First Printed Fish Books

Historians of natural science agree that the origins of biology, or the scientific discipline that deals with living beings, was essentially shaped by Aristotle (384–322 BC).[1] This applies especially to animal biology – generally referred to as zoology. Aristotle combined the speculative approach followed by most of the early Greek philosophers, whose thinking had focused on nature, with the empirical methods that were already used by some of them, developing further, and applying on a large scale, what he had adopted from his philosophical forerunners.

As the Roman scholar Pliny the Elder (23/24–79 AD) later described it, Aristotle had at his disposal 'some thousands of men in every region of Asia and Greece, comprising all those who followed the business of hunting, fowling, or fishing, or who had the care of parks, herds of cattle, the breeding of bees, fishponds, and aviaries'.[2] Like other parts of his works, Aristotle's biology and zoology never became really obsolete. His work was the beginning of a long tradition and was available to scholars in other regions of the world, even if the scope of what was still known of his writings and the intensity with which they were studied varied over time.[3]

1 See Bäumer Ä., *Geschichte der Biologie*. Vol. 1: *Biologie in der Antike bis zur Renaissance* (Frankfurt: 1991) 32–89 or Mayr E., *The growth of biological thought. Diversity, evolution, and inheritance* (Cambridge, MA – London: 1982) 87–94.

2 Pliny, *Naturalis historia* 8.17.44.

3 For an overview of how the knowledge and assessment of Aristotle's zoological writings developed, see, *inter alia*, Thompson D'Arcy W., *On Aristotle as a biologist with a prooemion on Herbert Spencer* (Oxford:1913); Lang H.S., "Aristotle and Darwin", *International Philosophical Quarterly* 23 (1983) 141–153; Gotthelf A., "Darwin on Aristotle", *Journal of the History of Biology* 32 (1999) 3–30; Gaukroger S., *The emergence of a scientific culture. Science and the shaping of modernity 1210–1685* (Oxford: 2006), chap. 2 and 3; Zimmermann H., "Die Bedeutung des Aristoteles vom Mittelalter bis zur Renaissance", *Forum Classicum* 53 (2010) 211–218; Leroi A.M., *The lagoon. How Aristotle invented science* (London: 2014) 353–358, 466.

© HOLGER FUNK, 2024 | DOI:10.1163/9789004681187_003

During the 13th century, the study of Aristotle was forbidden in Western Christianity for some years. However, the same century brought about the Latin translation of Aristotle's zoological works by Michael Scotus († 1235), translated from Arabic versions of these works (completed before 1220);[4] by William of Moerbeke († 1286), translated directly from the Greek (completed before 1260).[5] Shortly thereafter, Albertus Magnus († 1280) published a translation under the collective title *De animalibus*, which was printed for the first time in 1478. It was based on what Michael Scotus had provided and soon this version became authoritative within Western universities.[6]

Another important landmark was reached when George of Trebizond († 1472 or 1486)[7] completed his Latin translation of several of Aristotle's zoological works, again under the collective title *De animalibus* in 1450.[8] Even

---

Between 1495 and 1498 a five-volume edition of the works of Aristotle in the original Greek (*editio princeps*) was published by Aldus Manutius in Venice, see Sicherl M., *Handschriftliche Vorlagen der* Editio princeps *des Aristoteles* (Mainz: 1976); Grendler P.F., *The Universities of the Italian Renaissance* (Baltimore – London: 2002) 272, 274. The zoological writings are contained in volume 3 of 1497, see Anonymous, *Gesamtkatalog der Wiegendrucke*. Herausgegeben von der Kommission für den Gesamtkatalog der Wiegendrucke. Vol. 2 (Stuttgart – New York: 1968) 551–560 for details on the contents.

4 On Michael Scotus and his Aristotle translation, see Van Oppenraay A.M.I., "The Reception of Aristotle's History of Animals in the Marginalia of Some Latin Manuscripts of Michael Scot's Arabic-Latin Translation", *Early Science and Medicine* 8 (2003) 387–403 and Berger F., *Die Textgeschichte der* Historia animalium *des Aristoteles* (Wiesbaden: 2005) 52–54. Scotus applied a form of literal translation, as opposed to a so-called "analogous" translation aiming at the sense of a word or name that later became customary (Gaza, see below).

5 On Moerbeke's way of translation, see Berger, *Textgeschichte* 182, Beullens P., "Quelques observations sur la traduction de l'*Histoire des animaux* d'Aristote par Guillaume de Moerbeke", *Bulletin de Philosophie Médiévale* 34 (1992) 181–196 and idem, "Aristotle, his translators, and the formation of ichthyologic nomenclature", in: Goyens M. – De Leemans P. – Smets A. (eds.), *Science translated. Latin and vernacular translations of scientific treatises in medieval Europe* (Louvain: 2008) 105–122; Singer C., *Greek biology and Greek medicine* (Oxford: 1922) 155, 166, 169 manages to judge Moerbeke both critically and positively. Moerbeke has translated all zoological writings as well as several other works by Aristotle (listed by Schmitt C.B., *Aristotle and the Renaissance* (Cambridge: 1983) 166).

6 See Berger, *Textgeschichte*, 53 and the introduction and translation in the Albertus Magnus, *On Animals. A Medieval Summa Zoologica*, eds. K.F. Kitchell – I.M. Resnick (2 vols.) (Baltimore: 1999); finally Bäumer, *Biologie in der Antike* 136–159. Singer, *Greek biology* 73 noted that Albert, even though he knew Aristotle only in the Latin translation of Michael Scotus, productively continued his Greek protagonist's zoological studies, for example in his description of fish embryos.

7 On Trebizond, see Harlfinger D. (ed.), *Graecogermania. Griechischstudien deutscher Humanisten. Die Editionstätigkeit der Griechen in der italienischen Renaissance (1469–1523)* (Weinheim – New York: 1989) 10–13.

8 This translation remained unprinted and was of little importance and influence (Beullens P. – Gotthelf A., "Theodore Gaza's translation of Aristotle's *De Animalibus*: content, influence, and date", *Greek, Roman and Byzantine studies* 47 (2007) 469–513).

more influential was Theodore Gaza's († 1475)[9] Latin translation under the same title.[10] Both authors had translated directly from the Greek, but only Gaza's translation was printed. A reliable printed version of Aristotle's zoological works was now easily available for the first time in Latin, the language of Western scholars. During the next century, ichthyologists as Rondelet and Salviani used Gaza's translation in the Venice 1476 and Basel 1520 editions. A number of explanatory commentaries were also published at that time.

Aristotle's contributions to ichthyology, in particular, was the beginnings of a long tradition. Already Theophrastus, Aristotle's disciple, partner and successor as head of the Lyceum in Athens (better known for his botanical studies), wrote a book on fishes.[11]Given these favourable circumstances, it was somewhat upsetting for recent scholars that the renewed attention that was given to the zoological writings of Aristotle since the late fifteenth century appeared 'strangely uninterested'[12] in that part on which the Greek scholar had focused a considerable measure of his research as a naturalist, namely on the study of fishes and the marine fauna of the Aegean and the Black Sea.[13]

Aristotle mentions more than 100 different fish species by name, approximately one-fifth of all animals described by him.[14] But despite this rich ancient

9 On Gaza, see Harlfinger, *Graecogermania* 14–18.

10 For more on Gaza's way of translation in contrast to that of George of Trebizond, see Beullens, *Aristotle translators*.

11 The surviving fragments are documented by Sharples R.W. (ed.), *Theophrastus of Eresus, Commentary Volume 5: Sources on biology* (Leiden: 1995) 84–89. The second famous pupil who explored fishes, was Alexander the Great (Beullens, *Aristotle translators*). Lennox J.G., "The Disappearance of Aristotle's Biology: A Hellenistic Mystery", *Apeiron* 27 (1994) 7–24 has argued that in the Hellenistic world Aristotle's zoology, understood as a research project, had disappeared until Albertus Magnus's revitalisation, but regarding the study of fishes Polek N., "Die Fischkunde des Aristoteles und ihre Nachwirkung in der Literatur", in Hilberg I. – Jüthner J. (eds.), *Primitiae czernovicienses. Festgabe zur 50. Versammlung deutscher Philologen und Schulmänner in Graz* (Czernowitz: 1909) 35–57 conveys a different, less pessimistic view.

12 Nissen C., *Schöne Fischbücher. Kurze Geschichte der ichthyologischen Illustration. Bibliographie fischkundlicher Abbildungswerke* (Stuttgart:1951) 11; see also Hünemörder C., "Die Geschichte der Fischbücher von Aristoteles bis zum Ende des 17. Jahrhunderts", *Deutsches Schifffahrtsarchiv* 1 (1975) 185–200, here 193.

13 See Kullmann W., *Aristoteles als Naturwissenschaftler* (Berlin: 2014) 81–112 with a discussion of the (controversial) research literature on this topic.

14 The information on the number of fishes in Aristotle is strangely different. Beullens, *Aristotle translators* gives 100 as number, while Hünemörder, *Fischbücher* 186 once speaks of 'about 100' fish names, later (Hünemörder C., "Fische", *Der Neue Pauly* 4 (1998) 526) of 'about 133'; Polek, *Fischkunde*, in turn, speaks indefinitely of 'numerous passages' in the work of Aristotle. More recently, Leroi, *Lagoon* 391–392 presented a list of 43 Aristotelian fish names and their modern Linnaean equivalents, some of them only vaguely identified, whereas Ganias K. et al., "Aristotle as an ichthyologist: Exploring Aegean fish diversity

heritage, which included a series of other authors and even reasonably identifiable images of fishes and other marine creatures,[15] among the incunabula (books printed before 1500), only the *Hortus sanitatis* of 1491 had an illustrated section dedicated to fishes, which are, barely identifiable if at all [Fig. 2.1].[16] Significantly, in the printed edition of Albertus Magnus' *De animalibus* of 1478, a section on fishes was added but only much later in a German adaptation.[17]

Beside these encyclopaedic works with special chapters on fishes, there were specialist, practice-oriented books that focused on the work of professional fishermen and the required technical equipment (nets, fish traps, weirs etc.) rather than on specific fishes themselves. Their authors are unknown today, but they seem to have been successful because their books were imitated and translated, for example, the Dutch (Flemish) *Boecxken* ('Booklet') of 1506, which was published in German as *Büchlein* in 1511 [Fig. 2.2].[18]

2,400 years ago", *Fish and Fisheries* 18 (2017) 1–18 has a list of about 110 names (including marine mammals), several of them likewise only tentatively identified. Hünemörder, *Fischbücher* 186 also points out that Aristotle distinguished between fishes in a narrow sense (ἰχθύς) and other animals living in water (τὰ ἔνυδρα sc. ζῷα) and did not collapse them like his successors, especially at Renaissance times under the collective term *aquatilia*. Finally Thompson, *Glossary* provided a massive compendium of Aristotelian fishes in the broader sense including sponges, coelenterates (sea-anemones, corals, etc.), echinoderms (starfishes), mollusca (octopuses, cuttlefishes, shell-fishes, slugs, etc.) crustaceans and marine mammals (whales and dolphins); see also Strömberg R., *Studien zur Etymologie und Bildung der griechischen Fischnamen* (Göteborg: 1943) 19, who claims to know 800 ancient Greek fish names, which is a doubtfully high number.

15 Reproduced, for example, by Imhoof-Blumer F. and Keller O., *Tier- und Pflanzenbilder auf Münzen und Gemmen des Klassischen Altertums* (Leipzig: 1889) plate VIII, no. 23; Keller O., *Die antike Tierwelt*. Vol. 2 (Leipzig: 1913) figure 124 related to p. 393; 511; Singer, *Greek biology* 8; Sahrhage D. – Lundbeck J., *A history of fishing* (Berlin – Heidelberg – New York: 1992) 43–55, with images ranging from Minoan (Aegean Bronze Age, 1,500–1,400 BC) to Etruscan times (100 BC). It should be noted that most of this pictorial material did not become common knowledge until much later.

16 The anonymous *Hortus sanitatis* (or *Ortus sanitatis*), printed in 1491 in Mainz by Jacob Meydenbach, was previously often mistakenly attributed to Johann Wonnecke von Kaub or Johannes de Cuba (1430–1503), a German physician from the town of Kaub located on the right bank of the Rhine; fishes were dealt with in book 5. See also Jacquemard C. – Gauvin B. – Lucas-Avenel M.-A. (eds.), *Hortus sanitatis. Livre IV. Les poissons* (Caen: 2013); additionally Baumann B. – Baumann H., *Die Mainzer Kräuterbuch-Inkunabeln "Herbarius Moguntinus" (1484), "Gart der Gesundheit" (1485), "Hortus sanitatis" (1491)* (Stuttgart: 2010).

17 See the *Thierbuch Alberti Magni* (Frankfurt: 1545) book 3 "Von den Fischen".

18 On the different versions of the *Boecxken* or *Büchlein*, see Cockx-Indestege Elly, "Van een boekje om vogels en vissen te vangen naar een zeldzame Antwerpse postincunabel, nu in de Library of Congress te Washington", in: Van der Vekene E. (ed.), *Refugium animae bibliotheca. Festschrift für Albert Kolb* (Wiesbaden:1969) 109–138; Grimm H., "Neue Beiträge zur Fisch-Literatur des XV. bis XVII. Jahrhunderts und über deren Drucker und Buchführer". *Archiv für Geschichte des Buchwesens* 67 (1968) 2871–2887 and Zaunick R.,

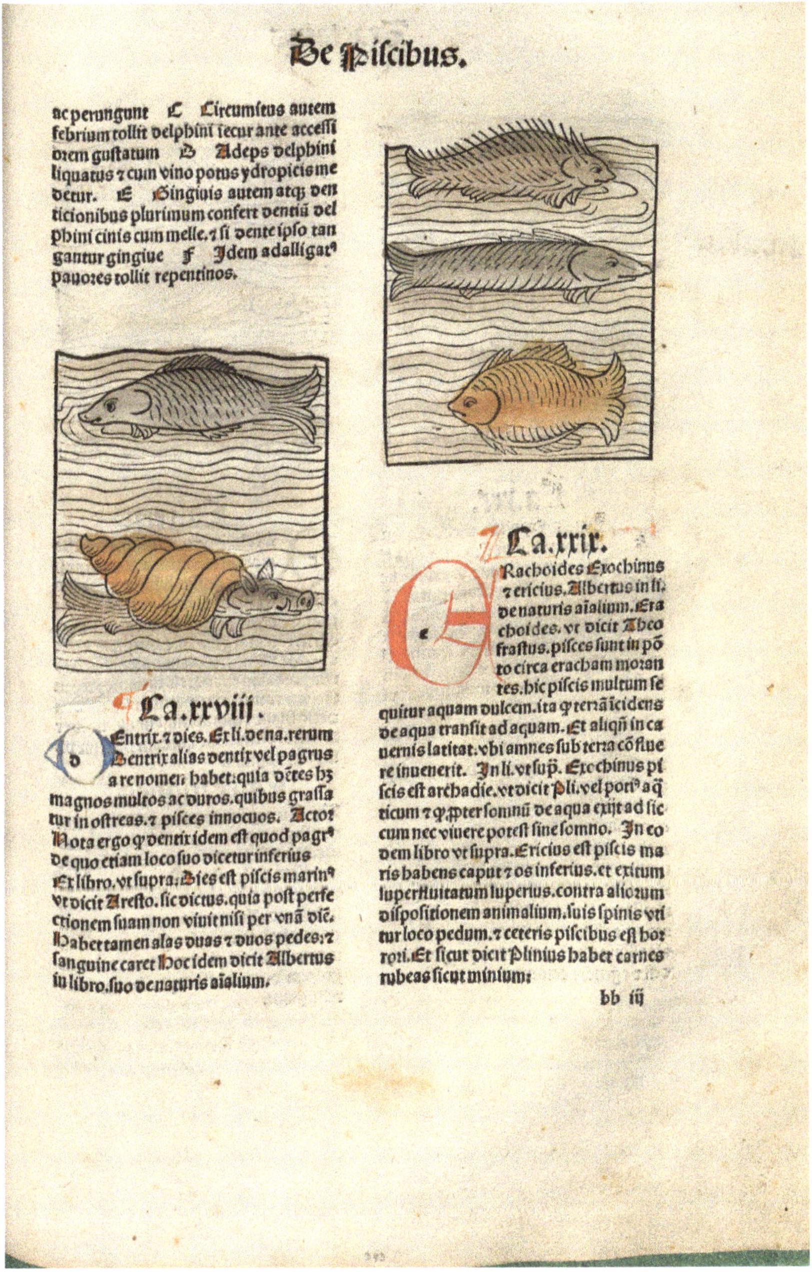

De piscibus.

ac perungunt C Circumfeus autem febrium tollit delphini iecur ante accessi onem gustatum D Adeps delphini liquatus ⁊ cum vino potus ydropicis me detur. E Gingiuis autem atq; den ticionibus plurimum confert dentiū del phini cinis cum melle. ⁊ si dente ipso tan gantur gingiue F Idem adalligatꝰ pauores tollit repentinos.

Ca.xxviii.

Dentrix. ⁊ dies. Ex li. de na. rerum Dentrix alias dentix vel pagrus a re nomen habet: quia dētes hʒ magnos multos ac duros. quibus grassa tur in ostreas. ⁊ pisces innocuos. Actor Nota ergo q̄ dentix idem est quod pagrꝰ de quo etiam loco suo dicetur inferius Ex libro. vt supra: Dies est piscis marinꝰ vt dicit Aresto. sic dictus. quia post perfe ctionem suam non viuit nisi per vnā diē. Habet tamen alas duas ⁊ duos pedes: ⁊ sanguine caret Hoc idem dicit Albertus in libro. suo de naturis aīalium.

Ca.xxix.

Eracoides Exochinus ⁊ ericius. Albertus in li. de naturis aīalium. Era choides. vt dicit Theo frastus. pisces sunt in pō to circa eracham moran tes. hic piscis multum se quitur aquam dulcem. ita q̄ terrā icidens de aqua transit ad aquam. Et aliqn̄ in ca uernis latitat. vbi amnes sub terra cō flue re inuenerit. In li. vt sup̄. Exochinus pi scis est archadie. vt dicit Pli. vel porꝰ aq̄ ticum ⁊ q̄ ꝓpter somnū de aqua exijt ad sic cum nec viuere potest sine somno. In eo dem libro vt supra. Ericius est piscis ma ris habens caput ⁊ os inferius. et exitum superfluitatum superius. contra aliorum dispositionem animalium. suis spinis vti tur loco pedum. ⁊ ceteris piscibus est hor rori. Et sicut dicit Plinius habet carnes rubeas sicut minium:

bb iij

FIGURE 2.1 Chapter 29 from the anonymous *Hortus sanitatis* (Mainz, Jacob Meydenbach: 1491), tract "De piscibus", fol. bbiij, showing, in the right column, *ericius*, the sea urchin (top), *erachoides*, the European weather loach (*Misgurnus fossilis* L.), and *exochinus* (unidentifiable, bottom)

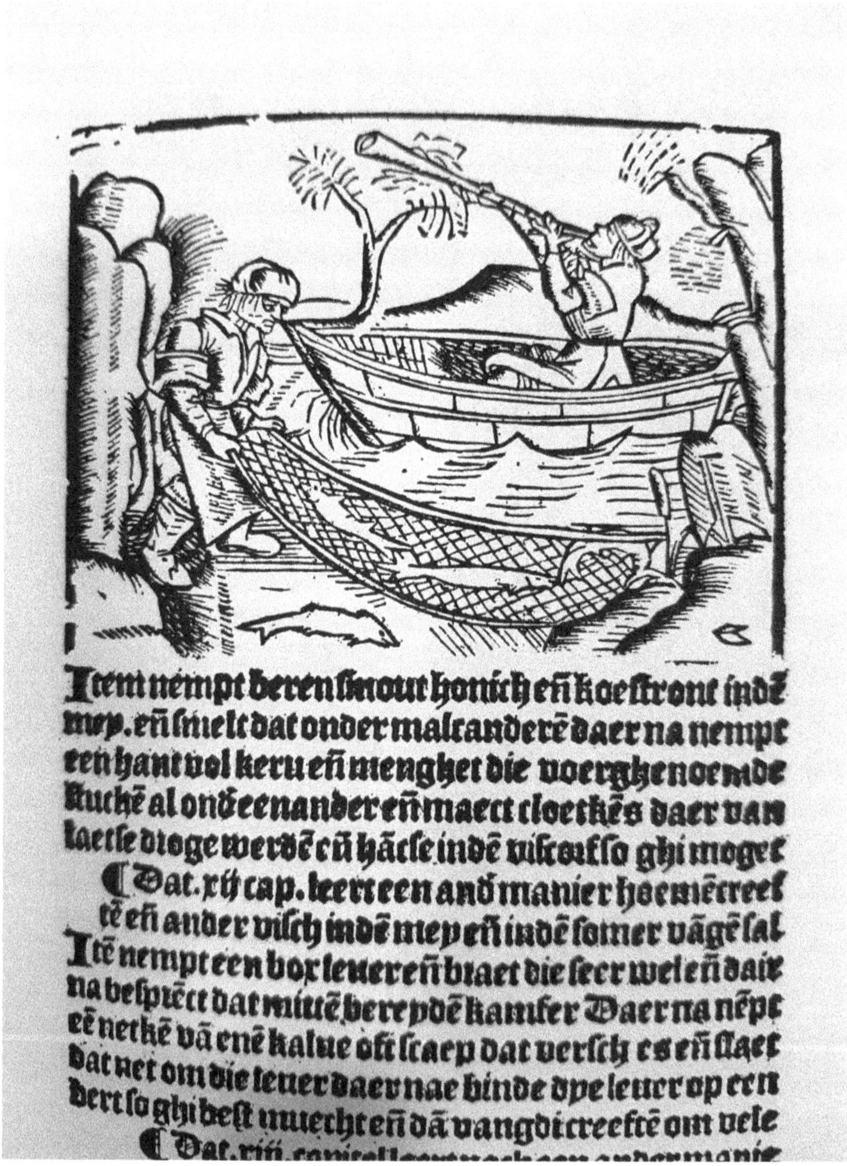

FIGURE 2.2 Fisherman with flute player. Anonymous, *Dit boecxken leert hoe men mach voghelen vanghen* [...] (Antwerp, Govert Back: 1506)

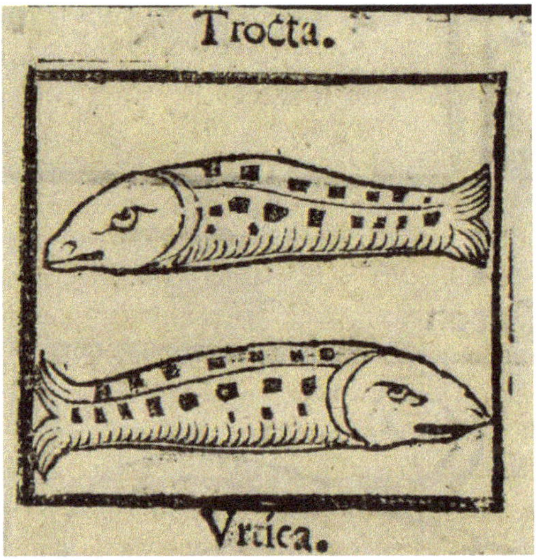

FIGURE 2.3
Brown trout (*Salmo trutta* L.), detail from Marschalk Nicolaus, *Historia aquatilium latine ac grece cum figuris* (Rostock, in edibus Thuriis: 1520), unpaginated appendix
COLLECTION BNU EN DÉPÔT À L'UNIVERSITÉ DE STRASBOURG. H 16.58

In 1520, the *Historia aquatilium* was printed [Fig. 2.3]. Its author was Nikolaus Marescalcus (Marschalk, *c.*1470–1525) from the University of Rostock on the Baltic Sea.[19] It was the first printed and illustrated book on fishes, rather than on fishing, which had been the subject of the *Boecxken* and the *Büchlein*. Marschalk was an outsider in a figurative and literal sense to whom we will return.

Das älteste deutsche Fischbüchlein vom Jahre 1498 und dessen Bedeutung für die spätere Literatur (Berlin: 1916).

19 See Marescalcus (Marschalk) Nicolaus, *Historia aquatilium latine ac grece cum figuris* (Rostock, in edibus Thuriis: 1520). The pictures were printed separately as early as 1517. – The lack of knowledge in ichthyological research about Marschalk and his work is remarkable: thus Bäumer Ä., *Geschichte der Biologie*. Vol. 2: *Zoologie der Renaissance – Renaissance der Zoologie* (Frankfurt: 1991) 346–347 translates some names of Marschalk's localities mistakenly (e.g., *Tyle* is not Tiel, a town in the middle of Holland, but the ancient name of the Shetland Islands), Glardon P., *L'histoire naturelle au XVIe siècle. Introduction, étude et édition critique de* La nature et diversité des poissons *de Pierre Belon (1555)* (Geneva: 2011) 54, who calls Marschalk a 'médecin méconnu' (in fact, he was neither unknown nor a physician, but a well-known jurist), finally, Zucker A., "Zoologie et philologie dans les grands traités ichtyologiques renaissants", *Kentron* 29 (2013) 135–173, here 140), who refers to a paper by Brian Ogilvie of 2005, who in turn owes his knowledge to a hint (not a publication) from Laurent Pinon. Solid information on Marschalk is provided only by Huber-Rebenich G., "Marschalk, Nikolaus", in Worstbrock F.J. (ed.), *Deutscher Humanismus 1480–1520. Verfasserlexikon*. Vol. 2. (Berlin – New York: 2013) 161–203 and Lisch G.C.F., "Buchdruckerei des Raths Dr. Nicolaus Marschalk". *Jahrbücher des Vereins für Mecklenburgische Geschichte und Altertumskunde* 4 (1839) 92–133.

The most popular fish book, however, of the 16th century and beyond, far ahead of the acclaimed works by Belon, Rondelet, Salviani or Gessner, which appeared three decades later, was published by the Italian Paulus Iovius (Paolo Giovio, 1483–1525) in 1524. Its title was *De Romanis piscibus*. Giovio had also planned to illustrate his book, but in the end, he failed with this aspect of his project.[20]

Following on from Marschalk's and Giovio's fish books were some more practice-oriented writings about fish but not all of these resulted, in printed books and, and those that were often lacked images, although they did contribute to the knowledge of fish. A well-known description of fish ponds, for example, came from the Bohemian Johannes Dubravius. It was printed in 1547.[21] Conrad Gessner later (1559) dedicated a separate treatise to this work.[22]

Aquaculture and fish ponds had existed since Roman times.[23] Later on, they were described in what is today France, Germany and Bohemia during the reign of Charlemagne at around 800. Fish culture was mostly practiced then in monasteries, where fish often served as substitute for meat, not only during Lent. The most eaten fish was the carp (*Cyprinus caprio*).[24]

Beginning in the 12th century, fish trade, export, and consumption expanded and evolved into a real industry, especially in the northern parts of Europe, along with carrier guilds and an efficient trading organization, known under

20 Giovio Paolo (Paulus Iovius), *De Romanis piscibus* (Rome, Francesco Minuzio Calvo: 1524). On Giovio's fishbook, see Baumann T. (ed. and transl.), *Pauli Iovii Novocomensis Medici De Romanis piscibus libellus ad Ludovicum Borbonium cardinalem amplissimum. Text, Übersetzung, Kommentar* (Unpublished doctoral thesis University of Mannheim, 1994), chap.v, ix, 122 and Beullens, *Aristotle translators*. A semi-critical edition of Giovio's work is available by Travi E. – Penco M. (eds.), *De piscibus Romanis*, in *Dialogi et descriptiones* (= *Pauli Iovii Opera* 9) (Rome: 1984) 3–64.

21 The treatise of Johannes Dubravius (Jan Skála z Doubravky, *c.*1486–1553) is available in translation, see Dubravius Johannes, *De piscinis et piscium qui in eis aluntur naturis libri quinque* (Breslau (Vratislaviae), Andreas Vinglerus: 1547) and Wüstner A. – Kollmann J. (eds. and transl.), *Buch von den Teichen und den Fischen, welche in denselben gezüchtet werden. In fünf Büchlein* (Vienna: 1906).

22 Gessner Conrad, *Iani Dubravii qui postea Olomucensis episcopus creatus est, De piscinis et piscium qui in eis aluntur naturis libri quinque* (Zurich, Andreas Gessner d. J.: 1559).

23 Higginbotham J., *Piscinae. Artificial fishponds in Roman Italy* (Chapel Hill – London: 1997).

24 On the importance of the common carp as edible and farmed fish, see Balon E.K., "The common carp, Cyprinus carpio: its wild origin, domestication in aquaculture, and selection as colored nishikigoi", *Guelph Ichthyology Reviews* 3 (1995) 1–55; Benoît P., "La carpe dans l'occident médiéval", in James-Raoul D. – Thomasset C. (eds.), *Dans l'eau, sous l'eau. Le monde aquatique au Moyen Âge* (Paris: 2002) 229–236; additionally Sahrhage – Lundbeck, *History of fishing* 50, 62–64.

its German name *Hansa* or *Hanse*.[25] Important in development of this industry was the construction of a new type of cargo ship, the Hanseatic cog, and other, even larger types beyond the mid-fifteenth century.[26] This made it possible not only to hunt large quantities of fish, but also to preserve and transport them. The herring (*Clupea harengus*), next to cod (*Gadus morhua*) was the the most targeted fish during these times. Gessner reports that unbelievable quantities were caught and consumed in boiled or smoked form, and it even seems that there were cases of overfishing.[27]

Of the large rivers, from an ichthyological perspective, the Elbe was most important, even more so than the Rhine or the Danube. Two descriptions of the Elbe fishes are worth highlighting, although neither of them was printed. Georg Handsch (1529–1578) published a large manuscript entitled "The Elbe fishing in Bohemia and Meissen" in two parts; the first part is a plagiarism of Dubravius' treatise, but the second part, written at an unknown time after his study of medicine in Padua, 1550 to 1553, is authentic and original.[28] In 1556, Johannes Kentmann (1518–1574) wrote an "Ichtyographia" (sic) of the Elbe, which he sent to his friend Gessner in Zurich a few years later.[29] Both works,

25 Dollinger P., *The German Hansa* (London – New York: 1999); Hünemörder C., "Fischfang und Fischkunde im Mittelalter", *Deutsches Schifffahrtsarchiv* 4 (1981) 183–190.

26 See Sahrhage – Lundbeck, *History of fishing* 70, 98; Bennema F.P. – Rijnsdorp A.D., "Fish abundance, fisheries, fish trade and consumption in sixteenth-century Netherlands as described by Adriaen Coenen", *Fisheries Research* 161 (2015) 384–399.

27 Gessner Conrad, *Historia animalium liber IV: qui est De piscium et aquatilium animantium natura* (Zurich, Christopher Froschauer: 1558) 487 mentions that incredibly huge amounts are caught, to the point that these fishes are reduced to a minimum (*copia incredibilis; ingens copia irretitur ... adeo imminuti sunt hi pisces*). I read this as an indication of overfishing, but the context is unclear; perhaps a fish epidemic was the true cause of the disappearance of herring populations near Heligoland. See also Hendrikx S., "Identification of herring species (Clupeidae) in Conrad Gessner's ichthyological works: A case study on taxonomy, nomenclature, and animal depiction in the sixteenth century", in Enenkel K.A.E. – Smith P.J. (eds.), *Zoology in early modern culture* (Leiden – Boston: 2014) 149–171, who tends to strain Gessner's endeavours, even if he was unable to carry out research of his own on the herring (how could he have done this in Switzerland?), but, as usual, only stringed together numerous foreign information.

28 Schubert O., *Georg Handsch von Limus'* Die Elbefischerei in Böhmen und Meißen (Prague: 1933), correcting Senfelder L., "Georg Handsch von Limus. Lebensbild eines Arztes aus dem XVI. Jahrhundert", *Wiener klinische Rundschau 1901*, 495–499, 514–516 and 533–535 about the authorship of the first part.

29 Hertel R., "Über die "Ichthyographie der Elbe" des Johannes Kentmann. Eine Studie über die ältesten sächsischen Fischfaunen (Pisces)", *Zoologische Abhandlungen. Staatliches Museum für Tierkunde in Dresden* 35 (1978) 75–100; Zaunick R., "Fragmente der ältesten sächsischen Fischfauna des Dr. Johannes Kentmann (1518–1574)", *Sitzungsberichte und Abhandlungen der Naturwissenschaftlichen Gesellschaft Isis* 1915: 15–36.

Handsch's as well as Kentmann's, are of interest today as a testimony to the former biodiversity in the Elbe.[30]

Among the large inland waters, Lake Constance played a special role. Here, it was more important than, for example, in fish ponds, to take into account ecological considerations to preserve the fish population. The best known work in this regard is the *Fischbuoch* by Gregor Mangolt (1498–*c*.1577) from 1557.[31] This book describes, in the format of a monthly calendar, 30 species of fish living in Lake Constance. Special emphasis is placed on spawning behaviour and diseases. Mangolt's book was printed under dubious circumstances, without the author's knowledge and permission, in Zurich by Andreas Gessner, a cousin of Conrad Gessner. Conrad Gessner had borrowed the manuscript from Mangolt in order 'to peruse it at home'. In the original manuscript still preserved today, a rather bewildered Mangolt notes that the text had been completely altered in sequence and illustrated with woodcuts [Fig. 2.4].[32]

What is described here, regarding fish literature on both sides of the Alps since the 1520s, shows that by the middle of the 16th century, special knowledge and, above all, originality was needed to stand out from already published books (Marschalk, Giovio, Dubravius, Mangolt) or other circulating manuscripts (Handsch, Kentmann) about fishes. Authentic, naturalistic pictures would be key here since they were absent from all the books up until this time.

In the middle of the century, this deficit of adequately illustrated books was resolved, when four monographs on fishes appeared almost simultaneously – each explicitly referring to Aristotle or Pliny. Two of these books were authored by Frenchmen Pierre Belon (1517–1564)[33] and Guillaume Rondelet (1507–1566),[34]

30 See for example Zarske A., "Autochthone Population oder Faunenverfälschung? Zum Fund der Nase (*Chondrostoma nasus* [L., 1758]) im oberen Elbegegbiet (*Teleostei: Cyprinidae*)", *Faunistische Abhandlungen. Staatliches Museum für Tierkunde* 20 (1996) 285–294 regarding *Chondrostoma nasus* L., a carp species.

31 Mangolt Gregor, *Fischbuoch. Von der natur und eigenschafft der vischen* (Zurich, Andreas Gessner: 1557).

32 For more details, see Violand G., "Historische Fischökologie des Bodensees. Das Fischbuch des Gregor Mangolt, 1557", *Österreichs Fischerei* 59 (2006) 169–171 and Hakelberg D., "Das Bodensee-Fischbuch von Gregor Mangolt in einem Basler Nachdruck von 1612", *Wolfenbütteler Barock-Nachrichten* 36 (2009) 107–114.

33 Belon's relevant works on fishes are *L'histoire naturelle des estranges poissons marins* (Paris, Regnauld Chaudiere: 1551) and *De aquatilibus libri duo* (Paris, Charles Estienne: 1553).

34 Rondelet's relevant works are *Libri de piscibus marinis* (Lyon: 1554) and *Universae aquatilium historiae pars altera* (Lyon, Macé Bonhomme: 1555), usually issued bound up in one volume with separate pagination (1554–1555); in 1558 an abbreviated French version of the Latin works appeared: *La premiere et la seconde partie de l'histoire entiere des poissons* (Lyon, Macé Bonhomme).

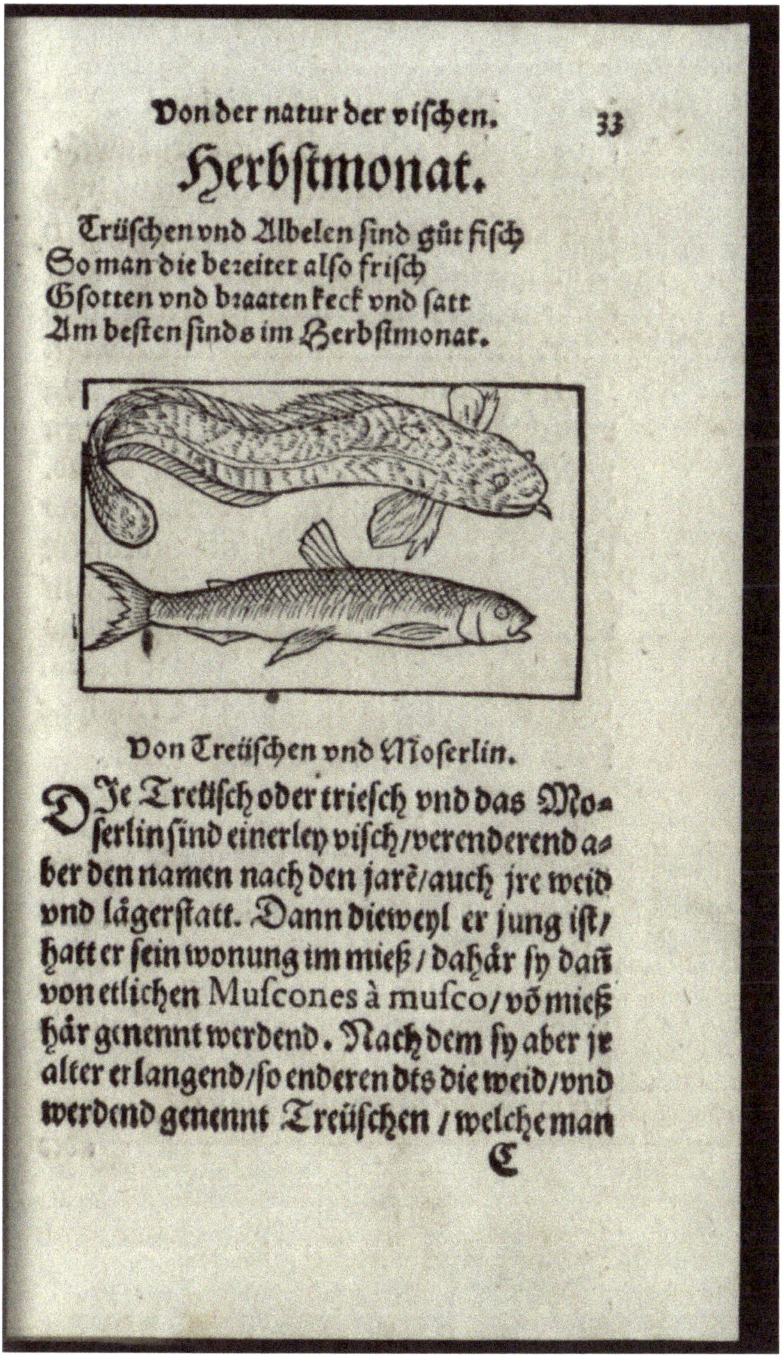
Von der natur der vischen. 33

Herbstmonat.

Trüschen vnd Albelen sind gůt fisch
So man die bereitet also frisch
Gsotten vnd braaten keck vnd satt
Am besten sinds im Herbstmonat.

Von Treüschen vnd Moserlin.

DJe Treüsch oder triesch vnd das Moserlin sind einerley visch/verenderend aber den namen nach den jarẽ/auch jre weid vnd lägerstatt. Dann dieweyl er jung ist/hatt er sein wonung im mieß/dahär sy dañ von etlichen Muscones à musco/võ mieß här genennt werdend. Nach dem sy aber je alter erlangend/so enderen sys die weid/vnd werdend genennt Treüschen/welche man

C

FIGURE 2.4 Fishes from Mangolt Gregor, *Fischbuoch. Von der natur und eigenschafft der vischen* (Zurich, Andreas Gessner: 1557) 33; on top, easily identifiable, a Burbot (*Lota lota* L.)

a third by the Italian Hippolito Salviani (1514–1572)[35] and finally Conrad Gessner (1516–1565) from Switzerland, who summed up but also extended the results of his predecessors.[36]

Between 1551 to 1560, the works of these scholars were published in such close succession that they overlapped each other, and, to make things really difficult, were issued in instalments (in part no longer extant), which were later bundled in separate editions or bound up in one book, but with different titles, separate page numbering and more than one date of publication. Additionally, there are several revisions and translations haunting bibliographical records until today.

All this led to considerable confusion at that time as well as in recent days because the authors occasionally refer to each other in their writings without specifying exactly which book or which instalment they are referring to. E.W. Gudger has tried to disentangle this tangle at least a bit. Another less successful attempt was made by Basford Dean.[37] Aside from these bibliographical difficulties, the works of these four authors are considered milestones in the history of ichthyology.[38] With them, realism finally began its triumphal march in fish images.[39] Examples from these four authors follow to substantiate

35 Salviani's work is entitled *Aquatilium animalium historiae liber primus, cum eorumdem formis, aere excusis* (Rome, "Printed in his own house": 1554 on the title page and 1557 and 1558 in two different colophons), see Mortimer R., *Catalogue of books and manuscripts. Part II: Italian 16th century books*. Vol. 2 (Cambridge, MA: 1974) 628–630. In 1559 *Aquatilium animalium historiae liber primus, cum eorumdem formis, aere excusis* (Rome) and in 1593 posthumously *Icones piscium* (Rome) solely illustrated editions without any texts came out, of which, in turn, pirated copies by the Meietti printing house in Venice were published in 1600 and 1602; the latter copy has survived, see Salviani, *Aquatilium animalium curendum* [sic] *formis* (Venice: 1602).

36 Gessner's fish-related works are *De piscibus et aquatilibus omnibus libelli III novi* (Zurich: 1556), on which see Bäumer Ä., "De piscibus et aquatilibus libelli III novi (Zürich 1556). Ein bisher unbeachtetes zoologisches Werk von Conrad Gesner", *Berichte zur Wissenschaftsgeschichte* 13 (1990) 177–181; then *Historia animalium liber IV* (1558) and *Nomenclator aquatilium animantium. Icones animalium aquatilium in mari et dulcibus aquis degentium* (Zurich: 1560).

37 Gudger E.W., "The five great naturalists of the sixteenth century: Belon, Rondelet, Salviani, Gesner and Aldrovandi. A chapter in the history of ichthyology", *Isis* 22 (1934) 21–40; Bashford D., *A bibliography of fishes*. Vol. 3 (New York: 1923) 216–217 on Belon, p. 309 on Rondelet and p. 312 on Salviani.

38 See Cuvier Georges – Valenciennes Achille, *Histoire naturelle des poissons. Planches* (22 vols.) (Paris: 1828–1849; vol. 1, 1828) 48–55; Bäumer, *Zoologie der Renaissance* 346–381; Nellen W. – Dulčić J., "Evolutionary steps in ichthyology and new challenges", *Acta Adriatica* 49 (2008) 201–232.

39 On realism in Italian Renaissance painting, see Blunt A., *Artistic Theory in Italy, 1450–1600* (Oxford: 1962).

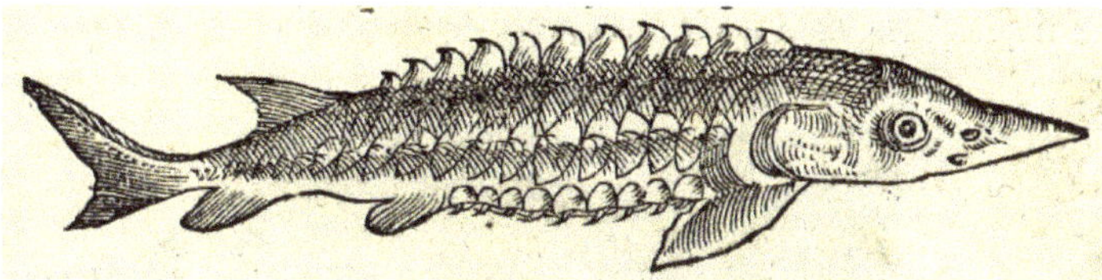

FIGURE 2.5A Sturgeon (*Acipenser sturio* L.) from Belon Pierre, *De aquatilibus libri duo* (Paris, Charles Estienne: 1553) 101
NIEDERSÄCHSISCHE STAATS- UND UNIVERSITÄTSBIBLIOTHEK GÖTTINGEN

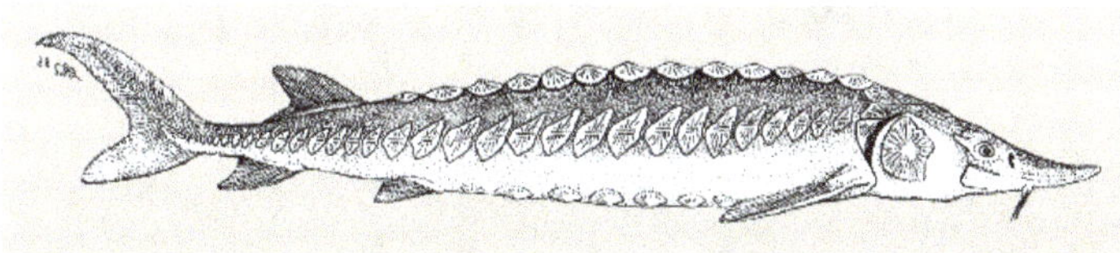

FIGURE 2.5B Sturgeon
ILLUSTRATION FROM FISHBASE/FAO

this statement [Figs. 2.5 to 2.8]. Our guiding criterion is the unambiguous recognition and identification of the respective fish, which is why a modern illustration (FishBase/FAO) is placed next to it.

Beside these masterworks, there were some other promising efforts that did not result in published books. Simone Porzio (Portius, 1496–1554), an Aristotelian philosopher, who lectured in Pisa from 1546 to 1552, produced a great study on fish entitled *De piscibus*, which he had begun already in his Neapolitan home region. The images of Porzio's project were commissioned by Grand Duke Cosimo I de' Medici and created by the famous painter Francesco Bacchiacca (1494–1557), so they must have been remarkable. Unfortunately, the work remained incomplete, only a manuscript fragment has survived.[40] Rondelet visited Porzio and mentions him in his fish book from 1554.

Gysbert van der Horst (Gysbertus Horstius, *c.*1991–*c.*1555), a Dutch physician living in Rome (where he also died), was collecting all kinds of fish specimens and had them drawn. Via an intermediary, a part of these drawings reached

40 Reproduced by Del Soldato E., "Un catalogo naufragato: il "De piscibus" di Simone Porzio", in De Bellis E. (ed.), *Aristotele e la tradizione aristotelica, Atti del convegno internazionale di studi, Lecce, 12–14 giugno 2008* (Lecce: 2008) 149–176; see also idem, *Simone Porzio. Un aristotelico tra natura e grazia* (Rome: 2010) 92–100.

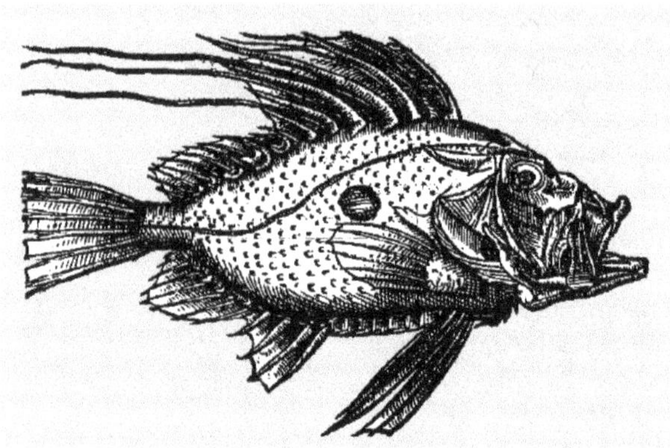

FIGURE 2.6A
John dory (*Zeus faber* L.) from Rondelet Guillaume, *Libri de piscibus marinis* (Lyon, Macé Bonhomme: 1554) 204
WIKIMEDIA COMMONS

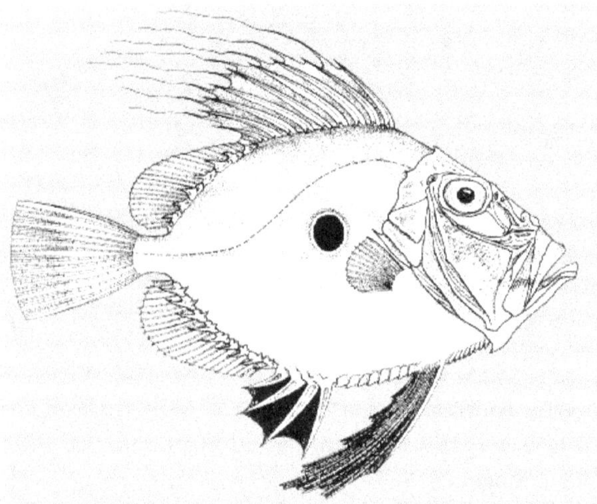

FIGURE 2.6B
John dory (*Zeus faber* L.).
ILLUSTRATION FROM FISHBASE/FAO

Conrad Gessner in Zurich, who used them in the fourth volume of his *Historia animalium* on fishes, which he published in 1558.[41]

About fifty years later, in the early 1600s, another important work on fish was completed. Its author was the Italian Ulisse Aldrovandi (1522–1605). His *De piscibus libri V* were posthumously edited in 1613 by Joannes Cornelis Wterweer (Johannes Cornelius Uterverius) and Geronimo Tamburin (Hieronymus

41 The whole story is told by Egmond F. – Kusukawa S., "Circulation of images and graphic practices in Renaissance natural history: the example of Conrad Gessner", *Gesnerus* 73 (2016) 29–72.

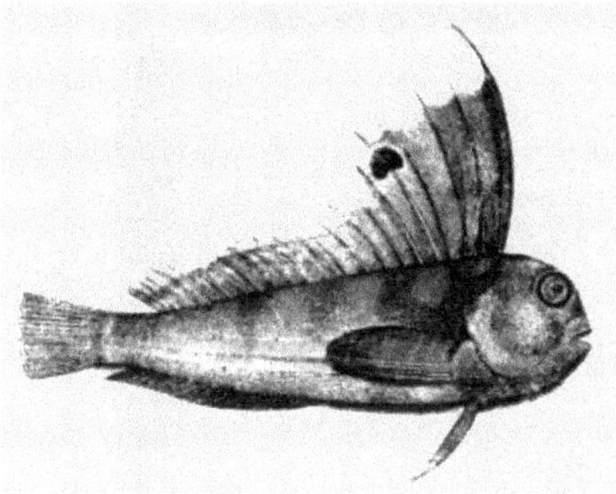

FIGURE 2.7A
Butterfly blenny (*Blennius ocellaris* L.) from Salviani Hippolito, *Aquatilium animalium historiae liber primus, cum eorumdem formis, aere excusis* (Rome, Hippolito Salviani: 1558) fol. 217r

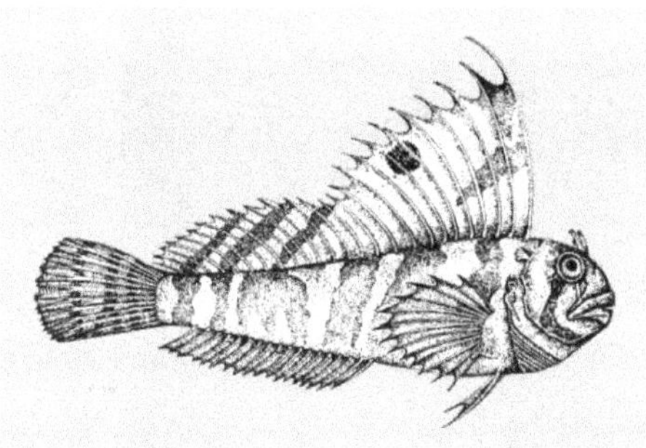

FIGURE 2.7B
Butterfly blenny from FishBase/FAO

Tamburinus). Like Gessner, Aldrovandi summed up and extended what Belon, Rondelet and Salviani had achieved previously.[42]

2 'True' Images or Portraits versus 'Dissimilar Similarity'

If we look at how fish literature developed over time, beginning with the works of the ancient Greeks and, above all, Aristotle, a significant change regarding the illustrations in fish books can be observed from the middle of the 16th century. This change is clearly visible in the overview of this progression that was provided in the first section of this article. It can be summarised as a transition from traditional to 'true' pictures.

42 Aldrovandi Ulisse, *De piscibus libri V. et De cetis lib. Unus*, ed. Ioannes Cornelius and Geronimo Tamburin (Bologna, Jo. Bapt. Bellagamba: 1613). On Aldrovandi's voluminous fish book, see Bäumer, *Zoologie der Renaissance* 96–102; on Aldrovandi's disciple Uterverius (or Uterwer), a Dutchman from Delft, see Richter C., "Hollandse vissenboeken. De onweerstaanbare kracht van de afbeelding", *Holland* 38 (2006) 161–176; Uterverius authored the preface to the reader.

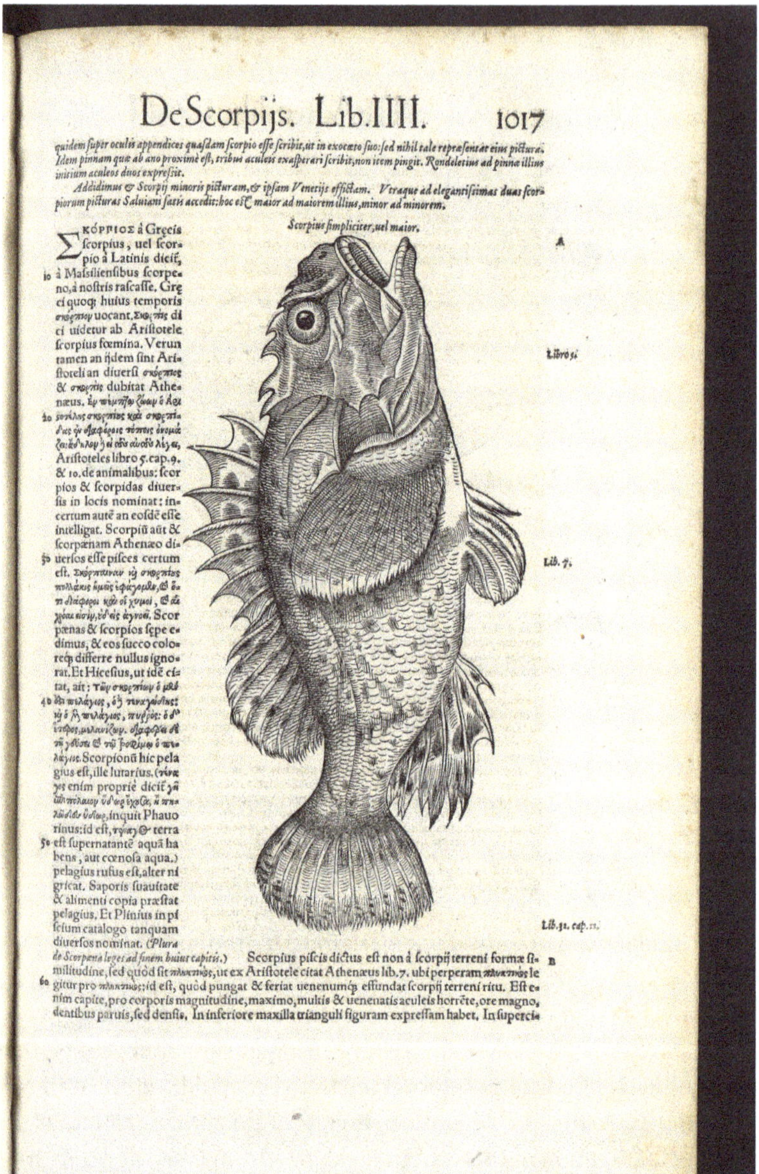

FIGURE 2.8A Red scorpionfish (*Scorpaena scrofa* L.) from Gessner Conrad, *Historia animalium liber IV: qui est De piscium et aquatilium animantium natura* (Zurich, Christopher Froschauer: 1558) 1017. Zentralbibliothek Zürich, NNN 48 | F, https://doi.org/10.3931/e-rara-16853 / Public Domain Mark

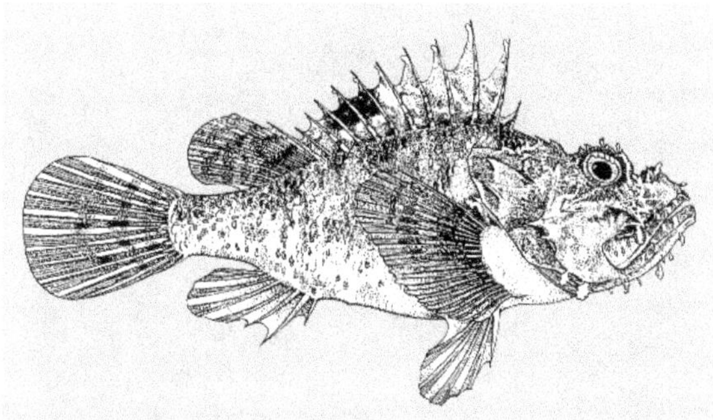

FIGURE 2.8B
Red scorpionfish from FishBase/FAO

According to the new approach, the author of a fish book and the illustrator who was employed to provide the pictures in the book had to fulfil some baseline requirements. They must have seen the fishes with their own eyes, which often made extensive traveling necessary and the fishes must have been available to them in a good state, that is still fresh, dried, or stuffed.[43] And beyond this, the ultimate aspiration was that the author had also dissected the fishes.

It was Pierre Belon, gifted with a keen sense for trends. who can be regarded as the first to have fulfilled, in a printed book, at least to some extent, this level the high requirements of personal observation and dissection as a basis for realistic or 'true' (vraie) fish pictures. Having made many observations in the Mediterranean, he announces already in the subtitle and again in the preface to his book from 1551 that he is going to offer 'true portraits' (vrais portraicts) or 'true pictures' (vraies peinctures) of fishes.

Typically, such a true portrait or picture[44] shows nothing but a specific single fish without any background, not even fishing equipment or habitat. Nor

43 On modes of preserving fishes in early and in more recent times, see Pinon L., "Clématite bleue contre poissons séchés: sept lettres inédites d'Ippolito Salviani à Ulisse Aldrovandi", *Mélanges de l'École française de Rome* 114 (2002) 477–492; Tosi A., "Acconciare, seccare, dipingere: pratiche di rappresentazione della natura tra le "spigolature" aldrovandiane", in Olmi G. – Simoni F. (eds.), *Ulisse Aldrovandi. Libri e immagini di Storia naturale nella prima Età moderna* (Bologna: 2018) 49–58; Davis P., "Collecting and preserving fishes: a historical perspective", in MacGregor A. (ed.), *Naturalists in the Field* (Leiden – Boston: 2018) 149–165 and Carusi C., "Salt and Fish Processing in the Ancient Mediterranean: A Brief Survey", *Journal of Maritime Archaeology* 13 (2018) 481–490. This point is of considerable importance with regard to the quality of a depicted fish, especially Salviani was attacked for that. To date, research has failed to provide accurate information on how fishes were kept fresh for more than a few days in the 16th century, especially in the hot Mediterranean climate (Peter Davis, Northumberland, by personal communication).

44 The equivalent term for 'Portrait' in contemporary German or Dutch (Flemish) fish books and especially in herbals was 'Contrafactur/Contrafayt/Contrefeyung' or 'conterfeytsel/

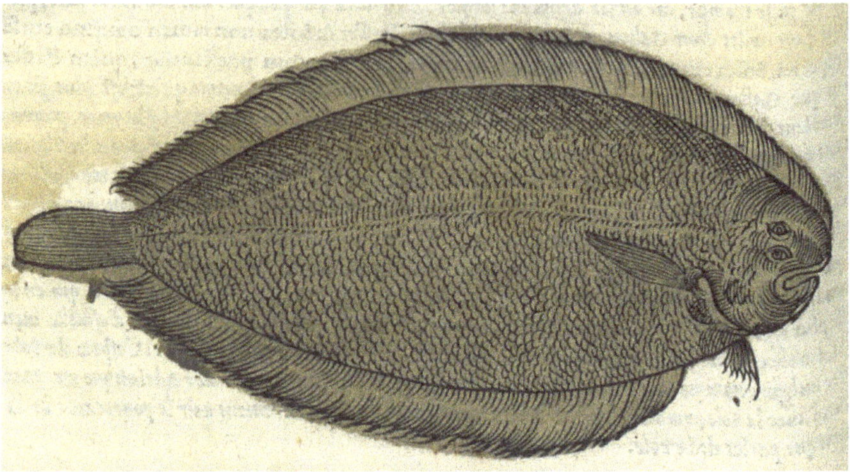

FIGURE 2.9 Common sole (*Solea solea* L., a marine fish) from Belon Pierre, *De aquatilibus libri duo* (Paris, Charles Estienne: 1553) 147
NIEDERSÄCHSISCHE STAATS- UND UNIVERSITÄTSBIBLIOTHEK GÖTTINGEN

are there any other requisites intended to convey further information or indicate or symbolise anything. A juxtaposition [Figs. 2.9–2.10] makes evident the difference between Belon's kind of fish portrait and the old traditions exemplified in the *Hortus sanitatis*, a work that was iconographically influential for a long time to come.

In Nikolaus Marschalk's *Historia aquatilium* of 1520, which has already been introduced here as the first printed and illustrated book ever that dealt with fishes only, the author had also claimed, in his dedication to Duke Albrecht of Mecklenburg, that as a young man he had been 'eyewitness' (*testis oculatus*) to the various fishes in the Atlantic Ocean and the North Sea. Yet Marschalk was not a doctor, but a lawyer, and his pictures seem to contradict his claims.

Most probably, Marschalk could not afford an artist of his own, so some of his pictures were adopted from others, or at least inspired by those that had

geconterfeyt', from Old German *conterfeit*, Old French *contrefait*, derived from Latin *contrafacere*, 'to recreate badly', 'to adulterate' (said of metals). However, while in French and English this pejorative meaning was retained ('counterfeit'/'contrefaçon'), in Germanic languages it dwindled away and a non-negative meaning remained, for example in *Hortus sanitatis* (Mainz:1491), in Albertus Magnus (Frankfurt: 1545) or in Otto Brunfels' *Contrafayt Kreüterbuch* (Strasbourg: 1532), Latin version *Herbarum vivae eicones* (Strasbourg: 1530), in Rembert Dodoens' *Den Nieuwen Herbarius* (Basel: 1545) and *Cruijdeboeck* (Antwerp: 1554). 'Portrait', in turn, is derived from Latin *protrahere*, to draw something out, namely the essential characteristics of an individual, see also Woodall J. (ed.), *Portraiture: Facing the Subject* (Manchester: 1997).

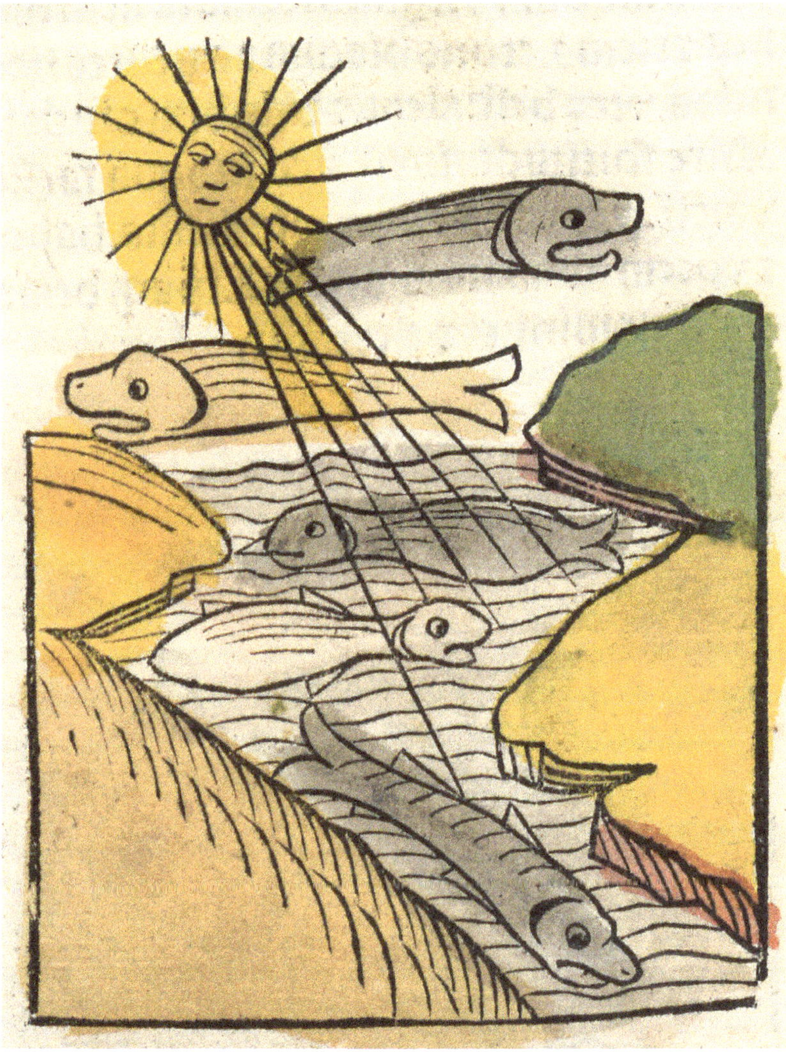

FIGURE 2.10 Common sole and burbot (*Lota lota* L., a river fish), showing habitat and the sun in allusion to their Latin names (*Solea* and *Solaris*) in *Hortus sanitatis* (Mainz, Jacob Meydenbach: 1491), tract "De piscibus", chap. 85
PROVIDED BY U.S. NATIONAL LIBRARY OF MEDICINE

appeared in the *Hortus sanitatis* in 1491,[45] which has also been mentioned already. The *Hortus sanitatis* still included a large number of pictures, in

45 See Timm W., "Die Holzschnitte zu Nikolaus Marschalks Historia aquatilium latine ac grece cum figuris, Rostock 1517–20", in Anonymous (ed.), *Festschrift Gottfried von Lücken* (Rostock 1968) 799–802.

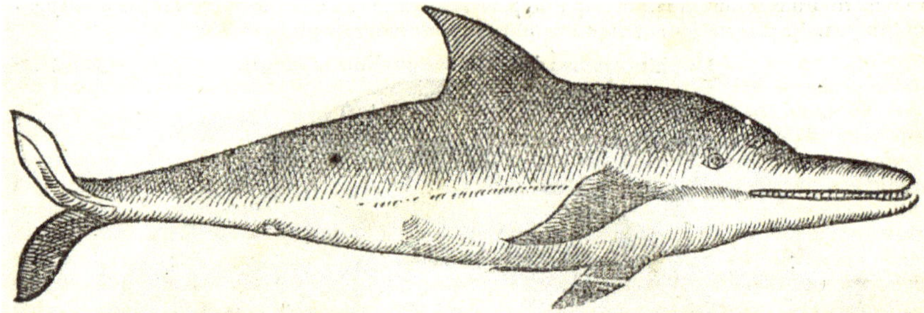

FIGURE 2.11 Dolphin (*Delphinus delphis* L.) in Belon Pierre, *L'histoire naturelle des estranges poissons marins* (Paris, Regnaud Chaudiere: 1551) fol. 28v and *De aquatilibus libri duo* (Paris, Charles Estienne: 1553) 9
NIEDERSÄCHSISCHE STAATS- UND UNIVERSITÄTSBIBLIOTHEK GÖTTINGEN

turn, either inspired or copied from even older books, such as the medieval bestiaries.[46] Gessner, therefore, dismissed Marschalk's pictures as absurd fantasy products (*picturae fictae et absurdae*).[47]

The difference between Belon's realistic images and the un-realistic (fictional) illustrations in the tradition of the *Hortus sanitatis* is striking. In Belon's book a dolphin, for example, looks like a real dolphin, not like a sea goddess or mermaid as in Marschalk's [Figs. 2.11 and 2.12].[48] And if an animal is fabulous, like Neptune's horse, this is expressly noted by Belon.[49]

---

46 On the pictorial (rather than scientific) importance of the medieval illustrated bestiaries for animal works of the Renaissance, see Hassig D., *Medieval bestiaries. Text, image, ideology* (Cambridge: 1995); Camille M., "Bestiary or Biology? Aristotle's Animals in Oxford, Merton College, MS 271", in Steel C. – Guldentops G. – Beullens P. (eds.), *Aristotle's Animals in the Middle Ages and Renaissance* (Louvain: 1999) 355–396; James-Raoul D., "Inventaire et écriture du monde aquatique dans les bestiaries", in James-Raoul D. – Thomasset C. (eds.), *Dans l'eau, sous l'eau. Le monde aquatique au Moyen Âge* (Paris: 2002) 175–226 and Beullens P., "Aristotle's Zoology in the Medieval World", in Boehrer B. – Hand M. – Massumi B. (eds.), *Animals, Animality, and Literature* (Cambridge: 2018) 29–42.

47 Gessner in the *Enumeratio authorum* of his *Historia animalium IV* of 1558: *picturis fictis et absurdis*. Nowadays such fantastic creatures would be characterised less pejoratively as a kind of 'hybrid knowledge': see Mackenzie L., "French early modern sea-monsters and modern identities, via Bruno Latour", in Cuneo P. (ed.), *Animals and early modern identity* (Farnham – Burlington: 2014) 329–349.

48 Fantastic representations in the *Hortus sanitatis* tradition were certainly not a serious point of attack. In the case of the dolphin, Belon (*L'histoire naturelle* 1551: Preface and *De aquatilibus* 1553: 9–11) said and showed that he intended to take action against faulty images on ancient and contemporary coins. For information on ancient dolphin images, see also Vidali S., *Archaische Delphindarstellungen* (Würzburg: 1997).

49 Belon, *De aquatilibus* 17.

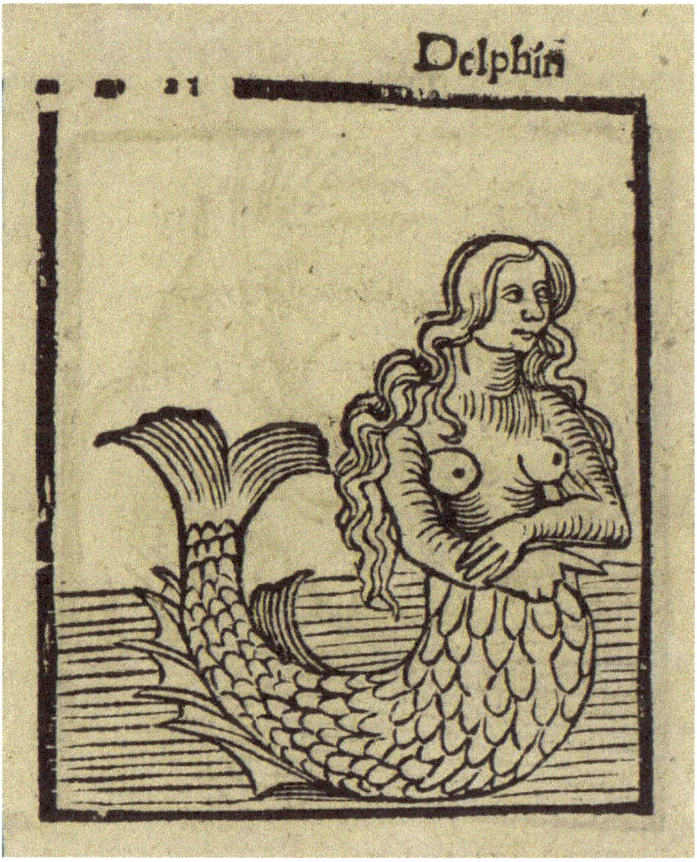

FIGURE 2.12 "Dolphin" in Marschalk Nicolaus, *Historia aquatilium latine ac grece cum figuris* (Rostock, in edibus Thuriis: 1520). No pagination
COLLECTION BNU EN DÉPÔT À L'UNIVERSITÉ DE STRASBOURG. H 16.583

1.1 *Botany in the Leading Role*

It's not a new insight that botany had acquired a leading role with regard to scientific innovation and progress in the 16th century; zoology was lagging at least twenty years behind botany in this period.[50] The authors of fish books could, therefore, build on the scientific approach that been developed by the so-called 'Fathers of Botany'.

50 Ogilvie B.W., *The Science of Describing: Natural History in Renaissance Europe* (Chicago – London: 2006) 49.

One of them was the German scholar Otto Brunfels († 1534). Already in his first publication, the *Herbarum vivae eicones* ('Lifelike herb pictures') of 1530, he had indicated the direction to take and provided the key concepts by declaring that botany was in a poor state and could hardly be taken seriously anymore. He claimed that to revive it, he had 'done away with the earlier, old-fashioned herbal books and published them anew equipped with pictures that are true to life (*vivae*) and artfully made like embroidery (*acupictae*). Furthermore, solid and reliable descriptions provided by the ancient original authors have been added. We have made an effort and taken care of both.'[51] This was a clear statement, and it almost went without saying that by old-fashioned herbal books the *Hortus sanitatis* of 1491 was meant in the first place, along with the German language *Gart der Gesundheit* ('Health Garden', this is also the meaning of the Latin title *Hortus sanitatis*) of 1485, which had served as a model in producing the *Hortus*.[52] Both books, the *Hortus* and the *Gart* were produced in numerous new editions, adaptations and translations and became the leading works of reference for medically oriented studies of nature until Brunfels and the other innovators appeared, who refreshed these studies with regard to descriptions and pictures.[53]

Detailed explanations of what the intention was in Brunfels's new herbal book were not needed.[54] The new key terms *verus* (true) and *vivus* (lifelike) immediately became clear.[55] The reference works that were in use at that time

51 In the original: '[…] abolitis prioribus ac veteribus Herbarijs, atque de novo vivis et acupictis imaginibus, editis. Deinde solidis ac firmis descriptionibus, ex priscis et autenticis autoribus prolatis, utrunque tentavimus, atque curavimus: Brunfels Otto, Herbarum vivae eicones ad naturae imitationem. Vol. 1. First edition (Strasbourg, Johann Schott: 1530) *Dedicatory letter to the Senate of Strasbourg*.

52 Green E.L., *Landmarks of botanical history*. Vol. 1 (Washington: 1906) 173.

53 See Rudolph P., *Im Garten der Gesundheit. Pflanzenbilder zwischen Natur, Kunst und Wissen in gedruckten Kräuterbüchern des 15. Jahrhunderts* (Vienna – Cologne – Weimar: 2020); Funk H., "Jan Černý's Knieha lekarska (1517): closing a gap in the history of printed illustrated herbals", *Archives of natural history* 42 (2015) 153–164.

54 See Brunfels Otto, *Novi Herbarii Tomus II* (Strasbourg: 1536) Appendix *De vera herbarum cognitione* 99. Only today's scientists have submitted a theoretical justification for such a claim, see e.g. De Angelis S., "Sehen mit dem physischen und dem geistigen Auge. Formen des Wissens, Vertrauens und Zeigens in Texten der frühneuzeitlichen Medizin", in Jaumann H. (ed.), *Diskurse der Gelehrtenkultur in der Frühen Neuzeit. Ein Handbuch* (Berlin – New York: 2011) 211–254.

55 See the impressive list of contemporary herbals with the respective titles in Kusukawa S., "Ad vivum Images and Knowledge of Nature in Early Modern Europe", in Balfe T. et al. (eds.), *Ad vivum? Visual Materials and the Vocabulary of Life-Likeness in Europe before 1800* (Leiden – Boston: 2019) 89–121.

included pictures and descriptions which did not serve their purpose. They were not just worthless, but could even be dangerous. This is obvious regarding plants that were above all studied and applied as medical remedies. There are plants, such as the autumn crocus (*Colchicum autumnale* L.), that can have a lethal outcome if falsely identified or administered using an improper dose. On the whole, it was important to thwart the efforts of 'Landtstreicher und Leutbescheißer' (vagabonds and quacks who rip off the people), as the Italo-German physician Johann Dantz (Dantzius) from Frankfurt dramatically expressed it in the preface to his herbal book in 1546.[56] This was no trivial matter, but a crucial element in terms of pharmaceutical control.[57]

The new fish books that were published in the middle of the 16th century had been written, like the new herbal books, by physicians or at least by medically educated scholars of natural things. And even though the medical benefit of fish is smaller than that for plants, it clearly existed.[58] In any case, it was at least mentioned by the authors in question. For example, wherever appropriate, Salviani included a section on potential medical benefits in his description of a particular fish.

It is, therefore, not surprising that the authors of the new fish books built on the scholarly standards that had by then been established through the works of their botanical forerunners. This meant that these authors tried to back up the fish descriptions they were providing in their books with 'true', 'lifelike'

56 See Dantz Johann, *Kreutter Buch. Des Hochberümpten Pedanij Dioscoridis Anazarbei, gründliche vnd gewisse beschreibung aller materien oder gezeugs der Artznei, das ist, nit alleyn der Kreutter, sonder auch alles, was sonst in der artznei gebreuchlich ist; in sechs Büchcren verfast* (Frankfurt, Jacob Cyriacus: 1546).

57 See Nauert C.G., "Humanists, Scientists, and Pliny: Changing Approaches to a Classical Author", *The American Historical Review* 84 (1979) 72–85 (here 83) or Watson G., *Theriac and Mithridatium. A study in therapeutics* (London: 1966) 102–104, 124 on the danger to life from incorrect dosages and on adulteration and quackery including criminal poisoning in Medieval and Renaissance times. See also Leu U.B., *Conrad Gessner (1516–1565). Universalgelehrter und Naturforscher der Renaissance* (Zurich: 2016) 299–302 on the fierce controversy between Mattioli versus Gessner and others about the identity of the plant 'Aconitum' (*Aconitum anthora* L., Yellow monkshood or Healing wolfsbane).

58 See Geoffroy Étienne François, *Suite de la Matiere medicale de M. Geoffroy. Par Mrs Arnault de Nobleville et Salerne*. Vol. 2: *Des poissons* (Paris, Desaint – Saillant – G. Cavelier – Le Prieur: 1756), who consistently directed the attention to the medicinal benefits of fish. A German translation of the work was published somewhat later: idem, *Fortsetzung der Abhandlung von der Materia Medica. Aus dem Französischen übersetzt. Von den Thieren*. Vol. 6 (Leipzig, Carl Ludwig Jacobi Wittwe: 1763).

and 'similar' (*similis / similitudo*) pictures.[59,60] They did this, however, without mentioning their forerunners at all. It was obviously sufficient for them to quote those key terms or make use of their meanings.

1.2 *Traditional Fish Pictures*

True pictures of fishes in the sense that this term had for Belon and other authors in the middle of the 16th century are hardly ever found in the traditional fish books and other literature on natural things. But why was this so? Why were traditional fish pictures usually not detailed and precise enough to allow the reader to identify any particular fish species or learn more about it?

There may be different reasons why a fish or any other object is not depicted in the traditional books in the ways that Belon and the others attempted to. Knowledge about fish may have been insufficient by these authors, and their illustration skills may have been poor, and, perhaps just as important, the purpose of showing fish in a book may not have been as clear to these authors as it was for Belon and his contemporaries.

How the purpose of a book may influence its illustrations can be seen, for example, in the above mentioned *Boecxken* ('Booklet'), which was printed in 1506. This book was written about how to catch fish, rather than how to identify various fish species. Accordingly, it provides descriptions and illustrations of fishing equipment, such as nets, traps and weirs. Regarding the fish itself, the author of this manual probably assumed that fishermen would know what

59 The similarity was measured on mostly clearly visible morphological features such as flowers, leaves and roots in plants and the number and position of the fins in fishes. Kyle S.R., *Medicine and humanism in late medieval Italy. The Carrara herbal in Padua.* (London – New York: 2017) showed that there were also other purposes, in particular to entertain the readers and keep them in suspense.

60 Such a similarity is also referred to by Brunfels in the subtitle of the work of 1530 as *imitatio naturae*. A sufficient similarity between a picture and what it represents can be considered to be a form of appropriateness with regard to subject matter (*Übereinstimmung mit der Sache*), also more shortly referred to as 'objectivity', in the sense that these terms have in modern theories of knowledge in science and the humanities. For a short overview of these theories, see Thiel C., "objektiv/Objektivität", in Mittelstraß J. (ed.), *Enzyklopädie Philosophie und Wissenschaftstheorie*. Vol. 6 (Stuttgart: 2016) 8–9, where it is also pointed out that the old meaning of objectivity differs considerably from the modern. While this meaning was still in use at the time when Renaissance scholars claimed to present "true pictures" of fishes, some authors have more recently introduced the term in its modern sense to the discussion about these pictures, see, for example, Daston L. – Galison P., *Objectivity* (New York: 2007), Egmond F., *Eye for detail. Images of plants and animals in art and science, 1500–1630* (London: 2017) 94, 134, or Balfe, *Ad vivum*.

fish they wanted to catch and how to identify the species. The fish shown in this book are, therefore, hardly ever identifiable in terms of species.

Another reason why fish would not be represented in a realistic manner by an illustration in a book is more complex. This may be related to symbolism and can be explained as follows. Perceiving a particular object as a symbol means perceiving it as something that stands for or *symbolises* something else. For example, a lamb symbolises innocence, a lion power, and so on. When an author describes or an illustrator depicts an object that is perceived by them as a symbol, the focus can be on this object or on what it stands for, or both. If the focus is not on the object, providing details about it becomes less important. As long as the illustration of a lamb in a book is somehow recognisable as a lamb, it may well serve the purpose of symbolizing innocence. Recognition may already be ensured by depicting this animal in a conventional, traditional manner. Adding more realistic traits to its illustration is not required, and may even distract from purpose. It seems that Christian symbolism, as it can be found in Western art, as well in the writings of Western philosophers and theologians over a long period of time, was mainly a symbolism of this kind, where the focus was on what an object stood for, rather than on the object itself. This explains how a tradition of depicting natural objects, among them fish, in a non-realistic manner could originate and remain.

This Christian tradition can be traced further back, however, even regarding the lack of interest in a realistic treatment of fish. It has been pointed out that the Bible does not report any names of fishes.[61] Thus, a folio from a medieval bestiary would, for example, show the sea with its teeming diversity, but not name any particular fish species or make it otherwise identifiable [Figs. 2.13 and 2.14].

This can be seen in the Harley Bestiary, a manuscript dating from ca. 1230–1240, and similarly, in the illustrations of a manuscript from 1465, rendering an epic named *The Salvation*, where fish are used to represent the element of 'water'. Even by the biblical 'whale' not the marine mammal in the modern zoological sense was meant, but like in the Hebrew original and its Greek

61 Pangritz W., *Das Tier in der Bibel* (Munich – Basel: 1963) 72; a medieval bestiary showing Adam naming the animals contains no fishes Hassig, *Bestiaries*, Fig. 10, see also James-Raoul, *Inventaire*. In the Book of Genesis (2:19–20) we read: 'So out of the ground the LORD God formed every beast of the field and every bird of the air, and brought them to the man to see what he would call them; and whatever the man called every living creature, that was its name. The man gave names to all cattle, and to the birds of the air, and to every beast of the field.' There is no talk of the animals of the water.

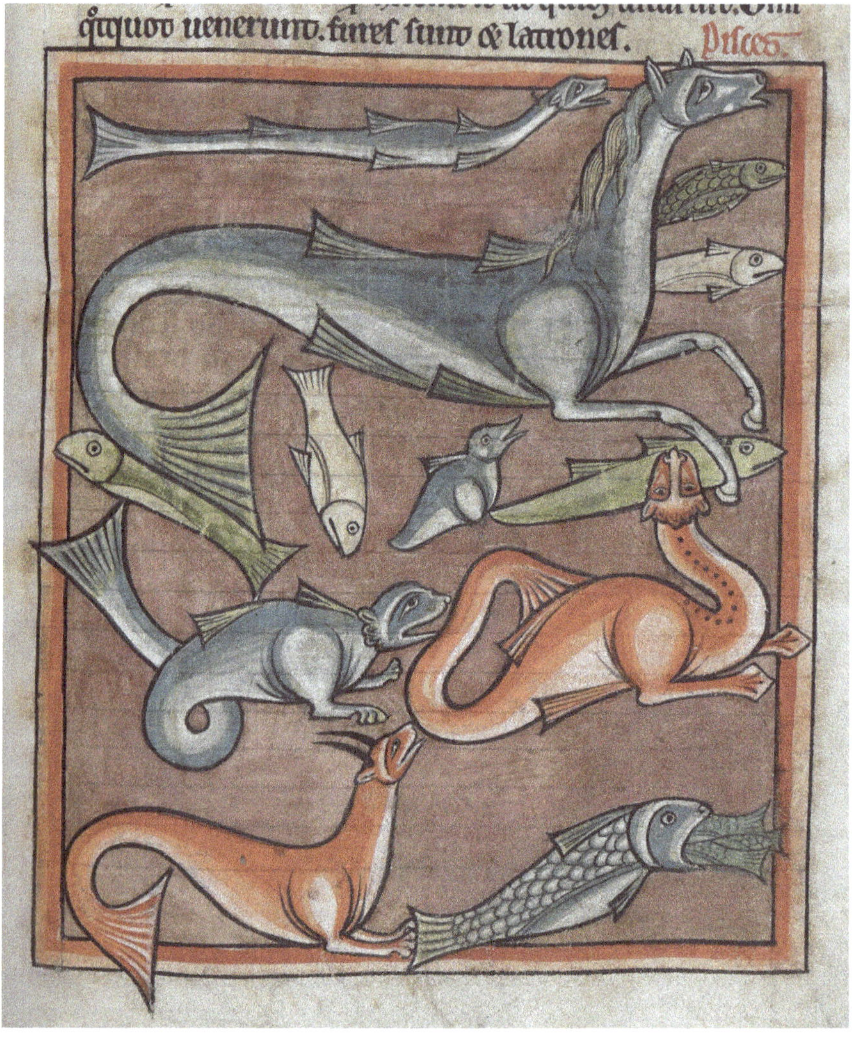

FIGURE 2.13 Marine fishes and hybrids in the Harley Bestiary MS 4751 (*c.*1230–1240)
COURTESY OF THE BRITISH LIBRARY, LONDON

(φάλλαινα) and Latin translations (*balaena*), to be any large fish or, even more generally, any sea monster (κῆτος, *cetus*) [Figs. 2.15 and 2.16].[62]

62 The term 'whale' (Jonah 1:17–2:10, Matthew 12:40) was introduced by Luther's and Tyndale's translations, respectively. See also Cambier H., "Un grand poisson qui pose question. La baleine au Moyen Âge", in Huber-Rebenich G. et al. (eds.), *Wasser in der mittelalterlichen Kultur* (Berlin: 2017) 532–541, ignoring, regrettably, ichthyological questions.

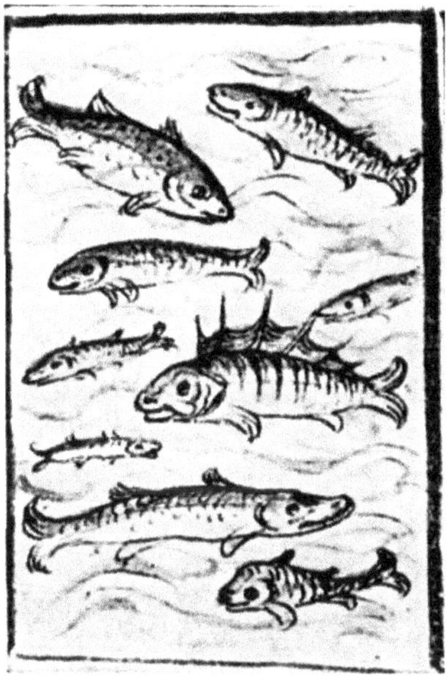

FIGURE 2.14
'Water', the third of the four elements in an epic entitled *Die Erlösung* ('*The Salvation*'), Nuremberg 1465; taken from in Exner M. et al., "Fisch I", *Reallexikon zur Deutschen Kunstgeschichte* IX (Rome: 1987) 18–88

When a fish was used in a painting or on a statue as a Saint's attribute ('Heiligenattribut'),[63] it served to help recognise who this particular saint was, whereas the fish itself remained unidentified. Overall, the extremely rich Christian fish imagery extended, based on the lecture of the Old and New Testament as well as on the writings of the patristic and later Christian theologians, to a large range of subjects, including baptism, the Eucharist, the Passion, resurrection and salvation.[64] For example, Christ could be represented as a fish (*piscis*) and his believers as little fishes (*pisculi*), all of them living in the water, which could be understood to symbolise the water of baptism. The early Christian philosopher and theologian Tertullian (after 150–after 220) expounded this in the following way: 'But we little fishes in the succession of our fish, Jesus Christ, are born in the water and by nothing else than dwelling in the water, we are in salvation.'[65]

63 See the numerous examples in Exner M. et al., "Fisch I", *Reallexikon zur Deutschen Kunstgeschichte* IX (1987) 18–88.

64 See the abundance of literary and pictorial evidence in Dölger F.J., *Ichthys* (5 vols.) (Rome: 1910–1957), Wehrhahn-Stauch L., "Christliche Fischsymbolik von den Anfängen bis zum hohen Mittelalter", *Zeitschrift für Kunstgeschichte* 35 (1972) 1–68 and Engemann J., "Fisch, Fischer, Fischfang", *Reallexikon für Antike und Christentum* 7 (1969): 959–1097.

65 Quoted from Wehrhahn-Stauch, *Fischsymbolik* 5, 61.

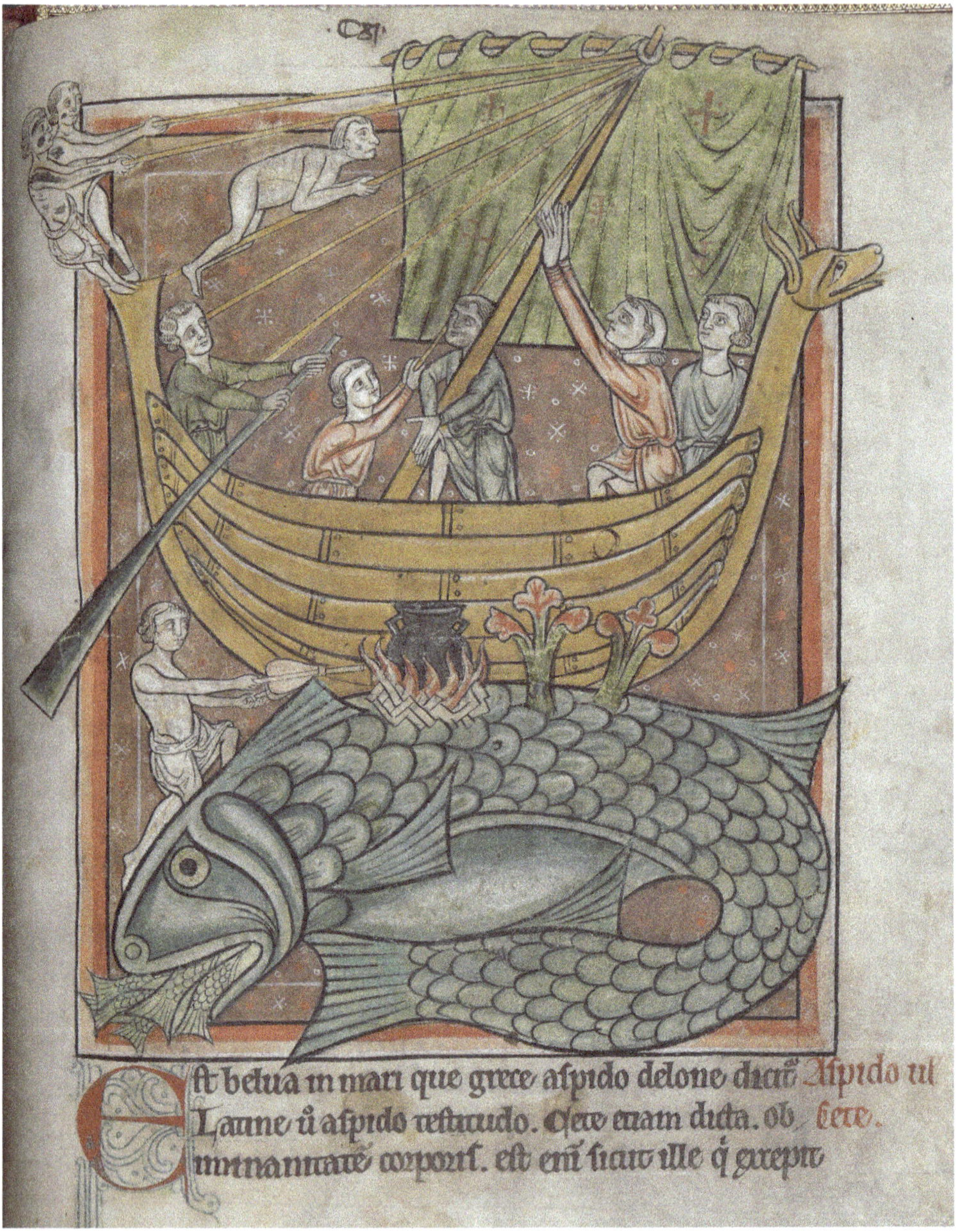

FIGURE 2.15 'Whale' in the Harley Bestiary MS 4751
COURTESY OF THE BRITISH LIBRARY, LONDON

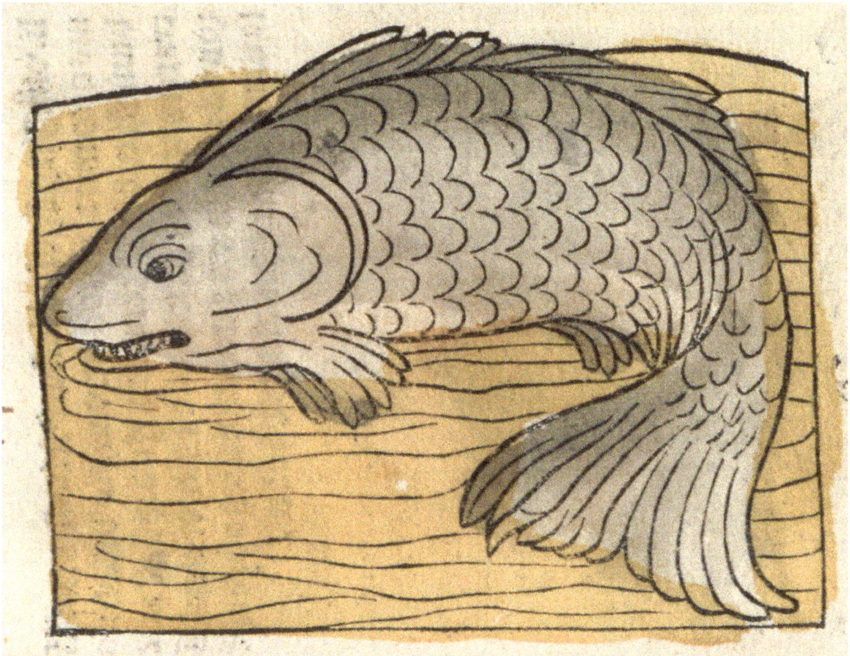

FIGURE 2.16 The same fish as in Fig. 2.16, called *Cetus* ('Large fish', 'Whale'), in *Hortus sanitatis* (Mainz, Jacob Meydenbach: 1491), tract "De piscibus", chap. 19
PROVIDED BY U.S. NATIONAL LIBRARY OF MEDICINE URL HTTPS://COLLECTIONS.NLM.NIH

Regarding its focus on what an object stands for, rather than on the object itself, this kind of symbolism aligns with a philosophical and theological doctrine that that we would like to call *figurative essentialism*. According to this doctrine, a distinction is to be made between *figura* and *essentia*. This means that the outward appearance and, consequently, the visual representation of any creature is in principle secondary to what it *essentially* is and, therefore, *figuratively* denotes.

Figurative essentialism is related to another doctrine, which has been summarised under the term 'dissimilar similarity' (*dissimilis similitudo*, ἀνόμοιος ὁμοιότης).[66] The term may seem paradoxical at first sight, but what is meant

66 Raised to doctrine at the Fourth Lateran Council (1215) but originating from centuries before, see Wehrhahn-Stauch, *Fischsymbolik* 4–7; Bader G, *Die Emergenz des Namens. Amnesie, Aphasie, Theologie* (Tübingen: 2006) 107; Dingel I. – Daugirdas K. (eds.), *Antitrinitarische Streitigkeiten. Die tritheistische Phase (1560–1568)* (Göttingen: 2013) 122.

is that an object that symbolises something else is similar to what it stands for, which enables it to serve as a symbol for it. For example, there is something perceived as powerful about a lion that enables it to symbolise power. On the other hand, the lion is an animal whereas power is a concept, so both are also dissimilar.

To get to the point: Figurative essentialism, as well as the concept of dissimilar similarity, rely on conventions for understanding the relationship between the figurative and essential value of an object as well as the similar and dissimilar traits that relate it to another object. For inventing any such relationship, individual ingenuity is required, bordering on arbitrariness. It is in any case, far away from what Belon and others had in mind when they were claiming to present in their books 'true' pictures of fishes.

Nevertheless, even these pioneers of realism in representing fishes were still familiar with the ideas and conventions of the traditional approach. At the end of his preface to the reader of the fish book published in 1558, Gessner writes (b3v): '[…] and, with nets and the hunting of unreasoning beasts abandoned, let us go forth as reasoning and spiritual hunters and fishers to the glory of God's eternal name.'[67] Gessner thereby refers to Christ's well-known sentence about the Apostles as fishers of humans.

In the dedication letters and brief prefaces of their fish books, also Belon, Rondelet and Salviani tell the reader about their intention to contribute to the glory of God, the Creator of all natural wonders. Since the addressees of these dedication letters were high and highest divine dignitaries, who excelled as patrons, one might be inclined to think that these references to the Christian sense of fish symbolism were merely conventional bows. It seems, however, that this was not so and that they were meant seriously, but it can neither be confirmed nor disproved, since the authors do not provide any further explanation.

Other purposes of these fish books, according to what their authors wrote in their prefaces and dedications were that they should serve to make the knowledge of the ancients regarding the medical benefits of fishes and their nutritional value useful to their contemporary and future readers, as well as to expand this knowledge.

Regarding the animal world, Aristotle was particularly important to the naturalists of the 16th century. This found expression in a series of extensive

67 The beginning of Gessner's wording is a mixture of Latin and Greek: 'relictisque retibus et venatione τῶν ἀλόγων, λογικοὶ καὶ πνευματικοὶ venatores ac piscatores', etc., the Greek part possibly being a quotation.

commentary volumes on the relevant Aristotelian writings.[68] However, these Aristotelian commentators were indulging in philological subtleties and overwhelmingly amounted to nothing more than pure bookish erudition, in particular regarding fishes.[69] None of them was of lasting effect, surviving only as archival fossils. Rondelet once remarked on these books that scholars 'trusted the ancients too much and considered it wrong to accuse them of error, and so they passed on to the remaining people the same opportunity to make mistakes that they had taken from others.'[70]

The naturalists of the 16th century, especially the fish experts among them, usually were not nearly the gifted draftsmen or knowledgeable theoreticians that Leonardo da Vinci or Albrecht Dürer were, perhaps with the exception of Gessner.[71] They were practitioners, who rushed into nearby as well as far-away nature, where they searched for and collected objects, comparing and dissecting what they had found, and also giving these objects names, mostly trying to follow the usage of ancient Rome and Greece. To their descriptions, they added illustrations, which they called 'true' (*verus*) or 'lifelike' (*ad vivum*).

What exactly they meant by these terms was not explained by them. From their point of view, the results obviously spoke for themselves. If we translate their dazzling key terms 'true' and 'lifelike' using the words 'realistic' or 'naturalistic', which are familiar to us, but were unknown within the Latin of their time, it becomes clear what the traditional illustrations they were opposed to were *not* in their eyes: not realistic, but rather fictitious and insufficient, in other words, they were not 'true',

This is, I argue, the spiritual and intellectual dimension of the situation that authors of fish books experienced during the middle of the 16th century. But behind all this, on closer inspection, also an emotional dimension of personal ambition and wounded vanity comes to the fore. This will be topical in what follows.

---

68 See Perfetti S., *Aristotle's zoology and its Renaissance commentators, 1521–1601* (Leuven: 2000).

69 In Augustinus Niphus' (Agostino Nifo, † 1538) *Expositiones in omnes Aristotelis libros* (Venice, Hieronymus Scotus: 1546), to mention an outstanding example, information on fishes are scattered (as in Aristotle) over hundreds of pages, of course without illustrations; on Niphus, see Perfetti, *Aristotle's zoology* 85–120.

70 Rondelet, *Libri de piscibus* 241.

71 On Leonardo da Vinci, Dürer and Gessner, see Zoller H., "Zum Wandel der Pflanzendarstellung während der Renaissance. Vom Beginn des 15. Jahrhunderts zu Leonardo da Vinci, Albrecht Dürer und Conrad Gessner", *Bauhinia* 9 (1989) 109–123.

3 Culture of Debate and Personal Animosities

Conrad Gessner furnished the fourth volume of his *Historiae animalium* on fishes and other aquatic animals (1558) with an extensive preface to the reader. Right at the beginning of this preface, Gessner quotes in Greek the last of a hexameter verse that remains untranslated without any information about author and work. As the only indication, Gessner had added the heading *Aemulatio bona et mala* (Good and bad competition) in the margin. The verse reads with translation:

ἀγαθὴ δ᾽ ἔρις ἥδε βροτοῖσιν (Noble competition is wholesome for men)

It can be assumed that most readers were neither familiar with the author, nor with a translation or the context. Gessner took the opportunity here to address a current conflict that weighed on him, shrouded in extremely erudite scholarship for experts. What was the context and what was this verse about?

The author of the cited verse was the Greek poet Hesiod (around 700 BC), and the quotation is taken from *Works and Days*, line 24, basically a didactic poem intended to instruct about agriculture. Before Hesiod turns to his basic topic, however, he takes the reader back in time to the birth of the gods, to theogony. Hesiod talks about *Eris* (Ἔρις), the goddess of discord and strife, who was a daughter of the Night (Νύξ, in Latin *Nox*), one of the goddesses who was born out of the primordial chaos. Eris is described as a 'wicked' (σχετλίη) woman, nothing but disastrous for mankind.

However, Hesiod places at her side – an innovation of his own – a good (ἀγαθή) goddess of competition and emulation, a daughter of the Night, too, bearing the same name as her sister. Now, while the first Eris separates people and sets them against each other, the second Eris encourages people to surpass each other in good deeds, useful for all.

Both Hesiod and Gessner had reasons to recall the difference between evil and useful rivalry. Hesiod wanted to admonish his brother Perses with his work to acquire wealth only through honest, rural work, not through intrigue and overreaching:

> Perses, attend, my just decrees observe,
> Nor from thy honest labour idly swerve;
> The love of strife, that joys in evils, shun;
> Nor to the forum from thy duty run.[72]

---

72 Lines 27–30 as translated by Thomas Cooke, *The Works of Hesiod* (London: 1743).

Gessner was also in an uncomfortable situation. In his work on fishes, he had relied heavily on direct or slightly modified citations from the fish books of Belon, Rondelet, and Salviani, which had recently appeared. Gessner appreciated all three authors very much and knew Belon and Rondelet personally.[73] He felt, therefore, compelled to comment on a public controversy that was simmering in particular between Rondelet and Salviani, and he did so by referring to the above cited line 24 from Hesiod, whereby he articulated his disapproval of 'publicly exposing science to ridicule by mutual wrangling and bad competition'.[74]

Gessner, an always friendly and balanced character, instead wished for a productive contest of arguments, ideas, and convincing creations, for 'if quarrelsome eagerness, reproaches and slander, ambition and self-love prevail all too much, this kind of competition displeases every honest man.'[75] As will be shown, exactly such an embarrassing bickering among educated men was the situation that prevailed between Rondelet and Salviani.

4 Rondelet's Attacks and Salviani's Defence

The French physician Guillaume Rondelet (1507–1566) – according to a biography written by his faithful but not uncritical student and successor as chancellor of the Faculty of Medicine at the University of Montpellier, Laurent Joubert – was a complex person. Nancy Siraisi[76] condensed some of Joubert's

73 Gessner had met Rondelet during his stay in Montpellier in 1540 (Fischer H., *Conrad Gessner (26. März 1516–13. Dezember 1565). Leben und Werk* (Zurich: 1966) 23), Belon visited Gessner in 1557 in Zurich (Barsi M., *L'énigme de la chronique de Pierre Belon* (Milan: 2001) 21; Glardon P., *Pierre Belon, L'histoire de la nature des oyseaux. Édition en fac-similé avec introduction et notes* (Geneva: 1997) LII). In a letter of October 15, 1556 to Leonhart Fuchs, Gessner, who was unaware of the details of the rivalry between Rondelet and Salviani, hastily sided with Rondelet. The letter is quoted by Leu, *Conrad Gessner* 185 and 276–277.

74 In the original: 'mutuis inter se rixis et vitiosis aemulantibus'.

75 Gessner had the misfortune to be involved at the same time in another spectacular case of insults and accusations raging between Melchior Wieland (Guilandinus), a German botanist who lived in Italy, and the famous Pietro Andrea Mattioli (Matthiolus) an irascible and contentious personality like Rondelet (Herrmann S., 'Ein Preuße in Venedig: Der Botaniker Melchior Wieland (ca. 1520–1589), Pionier der botanischen Feldforschung in der Levante', *Sudhoffs Archiv* 99 (2015) 1–14). In general, such public confrontation was an exception, as there was always the danger of attacking along with the opponent his high-ranking patron as well.

76 Siraisi N.G., *History, medicine, and the traditions of Renaissance learning* (Ann Arbor: 2007) 126.

assessments to form a revealing personality profile, showing that Rondelet was on the one hand a generous, hospitable and convivial man, but on the other hand careless, credulous and easily deceived. In addition, Rondelet was erratic in his scientific interests and literary activities, with the result that he barely finished anything ready for the press. It is well known that Carolus Clusius had to help him with the completion of his two fish books, but it is less known that another person, Johannes Molinaeus, from the Low Countries, had to help as well. And if all that was not enough, Rondelet was also reported to be irascible, which may explain why during the writing of his big fish book project, he started a quarrel not only with his Italian competitor Salviani but also with his other rival, his French compatriot Pierre Belon (1517–1564).

From his perspective, Rondelet had plenty of reason to be upset with Belon. Belon, as well as Rondelet, vied for the favour of their common patron, Cardinal François de Tournon (1489–1562),[77] and both worked on books about fishes. But while Rondelet was unable to publish his first volume until 1554 (as said, only with the help of others), Belon had already in 1551 and 1553 presented illustrated fish books in French and Latin. When finally, in 1554, the first volume of his *Libri de piscibus marinis* came out, Rondelet attacked Belon twice. In the preface, he ridiculed Belon's two fish books as 'booklets' (Belon's second book *De aquatilibus* of 1553 covered nearly 500 pages). Then again in the middle of the text, he contemptuously referred to Belon as a 'druggist' (Φαρμακοτρίβης, an allusion to the rumour that Belon had never formally completed his medical studies with a doctoral degree).[78] The key message of his attack was that Belon published pictures of fishes without mentioning that he, Rondelet, already had been working on this subject for some time. Belon responded to these taunts in the dedication to the Cardinal of Châtillon in his third fish book of 1555 (*La nature et diversité des poissons*), noting briefly and coolly that 'there has been no one who showed true pictures (of fishes) before us, nor of birds and snakes, which is enough to absolve us of all slander.'

Shortly after this unfriendliness towards Belon, the situation for Rondelet worsened. After spending more than 10 years working on his fish book at great personal expense,[79] he had to realise that his colleague and former friend (as he called him), Hippolito Salviani, was about to publish a fish book of his own with pictures made with a novel technique, pictures by which Salviani 'surpasses all by the accurate beauty of the copper engravings', as Gessner stated

77 On Tournon, see Huppert George, *The Style of Paris: Renaissance Origins of the French Enlightenment* (Bloomington: 1999) 1–6, 87.

78 Rondelet, *Libri de piscibus* fol. a5v and 423.

79 Joubert, *Gulielmus Rondeletius* 188–190.

in 1558.[80] Thus, not only was Belon chronologically ahead of Rondelet, boasting of more extensive expeditions than Rondelet had ever undertook, it was also no longer sufficient to produce good, nature-like fish images in the usual woodcut manner. Rondelet had 'only' the certainty of having provided *most* high-quality descriptions of fishes but not the *best* images artistically possible. In this unpleasant situation, which for Rondelet was apparently hard to bear, he decided in the second volume of his fish book (1555) to move on from malice and innuendo, as in the case of Belon, to open, personal hostility in order to present his own achievements in the proper light. In the following, we first let Rondelet speak, then proceed with Salviani's reply.

4.1 *Documents Part 1: Rondelet's Attacks*

4.1.1 Text 1: Rondelet, *Libri de piscibus marinis* (1554)

From the "Preface to the reader" ("Praefatio operis ad lectore", fol. a5v).

In his preface, Rondelet attacked contemporary authors of fish books without referring to anyone in particular. He accuses some of them of exploiting his writings for their own works, thus having committed plagiarism. But such alleged plagiarism has been carried out by them in such an inadequate manner, he says, their books provide no benefit to the reader at all.

> And also those who have recently published booklets on the same subject did not deter me from writing a history of. fishes.[81] Among them are some who, after inspecting my sheets, driven by the novelty of the matter and the beauty of the pictures and eager to snatch away the glory I deserve, cobbled together much of their writings from mine; but so unprepared in the matter to be treated that they only babble most disgracefully[82] when describing the parts of the fishes that are necessary for their understanding, and make grave mistakes in naming the different kinds of fishes as well as in providing quotes from ancient authors and carelessly adapting them to meanings of their own so that after reading their books the reader comes out not a bit wiser.

80 Gessner, *Historia animalium IV*, Preface, fol. b1r. That is, Belon (1551) and (1553) and the first instalments (?) of Salviani (1554–58). Salviani's book was not completed until 1557 and published as a whole, according to a second colophon, in 1558.

81 An allusion to the fish books of Belon and Salviani.

82 On Rondelet's term *hallucinati* (*exprimendis turpissime hallucinati sint*), see also Glardon (2011: 103–104).

4.1.2 Text 2: Rondelet, *Libri de piscibus marinis* (1554)

On the *Mugilis niger*,[83] chapter 6, book 15, pp. 423–424.

Rondelet's attacks continue later on in his fish hook of 1554, in the account of a particular fish, the *Mugilis niger*. Here he names individual authors and identifies others by giving personal details. Regarding the fish pictures in his books, he accuses two authors, Belon and Salviani, of copying them after obtaining them from him 'with lies and deceit'.

Before that, however, he praises two other authors, who behaved in a friendly and honourable manner towards him, Simone Porzio, who had sent him a specimen of the fish that this chapter is about, and Conrad Gessner, who sent him fishes from the Danube. Whereas these two are referred to by their names, Belon is just contemptuously called 'a druggist', as already mentioned above.

> This fish is unknown in our waters, and it is a rather rare fish that we present here. It is scaly, the appearance of the body very similar to the Mugilis, but completely dark, and has black stripes that stretch from the gills to the tail: I called it Mugilis niger for this reason. It has a strongly protruding lower jaw and, therefore, a wide gaping mouth. On its back it carries seven or eight separate spines, which are connected by no membrane, followed by a small fin.[84]
>
> This fish, given to me for drawing by Porzio[85] in Pisa, one of the most eminent philosophers who, as he combined an extraordinary erudition with a singular humanity, has created a collection not only including this fish, but also many others to take a look at;[86] that is why I always will be very grateful to him, for it is proper for a noble-minded and decent man to

83 Rondelet's fish is unidentifiable and does not matter in the subsequent attack on Salviani. The fish name *Mugil* (or *Mugilis*) was widely applied in ancient times, see Coney W.C., "Mulled thoughts: Mullus and Mugilis in Pliny's *Naturalis Historiae* and the *De Re Conquinaria* of 'Apicius'", *Pseudo-Dionysius* 18 (2016) 49–58.

84 The following sections are omitted in the French version: Rondelet, *Histoire des poisons* 326.

85 On Simone Porzio, see also the explanations above in the chapter *The Aristotelian legacy*. Porzio is also mentioned by Rondelet on p. 327. His *De piscibus* remained incomplete. A fragment has survived and is reproduced by Del Soldato *Catalogo naufragato*; see also eadem, *Simone Porzio* 92–100, Perfetti *Aristotle's zoology* 123–129 and Lavenia V., "Porzio, Simone", in *Dizionario Biografico degli Italiani*. Vol. 85 (Rome: 2016) 142–145. According to the report of a contemporary (Jacques-Auguste de Thou), Porzio was consumed with envy when in 1554 he realised that Rondelet had preempted him with the publication of his fish book, see Del Soldato, *Simone Porzio* 92.

86 The images of Porzio's *De piscibus* were made, commissioned by Grand Duke Cosimo I de' Medici from the famous painter Francesco Bacchiacca (1494–1557), and must have been remarkable.

> admit from whom he has received a benefaction. Of the same generosity towards me was the highly educated Gessner, who was anxious to send me fishes from the Danube, which he had collected with great zeal and effort, who can be sure that I will always think of him and be indebted whenever I can assist him with any service.[87] However, those were not of this attitude towards me who, after copying my fish pictures with lies and deceit, preferred to be guilty of the crime of plagiarism instead to acknowledge gratefully that they have received something from me.
>
> One of these people was a druggist with whom I worked to prepare medicines for the celebrated Cardinal François de Tournon.[88] When this man noticed that I often wrote about fishes and exchanged pictures of fishes with friends, he inquired about their names and picked out some pictures, and finally published them, some time after having been on a journey (which is something that I really very much approve of [424] if he had only combined a certain scholarship with his eagerness to see a lot of things), without letting me know and without mentioning my name even once.

After dealing with Belon, Rondelet takes on Salviani, again not introducing this author by name, but still clearly identifying him as the personal physician to the Cardinal who later became Pope Marcellus II. He blames Salviani not only for having stolen his own intellectual property, but also criticises his lifestyle. Rondelet concludes his attacks by again charging his rivals with having committed plagiarism while not having produced anything of value. He then asks the reader to judge his books in comparison to theirs. This is something Salviani would follow up on in his defence.

> In the same way, the personal physician of the Cardinal of the Holy Cross[89] was injurious to me, to whom I had left my sheets in Rome for inspection, when the physician, attracted by the novelty of the matter as

87 Gessner, in his *Historia animalium IV* of 1558, quoted verbatim numerous descriptions of Rondelet, including pictures, among them the first two paragraphs of the present description (1558: 653).

88 In the 1540s Rondelet was in the service of Cardinal François de Tournon (1489–1562) and undertook many diplomatic journeys with him.

89 This Cardinal is Marcellus Cervinus (Marcello Cervini, 1501–1555), later Pope Marcellus II; before his accession as Pope he had been Cardinal priest of the Basilica of the Holy Cross in Jerusalem. Salviani was the personal physician to three popes: Julius III, Marcellus II and Paul IV. On Pope Marcellus II, see Brunelli G., "Marcello II, papa", in *Dizionario Biografico degli Italiani*. Vol. 69 (Rome: 2007) 502–510.

well as by the variegated manner of depicting the fishes, suddenly gathered up for printing a great deal of my fish pictures as well as some others he had obtained incidentally, as I hear, and this at the Cardinal's expense. Is it surprising that Salviani is able to provide some commentaries of his own about fishes, for he has been travelling for long periods on different parts of the sea, has had the opportunity unremittingly to dissect fishes and inspect their parts, and to elaborate all this with commentaries and distinctions, but who spends whole days welcoming, receiving and escorting courtiers in order to earn a living?

These people tried, with the greatest zeal, to forestall me and did not want to admit honestly from whom they had obtained something in order to acquire a certain fame by a new treatment of the matter as well as by concealing my name. But they only achieved to produce blind puppies just like stray dogs. I say this to demonstrate the truth and to enable you to form a more reliable and better judgment of their and my writings, you should indeed believe, honourable reader.

4.2 *Documents Part II: Salviani's Defence and Counterattack*

In the 16th century, woodcuts dominated in illustrated books, while the seventeenth and eighteenth century has been called the age of copper engravings.[90] A remarkable exception to this general rule was Salviani, who was the first naturalist to use copper engravings in a book on natural science. He deserves the honour of being a true innovator and not – as commonly claimed[91] – Fabius Columna (Colonna), another Italian naturalist.[92] Actually, Columna, published as late as 1592 a work entitled *Phytobasanos* ('A painstaking inquiry into plants'), which included mainly images of plants, but also two images of a fish and a sea star.

Unlike Rondelet, Salviani was financially independent[93] and could afford to publish his fish images as copper engravings – a technique that was extremely

90 Nissen C., *Die zoologische Buchillustration. Ihre Bibliographie und Geschichte*. Vol. 2 (Stuttgart: 1978) 125–158.

91 For example Nissen C., *Die botanische Buchillustration. Ihre Geschichte und Bibliographie* (Stuttgart: 1966) 125–158 or Blunt W., *The art of botanical illustration* (Woodbridge: 1994) 99. Pinon L., *Livres de zoologie de la Renaissance: une anthologie (1450–1700)* (Paris: 1995) 95, however, is correct.

92 Columna (1567–1640) was a member of the *Accademia dei Lincei* in Naples, which was famous for its interest in technical novelties.

93 On Salviani's biography and earning capacity, see Jaitner-Hahner U., *Città di Castello nel Quattrocento e nel Cinquecento: economia, cultura e società* (Sansepolcro: 2020) and Andretta E., "Salviani, Ippolito", in *Dizionario Biografico degli Italiani*. Vol. 90 (Rome: 2017) 22–24; for Salviani's various medical activities, see Andretta E., *Roma medica. Anatomie d'un système médical au XVIe siècle* (Rome: 2011) 430–437. Salviani was initially supported

costly at the time, as each copperplate had to be integrated separately and could not be printed together with the text, as in the case of woodcuts.[94] Therefore, Salviani hired a skilled artist (*pictor*), Bernardus Aretinus, and print everything at his own expense. The images made such an impression on Antonio Lafreri (Antoine Lafréry, 1512–1577), a French cartographer and publisher active in Rome, that he in 1559 decided to publish with Salviani's approval a separate edition containing only the copperplates (but not all), so that other painters and illiterate people without education could also 'delight nonetheless in drawings engraved and printed in copper'.[95] In 1593, this text-free edition was published again, entitled *Eicones piscium*.

Salviani and Rondelet knew each other personally, and the Italian believed he was his French colleague's friend. However, it was Rondelet who started a severe controversy (instead of the reverse),[96] referring to Salviani not by name, yet unequivocally for insiders, as 'Medicus Cardinalis à Sancta Cruce', that is, the personal physician of the Cardinal of the Holy Cross, who was later to become Pope Marcellus II,[97] and accusing him of plagiarism. Salviani could not allow this to rest.[98]

4.2.1 Text 3: Salviani, *Aquatilium animalium historiae* (1558)
Historia octogesima septima: *De Callaria* (History no. 87: On the *Callarias*)[99]
fols. 231r–232r.

---

by his patron Marcellus Cervinus, both non-materially and financially, but later he was a wealthy man, who also had his own flourishing printing house in which he published his fish book.

94 See Griffiths A., *The Print Before Photography. An Introduction to European Printmaking, 1550–1820* (London: 2016).

95 Salviani, *Aquatilium animalium*, Preface.

96 As stated by Lewis G., "Clusius in Montpellier: A humanist education completed?", in Egmond – Hoftijzer – Visser (eds.), *Carolus Clusius. Towards a cultural history of a Renaissance naturalist* (Amsterdam: 2007) 65–98, here 84 and by Perfetti S., "Philosophers and animals in the Renaissance", in Boehrer B.T. – Kalof L. (eds.), *A cultural history of animals in the Renaissance* (Oxford, UK – New York: 2007) 147–164, here 161 and idem, *Animali pensati nella filosofia tra medioevo e prima età moderna* (Pisa: 2012) 125.

97 Rondelet (1554: 424), cf. my translation above (*Text 2*).

98 The public revocation of friendship, as it was done by Rondelet, was no trivial offense, it rather aimed at the moral discrediting and excommunication as a member of the world of honourable scholars (*Republic of Letters*). On the theory of friendship (*amicitia*) among sixteenth-century scholars, see Pinon, *Clématite* and Egmond F., "Clusius and friends: Cultures of exchange in the circles of European naturalists", in Egmond F. – Hoftijzer P. – Visser R. (eds.), *Carolus Clusius. Towards a cultural history of a Renaissance naturalist* (Amsterdam: 2007) 9–48, here 36–44; additionally see Pangle L., *Aristotle and the philosophy of friendship* (Cambridge: 2003).

99 The Cohen D.M. – Inada T. – Iwamoto T. – Scialabba N., fish is the Forkbeard (*Phycis phycis* Linnaeus, 1766), for ichthyological details see Cohen D.M., Inada T., Iwamoto T.,

Like Rondelet, Salviani uses his fish book to engage in the debate. In a chapter about a particular fish (the *Callarias*) he tells the reader how he obtained and read the book in which Rondelet had launched his attack. This caught him by surprise, Salviani says, since he had taken the French author to be a friend.

> When in writing these histories I had arrived at the 87th history, which is dedicated to the 93rd fish,[100] I got hold of Guillaume Rondelet's books on marine fishes, which had then been delivered for the first time to Rome and which I received with incredible affection and amazing benevolence as those of a man who, as I believed, was extremely experienced in fish matters and most friendly-minded towards me. And after interrupting entirely the task of writing also the very few of our stories that had remained, I immediately began to read these books with the greatest desire and incomparable pleasure, and did not stop until I had read them all carefully.
>
> After examining these books in a friendly and benevolent manner, I realised that my opinion had very much deceived me. For, as he renders some things, said with all due respect to him, unclearly, treats certain other things in a shortened manner, and most of them really falsely, he did not prove himself to be so excellent in this field as the opinion of himself exhibited in his ostentatious and boastful behaviour in the presence of almost all people. And what I find even more remarkable is that I should ever have believed, judging him in my own way, that he was a particularly good friend of mine. I realised that his mind is more hostile than friendly towards me, and this without any fault whatsoever on my side.

He then goes on to narrate how he and Rondelet had once met in Rome, where Salviani had introduced his supposed friend to his collection of anatomic showpieces. In return, Rondelet had shown him one of his illustrated fish books.

> For when, at the papal transition during which Julius III was elected Pope,[101] he had come to Rome as a travelling companion of the most

Scialabba N., *Gadiform fishes of the world (Order Gadiformes). An annotated and illustrated catalogue of cods, hakes, grenadiers and other gadiform fishes known to date* (Rome: 1990) 65–69. Its identity, however, is of little importance in the controversy.

100 That is, the *Callarias*, the subject of Salviani's 87. history. Its copper engraving (shown above) is prefixed to the history.

101 After the death of Paul III in a long conclave that began in November 1549, Julius III was elected pope in February 1550.

famous and honourable Cardinal de Tournon and found that I was not only skilled in dissecting bodies through much practice over a long period of time, but also preserved at my home certain most beautiful anatomic showpieces, he made an effort – in the same way as other learned men had wished before and afterwards to see our anatomic showpieces upon coming to Rome, not only Italians, but also Frenchmen, Germans, Hungarians[102] and Spaniards, due to their humanistic education and interest in the fine arts – to see our objects as well, driven by the same zeal and desire.

When, therefore, during the same interregnum,[103] Rondelet and three other highly learned French physicians, in the company of other venerable French Cardinals, had assembled in a friendly atmosphere of my house, I showed them excellently drawn single parts of the human body. Then I showed a human skeleton, constructed with such skill and care that it could be deconstructed into several pieces and kept in a cupboard of two cubits, and again, when needed, on the spot and quite easily could be put together from all parts and erected as a whole. Finally, we presented four-cubit-sized statues in which the position, origin, course, insertion of sinews, appearance, size, and natural colour of each of the muscles of the human body were represented extremely lifelike and distinct, according to the judgment of scholars, who had seen the same, an outstanding and useful work, created with the greatest labour of my own and at no small expense.

After having looked and weighed these things, Rondelet said: as you have done with dissecting bodies, I have spent many years with the exact knowledge of fishes, and I have faithfully painted pictures of almost all the fishes, which I will show you, if you like. And I answered him that nothing better could happen to me, for, as well as regarding flying and terrestrial beasts, I have always most enjoyed findings about aquatic animals, even though, distracted by anatomy, I have not devoted much effort to this highly distinguished field of philosophy.[104]

102 Salviani uses the name Pannonians (*Pannonij*) here, according to the name of the former Roman province of Pannonia. This province included much of the Danube basin, which was later on settled by the Hungarians.

103 The interregnum is the time of conclave after the death of Paul III from 1549 to 1550.

104 That is, *philosophia naturalis*. During the Renaissance, natural philosophy comprised what is now recognised as life sciences (including botany and zoology) and inanimate or physical sciences. Natural philosophy was regarded as a preliminary stage of medical studies, see Grendler, *Universities* 267–268.

When Salviani had looked at the fish pictures in Rondelet's book, so he lets the reader know, he was disappointed by their poor quality. He commissions an Italian artist with producing something better, which is in fact accomplished, and shows the results to Rondelet before the latter leaves Rome. In the following years, Salviani continues with his fish studies and eventually publishes an illustrated fish book of his own.

> But I added that since Marcellus Cervinus, the most honoured cardinal priest of the Titular Church of the 'Basilica the Holy Cross in Jerusalem',[105] my most benevolent lord, is exceedingly interested in all the fine arts and highly appreciates learned men, you could show the pictures, if you agree, to him and me at the same time, so that you would not only bias me even more towards you through this complacency, but will become better acquainted with this great man. Rondelet liked this idea and we determined a day. On the appointed day, Rondelet was kindly and cordially received by Cardinal Marcellus (in whom I had previously been an advocate of his merit and education), and he brought along a very large book full of numerous fish pictures which he showed one at a time to the Cardinal in my presence.
>
> These pictures were, however, besides not being painted with the natural colours, but merely with black ink, crude and [231v] inappropriate, so they did clearly not reflect in any way the appearance of the fishes they portrayed. So we were not a little surprised to have shown them by a man with such a big name. Wishing, therefore, to find out whether this had happened because of the difficulty of the subject or rather due to his negligence, I had the excellent painter, Bernardus Aretinus,[106] paint a mullet and a moray on the following day. When this was done as desired, I wanted him to paint two other fishes as well. And after a likewise rather outstanding work had successfully been accomplished, I hired, captivated by the elegance of the painting, this same Bernardus, having agreed

105 Marcellus Cervinus (Marcello Cervini, 1501–1555), later Pope Marcellus II; before his accession as Pope he had been Cardinal priest of the Basilica of the Holy Cross in Jerusalem in Rome (Salviani: "TT.S. ✠ Presbyter Cardinalis"); see also Brunelli, *Marcello II*.

106 The life data and more specific circumstances of the artist Bernardus Aretinus (Aretino) are completely unknown. Nissen's claim (*Zoologische Buchillustration* 117) that Nicolas Béatrizet (Beatricetto, *c.*1520–*c.*1560) in Rome made Salviani's copper-engravings is based on one single, obscure source from 1950. A highlight of confusion is provided by the Smithsonian Libraries, which identified Salviani's engraver Bernardus Aretinus with a medieval (!) theologian of the same name (also known as Bernard of Arezzo, † 1342), see https://www.biodiversitylibrary.org/bibliography/82337#/summary.

> on a monthly salary. He lived with me for two years and would still be living with me if he had not died prematurely after two years.

After providing the reader with this background, Salviani sets out to refute Rondelet's allegations one after another. First, he deals with having been blamed by Rondelet of interfering, through his fish studies, in the French scholar's business.

> All this was not unknown to Rondelet. In fact, when I showed him, before his departure from the city, twenty pictures of fishes that had already been drawn, he even admired their excellent quality. But when, after his departure, the number of drawings increased day by day, not only from indigenous people, but also from the help of friends, and from foreign fishes, the interest became stronger, and no longer satisfied with the pleasure of the mere pictures alone I also turned to tracking the ancient names of the fishes as well as their investigation in every other respect. And finally, things went so far that I did not shy away from making public those things on which I had been working day and night for a full seven years.[107]
>
> These are the facts, and we want to investigate unbiasedly whether Rondelet rightly or wrongly accuses me. In the sixth chapter of his fifteenth book on marine fishes, he says: 'In the same way, the personal physician of the Cardinal of the Holy Cross was injurious to me.'[108] Since he accuses me of having done something wrong with these first words, he says that in order not to appear as a slander, he will publicly announce what injustice he has suffered from me: 'After I had left to him my sheets in Rome for inspection, the physician, attracted by the novelty of the matter as well as by the variegated manner of depicting the fishes, suddenly gathered up for printing a great deal of my fish pictures as well as some others he had obtained incidentally, as I hear, and this at the Cardinal's expense.' But regarding the first, it is utterly ridiculous and childish to accuse my person of any injustice just because I should have begun, lured by the novelty of the matter, to deal with this same matter, which he had already been working on for a long time.

107 As mentioned, Salviani's fish book was completed in 1557 and finally appeared in 1558, but presumably pre-publications (instalments?) had been issued since 1554 which, however, have not survived and which Salviani did not connumerate.

108 This and the following quote are from Rondelet, *Libri de piscibus* 424, cf. my translation above.

> For even if it were so, who would therefore hold it against me, given that it has always been noble and worthy of every praise to burn for the passion of supporting the Republic of Letters? Or who will accuse me of injustice towards him, since if our elaborations are inferior to his, he will obviously earn a lot more of praise, but if ours prevail, he had better congratulate me for this than be angry with me, as much as the public benefit is to be preferred to the praise of one's own, where both cannot be achieved, by an honourable man?

Then he replies to Rondelet's claim that Salviani had committed plagiarism regarding his fish pictures. Salviani's line of defence is clear here and refers to the outstanding achievement in the history of ichthyology at that time: true pictures, the emergence of realistic fish images. The pictures in his own books are 'drawn superbly lifelike', Salviani says, whereas Rondelet's pictures are only 'crude illustrations'. There is, according to Salviani, really no indication that he might have 'stolen' anything from the French author.

> As for the next point: How wrong the claim is that I had stolen many of his fish images is not only evident from the fact that I have seen only once – as he himself knows and testifies – a larger quantity and, as it were, in passing, but can also be clearly recognised by comparing his pictures with mine. For no one will be so stupid that he will not notice very easily that our pictures are drawn superbly lifelike,[109] representing the fishes with such a great similarity[110] impossibly to be taken from his crude illustrations.

Rondelet had furthermore criticised Salviani's dependence on patronage and his courtier-like lifestyle. Salviani takes the opportunity to explain why Rondelet is wrong here, too.

> Finally, he adds: 'and this at the Cardinal's expense', but this does not amount to an injustice towards him, nor puts shame on me, but rather reflects his honesty. For if to be praised by a praised man is no unimportant praise, how much more is my honour enlarged by the fact that Marcellus Cervinus not only honoured me quite often, as many know, with his most honourable appreciation in front of a great number of people, but always helped me out as well in a most generous manner at

109 In the original: 'ad vivum egregie depictas'.

110 In the original: 'verissime ipsorum piscium similitudinem repraesentantes'.

> his own expense, regarding other domestic difficulties of mine, but also in our production of these illustrations and histories of fishes? He has deservedly been admired by all not only because of the most elevated dignity of being a Cardinal, but much more because of the purity of his life, the integrity of his mind, his observance of the Christian faith, his outstanding knowledge of all things, his benevolence towards all good and learned men, his compassion for the poor, and finally, because of his truly human attitude towards all men of whatever kind or rank. And as he was at last elected pope to the incredible joy of all, so he died prematurely and unexpectedly on the twenty-first day of his pontificate to the incomparable sorrow of almost the whole world.[111]

Salviani is confident that not only the pictures in his books, but also his descriptions of the different fishes clearly outdo what Rondelet has to offer. He even announces that he will publish another book shortly, in which he will deal in more detail with Rondelet's shortcomings regarding the study of fish.

> But as Rondelet vilified me with these words, so he persecutes me, not satisfied with his calumnies, to show his hostile attitude towards me even more, with injuries and maledictions, adding them to what he had already written [232r]: 'Is it surprising that Salviani is able to provide some commentaries of his own about fishes?'[112] But how bold this admiration from his side is, becomes clear through the matter itself. For as our fish pictures, in the opinion of all, surpass so much his own that they could or should in no way be compared to them, so we have no doubt that in the judgment of the scholars our history of fishes will prove to be more complete, better explained and more true. With the same boldness, he also says: 'Someone who has been travelling for long periods on different parts of the sea, who has had the opportunity to dissect fishes and inspect their parts, and to elaborate all this with commentaries and distinctions.' How much more efforts and diligence I have actually spent in doing all this, can also be clearly seen from our present reports; and in addition it

111 Salviani's portrait of Marcellus Cervinus was no fulsome praise for his patron, but corresponds to our present knowledge of Pope Marcellus II, see Brunelli, *Marcello II*. After the unexpected death of Pope Marcellus II, Salviani dedicated his fish book to his successor, Pope Paul IV; his unprinted dedication was published by Pollidorus Petrus (Pietro Pollidori), *De vita, gestis, et moribus Marcelli II. pontificis maximi commentarius* (Rome, Hiermonymus Mainardus: 1744) 85–87.

112 This and the following quotations are from Rondelet, *Libri de piscibus* 424 (cf. my translation above).

> will be very evident in our next book, which we will publish immediately after this issue, entitled *Critique of Guillaume Rondelet's books on fishes*.[113] For the errors that we have certainly noticed in many of his books could not be unfurled in our reports, partly because much of them had already been printed when his books reached us, partly because we considered it inappropriate to confuse our reports with rebuttals of Rondelet's mistakes and to increase their size; in the special book we have collected and refuted the mistakes. We have not done this out of malignity or envy, but rather in a friendly spirit, so that the truth may shine even brighter.

In contrast to what Rondelet holds against him, Salviani finally claims that he is not neglecting his duties as a scholar and physician in favour of courteous obligations, depending on other people's benevolence and support. As a renowned and valued physician, so the reader is told, he is well able to earn himself and his family a living. Then he returns to the subject matter of the current chapter in his book, the Callarias fish.

> So that it does not look in the end as if Rondelet had judged me by mistake, I quote the pretext for his opinion; he says: 'someone who spends whole days welcoming, receiving and escorting courtiers just in order to earn a living.' With these last words, he is guilty not only of insult and temerity, but also of mendacity. For besides the fact that this does not happen in Rome, where the equally highly regarded and venerable Cardinals, as well as the best and most distinguished Fathers of Christendom, are always accompanied by learned, noble, and honoured men, no one will be able to testify having seen me at any time, and even less during whole days, busy with such things. I am always fully occupied with my daily and nocturnal studies and visitations of the sick.
>
> And finally I am not someone who, if I am forced by Rondelet's impertinence to talk about myself, I implore you, honest reader, to take all of this not as complacency yet, would depend on earning himself some extra money, or who, if in need of anything, would have to obtain it by greeting and escorting. For in this city, I have such a good reputation, for which I thank God, as much as I can, that, as I have already been a full professor of practical medicine (as it is called) at this beneficial institution, the Gymnasium Romanum,[114] for many years and still continue to

113 This book has never been published.

114 *Gymnasium Romanum* is the old name of the University of Rome, the later *Sapienza*, see Schwarz B., *Kurienuniversität und stadtrömische Universität von ca. 1300 bis 1471* (Leiden: 2012) 218–220. Salviani taught practical medicine at this university from 1551 to 1568.

> do so, so some of the most well-known cardinals, many bishops and a great number of Roman and foreign noblemen will gladly call upon my services if they require medical help. As a result, I have enough income to feed my whole family honourably.
>
> But since I might have said more in defence of my honour against Rondelet's slanders than has been adequate, so let's drop this topic and return to the fish on our 93rd plate.

Reading Rondelet's and Salviani's texts, one gets the impression that, as in a court case, one word stands against another in the argument between the two opponents. It is difficult to form an objective, unbiased picture of the contrary statements, since there are no other testimonies, e.g., from eyewitnesses of the meeting between Rondelet and Salviani.

Instead of one-sided partisanship, I would like to try in the next section a simple approach that still promises more verifiability, which already Salviani had proposed: to contrast the fish images themselves, as they are available in their books. This comes down to a juxtaposition of Rondelet's woodcuts and Salviani's copper engravings. For reasons of space, such a comparison cannot be based on all or even many images of the two opponents. An exemplary comparison of two images representing the same species must suffice.

5 Rondelet's and Salviani's Pictures of the Flying Gurnard Compared

We would now like to put what Rondelet and Salviani are asserting, as well as the artistry of the draftsmen they employed, to the test and compare the pictures of a certain, outwardly not plain fish to see whether a suspicion of dependency (plagiarism) is justified and which picture is qualitatively more demanding. We have chosen the Flying gurnard (*Dactylopterus volitans* L.) and place Rondelet's woodcut by Georges Reverdy opposite to Salviani's copper-engraving by Bernhardus Aretinus [Figs. 2.17 and 2.18].

Rondelet's picture is not bad, in any case good enough to clearly identify the species. What one could criticise is that the two long spines in front of the dorsal fin are a bit sparse compared to those in Salviani's picture (red arrows). These two spines are important, they are the distinguishing mark of the Flying gurnard, as can be seen when comparing Salviani's picture with a modern drawing [Figs. 2.19 and 2.20].

Basically, the comparison of Rondelet's and Salviani's pictures shows, as every unbiased viewer will surely admit that Salviani's copperplate is not a plagiarism of Rondelet's woodcut, it's not even inspired by it. It is obviously more plastic, more filigree, in short overall better.

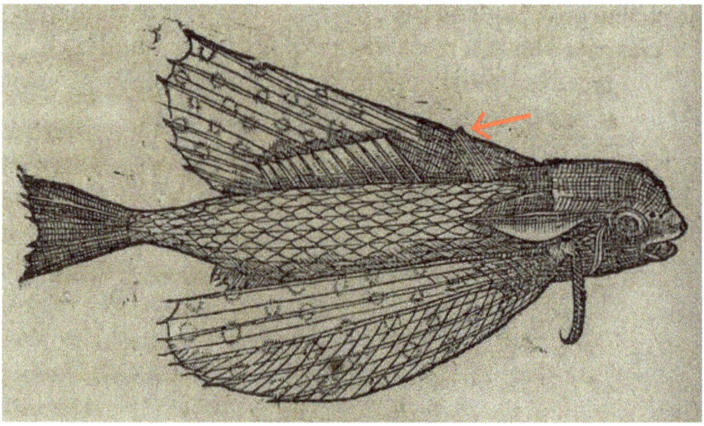

FIGURE 2.17 Flying gurnard erroneously called *Hirundo* ('Swallow') from Rondelet Guillaume, *Libri de piscibus marinis* (Lyon, Macé Bonhomme: 1554) 284. The name *Hirundo* was usually reserved for the Tropical two-wing flyingfish (*Exocoetus volitans* L.), even in Rondelet's time
BIBLIOTHÈQUE NATIONALE DE FRANCE

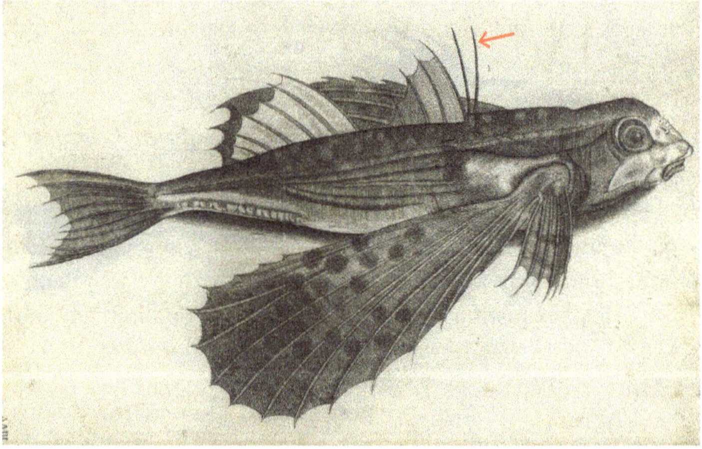

FIGURE 2.18 Flying gurnard called *Milvus* ('Hawk', 'Kite') from Salviani Hippolito, *Aquatilium animalium historiae liber primus, cum eorumdem formis, aere excusis* (Rome, Hippolito Salviani: 1558) fol. 187v. Also called *Milvus* by Belon, *De aquatilibus libri duo* (Paris, Charles Estienne: 1553) 197. Rondelet's *Milvus* in Rondelet, *Libri de piscibus marinis* (Lyon, Macé Bonhomme: 1554) 297, in turn, shows the Grey gurnard (*Eutrigla gurnardus* L.)
BIBLIOTHÈQUE NATIONALE DE FRANCE

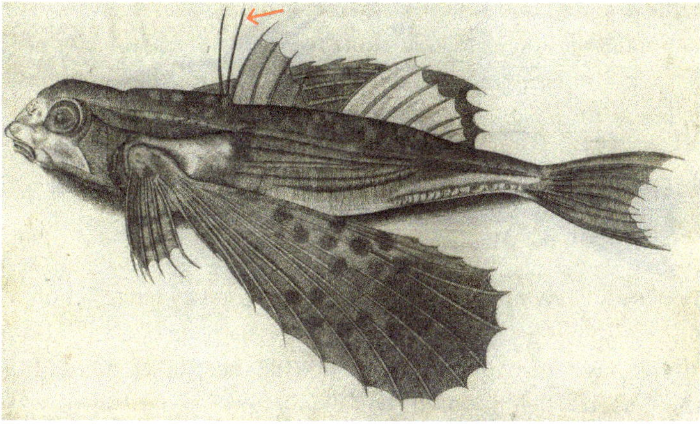

FIGURE 2.19 *Milvus* (mirrored) in Salviani Hippolito, *Aquatilium animalium historiae liber primus, cum eorumdem formis, aere excusis* (Rome, Hippolito Salviani: 1558) fol. 187v., showing Flying gurnard (*Dactylopterus volitans*)
BIBLIOTHÈQUE NATIONALE DE FRANCE

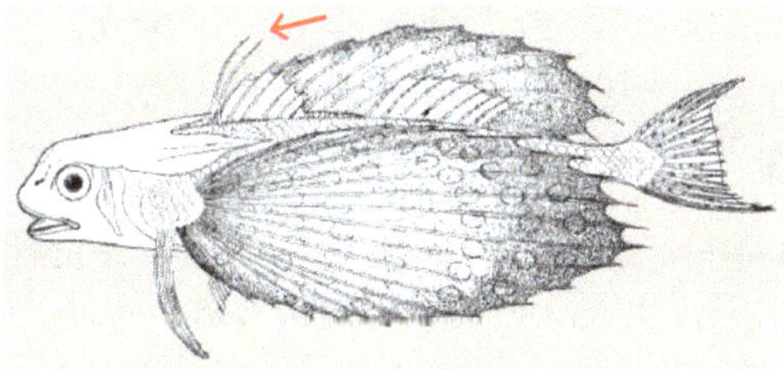

FIGURE 2.20
Flying gurnard (*Dactylopterus volitans*) in Šoljan T., *Fishes of the Adriatic* (Belgrade – Washington: 1963) 159

Regarding Salviani's engravings, we regretfully know no more than the name of the artist whom he had hired. It was, as we already mentioned, a certain Bernhardus Aretinus. We are better informed about the artist who was responsible for Rondelet's woodcuts. It was Georges Reverdy (?–1564/1565), who started as a woodcutter, but later also worked successfully as a copper-engraver.[115] Reverdy was, for sure, always a professional and not a bungler.

One would also have to take into account, of course, the general advantages of copper engravings over woodcuts. According to a standard work on techniques of graphics[116] these are among others:

115 Leutrat E., *Les débuts de la gravure sur cuivre en France: Lyon 1520–1565* (Geneva: 2007).
116 Koschatzky W., *Die Kunst der Graphik. Technik, Geschichte, Meisterwerke* (Munich: 1999).

- A more flexible layout of lines, allowing greater accuracy and richness in the reproduction of details and a greater variety of shapes
- The option of superimposing strokes, allowing smooth transitions as opposed to the typical hard contrast of light and dark in woodcuts
- A better rendering of surface qualities
- Sharper prints and copies.

6 Rondelet's and Salviani's Achievements Assessed by Later Scholars

Philippe Glardon[117] has endeavoured to arrive at a balanced assessment of the positions of the persons (Belon, Rondelet, Salviani and Gessner) involved in the debate that we have presented here. Other modern scholars have also tried to evaluate Rondelet's and Salviani's achievements.

On the whole, not many testimonies have survived that comment on the quality of Rondelet's fish images; fortunately, many more assessments of Salviani's pictures are available. We have already quoted what impression Salviani's copper engravings made on the publisher Lafreri in Rome, as well as Gessner's praise for the unsurpassed beauty of these images.[118] In addition to these immediate praises, similar ones were added over time, as by the ichthyologist Bashford Dean: Salviani's plates 'in accuracy and beauty surpass any figures published in the next 100 years'[119] or by historians such as Claus Nissen: Salviani's engravings 'far outstrip the poor characters of Rondelet'[120] or Laurent Pinon: Salviani is 'too little known and too often forgotten by historians of science, this magnificent work deserves to be studied more closely'.[121]

These acknowledgments can be contrasted with critical statements, all of which lack comprehensible justification. Thus the German zoologist Julius Victor Carus remarked (clearly adopted from Georges Cuvier, see below): 'The most important merit of Salviani lies in the technically beautiful execution of the illustrations, which, however, are not entirely useful with respect to natural history.'[122] Even more apodictically judged the historian Katharina Kolb,

117 Glardon P., "The relationship between text and illustration in mid-sixteenth-century natural history treatises", in Boehrer – Kalof (eds.), *A cultural history of animals in the Renaissance* 119–145, and idem, *Histoire naturelle* 95–109.

118 Gessner, *Historia animalium IV*, Preface, fol. b1r: 'iconum in aere expressarum accurata pulchritudine omnes superat.'

119 Dean, *Bibliography* 312.

120 Nissen, *Zoologische Buchillustration* 117.

121 Pinon, *Zoologie* 95.

122 Victor C.J., *Geschichte der Zoologie bis auf Johannes Müller und Charles Darwin* (Munich: 1872) 361.

claiming that Salviani's 'realistic aspect is sometimes deceptive' (examples would have been useful);[123] and similarly unsubstantiated decrees another historian, Gillian Lewis, praising 'Rondelet's decisive woodcuts of strongly individualized fishes' against 'Salviani's handsome, decorative and rather stylized pictures'[124] – again, as long as no examples are provided, this remains a blank, polemic assertion.

Neither the praise nor the criticism are really helpful or enlightening. The only expert who dealt more seriously with Rondelet's and Salviani's images was Georges Cuvier in the 19th century, even though in the end he remained in a peculiar manner ambivalent and irresolute.

Cuvier has dealt with Rondelet and Salviani twice. In the first volume of his monumental *Histoire naturelle des poissons* he writes (1828): 'The figures of Salviani are less numerous, but much more beautiful [than those of Belon and Rondelet], and engraved in intaglio, on a fairly large scale: *there are several that have not been surpassed in more recent works*' (here and below the emphasis in italics is mine).[125] A few years later, in the second volume of his *Histoire des sciences naturelles* (1841), Cuvier revises his earlier statement:

> If the characters of the fishes were sufficiently expressed, the work of Salviani would leave nothing to be desired. But for a painter to apply his talent perfectly to history, it is necessary for him to know for himself what to project; otherwise, it is indispensable that the naturalist who employs him should pay attention to the details which he must bring out. At the time of which we speak, no one thought that it would become important one day to count the rays of the fishes, the small serrations or spines that may exist in the bones of their heads; *these peculiarities are not sufficiently represented in the figures of Salviani. Apart from that, the whole is perfect, and these are the best designs we had until our time.*[126]

Comparing Salviani's achievements to Rondelet's, he asserts that Rondelet's woodcuts lack a bit of finesse ('manquant un peu de finesse'). Regarding the representation of crucial details (small serrations, spines, bones etc.), however, he believes that they are much better ('beaucoup mieux') than Salviani's copper-engravings.

123 Kolb K., *Graveurs, artistes & hommes de science. Essai sur les traités de poissons de la Renaissance* (Paris: 1996) 32.

124 Lewis, *Clusius* 84.

125 Cuvier Georges – Valenciennes Achille, *Histoire naturelle des poissons* (22 vols.) (Paris: 1828–1849) here vol. 1, 1828, 50–51.

126 Cuvier Georges, *Histoire des sciences naturelles depuis leur origine jusqu'à nos jours*. Vol. 2. (Paris: 1841) 74–75.

As Cuvier accuses Salviani of a lack of attention to detail, we would finally like to see whether he himself leaves nothing to be desired in terms of accuracy.

Cuvier has dealt extensively with the Dactylopteridae in the fourth chapter of his *Histoire naturelle des poissons* (1829).[127] He distinguishes between several species, one of which, named *Cephalacanthus spinarella* by Cuvier, is identical to Salviani's *Milvus*, i.e. to the Flying gurnard or *Dactylopterus volitans*.[128] As a comparison with Salviani's illustration of this species shows [Figs. 2.21 and 2.22], Cuvier lacks the two spines in front of the dorsal fin, the hallmark of this species; furthermore the typical triangular head shape is not reproduced, neither the distinctive fan-like pectoral fin [Fig. 2.22].

All of these features are perfectly rendered, however, in Cuvier's illustration of the Oriental flying gurnard, an extremely similar, close relative of the Flying gurnard [Fig. 2.22]. In contrast to this one, Cuvier's picture of the Flying gurnard [Fig. 2.21] can only be judged as complete failure – it's not a 'true' picture.[129] Here his criticism of Salviani falls back on himself.

It is obvious that Cuvier's requirements are well met by Salviani, occasionally even better than by Cuvier himself. This attention to detail can be found in nearly all of Salviani's images, which are regularly accompanied by a meticulous anatomical description. This impression is supported by another indicator: the number of pictures in Belon and Rondelet that cannot readily be identified (at least to family) is about 18%, while that in Salviani is zero.

In summary, according to our findings, two statements can be made:

1. In the controversy with Rondelet, Salviani credibly defended himself against the accusation of plagiarism by his French opponent.
2. Cuvier's criticism that Salviani's images lacked anatomical precision appears unwarranted when at least one of Cuvier's images (e.g., *Dactylopterus volitans*) is less accurate than Salviani's.

Finally, we would like to refer to Linnaeus, who valued Salviani and also defended him against Rondelet. When Linnaeus began to identify all known fishes and to give them unique names in the 10th and 12th editions of his *Systema naturae* (1758 and 1766), he referred to 80 of the 88 descriptions and

127 Cuvier – Valenciennes, *Histoire des poissons*, here vol. 4, 1829, 114–141.

128 Cuvier's name is listed here only as an obsolete synonym within today's valid nomenclature.

129 Reasons why a picture fails can be, for example, that the fish is (1) illustrated from a poorly preserved and/or damaged specimen; that (2) the illustration may depict a juvenile stage that is morphologically different from the adult; that the fish is (3) illustrated in a hurry by a ship naturalist who may not be experienced rendering fishes and thus is not mindful of the diagnostic significance of certain features.

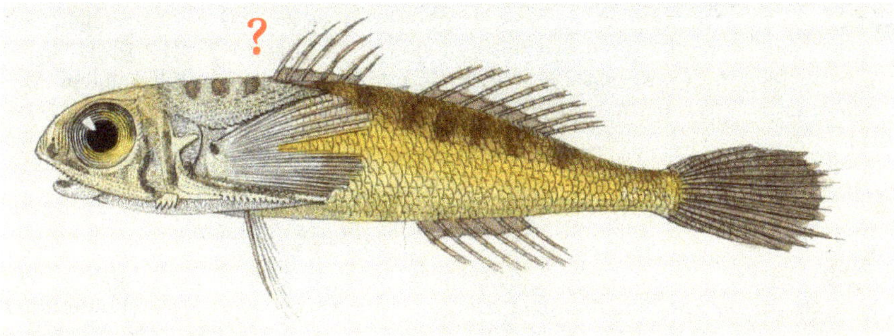

FIGURE 2.21 Cuvier's *Cephalacanthus spinarella*, showing the Flying gurnard (*Dactylopterus volitans*) in Cuvier Georges – Valenciennes Achille, *Histoire naturelle des poissons*, vol. 4 (Paris: 1829) 138, depicted on plate 73

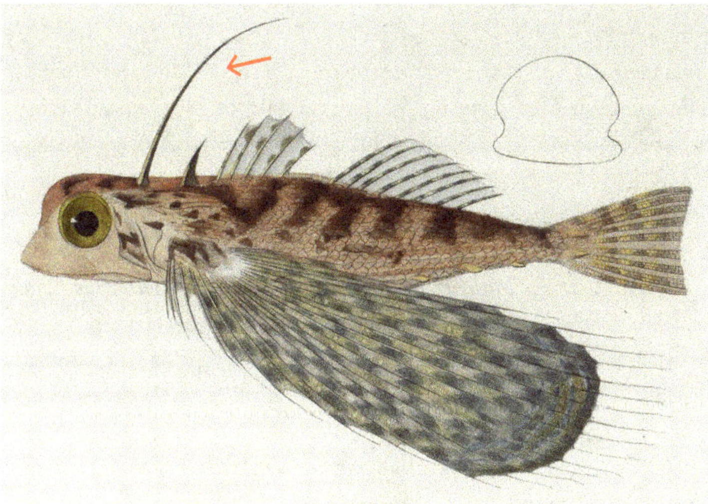

FIGURE 2.22 Cuvier's *Dactylopterus orientalis*, the Oriental flying gurnard (today renamed *Dactyloptena orientalis* Cuvier, 1829) in Cuvier Georges – Valenciennes Achille, *Histoire naturelle des poissons*, vol. 4 (Paris: 1829) 134, depicted on plate 76

images provided by Salviani. In doing so, Linnaeus relied on the preparatory work of Peter Artedi (1738), who had also consulted Salviani's images.[130]

130 Artedi Peter, *Ichthyologia, sive, opera omnia de piscibus*, ed. Carl Linnaeus (Leiden, Conrad Wishoff: 1738) 29, part *Bibliotheca ichthyologica*, has honoured Salviani's book as an outstanding ichthyological work. In addition, Linnaeus was of the opinion that Salviani had rightly defended himself against Rondelet's attacks.

Bibliography

Albertus Magnus, *On Animals. A Medieval Summa Zoologica*, eds. K.F. Kitchell – I.M. Resnick (2 vols.) (Baltimore: 1999).

Aldrovandi Ulisse, *De piscibus libri v. et De cetis lib. Unus*, ed. Ioannes Cornelius – Geronimo Tamburin (Bologna, Jo. Bapt. Bellagamba: 1613).

Andretta E., "Salviani, Ippolito", in *Dizionario Biografico degli Italiani*. Vol. 90 (Rome: 2017) 22–24.

Andretta E., *Roma medica. Anatomie d'un système médical au XVIe siècle* (Rome: 2011).

Anonymous, *Hortus sanitatis* (Mainz, Jacob Meydenbach: 1491).

Anonymous, *Dit boecxken leert hoe men mach voghelen vanghen metten handen end hoe mn mach visschen vanghen meten handen, ende oeck andersins*, Antwerp, Govert Back: 1506.

Artedi Peter, *Ichthyologia, sive, opera omnia de piscibus*, ed. Carl Linnaeus (Leiden, Conrad Wishoff: 1738).

Bader G., *Die Emergenz des Namens. Amnesie, Aphasie, Theologie* (Tübingen: 2006).

Balon E.K., "The common carp, Cyprinus carpio: its wild origin, domestication in aquaculture, and selection as colored nishikigoi", *Guelph Ichthyology Reviews* 3 (1995) 1–55.

Barsi M., *L'énigme de la chronique de Pierre Belon* (Milan: 2001).

Bashford D., *A bibliography of fishes*. Vol. 3 (New York: 1923) 216–217.

Baumann B. – Baumann H., *Die Mainzer Kräuterbuch-Inkunabeln "Herbarius Moguntinus" (1484), "Gart der Gesundheit" (1485), "Hortus Sanitatis" (1491)* (Stuttgart: 2010).

Bäumer Ä., "De piscibus et aquatilibus libelli III novi (Zürich 1556). Ein bisher unbeachtetes zoologisches Werk von Conrad Gesner", *Berichte zur Wissenschaftsgeschichte* 13 (1990) 177–181.

Bäumer Ä., *Geschichte der Biologie*. Vol. 1: *Biologie in der Antike bis zur Renaissance* (Frankfurt: 1991).

Bäumer Ä., *Geschichte der Biologie*. Vol. 2: *Zoologie der Renaissance – Renaissance der Zoologie* (Frankfurt: 1991).

Belon Pierre, *De aquatilibus libri duo* (Paris, Charles Estienne: 1553).

Belon Pierre, *L'histoire naturelle des estranges poissons marins* (Paris, Regnaud Chaudiere: 1551).

Bennema F.P. – Rijnsdorp A.D., "Fish abundance, fisheries, fish trade and consumption in sixteenth-century Netherlands as described by Adriaen Coenen", *Fisheries Research* 161 (2015) 384–399.

Benoît P., "La carpe dans l'occident médiéval", in James-Raoul D. – Thomasset C. (eds.), *Dans l'eau, sous l'eau. Le monde aquatique au Moyen Âge* (Paris: 2002) 229–236.

Berger F., *Die Textgeschichte der* Historia animalium *des Aristoteles* (Wiesbaden: 2005).

Beullens P. – Gotthelf A., "Theodore Gaza's translation of Aristotle's *De Animalibus*: content, influence, and date", *Greek, Roman and Byzantine studies* 47 (2007) 469–513.

Beullens P., "Aristotle, his translators, and the formation of ichthyologic nomenclature", in: Goyens M. – De Leemans P. – Smets A. (eds.), *Science translated. Latin and vernacular translations of scientific treatises in medieval Europe* (Louvain: 2008) 105–122.

Beullens P., "Aristotle's Zoology in the Medieval World", in Boehrer B. – Hand M. – Massumi B. (eds.), *Animals, Animality, and Literature* (Cambridge: 2018) 29–42.

Beullens P., "Quelques observations sur la traduction de l'*Histoire des animaux* d'Aristote par Guillaume de Moerbeke", *Bulletin de Philosophie Médiévale* 34 (1992) 181–196.

Blunt A., *Artistic Theory in Italy, 1450–1600* (Oxford: 1962).

Blunt W., *The art of botanical illustration* (Woodbridge: 1994).

Brunelli G., "Marcello II, papa", in *Dizionario Biografico degli Italiani*. Vol. 69 (Rome: 2007) 502–510.

Brunfels Otto, *Herbarum vivae eicones ad naturae imitationem*. Vol. 1. (Strasbourg, Johann Schott: 1530).

Brunfels Otto, *Novi Herbarii Tomus II* (Strasbourg, Johann Schott: 1536).

Cambier H., "Un grand poisson qui pose question. La baleine au Moyen Âge", in Huber-Rebenich G. et al. (eds.), *Wasser in der mittelalterlichen Kultur* (Berlin: 2017) 532–541.

Camille M., "Bestiary or Biology? Aristotle's Animals in Oxford, Merton College, MS 271", in Steel C. – Guldentops G. – Beullens P. (eds.), *Aristotle's Animals in the Middle Ages and Renaissance* (Louvain: 1999) 355–396.

Carusi C., "Salt and Fish Processing in the Ancient Mediterranean: A Brief Survey", *Journal of Maritime Archaeology* 13 (2018) 481–490.

Cockx-Indestege Elly, "Van een boekje om vogels en vissen te vangen naar een zeldzame Antwerpse postincunabel, nu in de Library of Congress te Washington", in: Van der Vekene E. (ed.), *Refugium animae bibliotheca. Festschrift für Albert Kolb* (Wiesbaden: 1969) 109–138.

Cohen D.M. – Inada T. – Iwamoto T. – Scialabba N., *Gadiform fishes of the world (Order Gadiformes). An annotated and illustrated catalogue of cods, hakes, grenadiers and other gadiform fishes known to date* (Rome: 1990).

Coney W.C., "Mulled thoughts: Mullus and Mugilis in Pliny's *Naturalis Historiae* and the *De Re Conquinaria* of 'Apicius'", *Pseudo-Dionysius* 18 (2016) 49–58.

Cooke Thomas (transl.), *The Works of Hesiod* (London, Printed for T. Longman, J. Osborn, S. Birt, and C. Hitch: 1743).

Cuvier Georges – Valenciennes Achille, *Histoire naturelle des poissons* (22 vols.) (Paris: 1828–1849).

Cuvier Georges, *Histoire des sciences naturelles depuis leur origine jusqu'à nos jours*. Vol. 2. (Paris: 1841).

Dantz Johann, *Kreutter Buch. Des Hochberümpten Pedanij Dioscoridis Anazarbei, gründliche vnd gewisse beschreibung aller materien oder gezeugs der Artznei, das ist, nit alleyn der Kreutter, sonder auch alles, was sonst in der artznei gebreuchlich ist; in sechs Bücheren verfast* (Frankfurt, Jacob Cyriacus: 1546).

Daston L. – Galison P., *Objectivity* (New York: 2007).

Davis P., "Collecting and preserving fishes: a historical perspective", in MacGregor A. (ed.), *Naturalists in the Field* (Leiden – Boston: 2018) 149–165.

De Angelis S., "Sehen mit dem physischen und dem geistigen Auge. Formen des Wissens, Vertrauens und Zeigens in Texten der frühneuzeitlichen Medizin", in Jaumann H. (ed.), *Diskurse der Gelehrtenkultur in der Frühen Neuzeit. Ein Handbuch* (Berlin – New York: 2011) 211–254.

Del Soldato E., *Simone Porzio. Un aristotelico tra natura e grazia* (Rome: 2010).

Del Soldato E., "Un catalogo naufragato: il "De piscibus" di Simone Porzio", in De Bellis E. (ed.), *Aristotele e la tradizione aristotelica, Atti del convegno internazionale di studi, Lecce, 12–14 giugno 2008* (Lecce: 2008) 149–176.

Dingel I. – Daugirdas K. (eds.), *Antitrinitarische Streitigkeiten. Die tritheistische Phase (1560–1568)* (Göttingen: 2013).

Dollinger P., *The German Hansa* (London – New York: 1999).

Dubravius Johannes, *De piscinis et piscium qui in eis aluntur naturis libri quinque* (Breslau (Vratislaviae), Andreas Vinglerus: 1547).

Egmond F. – Kusukawa S., "Circulation of images and graphic practices in Renaissance natural history: the example of Conrad Gessner", *Gesnerus* 73 (2016) 29–72.

Egmond F., "Clusius and friends: Cultures of exchange in the circles of European naturalists", in Egmond F. – Hoftijzer P. – Visser R. (eds.), *Carolus Clusius. Towards a cultural history of a Renaissance naturalist* (Amsterdam: 2007) 9–48.

Egmond F., *Eye for detail. Images of plants and animals in art and science, 1500–1630* (London: 2017).

Engemann J., "Fisch, Fischer, Fischfang", *Reallexikon für Antike und Christentum* 7 (1969): 959–1097.

Exner M. et al., "Fisch I", *Reallexikon zur Deutschen Kunstgeschichte* IX (1987) 18–88. Dölger F.J., *Ichthys* (5 vols.) (Rome: 1910–1957).

Fischer H., *Conrad Gessner (26. März 1516–13. Dezember 1565). Leben und Werk* (Zurich: 1966).

Funk H., "Jan Černý's Knieha lekarska (1517): closing a gap in the history of printed illustrated herbals", *Archives of natural history* 42 (2015) 153–164.

Ganias K. et al., "Aristotle as an ichthyologist: Exploring Aegean fish diversity 2,400 years ago", *Fish and Fisheries* 18 (2017) 1–18.

Geoffroy Étienne François, *Fortsetzung der Abhandlung von der Materia Medica. Aus dem Französischen übersetzt. Von den Thieren.* Vol. 6 (Leipzig, Carl Ludwig Jacobi Wittwe: 1763).

Geoffroy Étienne François, *Suite de la Matiere medicale de M. Geoffroy. Par Mrs Arnault de Nobleville et Salerne.* Vol. 2: *Des poissons* (Paris, Desaint – Saillant – G. Cavelier – Le Prieur: 1756).

Gessner Conrad, *De piscibus et aquatilibus omnibus libelli III novi* (Zurich, Christopher Froschauer: 1556).

Gessner Conrad, *Historia animalium liber IV: qui est De piscium et aquatilium animantium natura* (Zurich, Christopher Froschauer: 1558).

Gessner Conrad, *Iani Dubravii qui postea Olomucensis episcopus creatus est, De piscinis et piscium qui in eis aluntur naturis libri quinque* (Zurich, Andreas Gessner d. J.: 1559).

Gessner Conrad, *Nomenclator aquatilium animantium. Icones animalium aquatilium in mari et dulcibus aquis degentium* (Zurich, Christopher Froschauer: 1560).

Giovio, Paolo (Paulus Iovius), *De Romanis piscibus* (Rome, Francesco Minuzio Calvo: 1524). Baumann T. (ed. and transl.), *Pauli Iovii Novocomensis Medici De Romanis piscibus libellus ad Ludovicum Borbonium cardinalem amplissimum. Text, Übersetzung, Kommentar* (Unpublished doctoral thesis University of Mannheim, 1994).

Glardon P., "The relationship between text and illustration in mid-sixteenth-century natural history treatises", in Boehrer – Kalof (eds.), *A cultural history of animals in the Renaissance* (Oxford, UK – New York: 2007) 119–145.

Glardon P., *L'histoire naturelle au XVI*[e] *siècle. Introduction, étude et édition critique de* La nature et diversité des poissons *de Pierre Belon (1555)* (Geneva: 2011).

Glardon P., *Pierre Belon, L'histoire de la nature des oyseaux. Édition en fac-similé avec introduction et notes* (Geneva: 1997).

Gotthelf A., "Darwin on Aristotle", *Journal of the History of Biology* 32 (1999) 3–30.

Gaukroger S., *The emergence of a scientific culture. Science and the shaping of modernity 1210–1685* (Oxford: 2006).

Green E.L., *Landmarks of botanical history*. Vol. 1 (Washington: 1906).

Griffiths A., *The Print Before Photography. An Introduction to European Printmaking, 1550–1820* (London: 2016).

Grimm H., "Neue Beiträge zur Fisch-Literatur des XV. bis XVII. Jahrhunderts und über deren Drucker und Buchführer". *Archiv für Geschichte des Buchwesens* 67 (1968) 2871–2887.

Gudger E.W., "The five great naturalists of the sixteenth century: Belon, Rondelet, Salviani, Gesner and Aldrovandi. A chapter in the history of ichthyology", *Isis* 22 (1934) 21–40.

Hakelberg D., "Das Bodensee-Fischbuch von Gregor Mangolt in einem Basler Nachdruck von 1612", *Wolfenbütteler Barock-Nachrichten* 36 (2009) 107–114.

Harlfinger D. (ed.), *Graecogermania. Griechischstudien deutscher Humanisten. Die Editionstätigkeit der Griechen in der italienischen Renaissance (1469–1523)* (Weinheim – New York: 1989).

Hassig D., *Medieval bestiaries. Text, image, ideology* (Cambridge: 1995).

Hendrikx S., "Identification of herring species (Clupeidae) in Conrad Gessner's ichthyological works: A case study on taxonomy, nomenclature, and animal depiction in the sixteenth century", in Enenkel K.A.E. – Smith P.J. (eds.), *Zoology in early modern culture* (Leiden – Boston: 2014) 149–171.

Herrmann S., 'Ein Preuße in Venedig: Der Botaniker Melchior Wieland (ca. 1520–1589), Pionier der botanischen Feldforschung in der Levante', *Sudhoffs Archiv* 99 (2015) 1–14.

Hertel R., "Über die "Ichthyographie der Elbe" des Johannes Kentmann. Eine Studie über die ältesten sächsischen Fischfaunen (Pisces)", *Zoologische Abhandlungen. Staatliches Museum für Tierkunde in Dresden* 35 (1978) 75–100.

Higginbotham J., *Piscinae. Artificial fishponds in Roman Italy* (Chapel Hill – London: 1997).

Huber-Rebenich G., "Marschalk, Nikolaus", in Worstbrock F.J. (ed.), *Deutscher Humanismus 1480–1520. Verfasserlexikon*. Vol. 2. (Berlin – New York: 2013) 161–203.

Hünemörder C., "Die Geschichte der Fischbücher von Aristoteles bis zum Ende des 17. Jahrhunderts", *Deutsches Schifffahrtsarchiv* 1 (1975) 185–200.

Hünemörder C., "Fische", *Der Neue Pauly* 4 (1998) 526.

Hünemörder C., "Fischfang und Fischkunde im Mittelalter", *Deutsches Schifffahrtsarchiv* 4 (1981) 183–190.

Huppert George, *The Style of Paris: Renaissance Origins of the French Enlightenment* (Bloomington: 1999).

Imhoof-Blumer F. – Keller O., *Tier- und Pflanzenbilder auf Münzen und Gemmen des Klassischen Altertums* (Leipzig: 1889).

Jacquemard C. – Gauvin B. – Lucas-Avenel M.-A. (eds.), *Hortus sanitatis. Livre IV. Les poissons* (Caen: 2013).

Jaitner-Hahner U., *Città di Castello nel Quattrocento e nel Cinquecento: economia, cultura e società* (Sansepolcro: 2020).

James-Raoul D., "Inventaire et écriture du monde aquatique dans les bestiaries", in James-Raoul D. – Thomasset C. (eds.), *Dans l'eau, sous l'eau. Le monde aquatique au Moyen Âge* (Paris: 2002) 175–226.

Keller O., *Die antike Tierwelt*. Vol. 2 (Leipzig: 1913).

Kolb K., *Graveurs, artistes & hommes de science. Essai sur les traités de poissons de la Renaissance* (Paris: 1996).

Koschatzky W., *Die Kunst der Graphik. Technik, Geschichte, Meisterwerke* (Munich: 1999).

Kullmann W., *Aristoteles als Naturwissenschaftler* (Berlin: 2014).

Kusukawa S., "Ad vivum Images and Knowledge of Nature in Early Modern Europe", in Balfe T. et al. (eds.), *Ad vivum? Visual Materials and the Vocabulary of Life-Likeness in Europe before 1800* (Leiden – Boston: 2019) 89–121.

Kyle S.R., *Medicine and humanism in late medieval Italy. The Carrara herbal in Padua.* (London – New York: 2017).

Lang H.S., "Aristotle and Darwin", *International Philosophical Quarterly* 23 (1983) 141–153.

Lavenia V., "Porzio, Simone", in *Dizionario Biografico degli Italiani.* Vol. 85 (Rome: 2016) 142–145.

Lennox J.G., "The Disappearance of Aristotle's Biology: A Hellenistic Mystery", *Apeiron* 27 (1994) 7–24.

Leroi A.M., *The lagoon. How Aristotle invented science* (London: 2014).

Leu U.B., *Conrad Gessner (1516–1565). Universalgelehrter und Naturforscher der Renaissance* (Zurich: 2016).

Leutrat E., *Les débuts de la gravure sur cuivre en France: Lyon 1520–1565* (Geneva: 2007).

Lewis G., "Clusius in Montpellier: A humanist education completed?", in Egmond – Hoftijzer – Visser (eds.), *Carolus Clusius. Towards a cultural history of a Renaissance naturalist* (Amsterdam: 2007) 65–98.

Lisch G.C.F., "Buchdruckerei des Raths Dr. Nicolaus Marschalk". *Jahrbücher des Vereins für Mecklenburgische Geschichte und Altertumskunde* 4 (1839) 92–133.

Mackenzie L., "French early modern sea-monsters and modern identities, via Bruno Latour", in Cuneo P. (ed.), *Animals and early modern identity* (Farnham – Burlington: 2014) 329–349.

Mangolt Gregor, *Fischbuoch. Von der natur und eigenschafft der vischen* (Zurich, Andreas Gessner: 1557).

Marschalk Nicolaus, *Historia aquatilium latine ac grece cum figuris* (Rostock, in edibus Thuriis: 1520).

Mayr E., *The growth of biological thought. Diversity, evolution, and inheritance* (Cambridge, MA – London: 1982).

Mortimer R., *Catalogue of books and manuscripts. Part II: Italian 16th century books.* Vol. 2 (Cambridge, MA: 1974).

Nauert C.G., "Humanists, Scientists, and Pliny: Changing Approaches to a Classical Author", *The American Historical Review* 84 (1979) 72–85.

Nellen W. – Dulčić J., "Evolutionary steps in ichthyology and new challenges", *Acta Adriatica* 49 (2008) 201–232.

Niphus Augustinus, *Expositiones in omnes Aristotelis libros* (Venice, Hieronymus Scotus: 1546).

Nissen C., *Die botanische Buchillustration. Ihre Geschichte und Bibliographie* (Stuttgart: 1966).

Nissen C., *Die zoologische Buchillustration. Ihre Bibliographie und Geschichte*. Vol. 2 (Stuttgart: 1978) 125–158.

Nissen C., *Schöne Fischbücher. Kurze Geschichte der ichthyologischen Illustration. Bibliographie fischkundlicher Abbildungswerke* (Stuttgart: 1951).

Ogilvie B.W., *The Science of Describing: Natural History in Renaissance Europe* (Chicago – London: 2006).

Oppenraay A.M.I. van, "The Reception of Aristotle's History of Animals in the Marginalia of Some Latin Manuscripts of Michael Scot's Arabic-Latin Translation", *Early Science and Medicine* 8 (2003) 387–403.

Pangle L., *Aristotle and the philosophy of friendship* (Cambridge: 2003).

Pangritz W., *Das Tier in der Bibel* (Munich – Basel: 1963).

Perfetti S., "Philosophers and animals in the Renaissance", in Boehrer B.T. – Kalof L. (eds.), *A cultural history of animals in the Renaissance* (Oxford, UK – New York: 2007) 147–164.

Perfetti S., *Animali pensati nella filosofia tra medioevo e prima età moderna* (Pisa: 2012) 125.

Perfetti S., *Aristotle's zoology and its Renaissance commentators, 1521–1601* (Leuven: 2000).

Pinon L., "Clématite bleue contre poissons séchés: sept lettres inédites d'Ippolito Salviani à Ulisse Aldrovandi", *Mélanges de l'École française de Rome* 114 (2002) 477–492.

Pinon L., *Livres de zoologie de la Renaissance: une anthologie (1450–1700)* (Paris: 1995).

Polek N., "Die Fischkunde des Aristoteles und ihre Nachwirkung in der Literatur", in Hilberg I. – Jüthner J. (eds.), *Primitiae czernovicienses. Festgabe zur 50. Versammlung deutscher Philologen und Schulmänner in Graz* (Czernowitz: 1909) 35–57.

Pollidorus Petrus (Pietro Pollidori), *De vita, gestis, et moribus Marcelli II. pontificis maximi commentarius* (Rome, Hiermonymus Mainardus: 1744).

Richter C., "Hollandse vissenboeken. De onweerstaanbare kracht van de afbeelding", *Holland* 38 (2006) 161–176.

Rondelet Guillaume, *Libri de piscibus marinis* (Lyon, Macé Bonhomme: 1554).

Rondelet Guillaume, *Universae aquatilium historiae pars altera* (Lyon, Macé Bonhomme: 1555).

Rondelet Guillaume, *La premiere et la seconde partie de l'histoire entiere des poissons* (Lyon, Macé Bonhomme: 1558).

Rudolph P., *Im Garten der Gesundheit. Pflanzenbilder zwischen Natur, Kunst und Wissen in gedruckten Kräuterbüchern des 15. Jahrhunderts* (Vienna – Cologne – Weimar: 2020).

Sahrhage D. – Lundbeck J., *A history of fishing* (Berlin – Heidelberg – New York: 1992).

Salviani Hippolito, *Aquatilium animalium historiae liber primus, cum eorumdem formis, aere excusis* (Rome, Hippolito Salviani: 1554; other editions: Rome 1559; Rome 1593; Venice 1600; Venice 1602).

Schmitt C.B., *Aristotle and the Renaissance* (Cambridge: 1983) 166.

Schubert O., *Georg Handsch von Limus'* Die Elbefischerei in Böhmen und Meißen (Prague: 1933).

Schwarz B., *Kurienuniversität und stadtrömische Universität von ca. 1300 bis 1471* (Leiden: 2012).

Senfelder L., "Georg Handsch von Limus. Lebensbild eines Arztes aus dem XVI. Jahrhundert", *Wiener klinische Rundschau 1901* 495–535.

Sharples R.W. (ed.), *Theophrastus of Eresus, Commentary Volume 5: Sources on biology* (Leiden: 1995).

Sicherl M., *Handschriftliche Vorlagen der* Editio princeps *des Aristoteles* (Mainz: 1976).
Grendler P.F., *The Universities of the Italian Renaissance* (Baltimore – London: 2002).
Anonymous, *Gesamtkatalog der Wiegendrucke*. Herausgegeben von der Kommission für den Gesamtkatalog der Wiegendrucke. Vol. 2 (Stuttgart – New York: 1968).

Singer C., *Greek biology and Greek medicine* (Oxford: 1922).

Siraisi N.G., *History, medicine, and the traditions of Renaissance learning* (Ann Arbor: 2007).

Šoljan T., *Fishes of the Adriatic* (Belgrad – Washington: 1963).

Strömberg R., *Studien zur Etymologie und Bildung der griechischen Fischnamen* (Göteborg: 1943).

Thiel C., "objektiv/Objektivität", in Mittelstraß J. (ed.), *Enzyklopädie Philosophie und Wissenschaftstheorie*. Vol. 6 (Stuttgart: 2016) 8–9.

Thompson D'Arcy W., *On Aristotle as a biologist with a prooemion on Herbert Spencer* (Oxford:1913).

Timm W., "Die Holzschnitte zu Nikolaus Marschalks Historia aquatilium latine ac grece cum figuris, Rostock 1517–20", in Anonymous (ed.), *Festschrift Gottfried von Lücken* (Rostock 1968) 799–802.

Tosi A., "Acconciare, seccare, dipingere: pratiche di rappresentazione della natura tra le "spigolature" aldrovandiane", in Olmi G. – Simoni F. (eds.), *Ulisse Aldrovandi. Libri e immagini di Storia naturale nella prima Età moderna* (Bologna: 2018) 49–58.

Travi E. – Penco M. (eds.), *De piscibus Romanis*, in *Dialogi et descriptiones* (= *Pauli Iovii Opera* 9) (Rome: 1984) 3–64.

Victor C.J., *Geschichte der Zoologie bis auf Johannes Müller und Charles Darwin* (Munich: 1872).

Vidali S., *Archaische Delphindarstellungen* (Würzburg: 1997).

Violand G., "Historische Fischökologie des Bodensees. Das Fischbuch des Gregor Mangolt, 1557", *Österreichs Fischerei* 59 (2006) 169–171.

Watson G., *Theriac and Mithridatium. A study in therapeutics* (London: 1966).

Wehrhahn-Stauch L., "Christliche Fischsymbolik von den Anfängen bis zum hohen Mittelalter", *Zeitschrift für Kunstgeschichte* 35 (1972) 1–68.

Woodall J. (ed.), *Portraiture: Facing the Subject* (Manchester: 1997).

Wüstner A. – Kollmann J. (eds. and transl.), *Buch von den Teichen und den Fischen, welche in denselben gezüchtet werden. In fünf Büchlein* (Vienna: 1906).

Zarske A., "Autochthone Population oder Faunenverfälschung? Zum Fund der Nase (*Chondrostoma nasus* [L., 1758]) im oberen Elbegegbiet (*Teleostei*: *Cyprinidae*)", *Faunistische Abhandlungen. Staatliches Museum für Tierkunde* 20 (1996) 285–294.

Zaunick R., "Fragmente der ältesten sächsischen Fischfauna des Dr. Johannes Kentmann (1518–1574)", *Sitzungsberichte und Abhandlungen der Naturwissenschaftlichen Gesellschaft Isis* 1915, 15–36.

Zaunick R., *Das älteste deutsche Fischbüchlein vom Jahre 1498 und dessen Bedeutung für die spätere Literatur* (Berlin: 1916).

Zimmermann H., "Die Bedeutung des Aristoteles vom Mittelalter bis zur Renaissance", *Forum Classicum* 53 (2010) 211–218.

Zoller H., "Zum Wandel der Pflanzendarstellung während der Renaissance. Vom Beginn des 15. Jahrhunderts zu Leonardo da Vinci, Albrecht Dürer und Conrad Gessner", *Bauhinia* 9 (1989) 109–123.

Zucker A., "Zoologie et philologie dans les grands traités ichtyologiques renaissants", *Kentron* 29 (2013) 135–173.

CHAPTER 3

Beginnings of Ichthyological Natural History: Formal and Structural Questions

Philippe Glardon

In the project to draw a history of ichthyological knowledge, as in the history of science in general, the thread of events can be divided between moments of gestation and decisive steps, and is also marked by points of no return, such as the arrival of images in the treatises of the first half of the 16th century.

In this chapter I would like to look at what happened between two of these milestones: the translation of Aristotelian natural history works by Theodore Gaza and the publication of the great natural history treatises between 1540 and 1560, with a particular focus on the treatises on aquatic animals, the *aquatilia*.

When one follows a chronological perspective, the risk is, of course, reading or placing an a priori direction on efforts that one would like to see concordant and unidirectional. The risk and temptation are all the greater in the history of science. Thus, in the present case, we could see a movement away from the ancient texts, towards a form of pre-modernity. Of course, in the 18th century naturalists freed themselves from Aristotle, but this was not the case in the 16th century. And secondly, even if such an angle of view could be justified in the context of the study of the changes taking place in the discipline of natural sciences, this perspective would leave in the shade valuable information on the reflections carried out by the authors and on their hesitations and attempts to define their method.

Thus, in order to get from our point A (Theodore Gaza) to point B (edition of the great natural history treatises of the Renaissance), I propose defining five stages, which should help us understand this path followed by the naturalists, specifying that I will focus on stages II and III, *i.e.* the first attempts to match the old names with the ocular descriptions.

I. Restoration of the ancient text
II. Cutting or sequencing of the text, addition of consultation tools (indices)
III. Synthetic monographs or essays by extension from ancient texts
IV. Systematic confrontation of ancient text *vs* personal observations
V. Epistemological division toward new categories of knowledge

© PHILIPPE GLARDON, 2024 | DOI:10.1163/9789004681187_004

To sum it all up while avoiding the pitfall of anachronism in speaking of a linear overtaking of ancient texts during the 16th century, we could speak of "acculturation", a term defined as follows:

> Results from a multiplicity of micro-processes, invention, imitation, learning, or adaptation among thousands of interacting individuals and groups.[1]

Acculturation does not apply primarily to the ancient text but to nature itself, which will now be read through a different prism and then catalogued according to a new grid. Let us add that the five stages we have defined are necessarily schematic; they partly overlap and present geographical and chronological shifts. But our division should allow comparisons to be made and a general trend to be discerned in the efforts of naturalists. It should be relevant because naturalists are themselves very attentive to the approaches of their predecessors in developing their method. The best example of this is Conrad Gessner, whose prefaces contain abundant comments on the whole process that interests us.

Let us take up our plan at point 1, the restoration of ancient texts. The main contributor to this first stage is Theodore Gaza, from whom Gessner borrows a long extract of the preface to pay him a grateful tribute. It is very important for Gessner that this preface constitute a rhetorical performance defending the relevance of natural history, as Gessner highlights in a marginal note: 'How physicians turn to natural history and for what purpose'.[2]

Thanks to Gaza's plea, which defines a method by referring to Aristotle, natural history conquers its letters of nobility, even in the study of apparently insignificant beings. This shows a fact now well known, to which numerous re-editions of Gaza's translation attest: the Byzantine did pioneering work. It also illustrates the fact that the phases are intimately linked: here again Gaza was decisive in the search for equivalences or Latin transcriptions from the Greek, 'the most competent observer and imitator of the Plinian vocabulary of Greek equivalents of meaning'.[3] More broadly, we owe him the efforts to bring the texts of Aristotle and Pliny closer together, thanks to his close collaboration with Giovanni Andrea Bussi, at a time when the latter was working

---

1 https://www.cnrtl.fr, s.v. acculturation.

2 'In transferendis Graecorum sensibus Plinianorum verborum observator et imitator diligentissimus', Gessner Conrad, *Historiae animalium liber I. de quadrupedibus viviparis* (Zurich, Chistopher Froschauer: 1551), "Praefatio candidis lectoribus", fol. B3v.

3 Massario Francesco, *In nonum Plinii de naturali historia librum Castigationes et annotationes* (Paris, Michel Vascosan: 1542) fol. b7r.

on his edition of the *Natural History*, which was decisive in the history of the Plinian text.

The link established between works as different as Aristotle's *History of Animals* and Pliny's *Natural History* shows the early attempt to appropriate and transform the ancient texts, with the aim of developing a new key to understanding nature. Another very significant development on the Aristotelian source is the division into chapters that appears as early as the earliest editions of Gaza; the fact that there is no certainty that the division of the text is the work of Gaza himself is also important: it proves that there is a collective project to make adaptations to the ancient text, and that in the minds of its authors, this work is inseparable from the actual translation.[4] Thus, these first two phases closely complement each other. In the minds of the commentators, there can be no restoration without adaptation, already at the level of the translation. These first two stages, restoration of the ancient text, then cutting up or sequencing the text, with the addition of consultation tools (indices or chaptering), still concern the linear text, considered in its entirety.

Subsequently, Gessner notes several authors of editions of different construction and situated in the years 1520–1530: he even hails the first of them, Paolo Giovio, as a precursor: according to the Zurich physician, the first work that corresponds to our third phase is Giovio's *De romanis piscibus*, published in 1524 and reprinted several times until 1561, including an Italian version.[5] The form and content of this small work are probably the reason for its success. Gessner is quite clear on this point in his '*Praefatio candidis lectoribus*':

> Paulus Iovius was the first of our time who began to develop [*excolere*] history of fish and to manage to produce pictures of them, as he reports himself, even if I think that he didn't make them engrave for publication.[6]

4 On this point: Beullens P. – Gotthelf A., "Theodore Gaza's Translation of Aristotle's *De Animalibus*: Content, Influence, and Date", *Greek, Roman, and Byzantine Studies* 47 (2007) 469–513.

5 Giovio Paolo, *De Romanis Piscibus libellus ad Ludovicum Borbonium Cardinalem amplissimum* (Rome, F. Minitius Calvus: 1524 (April)); *De Romanis Piscibus libellus ad Ludovicum Borbonium Cardinalem amplissimum* (Rome, F. Minitius Calvus: 1524 (August)); *De piscibus marinis, lacustribus, fluviatilibus, item de testaceis ac salsamentis liber* (Rome, F. Minitius Calvus: 1527); *De Romanis Piscibus Libellus* [...] (Antwerp, Joannes Grapheus: 1528); *De Romanis Piscibus libellus* [...] (Basel, Hieronymus Froben: 1531); *Pauli Jovii de piscibus liber unus*, ed. Johannes Caesar (Strasbourg, Jacobus Cammerlander: 1534); *Libro di mons. Paolo Giovio de pesci romani*, transl. Carlo Zancaruolo (Venice, Gualtieri: 1560).

6 'Primus nostris temporibus Paulus Iovius, ut piscium historiam excolere coepit, ita et picturas eorum fieri curavit, ut ipse refert; quas tamen typis publicatas non puto.' (Gessner Conrad, *Historia animalium liber IIII, qui est de piscium et aquatilium animantium natura*,

In an almost identical sentence, he uses yet another verb to summarize Pierre Gilles' work:

> The Frenchman Pierre Gilles is in my opinion the first of our period after Paolo Giovio to have undertaken to illustrate [*illustrare*] natural history.[7]

What does Gessner mean by these two verbs, *excolere* and *illustrare*? A short survey of the Giovio book brings up some decisive elements. He proposes:
- To discuss the method and the difficulties.
- To compare about 40 Italian fish and their vernacular names with their ancient names.
- To report one's own observations and anecdotes or testimonies.
- To deal with medical and dietetic matters.

History is to be taken in its literal sense of inquiry, which mobilizes wide and complex competences, and Giovio's work, with its multifaceted approach, meets these requirements according to Gessner.

Another very important point is visible in the preface to *De romanis piscibus*: thanks to Giovio, natural history gained in prestige and became a literary genre of the court, an indispensable annotation for authors always in search of recognition and, above all, of funds to publish their works. The reading of this preface is enriching on this point. It is a real plea, written in a Latin imbued with Ciceronian terms, some examples of which are indicated by an asterisk*:

> You advised me, who can also impose it, O illustrious and highly respected Lord, to write down the subtle and erudite dialogue which took place on the subject of Roman fish, when Pope Clement received you at a convivial meal for recreation, and such was the case, in the company of Cardinal Johannes Lotharingus, who is well known for his intelligence. The subject is difficult to know and to treat, because of the almost infinite diversity of fish, which has tired me, as well as many others quite curious, who are lost in sterile efforts; this is because of the incredible disagreement between the authors, due to the multiplicity of languages, and which have thrown an impenetrable veil of darkness* over this kind of study. This is why it is necessary that the mind, shaken by the difficulty of the mission, should

[...] *continentur in hoc volumine Guilelmi Rondeletii* [...] *Petri Bellonii* [...] *de aquatilium singula scripta* (Zurich, Christopher Froschauer: 1558), "Ad candidum lectorem praefatio", fol. b2r.)

7 'Petrus Gillius Gallus, Primus (opinor) post Paolum Iovium nostro tempore piscium historiam illustrare coepit' (ibidem).

> not shrink back in fear, but should go deeper into the subject by devoting greater strength to method and judgement*.
>
> But your merit, your greatness of soul, the refinement of your mind, combined with your inclination for the best culture, are such that I prefer to neglect my honour and my small notoriety* [*gloriole*], even if they are the fruit of liberal studies, for the entire benefit of your most noble will. But you may judge afterwards whether, having undertaken a project of such high demands and requiring such intellectual faculties, I have yielded to a guilty audacity, or whether I will not be held responsible, so that I appear estimable and obliging*.
>
> Therefore, I wish this little work to be judged less severely by the more intransigent critics, since I have written it also in a spirit of levity and amusement, as it were in a bantering manner, while many have missed it at Saturnalia, dice games, and other such amusements, having recollected my former observations, and others which I have in part afterwards developed. This unfinished work, done in the leisure of a free man, will perhaps be the prelude to a complete and well-made work, if those who earn respect by letters and their labour, after having had it before their eyes, better examine what I shall have caught in my nets*, as it were, and after having been willing to give it more ornament by their more fruitful explanations*.
>
> I shall certainly be considered as one who has brought a favour, to you first of all, and to those to whom it will have pleased*; once I have recovered from this pleasant amusement, I shall resume my habits of nightly vigils devoted to furthering this history, more firmly and with more zeal.
>
> It is a fact that I shall shortly present to the public the first ten books of this laborious work, not without some aspiration to immortality, and moreover much increased and more abundant, if it should come to my attention that it has pleased you, as well as your King Francis, whom I have heard makes his pleasure of good letters.[8]

In the text itself, Giovio offers a combination of observations, erudite notations from ancient writings, and anecdotes borrowed from reliable witnesses who are often illustrious noblemen, court men, or ecclesiastics. This subtle alchemy gives a good idea of the courtly conversation that was popular at the time during hours of scholarly relaxation.

---

8 "Ad Lodovicum Borbonium cardinalem amplissimum" (Giovio, *De Romanis Piscibus libellus* (Basel: 1531), 3–5). For the Latin text, see the Annex.

The best definition of this activity appears in Pierre Belon's dedication of his *Histoire naturelle des estranges poissons marins* to "Monseigneur Monsieur le Reverendissime Cardinal de Chastillon, Liberal Mecenas des hommes studieus":

> [I] have described in our language the memorable things, and the riches of the earth of several strange countries where I have been, and the fertility of various seas, of which you have seen several portraits, of which it has pleased you to hear me speak; and knowing well that you have no greater pleasure than to employ the proper time in hearing things which are drawn from the intimate knowledge of natural history; and that you willingly give some hours of the day after meals, in discussing and hearing erudite talk which does not strain the mind.[9]

If the revival of botany in the Renaissance generally preceded the development of other fields of life because of its food and medical interest, ichthyology perhaps lends itself better to a culture that is both scholarly and entertaining: in addition to its food value, knowledge of aquatic world offers a taste for luxury and gastronomic refinement, and the innumerable and polymorphous people of marine monsters, an inexhaustible reservoir of anecdotes. Thus, in addition to having cleared the way for the new methodology in natural history, Giovio showed the way by bringing the discipline into the cultural sphere of the elites, a *sine qua non* condition for its economic and academic existence.

Gessner's very positive opinion of Giovio contrasts strikingly with his opinion on another author, Niklaus Marschalk, author of the *Historia aquatilium cum figuris*, published in Rostock in 1520, and underlines *a contrario* what Gessner appreciates in the former:

> Niklaus Marschalk's *Historia aquatilium* [...] was printed by himself in Rostock in 1520 in folio format; it is illustrated but with false and absurd images, such as are found in the books of Bartholomew the Englishman and other such inept writings on things of nature. It is a collection of

9 '[J'ai] descript en nostre langue, les choses memorables, et les richesses de la terre de plusieurs pays estranges ou j'ay esté, et la fertilité de diverses mers, dont vous avez veu plusieurs pourtraicts, et desquels il vous a pleu me ouir parler; et sachant bien que vous n'avez plus grand plaisir, que d'employer le temps convenable, à entendre les choses qui sont extraictes de l'intime cognoissance des histoires naturelles; et que donnez volontiers quelques heures du jour après les repas, à deviser et ouïr des propos d'érudition qui ne travaillent point l'esprit' (Belon Pierre, *L'histoire naturelle des estranges poissons marins* (Paris, Regnaud Chauldiere: 1551) fol. 2r).

> elements borrowed from the ancients, in alphabetical order, devoid of any personal contribution, which contains neither observations in kind, nor any Germanic names; which surprises me greatly, although the author boasts of having made long voyages on the seas. He also promises a zoography, a history of wild beasts and one on birds, which I do not believe were ever published.[10]

If we sum up Gessner's reproaches, Marschalk shows scarce, absurd, and mediaeval-like pictures, only collected facts without either any analysis or comment; he does not give any vernacular German name, and he promises never-published works.

As a matter of fact, when we look upon Marschalk's text itself, we can notice the difference from Giovio's text is strong. The first part of *Historia aquatilium* consists of 130 short chapters, presenting paraphrasis or summaries of antique fish's natural history (life, habitat, reproduction, feeding, etc.). The chapter devoted to fish names – "Unde imposita piscibus nomina" – consists of 10 lines, where Marschalk does not discuss any philological problem, nor does he address any difficulty regarding the link between Greek and Latin terminologies. The unique translator or commentator he mentions is Gaza – 'teste Theodoro' – without adding any comment or personal observation:

> According to M. Terentius Varro in Septumius, many names of fish are derived from those of land animals, such as anguilla, lingulaca, sudis, or from colours, such as asellus, umbra, turdus, or from the strength of animals, such as lupus, canicula, torpedo. Indeed, according to Pliny in Book IX of his *Natural History*, as we have written, all land animals have their equivalent in the sea. But their actual names are local, and they are foreign for others. Muraena, cybium, thynnus, melander, and uraeon are derived from the Greek; likewise, among shellfish, peloris, among oysters, echinus, and from the Greek still, polypus and hippopotamus. And from the vernacular, because of a similarity, syrene, pectunculi, ungues;

10 'Nicolai Marescalci [...] Historia aquatilium, impressa est Rostochii in aedibus ipsius, anno 1520 in folio, cum picturis sed fictis et absurdis, iisdem auf similis, quales libris Bartolomei Anglici et huis farinae scriptorum de rerum natura habentur, Sunt autem collectanea tantum ex authoribus ordine alphabetici congesta; proprium nihil, neque observatione ulla, neque nomen Germanicum ullum; quod hercle miror, cum de longinquis navigationibus suis per maria glorietur. Promittit et Zoographiam, et theriwn historiam, et ornithographiam, quae ipsum praestitisse non puto' (Gessner, *Historia animalium liber IIII*, "Enumeratio authorum qui de piscibus scripserunt, extantium et non extantium, veterum ac recentiorum" (fol. b4v)).

> for those which come on land, from Latin, mergus, rana and many others of this kind, but with new names.[11]

As can be seen from the original Latin text given in the notes, there are also transcription errors reminiscent of those of medieval copyists, or interpretations of previous editions like the Aldine but without any comment, which is, of course, unacceptable to humanists. Gessner's irritation could also be due to the inappropriate and, so to speak, usurped title of Marschalk's booklet. On the contrary, Giovio shows modesty in his methodology, although he applies it only to a few dozen species; he does open the way to a true natural history, while showing modesty towards the prestigious ancient predecessors, the 'Graeci authores qui naturae rerum abdita diligentissime perscrutati sunt', and Pliny, who depicted a 'tota natura ubique mirifice repraesentata'.[12]

The appearance of our third phase, manifested by the publication of monographs, is consecutive to this respectful attitude. Some passages are so complicated to elucidate that they lead the commentators to lengthy arguments that take on a form of autonomy, one might say by budding. Some of Giovio's chapters already show this tendency towards autonomy, presenting developments so important that they unbalance the work by their length. The chapter on the sturgeon (*Sturio*) is no less than 30 pages long out of a total of 139, *i.e.* almost a quarter of the entire work: it constitutes, so to speak, a work within a work. It testifies to the acute need of the naturalists to deepen their investigations, so that the formal unity of their work becomes secondary.

In some cases, this type of work can even take on a metaphorical dimension, when the *Systema naturae* set up by naturalists reflects the social hierarchy. Attention is focused on noble species, whose natural history is expanded, while insignificant species are treated much more briefly. Pierre Belon's *Histoire naturelle des estranges poissons marins* (1551), which incidentally still takes *De romanis piscibus* as its model, is in fact a direct tribute to François I. It focuses

11 'Unde imposita piscibus nomina. C. II. Piscium vocabula a terrestribus plaeraque (*sic*) sunt translata authore M. Terentio Varrone ad Septumium, ut Anguilla, lingua casudis (*sic*) [lingulaca, sudis], auta (*sic*) [aut a] coloribus asellus, umbra, turdus, aut ab animantium vi lupus, canicula, torpedo. Nam reperire omnia in mari, quae in terra animalia authore est C. Plinius libro historiae naturalis ut scripsimus nono. Vocabula vero ipsa vernacula partim sunt, partim peregrina. Muraena, cisibium (*sic*), thynnus, melandria ureus graece vocantur; in conchyliis e graecis peloris, in ostreis echinus, et e graecis polypus, hippopotamus. Vernacula ad similitudine, syrene, pectunculi, ungues, et qui in terram exeunt. Latina mergus, rana, et id genus caetera, multis vero noina (*sic*) [nomina] nova' (Marschalk Nikolaus, *Historia aquatilium latine ac grece cum figuris* (Rostock, Nikolaus Marschalk: 1517) fol. Aiv). Compare with Varro, *De lingua latina* V, 12, 77.

12 Giovio, *De romanis piscibus*, Chapter 1, fol. AA3r–v.

on the dolphin, which is the most important species of fish in the world. The other species are treated marginally and the whole structure of the work is subordinated to the centre of gravity constituted by the privileged status of the literally royal 'fish'.

> Now I have found a good opportunity to speak about the dolphin and other fish of its kind; I know well that it is a fish which holds the scepter of the sea and that it has been given the second rank in the French coats of arms and that it is the first after the fleurs-de-lis; I have decided to describe amply all the history which is appropriate to it, according to the particular observation of all its parts, both exterior and interior.[13]

Although he does not quote him, Belon seeks to follow the model of Giovio, who writes for a select audience; he writes a short work that is both scholarly and entertaining. The idea is understandable, especially since Belon is struggling to gain recognition as a scholar, after his skills as a Latinist and Hellenist were called into question. Nevertheless, this model of treatment of the subject is already obsolete. It was a time for great syntheses, as they had been appearing since the 1540s in the field of botany, headed by the German *De natura stirpium* by Leonhart Fuchs. This is undoubtedly why Gessner, who seeks to be moderate in his criticism, as we shall see later, mentions the work without any comment:

> From the same author, the book in French on foreign marine fish and the dolphin, printed at the same place [Paris] by Chaulderon, in -4°.[14]

It is interesting to note that this inequality of treatment will persist even in the great treatises with a global claim, those of Gessner or Aldrovandi, whose chapters devoted to "important" or "noble" species can extend over dozens of pages.[15]

---

13 'Maintenant que j'ay trouvé juste occasion de parler du Daulphin, et des autres poissons de son espèce; sachant bien qu'il soit un poisson qui tient le sceptre en mer, et qu'on luy ait donné le second lieu es armoiries en France et aussi qu'il soit en dignité le premier après les fleurs de lils; je suis mis en délibération de descrire amplement toute l'histoire qui luy convient, suivant une particulière observation de toutes ses parties, tant extérieures que interieures [...]' (Belon, *L'histoire naturelle des estranges poissons marins*, chapter 1, 4).

14 Gessner, *Historiae animalium liber IIII* fol. b4v, "Enumeratio authorum qui de piscibus scripserunt, extantium et non extantium, veterum ac recentiorum".

15 Noble species, because they are useful to man, such as the horse, or at the top of the animal hierarchy, such as the eagle.

But for the time being, the genre of textual commentary continues to grow. The *castigationes* on the entire work are becoming more extensive, as in the case of Sigismund Gelenius, and are thus finally gaining their autonomy by appearing in works published separately from the ancient source, by the pen of Francesco Massario or Ermolao Barbaro, for example.

This third stage – "synthetic monographs" or "essays" – therefore sees the critical apparatus gradually move away from the ancient text. In parallel, the commentary develops well beyond the strict search for lexical equivalents of species names.

Beatus Rhenanus, in a dedicace introducing Massarius's work, and Massarius himself, distinguish between the work of the *grammaticus*, which purifies language in general, and that of the *doctus cognitione linguarum*, whose mastery embraces ancient languages and vernacular correspondences, which are themselves useless without knowledge of the real species. It would seem that the "specialization" of the naturalist physician is emerging.[16]

> The Venetian Francescus Massarius certainly made every effort, since he testified that he had examined this sea on voyages of exploration, so that you know that Massarius did not only transcribe these elements using books, many of which are insufficient, but he tested them by eye examination and his own experiences.[17]

And the subtitle of his book also says it well: it is now a question of knowing the *aquatilia* themselves and no longer only of finding equivalences in the appellations, in order to purify the texts, even if this knowledge also includes references to ancient texts:

> Whoever you are, if you wish to study the nature of aquatic animals and gain a deeper knowledge of fish, buy this commentary by Massarius and read it. You will admire the labour and intelligence of a most brilliant

16 Massarius Franciscus, *In nonum Plinii de naturali historia librum castigationes et annotationes* (Paris, Michel Vascosan: 1542) fol. AA2v, Dedication to Doctor Balthazar Entzersberger.

17 'Equidem hanc piscium cognitionem adjuvare enixe studuit Franciscus Massarius Venetus, qui etiam ipsa marina navigando sese perlustrasse testatur, ut non librorum modo praesidiis ista tradidisse Massarum scias quae plerunque debilia sunt, verum ipsis oculis ac experimentis observasse' (Massarius, *In nonum Plinii de naturali historia librum castigationes et annotationes* fol. AA2r–v).

> man, who has long devoted the greatest pains to their study, for which scholars have greatly rejoiced.[18]

So, far from making Massarius into a somewhat timid scholar with only bookish knowledge, Rhenanus specifies that he himself sailed and had the fish he speaks of under his gaze. This precision on the activity of naturalists, which thus appears in the 1530s, will become an essential part of their task. Gessner emphasized this in connection with each author whose skills he praised.

Know-how anticipates the order of the matter. This was the case with Giovio, whose methodological preoccupations were hailed by all and were to become a school of thought for a long time, but whose work constitutes a disharmonious whole.

In the same vein, Giovio acknowledged that he could not follow his illustrious predecessor, Pliny, and be exhaustive in his descriptions of fish:

> And I will not imitate to the end Pliny himself, my compatriot who, in order to express in Latin the discoveries of the Greeks, has magnificently represented nature in its entirety, not only with regard to the fish of the Mediterranean and the rivers, but also the very monsters that crisscross the entire ocean.[19]

And Belon almost echoes Giovio's words:

> Among the Greek authors, Aristotle, Porphyry, and Aelian have written several books on the nature of animals; Oppian on fish; Nicander on snakes; Pliny, among the Latin authors, has indiscriminately described almost all of them, here and elsewhere, taking from the above-mentioned authors and others, who have long observed them; however, I have not hesitated to choose the only dolphin from among all those of which I was aware, by searching for them in their birthplace, and I have set aside the description and painting of them.[20]

---

18 'Quisquis de natura Aquatilium ac remotiore piscium cognitione edoceri cupis, hunc Massarii commentarium eme et lege. Admiraberis laborem ac ingenium hominis candidissimi, qui longe maximam operam in hiis indagandis, ut studiosi juvarentur, insumpsit'.

19 'Neque ipsum Plinium civem meum penitus imitabor, qui ut Graecorum inventa Romanis literis exprimeret, tota natura ubique mirifice repraesentata, non modo conclusi maris, ac fluminum pisces, verum etiam ispas beluas toto oceano fugientes persecutus est' (Giovio, *De romanis piscibus* 6).

20 'Combien que entre les autheurs Grecs, Aristote, Porphyre, et Elien aient escript plusieurs livres de la nature des animauls; Oppian, des poissons; Nicander, des serpents; et que

This attitude is undoubtedly marked by humility, but it opens the door to research that is still partial but thorough and in-depth, and that will go beyond the geographical and methodological framework of the models of Antiquity. What remains from now on is the task of shaping this emerging knowledge.

In his history of the progress made and still to be made, Gessner also cites Pierre Gilles, and his double work, which brings together a form of translation of Claudius Aelian's *De natura animalium* and a list of names of Marseilles fish, visibly inspired by *De romanis piscibus*:[21]

> The Frenchman Pierre Gilles, is the first of our time, I think, after Paolo Giovio, to have improved the history of fish, having published a small treatise of the French and Latin names of the fish of Marseilles, [...] in the format in 4°, by Sébastien Gryphe, in Lyon.[22]

But let there be no mistake, this is no longer a corrected and annotated edition. In his dedication to King François I, Pierre Gilles speaks of a *historiae naturalis nova periclitatio*, a new experiment in the natural history of all the animals of France, a term used by Cicero in *De natura deorum*. Gilles has indeed made a new attempt to adjust the ancient text, both in line with Giovio's model and with the intention of (re)thinking natural history. This appears in Gilles' complete title: *Historia de vi et natura animalium, per Petrum Gyllium tum ex Aeliano conversa, tum ex Porphyrio, Athenaeo, Heliodoro, Oppiano, tum ejusdem Gyllii marte, luculentis accessionibus aucta*.[23]

The chapters are made up of descriptive notes, devoted to species or groups of species, according to morphology or geography. But Gilles intersperses these

---

Pline entre les Latins, les ait indifferemment quasi touts recueillis ca et la, tant des dessus dicts, que de plusieurs autres autheurs, qui les avoient observez par long usage; toutesfois je n'ay laissé d'en elire le seul Daulphin entre touts ceuls dont j'ay eu la cognoissance, en les cherchant sur les lieux de leur naissance, duquel j'ay mis la description et peincture à part' (Belon, *L'histoire naturelle des estranges poissons marins*, "Preface", fol. A3r).

21 Gilles Pierre, *Ex Aeliani Historia per Petrum Gyllium Latini facti, itemque ex Porphyrio, Heliodoro, Oppiano, tum eodem Gyllio luculentis accessionibus aucti libri XVI De vi et natura animalium. Ejusdem Gyllii Liber unus, De Gallicis et Latinis nominibus piscium* (Lyon, Sebastien Gryphe: 1533).

22 'Petrus Gillius Gallus, primus, opinor, post Paulum Iovium nostro tempore, piscium historiam excoluit, libello De Gallicis et Latinis nominibus piscium Massiliensium aedito; et Aeliani de animalibus libris magna ex parte translatis, et accessionibus auctis; quod Sebastianus Gryphus excudit Lugduni in-4' (Gessner, *Historiae animalium liber IIII* fol. b4v, "Enumeratio authorum qui de piscibus scripserunt, extantium et non extantium, veterum ac recentiorum").

23 Gilles, *Ex Aeliani* 27.

with chapters on habits (food, habitat, reproduction, affinities and hatred between species, ways of catching species, etc.) or even more general groupings. The whole thing follows a generally decreasing order according to size, but it gives an impression of accumulation without any real order, which may make Gessner's irritation understandable. Concerning the *Aquatilia*, Gilles' chapters are:

- Whale and other *cetes* (following the example of Pliny and then Giovio, who place the largest *aquatilia* in first place)
- Tuna and catfish (*silurus*)
- *Scolopendra cetacea*
- Scaly fish
- *Molles pisces*
- Various fish
- Seal (*vitulus marinus*).
- Anglerfish (*rana piscatrix*) and other 'flatfish' (raies)
- Crustaceans
- Shellfish, sea urchins
- Fish division
- Nile fish
- Fish reproduction
- Fossils.[24]

24 As an example, here are the chapter titles only for the species descriptions, without general chapters, for books 10 to 12: 'Liber Decimus. De vi et natura animalium aquatilium. De balaena, de maximis cetis, de cetis et variis piscibus Taprobanae insulae, de cetaceo genere Gangis, De britannici ceti magnitudine, De duce cetacei generis, de physeteribus, de tritonibus, de delphinis, de cetis quae et rotae appellantur, de marinibus arietibus, de crocodilo et trochilo; Liber undecimus. De thynnis, de siluris piscibus, de caniculis marinis, de scolopendra et urtica, de scaro, de sargis, de ellope, de pisce aulopio, de acu, de exoceto, de castitate Aethnei piscis, de asello marino, de aspargis piscis, de aurata pisce, de callionymo pisce, de capitonibus maris Ionii, de marinis cicadis, de citharo pisce maris rubri, de denticibus piscibus, de marino dracone, de engraulis sive engrasicholis, de fluviatilibus equis, de glani, de marino grue, de pisce Hamerocita, de maris rubri Hygrophoenie pisce, de lacerto et charace et sagittario, de hyaena, trachuro, thunno, torpedine et pulmone piscibus, de hippocampo, de iulidibus piscibus, de lepore marino; liber duodecimus. De lolliginibus et sepiis, de pisce nuncupato luna, de lupo marino, de melanuro, de merula marina, de mullo pisce, de muribus marinis, de mustello et mustella, terrestri et marina, de myrone pisce, de orcyno, de orpho, de orphis sacris piscibus Myrensium, de ove et epatho psicibus marinibus, de oxyryncho maris rubri, de oxyryncho Nili pisce, de oxyrynchis Caspiis, de pagris et meotis piscibus Nili, de pardali pisce, de persaeo pisce maris rubri, de physa pisce, de physsalo pisce maris rubri, de pompylis piscibus, de remora, de scombris piscibus, de simia maris rubri, de thrissis aegyptiis, de thymo pisce, de torpedine, de piscibus vitulorum similibus, de marina vulpe, de mustello pisce, de

What are we to make of this recasting by Gilles? Certainly, Gessner acknowledges a certain posterity to Gilles. Guillaume Rondelet, he writes further on, followed his division in his own ichthyological writings. But Gessner adopts a much more severe tone on Gilles' contribution to natural history in his edition of Aelian's complete works. This time, the criticism is extensive and severe; we summarize it here:

- Gilles has neglected the order chosen by the author.
- He often deviated from the alphabetical order.
- He has not distinguished the species of the various genera.
- He has ruined the author's arrangement and the effect of the charm due to the variety and pleasure [of the writing].
- Likewise, he has butchered Aelian's associations justified by the quality or other common arguments between certain animals.
- He patched up the text.[25]

In fact, Gessner reproaches Gilles for having lacked both discernment and respect for the form of the text ('poemata elegantissime condata sunt'), the form which alone can have a heuristic value.

A final example for this stage shows better the direction in which naturalists should strive: reconciling the often concise references of the ancients, and the plurality of modern names: it is William Turner's small critical glossary, whose title is both modest and explicit:

> Avium praecipuarum, quarum apud Plinium et Aristotelem mentio est, brevis et succincta historia. Ex optimis quibusdam scriptoribus contexta, scholio illustrata et aucta. Adjectis nominibus Graecis, Germanicis et Britannicis.[26]

---

xiphia, et aliis piscibus Danubii, de xiphiae gladio, de marinis cantharis' (Gilles Pierre, *Ex Aeliani Historia* 283–386).

25 Gessner Conrad, *Claudii Aeliani Praenestini Pontificis et Sophistae, qui Romae sub Imperatore Antonino Pio vixit, Meliglosus aut Meliphthongus ab orationis suavitate cognominatus, opera, quae extant, omnia, Graece Latineque e regione, uti versa hac pagina commemorantur: partim nunc primum edita, partim multo quam antehac emendatiora in utra lingua, cura et opera Conradi Gesneri Tigurini* (Zurich, Gesneros fratres: 1556), "Epistola nuncupatoria", fol. a4v.

26 Turner William, *Avium praecipuarum, quarum apud Plinium et Aristotelem mentio est, brevis et succincta historia. Ex optimis quibusdam scriptoribus contexta, scholio illustrata et aucta. Adjectis nominibus Graecis, Germanicis et Britannicis* (Cologne, Johannes Gymnicus: 1544). Translation: 'A brief and succinct history of the principal birds, mentioned by Pliny and Aristotle, illustrated and enlarged by the commentary drawn from some of the best authors, together with their Greek, German and British names'.

Admittedly, the book lacks species descriptions and illustrations, but the intention is there. The door is now open for the last two phases: the systematic confrontation of ancient text *vs* personal observations, and epistemological division toward new categories of knowledge. Gessner sees them emerging, incompletely but promisingly, in two authors: Wotton and Belon, whom he had cited for his first work.

The approaches of these authors complement each other in a way. The first allows Gesner to note that there is now a unitary procedure for the analysis of ancient texts:

> Edward Wotton of Oxford edited a work titled *De differentiis animalium libri X*, printed in Paris by Vascosan in 1552 in folio format. Although he did not include any personal observations in his book and did not add anything to natural history in this respect, the work is nevertheless praiseworthy and worth reading: he has put in order many ancient writings and made them complement each other, so much so that they almost all seem to have been written by one author, with an equal and pure style; scholia and corrections have been made to many passages of the various authors, and finally, before coming to the explanation of the nature of the particular species, he has set forth in a very erudite manner explanations which may be called common and generic.[27]

The second, Pierre Belon, is cited for his observations, collected in France and during his travels:

> Belon is to be praised especially for having explored at length various little-known regions of Europe, Asia, and Africa, at great cost and risk to his life, on land and at sea, and thus made many discoveries unknown in our time and before. Thus, he has partly printed books about plants

27 'Edoardi Vuottoni oxionensis de Differentiis animalium libri X impressi sunt Lutetiae apud Vascosanum, anno 1552 in folio in quibus etiamsi suarum observationum quod ad historiam nihil adferat, neque novi aliquid doceat; laude tamen et lectione dignum est opus: quod pleraque veterum de Animalibus scripta ita digesserit ac inter se conciliarit, ut ab uno fere authore profecta videantur omnia, stylo satis aequabili et puro, scholiis etiam ac emendationibus in varios authorum locos adjectis; et quoque priusquam ad explicandas singulorum naturas accedat, quae communia et in genere dici poterant, doctissime exposuerit' (Gessner, *Historia animalium liber IIII* fol. b3v, "Enumeratio authorum qui de piscibus scripserunt, extantium et non extantium, veterum ac recentiorum"). Cf. Wotton Edward, *De differentiis animalium libri decem* (Paris, Michel Vascosan: 1552).

> and animals and many other things, and the publication of the rest is eagerly awaited.[28]

As we have seen, in his first ichthyological work, Belon still lacked methodological reference points. He claimed above all the rigour of his direct observations. He pointed out above all the repetition of erroneous information and not verified by eyewitnesses.

But very soon, in his second work on *aquatilia*, and then in the *Histoire des oyseaux*, Belon also engaged in a philological analysis and a comparison between the various ancient and modern sources on the one hand, and direct observations, his own or those of witnesses, on the other.[29] The influence of Gessner and the plural dimension that natural history took on in the mid-sixteenth century among naturalists in general can undoubtedly be seen here. However, Gessner does not welcome this development, as a discreet way of pointing out that Belon had not fulfilled his philological task.

Finally, it is Guillaume Rondelet who unified the two approaches:

> Rondelet devoted the greatest energy to deciphering the true and ancient names of aquatic animals and to writing descriptions of them; moreover, he also himself travelled through the Belgian and Italian regions; he added to this a great erudition and an uncommon ability to clarify obscure or doubtful passages in the authors.[30]

28 'In Bellonio hoc eximie laudandum, quod in diversis remotisque Europae, Asiae, et Africae regionibus peregrinatus, multo tempore, maximis laboribus, et discrimine vitae, per tot itinera et maria, mUlta huic nostro seculo et ante hoc pluribus incognitae prodidit; sicut in aliorum quoque animalium ac stirpium generem aliisque rebus multis, de quibus libri ejus partim aediti sunt, partim magno desiderio aedendi expetantur' (Gessner, *Historia animalium liber IIII*, "Ad candidum lectorem praefatio", fol. br).

29 Belon Pierre, *De aquatilibus Libri duo, cum ειconibus* [sic] *ad vivam ipsorum effigiem, quoad ejus fieri potuit, expressis* (Paris, Charles Estienne: 1553); idem, *Histoire de la nature des oyseaux avec leurs descriptions, et naïfs portraicts retirez du naturel: escrite en sept livres* (Paris, Guillaume Cavellat: 1555). Critical re-ed. Ph. Glardon (Geneva: 1997).

30 'Rondeletius diligentiae summae circa indaganda vera ac vetera piscium nomina, eorumque descriptiones (peregrinatus etiam ipse ad Belgas et Italos), variam eruditionem, et in explicandis dubiis obscurisque authorum locis haud vulgarem solertiam adjunxit' (Gessner, *Historia animalium liber IIII*, "Ad candidum lectorem praefatio", fol. br). Cf. Rondelet Guillaume, *Libri de piscibus marinis in quibus verae piscium effigies expressae sunt*, 2 vol., t. II: *Universae Aquatilium Historiae pars altera, cum veris ipsorum Imaginibus* (Lyon, Macé Bonhomme: 1554–1555); idem, *La premiere et la seconde partie de l'histoire entiere des poissons*, 2 vol. (Lyon, Macé Bonhomme: 1558).

Gessner was well aware that he also had to promote his work, and that criticism of the approach or content of works published before his own was therefore also aimed at competitors. Even though he does not shy away from promoting his own treatment of the subject, the concern to develop a general methodology in natural history is nonetheless very real for him; moreover, he feels himself to be the target of attacks, and responds to them in a long 'ad calumniatores' argument. In his epistle to the reader, he poses as a moderator and openly addresses the issue of the serious rivalries between Belon and Rondelet on the one hand, and Rondelet and Salviani[31] on the other, by emphasizing the positive aspects of the respective contributions of each. He undoubtedly did this in a spirit of appeasement, and his reputation, already well established at the time he wrote these lines, and the influence he exercised and would continue to exercise long after his death, clearly demonstrate that his work contributed, in the middle of the 16th century, to achieving a form of balance in the treatment of natural history.

The relationship with the ancient text had thus fundamentally changed; in a way, the relationship with the ancients had been appeased: the naturalists had established a methodology that clearly combined commentary and observation. The ancient text is respected, even if its contribution in terms of direct observation has become secondary. An essential dimension remained for the Renaissance scholar: the universal view of the ancients, which revealed the sacredness and harmony of nature, even in the form of the texts.

Naturalists have taken the measure of the ancient texts, and a distinction is now made between the form, content, and respective objectives of the two. However, a deep respect for the ancients remained, characteristic until the 17th century: knowledge of animals was a branch of natural history, whose primary purpose was to gain a better understanding of the divine work. This, in its immeasurable dimension, requires the pooling of everyone's skills and the ordering of all the knowledge available to educated men. In the quest for universality, the Aristotelian and Plinian approaches, like the texts of the ancient poets, remain keys to the sacred dimension of nature.

Annex

[Epistola] Pauli Jiovii novocomensis medici De romanis piscibus libellus ad Ludovicum Borbonium Cardinalem amplissimum.[32]

31 See the contribution of Holger Funk in this volume.

32 Giovio, *De romanis piscibus* (Basel, Froben: 1531) 3–5.

Suades Reverendissime ac Illustrissime domine, qui etiam jure optimo compellere potes, ut ea literis tradam, quae de Romanis piscibus erudite atque subtiliter fuere disputata, quum te, et Joannem Lotharingum praeclassimi ingenii Cardinalem Clemens Pontifex familiari convivio, veluti animum remissurus, hilariter excepisset. Res est cognitu tractatuque difficilis, cum propter infinitam fere naturae piscium varietatem, que me et plerosque alios admodum curiosos hactenus irrito labore fatigavit; tum propter incredibilem scriptorum discrepantiam, qui ex multiplici linguarum varietate perpetuas hujusmodi studiis tenebras offuderunt*. Quibus de causis necesse est, ut animus rei difficultate permotus totum hoc munus, quod majoribus doctrinae ex actiorisque judicii nervis esset extendum, haud mediocriter reformidet. Sed ea est dignitas, amplitudoque tua et morum suavitas cum singulari optimarum literarum cupiditate conjuncta, ut honoris mei vel gloriolae, siqua ingenuis studiis parta est, jactura plane facere, quam honestissimo desyderio tuo penitus deese malim. Verum tu postea judicaveris, an ego qui remtam difficilis argumenti et majorem omnino ingenii facultatem postulantis suscepi, impudentiae culpa vacaverim, quum ejus nominis veniam jam deprecer, ut officiosus et perhumanus appaream. Proinde hunc libellum ab acrioribus censoribus minus severe judicari velim, quando eum festiva quadam hilaritate veluti ludibundus, dum multi per haec Saturnalia, aleae, caeterisque voluptatibus vacarent, antiquis meis earum rerum observationibus memoriae repetitis, excogitaverim, et aliqua ex parte perfecerim. Eritque hic male feriati hominis ingenuus labor, fortasse praeludium justi, absolutique operis, si qui literis ac industria pollent, proposita oculis, et a me quasi per transennam commonstrata, accuratius contemplari, ac uberioribus interpretationibus illustrare voluerint. Ego certe hoc munere, et tibi in primis, et illis gratum fecisse videbor, animusque etiam meus tanquam amoeniore diverticulo recreatus ad institutas absolvendae historiae lucubrationes et firmior et alacrior revertetur. Exhibit enim in publicum propediem huijusmodi laborissimi operis prima Decas non sine aliqua spe immortalitis, et tum quidem multo auctior et ornatior, si eam tibi et Francisco Regi tuo, quem liberalibus studiis delectari audivimus, aliquando placuisse cognovero.

Vale. Ex Vaticano. Calendas Aprilis. M.D.XXIIII

Bibliography

Belon Pierre, *L'histoire naturelle des estranges poissons marins* (Paris, Regnaud Chauldiere: 1551).

Belon Pierre, *De aquatilibus Libri duo, cum* ειconibus [sic] *ad vivam ipsorum effigiem, quoad ejus fieri potuit, expressis* (Paris, Charles Estienne: 1553).

Belon Pierre, *Histoire de la nature des oyseaux avec leurs descriptions, et naïfs portraicts retirez du naturel: escrite en sept livres* (Paris, Guillaume Cavellat: 1555). Critical re-ed.: Ph. Glardon (Geneva: 1997).

Beullens P. – Gotthelf A., "Theodore Gaza's Translation of Aristotle's *De Animalibus*: Content, Influence, and Date", *Greek, Roman, and Byzantine Studies* 47 (2007) 469–513.

Gessner Conrad, *Historiae animalium liber I. de quadrupedibus viviparis* (Zurich, Chistopher Froschauer: 1551).

Gessner Conrad, *Claudii Aeliani Praenestini Pontificis et Sophistae, qui Romae sub Imperatore Antonino Pio vixit, Meliglosus aut Meliphthongus ab orationis suavitate cognominatus, opera, quae extant, omnia, Graece Latineque e regione, uti versa hac pagina commemorantur: partim nunc primum edita, partim multo quam antehac emendatiora in utra lingua, cura et opera Conradi Gesneri Tigurini* (Zurich, Gesneros fratres: 1556).

Gessner Conrad, *Historia animalium liber IIII, qui est de piscium et aquatilium animantium natura,* [...] *continentur in hoc volumine Guilelmi Rondeletii* [...] *Petri Bellonii* [...] *de aquatilium singula scripta* (Zurich, Christopher Froschauer: 1558).

Gilles Pierre, *Ex Aeliani Historia per Petrum Gyllium Latini facti, itemque ex Porphyrio, Heliodoro, Oppiano, tum eodem Gyllio luculentis accessionibus aucti libri XVI De vi et natura animalium. Ejusdem Gyllii Liber unus, De Gallicis et Latinis nominibus piscium* (Lyon, Sebastien Gryphe: 1533).

Giovio Paolo, *De Romanis Piscibus libellus ad Ludovicum Borbonium Cardinalem amplissimum* (Rome, F. Minitius Calvus: 1524 (April)); *De Romanis Piscibus libellus ad Ludovicum Borbonium Cardinalem amplissimum* (Rome, F. Minitius Calvus: 1524 (August)); *De piscibus marinis, lacustribus, fluviatilibus, item de testaceis ac salsamentis liber* (Rome, F. Minitius Calvus: 1527); *De Romanis Piscibus Libellus* [...] (Antwerp, Joannes Grapheus: 1528); *De Romanis Piscibus libellus* [...] (Basel, Hieronymus Froben: 1531); *Pauli Jovii de piscibus liber unus*, ed. Johannes Caesar (Strasbourg, Jacobus Cammerlander: 1534); *Libro di mons. Paolo Giovio de pesci romani*, transl. Carlo Zancaruolo (Venice, Gualtieri: 1560).

Marschalk Nikolaus, *Historia aquatilium latine ac grece cum figuris* (Rostock, Nikolaus Marschalk: 1517).

Massarius Franciscus, *In nonum Plinii de naturali historia librum castigationes et annotationes* (Paris, Michel Vascosan: 1542).

Rondelet Guillaume, *Libri de piscibus marinis in quibus verae piscium effigies expressae sunt*, 2 vol., t. II: *Universae Aquatilium Historiae pars altera, cum veris ipsorum Imaginibus* (Lyon, Macé Bonhomme: 1554–1555).

Rondelet Guillaume, *La premiere et la seconde partie de l'histoire entiere des poissons*, 2 vol. (Lyon, Macé Bonhomme: 1558).

Turner William, *Avium praecipuarum, quarum apud Plinium et Aristotelem mentio est, brevis et succincta historia. Ex optimis quibusdam scriptoribus contexta, scholio illustrata et aucta. Adjectis nominibus Graecis, Germanicis et Britannicis* (Cologne, Johannes Gymnicus: 1544).

Wotton Edward, *De differentiis animalium libri decem* (Paris, Michel Vascosan: 1552).

CHAPTER 4

The Many Names of Fish: Scientific and Poetic Fish Nomenclature in the Writings of Johann Fischart and Conrad Gessner

Tobias Bulang

Let me start by introducing a fish species, which might not be known in English. It is a fish that does not swim in any known waters, the so-called "Zwiebelfisch" – the term would literally be translated as "onionfish". The word belongs to a special technical terminology used by German book printers. "Zwiebelfisch" refers to a printed word that contains letters of different fonts; this can happen when a letter type is put into the wrong compartment of the letter case, or when certain letter types run out or are missing. To give an example:

Zwiebelfisch

One might note that the fonts of the first /ie/ and the /fis/ are different from the rest of the printed word. The page number 666 in a 16th-century German book about demons and witchcraft is a good example of a "Zwiebelfisch" [Fig. 4.1].[1]

When you look closer, you can see that the third number is a different font (a "Zwiebelfisch") – by the way, these are not three 'sixes', but three 'nines' placed upside down: The typesetter was superstitious and afraid of the devilish number mentioned in the Book of Revelation. And, he knew a few tricks of the trade to get out of the risky interaction with evil.

Printing workshops, which produced low-quality printed matter, were called "Zwiebelfischbuden" – "onionfish-shops". The word is a metaphor. But how does this metaphor work? Fish of lesser quality needs to be cooked with a lot of onions to cover the dingy taste; thus, at the fish market a "Zwiebelfisch" is a fish of lousy quality. And so is a printed volume without elegant typography at the book market.

How does this word relate to the phenomenon that it signifies? The relationship between words and things is always arbitrary, as modern linguistics has

1 Bodin Jean – Fischart Johann, *De magorum daemonomania. Vom außgelassnen wütigen Teufelsheer* (Strasbourg, Bernhard Jobin: 1586) 666.

DOI:10.1163/9789004681187_005

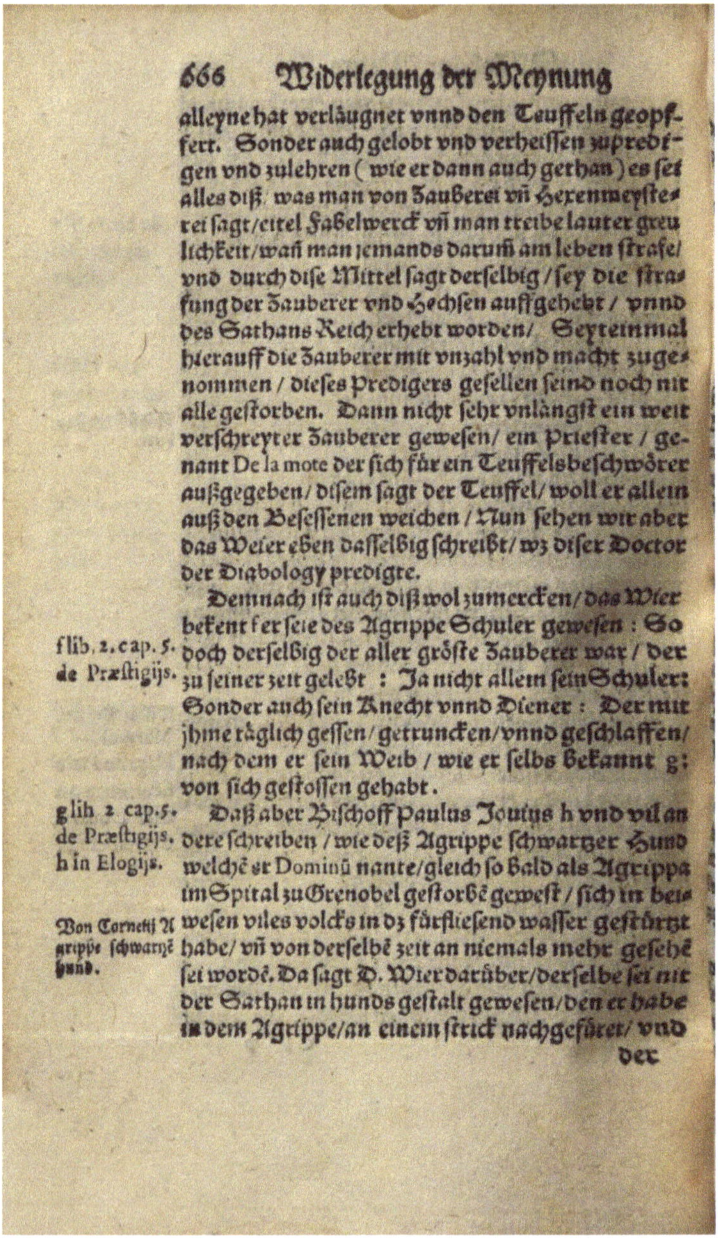

666 Widerlegung der Meynung

alleyne hat verläugnet vnnd den Teuffeln geopf-
fert. Sonder auch gelobt vnd verheissen zupredi-
gen vnd zulehren (wie er dann auch gethan) es sei
alles diß was man von Zauberei vñ Hexenmeyste-
rei sagt/eitel Fabelwerck vñ man treibe lauter greu
lichkeit/wañ man jemands darumb am leben strafe/
vnd durch dise Mittel sagt derselbig / sey die stra-
fung der Zauberer vnd Hechsen auffgehebt / vnnd
des Sathans Reich erhebt worden/ Seyteinmal
hierauff die Zauberer mit vnzahl vnd macht zuge-
nommen / dieses Predigers gesellen seind noch nit
alle gestorben. Dann nicht sehr vnlängst ein weit
verschreyter Zauberer gewesen/ ein Priester / ge-
nant De la mote der sich für ein Teuffelsbeschwörer
außgegeben/ disem sagt der Teuffel/ woll er allein
auß den Besessenen weichen / Nun sehen wir aber
das Weier eben dasselbig schreibt/ wz diser Doctor
der Diabology predigte.

Demnach ist auch diß wol zumercken/ das Wier
bekent f er seie des Agrippe Schuler gewesen : So
f lib. 2. cap. 5. de Præstigijs. doch derselbig der aller gröste Zauberer war / der
zu seiner zeit gelebt : Ja nicht allein sein Schuler:
Sonder auch sein Knecht vnnd Diener : Der mit
jhme täglich gessen/ getruncken/ vnnd geschlaffen/
nach dem er sein Weib / wie er selbs bekannt g:
von sich gestossen gehabt.

g lib 2 cap. 5. de Præstigijs. h in Elogijs. Daß aber Bischoff Paulus Jouius h vnd vil an-
dere schreiben / wie deß Agrippe schwartzer Hund
welchē er Dominū nante/ gleich so bald als Agrippa
im Spital zu Grenobel gestorbē gewest / sich in bei-
Von Cornelij Agrippe schwarzē hund. wesen viles volcks in dz fürfliesend wasser gestürtzt
habe/ vñ von derselbē zeit an niemals mehr gesehē
sei wordē. Da sagt D. Wier darüber/ derselbe sei nit
der Sathan in hunds gestalt gewesen/ den er habe
in dem Agrippe/ an einem strick nachgeführet/ vnd
der

FIGURE 4.1 Bodin Jean – Fischart Johann, *De magorum daemonomania. Vom außgelassnen wütigen Teufelsheer* (Strasbourg, Bernhard Jobin: 1586) 666 (detail)

taught us, or as Friedrich Nietzsche said: *all* words are nothing but metaphors.[2] In order to know how the word "Zwiebelfisch" relates to the phenomenon in a printed book, we need a specific cultural context, in this case culinary facts, institutions, issues of trade and craftsmanship and special technical jargon.

And this brings me to real fish. Fish do not tell us their names. The procedures that relate a concrete swimming or floating being to its name are complicated, sometimes involving translations from different languages, confusion with other species, different terms used by scholars or fishermen. Early modern ichthyology can be described as a major struggle to connect animals and words. As popular as 'animal studies' are in recent academic discourse, there should be more 'animal philology': we cannot speak scientifically about animals without considering the very nature of our language.

The starting point of my journey into the world of the philological aspects of ichthyology was a list of fish names in the first German translation of Rabelais' *Gargantua*. The Alsatian author Johann Fischart did not only translate Rabelais' text (first edition 1575), wherever possible he added long lists of German words, like lists of German sausages, cheeses, beers and wines, of guns and ammunition or of all the pumps used for mining. Grandgousier, the giant who is always hungry and thirsty, stows huge amounts of food in his storage rooms, including masses of fish. Fischart's German text has no equivalent in Rabelais' novel, it is an interpolation:

> Deßgleichen vergaß er sich auch nicht mit frischen Fischen, als allerhand Bratfischen vom Bodensee, Hausengalreien, gebratenen Forellen, Hausstockfischen, Doͤrren Posten, Proͤsem, *Stoͤren, scheiden, Rot Fohren, weiß Orffen, vnnd gel Haselnaschen, Raumen den Streydasgůtlein die Taschen. O kugelhaupt, gebachen Pirsching fuͤr die Pfaffen gut, gebraten Latfohren gut zum Salat, Miltzhåring gut zum sauren Kraut, gereuchert Rencken, blo Felchen, weiß vnnd gelb Gangfisch, Růdling, Kelchlin, Lauben,* Truschen, Ropelen ….[3]

This list is quite strange in many regards. Presented are not only the different common names of fish in the sea, but also the names of fish dishes prepared

2 See Nietzsche F., "Ueber Wahrheit und Lüge im aussermoralischen Sinne", in Friedrich Nietzsche. *Kritische Studienausgabe vol. 1: Die Geburt der Tragödie. Unzeitgemäße Betrachtungen I–IV. Nachgelassene Schriften 1870–1873*, eds. G. Colli – M. Montinari (Munich: 1999) 873–890, see: 878–879.

3 Fischart Johann, *Geschichtklitterung* (*Gargantua*). *Synoptischer Abdruck der Fassungen von 1575, 1582 und 1590. Mit drei Titelblättern und den Originalholzschnitten der Ausgabe von 1590 von Tobias Stimmer*, ed. H. Schnabel, 2 vols. (Halle a. S.: 1969) 82.

in the kitchen. The fish named by Fischart are supposed to be fish from the Bodensee, i.e. Lake Constance in southern Germany, Austria and Switzerland. And some of the fish from the list do indeed swim in that very body of water; however, some of the typical species found in Lake Constance are not mentioned. In 1546 Gregor Mangolt published a book about the fish in Lake Constance, and also Conrad Gessner often mentioned Lake Constance and specific fish found there in his list of vernacular fish names.[4] Among them are the eel, the carp as well as fish called "Inlancken" or "Blieck" – they do not appear in Fischart's list at all.

Another fishy thing about this fishy list is that it contains fish names which do not exist anywhere else: Species called "Haselnasche" or "Latfohre" are simply unknown.

Fischart did not make up that list. The lists in his novel, plus-texts he added to the Rabelais text, are excerpts from contemporary German publications. For a list of sausages, he used the *Nomenclator omnium rerum* by Hadrianus Junius; a list of cheeses was taken from Olaus Magnus' popular book about the northern countries; for other food items, Fischart replicated excerpts from Hieronymus Bock's *Teutsche Speiskammer*, and he copied a big collection of names of regional wines from Nicodemus Frischlin's *Carmen* on the royal wedding in Württemberg. Fischart used Leonhard Fronsberger's *Kriegsbuch* (guns) and Agricola's *Bergwerkbuch* (mining tools) for inventories of technical items. His translation of Rabelais with all the additional material may sometimes

4 Mangolt Gregor, *Fischbůch, Von der natur vnd eigenschafft der vischen* (Zurich, Andreas Gessner: 1557); see Hakelberg D., "Das Bodensee-Fischbuch von Gregor Mangolt in einem Basler Nachdruck von 1612", *Wolfenbütteler Barock-Nachrichten* 36 (2009) 107–114. Gessner's German list of fish names is part of his edition and commentary of (pseudo-)Ovid's *Halieutica*, which exists in two slightly different versions: *P. Ovidii Nasonis Helieuticon: hoc est, de piscibus libellus, multò quàm ante hac emendatior et scholijs illustratus. Emendantur et Plinij aliquot loca. Accedit Aquatilium animantium Enumeratio iuxta Plinium, emendata et explicata per Conradum Gesnerum, ordine alphabetico. Earvndem nomina Germanica eodem ordine* (Zurich, Andreas and Hans Gessner: 1556); *De piscibus et aquatilibus omnibus libelli III. novi. Authore Conrado Gesnero medico et philosophiae naturalis interprete in Schola Tigurina: I. Scholia et emendationes in Halieuticon P. Ouidij Nasonis. II. Aquatilium animantium enumeratio juxta Plinium, emendata et explicata serie literarum. III. Eorundem Nomenclator Germanicus longé copiosissimus. Et alia ad piscium historiam pertinentia* (Zurich, Andreas Gesner: 1556). The ichthyological catalogue "Teütsche nammen der Fischen vnd Wasserthieren" (97–279) is identical in both editions. A separate edition of only the catalogue with a remarkable introduction is by. Peters M. (ed.), *Konrad Gessner: Deutsche Namen der Fische und Wassertiere. Neudruck der Ausgabe Zürich 1556* (Aalen: 1974); cf. Leu U.B., *Conrad Gessner (1516–1565). Universalgelehrter und Naturforscher der Renaissance* (Zurich: 2016) 201f.

seem like an attempt to include all the books of his time in that one novel; his German translation is quite an encyclopaedic approach.[5]

It was not a fish book or a cook book from which Fischart took the fish names, but a poem by the famous German poet Hans Sachs from Nuremberg. The poem is called *The hundred and twenty-four fish and sea-mirabilia* and is dated to 6 May 1559.[6] In it, Hans Sachs claims that he received this information from an old fisherman, who told him about all the fish. He must have been a very, very old or very learned fisherman since he mentions Aristotle, Pliny, Isidor of Seville and Albertus Magnus as his references. Hans Sachs refers to the unknown fish first, excerpting from Pliny's *Historia naturalis*. Then he presents "the fish in our land" – in Germany.

Fischart copied the fish names in his list from this poem, even the sequence of the terms corresponds. Comparing the texts, one can also see how the "Haselnasche" came into being: Hans Sachs wrote:

> Hasseln vnd auch sampt den Aschen
> Raumen den Streußgůtlein die Taschen[7]

Fischart's "Haselnasche" originated from two fish species in Sachs' poem: the "Hasel" (*Leuciscus leuciscus*) and the "Äsche" (*Thymallus thymallus*). Thus, we can observe a miraculous multiplication of fish by making copy errors. The same is true for the enigmatic "Latfohre" in Fischart's list – in Sachs' poem it is the well-known fish named "Lachsfohre".

Some ethnobiological research into Sachs' poem proved that he mentioned species which can typically be found in lakes in the Bavarian Prealps[8] – Fischart turns these fish into fish from Lake Constance – which explains why some typical species are missing.

5 For cheese and sausages, cf. Bulang T., *Enzyklopädische Dichtungen. Fallstudien zu Wissen und Literatur in Spätmittelalter und früher Neuzeit* (Berlin: 2011) 346–348, 384–385. For wine, see Bulang T., "Spiele in Johann Fischarts *Geschichtklitterung*", in Jahn B. – Schilling M. (eds.), *Literatur und Spiel. Zur Poetologie literarischer Spielszenen* (Stuttgart: 2010) 45–69, see 52–53. For guns and mining tools, see Bulang, *Enzyklopädische Dichtungen* 350–351, 377–378.

6 Sachs Hans, *Die Hundert vnnd vier vnd zweintzig Fisch vnd Meerwunder mit jrer art. Das ander Buch: Sehr herrliche Schöne Artliche vnd gebundene Gedicht mancherley art* […] (Nuremberg, Christoph Heußler: 1560), "Das ander Theyl: von Tugent und Laster" fol. 90^r–92^r; Sachs Hans, "Die hundert vnd Vier vnd zweintzig Fisch vnd Meerwunder mit jrer art", in Sachs Hans, *Werke*, ed. A. von Keller, vol. 7. (Tubingen: 1873) 456–463. I will quote the text from 1560 and also refer to Keller's edition.

7 Sachs 1560, fol. CXI^v; Sachs 1873, 462, v. 1–10.

8 Gebhard T., "Das Spruchgedicht des Hans Sachs von den 124 Fischen", *Bayerisches Jahrbuch für Volkskunde* (Munich: 1985) 76–85.

How does this relate to the topic of the present volume? Both Hans Sachs and Fischart participated in the discourse of early modern ichthyology and turned the knowledge of their time into something like poetic energy. The work on the German names of things, including their relation to Latin terms and to other vernacular languages, is also very relevant to the field of natural history as well as ichthyology. This becomes evident in the works of Conrad Gessner (1516–1565). For his research on the nomenclature of fish, Gessner studied ancient texts about *aquatilia*; he extensively used, among other contemporary publications, the fish book by Guillaume Rondelet,[9] the Montpelier scholar who was among Gessner's correspondents. The depictions and descriptions of whales and sea-monsters in the popular book by Uppsala's archbishop Olaus Magnus about the northern lands were also of special interest to Gessner.[10] In addition to such bookish learning, Gessner received a wealth of information from his numerous correspondents – he corresponded with more than 500 scholars, pharmacists, doctors and others.[11] The starting point for Gessner's name-search were the ancient authors. However, they mainly addressed the *aquatilia* in the Mediterranean Sea, which do not necessarily correspond to the fauna north of the Alps. The writings of Aristotle, Pliny and Solinus were of major importance to the *aquatilia*. Didactic poetry, such as the *Halieutica* by Oppian, also had a profound influence.[12] And also Ovid played a major role, since Pliny mentioned Ovid's didactic poem about fish as a source in the 32nd book of his natural history. This poem has been known since the 9th or 10th century; but although it appears in many early modern Ovid editions, it is probably a medieval paraphrase of the fish Pliny listed in hexameters.[13] All these texts are not only considered by scholars for facts and dates, they are themself the object of major philological efforts in the 15th and 16th centuries, they are being commented, the manuscripts are being revisited, and the problem of how to

9 Rondelet Guillaume, *Libri de piscibus marinis in quibus verae piscium effigies expressae sunt* (Lyon, Mathias Bonhomme: 1554); Rondelet Guillaume, *Histoire entière des poissons* (Lyon, Mathias Bonhomme: 1558).

10 See the German translation: Magnus Olaus, *Beschreibüng allerley Gelegenheyte / Sitten / Gebräuchen und Gewonheyten / der Mitnächtigen Völcker* [...] (Strasbourg: Theodosius Rihel, 1567) 189ʳ–329ᵛ. For the tradition of fish books since Aristotle, see Hühnemörder Ch., "Die Geschichte der Fischbücher von Aristoteles bis zum Ende des 17. Jahrhunderts", *Deutsches Schiffahrtsarchiv* 1 (1975) 185–200.

11 See the table of Gessner's letters in Leu, *Conrad Gessner* 390–406. For Gessner's correspondence about fish names, see the introduction in Peters (ed.), *Konrad Gessner* 7–43, and Leu, *Conrad Gessner* 201–202.

12 Cf. Oppianus, *Halieutica*, transl. and ed. F. Fajen (Stuttgart – Leipzig: 1999).

13 Cf. *Der Neue Pauly*, vol. 5, 1998, s.v. "Helieuticon".

translate Greek and Latin names into vernacular ones becomes crucial.[14] The search for reference in this field is a huge project throughout Europe. Gessner himself edited and commented the pseudo-Ovidian *Halieutica*, which had already been published in several editions of Ovid's collected works after the *editio princeps* in Venice in 1534.[15] Gessner especially corrected the fish names according to other sources, added commentaries (scholia) and included an appendix with an enormous catalogue of German fish names.[16] As already mentioned, he also initiated the printing of the German book about the fish in Lake Constance, written by Gregor Mangolt.[17]

Gessner was not able to find all the German names of fish in the books of his day. The *copia verborum* of fish names, which he presents in his list, is the result of fieldwork and extensive correspondence. Gessner used contacts abroad to collect data and had friends send him inventories of locally used fish names. Gessner drew remarkably well, and he also systematically collected excellent drawings and paintings; this is evident in the original paintings used for the woodcuts of the *Historia animalium*, which were recently rediscovered by Florike Egmond.[18] He sent images to his correspondents and asked them to inquire of the local fishermen the names they used. Here, Gessner also played an important role in the lexicography of German dialects. The appendix in his pseudo-Ovidian *Halieutica* edition is followed by separate lists of fish that were caught in the Rhine river in Strasbourg and one list of fish in the Elbe river in Dresden. The latter list was composed and sent to Gessner by Johannes Kentmann, who produced a beautiful, handwritten herbal – the spectacular

14 Cf. Hendrikx S. – Smith P.J., "Ichthyology in Translation: Conrad Gessner's Fish Books", in Fournel J.-L. – Paccagnella I. (eds.), *Traduire – Tradurre – Translating. Vie des mots et voies des œuvres dans l'Europe de la Renaissance* (Geneva: 2022) 341–361.

15 The *Halieutica* can be found in the early modern editions of Ovid's works: e.g. Cologne, Martinus Gymnicus: 1545; Cologne, Walther Fabritius: 1554. They are also part of collections with classical texts about hunting (*cynegetica*): Augsburg, Steiner: 1534; Frankfurt am Main, Sigismund Feierabend: 1582. Modern edition: *Ovidii Haleutica, Gratii et Nemesiani Cynegetica* ex recessione Mauritii Hauptii accedent inedita Latine et tabula lithographica (Leipzig: 1838); Ovid, *Halieutiques*, transl. and ed. E. de Saint Denis (Paris: 1975). For Gessner's edition with *scholia*, see note 3. For Greek fish names, see Strömberg R., *Studien zur Etymologie und Bildung der griechischen Fischnamen* (Gohtenbug: 1943); see also Richmond J., *Chapters on Greek Fish-Lore* (Wiesbaden: 1973).

16 See note 3.

17 See note 3.

18 Egmond F. (ed.), *Conrad Gessners Thierbuch. Die Originalzeichnungen*, transl. G.M. Vorderobermeier (Darmstadt: 2018), see here the images of the album III C 22: "Tiere, die im Wasser leben" 35–219.

script can be viewed in Dresden's library.[19] To give an impression of the work on fish names and its epochal dimensions, I will present the Strasbourg list (the Rhine fish are of course also somewhat relevant to Lake Constance):

PISCES RHENI CIRCA ARgentinam. numero XXXIII.
Stőr.
Salm/ Forell.
Aal/ Lampred/ Berling oder Prick/ Nünaug/ Rufelck/ oder Rutt/ oder Trüsch. Koptt oder Gropp/ Steinbyß/ Grundel/ Schlye. Karpf/ Brachsme/ Meckel oder Blick. Meyfische oder Mañemer hengst/ die vß dem meer komen/ sunst werden die Hasele Meyfisch genennt. Rottel. Hecht. Aesch. Bersing. Kutt oder Goldfisch. Barb. Furn oder Alet. Naß. Schnottfisch oder Hasele. Elbele.
Rotteugel oder Schwal: sunst nennemn etlich ein Blieck ein Rotteugel. Lauck. Blieck. Kreß/ oder Gräßling. Stichling oder stahelfischle. Mylling oder orlē/ villiecht unsere Harlüchle/ and'schwo Pfrillen/ Pfålen/ Bintzbautten. Riemling/ villicht vnsere Bambele.

PISCES ALBIS, EX CATAtaloge quem Ioannes Kentmanus Dresdensis medicus nos dedit
I. Elbfische die auß der sehen komen/ und bleibē nicht/ nemen abe od' sterben.
Sthor/ Lachs/ Ziege/ wird sunst Goldfisch genennt/ kompt mitt dem Sthor in die Elbe/ wirdt mitt imm gefangen. Halbfisch. Lampreten. Nünaugen
II. Elbfische die auß den bechen in die Elbe komen/ gedien vnd nemen darin zů.
Steinbiß. Foren. Aeschen. Bapst/ der Mullen/ oder Kaulhaupt. Olruppen. Smerling. Steinsmerling. Elderitz. Stichling. Schlei. Aal. Beisecker.Welß.
III. Elbfisch die da schůppen haben.
Hecht. Karpf. Kaulbersing. Streiffbersing. Parme.
IIII. Elbfisch de gůt zubraten.
Jesen. Diebeln. Rappē. Blehen. Geusten Roteugel. Zorten oder Zerten.
V. Weißfisch.
Heseling. Grundling. Ockeln. Karas. Oberkottichen. Schneppelfischgen.
VI *Malostraca, & Ostracoderma*
Krebs. Schnecken.[20]

19 Not a print, but a manuscript with large illustrations, see Bürger T. (ed.). *Das Kräuterbuch des Johannes Kentmann von 1563. Mit einem Essay von Hansjochen Hancke und botanischen Erläuterungen von Marina Heilmeyer* (München et al.: 2004).

20 Gessner, *Teütsche nammen der Fischen vnd Wasserthieren* 267–268.

However, the plenitude of fish names and the number of creatures are not congruent. Fishermen used to name fish according to certain features, which might not be apparent in the first year of the fish, but in the third year. Therefore, Gessner also presented different names according to the age of the fish. A consideration of the constancy of species is not always relevant here. It was also not in the interest of the fishermen, who followed other practical principles for their nomenclature. Gessner's inventory is therefore influenced by geography, dialects and the age of the fish, making it the reason for an abundance of cross-references in his alphabetical list of names. Additionally, Gessner compared classical terms with the expressions in the vernacular languages, which makes room for observations in the field of etymology. It also shows how many gaps and blanks there still are. Thus, Gessner had to fill them, become a nomothete himself: He took expressions from other Germanic languages and created similar German words. In the beginning of his Latin edition of the *Historia Piscium* he mentions in an index of German names: "His non adiunximus quae ab authore conficta sunt. potest etiam fieri ut Anglica nonnulla sint admixta."[21] In any case, all the extensive work on the names of *aquatilia* finds its ultimate goal in the fourth volume of the *Historia animalium* from 1558. Over 1000 German and Dutch fish names are collected here, also the classical and various vernacular languages are taken into consideration, including Arabic and Hebrew, and sometimes Gessner even refers to hieroglyphs. (That the synonymy of ancient and vernacular fish names is all but clear nowadays is proven by Fritz Fajen's recent edition and translation of Oppian's *Halieutica*. Fajen notes that he was not able to identify all the names and that he simply transcribed the Greek names when in doubt.[22]) In its inventory of names, Gessner's enormous project connects the practices of editing classical texts, lexicography, correspondence, autopsy and comparative language studies; images and objects (skeletons, taxidermic fish, fossils) are also included here as are the exchange and trade of such items.[23] All these practises and procedures establish and affirm the reference of words. This project shows in an impressive way that the perception and understanding of the world, especially the collecting of empirical facts, is mediated by and intertwined in language and that philology therefore is, in the end, the science of the world itself. Without Gessner among others, the nomenclature of Linnaeus, which proved

---

21 Gessner Conrad, *Historia animalium liber IIII. qui est de piscium [et] aquatilium* [...] (Zurich, Christopher Froschauer: 1558). The indices are without pagination, see under: "[nomina] GERMANICA."

22 *Halieutica*, ed. Fajen (see note 11), 11, note 1.

23 See Leu, *Conrad Gessner* 190–194.

to be Ariadne's thread out of the labyrinth of words, words and words, could not have been developed at all.

All this is somehow also the background of the cultural archive – so to say – from which poetry takes its impulses. The fish list contributes to Rabelais' and Fischart's grotesque presentation of the gigantic heroes, who can eat the whole world. But there is more than just the literary aim in the German translation. Fischart attempts to prove in his novel that the German language is capable of competing with other languages, classical and vernacular, since it has the potential to name all things in the world.[24] This is how Fischart transfers the ideas of the *Pléiade* into German, and it is Fischart's patriotic aim when he creates long lists of things. For Johann Fischart the interesting aspects of early modern ichthyology are those which are relevant to the complete dictionary of the German language. And this is the reason why a literary scholar like me, who usually only cares about fish as a culinary item and is more interested in literary history and poetic form, sometimes also has to consider early modern ichthyology as important background for poetic imagination.[25]

Bibliography

Bodin Jean – Fischart Johann, *De magorum daemonomania. Vom außgelassnen wütigen Teufelsheer* (Strasbourg, Bernhard Jobin: 1586).

Brockstieger S., *Sprachpatriotismus und Wettstreit der Künste. Johann Fischart im Kontext der Offizin Bernhard Jobin* (Berlin – Boston: 2018).

Bulang T., "Spiele in Johann Fischarts *Geschichtklitterung*", in Jahn B. – Schilling M. (eds.), *Literatur und Spiel. Zur Poetologie literarischer Spielszenen* (Stuttgart: 2010) 45–69.

Bulang T., *Enzyklopädische Dichtungen. Fallstudien zu Wissen und Literatur in Spätmittelalter und früher Neuzeit* (Berlin: 2011).

Bürger T. (ed.). *Das Kräuterbuch des Johannes Kentmann von 1563. Mit einem Essay von Hansjochen Hancke und botanischen Erläuterungen von Marina Heilmeyer* (München et al.: 2004).

Egmond F. (ed.), *Conrad Gessners Thierbuch. Die Originalzeichnungen*, transl. G.M. Vorderobermeier (Darmstadt: 2018).

24 See Brockstieger S., *Sprachpatriotismus und Wettstreit der Künste. Johann Fischart im Kontext der Offizin Bernhard Jobin* (Berlin – Boston: 2018).

25 I want to thank Justine Gesell for the substantial improvement of the English version.

Fischart Johann, *Geschichtklitterung (Gargantua). Synoptischer Abdruck der Fassungen von 1575, 1582 und 1590. Mit drei Titelblättern und den Originalholzschnitten der Ausgabe von 1590 von Tobias Stimmer*, ed. H. Schnabel, 2 vols. (Halle a. S.: 1969).

Gebhard T., "Das Spruchgedicht des Hans Sachs von den 124 Fischen", *Bayerisches Jahrbuch für Volkskunde* (Munich: 1985) 76–85.

Gessner Conrad., *Historia animalium liber IIII. qui est de piscium [et] aquatilium [...]* (Zurich, Christopher Froschauer: 1558).

Hakelberg D., "Das Bodensee-Fischbuch von Gregor Mangolt in einem Basler Nachdruck von 1612", *Wolfenbütteler Barock-Nachrichten* 36 (2009) 107–114.

Hendrikx S. – Smith P.J., "Ichthyology in Translation: Conrad Gessner's Fish Books", in Fournel J.-L. – Paccagnella I. (eds.), *Traduire – Tradurre – Translating. Vie des mots et voies des œuvres dans l'Europe de la Renaissance* (Geneva: 2022) 341–361.

Hühnemörder Ch., "Die Geschichte der Fischbücher von Aristoteles bis zum Ende des 17. Jahrhunderts", *Deutsches Schiffahrtsarchiv* 1 (1975) 185–200.

Leu U.B., *Conrad Gessner (1516–1565). Universalgelehrter und Naturforscher der Renaissance* (Zurich: 2016).

Magnus Olaus, *Beschreibüng allerley Gelegenheyte/Sitten/Gebräuchen und Gewonheyten/ der Mitnächtigen Völcker [...]* (Strasbourg: Theodosius Rihel, 1567).

Mangolt Gregor, *Fischbůch, Von der natur vnd eigenschafft der vischen* (Zurich, Andreas Gessner: 1557).

Nietzsche F., "Ueber Wahrheit und Lüge im aussermoralischen Sinne", in Friedrich Nietzsche. *Kritische Studienausgabe vol. 1: Die Geburt der Tragödie. Unzeitgemäße Betrachtungen I–IV. Nachgelassene Schriften 1870–1873*, eds. G. Colli – M. Montinari (Munich: 1999) 873–890.

Oppianus, *Halieutica*, transl. and ed. F. Fajen (Stuttgart – Leipzig: 1999).

Ovidius (Pseudo-), *Helieuticon: hoc est, de piscibus libellus, multo quam ante hac emendatior et scholijs illustratus. Emendantur et Plinij aliquot loca. Accedit Aquatilium animantium Enumeratio iuxta Plinium, emendata et explicata per Conradum Gesnerum, ordine alphabetico. Earundem nomina Germanica eodem ordine* (Zurich, Andreas and Hans Gessner: 1556).

Ovidius (Pseudo-), *De piscibus et aquatilibus omnibus libelli III. novi. Authore Conrado Gesnero medico et philosophiae naturalis interprete in Schola Tigurina: I. Scholia et emendationes in Halieuticon P. Ouidij Nasonis. II. Aquatilium animantium enumeratio juxta Plinium, emendata et explicata serie literarum. III. Eorundem Nomenclator Germanicus longe copiosissimus. Et alia ad piscium historiam pertinentia* (Zurich, Andreas Gesner: 1556).

Peters M. (ed.), *Konrad Gessner: Deutsche Namen der Fische und Wassertiere. Neudruck der Ausgabe Zürich 1556* (Aalen: 1974).

Richmond J., *Chapters on Greek Fish-Lore* (Wiesbaden: 1973).

Rondelet Guillaume, *Histoire entière des poissons* (Lyon, Mathias Bonhomme: 1558).

Rondelet Guillaume, *Libri de piscibus marinis in quibus verae piscium effigies expressae sunt* (Lyon, Mathias Bonhomme: 1554).

Sachs Hans, "Die hundert vnd Vier vnd zweintzig Fisch vnd Meerwunder mit jrer art", in Sachs Hans, *Werke*, ed. A. von Keller, vol. 7. (Tubingen: 1873) 456–463.

Sachs Hans, *Die Hundert vnnd vier vnd zweintzig Fisch vnd Meerwunder mit jrer art. Das ander Buch: Sehr herrliche Schǒne Artliche vnd gebundene Gedicht mancherley art* [...] (Nuremberg, Christoph Heußler: 1560).

Strömberg R., *Studien zur Etymologie und Bildung der griechischen Fischnamen* (Gohtenbug: 1943).

CHAPTER 5

Aquatilia of Portugal in 1555–1556 According to Leonhardt Thurneysser zum Thurn

Bernardo Jerosch Herold and João Paulo S. Cabral

1 Introduction

The Swiss scholar Leonhardt Thurneysser zum Thurn (1531–1596) began to write about the natural history of Portugal in 1555, when he stayed in Lisbon as a guest of the royal chronicler and humanist Damião de Góis (1502–1574). Thurneysser's report is recorded in a manuscript, consisting of more than 300 pages with descriptions of plants and animals, which also includes other information, which he collected during his sojourn in Lisbon. Thirty-two folios of the manuscript deal mainly with fishes, mammals, crustaceans and molluscs.[1]

2 Who Was Leonhardt Thurneysser zum Thurn

Leonhardt Thurneysser zum Thurn [Fig. 5.1] was born on 22 July 1531 in Basel, the son of a goldsmith and was baptized in the Lutheran church.[2] As an apprentice in his father's shop, he developed an early interest in mineralogy and metallurgy. At the same time, he served as attendant for the professor of medicine Johannes Huber (1507–1571), collecting herbs and preparing medicines. On this occasion he became acquainted with the writings of the philosopher Paracelsus (Theophrastus von Hohenheim, 1493–1541), that influenced him profoundly throughout his life. From 1547 onward he undertook many travels, including a voyage to Lisbon in 1555; here he began to study the natural history of Portugal until his return home in 1556. In 1559 he started to work as a metallurgist in Tarrenz (Tyrol) and became owner of a mine. He

1 Ms. Germ. Fol. 97. Staatsbibliothek Berlin.

2 Moehsen Johann Carl Wilhelm, *Ein Beitrag zur Geschichte der Alchymie, wie auch der Wissenschaften und Künste in der Mark Brandenburg gegen Ende des sechszehnten Jahrhunderts*, (Berlin – Leipzig, George Jacob Decker: 1783); Spitzer G., … *und die Spree führt Gold: Leonhard Thurneysser zum Thurn, Astrologe – Alchimist – Arzt und Drucker im Berlin des 16. Jahrhunderts* (Wiesbaden: 1996) 16; Schumacher Y., *Leonhard Thurneysser, Arzt – Alchemist – Abenteurer* (Zurich: 2011) 33.

DOI:10.1163/9789004681187_006

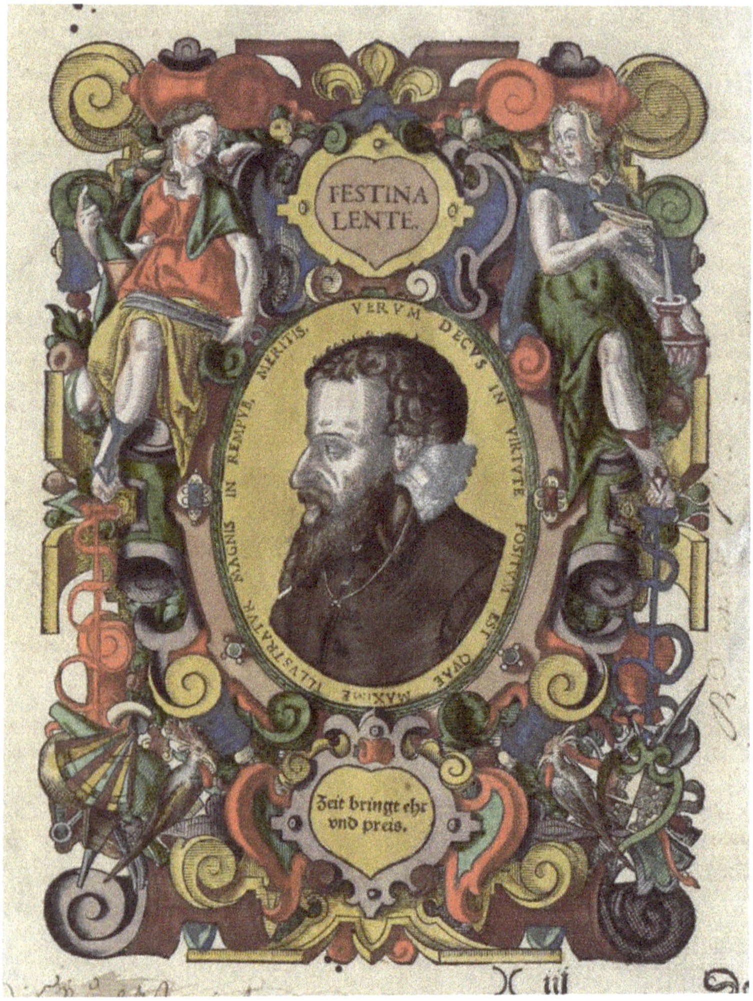

FIGURE 5.1 Leonhardt Thurneysser zum Thurn (1531–1596). Reproduced from Thurneysser Leonhardt, *Historia unnd Beschreibungen influentischer, elementischer und natürlicher Wirkungen aller heimischen und fremden Erdgewächssen*, 2nd edition (Cölln/Berlin, Johannes Gymnium: 1587)
COURTESY STAATSBIBLIOTHEK ZU BERLIN

became more and more well known as an apothecary, alchemist, astrologist and healer, drawing his wisdom from the experience he gained and observations that he made during his travels to many different countries. In spite of not having ever set foot into a university, his expertise in pharmacy, chemistry, metallurgy, botany, mathematics, astronomy and medicine was recognized by wealthy patricians and princes, such as Ferdinand II, Archduke of Further

Austria and Imperial Count of Tyrol (1529–1595) and his wife Philippine Welser (1527–1580) who came from a very wealthy family of Augsburg merchants and patricians. In 1561 Welser commissioned him to travel to many countries, including the Iberian Peninsula, North Africa and the Orient in order to collect minerals, plants and recipes for her curiosity cabinet. This included probably also a second trip to Portugal in 1562. As a result of his first voyage to Lisbon, Thurneysser wrote about his observations of the nature in Portugal, including a chapter on *aquatilia*,[3] and other information gathered, recorded in a manuscript, which was never published. In 1571 he moved to Berlin as the personal physician of John George, Elector of Brandenburg (1525–1598), who allowed him to use as his home, library and laboratory, as well as for his printing company, some buildings of a former Franciscan monastery, which later became known as the Grey Abbey. He became very wealthy by selling medicine of his own creation, as well as astrological calendars, horoscopes and talismans for protection against evil. He authored 39 books on a vast variety of subjects covering alchemy, astrology, hydrology, uroscopy, as well as multilingual glossaries, replies to defamations and many almanacs, as well as a book on herbs and their curative powers.[4] He also created a curiosity cabinet, a botanical garden and kept exotic animals.[5] Due to inheritance disputes he became unable to pay his debts and left Berlin 1584 for good and was baptized in the Catholic Church. He died impoverished in Cologne and was buried there 1596 in the Dominican Monastery of *Albertus Magnus*.[6]

3 Thurneysser's Travels to Lisbon

Thurneysser's burning curiosity for nature was certainly the main driving force for his many travels. During Spring 1555, Thurneysser was still working as a goldsmith and woodcut printer consecutively in Strasburg and Nurnberg. His

3 Fishes, mammals, crustaceans and molluscs.

4 Thurneysser L., *Historia sive plantarum omnium, tam domesticarum quam exoticarum* (Berlin, Michael Hentsken: 1578); ibid., *Historia unnd Beschreibungen influentischer, elementischer und natürlicher Wirkungen aller heimischen und fremden Erdgewächssen* (Berlin, Michael Hentsken: 1578; second edition Cölln/Berlin, Johannes Gymnium 1587).

5 Spitzer G., ... *und die Spree führt Gold: Leonhard Thurneysser zum Thurn, Astrologe – Alchimist – Arzt und Drucker im Berlin des 16. Jahrhunderts* (Wiesbaden: 1996); Schumacher Y., *Leonhard Thurneysser, Arzt – Alchemist – Abenteurer* (Zurich: 2011).

6 The monastery does not exist anymore. The location of the remains of Thurneysser is unknown but Albertus Magnus's remains rest after the lifting of the Dominican monastery (1804) since 15 November 1954 in the crypt of St. Andrew in Cologne (https://memim.com/albertus-magnus.html).

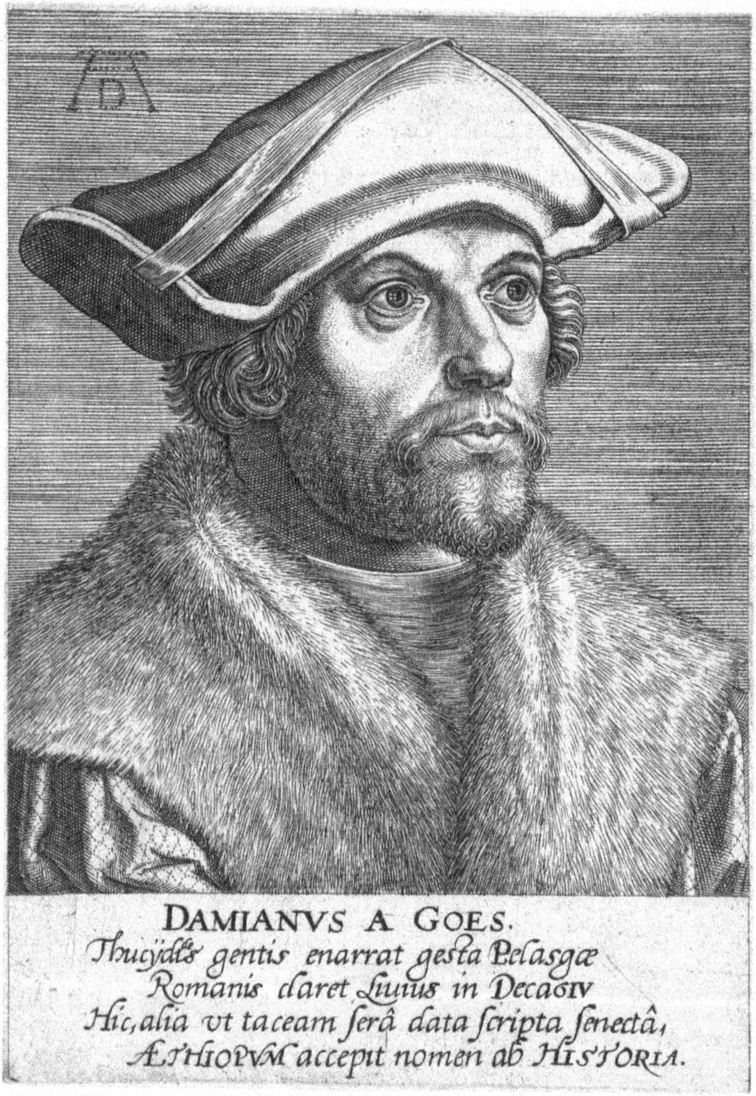

FIGURE 5.2 Damião de Góis (1502–1574). Rijksmuseum Amsterdam, RP-P-OB-6897
PUBLIC DOMAIN

arrival in Lisbon as early as solstice of the same year, as stated on the front page of his manuscript, would have only been possible if he travelled by sea from Antwerp, the main route of trade between Germany and Lisbon. The land route would have taken too long. Twice every year a convoy used to sail from the Portuguese factory in Antwerp to Lisbon, the first one departing by the end of May or early June, which took about two weeks to arrive in Lisbon in time for the midyear solstice.

After the closure of the Portuguese factory in Antwerp in 1549, the connection was continued mainly by vessels of the Hanseatic League.[7] In 1555, contemporaneously with the arrival of Thurneysser in Lisbon, an emissary of the senate of the Hanseatic city of Danzig, Heinrich Giese also arrived in Lisbon; there he is believed[8] to have met Damião de Góis [Fig. 5.2], Thurneysser's host. The existence of an acquaintance between Thurneysser and Giese offers a plausible explanation for a number of details in Thurneysser's manuscript, which were written in a way as to capture the attention of readers in Danzig. This points to the possibility of the merchants of Danzig having sponsored the voyage of Thurneysser. The important trade in salt from Portugal and timber and grain from Danzig going back to the Middle Ages had been boosted by the Portuguese overseas expansion movement with the purchase of Indian spices by Danzig merchants. This underpins the hypothesis that the first voyage to Lisbon of Thurneysser had been sponsored by commercial interests in Danzig. As noted earlier, the second trip to Portugal around 1562 was however probably sponsored by Archduke Ferdinand of Austria or his wife.

4 Relationship between Thurneysser and Damião de Góis

The front page of the first part of Thurneysser's manuscript concerning Portugal[9] [Fig. 5.3] tells us that he began to write in the house of the 'Nobleman and Lusitanian Knight Sir Damian de Goës by the solstice of 1555' and where

7 Marques A.H. de O., "Damião de Góis e os mercadores de Danzig", *Arquivo de Bibliografia Portuguesa* (Coimbra: 1959); Pohle J., *Os mercadores-banqueiros alemães e a Expansão Portuguesa no reinado de D. Manuel I* (Lisbon: 2017) (https://run.unl.pt/bitstream/10362/38843/2/MercadoresAlemaes.pdf); Arnold T. dos S., "Portugal and the Hanseatic League", in Lopes P.C. (ed.), *Portugal e a Europa nos séculos XV e XVI. Olhares, relações, identidade(s)* (Lisbon: 2019) 85–108.

8 Hirsch E.F., *Damião de Góis: The Life and Thought of a Portuguese Humanist* (The Hague: 1967).

9 Fol. 1r of *Ms. Germ. Fol. 97, Staatsbibliothek zu Berlin.*

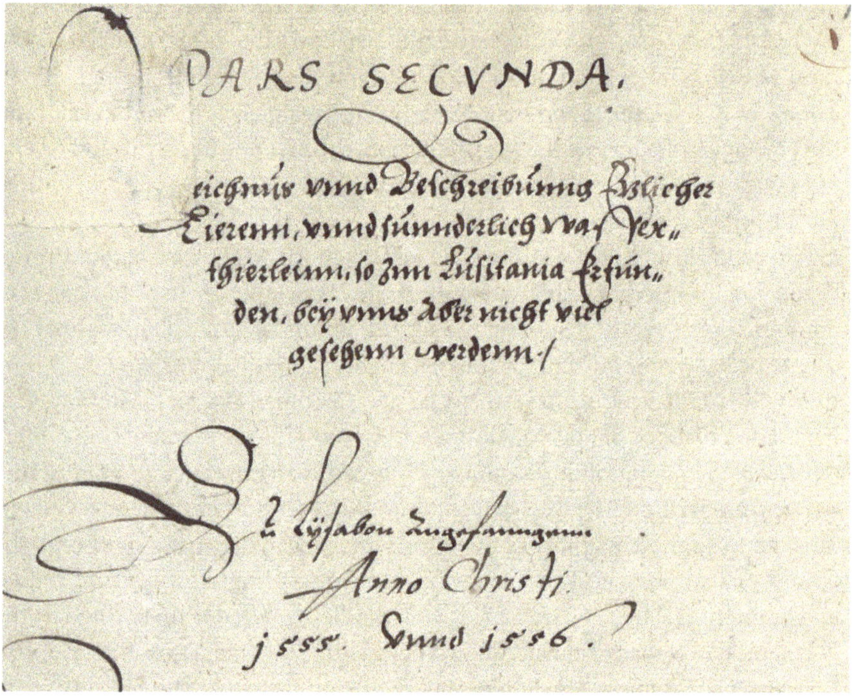
PARS SECVNDA.
Verzeichnis vnnd Beschreibung etlicher
Thieren, vnnd sunderlich wasser-
thierlein, so inn Lusitania gefun-
den, bey vns aber nicht viel
gesehen werden./
Zu Lissabon angefangen
Anno Christi.
1555. vnnd 1556.

FIGURE 5.3 Ms. Germ. Fol. 97. Front matter fol. 111r. Second Part. Index and Description of Various Animals especially small water animals found in Lusitania but not seen often at our home [country]. Begun in Lisbon Anno Christi 1555 and 1556
COURTESY STAATSBIBLIOTHEK ZU BERLIN

he remained until 1556. Damião de Góis, Portuguese humanist and diplomat,[10] was known to receive foreign visitors very often into his Lisbon home. This was common knowledge and held against him during his trial (1571–1572) by the inquisition tribunal. He was accused of heresy, because some of the humanist scholars, who he visited on his travels to Germany were leaders of the Reformation and because of his exchange of letters with them. Damião de Góis' friendship with several humanists in Basel, most of whom shared with Góis a friendship and admiration for Erasmus of Rotterdam[11] may have contributed to create a link, that led to Thurneysser being invited to his home by Góis. The latter had also cxchanged letters with the cosmographer Sebastian Münster of Basel (1488–1552), criticizing his negative evaluation of the economy and the population of Iberia, a region which Münster never had visited.

10 Hirsch, *Damião de Góis*.
11 Beau A.E., *As relações germânicas do humanismo de Damião de Góis* (Coimbra: 1941).

As a reply to this criticism, Damião de Góis published a description that highly praised the city of Lisbon.[12] An accurate description of the agriculture, and the flora and fauna of Portugal by Thurneysser could have been used by de Góis as an additional argument against Münster. In his manuscript, Thurneysser mentions Damião de Góis several times as a source of information or advice.

5 History, Structure and Main Contents of the Manuscript

The department of manuscripts of the Staatsbibliothek zu Berlin keeps a codex dated from the 16th Century[13] that is part of the estate left by Leonhardt Thurneysser zum Thurn. In an inventory published in 1925[14] it was named *Zur Naturbeschreibung von Portugal*, which one may translate as 'On the description of the nature of Portugal'. The existence of this manuscript was mentioned in several studies on Thurneysser,[15] Damião de Góis (1502–1574),[16] as well as on cultural relations between Portugal and Germany[17] or Switzerland.[18] None of these studies however was focused on the relevance of this manuscript to the early modern history of science. The Staatsbibliothek zu Berlin and the Centro Interuniversitário de História da Ciência e da Tecnologia (CIUHCT) agreed in 2013 to grant free access to a digitization of the complete codex in Early New High German to a team of researchers at the University of Lisbon.[19]

The contents of the parts of the manuscript concerning observations and information obtained in Portugal by Thurneysser were summarized in

12 Damião de Góis, *Vrbis Olisiponis Descriptio per Damianum Goem Equitem Lusitanum, in qua obiter tractantur nõ nulla de Indica nauigatione, per Graecos et Poenos et Lusitanos, diuersis temporibus inculcata* (Évora, Andreas Burgensis: 1554).

13 *Ms. Germ. Fol. 97*.

14 Degering H., *Kurzes Verzeichnis der germanischen Handschriften der Preussischen Staatsbibliothek* (*Berlin*), I: *Die Handschriften in Folioformat* (Leipzig: 1925).

15 Quelle O., "Leonhard Thurneysser zum Thurn", *Revista do Instituto de Cultura Alemã. Zeitschrift des Deutschen Kulturinstituts* 1/1 (1944) 99–102; ibid., "Deutsch-Portugiesische Kulturbeziehungen", *Zeitschrift für Politik* 34 (März/April 1944) 115–121; Spitzer, *… und die Spree führt Gold*; Schumacher, *Leonhard Thurneysser*.

16 Beau, *As relações germânicas do humanismo de Damião de Góis*; Hirsch, *Damião de Góis*.

17 Strasen E.A. – Gândara A., *Oito séculos de História Luso-Alemã* (Berlin: 1944).

18 Fischer B., *Dialogue Luso-Suisse: Essai d'une histoire des relations entre la Suisse et le Portugal du XVe siècle à la Convention de Stockholm de 1960* (Lisbon: 1960).

19 As a first result, its history and contents were briefly described in an article by Thomas Horst. Horst T., "A Rediscovered Manuscript about Portuguese Plants and Animals: Preliminary Observations", in Lopes M. dos S. – Leitão H. (eds.), *Renaissance Craftsmen and Humanistic Scholars. Circulation of Knowledge between Portugal and Germany* (Frankfurt/Main: 2016) 133–174.

Portuguese in a further study[20] and republished,[21] together with an annex with the transcription of the same parts. The codex has more than 900 pages with observations or information obtained in Portugal occupying four of the eleven parts of the codex (about 40% of the pages), namely

Part 1 (fol. 1r to 109v) on plants,
Part 2 (fol. 111r to 127v) on animals, including fishes, mammals, crustaceans and molluscs (*aquatilia*),
Part 3 (fol. 129r to 143v). *Miscellanea Lusitanica*, and
Part 4 (fol. 317r to 353v) continues the description of the natural history of Portugal.

The first three parts were a result from his first trip (1555/56), but the fourth part, however, presumably relates to his second trip (around 1562).

In spite of the manuscript not being in the original hand of Thurneysser, but merely a copy written presumably by his employee Adam Seidel from Basel, the notes on the margins of some pages, written in Thurneysser's own hand suggest his role in supervising the writing process of that copy. The fourth part, however, was written by yet a different hand.

Also on the front page of the first part[22] the reader is told that the manuscript contains an 'index and descriptions of a number of herbs, bushes and fruits that can be found mainly in Lusitania, which are rarely seen or not at all in our own country'. The front page of the second part[23] is identical with that of the first part, except that it tells that it contains an 'index and descriptions of a number of animals especially small aquatic animals that can be found mainly in Lusitania, which are rarely seen or not at all in our own country'.

Part 3 *Miscellanea Lusitanica* consists mainly of two chapters that deal with subjects not expected in a manuscript named 'On the Description of the Nature of Portugal'. One chapter with the title "Ritus Depositionis Mercatorum Norvegiæ" describes the rituals of initiation of newcomers to the Hanseatic factory of Bergen and the other describes the bodies of black Africans who

20 Herold B. – Horst T. – Leitão H., "A 'História Natural de Portugal' de Leonhardt Thurneysser zum Thurn, ca. 1555–1556", *Ágora. Estudos Clássicos em Debate* 19 (2017) 305–334.

21 Herold B. – Horst T. – Leitão H., A História Natural de Portugal *de Leonhardt Thurneysser zum Thurn, ca. 1555–1556*. [Including a] transcription of the Parts Concerning Portugal of the Manuscript Ascribed to Leonhardt Thurneysser zum Thurn, Ms. Germ. Fol. 97 of Berlin State Library (Berlin: 2019).

22 *Ms. Germ. Fol. 97, fol. 1r.*

23 *Ms. Germ. Fol. 97, fol. 111r.*

Thurneysser met in Lisbon. The latter has been reproduced, translated into English and commented on elsewhere.[24]

6 Language and Translation

The transcription of the original text in Early New High German can be downloaded from the digital library of the Lisbon Academy of Sciences.[25] The calligraphy of the original is in a 16th-century gothic cursive scholar's handwriting ('Gelehrtenschrift'), which highlights some words (mostly Latin or Portuguese expressions) by using italic letters of a slightly larger size.

In the next chapter, when translating entire phrases of the manuscript, the German original is shown as a footnote.

Regarding the names of the species, the taxonomic identification of fishes is based on Nobre,[26] Albuquerque,[27] Rodríguez de la Fuente,[28] Carneiro et al.,[29] Lloris[30] and the database FISHBASE,[31] that on molluscs on Macedo et al.[32] and the database WORMS,[33] that of crustaceans on Nobre,[34] Zariquiey Alvarez,[35]

---

24 Herold B.J. "The Diary of the Swiss Leonhardt Thurneysser and Black Africans in Renaissance Lisbon", *Renaissance Studies* 42 (2017) 463–488.

25 http://hdl.handle.net/10400.26/45218 last access 2023-09-27.

26 Nobre A., "Peixes de Água doce de Portugal", *Boletim do Ministério da Agricultura* XII, no. 2, série I (1932).

27 Albuquerque R.M., "Peixes de Portugal e Ilhas Adjacentes. Chaves para a sua determinação", *Portugaliae Acta Biologicae*, serie B, 5 (1954–1956).

28 Rodríguez de la Fuente F., *Cadernos de campo*. Lisbon: Artel. Colecção de 60 cadernos de campo sobre a fauna ibérica. Supervisão científica de Luís F. Matos, Cadernos de campo #18 and #32 for fishes, and #53 for crustaceans (Lisbon: 1984).

29 Carneiro M. – Martins R. – Landi M. – Costa F.O, "Updated Checklist of Marine Fishes (Chordata: Craniata) from Portugal and the Proposed Extension of the Portuguese Continental Shelf", *European Journal of Taxonomy* 73 (2014) 1–73.

30 Lloris, D., *Ictiofauna marina. Manual de identificación de los peces marinos de la Península Ibérica y Baleares* (Barcelona: 2015).

31 https://www.fishbase.se/search.php.

32 Macedo M.C.C. – Macedo M.I.C. – Borges J.P., *Conchas marinhas de Portugal* (Lisbon: 1999).

33 https://www.marinespecies.org/.

34 Nobre A., *Crustáceos decápodes de Portugal* (Oporto: 1931); Nobre A., *Crustáceos decápodes de Portugal*, second ed. (Barcelos: 1936).

35 Zariquiey Alvarez R., "Crustáceos decápodos ibéricos", *Investigación Pesquera* 32 (1968) 1–510.

Rodríguez de la Fuente, Saldanha,[36] Costello et al.,[37] Marco-Herrero et al.,[38] and that of marine mammals on Reis and Mathias,[39] and the database FAUNA EUROPAEA.[40] After each name used by Thurneysser and before the scientific name, the modern Portuguese common name is here represented in order to serve as a reference and comparison with the name used by Thurneysser in the manuscript.

7 About Fishes, Mammals, Crustaceans and Molluscs

Part 2 of the Thurneysser manuscript (fol. 111r to 127v) consists of the following chapters:

I – (fol. 112r to 112v) 'About little animals and fishes': Contains a list of the various qualities which would have to be reported in the descriptions of the various species.

II – (fol. 112v to 113r) 'Tenches': Contains remarks about tenches, beavers, crucian carps and crayfishes found in the waters of Gdansk.

III – (fol. 113r to 116v) 'Lusitanian lamprey'. Contains descriptions of two species of lamprey, probably the sea-lamprey (modern Portuguese common name: 'lampreia-de-mar', Lamprea Lvsitanica, *Petromyzon marinus* Linnaeus, 1758) and the river-lamprey ('Altera lamprea', Portuguese 'lampreia-de-rio', *Lampreta fluviatilis* (Linnaeus, 1758)). The description of 'Lamprea Lvsitanica' is very detailed and accurate: the body is 'like skye blue or grey blue, but not completely over the whole body, which in some places is, *interrumpiret* or spotted with white patches or blots which are rather large;[41] but their body is similar and identical with the eels. But they are generally one ell long and almost like a [Northern] pike [*Esox lucius* Linnaeus, 1758], which is also that long

36 Saldanha L., *Fauna submarina atlântica* (Lisbon: 1995).

37 Costello M.J. – Emblow C. – White R. (eds.), *European Register of Marine Species. A Check-list of the Marine Species in Europe and a Bibliography of Guides to Their Identification* (Paris: 2001).

38 Marco-Herrero E. – Abelló P. – Drake P. – García-Raso J.E. – González-Gordillo J.I. – Guerao G. – Palero F. – Cuesta J.A., "Annotated checklist of brachyuran crabs (Crustacea: Decapoda) of the Iberian Peninsula (SW Europe)", *Scientia Marina* 79 (2015) 243–256.

39 Reis M.S. – Mathias M.L., "Mamíferos", in Rodrigues P.D. – Crespo E.G. (eds.), *Portugal Moderno: Fauna, Enciclopédia Temática Pomo* (Lisbon: 1991) 160–191.

40 https://fauna-eu.org/.

41 wie liecht kemler, oder blaw aschenfarbig, aber doch nicht ganntz vnd gar, doch vber das ganntze Leib, den derselbig an etzlichenn Örtteren mit weissen Mackhlen oder Plackhenn *interrumpiret* oder befleckhet ist, wellche zimlich gros sein.

[and] thick, namely three fingers broad'.[42] Its mouth 'is garnished and *armiret* on both sides with many small rows of yellowish spikes which are not hard, but look like the sides of the top skin of a cow's tongue.'[43] The gill openings are described as being 'elongated little holes side by side at equal distances, through which it sucks the water in and returns it, when hanging by its mouth from the rocks, when sucking and *lambiert*'.[44] The skin is 'smooth and shining, without scales, just like the eels, but even smoother. It has a marvellous and soft flesh almost without little fish-bones or little bits of bones'.[45] Thurneysser gives an explanation for the fish's great flexibility: the fish has 'a *carthilaginisch* [cartilagineous] cord as a spine which is very flexible, this being the reason for the ready and fast agility of this fish, when it is caught and seized.'[46] The lamprey was 'caught mainly in January in running fresh waters'.[47] In fact, we know that the sea-lamprey lives in the sea until it reaches its adult stage. During the breeding season, generally in Spring, they go up the rivers and build a kind of nest where they lay their eggs. Juveniles live in fresh waters for up to 3–5 years, then descending rivers, growing in marine waters. According to Thurneysser, the fish was rare in the Lisbon markets, very appreciated and fetched high prices: 'along one whole year I have only seen once in the Lisbon market, and it was taken and delivered immediately or on the hour to the king. But it is offered and sold for a Portuguese *cruzado*'[48] (each gold *cruzado* weighted ca. 3.6 grams).

42 am Leibe aber ist es den Alen enlich vnnd gleich. Sie ist aber gemeiniglich einer Ellen lanng, vnnd fast wie ein Hecht, der auch so lang ist, dickhe, alls namlich dreyer zwerch Finnger dickh.

43 Es ist aber derselbigen Rach[en] zu beyden Seitten, mit vielen Reylein oder Rechen Zacherlein besetzet, vnnd *armiret*, wellche da gelblecht, vnnd nicht hardt, sunderen fast wie die stachlechtige Vberhaut der Ochsenn Zungen zu den Seitten anzusehen seindt.

44 lannglechtige Löcherlein, neben vnnd inn gleicher Weitte vonn einannder, durch wellchenn sie das Wasser zu sich zeüchte; vnnd auch widerumb vonn sich zeücht gibt, wann es mitt dem Munndt an den Steinklippen henngt vnnd dieselbige seügend vnnd *lambiert*.

45 Sie hat ein gar glatte vnnd glantzete Hautt, ohne Schuppenn od[er] Pflumen, gleich wie die Ale, aber sie seindt noch viel gletter. Sie hatt ein gar herrliches schon weiß, vnnd weich Fleisch, wellches kaum Gretlein od[er] Knochlein ... hatt.

46 ein *carthilaginis*ches [Thurneysser wrote ‚carthellagini' in italics, but the ending ‚sches' in German ‚Gelehrtenschrift', which was transcribed in roman characters] Kröspelein hatt zum Ruckhgrat, wellches gar brügig ist, doher dann auch die Vrsach der behennden vnnd geschwinnden Beweglgkheit dieses Fisches wol zu colligieren vnnd abzunehmen ist.

47 fürnehmlich im Jenner gefanngen, inn süssenn fliessenden Wasseren.

48 inn einem ganntzen Jar nur einmahl zu Lysabon auf dem Marckht gesehen hab, wellche doch strackhs oder vonn Stundt an zu dem Könnig gebracht oder getragen sein. Es wierdt aber alda eine umb einen portugallischen Kreüttzer feilgebotten vnnd verkaufft.

The description of the 'Altera lamprea' is also very fine and detailed. The body is blue but 'the belly is very white like with the eels.'[49] Comparing with the sea-lamprey, this lamprey has 'a more elongated and not very plump or round mouth'.[50] The dorsal fins (*Pinnas*) are 'firstly a bit smaller but then near the tail much larger and taller, continuing along the tail, although not in the same and equal height.'[51] It has 'a little broader tail, like the fish *Linguado* [sole] and under the belly it has no *Pinnas*'.[52] This lamprey was fished later, in February, and 'was not as expensive as the first [sea-lamprey].'[53] In fact, the river-lamprey has a way of life similar to that of the sea-lamprey, but it lives longer in fresh waters, climbs higher in rivers, and in the sea it does not go as far from the coast. The rise of rivers can occur in Spring or Autumn, or in both seasons. Here Thurneysser mentions the Spring rise. A curious method of conservation is also mentioned. The fish is cut into slices that are 'preserved in sand and not in salt, because in the sand they keep better and longer than in salt'.[54] The text refers to an illustration, which is lost.

Finally, it can be mentioned that the former species is more common in southern Europe, while the latter is more common in northern Europe.

IV – (fol. 116v to 117v) 'Seabreams'. Describes possibly five different species of seabreams: 'mvcharos' (mucharro, *Diplodus annularis* (Linnaeus, 1758)); 'sarves' (sargo, *Diplodus sargus* (Linnaeus, 1758)); 'goras' (goraz, *Pagellus bogaraveo* (Delaroche, 1809)); 'salem' (possibly salema, *Spondyliosoma cantharus* (Linnaeus, 1758)); 'vezugo' (besugo, *Pagellus acarne* (Risso, 1827)). Details about the body of these fishes are given in the text. Referring to the black spot or band near the tail fin of the bream, Thurneysser wrote that the 'sarves' on 'the rearmost part of the body at the beginning of the tail [...] have sometimes large and square shaped patches.'[55] and also 'long and black stripes on the body'.[56]

---

49 am Bauch aber ist sie gar weiß, wie die Ale.

50 ein lannglechtiges, vnd nicht sehr tranndes oder runndes Maul.

51 erstlich etwas geringens, darnach aber an Schwartze [meaning ‚am Schwantze'] vil grosser vnd höcher, wellches sie auch dem Schwantz lang streckht, wiewol nicht inn einem vnd gleicher grosser Höche.

52 ein wenig ein breitten Schwantz fast wie der Fisch *Linguado* zu vnnderst aber am Bauch hatt sie gar keine *Pinnas*.

53 nicht so teür wie die Erstenn verkaufft.

54 Sie werden aber inn dem Sand vnd nicht inn dem Saltz eingemacht; dann inn dem Sannd werden sie vil füeglicher, vnd lennger bewahret, dann inn dem Saltz.

55 fürnemlich am hinderst[en] Theil deß Leibes, zum oder am Anfanng deß Schwanntzes, so haben sie auch hin vnnd wider viel große schwartze, vnd vier eckhechtige Mackhlen am Leib.

56 lannge schwartze Striemen an dem Leib.

v – (fol. 117v to 118v) 'Gurnards'. Describes three or four different species of gurnards (with reference to a lost illustration), namely 'rvivo' (ruivo, a common name applied in Portuguese to several species of *Trigla* Linnaeus, 1758, in particular *Eutrigla gurnardus* (Linnaeus, 1758) and *Lepidotrigla cavillone* (Lacepède, 1801)) and 'cabra' (cabra, cabrinha or peixe-cabra, other species of this same genus, in particular *Trigla lyra* Linnaeus, 1758 and *Chelidonichthys obscurus* Bloch & Schneider, 1801).

vi – (fol. 118v–122v) 'Fishes in Lusitania especially those caught in the estuary of the river Tagus in Lisbon'; probably two different tuna species are described: 'tunn' (probably atum, *Thunnus thynnus* (Linnaeus, 1758)) and 'tunina' (possibly bonito, *Katsuwonus pelamis* (Linnaeus, 1758)). Thurneysser describes in detail the body of these two tuna species: 'Their heads are like black morions or closed helmets. Also like blackened and polished leather, which shines like a mirror.'[57] The 'teeth, which are not very large but sharp.'[58] 'Its body, belly and back are of the same thickness or a little larger than the head over a considerable length but nevertheless ending in a pointed tail, which at its end has not more than the thickness of a finger.'[59] 'On the back the color of its skin looks like black polished leather, the tail and its sides [are] a bit whitish and mainly towards the belly [here probably referring to *Thunnus thynnus* (Linnaeus, 1758)], but the belly is completely white [also probably referring to *Thunnus thynnus* (Linnaeus, 1758)].'[60] Referring to one of the two pointed dorsal fins, Thurneysser observed that 'On the back it has a raised horn, which looks more like a horn than as a *Pinna* or feather.'[61] Describing the small fins (finlets) that exist between the second dorsal fin and the caudal fin and between this the anal fin, Thurneysser mentions that 'on the back and under the belly many small *Pinnæ* or wings or feathers, which are of a yellow color'.[62]

57 Ire Heupten seindt einer schwartzen Sturmhauben, oder einem verschlossenen schwartzen Helm gleich. Item wie geschwertzt vnnd glatt gemacht Leder, wellches wie ein Spiegel glenntzet.

58 nicht sehr grosse aber doch scharpfe Zehne.

59 Sein Leib, Bauch, vnd Ruckhen seindt mit dem Haupt fast einerley grosse, oder ja ein wenig grosser inn zimlicher Lennge, ohne daß er sich enndtlich, inn einem gar spitzen Schwantz, wellcher am Ennde nicht viel vber eines Finngers dickh ist.

60 An dem Ruckhen ist an Farb sein Hautt anzusehen wie ein schwartz glentzet gemacht Leder, der Schwanntz vnnd seine Seitten etwas weißlechtig, vnnd sunderlich gegen den Bauch werts. Der Bauch aber ist im gar weiß,.

61 auf dem Ruckhenn hatt er ein einiges aufgerichtes Hornn, wellches mer ein Hornn, dann ein *Pinna* oder Feder anzusehen ist.

62 auf dem Ruckhen, vnd vnder dem Bauch vil kleiner *Pinnæ*, oder Fittigen oder Federen nacheinannder folgenn, wellche an Farb gelb, … sein.

Sturgeons (*Acipenser sturio* Linnaeus, 1758) are mentioned as extremely rare and expensive. This species is currently restricted in the western Atlantic to the Garonne River basin in France. In Portugal, the last specimens were observed in the 1980s in the lower Guadiana River. It is considered extinct in Portugal, with the last references of its reproduction being from the late 70's to the Douro River and from the beginning of the 80's to the Guadiana River.[63] Thurneysser mentions that in 1555 two very big sturgeons were caught and sold for a very high amount of money – 30–90 *cruzados*. Skates and rays – 'raiarvm' (raias, several species of *Raja* Linnaeus, 1758), plaices – 'solii' (solhas, *Pleuronectes platessa* Linnaeus, 1758, *Pleuronectes flesus* Linnaeus, 1758 and *Pleuronectes boscii* (Risso, 1810)), and wreckfish – 'cherne' (cherne, *Polyprion americanus* (Bloch & Schneider, 1801)) are also mentioned, as well as other fishes, namely 'coubra de mare' (cobra-de-mar, *Ophisurus serpens* (Linnaeus, 1758)) and 'bvdia' (possibly the bodião, *Labrus bergylta* Ascanius, 1767). This section has several references to lost illustrations.

VII – (fol. 122v to 123v) 'Sea mussels'. Different species of molluscs are mentioned, most of them by their Portuguese names with references to lost illustrations: 'brigigaun' (berbigão, *Cerastoderma edule* (Linnaeus, 1758) and *Cerastoderma glaucum* (Bruguière, 1789)); 'vieira' (vieira, *Pecten jacobeus* (Linnaeus, 1758) and *Pecten maximus* (Linnaeus, 1758); since Thurneysser says that he is referring to the 'Jacobi Muscheln', it is the first species, the pilgrim's to Santiago de Compostela shell); 'maxilhaon' (mexilhão, *Mytilus galloprovincialis* Lamarck, 1819); 'ostrae' (ostras, *Ostrea edulis* Linnaeus, 1758 and *Crassostrea angulata* (Lamarck, 1819) are, at present, the commonest species); *'conchæ margaritiferæ'* (Thurneysser might have been referring to the river mussel, *Margaritifera margaritifera* (Linnaeus, 1758), a species that is currently very threatened at a global and national level, classified as 'endangered' by the IUCN Red Book); 'caramunio' (caramujo is the Portuguese common name for rocky intertidal snails like *Littorina littorea* (Linnaeus, 1758) and *Monodonta lineata* (da Costa, 1778)); 'lapas' (lapas, currently four species occur, *Patella intermedia* Murray, 1897, *Patella rustica* Linnaeus, 1758, *Patella ulyssiponensis* Gmelin, 1791 and *Patella vulgata* Linnaeus, 1758).

VIII – (fol. 123v) 'Terrestrial mussels and trumpet snails': mentions only one land snail with a reference to a lost illustration.

IX – (fol. 124r) 'Crabs in Lusitania': mentions several crustaceans by their Portuguese names with references to lost illustrations: 'cangeios' (caranguejos, many species exist in the Portuguese coasts); 'centolas' (santolas, *Maja brachydactyla* Balss, 1922); 'lagonsta' (lagosta, *Palinurus elephas* (J.C. Fabricius, 1787)

63 Cabral M.J.M. (coord.), *Livro Vermelho dos Vertebrados de Portugal* (Lisbon: 2008).

and *Palinurus vulgaris* Latreille, 1803); 'lagostius' (possibly lagostim, *Nephrops norvegicus* (Linnaeus, 1758)); 'camerons' (camarões, namely *Palaemon serratus* (Pennant, 1777) and *Palaemon adspersus* Rathke, 1836).

X – (fol. 124r and 124v) '*Semianimalia* and reptiles'. Thurneysser mentions several 'half-alive and crawling sea creatures' namely 'Polypus maximus', 'Polypus medius' and 'Polypus minimus'. These names were used by Rondelet in his *Libri de piscibus marinis* (1554)[64] to describe several cephalopods: *Prima & secunda polyporum specie, Polypus octopus,*[65] is *Eledone cirrhosa* (Lamarck, 1798), today currently called in Portuguese 'polvo-do-alto'; *Tertia polyporum specie*[66] is *Octopus vulgaris* Cuvier, 1797, the 'polvo-comum'. The 'polypus' mentioned by Thurneysser are probably these two octopus species. This chapter also has several references to lost illustrations.

XI – (fol. 124v to 127v) 'Fishes in Portugal etc.': a list of several dozen names of fishes and marine mammals, most of which are in Portuguese, often with slightly erroneous spellings, probably due to having been recorded by somebody who does not know the Portuguese language after hearing these names mentioned in conversations with native fishermen or fishmongers. Each name has a reference to a lost illustration. The following fish can be mentioned, whose modern identification appears to be resonably secure (by the order they appear in the text, omitting those that had been referred to, before in the text): 'sardiniæ' (sardinha, *Sardina pilchardus* (Walbaum, 1792)); 'pataroxa' (pata-roxa, *Scyliorhinus canicula* (Linnaeus, 1758) and/or *Scyliorhinus stellaris* (Linnaeus, 1758)); 'sauelha' (savelha, *Alosa fallax* (Lacepède, 1803)); 'macarenæ sarda Anglica' (sarda, *Scomber scombrus* Linnaeus, 1758, an abundant species in the cold waters of northern Europe, perhaps hence the reference to England); 'chicharcino' (chicharro, *Trachurus trachurus* (Linnaeus, 1758) and *Trachurus picturatus* (Bowdich, 1825)); 'mugem' (mugem, *Mugil cephalus* Linnaeus, 1758, *Mugil curema* Valenciennes, 1836, *Chelon saliens* (Risso, 1810), *Chelon ramada* (Risso, 1827) and/or *Crenimugil seheli* (Forsskål, 1775)); 'bordalo' (bordalo is *Squalius alburnoides* (Steindachner, 1866), but could also refer to *Squalius cephalus* (Linnaeus, 1758) which, however, is also known for 'escalo'); 'robalo' (robalo, *Dicentrarchus labrax* (Linnaeus, 1758) and/or *Dicentrarchus punctatus* (Bloch, 1792)); 'chuopa' (choupa, *Spondyliosoma cantharus* (Linnaeus, 1758) and/or *Diplodus vulgaris* (Geoffroy Saint-Hilaire, 1817)); 'xyphion' (probably espadarte, *Xiphias gladius* Linnaeus, 1758, a common fish all around the world);

64 Rondelet Guillaume, *Libri de piscibus marinis, in quibus verae piscium effigies expressae sunt* (Lyon, Matthias Bonhomme: 1554).

65 Rondelet, *Libri de piscibus marinis* 513.

66 Rondelet, *Libri de piscibus marinis* 516.

'salmoneta' (salmonete, *Mullus surmuletus* Linnaeus, 1758 and *Mullus barbatus* Linnaeus, 1758); 'budiam' (bodião, there are several species with this common name, namely, *Labrus bergylta* Ascanius, 1767, *Labrus mixtus* Linnaeus, 1758 and *Symphodus bailloni* (Valenciennes, 1839)); 'cachuncho' (cachucho, *Dentex macrophthalmus* (Bloch, 1791)); 'pargo' (pargo, *Pagrus pagrus* (Linnaeus, 1758)); 'remora' (the fish described by Rondelet,[67] with this name is *Remora remora* (Linnaeus, 1758)); 'papagei' (papagaio, *Callanthias ruber* (Rafinesque, 1810)); 'truta' (there are currently two species in Portugal, *Salmo trutta* Linnaeus, 1758, an indigenous fish, and *Oncorhynchus mykiss* (Walbaum, 1792), the rainbow-trout, a fish originally from the USA and introduced into Portugal); 'abrontea' (abrotea, *Phycis blennoides* (Brünnich, 1768)); 'murea' (moreia, *Muraena helena* Linnaeus, 1758); 'murea altera' (the author may be referring to *Conger conger* (Linnaeus, 1758), a fish similar to the moray); 'eiros' (eiró, *Anguilla anguilla* (Linnaeus, 1758)); 'carapuos' (carapau, *Trachurus trachurus* (Linnaeus, 1758) and/or *Trachurus mediterraneus* (Steindachner, 1868)); 'pisce spanda' (probably peixe-espada, *Lepidopus caudatus* (Euphrasen, 1788)); 'rodoualho' (rodovalho, there are several species with this Portuguese common name, but the commonest is *Scophthalmus rhombus* (Linnaeus, 1758)); 'Linguadæ' (linguado, *Solea solea* (Linnaeus, 1758), today a very expensive fish); 'pastinacæ species duæ' (Rondelet described the pastinaca,[68] a name that Linnaeus will adopt for *Raja pastinaca* (current name *Dasyatis pastinaca* (Linnaeus, 1758)), the uge in modern Portuguese); 'pisce puta' (probably chaputa, *Brama brama* (Bonnaterre, 1788)); 'pampano' (pompano, *Stromateus fiatola* Linnaeus, 1758); 'douranda' (dourada, the commonest species is *Sparus aurata* Linnaeus, 1758); 'gallina' (galinha-do-mar, *Helicolenus dactylopterus* (Delaroche, 1809)); 'pisce porco' (peixe-porco, *Chilomycterus atinga* (Linnaeus, 1758)); 'cassaun' (cação, common name applied to species such as *Mustelus mustelus* (Linnaeus, 1758), *Mustelus asterias* Cloquet, 1819, *Galeorhinus galeus* (Linnaeus, 1758) and *Squalus acanthias* Linnaeus, 1758); 'lixa' (lixa, *Centrophorus squamosus* (Bonnaterre, 1788)); 'piscis caninus' (peixe-cão, *Bodianus scrofa* (Valenciennes, 1839)); 'pisce gallo' (peixe-galo, *Zeus faber* Linnaeus, 1758); 'barbo' (barbo, the European barbel is *Barbus barbus* (Linnaeus, 1758); in the Iberian Peninsula there is a barbel, *Barbus bocagei* Steindachner, 1864, which some consider as an autonomous species, others as a sub-species of *Barbus barbus*; in the Tagus and Guadiana rivers, there is another barbel, an endemic species, *Luciobarbus comizo* (Steindachner, 1864)); 'rana piscatrix' (Rondelet described a *Rana*

67 Rondelet, *Libri de piscibus marinis* 436.
68 Rondelet, *Libri de piscibus marinis* 331.

piscatrice,[69] called by Linnaeus as *Lophius piscatorius* Linnaeus, 1758, tamboril, in modern Portuguese); "coruina" (corvina, *Sciaena cirrosa* Linnaeus, 1758, and/or *Argyrosomus regius* (Asso, 1801)); 'bacalhao' (bacalhau, *Gadus morhua* Linnaeus, 1758); 'enxaroco' (enxarroco, *Zeus faber* Linnaeus, 1758).

Marine mammals are also mentioned by Thurneysser. 'Balena' and 'balena altera species', two cetaceans, possibly dolphins and/or small whales. 'Delphines', probably a dolphin; Rondelet, named a dolphin as 'delphino';[70] *Delphinus delphis* Linnaeus, 1758 is the commonest species in Portuguese waters. 'Phocae maris septentrionalis', possibly *Phocoena phocoena* (Linnaeus, 1758), the Portuguese 'bôto' or 'toninha', the smallest cetacean that occurs in waters of Continental Portugal.

XII – (fol. 127v) 'Sea monsters': six names of real or imagined *aquatilia*, four of which with references to lost illustrations. Many of the traditions of antiquity and the Middle Ages persisted into the Renaissance (and some to this day). Rondelet, of the purest Renaissance humanism, was also unable to 'emancipate' from some of these 'legends'. In his masterpiece, *Libri de piscibus marinis*, published in 1554, Rondelet described and drew (!) several of these sea monsters: "De monstro leonine";[71] "De pisce monachi habitu";[72] "De pisce episcope habitu";[73] "De nereide".[74] Thurneysser mentioned 'Monachus marinus' possibly Rondelet's "De pisce monachi habitu". Another master of the Renaissance, Conrad Gessner (1516–1565), in his *Historia animalium liber IV*,[75] described and drawn *Pan vel satyrus marinus*, probably Thurneysser's 'De Tritonibus segeues'.

XIII – (fol. 127v) 'Lusitanian ships and maritime gear': Only contains references to two lost illustrations.

Suprisingly, the first part of the manuscript, the one dealing with plants, contains in its last chapter[76] under the title "Proseves vel ungues marini" a very detailed description of goose barnacles. Since they grow on rocks in shallow seawater and are unable to move, Thurneysser, not knowing how they feed and reproduce themselves, considered them something marvelous and placed this chapter in the part of the manuscript dealing with plants in spite of concluding

69 Rondelet, *Libri de piscibus marinis* 363.
70 Rondelet, *Libri de piscibus marinis* 459.
71 Rondelet, *Libri de piscibus marinis* 491.
72 Rondelet, *Libri de piscibus marinis* 492.
73 Rondelet, *Libri de piscibus marinis* 494.
74 Rondelet, *Libri de piscibus marinis* 494.
75 Gessner Conrad, *Historiae animalium liber IIII qui est de piscium et aquatilium animantium natura* (Zurich, Christoph Froschauer: 1558) 1197.
76 Chapter number 32 (fol. 106r to 109v).

eventually that they are animals. This exemplifies how in early modern times the distinctions between plants and animals were difficult to draw (another example were corals, considered to be plants by early naturalists).

8 Conclusions and Outlook

This manuscript contains very important information on the aquatic and marine fauna of the Tagus estuary in 1555 with a large set of Portuguese common names, as well as descriptions, a few very detailed, of the morphology of some of the species. Unfortunately, the illustrations referred to in the text have been lost. To our knowledge, this is the first inventory of the aquatic and marine fauna of Continental Portugal.

Portuguese ichthyofauna and fisheries of the 18th century are reasonably known as the result of (1) inventories of the natural history (already according to the Linnaean nomenclature), carried out by the Italian naturalist Domingos Vandelli (1735–1816) hired by the Marquis of Pombal to renovate the University of Coimbra (the total number of fish species reported by Vandelli was 80), (2) reports of travelers who toured the country, visited the markets and saw what people ate, (3) menus of the Benedictine monasteries and (4) cookbook recipes.[77] All these sources of information agreed that our coasts were frequented by a high and diverse number of fishes and that many of these were captured, certainly for consumption by the populations. The situation for earlier periods is much less known, but this manuscript allows us to affirm that also in the 16th century the diversity of fish that existed on the Portuguese coasts and in the markets would have been high.

This high abundance and diversity of fish consumed by the people is also related to the Roman Catholic religious precepts that imposed abstinence from meat at certain times of the year, in which only fish were consumed as the main dish. Actually, Advent, Lent, and Fridays throughout the year, plus the eve of main holy days, corresponded to a very significant fraction of the annual calendar. In general terms, the privileged classes preferred meat and ate fish especially during periods of restriction. In contrast, the lower classes consumed mainly fish throughout the year. In certain social and religious groups

---

77 Cabral J.P.S., *A História Natural de Portugal em Domingos Vandelli* (Lisbon: 2018); Cabral, J.P.S., "Parte IV: Esboço biográfico de Domingos Vandelli. Flora e Fauna", in Cabral J.P.S. – Borges F.S. – Costa J.M.R. (eds.), *Um manuscrito inédito de Domingos Vandelli, Historia naturalis Olisiponensis (BPMPMs. 1127): leitura, interpretação e discussão* (Oporto: 2020).

there were other restrictions. Jews and New-Christians who maintained some of the Jewish practices, did not eat fish without scales, i.e., lampreys and chondrichthyans, such as skate.[78]

If we take into account Lacerda Lobo's history of Portuguese fisheries,[79] there are records of an abundance of fisheries that go back to at least to King João I (1357–1433). The quantity of fish was so great that it was exported to Castile and abroad. However, at the time that Lacerda Lobo was writing this memoir – 1812, its author considered that fishing had declined sharply and was in decay. Not for lack of fish, but because of fiscal, economic and financial reasons, including the lack of training of fishermen and deficient fishing equipment, the silting of many of the estuaries that made it difficult for fish to enter the bars and the mooring of boats, as well as the emigration of fishermen to the neighbouring country.

The current studies of the manuscript will centre on the tentative identification of the animals whose common names presented by Thurneysser are not similar to modern Portuguese names.

Acknowledgements

The authors wish to thank Dr. Gabriele Kaiser, librarian at the manuscript section of Staatsbibliothek zu Berlin for granting the access to a digitised version of Thurneysser's manuscript and Dr. Thomas Horst for cooperation in its transcription. We also thank for the support of Centro Interuniversitário de História da Ciência e da Tecnologia (CIUHCT) in Lisbon through Prof. Henrique Leitão. The authors are also grateful to Dr. Susan Turner (Brisbane) for assisting them with English language and scientific terminology.

Bibliography

Albuquerque R.M., "Peixes de Portugal e Ilhas Adjacentes. Chaves para a sua determinação", *Portugaliae Acta Biologicae*, serie B, 5 (1954–1956).

78 Rocha R., *A viagem dos sabores. Ensaio sobre a história da alimentação (séculos IX–XIX) seguido de 100 receitas em que vários mundos se encontram* (Lisbon: 1998); Braga I.D., "La place du poisson et des crustacés dans l'alimentation des Portugais (XVIe–XVIIIe siècle)", *Centre de Recherche d'Histoire Quantitative, Histoire maritime* 4 (2007) 241–250.

79 Lacerda Lobo C.B. de, "Sobre a decadencia das Pescarias em Portugal", *Memorias Economicas da Academia Real das Sciencias de Lisboa*. IV, Na Officina da mesma Academia (Lisbon: 1812) 312–383.

Arnold T. dos S., "Portugal and the Hanseatic League", in Lopes P.C. (ed.), *Portugal e a Europa nos séculos XV e XVI. Olhares, relações, identidade(s)* (Lisbon: 2019) 85–108.

Beau A.E., *As relações germânicas do humanismo de Damião de Góis* (Coimbra: 1941).

Braga I.D., "La place du poisson et des crustacés dans l'alimentation des Portugais (XVIe–XVIIIe siècle)", *Centre de Recherche d'Histoire Quantitative, Histoire maritime* 4 (2007) 241–250.

Cabral J.P.S., *A História Natural de Portugal em Domingos Vandelli* (Lisbon: 2018).

Cabral, J.P.S., "Parte IV: Esboço biográfico de Domingos Vandelli. Flora e Fauna", in Cabral J.P.S. – Borges F.S. – Costa J.M.R. (eds.), *Um manuscrito inédito de Domingos Vandelli, Historia naturalis Olisiponensis (BPMP Ms. 1127): leitura, interpretação e discussão* (Oporto: 2020).

Cabral M.J.M. (ed.), *Livro Vermelho dos Vertebrados de Portugal* (Lisbon: 2008).

Carneiro M. – Martins R. – Landi M. – Costa F.O., "Updated Checklist of Marine Fishes (Chordata: Craniata) from Portugal and the Proposed Extension of the Portuguese Continental Shelf", *European Journal of Taxonomy* 73 (2014) 1–73.

Costello M.J. – Emblow C. – White R. (eds.), *European Register of Marine Species. A Check-list of the Marine Species in Europe and a Bibliography of Guides to Their Identification* (Paris: 2001).

Damião de Góis, *Vrbis Olisiponis Descriptio per Damianum Goem Equitem Lusitanum, in qua obiter tractantur nõ nulla de Indica nauigatione, per Graecos et Poenos et Lusitanos, diuersis temporibus inculcata* (Évora, Andreas Burgensis: 1554).

Degering H., *Kurzes Verzeichnis der germanischen Handschriften der Preussischen Staatsbibliothek (Berlin)*, I: *Die Handschriften in Folioformat* (Leipzig: 1925).

Fischer B., *Dialogue Luso-Suisse: Essai d'une histoire des relations entre la Suisse et le Portugal du XVe siècle à la Convention de Stockholm de 1960* (Lisbon: 1960).

Gessner Conrad, *Historiae animalium liber IIII qui est de piscium et aquatilium animantium natura* (Zurich, Christoph Froschauer: 1558).

Herold B., "The Diary of the Swiss Leonhardt Thurneysser and Black Africans in Renaissance Lisbon", *Renaissance Studies* 42 (2017) 463–488.

Herold B. – Horst T. – Leitão H., "A 'História Natural de Portugal' de Leonhardt Thurneysser zum Thurn, ca. 1555–1556", *Ágora. Estudos Clássicos em Debate* 19 (2017) 305–334.

Herold B. – Horst T. – Leitão H., A História Natural de Portugal *de Leonhardt Thurneysser zum Thurn, ca. 1555–1556.* [Including a] transcription of the Parts Concerning Portugal of the Manuscript Ascribed to Leonhardt Thurneysser zum Thurn Ms. Germ. Fol. 97 of Berlin State Library (Berlin: 2019). Academia das Ciências de Lisboa, Biblioteca Digital http://hdl.handle.net/10400.26/45798 last access 2023-09-27.

Hirsch E.F., *Damião de Góis: The Life and Thought of a Portuguese Humanist* (The Hague: 1967).

Horst T., "A Rediscovered Manuscript about Portuguese Plants and Animals: Preliminary Observations", in Lopes M. dos S. – Leitão H. (eds.), *Renaissance Craftsmen and Humanistic Scholars. Circulation of Knowledge between Portugal and Germany* (Frankfurt/Main: 2016) 133–174.

Lacerda Lobo C.B. de, "Sobre a decadencia das Pescarias em Portugal", *Memorias Economicas da Academia Real das Sciencias de Lisboa*. IV, Na Officina da mesma Academia (Lisbon: 1812) 312–383.

Lloris, D., *Ictiofauna marina. Manual de identificación de los peces marinos de la Península Ibérica y Baleares* (Barcelona: 2015).

Macedo M.C.C. – Macedo M.I.C. – Borges J.P., *Conchas marinhas de Portugal* (Lisbon: 1999).

Marco-Herrero E. – Abelló P. – Drake P. – García-Raso J.E. – González-Gordillo J.I. – Guerao G. – Palero F. – Cuesta J.A., "Annotated checklist of brachyuran crabs (Crustacea: Decapoda) of the Iberian Peninsula (SW Europe)", *Scientia Marina* 79 (2015) 243–256.

Marques A.H. de O., "Damião de Góis e os mercadores de Danzig", *Arquivo de Bibliografia Portuguesa* (Coimbra: 1959).

Moehsen Johann Carl Wilhelm, *Ein Beitrag zur Geschichte der Alchymie, wie auch der Wissenschaften und Künste in der Mark Brandenburg gegen Ende des sechszehnten Jahrhunderts* (Berlin – Leipzig, George Jacob Decker: 1783).

Nobre A., *Crustáceos decápodes de Portugal* (Oporto: 1931).

Nobre A., "Peixes de água doce de Portugal", *Boletim do Ministério da Agricultura*, XIII, no. 2, série I (1932).

Nobre A., *Crustáceos decápodes de Portugal*, second ed. (Barcelos: 1936).

Pohle J., *Os mercadores-banqueiros alemães e a Expansão Portuguesa no reinado de D. Manuel I* (Lisbon: 2017). https://run.unl.pt/bitstream/10362/38843/2/Mercadores Alemaes.pdf.

Quelle O., "Leonhard Thurneysser zum Thurn", *Revista do Instituto de Cultura Alemã. Zeitschrift des Deutschen Kulturinstituts* 1/1 (1944) 99–102.

Quelle O., "Deutsch-Portugiesische Kulturbeziehungen", *Zeitschrift für Politik* 34 (März/April 1944) 115–121.

Reis M.S. – Mathias M.L., "Mamíferos", in Rodrigues P.D. – Crespo E.G. (eds.), *Portugal Moderno: Fauna, Enciclopédia Temática Pomo* (Lisbon: 1991) 160–191.

Rocha R., *A viagem dos sabores. Ensaio sobre a história da alimentação (séculos IX–XIX) seguido de 100 receitas em que vários mundos se encontram* (Lisbon: 1998).

Rodríguez de la Fuente F., *Cadernos de campo*. Lisbon: Artel. Colecção de 60 cadernos de campo sobre a fauna ibérica. Supervisão científica de Luís F. Matos, Cadernos de campo #18 and #32 for fishes, and #53 for crustaceans (Lisbon: 1984).

Rondelet Guillaume, *Libri de piscibus marinis, in quibus verae piscium effigies expressae sunt* (Lyon, Matthias Bonhomme: 1554).

Saldanha L., *Fauna submarina atlântica* (Lisbon: 1995).
Schumacher Y., *Leonhard Thurneysser, Arzt – Alchemist – Abenteurer* (Zurich: 2011).
Spitzer G., ... *und die Spree führt Gold: Leonhard Thurneysser zum Thurn, Astrologe – Alchimist – Arzt und Drucker im Berlin des 16. Jahrhunderts* (Wiesbaden: 1996).
Strasen E.A. – Gândara A., *Oito séculos de História Luso-Alemã* (Berlin: 1944).
Thurneysser Leonhardt, *Historia sive plantarum omnium, tam domesticarum quam exoticarum* (Berlin, Michael Hentsken: 1578).
Thurneysser Leonhardt, *Historia unnd Beschreibungen influentischer, elementischer und natürlicher Wirkungen aller heimischen und fremden Erdgewächssen* (Berlin, Michael Hentsken: 1578).
Thurneysser Leonhardt, *Historia unnd Beschreibungen influentischer, elementischer und natürlicher Wirkungen aller heimischen und fremden Erdgewächssen*, 2nd edition (Cölln/Berlin, Johannes Gymnium: 1587).
Zariquiey Alvarez R., "Crustáceos decápodos ibéricos", *Investigación Pesquera* 32 (1968) 1–510.

PART 2

Depicting

CHAPTER 6

Looking beyond the Margins of Print: Depicting Water Creatures in Europe, *c*.1500–1620

Florike Egmond

> Praeterea nihil e fossilibus vidi, quoniam tota hyeme
> Patavii delitui et bachanalia atque quadragesimam
> huc usque Venetiis transegi inquirendorum piscium gratia,
> quorum non minus iucunda quam herbarum est cognitio.
>
> By the way I saw nothing of minerals and stones because I passed
> the whole winter in Padua and spent Carnival and Lent
> between there and Venice for the sake of enquiring into fish.
> Finding out about fish is just as enjoyable as studying plants.[1]
>
> VALERIUS CORDUS to Agricola (from Venice, 20-04-1544)

1 Print and Drawing

Most histories of natural science tend to discuss the early-modern period mainly on the basis of printed works.[2] For example, the publications by Brunfels, Fuchs, Bock, and Mattioli of the 1530s–50s are seen as the founding

1 Valerius Cordus (Venice) to Georg Agricola, 20 April 1544, published in Wilisch Christian Gotthold, *Arcana bibliothecae Annabergensis in partes III divisa* (Leipzig, Lankischs Erben: 1730) 155–158; discussed in Horst U., "Die einzige erhaltene handschrift des Valerius Cordus, ein Brief aus seinem letzten Lebensjahr 1544", *Beiträge zur Geschichte der Pharmazie* 26.2 (1974) 9–14.

2 With many thanks to Giuseppe Olmi, Lucia Tongiorgi, Paul Smith, Marlise Rijks, Elisa Andretta, José Pardo Tomás, Sarah Kyle, María M. Carrión, Arthur MacGregor, Sachiko Kusukawa, Silva Dobalová, Cesare Pastorino, Stefano L'Occaso, Emma Sallent, Giulia Simonini, and especially to Peter Mason, who read several versions and translated all Latin quotations, and to Holger Funk, who generously shared his published and unpublished research, and commented on two versions of this text. The research for this article was done as part of the NWO-funded projects *Re-Reading the Book of Nature* and *A New History of Fish*, both at Leiden University.

 | DOI:10.1163/9789004681187_007

works of botany, in which their humanistically and often medically trained authors incorporated and critically discussed knowledge from antiquity while expanding the corpus of known plants by means of new observations, descriptions and identifications. Similarly, the illustrated publications of the 1550s by the Frenchmen Pierre Belon and Guillaume Rondelet, the Italian Ippolito Salviani and the Swiss Conrad Gessner are regarded as the founding canon of ichthyology.

This historiographical focus on print has created its own chronology and geography, which have helped to define the boundaries of what has come to be regarded as natural science. I suggest that the current chronology of natural history is limping on the single leg of mainly printed works, the analysis of which has, until recently, largely focused on their texts. This situation needs to be redressed by the inclusion of a variety of other sources and the consideration of their implications for the chronology, geography and ultimately definition of natural science itself. They are particularly essential if we want to get a clearer idea of how the wider domain of knowledge making about nature relates to the formation of natural science as a scientific discipline, and to the ways in which practice-based knowledge entered into the wider corpus of visual science. Those 'other' sources should include non-printed visual material, decorations, and many objects in collections, but that range lies beyond the scope of the present article.

Here I will mainly attempt to reconnect the European corpus of illustrated printed works on *aquatilia* with the large collections of generally coloured drawings of water creatures. There are (or were) thousands of original 16th-century drawings (often incorporated in albums or paper collections) depicting a wide range of aquatic creatures, including fish, molluscs, shells, marine mammals, some sea 'monsters', and an occasional sea bird and crocodile. Much research has been done on individual collections of animal drawings and albums during the last thirty or so years. However, no up to date survey exists to my knowledge that discusses their chronology, geographical patterns of collecting, or the possible functions of these drawings.[3] The following text consists therefore of

3 Among the main surveys focusing on *printed* illustrated works are Gudger E.W., "The five great naturalists of the sixteenth century: Belon, Rondelet, Salviani, Gesner and Aldrovandi: a chapter in the history of Ichthyology", *Isis. International review devoted to the history of Science and Civilization* 22 (1934) 21–40; Idem, "Beginnings of fish teratology, 1555–1642. Belon, Rondelet, Gesner and Aldrovandi, the fathers of ichthyology, the first to figure abnormal fishes", *Scientific Monthly* 43 (1936) 252–261; Nissen C., *Schöne Fischbücher. Kurze Geschichte der ichthyologischen Illustration* (Stuttgart: 1951); Idem, *Die zoologische Buchillustration. Ihre Bibliographie und Geschichte*. Band II (Stuttgart: 1978); Bäumer A., *Geschichte der Biologie. Zoologie der Renaissance – Renaissance der Zoologie Band 2* (Frankfurt am Main: 1991); Pinon L., *Livres de zoologie de la Renaissance, une anthologie (1450–1700)* (Paris: 1995); Kolb K., *Graveurs Artistes et hommes de science: Essai sur les traités de poissons de la Renaissance*

a chronological synopsis of our present knowledge about *printed* illustrated works on *aquatilia*, and goes on to a more detailed survey of *aquatilia* drawings and albums between circa 1500 and *c.*1615, with special attention to Italy and the rather underexplored first half of the 16th century.[4]

Will this non-printed visual material adjust our notions of the chronology, geography and nature of a scientific interest in the aquatic world in early-modern Europe? And can it throw light on the as yet insufficiently understood phenomenon that *aquatilia* (apparently) were the first of all animal categories to be intensively studied in the early modern period?[5] Only the interest in the medicinally much more relevant plants manifested itself earlier, in printed works and before the age of print in numerous manuscript herbals.

My focus on the visual domain – in print and in drawing, but always closely connected with the *study* of nature – is inspired by the often-noted visual turn of 16th-century natural history itself, and by the fact that printed illustrations were always linked to drawings. The latter may need some explanation. In the early modern period nearly all printed illustrations (generally woodcuts) went back to model drawings. Many of those were destroyed in the process of transferring the image to the woodblock. Of course, the fact that prints were based on drawings by no means implies that all drawings were created as models for print. Even when a drawing was indeed used as a model for print, it does not necessarily follow that the collector or painter had intended or made it for that purpose, as we will see below. Links between drawings and printed illustrations could, moreover, be far more complex than a simple connection between

(Paris: 1996); Nellen W. – Dulčić J., "Evolutionary steps in ichthyology and new challenges", *Acta Adriatica* 49 (2008) 201–232; and Zucker A., "Zoologie et philologie dans les grands traités ichtyologiques renaissants", *Kentron. Revue pluridisciplinaire du monde antique* 29 (2013) 135–174. Cf. Minonzio F., "Diffrazioni pliniane prima di Belon (1553): descrizione e classificazione di pesci in Paolo Giovio, Francesco Massari e Simone Porzio", in Maraglino V. (ed.), *Scienza antica in età moderna* (Bari: 2012) 401–439. For surveys with some attention to *non-printed* images, see Hünemörder C., "Die Geschichte der Fischbücher von Aristoteles bis zum ende des 17. Jahrhunderts", *Deutsches Schiffahrtsarchiv* 1 (1975) 185–200, on pre-Renaissance books; and Nissen, *Schöne Fischbücher*.

4 Obviously, I cannot claim completeness. I will discuss neither the (mainly earlier) illuminated manuscripts and incunables that do not systematically treat *aquatilia*. Nor will I go into important but generally *non*-illustrated material that contains early modern fish knowledge, such as culinary books and texts that deal with health, agriculture, hunting and fishing. A good survey of sixteenth-century fishing in the Mediterranean is Nicolò M.L. De, *Il Mediterraneo nel Cinquecento tra antiche e nuove maniere di pescare* (Pesaro: 2011).

5 I refer here to studies in the domain of natural history, not to discussions of animals involved in the hunt or to farm animals. Gudger, "The five great naturalists" points out how fish became the first animals to be studied, but does not explain why. On zoology before the 1550s, see esp. Perfetti S., *Aristotle's Zoology and its Renaissance Commentators (1521–1601)* (Leuven: 2000); and Minonzio, "Diffrazioni pliniane".

model drawing and printed illustration. Sometimes drawings were even copied after prints. Those complex relations between painted and printed images of naturalia, which can differ per image and even within one paper collection, make it all the more important not to separate them.

In what follows, I will pay attention to the chronology, geography, collectors and in so far as possible quantities of *aquatilia* drawings. If information is available, I will discuss modes of collecting and the functions and uses of the drawings – placing this essay also in the context of the history of visual collections. I will not analyse painterly styles, however, nor go into fish identifications according to modern criteria.

2 A Printed Visual Corpus in Development

2.1 *The Printed Works before the 1550s*

Even if we limit ourselves for the moment to the domain of print, the older historiography is in need of some nuances. While the canonical publications on *aquatilia* belong to the 1550s, specialized books on this subject began to appear much earlier. Intriguingly, the first of these was published far away from the core areas of European printing at the time. In 1517–20 the jurist-humanist and printer Nicolaus Marschalk in Rostock published a 115-page book in Latin in which he first describes fish thematically and then discusses them alphabetically per species. Although he refers to his own experience in crossing the seas, most of his descriptions actually go back to combinations of classical (and thus Mediterranean) sources. The rather rough woodcut illustrations that conclude his work are hardly connected with the text: they belong to the late medieval tradition of bestiaries and books of health (going back in particular to the *Hortus Sanitatis* of the early 1490s), and include 'monsters', such as sea swine, mermaids, and dragons. Visually and in terms of approach this publication stands on the edge between classical tradition and new observational science, between late medieval imaginary and erudite humanism. Its influence was probably only regional, and the work was soon forgotten. Yet, together with a contemporaneous 100-page unillustrated publication on serpents and medicine by Niccolò Leoniceno (1518) it is the first printed work in Europe that attempts to bring together knowledge about a whole category of animals. Marschalk also speaks explicitly of *ichtyographia*.[6]

6 Marschalk [Marescalcus] Nikolaus, *Historia Aquatilium latine ac grece cum figuris* (Rostock, Nikolaus Marschalk: 1517–1520); the impressum states 1520, but the printing began in 1517. Not all exemplars have the illustrations. See Haye T., "Notizen zu Nicolaus Marschalk", *Daphnis* 23 (1994) 205–236; Huber-Rebenich G., "Marschalk, Nikolaus", in Worstbrock F.J.

From that moment on publications about *aquatilia* never stop. The publication order of the following works indicates that the cultural interest in *aquatilia* grew steadily and did not suddenly explode in the 1550s. In 1524 the Vatican church official and Papal advisor Paolo Giovio published a circa 125-page unillustrated work on fish in Latin in which he critically discussed fish names in sources from antiquity (among the most important were Oppian's *Halieutica*, Aelian's work on animals, and Pliny's *Naturalis Historia*). Giovio also attempted to match fish names and descriptions with what he knew about aquatic life from practice and contemporary sources. The work was written on the suggestion of Cardinal Francesco Ludovico di Borbone, and to a large extent the result of the many feasts and banquets at the Papal court and Giovio's visits to the Roman fish market. In contrast with Marschalk's publication, Giovio's work proved so popular that it was reprinted (in Latin) until the 1730s in various European countries; an Italian edition appeared already in 1560.[7] Importantly, Giovio had intended to illustrate his work, but did not receive the necessary funding from his noble patron.[8]

Fish studies almost immediately also involved travel and personal observation. In the course of the 1520s, the Venetian Francesco Massari travelled to Greece and Constantinople, partly for his fish research. He left important, but unillustrated comments on the 9th book of Pliny's *Naturalis Historia* with abundant information especially about vernacular Venetian and Greek fish terminology. Written during the 1520s, his work was only published in 1537.[9]

(ed.), *Deutscher Humanismus 1480–1520. Verfasserlexikon, Vol. 2* (Berlin – New York: 2013) 161–203; and Timm W., "Die Holzschnitte zu Nikolaus Marschalks *Historia aquatilium latine ac grece cum figuris*. Rostock 1517–20", *Wisschenschaftliche Zeitschrift der Universität Rostock – Gesellschafts- und Sprachwissenschaftliche Reihe* 17.9/10 (1968) 799–802. Leoniceno Niccolò, *De serpentibus opus singulare ac exactissimum* (Bologna, Giovanni Antonio de' Benedetti the younger: 1518).

7 Giovio Paolo, *De romanis piscibus libellus* (Rome, Francesco Minuzio Calvo: 1524). On Giovio, see esp. Baumann T., *Pauli Iovii Novocomensis Medici De Romanis piscibus libellus ad Ludovicum Borbonium cardinalem amplissimum. Text, Übersetzung, Kommentar* (Ph.D. dissertation, Universität of Mannheim: 1994); and Minonzio, "Diffrazioni pliniane". For a biography, see Zimmermann T.C.P., *Paolo Giovio* (Princeton: 1995).

8 Giovio's passages concerning the swordfish and the melanurus refer to intended images. I thank Holger Funk for pointing this out. Minonzio, "Diffrazioni pliniane" 426, mentions a unique exemplar of Giovio's first edition in an Italian private collection that is both annotated and illustrated in the margins with 'hundreds of drawings, in pen and black or sepia ink, of fish that are depicted with extraordinary and even anatomical precision'; these drawings are based on Rondelet's (later) illustrations. I have no further information about it.

9 Massari Franciscus, *Veneti in nonum Plinii De Naturalis Historia librum castigationes et annotationes* (Basel, Officina Frobeniana: 1537). See on this Folena G., "Per la storia dell'ittionomia volgare", in Olmi G. – Tongiorgi Tomasi L., *De piscibus. La bottega artistica di Ulisse Aldrovandi e l'immagine naturalistica* (Rome: 1993) 113–150, here 130 and note 34; and Minonzio, "Diffrazioni pliniane" 427–432.

One of the most interesting early publications on (Mediterranean) fish appeared a decade after Giovio's first edition: a Latin edition of Aelian by the French naturalist Pierre Gilles (1535). A process familiar from the herbals is also clearly visible in Gilles's works: newly gathered information clustered around and soon began to take over from the core of classical knowledge. Gilles's 1535 edition presents much information about Mediterranean fish and their local names, based on his own observations from Marseille to Liguria. And his successive editions contain additional information on *aquatilia* from Rome, Naples, the Adriatic, Greece, and Constantinople.[10] Gilles not only integrated practice-based knowledge; he also explicitly acknowledges it in his references to fishermen-informants.

Gilles's widely read work triggered further investigations into aquatic life. One of these is a short, unillustrated Latin treatise by the Sicilian monk, mathematician and astronomer Francesco Maurolico, written (1543) in the form of a letter to Gilles. Maurolico aimed to provide the latter with more information about marine and fresh water *aquatilia*, fishing techniques, and marine monsters from Sicily. Maurolico's treatise remained unpublished until 1807, but it throws some light on why there was such a great interest in the world of water creatures in his time. Maurolico states that neither classical authors nor modern naturalists provided sufficient detail about *aquatilia* to facilitate their identification and further discussion. The domain of water, he continues, appears filled with a riches of life at least as varied and fascinating as that on land, and potentially more so, since it is inscrutable and largely invisible. Maurolico also points to the need for clear descriptions, since fish names differ per region or even location. He thus directly links the lack of uniformity in naming with the need for detailed description.[11]

Strikingly, *none* of the works printed before 1550 has any illustrations, apart from Marschalk with its bestiary-based images. This group is larger than we might think. Besides the ones discussed above, it also includes Figulus (1540) on Ausonius's fourth-century poem concerning the river Moselle and its fish; Estienne (1544) on the names of all kinds of naturalia; and Dubravius (1547)

10 Gilles Pierre, *Ex Aeliani historia per Petrum Gyllium latini facti, itemque ex Porphyrio, Heliodoro, Oppiano, tum eodem Gyllio luculentis accessionibus aucti libri XVI* (Lyon, Sebastian Gryphius: 1535); Idem, *De Bosporo Thracio Libri III* (Lyon, Guillaume Roville: 1561); Idem, *Aeliani De historia animalium libri XVII* (Lyon, Guillaume Roville: 1565).

11 Maurolicus Franciscus, *Tractatus per epistolam Francisci Maurolyci ad Petrum Gillium de piscibus siculis*, Messina, 1543, manuscript, Florence, Biblioteca Nazionale; it is published in Sestini D., *Viaggi e Opuscoli diversi* (Berlin: 1807). For a discussion, see Castelli D., "Tra ricerca empirica e osservazione scientifica: gli studi ittiologici di Simone Porzio", *Archives Internationales d'Histoire des Sciences* 57 (2007) 105–123, here 111.

on freshwater fish and fish ponds.[12] The situation changed dramatically in the 1550s. Not the interest in *aquatilia* suddenly exploded, therefore, but their visual representation.

2.2 *The Visual Revolution in Print*

In the 1550s one printed publication on *aquatilia* swiftly followed another: all are illustrated, and the number of illustrations and of species discussed increased rapidly. All of the publications now regarded as canonical works that mark the birth of *ichthyology* as a scientific discipline were published in this decade. The first of this series, the French edition of Belon (1551) has only 20 illustrated pages, but his much-expanded *La nature et diversité des poissons* of 1555 already contains 185 pages with illustrations. Rondelet's well over 600-page work on marine fish of 1554 counts almost 220 pages with woodcuts, often with more than one image per page. Its second part (1555) has some 130 illustrated pages which include shells as well as amphibians and fish. Salviani (1554–1558) has 81 pages with copper engravings; several show more than one fish.[13] Gessner's massive, almost 1300-page *Historia Animalium* IV (1558), finally, has 507 pages with woodcuts of aquatic creatures. Still within the same decade, water creatures gained an important place in two further publications. Olaus Magnus' famous work on life in Scandinavia (Rome, 1555) is illustrated throughout and has a considerable section on fish and fishing. More surprisingly, in 1554 the Italian naturalist Pietro Andrea Mattioli (1501–1578) first introduced illustrations of animals, especially marine ones, in his famous and mainly botanical *Commentarii* on Dioscorides.[14]

How crucially important illustrations became in this decade is also clear from much smaller illustrated works with a regional focus. A significant example is the modest work about fish and fishing in Lake Constance by Gregor Mangolt (1498–*c.*1584), a German Protestant preacher and bookseller who worked for some time as a corrector for the Froschauer publishing house

12 Figulus Carolus, *ΙΧΘΥΟΛΟΓΙΑ, seu Dialogus de piscibus* (Cologne, Eucharius Cervicornus: 1540); Estienne Charles, *De Latinis et Graecis nominibus arborum, fruticum, herbarum, piscium et avium liber* (Paris, Charles Estienne: 1544); and Dubravius Jan, *De piscinis* (Breslau, Andreas Vinglerus: 1547).

13 See Rondelet Guillaume, *Libri de piscibus marinis, in quibus verae piscium effigies expressae sunt* (Lyon, Macé Bonhomme: 1554); and Idem, *Universae aquatilium historiae pars altera, cum veris ipsorum imaginibus* (Lyon, Macé Bonhomme: 1555). The numbering of pages and images in Salviani is rather chaotic: 81 is the total of illustrated pages, not of species, and not including the title page.

14 In Magnus Olaus, *Historia de Gentibus Septentrionalibus* (Rome, Giovanni Maria Viotti: 1555) see esp. XX and XXI, *De piscibus* and *De piscibus monstrosis* (697–778). On further Mattioli editions with fish images, see below.

in Zurich. As Hakelberg has shown, Mangolt probably wrote the text of his fish book before *c.*1548. He focused on fish names and local fish expertise, and had no illustrations. Only in the mid 1550s, the Swiss publisher Andreas Gessner decided to print Mangolt's work on the instigation of Conrad Gessner (Andreas's relative). Without ever consulting the author, the publisher inserted some 24 fish images (grouped in twelve woodcuts) which he took from old calendar illustrations. He also reorganized the text according to the months and inserted some short calendar poems as well as a whole section on fishing techniques, which goes back to a late 15th-century text.[15] With a definite eye for the market this publisher thus created a bricolage, spicing up an unillustrated but recent treatise with texts and images that were old at the time. There was no intrinsic connection between image and original text. In fact, they even belonged to distinct genres.

The adoption of a presentational format in which illustrations – whatever their quality – became an absolutely essential ingredient was a real visual turn. Only three relevant works of the 1550s are unillustrated. In two cases that absence is significant.[16] The English physician-naturalist Edward Wotton's strongly Aristotelian work of 1552 excerpts and discusses the classical sources on various categories of animals. The Italian polymath Girolamo Cardano too wrote about *aquatilia* (1557) in a far wider context, and his discussion moves

15 Mangolt Georg, *Fischbuch. Von der Natur und Eigenschafft der vischen* (Zurich, Andreas and Jakob Gessner: 1557). See Hakelberg D., "Das Bodensee-Fischbuch von Gregor Mangolt in einem Basler Nachdruck von 1612", *Wolfenbütteler Barock-Nachrichten* 36 (2009) 107–114; cf. Violand G., "Historische Fischökologie des Bodensees. Das Fischbuch des Gregor Mangolt, 1557", *Österreichs Fischerei* 59 (2006) 169–171. Both illustrations and short poems occur only in the 1557 edition of Mangolt. A few woodcut illustrations can also be found in some late fifteenth to early sixteenth-century German and Netherlandish treatises on fishing. See Zaunick R., "Fragmente der ältesten sächsischen Fischfauna des Dr. Johannes Kentmann (1518–1574)", *Sitzungsberichte und Abhandlungen der Naturwissenschaftlichen Gesellschaft Isis* (1915) 15–36; and Cockx-Indestege E., "Van een boekje om vogels en vissen te vangen naar een zeldzame Antwerpse postincunabel, nu in de Library of Congress te Washington", in Van der Vekene E. (ed.), *Refugium Animae Bibliotheca. Festschrift Albert Kolb* (Wiesbaden: 1969) 109–138.

16 Gessner Conrad, *De piscibus et aquatilibus libelli III novi* (Zurich, A. Gesner: 1556) on fish names was intended as a forerunner of his massive *Historia animalium, Liber IIII, qui est de piscium et aquatilium animantium natura* (Zurich, Froschauer: 1558), so there are practical reasons for the former's lack of illustrations. On Gessner's printed works on fish and his translations of works by others on fish, Hendrikx S. – Smith P.J., "Ichthyology in Translation: Conrad Gessner's Fish Books", in Fournel J.-L. – Paccagnella I. (eds.), *Traduire – Tradurre – Translating. Vie des mots et voies des oeuvres dans l'Europe de la Renaissance*, (Geneva: 2022) 341–361.

from general characteristics of aquatic creatures to more specific ones.[17] These were general analytic works on nature and natural philosophy, therefore, in which illustrations of specific species were far less functional than in surveys of fish. The enormous emphasis on illustrations is not the only characteristic that distinguishes the *aquatilia* publications of the 1550s from those of the preceding four decades, however. The 1550s publication wave also shows increasing attention to region-specific information and local knowledge; a partial shift to vernacular languages; and a more wide-ranging (encyclopaedic) effort to include as many species as possible. After botanical works, *aquatilia* publications were indeed forerunners in this respect.

All these changes together and especially the rapidly increasing number of illustrations in the 1550s indicate that a genre of natural history encyclopaedia was finding its form. In that genre, illustrations and textual description *together* served the essential purposes of identification in a situation of confused naming and huge and partly invisible natural variety.[18] The visual turn of 16th-century natural history publications with their unprecedented numbers of illustrations is more, therefore, than mere evidence of acute commercial awareness on the part of the publishers – though there is no doubt that the latter pushed for images and realized that illustrated works sold better than non-illustrated ones. The turn to visual information was intrinsically connected with the function and identity of the genre as a whole, and with the functions of illustrations within these publications.

The influence of *aquatilia* publications of this type was lasting, but the actual 16th-century publication wave was relatively short-lived. Further editions of Gessner's major fish work of 1558 with newly added illustrations appeared in 1560 and 1563.[19] In the 1562 Latin edition and especially in the Valgrisi-editions of 1565 and 1568 Mattioli published expanded and revised sections on animals with more, newly executed and larger woodcut illustrations in his work on

17 Wotton Edward, *De differentiis animalium libri decem* (Paris, Vascosan: 1552). And Cardano Girolamo, *De rerum varietate libri XVII* (Basel, Henricus Petri: 1557).

18 As already pointed out by Gessner (see esp. Folena, "Per la storia" 133–34), the naming confusion was enormous. Identification was not the only function, of course, of the illustrations. See on the visual turn and the functions of illustrations, esp. Pinon, *Livres de zoologie* 24; and Smith P.J. – Trijp D. van, "Dynamiques européennes de l'humanisme érudit dans l'histoire naturelle. Le cas de l'ichtyologie, de Belon, Rondelet et Gessner à Willughby et Ray", in Crouzet D. – Crouzet-Pavan E. – Desan P. – Revest C. (eds.), *L'humanisme à l'épreuve de l'Europe (XVe–XVIe siècle). Histoire d'une transmutation culturelle* (Ceyzérieu: 2019) 167–181.

19 Gessner Conrad, *Nomenclator Aquatilium animantium. Icones animalium aquatilium* (Zurich, Christopher Froschauer: 1560); and Gessner Conrad – Forer Cunrat, *Fischbuch* (Zurich, Christopher Froschauer: 1563).

Dioscorides: his animal section starts with water animals and he pays special attention to the sea creatures of the Adriatic [Fig. 6.1].[20] And an influential new German translation by Johannes Heyden of the popular books VII–XI of Pliny's *Naturalis Historia* appeared in 1565 in Frankfurt am Main, on the initiative of the publisher Sigmund Feyerabend. It had some 200 newly cut illustrations by the famous Jost Amman and Virgil Solis, including a large number of fish and sea monsters. Feyerabend had obviously been inspired by the success of the 1550s wave of naturalia publications, and his commercial instincts proved correct. This German edition was reprinted at least eight times until 1651.[21]

No major new works on fish were published in Europe during the rest of the 16th century.[22] Obviously, this does not imply that research had ended or that interest petered out. Some of the works discussed earlier continued to be reprinted. And one important work was still in preparation: Ulisse Aldrovandi's huge volume on *aquatilia* with five large sections on fish and one on whales (*ceti*) only came out posthumously, in 1613, edited by Aldrovandi's former student and successor, the Dutchman Johan Cornelis Uterweer (Uterverius). Of its well over 700 pages, some 340 pages are illustrated, often with more than one woodcut or with full page images. It was based on almost half a century of Aldrovandi's collecting information, dried fish, and fish drawings, and its origins went back as far as the years 1549–51, as will be discussed below. In terms of its conception, character and formative decades Aldrovandi's work

20 The first Latin edition of Mattioli with animal figures contains a limited number of small-sized woodcuts: Mattioli Pietro Andrea, *Commentarii in libros sex Pedacii Dioscoridis Anazarbei, De Materia Medica* (Venice, Vicenzo Valgrisi: 1554). Their number but not their size expands slightly in the 1562-edition: Mattioli Pietro Andrea, *Commentarii denuo aucti, in libros sex Pedacii Dioscoridis Anazarbei de medica materia. Adiectis quamplurimis plantarum, & animalium imaginibus, quae in prioribus editionibus non habentur* (Lyon, Gabriel Coterius: 1562). But a much larger number of newly designed and bigger woodcuts of water creatures (modelled on the older images but much more finely executed) first appears in the 1565-edition: Mattioli Pietro Andrea, *Commentarii in sex libros Pedacii Dioscoridis Anazarbei De medica materia* (Venice, Vicenzo Valgrisi: 1565). On the painter see further below. The German Handsch edition of 1563 does not contain this animal section, in so far as I have been able to see from editions available online: Mattioli Pietro Andrea – Handsch Georg, *New Kreüterbuch: Mit den allerschönsten vnd artlichsten Figuren aller Gewechß, dergleichen vormals in keiner sprach nie an tag kommen* (Prague, Melantrich von Auentin and Vicenzo Valgrisi: 1563).

21 Heyden Johann, *Caij Plinij Secundi, Des fürtrefflichen* [...] *Philosophi, Bücher vnd Schrifften, von der Natur* (Frankfurt am Main, Sigmund Feyerabend: 1565).

22 Two smaller publications from Germany discuss fish of the river Elbe: Fabricius Georg, *Rerum Misnicarum libri VII* (Leipzig, Ernst Vögelin: 1569); and Albinus Petrus, *Commentarius novus de Mysnia, Oder Newe Meysnische Chronica* (Wittenberg, Hans Lufft: 1580).

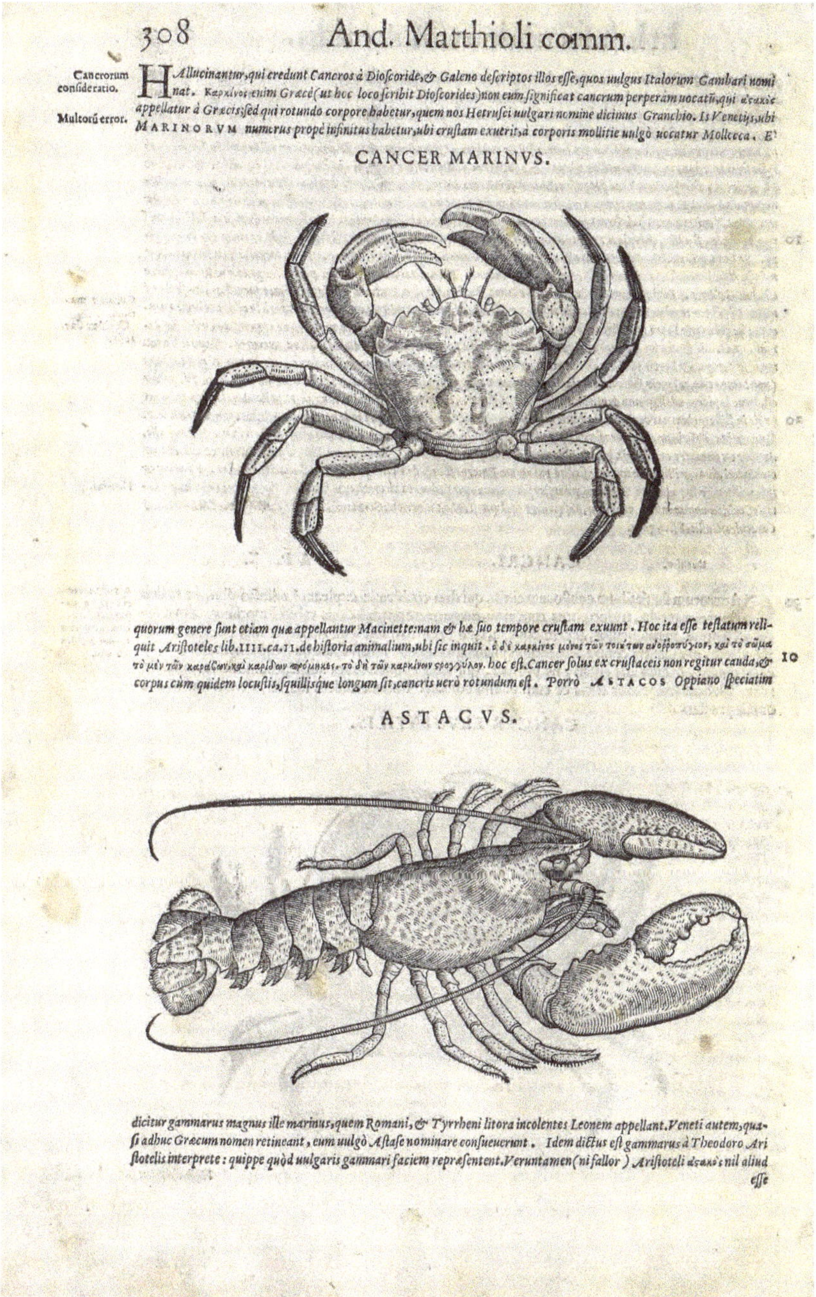

308 And. Matthioli comm.

Cancrorum consideratio. Multorũ error.

HAllucinantur, qui credunt Cancros à Dioscoride, & Galeno descriptos illos esse, quos uulgus Italorum Gambari nominat. Καρκίνος enim Græcè (ut hoc loco scribit Dioscorides) non eum significat cancrum perperam uocatũ, qui ἀστακὸς appellatur à Græcis; sed qui rotundo corpore habetur, quem nos Hetrusci uulgari nomine dicimus Granchio. Is Venetijs, ubi MARINORVM numerus propè infinitus habetur, ubi crustam exuerit, à corporis mollitie uulgò uocatur Molleca. E

CANCER MARINVS.

quorum genere sunt etiam quæ appellantur Macinette: nam & hæ suo tempore crustam exuunt. Hoc ita esse testatum reliquit Aristoteles lib. IIII. ca. II. de historia animalium, ubi sic inquit. ὁ δὲ καρκίνος μόνος τῶν τοιούτων ἀνορροπύγιον, καὶ τὸ σῶμα τὸ μὲν τῶν καράβων, καὶ καρίδων πρόμηκες, τὸ δὴ τῶν καρκίνων στρογγύλον. hoc est. Cancer solus ex crustaceis non regitur cauda, & corpus cùm quidem locustis, squillisque longum sit, cancris uerò rotundum est. Porrò ASTACOS Oppiano speciatim

ASTACVS.

dicitur gammarus magnus ille marinus, quem Romani, & Tyrrheni litora incolentes Leonem appellant. Veneti autem, quasi adhuc Græcum nomen retineant, eum uulgò Astase nominare consueuerunt. Idem dictus est gammarus à Theodoro Aristotelis interprete: quippe quòd uulgaris gammari faciem repræsentent. Veruntamen (ni fallor) Aristoteli ἀστακὸς nil aliud esse

FIGURE 6.1 Crab and crayfish. Woodcut, probably based on a drawing by Giorgio Liberale da Udine, in Mattioli Pietro Andrea, *Commentarii in sex libros Pedacii Dioscoridis Anazarbei De medica materia* (Venice, Vincenzo Valgrisi: 1565) 308 PUBLIC DOMAIN: HTTPS://WWW.BIODIVERSITYLIBRARY.ORG/ITEM/125718#PAGE/484/MODE/1UP

on *aquatilia* is thus a very late manifestation of the big wave of the 1550s rather than an example of seventeenth-century natural science.[23]

In the domain of print, copper engravings of *aquatilia* began to play a more important role in the final years of 16th century, in particular in the Low Countries.[24] More or less contemporaneously with the emergence of plant and animal still lifes and market scenes in oil, smallish groups of copper engravings began to appear. They hover on the edge between artistic models or scenes and visual natural science.[25] Nicolaes de Bruyn's (undated) group of 13 copper engravings of *aquatilia*, for instance, was explicitly intended as a model book for painters and sculptors.[26] Adriaen Collaert created (from *c.*1590) a series of engravings that depict various fish species in a setting of beach and coastal

23 Aldrovandi Ulisse – Uterverius Johannes Cornelius, *De piscibus libri v et De cetis liber unus* (Bologna, Hieronymus Tamburinus: 1613). This also applies to the unillustrated Schwenckfeld Caspar, *Theriotropheum Silesiae* (Liegnitz, David Albertus: 1603) which went back to research done before 1560.

24 Unlike woodcuts, and because of different printing techniques, copper engravings are not integrated in the textual pages, but printed on separate sheets.

25 On artistic developments in particular in Antwerp during the late sixteenth century with attention to representations of aquatilia, see Göttler C., "The place of the 'Exotic' in seventeenth-century Antwerp", in Schrader S. (ed.), *Looking East. Rubens's encounter with Asia* (Los Angeles: 2013) 88–107; Rijks M., "A painter, a collector, and a horseshoe crab. Connoisseurs of art and nature in early modern Antwerp", *Journal of the History of Collections* 31.2 (2019) 343–361; Idem, "'Unusual Excrescences of Nature': Collected Coral and the Study of Petrified Luxury in Early Modern Antwerp", *Dutch Crossing* 43.2 (2019) 127–156; Idem, "Scales, skins, and carapaces in Antwerp collections", in Bol M. – Spary E. (eds.), "Scales, Skins, and Carapaces in Antwerp Collections", in Bol M. – Spary E. (eds.), *The Matter of Mimesis: Studies on Mimesis and Materials in Nature, Art and Science* (Leiden – Boston: 2023) 295–320; and her contribution to the present volume. See also the two notes below. On seventeenth-century shell collecting and painting, see Leonhard K., "Shell collecting. On 17th-century conchology, curiosity cabinets, and still life painting", in Enenkel K. – Smith P.J. (eds.), *Early Modern Zoology. The construction of animals in science, literature and the visual arts*, Intersections 7 (Leiden – Boston: 2007) 177–214.

26 Bruyn Nicolaes de, *Libellus varia genera piscium complectens, pictoribus, sculptoribus utilis et necessarius* (Amsterdam, no publisher: year uncertain). This fish series has been dated to either the late 1590s or *c.*1630; Rikken argues for the 1590s: Rikken M., *Dieren verbeeld. Diervoorstellingen in tekeningen, prenten en schilderijen door kunstenaars uit de Zuidelijke Nederlanden tussen 1550 en 1630* (Leiden: 2016), here 66. On Italian fish still lifes, see Fratta V. Di, "De Aquatilibus. Dall'illustrazione scientifica alla nascita e sviluppo del genere della natura morta dei pesci nel XVII secolo", in [no editor] *Per la conoscenza dei beni Culturali, VI* [Ricerche del Dottorato in Metodologie conoscitive per la Conservazione e la Valorazione dei Beni Culturali, Università degli Studi di Napoli] (Naples: 2016) 57–72. Examples of Italian fish market scenes are by Vincenzo Campi in Venice (late 1570s), and a scene at a fish monger's (1580s) by Bartolomeo Passerotti from Bologna.

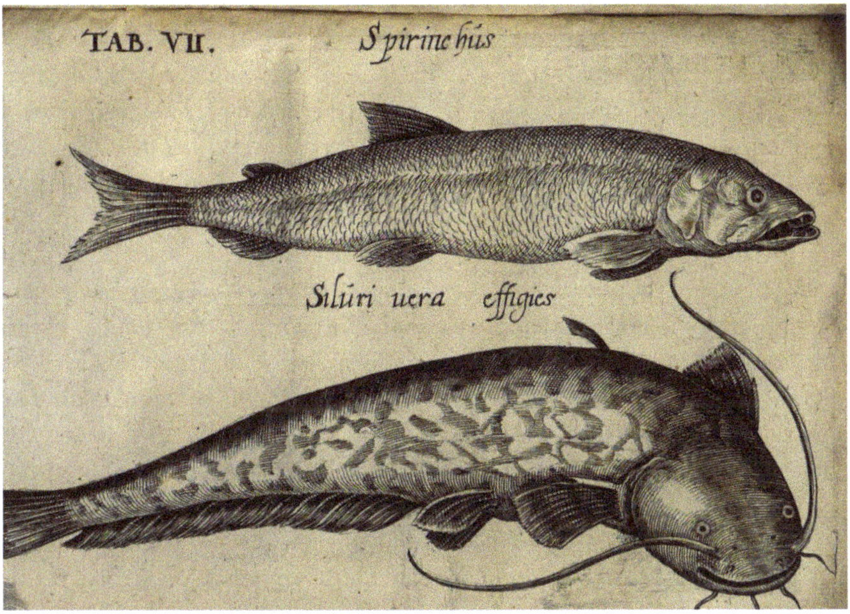

FIGURE 6.2 Spirinchus and silurus. Woodcut in Schonefeld Stephan von, *Ichthyologia et nomenclaturae animalium marinorum, fluviatilium, lacustrium* (Hamburg, Bibliopolis Heringianus: 1624), Table VII
PUBLIC DOMAIN: HTTPS://WWW.DIGITALE-SAMMLUNGEN.DE/DE/VIEW /BSB11220552?PAGE=5

landscapes.[27] The Dutch printer-engraver Crispijn de Passe produced (1620s?) a small group of copper engravings in which each page shows a handful of *aquatilia* draped artistically against a watery background. Engravings of aquatic creatures also embellish a small section on fish in Colonna's publication on plants of 1592 (with images based on Salviani), the published description of the Calzolari museum in Verona (1622); and seven fold-out copperplates conclude Schonefeld's 100-page work (1624) on the *aquatilia* of the North-German coastal zones [Fig. 6.2].[28]

27 Collaert Adriaen, *Piscium vivae icones. In aes incise et editae ab Adriano Collardo* (Antwerp, no publisher: c.1598). See Rikken, *Dieren verbeeld* 63–65.

28 See Passe Crispijn de, *Piscium vivae Icones* (Utrecht, no publisher: 1620?); Colonna Fabio, Φυτοβασανος (Phytobasanos) *sive plantarum aliquot historia, part Piscium aliquot plantarumque novarum historia* (Naples, Horatio Salviani: 1592); Ceruti Benedetto, *Musaeum Franc. Calceolari iun. Veronensis* (Verona, Angelus Tamus: 1622); and Schonefeld Stephan von, *Ichthyologia et nomenclaturae animalium marinorum, fluviatilium, lacustrium, quae In florentissimis ducatibus Slesvici et Holsatiae et celeberrimo Emporio Hamburgo occurrunt triviales* (Hamburg, Bibliopolis Heringianus: 1624).

All of the illustrations in the 1550s-wave of publications aimed to represent *aquatilia* accurately, true to life. Most authors valued empirical observation. Ideally this would be direct, personal observation of the live or dead fish by the naturalist and painter. In practice it often meant that at least someone had seen the fish and had it portrayed, after which a copy, or a copy of a copy, of the image had been sent to the naturalist. Interestingly, only very few of the printed images of the 1550s were copied after older woodcuts, such as the relatively well-known section on fish in the German *Hortus Sanitatus* (1491).[29] The vast majority of the illustrations of the long decade 1550–1563 was newly made. Something quite amazing happened, therefore. Although the genre of natural history publications in general and the interest in fish in particular developed fairly gradually from about 1500, a large, almost full-blown visual corpus of printed *aquatilia*-images emerged out of nothing in less than a decade in the mid-16th century. The influence of that corpus can be traced until far into the eighteenth century.

There is a further curious aspect to this corpus. Illustrations that appeared within the same decade are totally dissimilar in technical quality and artistic level, and it is impossible to discern any kind of artistic development. Right in the middle of the 1550s-wave, Salviani's Roman work on fish appeared as the first and virtually only publication in the whole of 16th-century natural history to be illustrated with copper engravings – a more refined and vastly more expensive type of illustration than woodcuts. Artistically, these engravings belong to the top; they are equalled only in the eighteenth century, if ever. That top was reached, therefore, within the first five years of the publishing pattern described here, and Salviani's engravings are contemporaneous with both excellent *and* extremely clumsy woodcuts.

This situation sounds unreal, which suggests that we are looking in the wrong way. First, I propose that there is no point in comparing the incomparable and trying to place this heterogeneous material in a line of qualitative visual development: Salviani's engravings aimed for excellence and were sponsored both financially and in terms of information by a Roman cardinal. They cannot be put on a level with works produced (often for different purposes and a different public) by image makers who had fewer funds, less access to information, less sophisticated printers, and probably fewer skills. Secondly, a very large part of the visual story is missing. In what follows we will look for it, outside the margins of print, and paying most attention to the crucial decades c.1530–1570. In that period, we may expect the most interesting information

29 For an edition with comments of its fish sections, see Jacquemard C. – Gauvin B. – Lucas-Avenel M. (eds.), *Hortus Sanitatis. Livre IV: Les Poissons* (Caen: 2013).

concerning the character, motivation, and background of the collecting of *aquatilia* drawings, and thus about the formation of the new visual corpus.

3 The Early Drawing Collections

3.1 *Early Italian Drawing Collections Begun in the 1520s–1540s*

The domain of 16th-century naturalia drawings is at the same time wider, more diffuse, and richer in terms of colours and styles than that of printed illustrations. It is also much less certain in terms of dating and chronology, since collections and albums of drawings were often created over several decades and can contain items made by (generally anonymous) painters from several countries in various periods.[30] Those uncertainties have undoubtedly contributed to a certain neglect of this material by historians, but a tacit assumption that drawings lost their usefulness and relevance once printed works with illustrations took over may have been as important. What follows should therefore not only inform us about the existence of naturalia drawings and albums, but also confirm the continued relevance of drawings in the age of print.

First of all, the almost complete absence of such a visual corpus for the 15th century is striking. It needs no arguing that lifelike representations of animals can be found in Europe ever since Greek and Roman antiquity, and that the fourteenth-century Giovannino de' Grassi, the early 15th-century Pisanello, and numerous illuminators of the 15th century – to name but a few examples – were more than capable of naturalistically depicting animals.[31] Some drawings of *aquatilia* figure, of course, in medieval *bestiaria*, various 15th century manuscript books of health and herbals, and the illuminations of Pliny's *Naturalis Historia*. But no codices, albums or collections of drawings from this period exist to my knowledge that are devoted to fish or contain a large visual corpus of recognizable *aquatilia*. In contrast, plants are depicted by the hundreds in 15th-century herbals.

Something changed around 1500. An example are the eight beautifully painted fish species that appear on the *Fischereipatent* of 1506 (a legislative document concerning fishing in the Danube) of the Habsburg Emperor Maximilian I, who was known for his interest in naturalia and his passion

30 My focus remains on *groups* of drawings that can be regarded as evidence of an interest in living nature. It therefore excludes single items, mainly emblematic and heraldic ones, and purely decorative images.

31 The point has been made before, e. g. in Nissen, *Schöne Fischbücher* 10–12, but the implications have as yet not been sufficiently investigated.

for hunting and fishing. The illustrated Tiroler *Fischereibuch* of 1504, again for Maximilian I, shows scenes of fishing with nets and rods, but no detailed representations of fish species.[32] Although these sources are not directly connected with the study of nature, and the tradition of local *Fischereiordnungen* in Bavaria went back much further in time, these particular sources are visually interesting. The 1506-drawings emphasize that extremely lifelike images of living creatures were produced at the very start of the 16th century – and not only in the workshops of contemporary masters such as Dürer, Da Vinci and Carpaccio.

The real boom of *aquatilia* drawings started in the 1530s and early 1540s, some ten to fifteen years before the major wave of the illustrated printed works discussed above.

The earliest evidence that points to the presence of actual fish in collections comes from Mantua and Venice; it is not precisely dated, but can be traced to before 1530 and most likely after 1500. In her collection of rarities in the Gonzaga palace in Mantua, Isabella d'Este (1474–1539) had not only narwal teeth (then classified as unicorn horns), but also other teeth of fish as well as corals. Her son, Duke Federico II Gonzaga (1500–1540), expanded this collection: an inventory of 1540 refers to his *museo naturalistico* and describes how sea fish, including five *colombi di mare* (a name used for rays, which in this case probably had been dried to look like small sea dragons) and 'other marine monsters' were hanging in his study between portraits. There were also several crocodiles, the skin of a large sea fish, the sword of a sword fish, teeth of fish, and eleven seasnails.[33] Similar evidence comes from Venice: an inventory made in 1532 of the enormous collection of antiquities, medals and coins, paintings, sculptures et cetera of Andrea Odoni – who belonged to an extremely rich family originally from Milan – also included several crabs, petrified serpents, a dried chameleon, and bizarre fish. These may have come from an even earlier collection owned by the Venetian Francesco Zio (also Giglio).[34]

---

32 These sources and the possible painters are discussed in Scheichl A., "Wer war(en) Jorg Kolderer, Innsbrucker Hofmaler und Tiroler Baumeister", in Sternath M.L. – Michel E. (eds.), *Emperor Maximilian and the age of Dürer*, exh. cat. Vienna (Munich: 2012) 81–90, 181–182, 315–319. For an edition of the *Fischereibuch*, see Hochleithner M. – Hohenleiter W., *Das Fischereibuch Maximilians I. – Faksimile und Transkription* (Kitzbühel: 2013).

33 Franchini D.D. – Margonari R. – Olmi G. – Signorini R. – Zanca A. – Tellini Perina C., *La scienza a corte. Collezionismo eclettico natura e immagine a Mantova fra Rinascimento e Manierismo* (Rome: 1979), here 87–89.

34 On these Venetian collections, see Morelli J., *Notizia d'opere di disegno, pubblicata e illustrata da D. Jacopo Morelli, Seconda edizione, riveduta ed aumentata per cura di Gustavo Frizzoni* (Bologna: 1884), here 152–159 and 176–177.

These were fishy objects in what looks like early manifestations of curiosity collections. The very earliest reference to an album specifically with fish drawings comes from this context of rich Venetian collectors too. Gabriele Vendramin (1484–1552) was a notable patron of artists – he commissioned Giorgione's *The Tempest* – and the owner of one of the most important collections in this city. His collection included (by 1530) a book with animal drawings; one with coloured bird drawings; and a small group of albums with drawings made by a certain Pre Vido Celere (or: Frate Guido Celere) that consisted of one book in quarto of birds; two books in quarto of fish; and two books in quarto of Roman antiquities.[35]

Of course, it makes sense that naturalists-authors would have begun to collect naturalia drawings as part of the preparatory research for their planned publications, and thus at least months but more probably years before those were printed. It took time and effort to find decent painters and obtain drawings that met the requirements of scientific naturalism.[36] Even more time was needed to obtain drawings from contacts in other regions or countries. In fact, Belon, Rondelet, Gessner and Salviani are all known to have collected fish drawings at least several years before their respective publications came out in the 1550s.[37] But were all those drawings actually made for print?

A closer look at five of the earliest still traceable collections of *aquatilia* drawings can help us understand what kind of interest in living nature they served or expressed, and clarifies to what extent these drawings were connected with publication projects or served other purposes. All originated in the 1540s. Three of these five collections are no longer extant, but we can trace their origins and sometimes their contours. Two further and still extant ones

35 For a brief reference, see Hochmann M., "Plinio Scarpelli, pittore di Daniele Barbaro e dei Grimani di Santa Maria Formosa", *Arte Veneta* 67 (2010) 43–53, here 44 and note 7. On the Vendramin collection and these albums see Morelli, *Notizia d'opere* 214–222. The huge Vendramin inventory was published by Rava A., "Il 'Camerino delli antigaglie' di Gabriele Vendramin", *Nuovo Archivio Veneto* 39 (1920) 155–181, here 168. Celere was a *miniator* and possibly also goldsmith, born around 1470, and active in Venice in the 1510s; he may have been part of the circle of Giorgione. See Ludwig G., "Archivalische Beiträge zur Geschichte der venezianischen Malerei", *Jahrbuch der Königlich Preussischen Kunstsammlungen* 24 Beiheft (1903) 1–109, here 42–44. On early (fish) collecting in Venice, see also Schmitter M., "'Virtuous Riches': The Bricolage of Cittadini Identities in Early-Sixteenth-Century Venice", *Renaissance Quarterly* 57.3 (2004) 908–969.

36 The term *wissenschaftliche Naturalismus* goes back to Kris E., "Georg Hoefnagel und der wissenschaftliche Naturalismus", in Weixlgärtner A. – Planiscig L. (eds.), *Festschrift für Julius Schlosser zum 60. Geburtstag* (Zurich: 1927) 243–253.

37 For the drawings of Gessner and Salviani see below. Rondelet's original drawings (see also below) seem to have disappeared. Christine Kleiter (Florence) is researching Belon and drawings connected with him.

continued to expand during the 1550s and early 1560s. All are closely linked with Italy and the Mediterranean.

An almost completely forgotten image collection was created by the German physician-botanist Valerius Cordus (1515–1544).[38] Its former existence is well hidden in a few sentences in letters by Conrad Gessner that date from some twenty years after Cordus's death. Gessner published several works by Cordus, who died very young of fevers in September 1544 in Rome in the course of a *peregrinatio medica* undertaken in Italy with a group of young fellow physicians and medical students from north of the Alps. Cordus was by no means only interested in plants. He also investigated minerals and ores during his travels in the Alps and Dolomites in the winter and spring of 1543–1544. In this same period, he also discovered the pleasures of studying aquatic creatures, as the quotation at the start of this essay indicates. Those investigations took him from Padua to Venice – a key town for the investigation (and eating) of fish from the Adriatic. Gessner himself had spent some weeks in Venice almost exactly one year earlier (summer 1543), where he examined Greek manuscripts in the famous library of Diego Hurtado de Mendoza and fish on the Venetian fish market.[39]

Cordus systematically collected information about fish during his stay in the Veneto. A series of very accurate descriptions of 66 different marine fish put together in Venice by Cordus, as well as a group of big and beautiful images of *aquatilia* that Cordus had had made during his stay in Italy in 1544 ended up with Gessner. The fact that Gessner was still thinking of using Cordus's material in 1563 and 1565 suggests that it only reached Gessner *after* the latter's major fish publications in Latin (1558, 1560) had been published.[40] For

38 It is mentioned, though, by Glardon P., *L'histoire naturelle au XVIe siècle. Introduction, étude et édition critique de* La nature et diversité des poissons *de Pierre Belon (1555)* (Geneva: 2011), here 52; and discussed briefly in Horst, "Die einzige".

39 See Nelles P., "Conrad Gessner and the Mobility of the Book: Zurich, Frankfurt, Venice (1543)", in Bellingradt D. – Nelles P. – Salman J. (eds.), *Books in Motion in Early Modern Europe. Beyond Production, Circulation and Consumption* (Cham: 2017) 39–66; and Leu U., *Conrad Gessner (1516–1565). Universalgelehrter und Naturforscher der Renaissance* (Zurich: 2016), here esp. 140.

40 See Dilg P., "Über die Schrift 'De halosantho seu spermate ceti vulgo dicto' des Valerius Cordus (1515–1544). Ein Beitrag zur möglichen Kenntnis des Walrats in der Antike", in Friedrich C. – Telle J. (eds.), *Pharmazie in Geschichte und Gegenwart. Festgabe für Wolf-Dieter Müller-Jahncke zum 65. Geburtstag* (Stuttgart: 2009) 111–132, esp. 118. Cordus's fish collecting (cf. the quotation opening this essay) is evident from Gessner's letters of August 1563 to Crato von Krafftheim; see Huth M. – Walter T., "Der Briefwechsel zwischen Conrad Gessner und Johannes Crato von Krafftheim", in Leu U. – Opitz P. (eds.), *Conrad Gessner (1516–1565). Die Renaissance der Wissenschaften / The Renaissance of Learning* (Berlin: 2019) 377–404. Cf. Gessner Conrad – Wolf Caspar, *Epistolarum medicinalium*

present purposes it is more important, however, to point to the early dating and systematic nature of Cordus's *aquatilia* collecting, to the central role of Venice, and to how images and verbal descriptions already belonged indelibly together as a research apparatus in the 1540s, following the model of botanical collecting and describing.

A second lost collection of fish drawings is connected with the Medici court in Florence. Duke Cosimo I de' Medici invited the Neapolitan physician and philosopher Simone Porzio (1497–1554) to lecture in medicine at the University of Pisa, and Porzio wrote several of his most important works during his years in Tuscany (*c.*1545–1553/4). It may indeed have been Duke Cosimo I himself, a passionate collector of and expert on naturalia, who commissioned Porzio's short treatise on fish (*Tractatus de piscibus*, datable to *c.*1548/49–1553), which remained unfinished and unpublished.[41] Some letters from Porzio to Cosimo I certainly confirm their shared interest in *aquatilia*. One (1549) contains a fairly accurate description of a Mediterranean monk seal. In another (1550) Porzio describes rare fish that have reached him from La Spezia (north of Pisa) and sends Cosimo their drawings, which had been made for him by the Medicean court painter Francesco Bachiacca – who is also known for his (now badly decayed) wall paintings of plants in Cosimo's studiolo in Palazzo Vecchio in Florence. Porzio was, therefore, collecting visual documentation to accompany his fish treatise in manuscript. The whole may have eventually been intended for publication, but there is no doubt that the fish drawings and the dried fish themselves first of all served study and entertainment. In fact, Porzio writes (Sept. 1550) that the Cardinal of Santa Croce had asked him for rare fish because they helped to dispel his melancholic moods.[42]

Conradi Gesneri [...] *libri III* (Zurich, Froschauer: 1577) fols. 12v–13, cf. fols. 6–7; and Gessner's dedicatory letter in: Cordus Valerius – Gessner Conrad, *Valerii Cordi Simesusii De Halosantho* (Zurich, Froschauer: 1566 = 1565) a4. The images that Cordus collected probably followed the same route after his death as his manuscripts. Blair mentions that those were handed down by the uncle of the deceased to Ioannes Placotomus and on to Gessner, who published (parts of) them in 1561 in Strasbourg; see Blair A., "Humanism and printing in the work of Conrad Gessner", *Renaissance Quarterly* 70 (2017) 1–43, here 34. Cf. Horst, "Die einzige" 12. In so far as known, Gessner never published any of the Cordus *aquatilia* drawings; thus far none have been identified.

41 Porzio and his manuscript treatise have been discussed by Soldato E. Del, "Un catalogo naufragato: il *De piscibus* di Simone Porzio", in Bellis E. de (ed.), *Aristotele e la tradizione aristotelica. Nuove tematiche per il turismo culturale* (Soveria Mannelli: 2008) 149–176; Idem, *Simone Porzio. Un aristotelico tra natura e grazia* (Rome: 2010), esp. 92–99; Castelli, "Tra ricerca empirica"; and Minonzio, "Diffrazioni pliniane", on whose work this paragraph is based.

42 Castelli, "Tra ricerca empirica" 116–119, with quotations from these letters. See for Bachiacca and Cosimo's studiolo esp. Vossilla F., "Cosimo I, lo scrittoio del Bachiacca,

That remark is not merely evidence of the pleasure that naturalia offered. This cardinal was none other than the erudite humanist Marcello Cervini (1501–1555), collector of rare editions and Greek manuscripts, key member of the Roman *Accademia Vitruviana*, and Cardinal Supervisor of the Vatican Library. He became Pope (Marcellus II) in 1555, but died within a few weeks after his election. Cardinal Cervini was personally connected with Hurtado de Mendoza, in whose Venetian library Gessner had studied. Cervini was also a key figure in Salviani's project to publish a great work on Roman fish. Salviani's model drawings are the third example of an early but lost collection of *aquatilia* images of which we can trace some contours. As Salviani wrote, Cardinal Cervini 'mi consigliaste che io facessi dipingere e incidere in rame le immagini di tutti i pesci che mi fosser venuti alle mani' (advised me that I should have images made and cut in copper plate of all fish that I could lay hands on). Cervini also provided Salviani with the necessary funding to cover the great expense of these engravings. Salviani acknowledges Cervini's generous support more than once in his fish work, and in the first dedication of that same work to Cervini – then still a cardinal – which was removed when Cervini died before the printing was completed.[43]

Cervini's role went even further than this crucial sponsorship. In 1549–50 and perhaps earlier, he actively helped Salviani obtain drawings and information from both Italy and abroad concerning the *aquatilia* that the latter wanted to include in his book. As Salviani writes: Cervini,

> col vostro eloquente parlare e col vostro esempio eccitando altri tra' cardinali a far lo stesso, come anche facendo che a spese vostre molte sorte di pesci a noi sconosciute, e senza le quali imperfetta sarebbe stata questa mia storia, venissero esattamente dipinte dalla Francia, dall'Allemagna, dal Portogallo, dalla Brettagna e per fin dalla Grecia.

una carcassa di capodoglio e la filosofia naturale", *Mitteilungen des Kunsthistorischen Institutes in Florenz* 37.2/3 (1993) 381–395. On Bachiacca and the fish drawings, see Garbari F. – Tongiorgi Tomasi L. – Tosi A., *Giardino dei Semplici / Garden of Simples* (Pisa: 2002), 154–156; Else F.M., "Globefish, Sturgeon and Trout: Duke Cosimo I de' Medici, Bachiacca and the Consuming Culture of Fish", *Medicea* 9 (2011) 20–29; and Minonzio, "Diffrazioni pliniane" 437.

43 Quoted in Tiraboschi Girolamo, *Storia della Letteratura Italiana*, Vol. VII, part 2 (Florence, Molini: 1810; 1st edition 1772), here 292–293, who also quotes the first dedication (to Cervini). The later dedication was to the new Pope Paul IV, elected in 1555. On Salviani's sponsoring, see Pinon L., "Clématite bleue contre poissons séchés: Sept lettres inédites d'Ippolito Salviani à Ulisse Aldrovandi", *Mélanges de l'École Française de Rome* 114 (2002) 477–492. Blair, "Humanism and printing" 28, shows that Gessner was slightly envious of Salviani's sponsoring by cardinals.

> by your eloquent conversation and example stimulating other cardinals to do the same, as well as taking care that at your expense many kinds of fish that were unknown to us and without which my history would have been imperfect, would be accurately depicted from France, Germany, Portugal, Bretagne and even from Greece.[44]

In precisely the same years when this melancholic cardinal received Porzio's dried fish, Cervini thus used church diplomatic and personal contacts in order to obtain coloured drawings of fish from abroad that were inaccessible to Salviani: the apostolic nuncio in Lisbon, for instance, sent Cervini twelve drawings of Atlantic fish for Salviani. At least a part of the model drawings for Salviani's work, therefore, originated outside Italy and was made in different countries and by different painters.[45]

Salviani himself throws some further light on the relations between patrons-collectors and naturalists, and on the importance of the naturalia drawings themselves. Four interesting points emerge. First of all, Salviani (a physician and anatomist by profession) declares that he loved the study of aquatic creatures even better than that of birds and land animals. Second, he describes how his collecting of fish images triggered his deeper interest in their names, identifications, and characteristics – a remarkable reversal of what we usually hear about the humanist interest in nature, which was supposed to arise mainly from (classical) texts and then move on to the visual domain and direct observation:

> Ita animus ardentior factus, neque sola nudarum picturarum voluptate contentus, priscis etiam nominibus piscium indagandis, atque reliquae eorum cognitioni sese applicuit.

> I became more eager and not content with just the pleasure of the paintings alone, I applied myself to the study of the ancient names of the fishes and other knowledge about them.[46]

44 Quoted in Tiraboschi, *Storia* 292–293.

45 See on Cervini as sponsor and his involvement with Portugal esp. Andretta E., *Roma Medica. Anatomie d'un système médical au XVIe siècle* (Rome: 2011), here 430–436; and also Andretta E. – Pardo Tomás J., "Books, plants, herbaria: Diego Hurtado de Mendoza and his circle in Italy (1539–1554)", *History of Science* 58 (2020) 3–27. Nissen, *Die zoologische Buchillustration* 117, has pointed out that the model drawing for Salviani's Mola mola illustration had come from Luca Ghini in Pisa, and that of a burbot from Andreas Masius in Brussels. See Salviani Ippolito, *Aquatilium animalium Historiae liber primus. Cum eorundem formis, aere excusis* (Rome, Ippolito Salviani: 1554–1558) 155–155v.

46 Salviani, *Aquatilium* fol. 231v.

While discussing his controversy with Rondelet which centred on accusations of copying model drawings, Salviani makes further interesting remarks. He states that he visited his patron Cardinal Cervini together with Rondelet in 1549–50, some four to seven years, therefore, before the printing of Rondelet's big works on fish (1554, 1555, and 1558) *and* of Salviani's own work of 1554–58. During this visit Rondelet had shown the cardinal *his* book of fish drawings, which were only in black and white and definitely crude according to Salviani. The latter argues that he would never have copied or taken images from Rondelet, since anyone could see how inferior the quality of Rondelet's woodcuts was to his own illustrations:

> Quae quidem praeter id quod non nativis coloribus, sed solo atramento erant pictae; rudes etiam et ineptae erant adeo, ut nullo prorsus pacto eorum piscium, quorum erant icones, faciem repraesentarent.
>
> Which apart from not being painted in the natural colours but only in black, were also so crude and absurd that they in no way conveyed the appearance of the fish of which they were images.[47]

The point here is not the controversy itself, of course, but the fact that such original drawings – the eventual models for print – were clearly held in great value, plus the fact that Rondelet carried his model drawings with him on his travels. Perhaps they also served as a kind of visual catalogue with which to compare new fish that would still need depicting.

Finally, shortly after this visit to Cervini in 1549–50, Salviani hired Bernardo Aretino (from Arezzo) to paint fish for him: Bernardo lived for two years in Salviani's house, was paid by the latter on a monthly basis, and left under a shadow.[48] Salviani emphasizes that Bernardo's drawings, of which some twenty were seen by Rondelet before the latter left Rome, were of top quality: 'Nostras ad vivum egregie depictas, et quam verissime ipsorum piscium

47 Salviani, *Aquatilium* fol. 231–231v, also for the quotation. Cervini, Rondelet and the latter's patron, Cardinal de Tournon, were in Rome during the conclave of 1549–50. On Rondelet and the limited information concerning his model drawings, painter and illustration format, see Kolb, *Graveurs artistes et hommes de science* 96; and Lewis G., "The debt of John Ray and Martin Lister to Guillaume Rondelet of Montpellier", *Notes and Records of the Royal Society* 66 (2012) 323–339. See on Salviani and Rondelet especially the contribution by Holger Funk in the present volume.

48 On Bernardo, see Salviani, *Aquatilium* fol. 231v; and Tosi A., "Acconciare, seccare, dipingere: pratiche di rappresentazione della natura tra le "spigolature" aldrovandiane", in Olmi G. – Simoni F. (eds.), *Ulisse Aldrovandi. Libri e immagini di Storia naturale nella prima Età moderna* (Bologna: 2018) 49–58, here 56.

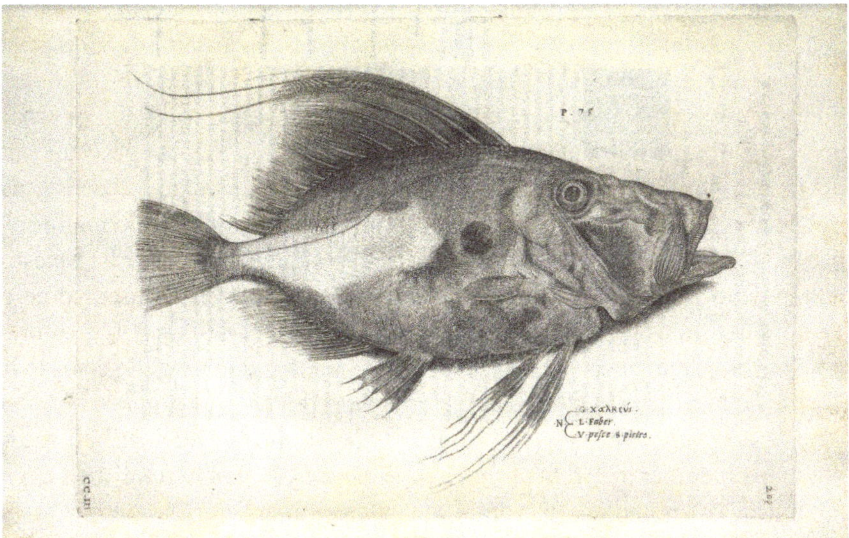

FIGURE 6.3 St Peter's fish. Copper engraving in Salviani Ippolito, *Aquatilium animalium Historiae liber primus. Cum eorundem formis, aere excusis* (Rome, Ippolito Salviani: 1554–1558)
PUBLIC DOMAIN: HTTPS://WWW.BIODIVERSITYLIBRARY.ORG/BIBLIOGRAPHY/82337

similitudinem repraesentantes' (Ours are splendidly painted ad vivum and give the likeness of the fish themselves very truthfully).[49] Behind the consistently beautiful and accurate printed copper engravings in Salviani's work lies a long and intricate story, therefore, of model drawings, painters, engravers, sponsors and collectors [Fig. 6.3]. The fish drawings made for Salviani in Rome – mostly by Bernardo Aretino – were definitely not his only visual stock, since he received further drawings from various European countries and Lisbon via Cardinal Cervini and from (or via) several other Italian collectors and naturalists. These drawings cannot possibly have been uniform in style and quality. It seems likely, therefore, that Salviani's engraver must have contributed much to the fairly consistent style throughout most of the illustrations that were eventually printed.[50]

49 Salviani, *Aquatilium* fol. 231v.
50 It seems unlikely to me that Salviani's engraver and painter were one and the same person, as Holger Funk proposes (personal communication). A small handful of fish drawings in Cassiano dal Pozzo's seventeenth-century collection are copied after either Salviani's printed illustrations or (given their reversal) after his original drawings; see McBurney H. – Rolfe I. – Napoleone C. – Findlen P. et al., *The Paper Museum of Cassiano*

Some drawings in Salviani's image collection had been gifts from the Venetian patrician, church official and humanist Daniele Barbaro (1514–1570), poet, expert on architecture, optics and mathematics as well as one of the key founders of the botanical garden in Padua. Salviani acknowledged his visual debt to Barbaro's image collection when he wrote in 1558: 'Dell' pesci depenti da messer Plinio, che stà con Monsignore Daniello Barbaro, io ne so informatissimo, et me ne so anco serviti di alcuni, perché Monsignore è molto mio patrone' (I am extremely well informed about the fish painted by Maestro Plinio, who is with Monsignore Daniele Barbaro, and I have used some of them, since Monsignore is very much my patron).[51] Barbaro was the creator of the fourth systematic collection of *aquatilia* images pre-dating the 1550s to be discussed here. He was nominated Patriarch of the town of Aquileia between Trieste and Venice in 1550, but is better known as humanist-author, young founding member in Padua of the *Accademia degli Infiammati* (created 1540), diplomat, and friend and patron of the painter Paolo Veronese and the architect Andrea Palladio.[52] His collection of fish drawings was known as the *Libro dei Pesci del Patriarcha.*

More is known about Barbaro's images than about any other of these early lost collections, thanks to the fact that he let naturalists such as Aldrovandi, Belon and Salviani borrow his drawings and copy them.[53] The collection probably took the shape of an expanding set of loose sheets with drawings. It may have been bound at some later stage. Thanks to the testimony of Belon, who met Barbaro while the latter was Venetian ambassador at the English court in 1549–51, we know that the collection was already sizeable at this time; that Barbaro employed a painter, Maestro Plinio, for eight years at his own expense;

---

dal Pozzo, Series B (parts 4 and 5). Natural History: Birds, Other Animals and Natural Curiosities, 2 vols. (Turnhout – London: 2017), here vol. 2, 526–588.

51 Letter from Salviani to Aldrovandi, 15 October 1558, published in Pinon, "Clématite bleue" 490–491; cf. Fantuzzi Giovanni, *Memorie della vita di Ulisse Aldrovandi* (Bologna, Lelio dalla Volpe: 1774) 218–219.

52 See Azzi Visentini M., *L'Orto Botanico di Padova e i giardini del Rinascimento* (Milan: 1984) 159–165. The principal works on Barbaro contain little information about his interest in naturalia and none about the fish images. See Laven P.J., *Daniele Barbaro. Patriarch-Elect of Aquileia with special reference to his circle of scholars and to his literary achievement* (Ph.D. dissertation, University of London: 1957); Marcon S. – Moretti L. (eds.), *Daniele Barbaro 1514–70. Letteratura, scienza e arti nella Venezia del Rinascimento* [exh. cat. Venice] (Venice: 2015); and Lemerle F. – Zara V. – Caye P. – Moretti L. (eds.), *Daniele Barbaro 1514–1570: Vénitien, patricien, humaniste* (Turnhout: 2017).

53 For a further discussion of Barbaro's fish book, see Egmond F., "Visual immersion: Daniele Barbaro's fish album and the wave of interest in aquatic creatures in mid sixteenth-century Europe", *Notes and Records of the Royal Society* (published online May 2022, doi:10.1098/rsnr.2021.0084).

and that this painter accompanied him also during his stay in England.[54] Belon reports that Barbaro had instructed Maestro Plinio to use most of his time during his eight-year employment to depict all sorts of fish, not only from the Adriatic, but also from the Mediterranean, the Black Sea, and the rivers and lakes of Italy. He had done his work so well, that Barbaro possessed (in 1550):

> le portraict contrefaict au naturel des vives images non seulement de ceuls qui ont estés apportez au marché ou es poissoneries de Venise, mais aussi des autres qui luy ont estés singulierement envoiez des ports et plages d'Esclavonie; lesquelles peinctures sont beaucoup plus de trois cents de compte faicts.
>
> the lifelike portrait in accurate images not only of those brought to the market or fishmongers of Venice, but also others that have been sent to him one by one from the ports and the beaches of Slavonia. There are many more than three hundred of these paintings all added up.[55]

Barbaro clearly also carried this visual material with him during his longer stay abroad in England. This suggests that he needed it to check whether a fish observed there should be newly portrayed or was already present in his collection. In England, Maestro Plinio did not paint only fish. He may actually have worked as a visual documentalist for Barbaro: 'luy [i.e. Barbaro] qui est prudent et diligent inquisiteur des haults faits de l'Eternel, ne voulant rien laisser en arriere, avoit un peintre avec luy, pour luy representer ce qu'il trouvoit digne' (he who is a judicious and diligent investigator of the wonders of Creation, wished to leave nothing undone and had a painter with him to depict for him what he found worthwhile).[56]

None of the original 300 or more images in Barbaro's fish book seem to have survived.[57] As already explored by the art historian Michel Hochmann, we can

54 Barbaro was in England between circa July 1549 and early spring 1551. For further details on the Belon-Barbaro meeting, see Glardon, *L'histoire naturelle* 49–50; of an earlier meeting (1545) in Venice that Glardon mentions (189) I have not been able to find further information.

55 Belon Pierre, *L'Histoire naturelle des estranges poissons marins, avec la vraie peincture et description du daulphin, et de plusieurs autres de son espece* (Paris, Regnaud Chaudière: 1551) 6v–7r.

56 Belon Pierre, *L'Histoire de la nature des oyseaux, avec leurs descriptions et naïfs portraicts retirez du naturel, escrite en sept livres* (Paris, Guillaume Cavellat: 1555) 239.

57 Tosi, "Acconciare" 56, refers to an article published in 1868 by the Venetian fish expert Nardo in which the latter says that he was in the possession of a possibly sixteenth-century fish album with some 150 painted fish without text which he had bought in 1856 from the

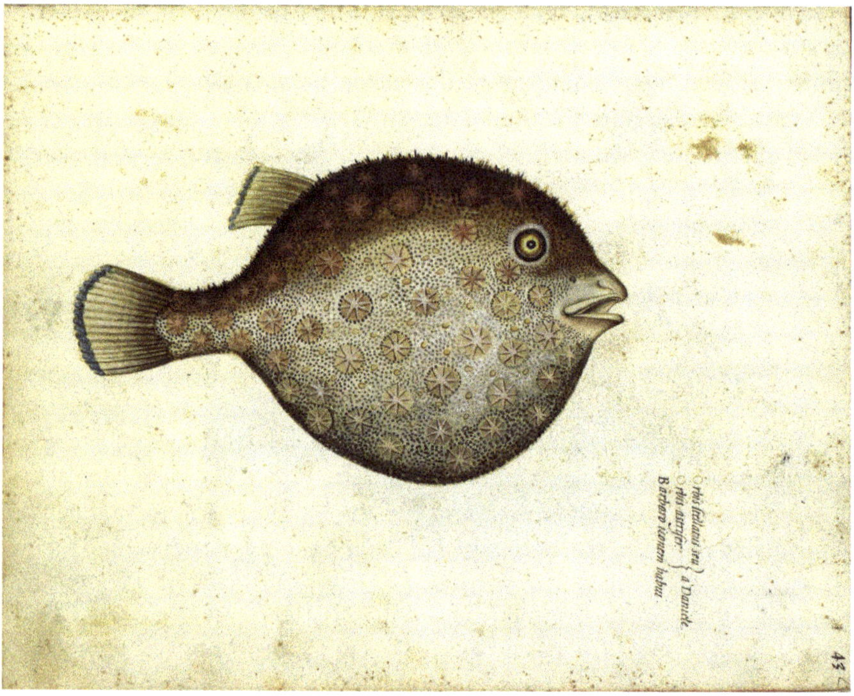

FIGURE 6.4 Orbis stellatus in the Aldrovandi collection, second half 16th century, copied after Daniele Barbaro's fish book. Aldrovandi, *Tavole*, vol. 4, fol. 43

only surmise what they looked like by inspecting a limited number of illustrations in the printed works of Belon and Salviani that were indirectly based on Maestro Plinio's drawings.[58] The Aldrovandi collection also contains some coloured (copied) drawings that are explicitly based on the Barbaro collection [Fig. 6.4]. Aldrovandi borrowed that collection in the course of the 1550s and again in the 1560s, and referred to it as *Ex Patriarcha De piscibus*. He not only

Padovan bookshop Zambeccari. Nardo wondered whether it could be Barbaro's fishbook; see Nardo D., "Brevi cenni storici sui progressi dell'Adriatica Fauna da Oppiano fino a' di' nostri", *Commentario della fauna, flora e gea del Veneto e del Trentino* 3.1 (1868) 125–126. As Tosi relates, no further information is available about this album after Nardo left it to his daughter.

58 Barbaro permitted his painter Plinio to copy a selection of his fish images for Belon while both men were in England in 1550. See Barsi M., *L'énigme de la chronique de Pierre Belon. Avec édition critique du manuscrit Arsenal 4651* (Milan: 2001) 134; Belon Pierre, *La nature et diversité des poissons avec leurs pourtraicts, représentez au plus près du naturel* (Paris, Charles Estienne: 1555) 131, 245; and Belon, *L'Histoire de la nature des oyseaux* 239.

copied images but also a list of some 100 fish names (partly in Italian, partly in Latin), which gives us some idea of the range of fish documented by Barbaro and his painter. That list includes numerous common Mediterranean fish, from mullus and donzella to sparus and fragolin, but also several fish described as 'di Fiandra' (from Flanders), the orca, the whale (*balena*), trout from Lake Garda, and a 'rocco d'Inghilterra', which Barbaro perhaps added to his album while in England, just like the fish 'sandilz' (a rendering of the English *sandeels*). Aldrovandi copied the 'sandilz' image from an image that Salviani had in his turn received from Barbaro.[59]

Finally, the Barbaro collection is one of the few examples in which at least something is known about the principal painter. Hochmann has identified Maestro Plinio as the painter Magister Plinius – son of Francesco Scarpelli from Tolentino in the Marche – who acted as a witness to the signing of certain notarial deeds (1546–48) by Barbaro's friends and kinsmen from the powerful Venetian Grimani family, and to a deed (1566) that involved the painter Camillo Mantovano. Earlier notarial records (1540) from Mantua further show that this Francesco Scarpelli, Plinio's probable father, had acted as a witness to a payment for a house in Mantua by the painter Giulio Romano. Barbaro had thus chosen to hire a professional but non-famous painter possibly connected with the artists active in both Mantua (the Gonzaga court) and Venice.[60]

---

59 On the sandils, see Salviani, *Aquatilium* 70v. Aldrovandi's undated list entitled 'Ex Patriarcha De Piscibus [...] secundum numerum [...]' must be from after 1550 (when Barbaro was appointed Patriarch); see MS Aldrovandi 136, t. V, University Library Bologna. See on this image copying Barbaro-Salviani-Aldrovandi the excellent Hochmann, "Plinio Scarpelli" 43–53, with a complete transcription of Barbaro's fish list as copied by Aldrovandi. And see Sallent del Colombo E., "Natural History Illustration between Bologna and Valencia: The Aldrovandi-Pomar Case", *Early Science and Medicine* 21 (2016) 182–213, esp. 193; and Sallent Del Colombo E. – Pardo Tomás J., "Materiali aldrovandiani in Spagna: l'enigmatico caso del *Códice Pomar*", in Olmi – Simoni (eds.), *Ulisse Aldrovandi* 37–48. I thank Emma Sallent for giving me photos and her transcription of this original document, and Giuseppe Olmi for photos of further Aldrovandi manuscript material connected with fishes. See also the correspondence between Aldrovandi and Salviani in 1557–58, as discussed in Pinon, "Clématite bleue" 482, 485, 489–490, in which Salviani says he has used *some* (alcuni) as models for his own illustrations.

60 Hochmann, "Plinio Scarpelli" 43–44. See the notarial records of Mantua online (Febr. 2022): http://banchedatigonzaga.centropalazzote.it/giulioromano/index.php?page=Pdf&scheda=1267, which is ASMN, Archivio Notarile, Notary Giovanni Giacomo Calzoni (4 November 1540). Francesco Scarpelli is mentioned there as son of Giovanni (de) Scarpelli from the quarter Cornu, which seems to imply that Giovanni was or already had been living in Mantua. See more generally on Venetian collecting and *naturalia*, Tartaglini, Barbaro and Aldrovandi: Hochmann M. – Lauber R. – Mason S. (eds.), *Il collezionismo d'arte a Venezia. Dalle origini al Cinquecento* (Venice: 2008), here esp. pp. 3–39.

Given its content, size, geographical range, and the role of the painter, Barbaro's *Libro dei Pesci del Patriarcha* was not merely an interesting group of *aquatilia* drawings, but an important research collection that aimed to provide a visual survey of fresh water and marine creatures of Italy, and the marine ones of the Mediterranean and Black Sea. Barbaro was purposely creating a visual corpus of aquatic life, and as eminent classicist must have been perfectly aware of the lack of such material in the classical sources. Barbaro certainly did *not* intend to publish these aquatic drawings himself. As befitted an erudite patrician and patron of arts and sciences, he lent his visual material to naturalists for use in *their* publications. Since much of this image collection was in existence by 1550, Maestro Plinio must have begun painting in the 1540s, precisely in the period when Barbaro was closely involved in the creation of the hortus in Padua. One of the purposes of that garden was to bring together a wide range of plant species, and it seems not too farfetched to interpret Barbaro's aquatic and botanical projects as linked by similar aims and approaches.[61]

Scanty though the information about nearly all of these vanished early collections is, the web of image exchanges thickens, and we have begun to discern a whole world of erudite, humanistically trained, expert naturalia and antiquity collectors in Italy during the late 1530s, 1540s and 1550s who commissioned and exchanged naturalia images. They were members of learned academies, often occupied high functions in the church hierarchy, and were actively involved in the patronage, promotion and sponsoring of publication projects. Cardinal Cervini in particular appears to have been a key figure. We have already seen how he was linked with Porzio and Salviani. There was a personal connection between Cervini and the 13 years younger Barbaro as well. In 1551, for instance, Barbaro sent fish drawings from Venice to both Cervini and Salviani.[62] Cervini maintained good connections with humanist circles of the Veneto, had himself studied in Padua, and was a friend of one of Barbaro's teachers.[63] Cardinal Cervini thus appears to form a link – as sponsor, background figure, and source

61 Hochmann, "Plinio Scarpelli" 44, dates the making of Barbaro's fish book to *c.*1542–50. On the creation of the Padua hortus and Barbaro's role in it, see Azzi Visentini, *L'Orto Botanico*; and Laven, *Daniele Barbaro*. Barbaro (born 1514) and Cordus (born 1515) would probably have met while the latter spent much time in both Padua and Venice in early 1544.

62 See Olmi G., *L'Inventario del mondo. Catalogazione della natura e luoghi del sapere nella prima età moderna* (Bologna: 1992) 229; Andretta, *Roma Medica* 431–437; Andretta – Pardo, "Books, plants, herbaria". On the Barbaro family and collecting, see Howard D., "I Barbaro come collezionisti rinascimentali", in Hochmann – Lauber – Mason (eds.), *Il collezionismo* 193–205.

63 Laven, *Daniele Barbaro* 22–23.

of inspiration and information – between the creators of three of the early Italian collections of fish drawings: Porzio, Salviani, and Barbaro.

Although collectors could be rivals, there was considerable collaboration too, and joint efforts certainly furthered the creation of a new visual corpus of *aquatilia* drawings that served entertainment, documentation, nature research, fashion, self-presentation, and in some cases eventually publication. The French naturalists Belon and Rondelet were engaged in the same pursuit in the same period. We know very little so far about what may have been the fate of their collections of drawings, but it is clear that both had Cardinal-patrons and that Belon and Rondelet as well as their patrons had close links with Italy and with Italian naturalists.

3.2 *Gessner's and Kentmann's Collections and Their Italian Connections*

Two interconnected and still partly extant collections of *aquatilia* drawings that likewise originated in the 1540s and early 1550s are those of Conrad Gessner in Zurich and of his friend, the German physician-naturalist Johannes Kentmann (1518–1574), who spent most of his life in Meissen and Torgau to the east of Leipzig. Both collections are large.

Kentmann's collection comprises some 120 folios with coloured drawings of marine and freshwater *aquatilia*; many sheets show more than one fish. He collected *aquatilia* drawings from at least 1549 to far into the 1560s [Fig. 6.5].[64] Kentmann sent many of his drawings to Gessner, who copied and used a number of them for the woodcuts in his *Historia Animalium* IV (1558). But Kentmann also borrowed from Gessner, while both copied drawings from a third party as well. The Kentmann and Gessner drawing collections therefore contain a considerable number of not-quite-identical copies or replicas. Kentmann personally drew many plants during his travels in Italy, and he may also have painted (or copied) various fish images both in Germany and in Italy. Undoubtedly the drawings and descriptions of a number of fish from the river Elbe in his album are based on his own experience.[65]

---

64 See List of Codices. The fish drawings are bound together with Kentmann's plant drawings and with his son Theophilus' later nature prints. The year 1549 on the section with the *aquatilia* drawings (entitled *Animalium Aquatilium in mari et dulcibus aquis degentium, Icones ad vivum expressae, 1549*) probably indicates not the start of collecting, but the year when he began to put the drawings together.

65 The Codex Kentmann also includes a multipage manuscript map that outlines the riverbed of the Elbe from start to finish, with short descriptions in German of the various adjacent towns. On Kentmann, see Kusukawa S., "Image, Text and Observation: The Codex Kentmanus", *Early Science & Medicine* 14.4 (2009) 445–475. On Kentmann and fish in Germany, see Zaunick, "Fragmente"; and Hertel R., "Über die 'Ichtyographie der Elbe' des Johannes Kentmann. Eine Studie über die ältesten sächsichen Fischfaunen

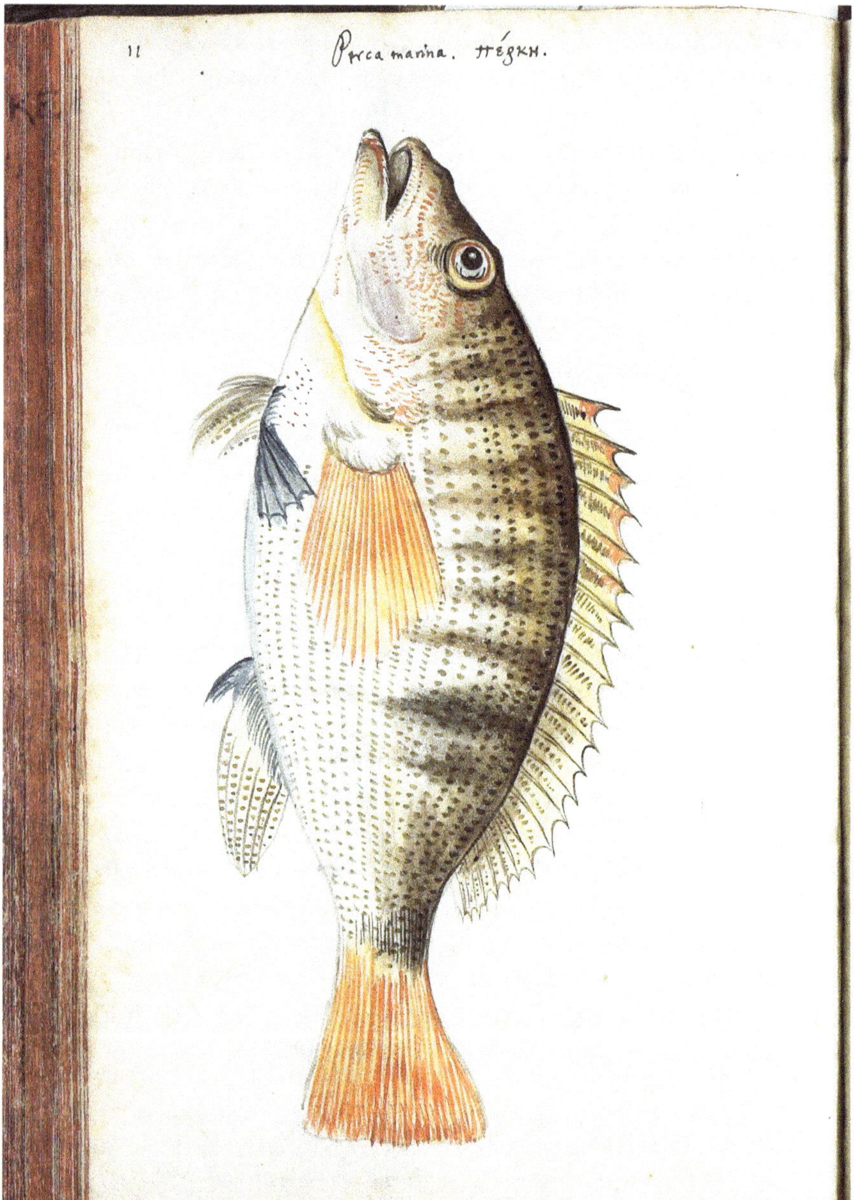

FIGURE 6.5 Perca marina. Drawing in the Kentmann Codex, middle to late 16th century, Meissen, Thorgau, Fol 323, fol. 11

Gessner's collection of *aquatilia* drawings was even larger. The model drawings of 159 of Gessner's aquatic illustrations (not counting shells) have been identified in the Gessner-Platter albums, rediscovered in 2012 in Amsterdam, and he must have owned at least dozens more.[66] Gessner was already busy obtaining fish drawings during his stay in Venice in 1543, where he incidentally also met Pierre Gilles. He continued collecting until his death in 1565.[67] It may be surprising that two men whose home regions could hardly have been further removed from the seas and who were no great travellers, chose to study fish at all. It must have been precisely their lack of personal access to the marine world, however, that made these drawings as 'placeholders' for the actual naturalia crucially important as research documentation. *Aquatilia* are – besides plants – by far the largest group in both their extant collections, which once more underpins the intriguing parallels between botany and fish studies.

Thanks especially to Sachiko Kusukawa's research much is known about *how* Gessner collected animal drawings as part of the investigations for his *Historia Animalium*, and about the crucial role of his wide European network.[68] Much of his visual fish material arrived from and via friends. In 1547, for instance, Gessner's friend the theologian Jean Ribit sent him a letter from Lausanne with a drawing of a troutlike fish (locally known as *amble* or *omble*) from Lake Geneva. It was drawn in Ribit's presence by first laying the fish on paper to draw its outline and then painting it as exactly as possible so that Ribit was satisfied.[69] In 1556 and 1557 Gessner sent specific requests to experts in his

(Pisces)", *Zoologische Abhandlungen Staatliches Museum für Tierkunde in Dresden* 35.5 (1978) 75–100. On the Kentmann Gessner exchanges of fish images, see Egmond F. – Kusukawa S., "Circulation of images and graphic practices in Renaissance natural history: the example of Conrad Gessner", *Gesnerus* 73 (2016) 29–72; and Egmond F. – Kusukawa S., "Gessner's fish: images as objects", in Leu – Opitz (eds.), *Conrad Gessner* 581–605. On the links between Gessner, Rondelet, Belon and Salviani, and their publications on fish, see esp. Glardon, *L'histoire naturelle*.

66 See Egmond F., "A collection within a collection. Rediscovered animal drawings from the collections of Conrad Gessner and Felix Platter", *Journal of the History of Collections* 25.2 (2013) 149–170; Egmond F. (ed.), *Conrad Gessners Thierbuch. Die Originalzeichnungen* (Darmstadt: 2018).

67 In 1563–65 Gessner still intended to publish Cordus' fish. Cf. Blair, "Humanism and printing" 14–16, with several examples of his fish collecting. For Gessner and Venice, see Egmond – Kusukawa, "Gessner's fish".

68 See esp. Kusukawa S., "The sources of Gessner's pictures for the *Historia animalium*", *Annals of Science* 67.3 (2010) 303–328; and Leu, *Conrad Gessner*. On Salviani's problems in obtaining fish species not availabe in Central Italy, see Pinon, "Clématite Bleue" 486.

69 Jean Ribit (Ribittus), manuscript letter to Conrad Gessner, Latin, 1 Oct. 1547, Paris, Bibliothèque Nationale de France, Ms 8641 [accessible via www.aerztebriefe.de/id/00012298].

network for drawings of various species of Danube fish, such as *silurus* and *jentling*.[70] And in 1556 he dedicated part of his *De piscibus* to a man he knew only by reputation – Burchard Mythobius in the far north of Germany – precisely in the hope that the latter would send him drawings of fish from the Baltic.[71]

Many of Gessner's early images must have first served as a visual research collection, as we can infer from his writings; only years later some of these became the models for the woodcuts in his *Historia Animalium* IV of 1558. The detailed visual comparisons between images to which Gessner refers frequently may have helped him to obtain a better idea of the usual shape and appearance of a species – a fishy equivalent of the ideal or generic image of a plant – and to distinguish local varieties and different species [Fig. 6.6].[72] Such comparisons certainly led Gessner to reject some drawings in favour of others in terms of accuracy rather than purely painterly qualities. Unlike Gessner, Kentmann seems to have never considered publishing about either plants or *aquatilia*. His drawings appear to have served purely as study material. But the drawings were also a source of pleasure for both men, and formed the substance of gifts that cemented their friendship and helped to establish new bonds with other naturalists and *curiosi*.

For the landlocked Gessner and Kentmann their links with Italy – which were forged during their formative years as physicians – may well have been crucial in stimulating an interest in the vast variety of *aquatilia*. Gessner personally acquired his first large cluster of *aquatilia* drawings in Venice in 1543, while several further important groups of fish drawings came into his possession via Italy. A large but relatively badly documented group consists of some forty-five drawings of *aquatilia* made in Venice that survive in the Gessner-Platter album in Amsterdam and were used by Gessner as *Vorlagen*

70 See Gessner's letters to Caspar von Nidbruck on fish of the Danube, naming the species, *'Schaid vel Schaidle, Schilln, Zinne vel Zindel, Jentling'*: two manuscript letters, 2 Febr. 1556 and 12 Jan. 1557, Latin, both Cod. 9737 k in Österreichische Nationalbibliothek, Vienna [accessible via www.aerztebriefe.de/id/00003401 and www.aerztebriefe.de/id/00003402]. On the English naturalist William Turner's letter (1557) about fish to Gessner, see Wheeler A. – Davis P.S. – Lazenby E., "William Turner's (c 1508–1568) notes on fishes in his letter to Conrad Gessner", *Archives of natural history* 13 (1986) 291–305.

71 As argued by Blair, "Humanism and printing" 21. In a letter to the theologian-historian David Kochhafe (Chytraeus) in Rostock, Gessner asked specifically for descriptions, drawings and names of fish in the Baltic. See Gessner to Kochhafe (Chytraeus), Latin, date uncertain (1563?), Staats- und Universitätsbibliothek Bremen, Msa 0009, nr. 9 [accessible via www.aerztebriefe.de/id/00034444].

72 On generic versus specific representations of plants and animals, see Kusukawa S., *Picturing the Book of Nature. Image, Text and Argument in Sixteenth-Century Human Anatomy and Medical Botany* (Chicago – London: 2012); and Egmond F., *Eye for Detail: Images of Plants and Animals in Art and Science, 1500–1630* (London: 2017).

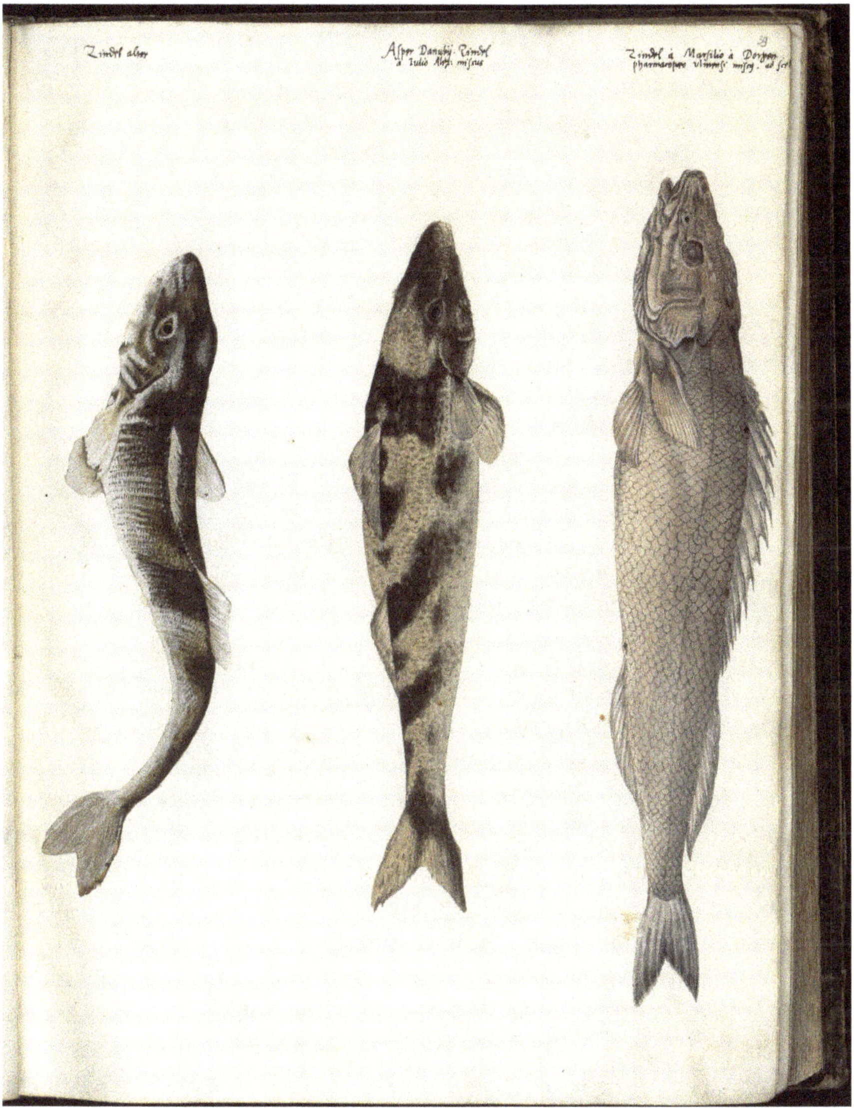

FIGURE 6.6 Three river fish. Coloured drawings in the Gessner-Platter Animal Albums, collected by Gessner in Zurich, *c.*1540s–65, Album III C 22, fol. 39

for his printed illustrations. Nearly all of these drawings are undated, and no names of painters or donors have so far been securely identified in either Gessner's printed texts or the annotation of the drawings.[73] Perhaps this is the material that Gessner acquired himself in Venice.

73 For a detailed discussion, see Egmond – Kusukawa, "Gessner's fish".

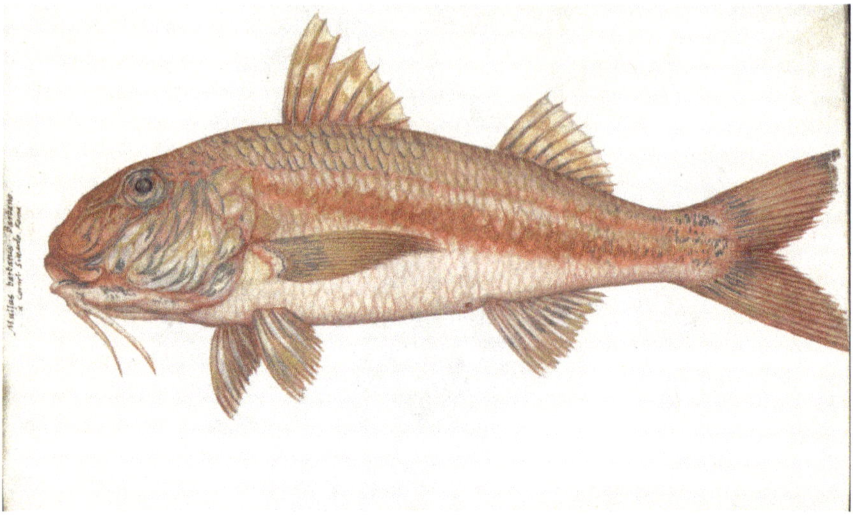

FIGURE 6.7 Mullus barbatus. Coloured drawing sent by Horstius in Rome to Gessner in Zurich, *c.*1550s. In the Gessner-Platter Animal Albums, Album III C 22, fol. 88 © AMSTERDAM, UNIVERSITY LIBRARY

A much better documented cluster of 37 *aquatilia* drawings in Gessner's collection originated with the Dutch-Roman physician and naturalist Gysbert Horst (also Gisbertus Horstius, from Amsterdam, *c.*1491–1555/1556). Given their diverse styles they were probably copies of drawings in Horstius's personal collection, or drawings freshly commissioned by Horstius from some painterly workshop in Rome [Fig. 6.7]. Horstius was not merely a provider of images to Gessner, but a Roman hub in an international exchange network of naturalists interested in *aquatilia,* plant medicine and minerals. It stretched across the Alps into Germany and the Low Countries. He had an international reputation as expert on fish, serpents, poisons, and exotic plants and was very well connected in the Roman world too. Horstius practised during the 1540s–50s as senior physician at the Hospital of Santa Maria della Consolazione next to the Forum Romanum and was in touch with the top physicians of Rome and the Vatican of the 1540s–50s. Several North-European physicians on their Italian *peregrinatio* sought him out for his medical and naturalistic expertise since at least the very early 1540s. And Horstius personally exchanged fish information with Belon and Rondelet, and studied fish on the Roman fish market together with Salviani at the time when the latter was preparing his big publication sponsored by Cardinal Cervini.[74] It is hard to believe that Horstius would not

74 Horst is discussed more fully in Egmond – Kusukawa, "Circulation of images" and "Gessner's fish".

also have met Aldrovandi, Giovio and Gilles during their respective stays in Rome between about 1549 and 1551.[75]

Gessner's friend Kentmann, who was also in personal touch with Belon, discussed plants and probably fish with Horstius, and personally painted some plants and trees in Horstius' Roman garden during his stay in Rome in 1549. Since Horstius transmitted his fish drawings to Gessner in Switzerland via yet another German physician on *peregrinatio*, Cornelis Sittardus (one of Valerius Cordus's companions on the fatal trip of 1544), and tutored the Dutch physician Pieter van Foreest (another companion of Cordus), we can discern the contours of an international cluster of naturalists and fish experts active in or connected with Rome. That cluster linked Central Italy with Southern France, Switzerland, Germany and the Low Countries, connecting Europe south and north of the Alps.[76]

In fact, in this brief period Rome and the Roman fish market were a special site of knowledge with a formative influence on a whole generation of famous fish naturalists. Rondelet was in Rome with his patron Cardinal de Tournon for the conclave of 1550. Salviani and Horstius lived permanently in Rome, and while Rondelet managed to fall out with Salviani over image copying, he made friends with the young Aldrovandi, who was in Rome to be investigated by the Inquisition on the suspicion of heresy. Rondelet and Aldrovandi regularly visited the Roman fish market together in order to 'osservare i pesci più rari, che capitavano, e studiarne le più rimarchevoli strutture e proprietà' (observe the rarest fish that could be found there, and to study their most notable physical characteristics and properties).[77] Rondelet also often discussed fish with Giovio.

The stay in Rome was certainly crucial to Aldrovandi's scientific development. His eighteenth-century biographer Fantuzzi emphasizes that the erudite company of the fish experts mentioned above stimulated Aldrovandi to extend his interest from the antiquities that could be seen locally to naturalia – first of all to fish, only later to plants. The collection of fish and presumably fish images in Rome in these years constituted the very beginning of Aldrovandi's naturalia collection.[78] It is yet another example of the primacy of fish and

---

75 See Castelli, "Tra ricerca empirica" 106, for the Rondelet certainty and the Giovio suggestion.

76 Cordus and Sittardus observed aquatic animals together during this Italian trip. See for more information, Egmond – Kusukawa, "Circulation of images" and "Gessner's fish", esp. 298–299. The role of Rome as a centre for fish information was already pointed out by Pinon, "Clématite bleue" 484.

77 Fantuzzi, *Memorie della vita* 13.

78 Ibidem. See: Aldrovandi Ulisse, *Le antichità de la città di Roma* (Venice, Giordano Ziletti: 1556).

botanical studies in the larger field of naturalia research and collecting. What is more, that connection was also made explicit by Aldrovandi, as Tosi argues:

> Era Aldrovandi a suggerire agli allievi l'indagine botanica come esercizio prioritario nello studio della realtà e della storia naturale, a indicare la 'pescharia' come luogo privilegiato della curiosità, palestra formativa come lo era stata per lui ai tempi del soggiorno romano.
>
> It was Aldrovandi who advised his pupils to regard botanical research as leading branch of the study of existence and natural history, and to indicate the fish market as privileged site of curiosity, a training ground, as it had been for him during his stay in Rome.[79]

3.3 *Naturalism, Observation and Preservation*

The collections of *aquatilia* drawings discussed so far preceded the illustrated publication wave of the 1550s, but mainly originated *after* the series of unillustrated 1520s–30s publications on *aquatilia*, which had been a largely textual continuation of a tradition rooted in classical natural history. Practical experience and personal observation were creeping in, and words were no longer enough. There was nothing fortuitous about this timing. The early collections of *aquatilia* drawings of the late 1530s–early 1540s fit in extremely closely with the earliest traceable phase of botanical and marine fieldwork by European naturalists in Germany, Italy, South-Eastern Europe and the Middle East (1530s–early 1540s); with the dating of the oldest extant herbaria containing dried plants; and with the creation of the earliest university botanical gardens in Italy in the mid 1540s and that of many private collector's gardens in various European countries in the same decade.[80] Such a close correlation suggests that we are looking here at a really new and many-stranded development, and not merely at a phenomenon of which we cannot fathom earlier phases

79 Tosi, "Acconciare" 54.

80 For a survey of sixteenth-century botanical fieldwork, see Egmond F., "Into the wild. Botanical fieldwork in the sixteenth century", in MacGregor A. (ed.), *Naturalists in the Field. Collecting, recording and preserving the natural world from the fifteenth to the twenty-first century* (Leiden – Boston: 2018) 166–211. On aquatic fieldwork and direct observation, especially by Belon and Rondelet, see Perfetti S., "Aquatilium historiae. Epistemologia aristotelica e osservazioni sul campo nelle monografie ittiologiche di Rondelet e Belon (XVI secolo)", in Geruzzi S. (ed.), *Uomini, demoni, santi e animali tra medioevo ed età moderna* (Pisa – Rome: 2010) 169–183. For a recent survey of herbaria with dried plants, see Thijsse G., "'Tusschen Pampier Geleyt.' Ontstaan, verspreiding en gebruik van de vroegste herbaria", in IJpelaar L. – Chavannes-Mazel C.A. (eds.), *De Groene Middeleeuwen. Duizend jaar gebruik van planten (600–1600)* (Eindhoven: 2016) 64–93.

for lack of sources. Geographically, the Mediterranean was the core area with centres of image production and distribution in Venice and the Veneto, Rome, almost certainly in Montpellier, and possibly in Lyon.[81] But interest in *aquatilia* rippled northwards – following the return paths of the medical *peregrinatio* by north Europeans and of their journeys in the service of diplomacy and espionage.

The evidence of the early image collections also points to the crucial role of coloured drawings as placeholders. Fish posed exceptional problems of preservation. Only few of their body parts were hard enough to be dried. In the 1540s-circles of Salviani and Porzio, drying techniques similar to those developed for plants were certainly used to preserve fish skins.[82] But the dried material soon became brittle, as we can see on a few exceptional pages in the German Ratzenberger plant herbarium (1590s) that present dried ray's eggs, parts of shells, and even fragments of a cuttlefish bone.[83] Once out of the water, fish also quickly lose their distinctive colours, and this is not only aesthetically relevant. Colours and shine are as important to correct identification – and therefore fish study – as shape, size and build. While plants posed the painterly-scientific challenge of how to render their various stages of growth in one representation, the major difficulty with (most) *aquatilia* was how to depict them as colourful and shiny as they looked when alive.

Two letters show how clearly contemporary collectors-naturalists were aware of these problems, and how they discussed them. The first dates from 1552, right in the middle of the wave of fish interest, and shows us the young Carolus Clusius in the rather unusual role of a dried fish collector in Montpellier, where he was staying in Rondelet's house. The young German physician Lorenz Gryll, who had stayed earlier with Rondelet, wrote to Clusius from Paris, asking him for some thirty species (specified in a list) of dried Mediterranean fish that were unfamiliar to the Germans:

81 To what extent Paris was relevant as place of origin for Belon's visual documentation must remain open here.

82 On fish preservation in history, see Davis P., "Collecting and preserving fishes: a historical perspective", in MacGregor (ed.), *Naturalists in the Field* 149–165; and on techniques of drying fish skins et cetera, see Pinon, "Clématite bleue" 485. Alcohol was not regularly used at the time.

83 On this herbarium see Zahn G., "Das Herbar des Dr. Caspar Ratzenberger (1598) in der Herzoglichen Bibliothek zu Gotha", *Mitteilungen des Thüringischen Botanischen Vereins* N.F. 16 (1902) 50–121; and Schaffrath U., "Läuse, Muscheln und Tabak – Das Herbar Ratzenberger", *Philippia. Abhandlungen und Berichte aus dem Naturkundemuseum im Ottoneum zu Kassel* 15.3 (2012) 191–214.

> […] ineunte vere si probe siccati sint pisces, de iis transmittendis consilium non deerit. Seliges autem corpora piscium nec grandia nimis nec minutissima, sed mediae magnitudinis. Uteris tamen tuo arbitrio hac in re quod quidam pisces minuti plane alii sunt, in plerisque autem eadem figura constat quaecumque sit aetas. Dubii autem et qui aridi formam mutant schedulis appensis erunt notandi. Potes hoc indice iuvari, quamvis plures alii tibi succurrere possint. Scis autem quosdam plane integros siccari posse, alios autem exemptis interaneis palea replendos.
>
> By early spring, if the fish are properly dried there will be no lack of advice about their dispatch. Choose the bodies of fish that are neither too large nor too small but of medium size. Use your judgment in this matter for some fish when they are small are completely different, while in most cases they retain the same shape whatever their age. Uncertain ones and those which change their shape when dried will have to be noted in appended labels. You can make use of this list although several others may occur to you. You know that some can be dried as they are, while others have to be filled with straw after their internal organs have been removed.[84]

A second letter, sent from Crete by the Italian naturalist-physician and antiquarian Onorio Belli (1596) shows clear awareness of the relevance of colour to identification:

> Hunc[!] scaros, sicut postulas siccos accepies: qui nihil cum recent[e] captis conveniunt, nam colores quibus insigniti a natura sunt omino perierunt. Scari hi duorum sunt generum. Maiores iconem tibi missum aemulantur. Minores vero colore rubri et melli sunt, quibusdam extremitatibus cerulei, ut mirum immodum oculis gratissimi videantur. [….] Figuram, quam Bellonius ad calcem octavi capitis primi observationum libri exhibit, mala est: neq[ue] scarum exhibit: ut ex pisce sicco, et icone missa facillime percipies.
>
> Here are the dried *scari*, as requested. They are nothing like freshly caught ones, because their natural colouring has been completely lost. These *scari* are of two sorts. The larger ones match the image sent to you; the smaller ones are naturally red and honey-coloured with blue

84 Manuscript letter, Lorenz Gryll (Laurentius Gryllus) to Clusius, 21 August 1552, Paris to Montpellier, Leiden University Library, VUL 101.

> extremities. […] The image which Belon shows […] is bad: it is not even a *scarus*, as you can easily see from the dried exemplar and from the image sent to you.[85]

Some experimented with special preservative liquids, as is documented for Grand Duke Francesco I de' Medici and his painter Jacopo Ligozzi in the late 1570s.[86] But the more obvious solution was to make drawings as quickly as possible after fish had been caught, and to consult persons who had actually seen them alive or regularly caught or sold them.

Few early modern naturalists actually observed fish alive, out at sea, on the beaches, or immediately after they had been caught. Belon on his Mediterranean voyages, Mattioli during his coastal explorations along the Adriatic, and the Dutch wholesale merchant and specialist in fish Adriaen Coenen (see below) are a few examples. Mattioli, for instance, explicitly refers to his personal, sensorial experience: he had seen many gobio fish in the lakes and rivers of the Trentino; suffered the stings of jellyfish; and had seen on the beaches of Istria how a stingray smashed its hook deep into the wood of a fishing boat.[87] Belon too emphasized his personal observation when stating that he had often travelled by the night boat from Padua on Thursday evening down the river Brenta, arriving next morning in Venice, where he remained all day to observe fish that had been brought to the famous Venetian fish market. He had also stayed for the whole of Lent in Venice and interviewed fishermen.[88]

Fish markets and ports, especially at Lent, offered some of the best sites to observe common and unusual fish that at least retained a little of their colours and shine. They became even more important when fishmongers and fishermen knew that a naturalist was searching for special and rare items, and probably willing to pay for them. That was certainly the case with Horstius and Salviani in Rome. The former received information about a special catch from

85 Letter, Onorio Belli to the Swiss philologist-physician Jacob Zwinger in Basel, 15 August 1596, published in Beschi L., *Onorio Belli Accademico Olimpico. Scritti d' Antiquaria e Botanica (1586–1602)* (Rome: 2000) 186–187.

86 See Luca M.E. De – Faietti M. (eds.), *Jacopo Ligozzi "Altro Apelle"* (Florence: 2014) 36–37; and Cecchi A. – Conigliello L. – Faietti M. (eds.), *Jacopo Ligozzi "pittore universalissimo"* (Livorno: 2014) 48–49.

87 Mattioli Pietro Andrea, *I discorsi di M. Pietro Andrea Matthioli sanese, medico cesareo, et del serenissimo principe Ferdinando archiduca d'Austria & c. nelli sei libri di Pedacio Dioscoride Anazarbeo della materia medicinale* (Venice, Valgrisi: 1568) 360 (gobio), 365 (jellyfish), 347–48 (stingray). Concerning personal observation, Mattioli praises Salviani (in the same edition 339, 350, 357–358), but criticizes Belon (351).

88 Belon, *L'Histoire naturelle* 6v. On Lent and especially well stocked fish markets, cf. Pinon, "Clématite bleue" 486.

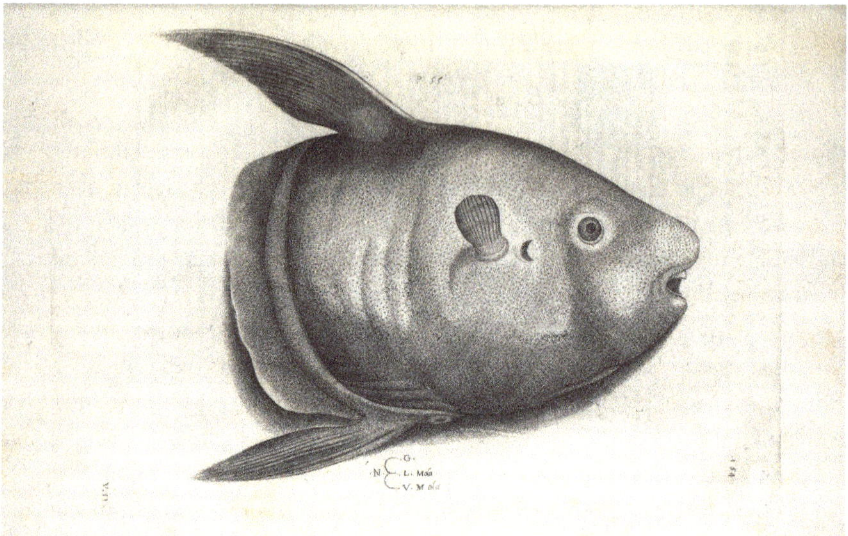

FIGURE 6.8 *Mola mola*. Copper engraving in: Salviani Ippolito, *Aquatilium animalium Historiae liber primus. Cum eorundem formis, aere excusis* (Rome, Ippolito Salviani: 1554–1558)
PUBLIC DOMAIN: HTTPS://WWW.BIODIVERSITYLIBRARY.ORG/BIBLIOGRAPHY/82337

the port of Civitavecchia some 80 kilometres northwest of Rome. And Salviani describes how fishmongers at the Roman fish market (located since the Middle Ages in the Portico d'Ottavia that dates back to the second century BC) often sent him messages to come and inspect rarities. On one such occasion, Salviani rushed to the fish market and found a large and rare *Mola mola* (sunfish), which he recognized because the naturalist Luca Ghini had earlier sent him a drawing of this species. Salviani had the *Mola mola* transported to his house, where many erudite friends who shared his interest in naturalia admired it. He personally dissected the fish so that they could inspect its inner structure and organs, and corroborate the correctness of his drawing. The drawing of this *Mola mola* served as the model for his printed illustration [Fig. 6.8].[89]

The growing textual emphasis on personal observation and the matching visual emphasis on naturalism and *ad vivum* depicting formed part of a rhetorical strategy that underpinned reliability by pointing to *experientia* – a well-known technique by this time that was also used extensively in botanical

89 Salviani, *Aquatilium* fols. 155r–v.

and anatomical studies.[90] But even if some of these claims were vastly exaggerated, all naturalists of this period did indeed believe that personal observation was important and that visual representations should be as lifelike as possible in the service of accurate identification. If they had not attached great value to that match between a representation and its subject, men like Barbaro and Rondelet would have hardly carried their drawings collections with them on their international journeys; nor would many naturalists have used their international networks to obtain more accurate drawings. In fact, there would not have been much reason for field observations at all.

The history of the image collections discussed above demonstrates that many drawings – and even some whole collections – were created without any intention to print this material on the part of either painter or collector. Large numbers of drawings were first and foremost intended as study material. Eventually, and in only in some cases, such image collections could also serve as a fund from which to select models for printed illustrations. The whole of the corpus of *aquatilia* newly drawn in the 1540s–50s attempted a lifelike visual rendering of naturalia. That visual format mainly served identification. On many drawings the names (or short descriptions) of the *aquatilia* are written on the page close to the drawing – again following the model of plant studies – in an attempt to link the images to fish names used in classical sources, 16th-century vernacular languages and local dialects.[91] The drawings also played an important part in the gift exchanges and construction of friendship networks among naturalists and collectors. And they provided pleasure to collectors, viewers, the naturalists themselves and presumably the painters. They had distinct lives, therefore, outside the domain of print.

4 Fish Move into Fashion, 1540s–1580s

4.1 *Arts and Sciences*

Looking back from the later 16th century, it is hard to avoid the impression that the wave of the 1540s–1550s enhanced – and probably was itself part of – an

90 See Egmond, *Eye for Detail*; Egmond F., "The Ad Vivum Conundrum: Eye witnessing and the Artful Representation of Naturalia in Sixteenth-Century Natural Science", in Schmiedel I. – Oy-Marra E. (eds.), *Zeigen – Überzeugen – Beweisen. Methoden der Wissensproduktion in Kunstliteratur, Kennerschaft und Sammlungspraxis der Frühen Neuzeit* (Mainz: 2020) 33–62; and Smith – Van Trijp, "Dynamiques européennes".

91 See Glardon P., "La terminologie botanique dans le *De historia stirpium* de Leonhart Fuchs (1542) et ses premières traductions françaises", *Seizième Siècle* 8 (2012) 57–74, on parallels between the naming and ordering of plants and fish.

impressive *aquatilia* fashion that fanned out from Italy and France to the rest of Europe and lasted until well into the seventeenth century. By fashion I mean a fascination that spread very widely through European society and manifested itself in scientific research as much as in decorative arts, literature, folk songs, and painting. In spite of the general popularity of living nature and natural history in this period, no such *wave* seems to have occurred with respect to other living creatures. The interest in plants showed a more continuous and steady growth. Birds followed almost immediately in the slipstream of the *aquatilia*. A fascination with insects came into its own in the later decades of the 16th century. But water and fish dominated the mid 16th century and remained *en vogue* for a very long time. Here, courtly art and artful science found a perfect meeting point: aristocratic collecting and self-fashioning fused with elite patronage of naturalists and painters in setting visual trends and scientific passions.

Outside the domain of coloured drawings and printed illustrations this fascination can be traced in a series of phenomena that appear to culminate in the 1550s–1560s as well. These deserve to be mentioned briefly here and should be studied together, precisely because they incorporate and link up with the 'scientific' information, while fitting into the yet larger context of the partly classically inspired fascination with water, fountains, baths, river gods, cascades, shells, grottoes, and their Mannerist representations and designs.

A cluster of hard to classify texts incorporate both classical fish information and new, 16th-century study of the aquatic world. The best-known example is probably *De natura aquatilium carmen* (1558) by the French physician François Boussuet (1522–1572). The illustrations and the textual information of this work of epigrammatic verses in Latin about fish are largely based on Rondelet's *Libri de piscibus marinis*, which had been published a few years earlier (1554–55) by the very same publisher in Lyon.[92] All other examples in this genre known to me are *un*-illustrated. They include a Latin dialogue published in 1554 by the German physician Johannes Lange (1485–1565) in which two German students discuss fish on the Venetian fish market, and the better known and much longer vernacular poems (*rime pescatorie*) published in the 1550s in that same city, and frequently reprinted.[93]

92 Boussuet François, *De natura aquatilium carmen, in universam Gulielmi Rondeletii* (Lyon, Macé Bonhomme: 1558). Cf. Rondelet, *Libri de piscibus marinis*. See on Boussuet, Hendrikx S.M. – Smith P.J., "Connaissances ichtyologiques sous forme emblématique: le cas du *sargus*", *RursuSpicae* 4 (2022) 1–26 (https://journals.openedition.org/rursuspicae/2258).

93 Lange Johannes, *Medicinalium epistolarum miscellanea* (Basel, Johannes Oporinus: 1554), where it is followed by another letter on river and lake fish and how to cook them. For

An even more intriguing group of texts belongs specifically to the area of Naples. These *egloghe pescatorie* consist of highly erudite, book-length poems in Latin that focus almost completely on fish and fishermen – transposing the format of Virgil's bucolic egloghe to the domain of the sea. Their roots reach back to the Neapolitan humanist and poet Jacopo Sannazaro (1458–1530), author of *Arcadia* and of five *Eclogae piscatoriae* that he himself published in 1526. Two of the most famous *egloghe pescatorie* explicitly link the findings of contemporary, 16th-century natural science with the classical tradition: *Egloghe pescatorie* by the Neapolitan poet Ber(n)ardino Rota (1560) and *Mergellina. Egloghe piscatorie* (1598) by the Neapolitan theologian, poet and archaeologist Giulio Cesare Capaccio. They bridge the genres and have been studied by Caracciolo not only as a literary genre with classical roots, but also in the context of natural science and the *Kunst- und Wunderkammern* tradition. The publication years of these dialogues and poems are as significant in the present context as the fact that many originated in the two principal Italian ports.[94]

A clear manifestation of how the fascination with the world of water entered the *decorative arts* can be seen in the frescoes (and some tapestries) that decorate vaulted ceilings and more occasionally walls in Italian villas and urban palaces. Many of these show mythological and classicizing scenes with often

---

an example of the *rime pescatorie*, see Calmo Andrea, *Le bizzarre, faconde, et ingegnose rime pescatorie, Nelle quali si contengono Sonetti, Stanze, Capitoli, Madrigali, Epitafij, Disperate, e Canzoni* (Venice, Iseppo Foresto: 1557). On their linguistic importance and for further printing information see Folena, "Per la storia" 120–121. Ichthyological knowledge also entered into French literary works, most famously those by Rabelais and Guillaume du Bartas. See esp. Smith P.J., *Tussen taal en natuur: de poëzie van Du Bartas* (Leiden: 2020); Idem, "Ronsard and Du Bartas in the Low Countries: Evidence from Early Modern Dutch Private Libraries and a *Vanitas* Still-Life by Edwaert Collier (ca. 1664)", in Pouey-Mounou A.P. – Smith P.J. (eds.), *Ronsard and Du Bartas in Early Modern Europe* (Leiden – Boston: 2021) 207–228; Idem, "Ichthyological Topics of the European Reception of Du Bartas" in Pouey-Mounou – Smith (eds.), *Ronsard and Du Bartas* 55–279; Idem, "Rabelais ichtyologue", in Garnier I. – La Charité C. – Menini R. – Pouey-Mounou A.-P. – Réach-Ngô A. – Tran T. – Viet N. (eds.), *Narrations fabuleuses. Mélanges en l'honneur de Mireille Huchon* (Paris: 2022) 439–452.

94 Capaccio Giulio Cesare, *Mergellina. Egloghe piscatorie di Giulio Cesare Capaccio napolitano* (Venice, Eredi di Melchior Sessa: 1598). Rota Bernardino, *Egloghe pescatorie* (Naples, Giovan Maria Scotto: 1560). A further example is Regio Paolo, *Siracusa Pescatoria* (Naples, Giovanni de Boy: 1569). For excellent publications around this theme, see Caracciolo D., "L'enciclopedia ittiologica di Giulio Cesare Capaccio ed i suoi rapporti con il collezionismo Napoletano", *Napoli Nobilissima* serie 5.x (2009) 3–20; Idem, "Per una Wunderkammer letteraria, Mergellina, la 'fatica marittima' di Giulio Cesare Capaccio", *Annali di critica d'arte* 5 (2009) 33–80; and Mauriello A., "La 'Siracusa' di Paolo Regio e la tradizione letteraria napoletana tra primo e secondo Cinquecento", *Studi rinascimentali* 6 (2008) 91–97.

fantastic water creatures and grotesque fishes, but they also contain lifelike, identifiable *aquatilia*. That distinction is not only a modern one. In the early 1580s, as discussed by Acciarino, the Bolognese naturalist Aldrovandi emphasized the difference between grotesque and lifelike representations: the latter could serve scientific purposes since they aimed at lifelike representation, while the former had no immediate connection with nature but with fantasy.[95]

Interestingly, the prominence of lifelike fish on frescoes in Italian palazzi and villas chronologically peaks once again between the early 1550s and the mid 1570s. Examples include the vaulted ceiling in Palazzo Chiericati in Vicenza (designed by Palladio and built from 1551) with numerous lifelike fish – from tortoises to eels, crabs, lobsters, flat fish, sturgeon, and red gurnard – among the grotesques painted in 1557–58 by Eliodoro Forbicini from Verona, who appears to have worked in other Palladian villas as well; the vaulted ceiling of the entrance to Palazzo Vitelli (constructed *c.*1540s–50s) in Città di Castello not far from Perugia; ceiling frescoes dated to 1558–61 in the *Sala dei Pesci* (also of Neptune) in the Palazzo Ducale of the Gonzaga in Mantua [Fig. 6.9]; and ceiling frescoes in the famous Villa d'Este in Tivoli near Rome, built for Cardinal Ippolito II d'Este from *c.*1561.[96]

Ceiling and wall frescoes in the *Stanza degli Elementi* of Palazzo Firenze in Rome are an even more impressive example. There is hardly a trace of the grotesque here, and the symbolic grouping of living creatures according to their element goes hand in hand with a highly naturalistic representation. It is easy to recognize several types of blowfish, various flatfish, crabs and lobsters, many shells, and a curled-up eel that looks as if it comes straight out of either Gessner or Salviani; in the fish festoons there may even be the face of a sunfish. As further visual references among the animals (a rhinoceros, a chameleon) show, these frescoes do indeed incorporate contemporary 'scientific' and published knowledge. They were painted in the 1570s by Jacopo Zucchi for Ferdinando de' Medici, at that time Cardinal and later successor of his older brother Grand Duke Francesco de' Medici [Fig. 6.10].[97]

---

95 See Acciarino D., "Antipoetica delle grottesche: le *Lettere sulla pittura* di Ulisse Aldrovandi", *Schede Umanistiche. Rivista annuale dell'Archivio Umanistico Rinascimentale Bolognese* N.S. XXX (2016) 169–196; more in general on grotesques, Morel P., *Les Grotesques. Les figures de l'imaginaire dans la peinture italienne de la fin de la Renaissance* (Paris: 1997). On the Medici tapestries with lifelike fish, see Else, "Globefish".

96 I have not made a systematic study of such frescoes in Italy. Apart from the Vatican Loggias of 1519, I do not know of other examples of *aquatilia* prominence on frescoes that date from well before the 1540s. Nor have I so far found examples for the very *late* sixteenth century.

97 Palazzo Firenze is in the Campo Marzio. Ferdinando de' Medici kept a menagerie with wild animals in the gardens with exotic plants at Villa Medici in Rome, which has a garden

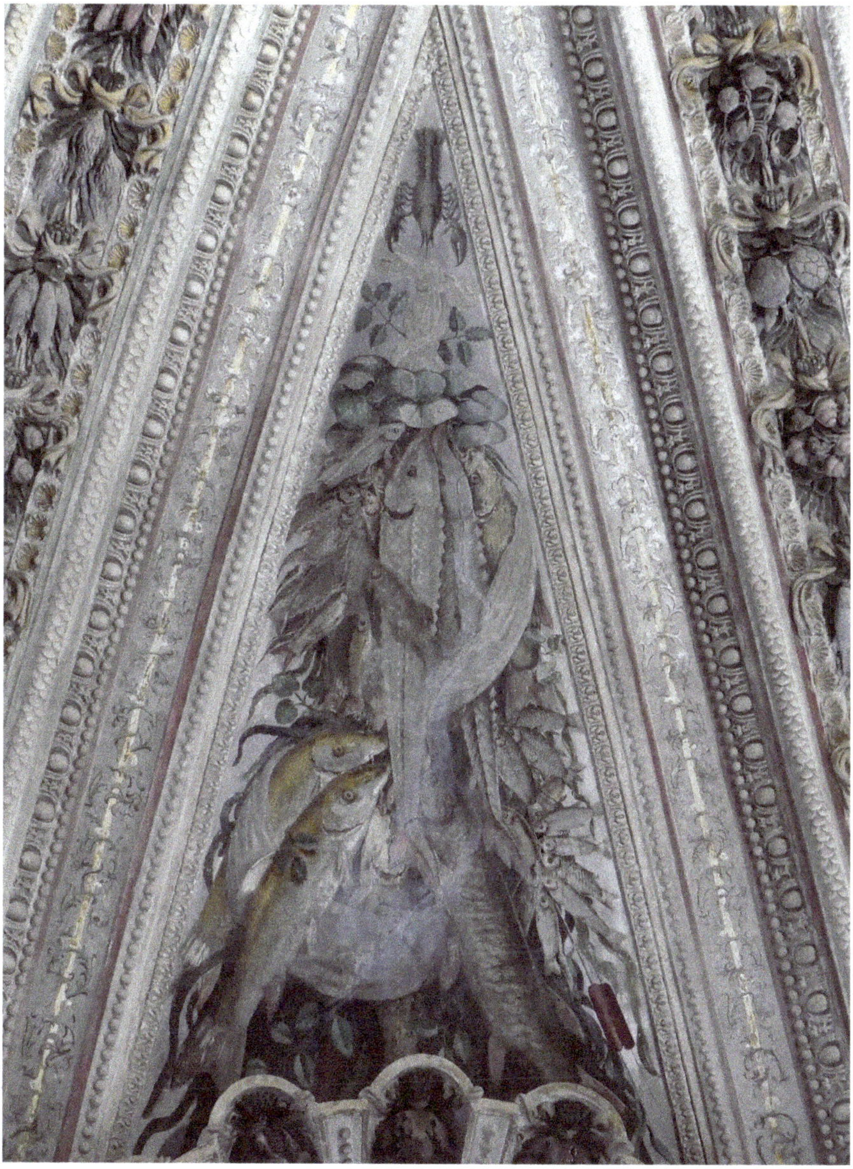

FIGURE 6.9 Detail of the ceiling frescoes and stucco of *aquatilia* on the ceiling of the Sala dei Pesci/di Nettuno in the Palazzo Ducale, Mantua, 1558–61
PHOTO BY THE AUTHOR

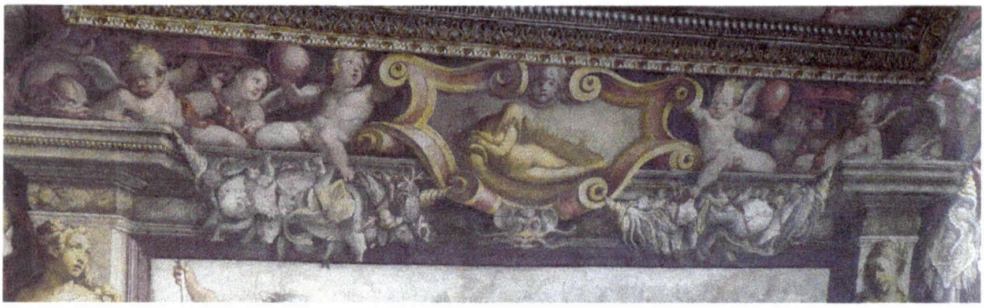

FIGURE 6.10 Frescoes painted by Jacopo Zucchi for Cardinal Ferdinando de' Medici in Palazzo Firenze, Rome, 1570s
PHOTO BY THE AUTHOR

The very earliest and highly influential examples of lifelike decorative fish date from much earlier, however, and can be found among the grotesque decorations of Raphael's Loggias in the Roman Vatican (*c.*1519). They have been intensively studied and include many zoologically identifiable painted fish, birds (including a very early American colibri) and other animals, in particular on the pilasters.[98] The influence of these particular fish representations can be traced, for instance, in illuminated decorations on several folios of the Book of Hours painted (1530s) by the Papal illuminator Vincenzo Raimondi in Rome for Eleonora Gonzaga, wife of Francesco Maria I, Duke of Urbino.[99] And they are directly linked as well with fresco decorations in Palazzo Grimani in Venice. In fact, the same painter, Giovanni da Udine, who was responsible for the pilaster decorations in the Vatican Loggia's of *c.*1519 also designed fresco decorations in the Venetian Grimani palace twenty years later (*c.*1537–39). Two wall frescoes in the *Stanza di Psyche* of Palazzo Grimani with highly naturalistic frescoes of birds and fish that hang from candelabra-like painted shapes closely resemble those in the Vatican.[100] The Grimani context is particularly significant. Several members of the Grimani family were among the great antiquities collectors

house with frescoed plants. See on this Tice L.N., "Collecting in the garden. Inventories of *casini* in late sixteenth- and early seventeenth-century Rome", *Journal of the History of Collections* 23.2 (2011) 315–331.

98 See Caneva G. – Carpanetto G. (eds.), *Raffaello e l'Immagine della Natura. La raffigurazione del mondo naturale nelle decorazioni delle Logge Vaticane* (Milan – Rome: 2010), which includes zoological identification of all depicted naturalia.

99 As argued by Alexander J.J.G. (ed.), *The Painted Page: Italian Renaissance Book Illumination, 1450–1550* (Munich: 1994) 243–244. Eleonora's Book of Hours is Bodleian Library MS Douce 29, see esp. folios 110v–111 and 131v–132.

100 These frescoes were most probably designed by Da Udine and actually painted by his much younger assistant Camillo Capelli (also: Camillo Mantovano, died 1568), who had a special aptitude for depicting naturalia.

of the Veneto, while Giovanni (1506–1593), who in 1558 became the sole proprietor of the Grimani Palace, was not only a collector and bishop, but also Patriarch of Aquileia. He was both a kinsman of Daniele Barbaro and the latter's immediate predecessor as Patriarch [Fig. 6.11].[101]

Live fish also played a part in mannerist display, decoration and entertainment. At the very end of the 16th century, Archduke Ferdinand II of Tyrol personally designed a circular fish pond protected by a pagoda-like roof. It was planned with a central viewpoint that could be reached by a footbridge. From there the viewer would be able to look at all segments of the pond: each should contain different fish species, alternating exotic and European ones.[102] An even more ingenious use of live fish was made in the famous Big Grotto of the Florentine Boboli gardens, created between 1583 and 1593 by Bernardo Buontalenti for the Medici. It foreshadows the tunnel-under-water experience in modern aquariums. Originally, the opening in the ceiling of this Big Grotto was closed off with a large glass slab that let in light and also served as the transparent bottom of an aquarium-pond. Fish thus swam above the heads of the grotto's visitors, and shadows cast by both fish and water created enchanting underwater effects on the walls of the grotto below.[103]

Yet further evidence of the spreading fascination with *aquatilia* in the 16th century occurs in manuscript illumination. Three diverse pre-16th century texts about the animal world that originally had no illustrations at all were decorated with coloured illuminations of fish (and other animals) between about 1550 and 1590 – decades to centuries after the texts themselves originated. One of these is a work about the animal world written in the mid-15th century by the humanist Pietro Candido Decembrio (1399–1477) for the ruling Gonzaga family in Mantua. As argued by Pyle, more than a century later an unknown painter inserted finely painted, coloured images of animals in the

101 See Bristot A. (ed.), *Palazzo Grimani a Santa Maria Formosa. Storia, arte, restauri* (Venice: 2008) 72–77, 91, 99. Between the early 1530s and the late 1560s the Grimani palace was expanded and remodelled. During the remodellations of the 1560s Camillo Mantovano created a magnificent frescoed ceiling full of plants and birds in the *Stanza dei fogliami*, and another frescoed ceiling with hanging fish (some depicted in nets) and birds that was finished in 1567. Mantovano probably worked with a team. See also Hochmann M., "La famiglia Grimani", in Hochmann – Lauber – Mason (eds.), *Il collezionismo* 206–223; and Idem, "Plinio Scarpelli".

102 See Dobalová S., "Erzherzog Ferdinand II. von Habsburg, das Lusthaus Belvedere und die Fischbehälter im Königlichen Garten der Prager Burg", *Die Gartenkunst* 20 [Beilage zu Heft 2] (2008) 11–18, here 15–16. The structure was never built.

103 See Heikamp D., "La Grotta Grande del Giardino di Boboli", in Capecchi G. (ed.), *Palazzo Pitti: la reggia rivelata* (Florence: 2003) 446–474, here 446–447. The 'aquarium' only functioned briefly.

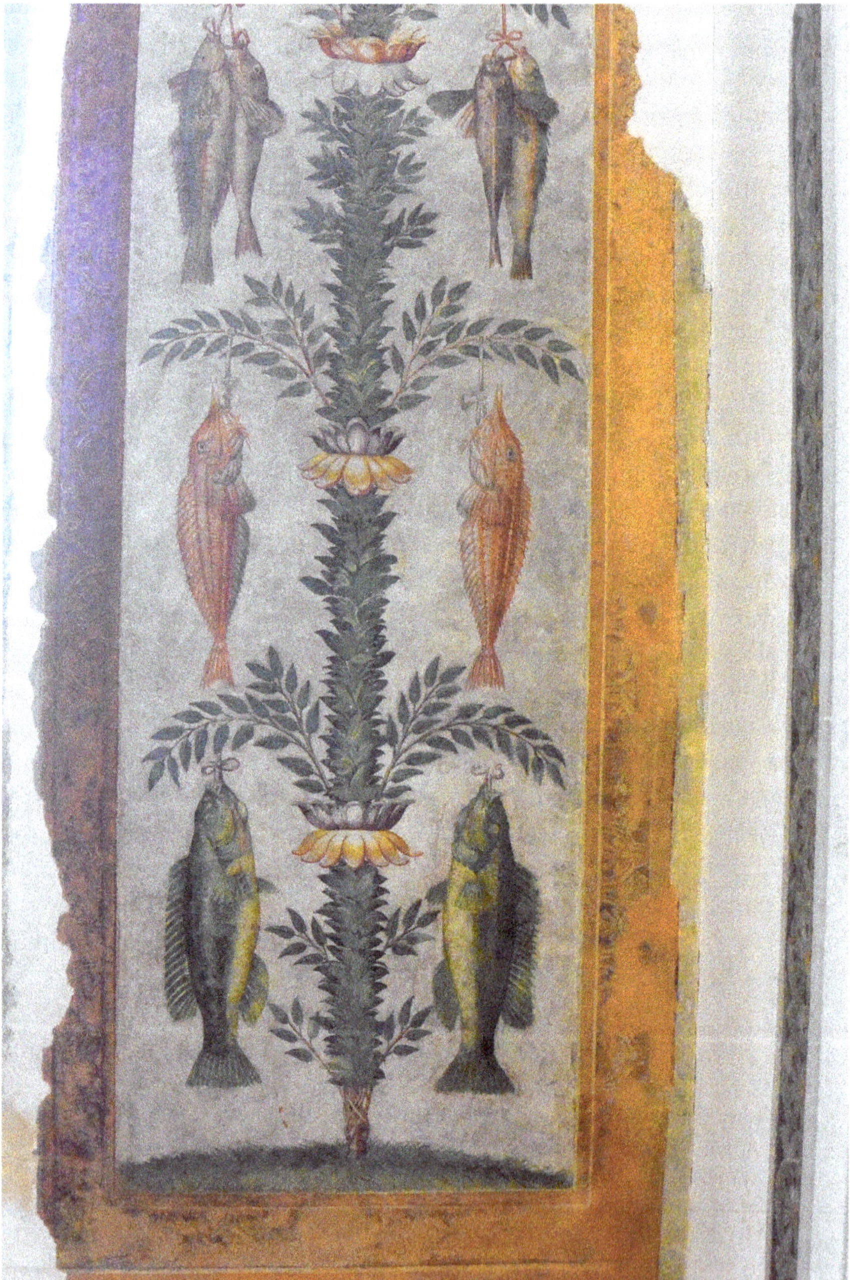

FIGURE 6.11 Wall frescoes by Camillo Mantovano in the *Stanza di Psyche* of Palazzo Grimani, *Venice*, detail
PHOTO BY THE AUTHOR

bottom margins of most pages of its 15th-century exemplar on parchment now in the Vatican Library. Almost certainly an elite collector commissioned these illuminations. They include a large number of *aquatilia*. Since many of these illuminations are directly based on Gessner's printed illustrations, they must date from after 1558.[104] The fact that this happened at all tells us that the late 16th-century owner of the Candidus Codex felt the absence of images so strongly that he or she thought it worthwhile to have them added to this valuable manuscript.

Handpainted illumination in colour was, in fact, by no means outmoded and illuminated manuscripts remained highly valued and precious in the mid to late 16th century as one glance at the work of famous illuminators such as Giulio Clovio (active 1530s–70s in Venice, Mantua, Florence, and Rome) and Georg Hoefnagel (active at the Habsburg courts of Prague and Vienna around 1600) demonstrates. Elite collectors, in fact, *newly* commissioned works by both illuminators and calligraphers. Those works included old and new texts about nature. In the mid-16th century the Cretan calligrapher Angelos Vergekios, who arrived in France via Venice and worked at the French royal court from *c.*1539 until his death in 1569, produced at least 11 copies of an even older text than the Candidus one. This was a 2000-lines long poem about the natural world, *De animalium proprietate*, by the Byzantine poet Manuel Phyles (*c.*1275–*c.*1345). Vergekios's manuscript copies are illustrated with coloured animal drawings by an unknown painter. The exemplars differ slightly in both quantity and characteristics: the Vatican exemplar of 1560 has 109 illuminations on 60 folia, the Bodleian of 1564 has 106 illustrations on 62 folia, for instance. There is a notable presence of *aquatilia*, and again many of these illuminations are based on *printed* zoological illustrations of the 1550s–60s.[105]

A third, far less investigated and equally intriguing example concerns a unique incunable of Pliny, printed in 1481 in Parma: besides the initial and incomplete illumination that presumably dates from the 1480s, a considerable

104 See the contribution by Cynthia Pyle in the present volume, and Pyle C., *Das Tierbuch des Petrus Candidus. Codex Urbinas Latinus 276. Eine Einführung* (Zurich: 1984), with details on the possible painter 85, 98–99; and Idem, "The Art and Science of Renaissance Natural History: Thomas of Cantimpré, Pier Candido Decembrio, Conrad Gessner and Teodoro Ghisi in Vatican Library MS Urb. lat. 276", *Viator* 27 (1996) 265–321.

105 For a detailed discussion of the Philes-Vergekios manuscripts and their textual and visual traditions, see Peers G., "Thinking with Animals: Byzantine Natural History in Sixteenth-Century France", *Bibliothèque d'Humanisme et Renaissance* 68 (2006) 457–484; and Idem, "Forging Byzantine animals: Manuel Philes in Renaissance France", *Rivista di Studi Bizantini e Neoellenici* N.S. 49 (2012–2013) 79–103. The manuscript in the Vatican Library was a state gift from the French King François II to the Duke of Savoy. See List of Codices for the British Library exemplar.

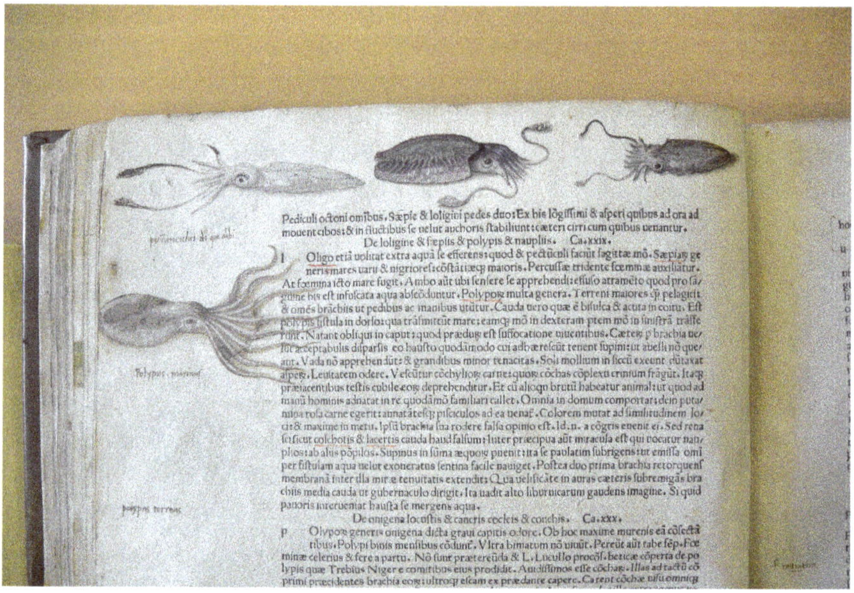

FIGURE 6.12 Detail of the decorations in Pliny, *Naturalis Historia* (Parma, Andreas Portilia: 1481), Parma, Biblioteca Palatinense, Inc. 1158, fol. 74v

number of extremely finely painted watercolours of *aquatilia*, a few birds, and some fruits appears in the margins of its later pages [Fig. 6.12]. Even less is known in this case about the number and identity of the painters involved, who commissioned the drawings, and in which year (or periods) they were inserted, except that one image dates from after 1515 and one textual annotation must post-date 1523. So far, no models in either print or drawing have been discovered for the extremely lifelike coloured fish, fruit and bird drawings. Stylistically they seem to belong to the 1570s–1610s.[106] If, however, the *aquatilia* illuminations in this Parma Pliny would turn out to have been made in the 1530s–40s, they should be rated among the most sophisticated early images of naturalia in Europe.

In each of these three very different codices the illuminations are directly related to the text and literally illustrate it, but they remain separate from it

106 Pliny, *Naturalis Historia* (Parma, Andreas Portilia: 1481), with later drawings in the margins, Parma, Biblioteca Palatinense, Inc. 1158. Most of its illuminations have never been published. The only detailed analysis so far is Walter H., "An illustrated incunable of Pliny's Natural History in the Biblioteca Palatina, Parma", *Journal of the Warburg and Courtauld Institutes* 53 (1990) 208–216. I hope to publish its illuminations.

and from each other. Nearly all are painted on the whitish background of the page, sometimes with a *trompe-l'oeil* shadow to give them further relief, and they are unaccompanied by any other figures or decorative patterns. These were illustrations *added* to texts at a time when the visual turn was enforcing itself. Their functions were clearly different from those of the image collections discussed earlier. On the one hand these prized miniatures embellished texts about nature and turned precious manuscripts into even more valued (and valuable) exemplars for wealthy collectors. On the other hand, precisely by modelling themselves on printed scientific illustrations, these hand-painted illuminations demonstrated that the collector's knowledge of living nature was state-of-the-art.

5 Drawing Collections after the Great Wave of the 1550s

5.1 *New Questions*

It is impossible to discuss the post-1550s history of *aquatilia* drawings in the same way as that of the earlier decades because the situation was profoundly different. Before the 1550s basically *no* printed works with large numbers of *aquatilia* images existed in Europe. A new visual corpus consisting of very large numbers of mostly coloured drawings was created in the 1530s–50s; parts of that corpus entered into print. By about 1560, in contrast, thousands of *aquatilia* images – mainly in black-and-white – were accessible in print, while numerous coloured drawings circulated via repeated copying among expert collectors and naturalists.

That raises several questions concerning the making and collecting of naturalia drawings in the context of the study of nature *after* c.1560. What exactly was the point of commissioning and collecting newly made drawings? Was the visual corpus in print not enough? Was colour so essential to *aquatilia* identification that naturalists and collectors continued to need drawings? Did the post-1560 drawings collections mainly focus on new fish species that had not yet been visually documented, or that had been unclearly identified? Or were the post-1560 drawings better, in the sense of scientifically more accurate and truer to life? Were they less or more closely connected to printing purposes than before? And who were the principal collectors?

The question about the contents of the later image collections is easily answered. They do indeed include 'new' fish species, but all later collections – even those begun around 1590–1600 – continue to depict (both in copied and newly drawn images) an already well-known range of fish species. None of them contains only new species. Repetition of the most common species is the rule. If a quantitative comparison of species in all these image collections would ever be undertaken, it might even indicate a particular set of canonical species that could well go back to the influence of classical textual sources. Furthermore, the fact that hardly any new printed works about *aquatilia* appeared after *c.*1560 already suggests that the drawings of the decades *c.*1560–1610 were usually *not* made or collected as models for eventual printed and illustrated works – always with the exception of Aldrovandi.

A good example from the late 16th century is the huge collection of naturalia drawings created by the physician-naturalist Felix Platter in Basel. He had no intention to publish them. Both his drawings and the dried plants of his herbarium served as a research collection, a visual and wide-ranging database on living nature, in which he incorporated whole earlier collections created by and for others: the most famous one consists of Gessner's model drawings of animals that dated back to *c.*1545–1565. The *aquatilia* drawings that Platter collected personally follow these earlier patterns perfectly. Like Gessner, Platter's collection method of the 1580s–1600s was encyclopaedic: his range was very wide, and he did not focus on new, additional species. Though painterly styles differ, the visual formats are identical. And both Gessner and Platter collected important clusters of *aquatilia* drawings in or from Venice.[107]

5.2 *North-Italian* Aquatilia *Drawings, c.1560s–c.1580s*

Three clusters of *aquatilia* drawings from the period *c.*1560–*c.*1585 help us to better understand shifts in how image collecting of naturalia was connected with changing knowledge about aquatic nature; who collected them; and what the functions of their drawings may have been.

The largest of these interconnected clusters is also the most complex one. It comprises hundreds of *aquatilia* drawings in Aldrovandi's massive collection in Bologna; a (now lost) album of fish drawings commissioned by the Venetian herbalist cum barber-surgeon and collector Leone Tartaglini (died *c.*1576); and a beautiful oblong codex on parchment in the Habsburg collections. That codex (Cod. Min. 83) has only 20 folios, but each of those shows up to five water creatures. The link between these three sets of images is the Medici court painter Jacopo Ligozzi (1547–1627), who is famous for his extremely

107 For further discussion, see Egmond – Kusukawa, "Circulation of images" and "Gessner's fish"; and Egmond, *Eye for Detail* and *Conrad Gessners Thierbuch*. On Platter's herbarium and its functions, see Dauwalder L. – Lienhard L. (eds.), *Das Herbarium des Felix Platter. Die älteste wissenschaftliche Pflanzensammlung der Schweiz* (Bern: 2016); and Benkert D., *Ökonomien botanischen Wissens. Praktiken der Gelehrsamkeit in Basel um 1600* (Basel: 2020) 60–98.

finely painted and naturalistic drawings of plants and animals.[108] Ligozzi is the probable painter of this parchment fish codex. He also made fish drawings for Tartaglini in Venice, some of whose visual material entered Aldrovandi's collection via copying. And Ligozzi both made and copied naturalia drawings directly for Aldrovandi, for whom he sometimes also copied his own drawings made for the Medici. The latter were both Ligozzi's and Aldrovandi's patrons during a large part of the 1570s–80s.[109]

By the late 16th century Aldrovandi's collection of naturalia drawings (*Tavole*) had grown to be probably the largest one in Europe. The extant part consists of 10 volumes, some in two parts, each of which contains between 100 and almost 300 figures; there are at least 250 figures of *aquatilia* [Fig. 6.13].[110] The roots of his image collection go back to Aldrovandi's years in Rome (1549–50), as discussed earlier. His entire collection reflects half a century of nature research (*c.*1550–*c.*1600), but since most of the drawings cannot be individually dated, it is impossible to reconstruct a clear chronology of his collecting. Olmi and Tongiorgi Tomasi in particular have shown that Aldrovandi employed a 'bottega' of painters in the course of these 50 years. There was a clear division of tasks. In terms of the production process, painters of documentary drawings and of model drawings for print came first. They were followed by those who transferred selected drawings to woodblocks that would then be handed over to the woodcutters. Aldrovandi's painters did not all work at the same time: some of them coincided in the course of this long period. Some painted large quantities of images for him, others only a few. Furthermore, painters might work on drawings years and even decades before these would be printed, and

108 See List of Codices. As first shown by Conigliello L., "Pesci, crostacei e un'iguana per l'imperatore Rodolfo II", *Paragone. Arte* 42. 493–495 (1991) 22–29.

109 The majority of Ligozzi's naturalia drawings (some 78 of plants and 65 of animals) are kept in the *Gabinetto dei disegni e stampe* of the Uffizi in Florence. On Ligozzi, see De Luca – Faietti (eds.), *Jacopo Ligozzi*; Cecchi – Conigliello – Faietti (eds.), *Jacopo Ligozzi*; and Groom A., *Exotic Animals in the Art and Culture of the Medici Court in Florence* (Leiden: 2019). On Aldrovandi and the Medici, see Olmi G., "Bologna nel secolo XVI: una capitale europea della ricerca naturalistica", in Frommel S. (ed.), *Crocevia e capitale della migrazione artistica. Forestieri a Bologna e bolognesi nel mondo (secoli XV–XVI)* (Bologna: 2010) 61–80; and Tosi, "Acconciare".

110 Most of these can be found in volumes IV, V, VI and VII of Aldrovandi's *Tavole* (see List of Codices). See on these Alessandrini A. – Ceregato A. (eds.), *Natura Picta. Ulisse Aldrovandi* (Bologna: 2007); and Tongiorgi P., "Dalle profondità dei mari del XVI secolo", in Alessandrini – Ceregato (eds.), *Natura Picta* 89–93, with identifications of Aldrovandi's depicted *aquatilia*.

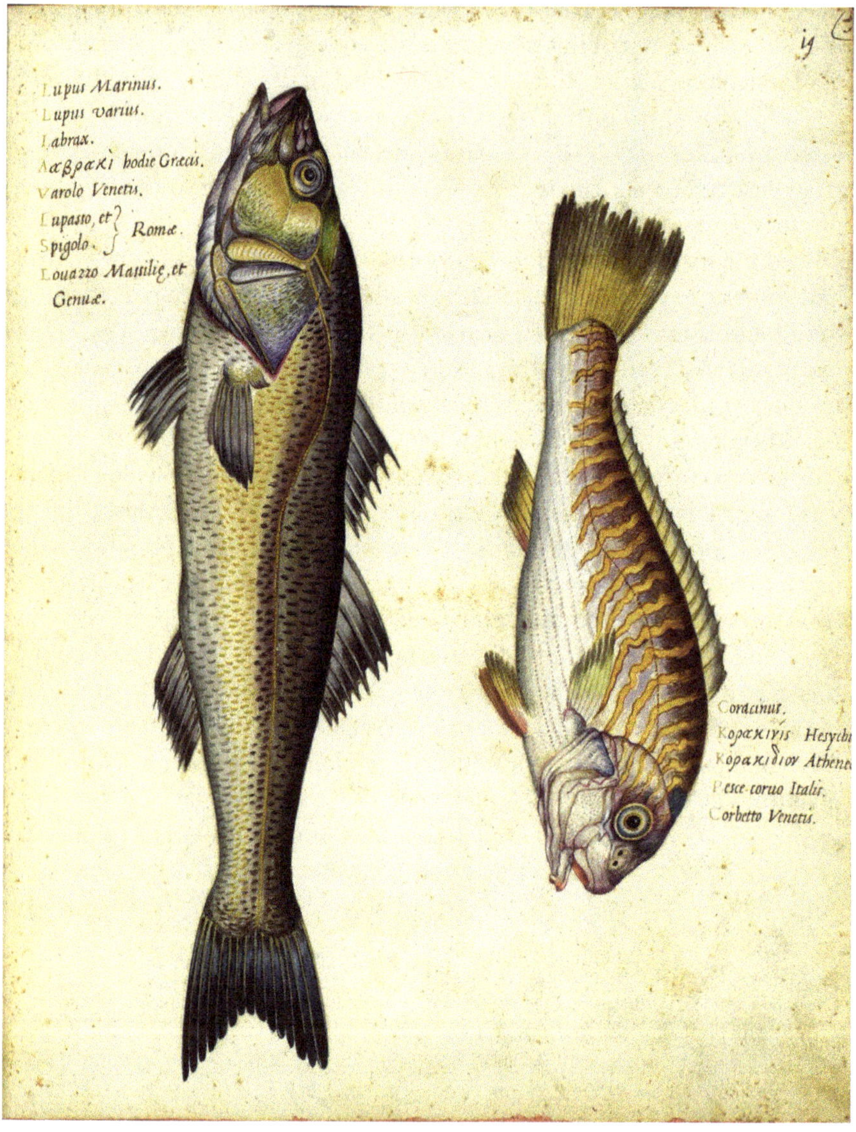

FIGURE 6.13 Lupus marinus, and coracinus in the Aldrovandi collection, second half 16th century. Aldrovandi, *Tavole*, vol. 4, fol. 19
© BOLOGNA, BIBLIOTECA UNIVERSITARIA

may not always have known whether their drawings would serve only research or also print.[111]

Ligozzi was certainly the most famous painter connected with Aldrovandi (from 1577 on), but his total contribution to the latter's image collection was modest in a quantitative sense: no more than about 30 drawings of naturalia can securely be attributed to him, as Olmi argues. By far the larger part of Aldrovandi's animal drawings was painted by Giovanni de' Neri, whom Aldrovandi employed in his home town Bologna and paid for nearly thirty years, *c.*1558–*c.*1590. From about 1558 to 1564 Neri may even have been Aldrovandi's only painter. Neri's high production rate is reflected in the unimpressive artistic quality of many of his drawings, which clashes especially with those by Ligozzi.[112]

Aldrovandi's collection consisted not only of *newly* drawn images, however. Like Gessner, Salviani and many other naturalists, he also incorporated drawings sent him by friends. These had been made elsewhere and by other painters, therefore. But more importantly, Aldrovandi's own painters often copied drawings in the collections of others or made duplicate images of their own drawings. We have already seen that Aldrovandi had a whole cluster of Daniele Barbaro's Venetian fish drawings copied in the mid 1550s; he borrowed the latter's fish book again in the late 1560s.

Not much hard evidence is available about the album or book with fish drawings on parchment commissioned by Leone Tartaglini in Venice, but it is possible to sketch a few contours and exclude some possibilities. While it is very likely that Tartaglini had some images copied from the older Barbaro album, the Barbaro and Tartaglini fish albums – both lost, both Venetian – were two distinct image collections with different chronological origins, different painters, and very probably a different materiality and functions. First of all, they were two distinct objects: in October 1571 Aldrovandi saw not only Tartaglini and his collection in Venice, but also admired the late Daniele Barbaro's fish book in a palazzo of the Barbaro family.[113] Barbaro's fish book was already size-

111 See Olmi, *L'Inventario del mondo* 64–85; and Olmi – Simoni (eds.), *Ulisse Aldrovandi*; as well as Tosi, "Acconciare". Some of the earliest information about Aldrovandi's painters goes back to Fantuzzi, *Memorie della vita* 89–90. On the Codex Pomar and copying Aldrovandi's images after his death, see Sallent del Colombo, "Natural History Illustration" and Sallent Del Colombo – Pardo Tomás, "Materiali aldrovandiani".

112 Olmi, *L'Inventario del mondo* 64–70.

113 See Conigliello, "Pesci, crostacei" 25; and Hochmann, "Plinio Scarpelli" 47, 52 note 19. Tosi A., "Contrivances of art. The power of imagery in the early modern culture of curiosity", in Beretta M. – Conforti M. (eds.), *Fakes. Hoaxes, Counterfeits and deception in early*

able by 1550. Tartaglini's collecting of fish drawings seems to have belonged rather to the 1560s–70s, though an earlier start cannot be excluded. The fish album was only one of Tartaglini's 'painted books' with naturalia drawings, moreover: others depicted birds and plants. Finally, the contrasting ways in which Barbaro and Tartaglini treated their albums offers some important clues. As we have seen, Barbaro travelled abroad carrying around his large *aquatilia* image collection of at least 300 figures, and lent his albums repeatedly to Aldrovandi. This was a research collection for Barbaro's own delight that could also be lent to naturalists who enjoyed his patronage. Whatever their scientific quality, it seems unlikely that these were luxury drawings on expensive parchment. Tartaglini's naturalia drawings, on the other hand, were definitely painted on parchment. And his last will of 1576 indicates that he regarded his parchment books of bird, plant and fish drawings as so precious that he kept them in a safe place, so that no one would touch them but himself and the few persons to whom he gave access. He certainly did not let them out of his sight.[114]

The intriguing Tartaglini, who originally did not come from Venice but from a small town in eastern Tuscany, combined a number of professional activities. He was an herbalist with his own stall (*banco*) on the central Venetian Piazza San Marco, a *cerusico* (barber-surgeon), a collector of naturalia, antiquities and coins, who also owned exotic animals such as a live monkey and a (dead) bird of paradise. Tartaglini dried and sold many fish, including fake dragons, and he had a garden on Murano where he grew herbs and special plants that interested collectors. The rarities he owned included plant drawings on parchment by Ligozzi, marble antiquities, seeds, minerals, and an exotic fish skin. He was both an expert and a commercial entrepreneur who operated in the world of antiquities and naturalia collectors, as Paula Findlen has pointed out, calling him 'an artisan of nature and a vendor of secrets'. He may well have acted as agent and dealer not merely in the naturalia themselves, but also in the costly images of naturalia on parchment. His last will states that his precious

modern Science (Sagamore Beach: 2014) 153–175, here 167–168, discusses Tartaglini's collecting in the 1560s–1570s and the copying of his images for Aldrovandi by the latter's painter Giovanni de' Neri.

114 The testament is printed in Brusegan M. (ed.), *Leone Tartaglini, Opera nuova nella quale se contiene la natura dil sonno cioe come lhuomo debbe dormire per mantenersi sano con alchuni bellissimi & utilissimi secreti medicinali, Venetia 1551* (Fermo: 1996). Cf. Hochmann, "Plinio Scarpelli" 49; Tosi A. (ed.), *Ulisse Aldrovandi e la Toscana. Carteggio e testimonianze documentarie* (Florence: 1989) 17–18, 126–128 on contacts Tartaglini-Fulcheri-Aldrovandi and a copied Tartaglini fish list that should reach Aldrovandi in 1569 from Fulcheri; cf. Idem, "Acconciare" 55–57.

parchment albums of birds, plants and fish should be offered for 70 scudi to the bishop of Verona (Agostino Valier), who had already admired them earlier.[115]

It seems likely that more than just one painter was involved in the making of Tartaglini's three albums during the 1560s–70s. Jacopo Ligozzi from Verona, who spent some time in Venice in the 1570s, certainly painted fish and plants for Tartaglini, and there was also a definite link with Francesco Ligozzi, the slightly less talented brother of Jacopo.[116]

Jacopo Ligozzi's involvement in the making of the Tartaglini albums might even help explain how Ligozzi gained enough of a reputation to be appointed court painter by the Medici in 1577 – which itself could throw some further light on the dating of other naturalia drawings. Among those are the 20 folios with water creatures of the abovementioned Viennese Codex Miniatus 83, which has formed part of the imperial Habsburg collections since the years of Rudolph II. Conigliello has pointed out that its large format parchment folios and the very fine painterly style with highly realistic details, some fantastic fish, and mannerist traits are reminiscent of Jacopo Ligozzi. She also argues that most of the depicted *aquatilia* are Mediterranean, and suggests that several animals seem painted from life, but that others definitely go back to printed sources, which range from Belon's editions of 1551 and 1553 to Salviani's work of the mid 1550s. There are close visual links as well with some of Aldrovandi's images. One of the Codex's two drawings of sunfish (*Mola mola*), for instance, goes back straight to Salviani's sunfish. The filiation of its second *Mola mola* leads us into yet another loop: it is based on Aldrovandi's *Mola mola* that itself was copied after Daniele Barbaro's fish book [Fig. 6.14].[117]

Many questions about Cod. Min. 83 still remain open, however, and its drawings have not yet been systematically compared with various other image collections. It seems more than likely that Jacopo Ligozzi was involved, while

115 See Findlen P., "The Market and the World. Science, Culture and Collecting in the Venetian Republic", in Aikema B. (ed.), *Il collezionismo a Venezia e nel Veneto ai tempi della Serenissima* (Venice: 2005) 55–68, here 61–62 (also for the quotation); Brusegan (ed.), *Leone Tartaglini* 20–26, 30; and for more details Tosi (ed.), *Ulisse Aldrovandi e la Toscana* 17–18; and Idem "Contrivances of art" 167–169.

116 On Jacopo and Francesco Ligozzi and Tartaglini, see Conigliello, "Pesci, crostacei" 23–26; cf. Tosi, "Contrivances of art" 170; Hochmann, "Plinio Scarpelli" 49; and Brusegan (ed.), *Leone Tartaglini* 23. Olmi, *L'Inventario del mondo* 83–84, mentions that Jacopo Ligozzi's brother Francesco also worked in Venice for Aldrovandi, painting birds, and should not be confused with Francesco Mercurio di Ligozzi, a cousin of Jacopo, who also painted naturalia (possibly esp. exotica) and was in contact with Aldrovandi *c.*1590–92.

117 Conigliello, "Pesci, crostacei" 22–26. The same filiation (Barbaro – Aldrovandi – Cod. Min. 83) also applies to the Orbis stellatus and the Clupea.

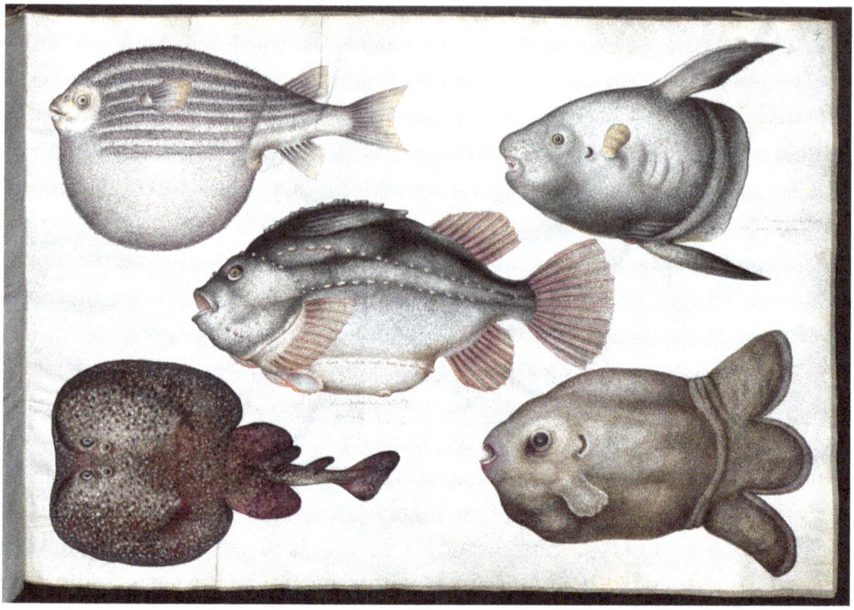

FIGURE 6.14 Parchment folio with fish drawings, including a *Mola mola* probably copied after Salviani (cf. Fig. 6.8.). Undated, attributed to Jacopo Ligozzi. Cod. Min. 83, fol. 7

new research by Concin indicates that he made this album (probably 1577–78) while in the service of the Medici, and that it was part of a lavish set of diplomatic gifts in 1578 from Francesco I de' Medici to Rudolph II.[118] It was certainly possible for a painter to work under both Habsburg and Medici patronage – though usually not at the same time. The miniaturist Daniel Froeschl (*c.*1572–1613), for instance, worked at Rudolph II's court in Prague and was responsible for an early seventeenth-century court inventory of Habsburg albums that links Ligozzi's name with Cod. Min. 83. Froeschl the painter also made at least some naturalia drawings in a heterogeneous and undated codex in Pisa: some of its circa 25 pages with *aquatilia* drawings are clearly linked

118 Concin A., "Splendid Gifts and a Florentine Architect for Emperor Rudolf II: Antonio Lupicini at the Imperial Court in Prague (1578–1580)", *Studia Rudolphina* 20 (2020) 24–49, here esp. 34–40.

with Habsburg codices while others are connected visually with Aldrovandi drawings [Fig. 6.15].[119]

The group of Habsburg codices with naturalia drawings inventorized by Froeschl comprises two further sets of drawings that focus completely on *aquatilia*: a coarsely painted and undated large album, and a stunningly beautiful (unbound) series of drawings on very large parchment sheets.[120] About the former album, Cod. Ser. 2648, very little is known. It supposedly dates from the second half of the 16th century, has almost 60 folios, and contains a large number of drawings of marine *aquatilia* accompanied by their names in Italian.[121] The beautiful parchment codex of loose sheets (Cod. Ser. 2669) is a very different case. It once more highlights a close collaboration between a naturalist-physician – Pietro Andrea Mattioli – and a painter: Giorgio Liberale da Udine (*c.*1527–1579). Both were Italians; both worked mainly under the patronage of the Austrian Habsburgs.

Mattioli's European fame as a naturalist rests on his work on plants, and in particular on his numerous, ever expanding and permanently updated editions of Dioscorides. While printed illustrations began to appear in the 1554-edition, as discussed earlier, increasing numbers of large fish and animal illustrations can be found in the Latin editions of 1563 and 1565 published by Valgrisi in Venice. Giorgio Liberale da Udine was the painter of by far the larger part of all these images – plants and animals. His collaboration with Mattioli started in 1553. Mattioli was not exaggerating when he called Liberale his personal painter if we inspect the quantities and quality of the woodcut illustrations and try to imagine the amount of time that Liberale must have spent on the model drawings. The 1554-Mattioli edition contains some 600 small plant and animal drawings. By 1565 there are some 900 large woodcuts of plants and some 100 of animals.[122]

Obviously, the chronology of this collaboration is significant. It indicates that Mattioli and his publishers were highly sensitive to the visual turn in

119 See List of Codices. Much of the Codex Froeschl-De Bruyn seems to date from after 1577; the fish section starts on page 209. See Garbari – Tongiorgi Tomasi – Tosi, *Giardino dei Semplici*. On Froeschl's connections with both Pisa and Prague, see Olmi, *L'Inventario del mondo* 86.

120 See List of Codices. On the whole group, see Irblich E. (ed.), *Thesaurus Austriacus: Europas Glanz im Spiegel der Buchkunst: Handschriften und Kunstalben von 800 bis 1600* (Vienna: 1996); and Weiler C. (ed.), *Von Fischen, Vögeln und Reptilen. Meisterwerke aus den kaiserlichen Sammlungen* (Vienna: 2011) with further references.

121 See Weiler (ed.), *Von Fischen, Vögeln* 187–188.

122 On Liberale's role as Mattioli's personal painter, see Tongiorgi Tomasi L., "Il problema delle immagini nei 'Commentarii'", in Ferri S. (ed.), *Pietro Andrea Mattioli: Siena, 1501 – Trento, 1578: la vita e le opere* (Perugia: 1997) 369–376, here 369–371.

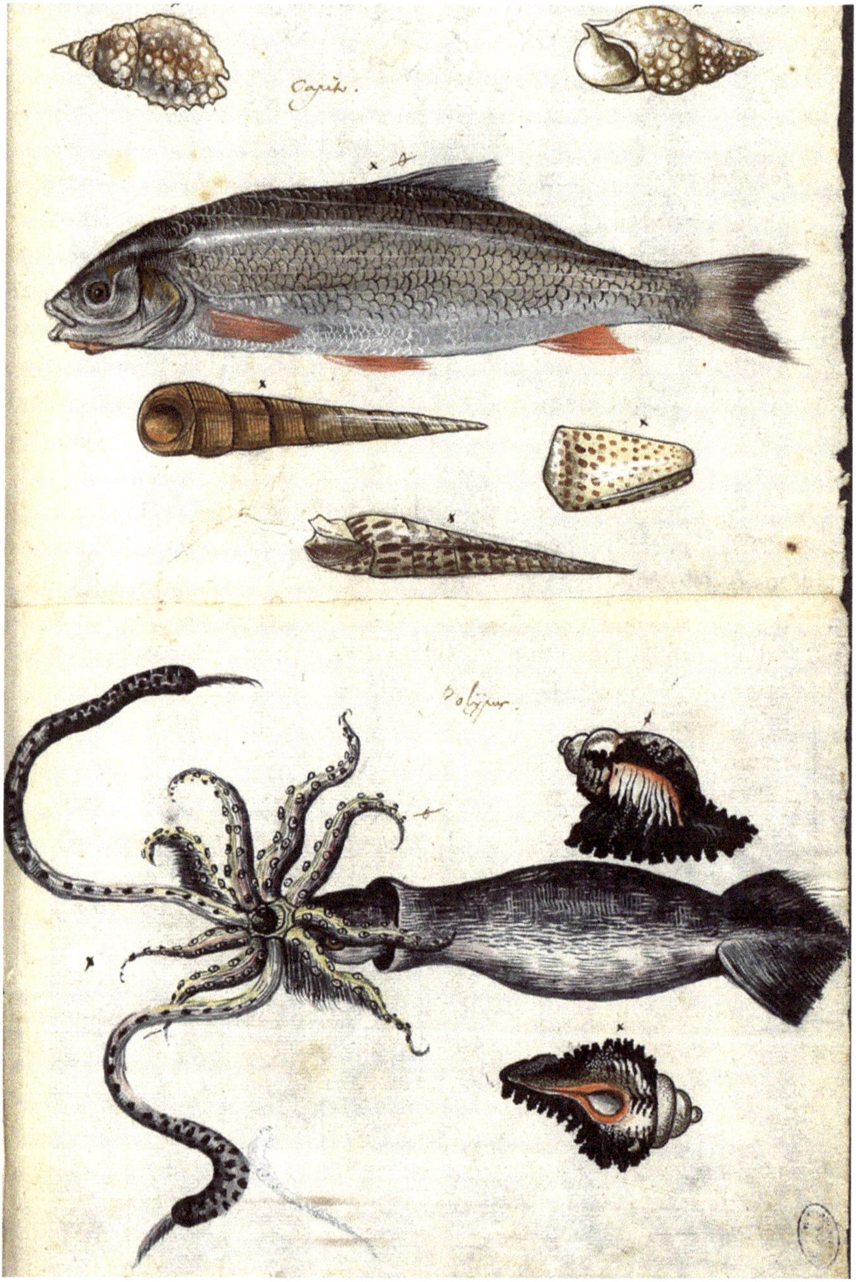

FIGURE 6.15 Various *aquatilia* in the Codex Daniel Froeschl – Nicolaes De Bruyn, MS 514, fol. 272, *c.*1570–1620s
© PISA, BIBLIOTECA UNIVERSITARIA

the early 1550s. It also suggests that Mattioli was well aware in 1553–54 of the fish works that fellow-naturalists and rivals were publishing and preparing. Liberale's drawings made between *c.*1553 and *c.*1565 for Mattioli were, therefore, intended from the start for print – not for collecting, nor for display, nor exclusively as research material. Presumably for this reason his original model drawings do not survive: they may have been destroyed during the printing process. Other originals by Liberale do still exist, however.

In early 1555, Mattioli moved from Gorizia (north of Trieste) to Prague, where he became court-physician and naturalist of Archduke Ferdinand II, younger brother of Emperor Maximilian II, governor of Bohemia (1547–67), and ruler of Tyrol from 1564. Liberale da Udine appears to have followed Mattioli soon afterwards to Prague, at least for the duration of his work on the large format book illustrations. Ferdinand II extended his patronage to Liberale, and *c.*1562 commissioned him to create what can only be called a visual fauna of the Adriatic Sea on a princely scale [Fig. 6.16]. The result was the very large format Cod. Ser. 2669 that consists of 100 oblong folios of calf parchment, measuring 88 × 64 cm. They are painted in incredibly fine brushwork, mostly on both sides. There were many delays, however. Nine years after the initial contract, which promised Liberale 200 florins per year, the conditions were changed to payment only for what Liberale actually delivered. The collection of painted parchment sheets thus grew slowly in the course of decades. In 1576–77 Liberale moved to Innsbruck with his family, where Ferdinand II had chosen Schloss Ambras as his new seat as ruler of Tyrol. Liberale eventually died *c.*1580 with his fish codex not quite completed.[123]

The Liberale codex shows marine fish, mammals, molluscs, some serpents, reptiles, and insects. Most or all of the *aquatilia* depicted belong to the Adriatic Sea or more generally to the Mediterranean. Most sheets represent several *aquatilia*: from one to as many as two dozen in the cases of small crabs, shells, and the insects. Some pages are unfinished with only the fine outline of the fish sketched. Many completed sheets show not merely the naturalia, however, but

123 See List of Codices. The best biographical information on Liberale is Coronini Cronberg G., "Giorgio Liberale e i suoi fratelli", in Brussich G. – Del Fabbro I. (eds.), *Studi di storia dell'arte in onore di Antonio Morassi* (Venice: 1971) 85–96. See further Tongiorgi Tomasi, "Il problema delle immagini"; and Mazal O., "Tierbilder aus der Ambraser Kunst und Wunderkammer Erzherzorg Ferdinands von Tirol (Cod. Ser. N. 2669 der Österreichische Nationalbibliothek)", *Codices Manuscripti* 8 (1982) 12–38, here 19–21, which is still fundamental on this codex; he reports that Liberale also received a commission to depict birds from Ferdinand II in 1577. Cf. Weiler (ed.), *Von Fischen, Vögeln*. On Mattioli and Prague, see Bohatcová M., "Prager Drucke der Werke Pierandrea Mattiolis aus den Jahren 1558–1602", *Gutenberg Jahrbuch* 60 (1985) 167–185.

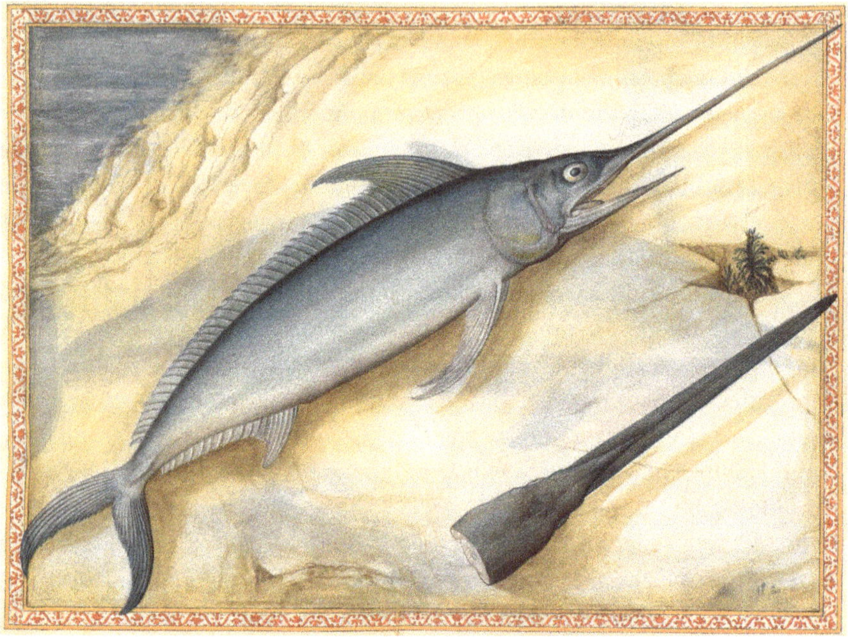

FIGURE 6.16 Swordfish in a parchment codex, painted by Giorgio Liberale da Udine for Archduke Ferdinand II of Tirol, *c.*1562–80, Gorizia and Innsbruck. Cod. ser. 2669, fol. 51v

a composition with an invented or sketched background of watery environs and rocky or sandy surroundings, and even some coastal scenes. These backgrounds are often evocative and range from the almost completely abstract to the almost completely naturalistic. In nearly all cases the naturalia are depicted 'on top' of these backgrounds rather than integrated into a composition, and their poses and grouping per class of animals belong explicitly to the genre of scientific representation. Yet, many sheets also have a decorative frame of *trompe-l'oeil* painted gems, jewels, pearls and scrolls, that belongs very much to the long tradition of manuscript illumination. The quality of the large format parchments sheets, the painterly style in extremely fine detail, and these precious decorations place this whole image collection in the sphere of courtly collecting, rarity, and treasure.[124]

The large format of Liberale's parchment sheets is as nothing compared to those of a little known Piedmontese fish album, *Album dei Pesci*, in Turin. That album contains some 75 paper folios on which cut-out, brightly coloured

124 See Mazal, "Tierbilder" 15–16.

drawings (also on paper) are pasted of fish and other *aquatilia*, such as whales, seals, sea urchins, jelly fish, and lobsters. They represent mostly Mediterranean and some fresh water *aquatilia*, besides a few reptiles. The animals are accompanied by their handwritten names in Italian, some in the dialect of Nice, sometimes in Latin, and often (also) in Portuguese. The truly stunning aspect of this album is the size of many images, which has caused them to be folded several times to fit into the album. Many fish are depicted close to life size. Several drawings are more than 1 meter long; one reaches almost 3 meters. Conte suggests that up to four painterly hands can be distinguished of different artistic quality; the quality of most drawings is good, though not quite at the Ligozzi level. The drawings probably date to the years between about 1580 and 1624. They were thus made during the rule of – and possibly for – Duke Carlo Emanuele I of Savoy (ruled 1580–1630), whose interest extended not only to warfare but also to the Aldrovandi and Medici naturalia collections. Conte has suggested links between some of these fish drawings and patterns in the mosaic floor of this Duke's (no longer extant) Galleria Grande in Turin, which could point to the drawings as either models for or copies after that floor. They might even have had an explanatory function of the 'marine encyclopaedia' of the floor mosaic.[125] A link that has not been explored to my knowledge is that with Liberale da Udine: a number of the drawings in the Turin Codex show remarkable similarities with Liberale's drawings in the parchment codex discussed above, for which Liberale also made use of his own model drawings for Mattioli, as we have just seen. There are not merely stylistic similarities, but several *aquatilia* are also depicted both from above and from below, as in the Liberale album. Could the Turin drawings be connected somehow with Liberale da Udine?

In Liberale we thus find a painter who created *aquatilia* drawings first of all as models explicitly intended for the illustration of Mattioli's printed work on natural history, and who then transferred the visual format of scientific illustration (but not the painterly style) into the very different domain – and materiality – of precious illuminated manuscripts made for the Habsburg court,

125 See List of Codices. With thanks to Giulia Simonini for drawing my attention to this codex. The Galleria Grande (destroyed by fire in 1659) connected Palazzo Madama and the Palazzo Reale in Turin. Duke Carlo Emanuele I commissioned its decorations, starting in 1587. See Conte F., "Intorno alla Grande Galleria di Carlo Emanuele I: inventari del sapere a Torino tra Cinque e Seicento", in Cioffi R. – Sconamiglio O. (eds.), *Mosaico. Temi e metodi d'arte e critica per Gianni Carlo Sciolla* (Naples: 2012) 197–206, esp. 200; and see Bava A.M. – Pagella E. (eds.), *Le meraviglie del mondo. Le collezioni di Carlo Emanuele I di Savoia*, exh. cat. Turin (Genua: 2016), here 137–139. I have not been able to see the original album (though restored, the drawings are very fragile) or to consult the older literature.

where it would have served courtly display and study purposes. Interestingly, his presentations in the albums are more innovative than those made for printing purposes. For Mattioli, who knew many Adriatic fish by personal observation, Liberale painted a relatively small number of *aquatilia*. As illustrator he remained tied to the textual discussions of *aquatilia* by Dioscorides and Mattioli, and he focused completely on the fish. Ferdinand II's commission, on the other hand, allowed him to create a perhaps incomplete, but certainly impressive visual aquatic fauna, in which he often combines groups of animals on one folio and inserts painted backgrounds. We cannot tell how many of those drawings were based on direct personal observation by Liberale, but it is surely significant that he spent most of his life close to the Adriatic Sea, in Gorizia.

Liberale's codex for Ferdinand II may well be the very first visual fauna created in Europe. Again, the chronology seems significant. Its production between 1562 and *c.*1580 coincides with that of two works about plants that aimed at similar regional surveys (in both text and image): Carolus Clusius' Spanish and Pannonian floras. Clusius researched the former in Iberia in 1564–65 and published it in 1576; the latter was researched *c.*1578–81 from the Viennese and Hungarian courts and appeared in 1583.[126] One mid 16th-century Portuguese-German manuscript that is currently being analysed may be an even earlier example of this same trend. It is the *História Natural de Portugal* written in Lisbon (during 1555–56) in the house of the Portuguese humanist and philosopher Damião de Góis by the Swiss polymath, alchemist-astrologist-physician-apothecary Leonhard Thurneysser zum Thurn (1531–1596). The latter was, incidentally, an almost exact contemporary and fellow citizen of the naturalist and collector Felix Platter from Basel. This manuscript in German contains a long section on plants observed in Portugal that seemed different from those north of the Alps, plus a shorter section of some 18 folios with descriptions of animals, in particular aquatic ones, that could be found in Portugal.[127] The regional focus of these works and their attempts to provide a survey of a certain

126 See Clusius Carolus, *Rariorum aliquot stirpium, per Hispanias observatarum historia* (Antwerp, Christopher Plantin: 1576); and Idem, *Rariorum aliquot stirpium, per Pannoniam, Austriam, & vicinas quasdam provincias observatarum historia* (Antwerp, Christopher Plantin: 1583). Perhaps the following (in so far as I know *non*-illustrated) manuscript treatise on fish and fishing in the river Elbe points in a similar direction: Handsch von Limus, Georg, *Die Elbefischerei in Böhmen und Meißen*, ed. O. Schubert (Prague: 1933); the original text dates from not before 1547. Handsch too worked for the Innsbruck-Ambras Habsburgs.

127 On this manuscript (Staatsbibliothek Berlin, Ms. Germ. Fol. 97), see Herold B.J. – Horst T. – Leitão H., *A História Natural de Portugal de Leonhard Thurneysser zum Thurn, ca. 1555–1556* (Lisbon: 2019), and Herold's and J.P.S. Cabral's contribution in the present volume. The

category of naturalia seem to indicate a new development in nature studies of the second half of the 16th century. After the vast encyclopaedic works and collections produced during the 1540s–50s that aimed to comprise more or less everything, some naturalists had begun to realize that focus and specialization could contribute to exhaustive surveys.

In spite of the numerous uncertainties regarding the *aquatilia* codices discussed in this section it is clear that these image collections are linked by lines of filiation and copying. Most originated in northern Italy, but there are links to Bologna, Florence and Pisa southwards, and to the main Habsburg centres in Prague, Innsbruck and Vienna northwards. In terms of function these codices were extremely different, however. Aldrovandi's vast visual corpus at first mainly served study and research purposes; he intended to publish, however, and selected images were eventually used in Aldrovandi's (mostly posthumous) publications. His visual documentation helped to expand the number of known species, but was based on the same notion of describing, inventorizing and depicting nature that had inspired the publications of the 1550s. Felix Platter's herbarium and compound image collection was a display and visual research collection for himself and for friends-colleagues. The lost Tartaglini parchment albums and the Habsburg Cod. Min. 83 did not serve research purposes, but embodied expert knowledge and showed artistic top quality. Perhaps they were made with the elite collectors in mind who eventually came to own them, for knowledge and delight.[128] There is no better example than Liberale da Udine's Cod. Ser. 2669, however, to demonstrate that no rigid divide separated scientific drawings from images for delight and display. Fish drawings by Liberale served as illustrations in Mattioli's scientific work on nature. Given backgrounds, a decorative frame and a new materiality in the form of precious parchment, the same drawings were perfectly suitable at court.

5.3 *Moving North of the Alps*, c.1565–1600

The geographical focus of this survey has so far been mainly on Italy and to a much lesser extent on France, since no collections of *aquatilia* drawings from

extant exemplar does not contain the original drawings. I thank Bernardo Herold for sharing his full transcription of the naturalia lists in the manuscript.

128 On Rudolph II as expert and collector and the Prague court as a cultural and scientific centre, see esp. Bukovinská B., "The known and unknown Kunstkammer of Rudolf II", in Schramm H. – Schwarte L. – Lazardzig J. (eds.), *Collection, laboratory, theater: Scenes of knowledge in the 17th century* (New York: 2005) 199–227; and Idem, "Die Kunstkammer Rudolfs II – Entstehung, Niedergang, Wiederentdeckung", in Haag S. – Kirchweger F. – Rainer P. (eds.), *Das Haus Habsburg und die Welt der fürstlichen Kunstkammern im 16. und 17. Jahrhundert* (Vienna: 2015) 229–252.

before 1560 are known to me that originated anywhere else in Europe. In this last section we will look northwards, starting from Venice and the Habsburg court cities in Central Europe.

Of the many codices with painted naturalia that circulated at the Habsburg courts in Vienna, Prague and Innsbruck between about 1575 and 1615 some were newly created at court by painters such as Arcimboldo, Daniel Froeschl, and Georg Hoefnagel. Occasionally they even portrayed animals that actually lived in the court menageries. Other albums, such as Liberale's codex, though commissioned by the Habsburgs, were painted elsewhere. Court painters also copied drawings from other codices, scrapbook collections, model books, and from print. Even within the courtly context a major amount of image re-cycling and copying was thus going on, while many painters active at these courts were, moreover, not limited to the Habsburg circles and cities. They operated in more than one city or at more than one court. They travelled, and their images travelled even more.

The patterns of these movements are multidirectional. We have already seen how images entered the Habsburg court circles from Italy. But naturalia drawings also arrived there from the Low Countries, which – in spite of the Dutch Revolt – were still partly under Habsburg rule. Before 1610 (probably before 1600) Rudolph II acquired, for instance, three albums with animal drawings painted on parchment by the Flemish painter Hans Bol (1534–1593): one with four-footed animals, one of birds and one of *aquatilia*. The latter contains some 70 illustrated folios. Bol painted these albums in Antwerp, starting possibly as early as 1572. As Rikken has shown, once the albums reached Prague, their motifs entered via copying into animal codices that originated at the Prague court itself.[129] Copied images of *aquatilia* also travelled outwards from the Habsburg courts, to the German-speaking countries, presumably to parts of Eastern Europe, and certainly to the Low Countries and Italy.[130] As Olmi has shown, for instance, the Italian court physician Francesco Padovani in Prague provided Aldrovandi in Bologna with animal drawings 'ad vivum

129 See Rikken, *Dieren verbeeld* 38–39, 115–116. See List of Codices. The catalogue of the Royal Library Copenhagen dates the Bol albums to the early 1590s.

130 I know of no German *aquatilia* albums from the late sixteenth or early years of the seventeenth century – that is from before the special and original albums of river and lake fish drawings (as well as other animals) by the Strasbourg fisherman-naturalist Leonhard Baldner (*c*.1653–1666). The Nuremberg album of animal drawings by (and collected by) Lazarus Röting (*c*.1580–1614) contains few *aquatilia* drawings, which seem to be mainly based on printed illustrations by Gessner; see Hackethal S., "Das Theatrum Naturae des Michael Rötenbeck. Unbekannte Naturstudien 100 Jahre nach Dürer", in Damaschun F. et al. (eds.), *Klasse Ordnung Art. 200 Jahre Museum für Naturkunde* (Rangsdorf: 2010) 70–75. For the albums, see List of Codices.

coloribus delineatas' (painted *ad vivum* in colour) in Prague; some of those may have been made by or copied after Arcimboldo.[131] The Habsburg courts functioned, therefore, as centres of naturalia image production, 'engines' of circulation, and crossroads in the transit of images between southern and northern Europe. That should make us think anew about the effects and reach that drawings could have – even when made for the eyes of a limited and courtly public. In this context the distinction between a newly made drawing (even one based on direct observation of an animal) and its multiple copies may well have been irrelevant.

There was a conspicuous presence of painters and scholars from the Low Countries at the Habsburg courts between about 1575 and 1610. Some of them were known for a strong interest in naturalia that could even amount to specialization. Father (Georg) and son (Jacob) Hoefnagel from Antwerp were particularly famous for their miniatures on both parchment and paper of shells, insects, other small animals and flowers. Roelant Saverij – a former pupil of Hans Bol in Antwerp – worked at the Prague court from 1603 and produced various well-known animal paintings. Dirck de Quade van Ravesteijn (*c.*1565/70–1618) from The Hague(?) was the principal painter of the albums known as Rudolph II's 'Bestiary', in which he used various motifs from Hans Bol's albums; he worked in Prague from 1588 until his death. And Anselmus de Boodt (1550–1632) from Bruges acted as Rudolph's court physician, specialist in precious stones and their alchemical-medicinal uses, and painter of naturalia.[132]

Although *aquatilia* are present in the albums and works by these predominantly Southern Netherlandish painters, they rarely occupy an important place. There are relatively few in the works of the Hoefnagels, while the small number of *aquatilia* images in Quade van Ravesteijn's 'Bestiary' mainly represent *Kunstkammer* objects of marine origin (sea urchins, corals, shells, dried specimens and dragons made of dried rays) and not a marine fauna.[133] These image collections did not aim to present encyclopaedic completeness, therefore. Anselmus de Boodt's albums made for Rudolph II, were different and cover as wide a range of animals as possible. In this large collection water creatures form an important segment of some 70 folios, mostly with one or two figures. De Boodt, who spent the years 1587–1614 mainly in Prague, was both

131 Quoted and discussed in Olmi, "Bologna nel secolo XVI" 69–70.

132 See Rikken, *Dieren verbeeld* esp. 117, 126–127. Cf. Weiler (ed.), *Von Fischen, Vögeln*.

133 See List of Codices, Cod. Min. 129–130. See Haupt H. – Vignau-Wilberg T. – Irblich E. – Staudinger M., *Le bestiaire de Rodolphe II. Cod. Min. 129 et 130 de la Bibliothèque nationale d'Autriche* (Paris: 1990). On the Hoefnagels, see Vignau-Wilberg T., *Joris and Jacob Hoefnagel. Art and Science around 1600* (Berlin: 2017).

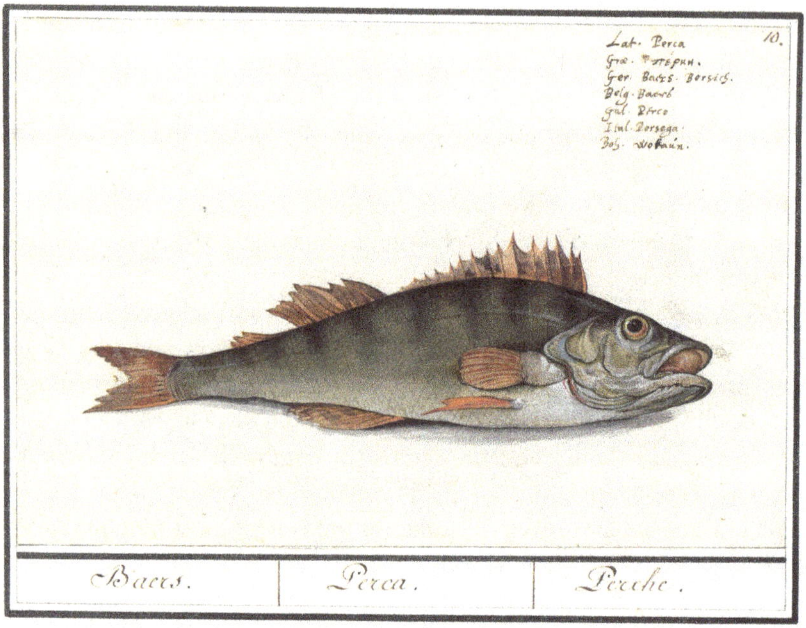

FIGURE 6.17 Perch in the albums of Anselmus de Boodt made for Emperor Rudolph II, vol. VI, fol. 42, Prague, late 16th–early 17th century
© AMSTERDAM, RIJKSMUSEUM

painter and compiler of these albums: he personally painted numerous animals, but also copied images that circulated at court and even included various animal drawings painted by others [Fig. 6.17].[134]

Chronologically the involvement of painters from the Low Countries in the animal painting of the Habsburg courts appears to belong mainly to the years *after* 1590 or even 1600. The Flemish painters did not turn to naturalia painting once they arrived at the Habsburg courts, however; they were already expert in these subjects, and the tradition of animal drawing *in* the Low Countries (especially the Antwerp region) went back well before the 1590s–1600s, as was already clear from the example of Hans Bol. Nor did the Netherlandish painters lose touch with that tradition while in Prague or Vienna. Some indeed returned from Prague to the Low Countries at the end of their employment at court or during the last years of their lives.

134 See List of Codices. De Boodt certainly included drawings by Elias Verhulst from Mechelen. See Maselis M.C – Balis A. – Marijnissen R.H., *The albums of Anselmus de Boodt (1550–1632): natural history painting at the court of Rudolph II in Prague* (Ramsen: 1999).

The question thus emerges of when a Netherlandish tradition of animal drawing originated, and of whether it did so independently from the Italian one that went back at least as far as the 1530s–40s. Or was it an offshoot of that Italian tradition – perhaps via as yet unexplored contacts between the Habsburg courts of Vienna and Brussels in the 1550s–70s, via the even earlier Dürer connection, or via direct contacts between Italy and the Low Countries (especially the major ports of Venice and Antwerp)? This question requires a wide-ranging discussion of European animal drawing and painting that goes far beyond the scope of the present essay, but at least some of the evidence from the Low Countries points in the direction of diffusion rather than an independent development.

Rikken has shown that the earliest known, quite small, groups of animal drawings by painters from the Low Countries were collected in albums – like their southern counterparts. The principal painters involved were Lambert Lombard (1550s–70s), Hans Bol (mainly 1570s–80s), and Hans Verhagen (post 1547 and mainly 1560s).[135] Chronologically, therefore, none of the extant animal drawings from the Low Countries seems to date back further than the late 1550s or more probably the 1560s. By the 1560s, major collections existed in the Habsburg Netherlands and included marine creatures. The collection of the Spanish humanist and orientalist Benito Arias Montano, who lived in Antwerp during the years 1568–79 and was closely linked with the circles of Ortelius and Plantin, even included a section called 'The Sea' which probably included marine material collected in Antwerp.[136] It is likely, as we will see below, that some of these collections included albums with naturalia drawings, as was also the case in the Habsburg and many Italian cases. As Rijks has discovered, a fascinating but lost Antwerp album with either drawings or engravings of *aquatilia* is mentioned in the inventory (1625) of a Antwerp collector, the notary Gillis de Kimpe, and may date from the 16th century.[137] The link between collections and albums with naturalia drawings is far less clear, however, for the Low Countries than for the Viennese-Prague collections or for some of the Italian ones. Animal painting in the Low Countries fairly quickly

135 See Rikken, *Dieren verbeeld* 25–33. On *aquatilia* collecting as objects in Antwerp, see esp. Rijks, "A painter, a collector" and "'Unusual Excrescences of Nature'", and her contribution in the present volume.

136 See Göttler C., "Extraordinary Things: 'Idols from India' and the Visual Discernment of Space and Time, circa 1600", in Göttler C. – Mochizuki, M.M. (eds.), *The Nomadic Object. The Challenge of World for Early Modern Religious Art* (Leiden – Boston: 2018) 37–73, here 44; and for a list of the collection, Salazar A., "Arias Montano y Pedro de Valencia", *Revista de estudios extremeños* 15 (1959), 475–493, here 490–491.

137 See Rijks, "A painter, a collector" 346 for details about this collector and the inventory.

developed different modes and formats, moreover. During the last quarter of the 16th century the emphasis shifted away from painted albums to print series (Abraham and Nicolaes De Bruyn; Adriaen Collaert; Marcus Gheeraerts; Jacob Hoefnagel) and to oil paintings (from *c.*1590).[138]

In this context the mainly unpublished and still hardly studied animal drawings of superb quality in the *Libri Picturati* volumes A16 (*aquatilia* and four-footed animals) and A17 (birds) may turn out to be crucial to the understanding of how animal painting developed in the Low Countries. The majority of those drawings can be provisionally dated on the basis of watermarks to the 1560s, just like the oldest core of plant drawings in the same collection. All of those oldest drawings were commissioned by the nobleman Charles de Saint Omer (1533–1569): after a military career he retired on account of ill health, at the age of about 30, to a castle in the environs of Bruges. His years of collecting naturalia and their drawings can thus be pinpointed to a relatively short period, between *c.*1555 and 1569. After *c.*1595 a new owner of the Saint Omer albums – Prince Charles of Arenberg, who was an expert naturalist himself – added a considerable number of further plant drawings.[139]

The main body of animal drawings in the Libri Picturati depicts birds: volume A17 contains 127 painted pages, for the smaller songbirds with up to 6 figures per page. The water creatures are the next important group (41 painted pages, mostly with one or two creatures per folio), while the four-footed animals occupy 23 painted pages (mostly with one animal per page). While many depicted bird species are typical of the coastal wetlands near Bruges and along the North Sea, a regional focus is much harder to establish for the aquatic creatures. The seal, jellyfish, oysters, and a rarely depicted sea mouse are associated with the Atlantic or North Sea, but many other *aquatilia* (e.g. sunfish, St Peter's fish, saw of the sawfish, lamprey, sturgeon, needlefish) occur in both these northern seas and the Mediterranean. Their images are also commonly found in Italian fish image collections [Fig. 6.18].

Although some animals have most probably been drawn from direct observation, there is at least one definite case of copying among the *aquatilia*

138 See Rikken, *Dieren verbeeld* 14 (print series) and 3–4 (on oil paintings); and Rijks in the present volume.

139 See List of Codices. The botanical drawings in these Libri Picturati have been fully published in Koning J. De – Uffelen G. van – Zemanek A. – Zemanek B. (eds.), *Drawn after nature. The complete botanical watercolours of the 16th-century Libri Picturati* (Zeist: 2008). The only *aquatilia* drawings published so far can be found in Egmond F., "Curious fish: connections between some sixteenth-century watercolours and prints", in Enenkel K.A.E. – Smith P.J. (eds.), *Early Modern Zoology. The Construction of Animals in Science, Literature and the Visual Arts* (Leiden – Boston: 2007) 245–272, and Idem, *Eye for Detail.*

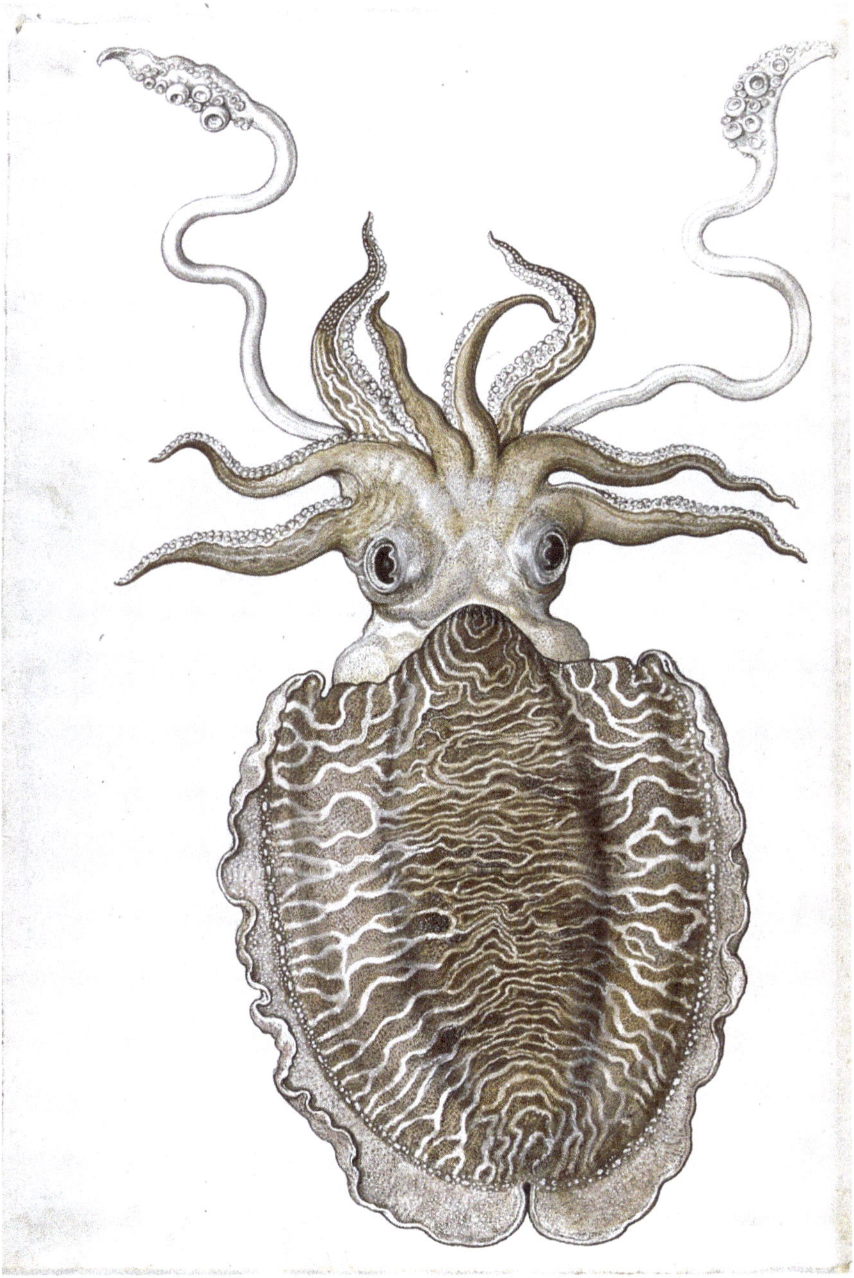

FIGURE 6.18 Squid in the Flemish *Libri Picturati* by an anonymous painter, *c.*1560s, made near Bruges for Charles de St Omer. *Libri Picturati*, A16, fol. 20

drawings in the *Libri Picturati*. It concerns the drawing of a 'tattooed' tuna fish with strange markings on its skin that look like boats and men of war; this fish had been caught in the Mediterranean. The *Libri Picturati* drawing copies a printed pamphlet that circulated in various versions in Europe in the 1560s.[140] Much more significant cases of copying – but in the other direction – occur among the birds (4) and four-footed beasts (1 llama). These provide direct visual links between the Bruges *Libri Picturati* and the much later Prague albums of Anselmus de Boodt from Bruges.[141] Here, style, detail, watermarks and the animals depicted all indicate that the *Libri Picturati* drawings of the 1560s were the models for the far less sophisticated ones that De Boodt produced in Prague for Rudolph II. Unlike a considerable number of the plant drawings in the *Libri Picturati* volumes, the animal drawings have never served as model for printed illustrations, and it is unclear whether Saint Omer ever intended them for that use.[142]

Charles de Saint Omer certainly commissioned the Bruges painter Jacques van den Corenhuyze to depict some plants, and the monogram JC (for that same painter?) figures on several of the bird drawings. Peeter van der Borcht from Malines, who also made a large number of model drawings for woodcuts in various Plantin editions, botanical and not, likewise created drawings for the *Libri Picturati*. But like Daniele Barbaro in the 1540s, Saint Omer also employed several painters on a more permanent basis at his court. The Dutch fish expert Adriaen Coenen reported in 1565 that Saint Omer was 'een groot beminder van selsame vremde dinghe van visschen ende vogelen. Hij hadde twee scilders die hem alle dagen selsame vogelen ende visschen uutscilderde dese scilders hadde hij al binnen zijn hof' (a great lover of rare and curious fish and birds. He had two painters who daily painted rare birds and fish for him, these painters were at his court).[143] The fact that Saint Omer collected these drawings in albums, the encyclopaedic range of his plant and bird drawings, and his employment of painters specifically for this purpose all suggest that this Flemish image collection of the early 1560s was inspired by models of image collecting from abroad. Given both Saint Omer's social position in the very top of the old nobility of the Habsburg Low Countries, and his friendship with humanists and erudite collectors in the Bruges area, he must have been aware of such traditions in the Habsburg world and Italy.[144]

---

140 See Egmond, "Curious fish".

141 This concerns two types of gulls, one red-breasted goose, a bird of prey, and a llama.

142 *Plant* drawings in the *Libri Picturati* did serve as models for the printed illustrations in several works by Clusius.

143 Coenen, *Visboeck* 193v = new 200v.

144 See Egmond F., "Clusius, Cluyt, Saint Omer. The origins of the sixteenth-century botanical and zoological watercolours in Libri Picturati A. 16–30", *Nuncius* 20 (2005) 11–67.

Two of the most intriguing and charming extant *aquatilia* albums created in 16th-century Europe largely support this tentative reconstruction of patterns in naturalia image collecting in northern Europe, but also raise new questions [Fig. 6.19]. The *Visboock* (*Fish Book*, 1577–81) and the *Walvisboock* (*Whale Book*, *c.*1584–85) are bound manuscript albums full of illustrations in water colour devoted to marine creatures – though there is some space for insects, sea birds, and natural phenomena in the skies. Together with an unfinished fragment of a third painted manuscript on fish (*c.*1586–87) and a now lost earlier fish album, the *Fish Book* and the *Whale Book* were the results of a lifelong fascination with the underwater world of the Dutch fish merchant and expert Adriaen Coenen (1514–1587). Coenen not only wrote but also painted these colourful encyclopaedic manuscripts.[145]

Coenen's albums are highly unusual for a number of reasons that include their very wide-ranging contents; the combination of book learning and practice-based knowledge; Coenen's social and educational background; and the wealth of information that he provides about how he collected information about marine life. Coenen was born in Scheveningen, a fishing village a few kilometres from The Hague, as the son of a fisherman. He worked his way up to become a well-to-do and respected local notable and wholesale merchant in fish who traded long distance with the German hinterland and the region of Bruges, Mechelen and Antwerp. Coenen had no formal education beyond the local elementary school and never learnt Latin, though he could speak and read German and French, and perhaps some English. Coenen's richly illustrated manuscripts are the only known extant encyclopaedic albums of animal drawings in 16th-century Europe created by a man from a lower to lower middle-class background. The texts of Coenen's manuscripts reveal much of his approach to nature, marine life, fishing, and nature study. He also discusses the making of his albums, repeatedly mentions and demonstrates his critical attitude to his sources, and shows how he modestly but systematically privileges personal experience and observation of natural phenomena above book learning – even when this means calling in doubt the greatest authorities of antiquity and of his own time, from Aristotle to Pliny and from Gessner to Belon and Rondelet.[146]

145 See List of Codices. Coenen's earliest fishbook is lost: he personally gave it to Prince William of Orange, who still had it *c.*1578. See on Coenen and his work, Egmond F. – Mason P. (eds.), *The Whale Book. Whales and other marine animals as described by Adriaen Coenen in 1585* (London: 2003); Egmond F., *Het Visboek. De wereld volgens Adriaen Coenen* (Zutphen: 2005); and Bennema F.P. – Rijnsdorp A.D., "Fish abundance, fisheries, fish trade and consumption in sixteenth-century Netherlands as described by Adriaen Coenen", *Fisheries Research* 161 (2015) 384–399.

146 On Coenen's critical attitude, see Egmond, *Het Visboek*, and Idem, *Eye for Detail*.

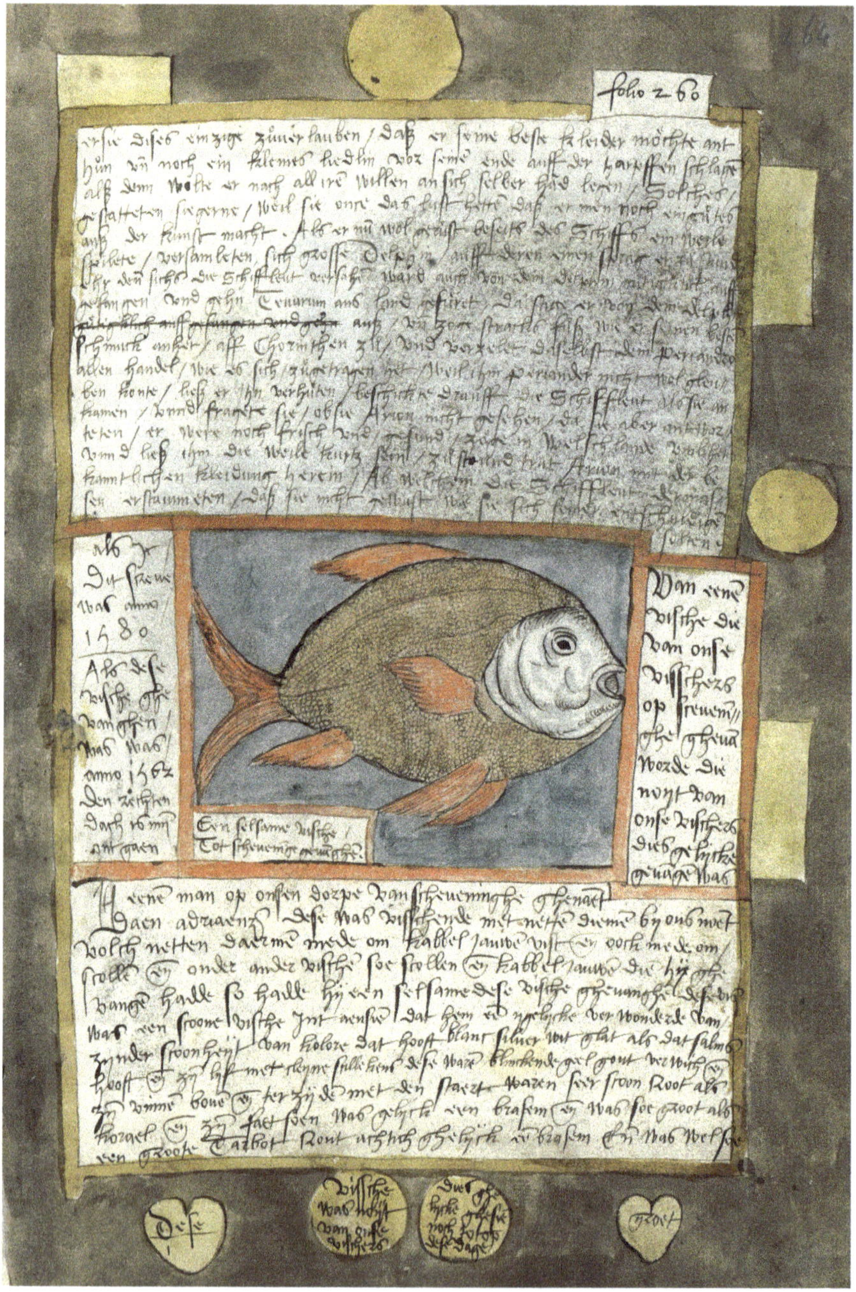

FIGURE 6.19 A rare fish (probably *Lampris guttatus*) depicted and described in Adriaen Coenen's *Visboeck*, 1577–81, Holland, Ms 78 E 54, fol 264r (old nr. 260) © THE HAGUE, KONINKLIJKE BIBLIOTHEEK

Coenen did not intend his fish albums for print. They were inspired, as he explicitly states, by a lifelong curiosity about living nature that mixed scientific interest (the need to know) with his vision of nature study as honouring God. He intended his albums as a repository of knowledge and information, made and consulted for personal enjoyment and sociability. But he also used them for commercial purposes (showing them on the local fair for payment), and for self-presentation as an expert in marine matters, which promoted his access to higher social circles. Probably shortly before 1576–78 he used his very first fish album – now lost – as a gift to Prince William of Orange, at that moment leader of the Dutch Revolt, in order to obtain a personal privilege and (unsuccessfully) an appointment for his young son as a page at the Prince's court. Though Coenen the painter was no great artist, his albums were seen as valuable objects in his own time.[147]

A considerable quantity of textual information and an even larger percentage of the visual information in Coenen's fish albums had come *from* print in the first place. His albums are, in part, a massive compilation and amalgam from the major published and illustrated works on *aquatilia* that he could consult.[148] For textual information he was obviously limited to vernacular editions, but among his main visual sources were most of the famous fish publications of the 1550s (e. g. Belon, Rondelet, Gessner, Olaus Magnus, the German Heyden edition of Pliny) besides older chronicles and contemporary illustrated pamphlets. The albums offer much more than excerpts from printed works, however. Per fish or group of fishes, and where applicable, Coenen inserted his own lifelong practice- and observation-based experience of marine life in the Atlantic and North Sea in text and sometimes in image. Moreover, important sections on North Sea fishing, herring, cod, plaice, salmon and so on, and on beached whales and other curious fish in the Channel are almost exclusively based on his own observation and on that of the local fishermen of Scheveningen. As Coenen proudly states, he generally names his sources. Often, he indeed visually separates copied or paraphrased texts from the ones based on his own experience, which are usually written in the first person. In the drawings such distinctions are of course harder to trace.

If we consider Coenen's albums, created in The Hague, Scheveningen and Leiden, in the context of the question of how Netherlandish animalia drawings originated, the questions of dating the origins of his interest and collecting,

147 For more details, see Egmond, *Het Visboek*.

148 Coenen's text is mainly in Dutch, but he also quotes long passages in German and French, and some shorter phrases in Latin.

and his sources and cultural models are of fundamental importance. Mainly thanks to clues in Coenen's own texts, we can partially trace both, although that reconstruction is complicated by the fact that Coenen's life bridged not merely the Reformation but also the Dutch Revolt against Spanish Habsburg rule. Coenen was born as a Catholic inhabitant of the united and Habsburg-ruled Low Countries. By the last two decades of his life, he had become a (moderate) Protestant and found himself living in the newly proclaimed Dutch Republic. These big events were of direct relevance to Coenen's activities as a collector and student of marine life.

A major cultural influence on Coenen's collecting and his main access to the publications that he used as visual and textual sources can be found in the circles of top Habsburg functionaries in the Northern Netherlands. The central figure was Cornelis Suys (1514–1580), President of the Court of Holland (1559–72), who shared Coenen's passion for marine natural history, certainly lent Coenen Rondelet's work on fish, and regularly received Coenen at his table in The Hague.[149] Via the meals at Suys's house in The Hague Coenen met a number of high Habsburg officials, nearly all staunch Catholics with international networks that stretched from Portugal to Italy, but also to Ortelius's circle in Antwerp and Charles de Saint Omer in Bruges, the man who commissioned the watercolours of the *Libri Picturati* discussed earlier. It was most probably via Suys's connections that Coenen himself came into personal contact with Saint Omer during the early 1560s.[150]

The cumulative evidence of Adriaen Coenen's contacts with these elite circles between about 1550 and 1565 helps us understand where the cultural models for Adriaen Coenen's fish albums came from. The men in these circles – whether old aristocracy, newer *noblesse de robe*, or humanists and local notables – were Catholics and nearly all remained so. Most were directly linked to the courtly Habsburg elite, either in Brussels and Mechelen or in The Hague. Nearly all disappeared from public life or even from the Northern Netherlands once the Revolt gained momentum, especially after the iconoclasm of 1566. Coenen remained in Holland, and became a staunch supporter of Prince William of Orange and the Revolt. These different choices in the years 1565–72 appear to have ended most of his contacts with the extended Suys

149 This was the highest judicial institution (and only appeal court) in the key provinces Holland and Zealand of the Northern Netherlands.

150 On the Suys and Saint Omer connections, see Egmond, "Clusius, Cluyt, Saint Omer"; research on these networks is ongoing. In 1566, Saint Omer lived for several months in The Hague in connection with an important court case at the Court of Holland (presided by Suys). Coenen ate several times at Saint Omer's table. See Coenen, *Visboeck*, 104v = new 113v.

circle. Most but not all. In the last years of his life Coenen moved to Leiden, where he found himself living almost next door to the old Rembert Dodoens (Dodonaeus; 1517–1585), one of the most famous naturalists in Europe, and a recently appointed professor (1582) at Leiden University. Dodoens had spent much of his life at Mechelen, and a number of years as court physician at the Habsburg court in Vienna. And from his personal collection a small collection of fish drawings and sketches have recently emerged that include a few Coenen drawings, made undoubtedly by Coenen himself for his near neighbour [Fig. 6.20].[151]

Thus far, the information concerning both Saint Omer's *Libri Picturati* and Coenen's series of fish albums – the only larger sets of *aquatilia* drawings from the Low Countries in this period – underpins the tentative conclusion that animal painting in the Low Countries was not an independent development, but a subsidiary one that derived from international (especially Habsburg) court culture. There is some evidence, however, that challenges this thesis, and it again comes from Coenen himself. It reopens to some extent the debate about the chronological and geographical development of nature collecting and painting in Europe.

Coenen himself explicitly places the early roots of his interest in 'curious' fish in his youth and in practice, when he dried and dissected thousands of rays and inspected rare *aquatilia* while working as assistant to the Scheveningen fish auctioneer. From a very early date he used to take notes and make drawings in what he calls 'mijn memorijboockxken daer Ic ander zelsame visgen in geteijkent hadde die bij mijnen tijden ander zee gecomen waren' (my memory booklet in which I had also drawn other rare fish that had arrived in the sea in my time).[152] As early as 1545–1546 Coenen was clearly already known as a man with a pronounced interest in rare and unusual fish who had some of these depicted by a local painter. These are the same years when Valerius Cordus and Daniele Barbaro were creating two of the very early fish image collections that we know of in Italy. Internal evidence to his manuscripts indicates that Coenen must have begun his first (lost) fish book already in the mid to late 1550s. His interest in marine life thus originated *before* any of the great fish works of the 1550s had been published. Even if the format of Coenen's fish albums was modelled on encyclopaedic printed works by learned 'fish describers' as he called them, while his interest was definitely shaped by Southern

151 This small collection of fish and a few plant drawings was personally annotated by Dodonaeus; it arrived in the collection of the Rijksmuseum from the private Van Regteren Altena collection. See List of Codices.

152 Coenen, *Visboeck* 405 = new 407.

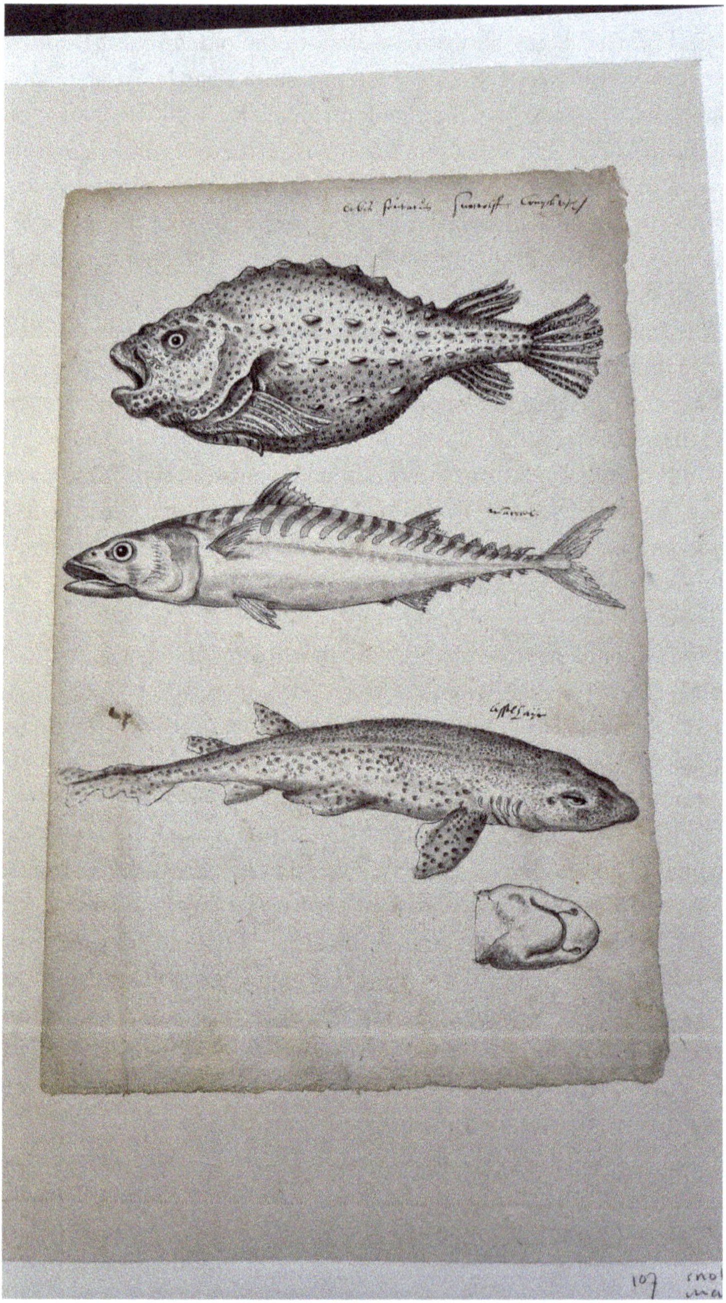

FIGURE 6.20 Fish drawings from a set of loose drawings that belonged to Rembertus Dodonaeus and included drawings by Adriaen Coenen, before 1585. Formerly Collection Van Regteren Altena, fol. 107a

Netherlandish courtly and learned traditions of collecting and studying living nature, his very early interest in marine life chronologically developed in parallel with that of Daniele Barbaro and Leone Tartaglini in Venice, Cardinal Cervini and Ippolito Salviani in Rome, and Conrad Gessner, Pierre Belon and Guillaume Rondelet in Switzerland and France.

That finding should make us think twice about the possible divergence between histories of cultural formats (books, albums, practices of collecting and display) and those of interests and fascinations. Depending on our perspective, Coenen can be any of three things. An original and highly individual figure who operated at the geographical margins of the cultural influence of heartland Europe and whose albums reflect rather than radiate. A bridge figure who reflects and radiates at the same time; who embodies the social and geographical transmission and elaboration of new interests in living nature (and marine creatures in particular) and of a critical attitude with respect to both classical and contemporary authority. Or a key witness to the meeting points of largely practice-based popular knowledge, book learning and collecting practices.

6 Conclusion

Does the re-integration of this large amount of non-printed visual material into the history of printed works indeed adjust our notions of the geography, chronology, and nature of a scientific interest in the aquatic world in early-modern Europe?

Geographically it defines the boundaries of the main zones of fish studies in Europe more sharply than before, in part because we can trace patterns of copying, exchanging and collecting images before print. Italy was of crucial importance to visual fish research during the whole of the 16th century – (even) more so than for plant or bird studies. Southern France and Switzerland were two further key zones in the early and mid-16th century. Germany is, so far, surprisingly underrepresented in terms of known image collections, with only Cordus and Kentmann. The activities of both men as fish image collectors should, moreover, be largely understood as a result of their life and travel in Italy. Italian visual traditions also profoundly influenced Gessner's fish image collection, which contained large clusters of Roman and Venetian material, while he further re-used much of Rondelet's image material. Until the 1560s, the geographical emphasis is strongly on the Mediterranean, therefore. More research concerning Naples and Sicily may underpin this even further. During and after the 1560s–70s the Habsburg centres of Vienna, Prague, Innsbruck, Mechelen, Brussels and to some extent The Hague become prominent, while

Swiss collectors continue to accumulate fish images until the mid 1610s. Unsurprisingly, two principal European ports and their environs – Venice (during the whole century) and Antwerp (mainly after 1560–70) – turn out to be especially important as sources of *aquatilia* information and drawings. Why this does not seem to be the case for Lisbon is an open question.

There are some striking gaps as well. Scandinavia, most of eastern Europe, Great Britain, Spain and even Portugal seem to have produced no *aquatilia* image collections. New research may fill in some of these gaps. For Spain and Great Britain this absence fits in, however, with a general lack of visual *naturalia* sources for most of the 16th century.

Chronologically, the non-printed visual material adds many nuances and provides a much richer pattern. It suggests that the phenomenon of *aquatilia* image collecting was fairly circumscribed in time: it may have started around 1500, certainly became more prominent around 1530s and peaked very quickly, in the 1540s–1560s, in terms of both quantity and quality. The wave of printed fish publications and albums with drawings of *aquatilia* of those decades coincided with a prominent presence of *aquatilia* on frescoes, in illuminated manuscripts, in various decorative arts, and in specially devised literary genres that incorporated new scientific knowledge. The production of *aquatilia* drawings and albums continued until *c.*1600–1615. Between *c.*1590 and 1620 other visual genres emerged, such as fish print series, market paintings and fish still lifes. In the eighteenth century albums with naturalia drawings returned in major way, but often with different functions, modalities and visual formats.

Two distinct phases can be distinguished within the period discussed here. The first one (*c.*1530–*c.*1560 – though Aldrovandi continued until *c.*1600) saw the creation of a completely new visual corpus in the service of the major collective effort to identify as many aquatic creatures as possible, and to connect images to the names that circulated in the classical literature (esp. Aristotle, Pliny, Oppian) and in the vernacular languages and dialects. This effort was only in part geared towards print: most early image collections started out as research collections, fast growing visual databases with fish names written next to the drawings that aimed to comprise as many *aquatilia* as possible. Some of these study collections with encyclopaedic aims also served as visual funds for the selection of illustrations for printed works. But even if such publications eventually appeared (Gessner, Aldrovandi, Belon, Rondelet), authors often worked with the visual material they had at hand. Only in a few cases, the painters had been instructed beforehand to create images suitable for print.

It was generally the collector who *created* the image collection and brought together drawings by many different painters, sometimes from different countries, and very often including copies of drawings in other collections. The image collections of naturalists like Cordus, Gessner and Aldrovandi almost exclusively served their research and were used as a fund of exchange items that strengthened their bonds of friendship and collegiality with other naturalists. In commissioning and sponsoring image collections, Italian aristocrats, patricians, high officials and especially cardinals played a crucial role. As expert patrons-collectors they directly and indirectly stimulated research by naturalists. Virtually all of the major patron-collectors regarded and used their image collections also as objects of display and self-presentation as erudite connoisseurs. The drawings must have been a source of entertainment, pleasure and conviviality for all.

During the second phase, from about 1565, the visual corpus newly created in the 1530s–60s had either become accessible in print (mainly in black-and-white) or circulated as coloured drawings (via copying) among naturalists and collectors. The collections of drawings that originated after *c.*1565 – again with the exception of Aldrovandi, who continued to operate along encyclopaedic lines – are virtually unconnected with publication projects. In fact, there hardly were any new publications on fish in the making in Europe at this time. While these later collections of *aquatilia* drawings thus moved away from the domain of print, they continued to function as research collections and visual depositories. Coloured drawings by no means lost their functions because of print or the availability of printed illustrations. Drawings continued to be made as substitutes for the object, placeholders for the dead fish that had lost both colour and shine, and to function in ways that neither the real thing nor the printed illustration could accomplish. Platter's large albums with fish images probably served as a visual database for comparative research, just like his herbarium. These later collections by no means focused only on new fish species, although some of course included 'new' fish. Adriaen Coenen's fish books, for instance, re-elaborated the older printed literature, but at the same time added many North Sea fishes that he had personally observed. Many later albums, in fact, continued to repeat a number of known species, almost a fish canon.

At the same time, virtually all later image collections had strong representative and display functions, whether they were made for the Habsburg rulers, the Medici Grand Dukes, the Duke of Savoy, the Flemish nobleman Charles de Saint Omer, or the fish merchant Adriaen Coenen. The albums themselves – such as Liberale da Udine's codex with Adriatic fish, Saint Omer's *Libri Picturati*, the Turin fish codex with its huge fold-outs, and the parchment albums of Tartaglini in Venice – were objects of display and formed part of a culture in which elite connoisseurship of nature went hand in hand with an emphasis on curiosity and preciosity. That function had been present from the start, but it seems to have grown stronger in the later 16th century. Some of the later image collections also show signs of increasing specialization, such as a focus on the marine fauna of the Adriatic (Liberale da Udine), special interest in the North Sea (Coenen), or increasing attention to freshwater fish (Platter).

Whether there was a visual trend towards greater scientific accuracy must remain an open question here: it can only be answered by an expert ichthyologist who is willing to avoid an anachronistic comparison with present day standards and can take into account the changes over time in what scientific accuracy actually meant. In terms of attention to detail and lifelikeness it is impossible to argue for any chronological development at all. Top quality representations can be found both in the (un-dated) marginal illuminations of the Parma Pliny, Salviani's engravings of the 1550s, the drawings in the Southern Netherlandish Libri Picturati (1560s), Ligozzi's drawings of the 1570s–80s, and Liberale da Udine's drawings of *c.*1560–80.

All of this material confirms that *aquatilia* were indeed the first of all animal categories to be intensively studied and depicted in the early modern period, and that fish studies directly followed the model of plant studies in terms of depicting, image collecting and research methods. Of course, many experts, such as Cordus, Barbaro, Aldrovandi, and Mattioli studied both plants and *aquatilia*. None of this sufficiently explains, however, *why* fish studies were the first. A changing food culture among the elite, with a new emphasis on fish in the circles of Italian cardinals in particular, as suggested by Grieco, may have played a part, but it can hardly have been the only factor.[153] After all, fish and other water creatures had for many centuries formed a significant source of food for a large part of the non-elite in Europe, an importance reflected in the huge variety of local and dialectal names for fishes. Most probably the fascination of an enormously varied and yet largely invisible underwater world formed a strong incentive to study *aquatilia*, as the Sicilian Maurolico suggested. The great variety of local fish names (that persists to this day) and the resulting complications for their identification surely stimulated fish research and depicting at the time. But the existence of an influential classical corpus of Mediterranean fish descriptions in the works of Aelian, Oppian and Pliny definitely also triggered an early focus on fish. Perhaps, we should look at the wider picture, however. The fact that a huge wave in fish interest can be traced in a variety of domains – from decorative arts to drawings and from illuminated manuscripts to literature – itself suggests that scientific studies simply

153 See Grieco A.J., "Fiordiano Malatesta da Rimini e i trattati di ittiologia della metà del Cinquecento", in Laurioux B. – Moulinier-Brogi L. (eds.), *Scrivere il Medioevo. Lo spazio, la santità, il cibo* (Rome: 2001) 305–318; cf. Folena, "Per la storia".

followed (and in their turn enhanced) a far more general cultural phenomenon: the fashion of water, that can also be traced in garden design, grottoes, fountains, grotesques, and much more.

Finally, the case of Adriaen Coenen and his long involvement with fish investigations between about 1545 and 1580 subtly undermines the chronology presented above. It indicates that we have investigated the history of a particular type of object (drawings, albums and printed illustrations) as a cultural manifestation of a profound interest in aquatic nature. Per definition that type of object belonged to the domain of the intellectual and social elites. But Coenen's interest in aquatic naturalia predated that of many learned naturalists and collectors in Europe, and seems to have originated in his practical experience. And that shows that the histories of genres and formats are never sufficient to fully explore the history of interests and fascinations.

Bibliography

List of Codices / Aquatilia Drawings

(with url if online accessible; last consulted Sept. 2023)

Album dei Pesci, anonymous, probably Piedmontese, *c.*1580–1624, 75 folios with aquatic creatures, height 55 cm, width 1 meter 17 cm. Turin, Biblioteca Reale, Ms 14491 D.C. (*Colloc. Varia* 230) 14688 D.C. [https://museireali.beniculturali.it/catalogo-on-line/#/cerca?text=pesci&size=80&page=0].

Aldrovandi, Ulisse, *Tavole*, Aldrovandi as naturalist-collector; various painters, *c.*1560–1610, Bologna. 10 vols, with some 2,500 images; animals, plants, fruit and some human beings and stones. Bologna, Biblioteca Universitaria. [http://aldrovandi.dfc.unibo.it/pinakesweb/main.asp].

Baldner, Leonhard, *Vogel- Fisch- und Thierbuch*, Strasbourg, *c.*1653–1666. Kassel, University Library, 2° Ms. phys. et hist. nat. 3. [https://orka.bibliothek.uni-kassel.de/viewer/image/1343227732549/1/#topDocAnchor].

Bol, Hans, Albums, drawings of fourfooted animals, aquatic creatures, and birds. Aquatic album: *Icones Quorundum Piscem nativam formam referentes*. Copenhagen, Royal Library, GKS 3471.

Boodt, Anselmus de, Albums, drawings of animals and plants, made for Emperor Rudolph II; painters De Boodt, Elias Verhulst, and others. Prague, late 16th-early 17th century. 11 vols with some 800 illustrated folios. Amsterdam, Rijksmuseum. For the fish, see:

[https://www.rijksmuseum.nl/en/search/objects?set=RP-T-BR-2017-1-6#/RP-T-BR-2017-1-6,0].

Cod. Min. 83, album with drawings of *aquatilia*. 20 folios with multiple images per folio. Prague?, late 16th century? Attributed to Hoefnagel or Ligozzi. Vienna, Österreichische Nationalbibliothek.

Cod. min. 129 and 130, also known as *Rudolf II's Bestiary*, codex in 2 volumes, various painters, especially Dirck de Quade van Ravesteijn, for Emperor Rudolf II, *c.*1600–1612, Prague. C. 90 folios per volume. Vienna, Österreichische Nationalbibliothek. *Aquatilia* in Cod. min. 129.

Cod. Ser. 2648, album with crudely drawn *aquatilia*. 59 folios, probably late 16th century, Italian? Vienna, Österreichische Nationalbibliothek.

Coenen, Adriaen, *Visboeck*. Written and painted by Adriaen Coenen, 1577–81, Holland. C. 400 folios; mainly aquatic animals. The Hague, Koninklijke Bibliotheek, Ms 78 E 54.
[https://www.kb.nl/ontdekken-bewonderen/topstukken/visboeck].

Coenen, Adriaen, *Walvisboeck*. Written and painted by Adriaen Coenen, *c.*1584–85, Holland. C. 125 folios; aquatic animals. Antwerp, Erfgoedbibliotheek Hendrik Conscience, EHC 707256.
[https://consciencebibliotheek.be/nl/pagina/walvisboek].

Coenen, Adriaen, *Haringkoningboek*. Written and painted by Adriaen Coenen, *c.*1586–87. C. 25 folios, unfinished. Cologne, Historisches Archiv, Ms 296.

Dodonaeus, Rembertus, loose drawings of plants and animals that belonged to Dodonaeus, before 1585. Rijksmuseum Amsterdam, former collection Van Regteren Altena.
[https://www.rijksmuseum.nl/en/search?q=dodonaeus&v=&s=&ondisplay=False&ii=0&p=1].

Froeschl, Daniel – Nicolaes De Bruyn, Codex with animal drawings, *aquatilia* starting at p. 209, *c.*1570–1620s. Pisa, University Library, Ms 514.
[https://www.internetculturale.it/it/16/search?q=Froeschl&instance=magindice].

Gessner-Platter Animal Albums, collector Felix Platter, including the original drawings collected by Conrad Gessner and used as models for the illustrations in the latter's *Historia Animalium*. C. 1565–1615 for Platter, Basel; 1540s–65 for Gessner, Zurich. 2 volumes: I (III C 22), some 230 folios with aquatic creatures; II (III C 23), some 150 folios with beasts, reptiles, amphibians and insects. Amsterdam, University Library Ms III C 22 and III C 23.
Beasts, insects etc.:
https://www.flickr.com/photos/bijzonderecollectiesuva/sets/72157632809370911/.
Aquatilia:
https://www.flickr.com/photos/bijzonderecollectiesuva/sets/72157632814218926/.

Kentmann Codex, by Johannes Kentmann and his son Theophilus Kentmann. Middle to late 16th century, Meissen, Thorgau. Johannes' section contains *c.*120 folios with *aquatilia*. Weimar, Herzogin Anna Amalia Bibliothek, Fol 323. https://haab-digital.klassik-stiftung.de/viewer/image/942369343/2/LOG-0000/.

Liberale da Udine Codex (Cod. Ser. 2669), painted by Giorgio Liberale da Udine for Archduke Ferdinand II of Tirol, *c.*1562–80, Gorizia and Innsbruck, 100 folios. Vienna, Österreichische Nationalbibliothek. [https://onb.digital/search/320067].

Libri Picturati A16–30, collector Charles de Saint Omer (1560s), with some additions (*c.*1598–1620) by Charles d'Arenberg; various Flemish painters, including Jacques van den Corenhuyse. Bruges, Antwerp, Malines. 15 vols: 2 vols animals (A16–17, some 170 sheets); and 13 vols plants (A18–30, some 1,150 illustrated sheets). Kraków, Jagiellonian Library.

Ligozzi, Jacopo, naturalia drawings, some 78 of plants and 65 of animals, loose sheets, mainly kept in the Gabinetto dei disegni e stampe, Uffizi, Florence.

Philes, Manuel – Angelos Vergekios, *De animalium proprietate*. London, British Library, MS Burney 97. Dated between *c.*1540 and 1575. Manuscript by Vergekios; painter of animals unknown.

Pliny, *Naturalis Historia* (Parma, Andreas Portilia: 1481). Parma, Biblioteca Palatinense, Inc. 1158 (numerous aquarel drawings).

Röting, Lazarus, Album, *Theatrum Naturae von Michael Rötenbeck*. Drawings mainly by painter-naturalist Lazarus Röting, Nuremberg, *c.*1580–1614; bound into an album (1615) by his nephew Michael Rötenbeck. Volume with *c.*190 folios with drawings of animals (predominantly birds), some *aquatilia*, stones, and plants. Berlin, Museum für Naturkunde.

Sources (*Printed Works before 1800*)

Albinus Petrus, *Commentarius novus de Mysnia, Oder Newe Meysnische Chronica* (Wittenberg, Hans Lufft: 1580).

Aldrovandi Ulisse – Uterverius Johannes Cornelius, *De piscibus libri V et De cetis liber unus* (Bologna, Hieronymus Tamburinus: 1613).

Aldrovandi Ulisse, *Le antichità de la città di Roma* (Venice, Giordano Ziletti: 1556).

Belon Pierre, *L'Histoire naturelle des estranges poissons marins, avec la vraie peincture et description du daulphin, et de plusieurs autres de son espece* (Paris, Regnaud Chaudière: 1551).

Belon Pierre, *La nature et diversité des poissons avec leurs pourtraicts, représentez au plus près du naturel* (Paris, Charles Estienne: 1555).

Belon Pierre, *L'Histoire de la nature des oyseaux, avec leurs descriptions et naïfs portraicts retirez du naturel, escrite en sept livres* (Paris, Guillaume Cavellat: 1555).

Boussuet François, *De natura aquatilium carmen, in universam Gulielmi Rondeletii* (Lyon, Macé Bonhomme: 1558).

Bruyn Nicolaes de, *Libellus varia genera piscium complectens, pictoribus, sculptoribus utilis et necessarius* (Amsterdam, no publisher: year uncertain).

Calmo Andrea, *Le bizzarre, faconde, et ingegnose rime pescatorie, Nelle quali si contengono Sonetti, Stanze, Capitoli, Madrigali, Epitafij, Disperate, e Canzoni* (Venice, Iseppo Foresto: 1557).

Capaccio Giulio Cesare, *Mergellina. Egloghe piscatorie di Giulio Cesare Capaccio napolitano* (Venice, Eredi di Melchior Sessa: 1598).

Cardano Girolamo, *De rerum varietate libri XVII* (Basel, Henricus Petri: 1557).

Ceruti Benedetto, *Musaeum Franc. Calceolari iun. Veronensis* (Verona, Angelus Tamus: 1622).

Clusius Carolus, *Rariorum aliquot stirpium, per Hispanias observatarum historia* (Antwerp, Plantin: 1576).

Clusius Carolus, *Rariorum aliquot stirpium, per Pannoniam, Austriam, et vicinas quasdam provincias observatarum historia* (Antwerp, Plantin: 1583).

Collaert Adriaen, *Piscium vivae icones. In aes incise et editae ab Adriano Collardo* (Antwerp, no publisher: *c.*1598).

Colonna Fabio, Φυτοβασανος [Phytobasanos] *sive plantarum aliquot historia, part Piscium aliquot plantarumque novarum historia* (Naples, Horatio Salviani: 1592).

Cordus Valerius – Gessner Conrad, *Valerii Cordi Simesusii De Halosantho* (Zurich, Christopher Froschauer: 1566 = 1565).

Dubravius Jan, *De piscinis* (Breslau, Andreas Vinglerus: 1547).

Estienne Charles, *De Latinis et Graecis nominibus arborum, fruticum, herbarum, piscium et avium liber* (Paris, Charles Estienne: 1544).

Fabricius Georg, *Rerum Misnicarum libri VII* (Leipzig, Ernst Vögelin: 1569).

Fantuzzi Giovanni, *Memorie della vita di Ulisse Aldrovandi* (Bologna, Lelio dalla Volpe: 1774).

Figulus Carolus, *ΙΧΘΥΟΛΟΓΙΑ, seu Dialogus de piscibus* (Cologne, Eucharius Cervicornus: 1540).

Gessner Conrad, *De piscibus et aquatilibus libelli III novi* (Zurich, Andreas Gessner: 1556).

Gessner Conrad, *Historiae Animalium, Liber IIII, qui est de Piscium & Aquatilium animantium natura* (Zurich, Christopher Froschauer: 1558).

Gessner Conrad, *Nomenclator Aquatilium animantium. Icones animalium aquatilium* (Zurich, Christopher Froschauer: 1560).

Gessner Conrad – Forer Cunrat, *Fischbuch* (Zurich, Christopher Froschauer: 1563).

Gessner Conrad – Wolf Caspar, *Epistolarum medicinalium Conradi Gesneri* [...] *libri III* (Zurich, Christopher Froschauer: 1577).

Gilles Pierre, *Ex Aeliani historia per Petrum Gyllium latini facti, itemque ex Porphyrio, Heliodoro, Oppiano, tum eodem Gyllio luculentis accessionibus aucti libri XVI* (Lyon, Sebastian Gryphius: 1535).

Gilles Pierre, *De Bosporo Thracio Libri III* (Lyon, Guillaume Roville: 1561).

Gilles Pierre, *Aeliani De historia animalium libri XVII* (Lyon, Guillaume Roville: 1565).

Giovio Paolo, *De romanis piscibus libellus* (Rome, Francesco Minuzio Calvo: 1524).

Heyden Johann, *Caij Plinij Secundi, Des fürtrefflichen* [...] *Philosophi, Bücher vnd Schrifften, von der Natur* (Frankfurt am Main, Sigmund Feyerabend: 1565).

Lange Johannes, *Medicinalium epistolarum miscellanea* (Basel, Johannes Oporinus: 1554).

Leoniceno Niccolò, *De serpentibus opus singulare ac exactissimum* (Bologna, Giovanni Antonio de' Benedetti the younger: 1518).

Mangolt Georg, *Fischbuch. Von der Natur und Eigenschafft der vischen* (Zurich, Andreas and Jakob Gessner: 1557).

Marschalk Nikolaus [Marescalcus], *Historia Aquatilium latine ac grece cum figuris* (Rostock, Nikolaus Marschalk: 1517–1520).

Massari Franciscus, *Veneti in nonum Plinii De Naturalis Historia librum castigationes et annotationes* (Basel, Officina Frobeniana: 1537).

Mattioli Pietro Andrea, *Commentarii in libros sex Pedacii Dioscoridis Anazarbei, De Materia Medica* (Venice, Vincenzo Valgrisi: 1554).

Mattioli Pietro Andrea, *Commentarii denuo aucti, in libros sex Pedacii Dioscoridis Anazarbei de medica materia. Adiectis quamplurimis plantarum, & animalium imaginibus, quae in prioribus editionibus non habentur* (Lyon, Gabriel Coterius: 1562).

Mattioli Pietro Andrea, *Commentarii in sex libros Pedacii Dioscoridis Anazarbei De medica materia* (Venice, Vincenzo Valgrisi: 1565).

Mattioli Pietro Andrea, *I discorsi di M. Pietro Andrea Matthioli sanese, medico cesareo, et del serenissimo principe Ferdinando archiduca d'Austria & c. nelli sei libri di Pedacio Dioscoride Anazarbeo della materia medicinale* (Venice, Vincenzo Valgrisi: 1568).

Mattioli Pietro Andrea – Handsch Georg, *New Kreüterbuch: Mit den allerschönsten vnd artlichsten Figuren aller Gewechß, dergleichen vormals in keiner sprach nie an tag kommen* (Prague, Melantrich von Auentin and Vicenzo Valgrisi: 1563).

Magnus Olaus, *Historia de Gentibus Septentrionalibus* (Rome, Giovanni Maria Viotti: 1555).

Passe Crispijn de, *Piscium vivae Icones* (Utrecht, no publisher: 1620?).

Pliny, *Naturalis Historia* (Parma, Andreas Portilia: 1481).

Regio Paolo, *Siracusa Pescatoria* (Naples, Giovanni de Boy: 1569).

Rondelet Guillaume, *Universae aquatilium historiae pars altera, cum veris ipsorum imaginibus* (Lyon, Macé Bonhomme: 1555).

Rondelet Guillaume, *Libri de piscibus marinis, in quibus verae piscium effigies expressae sunt* (Lyon, Macé Bonhomme: 1554).

Rota Bernardino, *Egloghe pescatorie* (Naples, Giovan Maria Scotto: 1560).
Salviani Ippolito, *Aquatilium animalium Historiae liber primus. Cum eorundem formis, aere excusis* (Rome, Ippolito Salviani: 1554–1558).
Sannazaro Jacopo, *Eclogae Piscatoriae* (Naples, Jacopo Sannazaro, 1526).
Schonefeld Stephan von, *Ichthyologia et nomenclaturae animalium marinorum, fluviatilium, lacustrium, quae In florentissimis ducatibus Slesvici et Holsatiae et celeberrimo Emporio Hamburgo occurrunt triviales* (Hamburg, Bibliopolis Heringianus: 1624).
Schwenckfeld Caspar, *Theriotropheum Silesiae* (Liegnitz, David Albertus: 1603).
Tiraboschi Girolamo, *Storia della Letteratura Italiana*, Vol. VII, part 2 (Florence, Molini: 1810; 1st edition 1772).
Wilisch Christian Gotthold, *Arcana bibliothecae Annabergensis in partes III divisa* (Leipzig, Friedrich Lankischs Erben: 1730).
Wotton Edward, *De differentiis animalium libri decem* (Paris, Michel de Vascosan: 1552).

Works Printed after 1800

Acciarino D., "Antipoetica delle grottesche: le *Lettere sulla pittura* di Ulisse Aldrovandi", *Schede Umanistiche. Rivista annuale dell'Archivio Umanistico Rinascimentale Bolognese*, N.S. 30 (2016) 169–196.
Alessandrini A. – Ceregato A. (eds.), *Natura Picta. Ulisse Aldrovandi* (Bologna: 2007).
Alexander J.J.G. (ed.), *The Painted Page: Italian Renaissance Book Illumination, 1450–1550* (Munich: 1994).
Andretta E., *Roma Medica. Anatomie d'un système médical au XVIe siècle* (Rome: 2011).
Andretta E. – Pardo Tomás J., "Books, plants, herbaria: Diego Hurtado de Mendoza and his circle in Italy (1539–1554)", *History of Science* 58 (2020) 3–27.
Azzi Visentini M., *L'Orto Botanico di Padova e i giardini del Rinascimento* (Milan: 1984).
Barsi M., *L'énigme de la chronique de Pierre Belon. Avec édition critique du manuscrit Arsenal 4651* (Milan: 2001).
Baumann T., *Pauli Iovii Novocomensis Medici De Romanis piscibus libellus ad Ludovicum Borbonium cardinalem amplissimum. Text, Übersetzung, Kommentar* (Ph.D. dissertation, University of Mannheim: 1994).
Bäumer A., *Geschichte der Biologie. Zoologie der Renaissance – Renaissance der Zoologie Band 2* (Frankfurt am Main: 1991).
Bava A.M. – Pagella E. (eds.), *Le meraviglie del mondo. Le collezioni di Carlo Emanuele I di Savoia*, exh. cat. Turin (Genua: 2016).
Benkert D., *Ökonomien botanischen Wissens. Praktiken der Gelehrsamkeit in Basel um 1600* (Basel: 2020).
Bennema F.P. – Rijnsdorp A.D., "Fish abundance, fisheries, fish trade and consumption in sixteenth-century Netherlands as described by Adriaen Coenen", *Fisheries Research* 161 (2015) 384–399.

Beschi L., *Onorio Belli Accademico Olimpico. Scritti d'Antiquaria e Botanica (1586–1602)* (Rome: 2000).

Blair, A., "Humanism and printing in the work of Conrad Gessner", *Renaissance Quarterly* 70 (2017) 1–43.

Bohatcová M., "Prager Drucke der Werke Pierandrea Mattiolis aus den Jahren 1558–1602", *Gutenberg Jahrbuch* 60 (1985) 167–185.

Bristot A. (ed.), *Palazzo Grimani a Santa Maria Formosa. Storia, arte, restauri* (Venice: 2008).

Brusegan M. (ed.), *Leone Tartaglini, Opera nuova nella quale se contiene la natura dil sonno cioe come lhuomo debbe dormire per mantenersi sano con alchuni bellissimi & utilissimi secreti medicinali, Venetia 1551* (Fermo: 1996).

Bukovinská B., "The known and unknown Kunstkammer of Rudolf II", in Schramm H. – Schwarte L. – Lazardzig J. (eds.), *Collection, laboratory, theater: Scenes of knowledge in the 17th century* (New York: 2005) 199–227.

Bukovinská B., "Die Kunstkammer Rudolfs II – Entstehung, Niedergang, Wiederentdeckung", in Haag S. – Kirchweger F. – Rainer P. (eds.), *Das Haus Habsburg und die Welt der fürstlichen Kunstkammern im 16. und 17. Jahrhundert* (Vienna: 2015) 229–252.

Caneva G. – Carpanetto G. (eds.), *Raffaello e l'Immagine della Natura. La raffigurazione del mondo naturale nelle decorazioni delle Logge Vaticane* (Milan – Rome: 2010).

Caracciolo D., "L'enciclopedia ittiologica di Giulio Cesare Capaccio ed i suoi rapporti con il collezionismo Napoletano", *Napoli Nobilissima* serie 5.X (2009) 3–20.

Caracciolo D., "Per una Wunderkammer letteraria, Mergellina, la 'fatica marittima' di Giulio Cesare Capaccio", *Annali di critica d'arte* 5 (2009) 33–80.

Castelli D., "Tra ricerca empirica e osservazione scientifica: gli studi ittiologici di Simone Porzio", *Archives Internationales d'Histoire des Sciences* 57 (2007) 105–123.

Cecchi A. – Conigliello L. – Faietti M. (eds.), *Jacopo Ligozzi "pittore universalissimo"* (Livorno: 2014).

Cockx-Indestege E., "Van een boekje om vogels en vissen te vangen naar een zeldzame Antwerpse postincunabel, nu in de Library of Congress te Washington", in Van der Vekene E. (ed.), *Refugium Animae Bibliotheca. Festschrift Albert Kolb* (Wiesbaden: 1969) 109–138.

Concin, A., "Splendid Gifts and a Florentine Architect for Emperor Rudolf II: Antonio Lupicini at the Imperial Court in Prague (1578–1580)", *Studia Rudolphina* 20 (2020) 24–49.

Conigliello L., "Pesci, crostacei e un'iguana per l'imperatore Rodolfo II", *Paragone. Arte* 42, 493–495 (1991) 22–29.

Conte F., "Intorno alla Grande Galleria di Carlo Emanuele I: inventari del sapere a Torino tra Cinque e Seicento", in Cioffi R. – Sconamiglio O. (eds.), *Mosaico. Temi e metodi d'arte e critica per Gianni Carlo Sciolla* (Naples: 2012) 197–206.

Coronini Cronberg G., "Giorgio Liberale e i suoi fratelli", in Brussich G. – Del Fabbro I. (eds.), *Studi di storia dell'arte in onore di Antonio Morassi* (Venice: 1971) 85–96.

Dauwalder L. – Lienhard L. (eds.), *Das Herbarium des Felix Platter. Die älteste wissenschaftliche Pflanzensammlung der Schweiz* (Bern: 2016).

Davis P., "Collecting and preserving fishes: a historical perspective", in MacGregor A. (ed.), *Naturalists in the Field. Collecting, recording and preserving the natural world from the 15th to the twenty-first century* (Leiden – Boston: 2018) 149–165.

Dilg P., "Über die Schrift 'De halosantho seu spermate ceti vulgo dicto' des Valerius Cordus (1515–1544). Ein Beitrag zur möglichen Kenntnis des Walrats in der Antike", in Friedrich C. – Telle J. (eds.), *Pharmazie in Geschichte und Gegenwart. Festgabe für Wolf-Dieter Müller-Jahncke zum 65. Geburtstag* (Stuttgart: 2009) 111–132.

Dobalová S., "Erzherzog Ferdinand II. von Habsburg, das Lusthaus Belvedere und die Fischbehälter im Königlichen Garten der Prager Burg", *Die Gartenkunst* 20 [Beilage zu Heft 2] (2008) 11–18.

Egmond F., "Clusius, Cluyt, Saint Omer. The origins of the sixteenth-century botanical and zoological watercolours in Libri Picturati A. 16–30", *Nuncius* 20 (2005) 11–67.

Egmond F., *Het Visboek. De wereld volgens Adriaen Coenen* (Zutphen: 2005).

Egmond F., "Curious fish: connections between some sixteenth-century watercolours and prints", in Enenkel K. – Smith P. (eds.), *Early Modern Zoology. The Construction of Animals in Science, Literature and the Visual Arts* (Leiden – Boston: 2007) 245–272.

Egmond F., "A collection within a collection. Rediscovered animal drawings from the collections of Conrad Gessner and Felix Platter", *Journal of the History of Collections* 25.2 (2013) 149–170.

Egmond F., *Eye for Detail: Images of Plants and Animals in Art and Science, 1500–1630* (London: 2017).

Egmond F. (ed.), *Conrad Gessners Thierbuch. Die Originalzeichnungen* (Darmstadt: 2018).

Egmond F., "Into the wild. Botanical fieldwork in the sixteenth century", in MacGregor A. (ed.), *Naturalists in the Field. Collecting, recording and preserving the natural world from the 15th to the twenty-first century* (Leiden – Boston: 2018) 166–211.

Egmond F., "The Ad Vivum Conundrum: Eye witnessing and the Artful Representation of Naturalia in Sixteenth-Century Natural Science", in Schmiedel I. – Oy-Marra E. (eds.), *Zeigen – Überzeugen – Beweisen. Methoden der Wissensproduktion in Kunstliteratur, Kennerschaft und Sammlungspraxis der Frühen Neuzeit* (Mainz: 2020) 33–62.

Egmond F., "Visual immersion: Daniele Barbaro's fish album and the wave of interest in aquatic creatures in mid sixteenth-century Europe", *Notes and Records of the Royal Society* (published online May 2022, doi:10.1098/rsnr.2021.0084).

Egmond F. – Mason P. (eds.), *The Whale Book. Whales and other marine animals as described by Adriaen Coenen in 1585* (London: 2003).

Egmond F. – Kusukawa S., "Circulation of images and graphic practices in Renaissance natural history: the example of Conrad Gessner", *Gesnerus* 73 (2016) 29–72.

Egmond F. – Kusukawa S., "Gessner's fish: images as objects", in Leu U. – Opitz P. (eds.), *Conrad Gessner (1516–1565). Die Renaissance der Wissenschaften / The Renaissance of Learning* (Berlin: 2019) 581–605.

Else, F.M., "Globefish, Sturgeon and Trout: Duke Cosimo I de' Medici, Bachiacca and the Consuming Culture of Fish", *Medicea* 9 (2011) 20–29.

Findlen P., "The Market and the World. Science, Culture and Collecting in the Venetian Republic", in Aikema B. (ed.), *Il collezionismo a Venezia e nel Veneto ai tempi della Serenissima* (Venice: 2005) 55–68.

Folena G., "Per la storia dell'ittionomia volgare", in Olmi G. – Tongiorgi Tomasi L., *De piscibus. La bottega artistica di Ulisse Aldrovandi e l'immagine naturalistica* (Rome: 1993) 113–150.

Franchini D.A. – Margonari R. – Olmi G. – Signorini R. – Zanca A. – Tellini Perina C., *La scienza a corte. Collezionismo eclettico natura e immagine a Mantova fra Rinascimento e Manierismo* (Rome: 1979).

Fratta V. Di, "De Aquatilibus. Dall'illustrazione scientifica alla nascita e sviluppo del genere della natura morta dei pesci nel XVII secolo", in [no editor] *Per la conoscenza dei beni Culturali, VI* [Ricerche del Dottorato in Metodologie conoscitive per la Conservazione e la Valorazione dei Beni Culturali, Università degli Studi di Napoli] (Naples: 2016) 57–72.

Garbari F. – Tongiorgi Tomasi L. – Tosi A., *Giardino dei Semplici / Garden of Simples* (Pisa: 2002).

Glardon P., *L'histoire naturelle au XVIe siècle. Introduction, étude et édition critique de* La nature et diversité des poissons *de Pierre Belon (1555)* (Geneva: 2011).

Glardon P., La terminologie botanique dans le *De historia stirpium* de Leonhart Fuchs (1542) et ses premières traductions françaises, *Seizième Siècle* 8 (2012) 57–74.

Göttler C., "The place of the 'Exotic' in seventeenth-century Antwerp", in Schrader S. (ed.), *Looking East. Rubens's encounter with Asia* (Los Angeles: 2013) 88–107.

Göttler C., "Extraordinary Things: 'Idols from India' and the Visual Discernment of Space and Time, circa 1600", in Göttler C. – Mochizuki, M.M. (eds.), *The Nomadic Object The Challenge of World for Early Modern Religious Art* (Leiden – Boston: 2018) 37–73.

Grieco A.J., "Fiordiano Malatesta da Rimini e i trattati di ittiologia della metà del Cinquecento", in Laurioux B. – Moulinier-Brogi L. (eds.), *Scrivere il Medioevo. Lo spazio, la santità, il cibo* (Rome: 2001) 305–318.

Groom A., *Exotic Animals in the Art and Culture of the Medici Court in Florence* (Leiden: 2019).

Gudger E.W., "The five great naturalists of the sixteenth century: Belon, Rondelet, Salviani, Gesner and Aldrovandi: a chapter in the history of Ichthyology", *Isis. International review devoted to the history of Science and Civilization* 22 (1934) 21–40.

Gudger E.W., "Beginnings of fish teratology, 1555–1642. Belon, Rondelet, Gesner and Aldrovandi, the fathers of ichthyology, the first to figure abnormal fishes", *Scientific Monthly* 43 (1936) 252–261.

Hackethal S., "Das THEATRUM NATURAE des Michael Rötenbeck. Unbekannte Naturstudien 100 Jahre nach Dürer", in Damaschun F. et al. (eds.), *Klasse Ordnung Art. 200 Jahre Museum für Naturkunde* (Rangsdorf: 2010) 70–75.

Hakelberg D., "Das Bodensee-Fischbuch von Gregor Mangolt in einem Basler Nachdruck von 1612", *Wolfenbütteler Barock-Nachrichten* 36 (2009) 107–114.

Handsch von Limus, Georg, *Die Elbefischerei in Böhmen und Meißen*, ed. O. Schubert (Prague: 1933).

Haupt H. – Vignau-Wilberg T. – Irblich E. – Staudinger M., *Le bestiaire de Rodolphe II. Cod. Min. 129 et 130 de la Bibliothèque nationale d'Autriche* (Paris: 1990).

Haye T., "Notizen zu Nicolaus Marschalk", *Daphnis* 23 (1994) 205–236.

Heikamp D., "La Grotta Grande del Giardino di Boboli", in Capecchi G. (ed.), *Palazzo Pitti: la reggia rivelata* (Florence: 2003) 446–474.

Hendrikx S. – Smith P.J., "Ichthyology in Translation: Conrad Gessner's Fish Books", in Fournel J.-L. – Paccagnella I. (eds.), *Traduire – Tradurre – Translating. Vie des mots et voies des oeuvres dans l'Europe de la Renaissance*, (Geneva: 2022) 341–361.

Hendrikx S.M. – Smith P.J., "Connaissances ichtyologiques sous forme emblématique: le cas du *sargus*", *RursuSpicae* 4 (2022) 1–26 (https://journals.openedition.org/rursuspicae/2258).

Herold B.J. – Horst T. – Leitão H., *A História Natural de Portugal de Leonhard Thurneysser zum Thurn, ca. 1555–1556* (Lisbon: 2019).

Hertel R., "Über die 'Ichtyographie der Elbe' des Johannes Kentmann. Eine Studie über die ältesten sächsichen Fischfaunen (Pisces)", *Zoologische Abhandlungen Staatliches Museum für Tierkunde in Dresden* 35.5 (1978) 75–100.

Hochleithner M. – Hohenleiter W., *Das Fischereibuch Maximilians I. – Faksimile und Transkription* (Kitzbühel: 2013).

Hochmann M., "Le collezioni veneziane nel Rinascimento: storia e storiagrafia", in Hochmann M. – Lauber R. – Mason S. (eds.), *Il collezionismo d'arte a Venezia. Dalle origini al Cinquecento* (Venice: 2008) 3–39.

Hochmann M., "La famiglia Grimani", in Hochmann M. – Lauber R. – Mason S. (eds.), *Il collezionismo d'arte a Venezia. Dalle origini al Cinquecento* (Venice: 2008) 206–223.

Hochmann M., "Plinio Scarpelli, pittore di Daniele Barbaro e dei Grimani di Santa Maria Formosa", *Arte Veneta* 67 (2010) 43–53.

Horst U., "Die einzige erhaltene handschrift des Valerius Cordus, ein Brief aus seinem letzten Lebensjahr 1544", *Beiträge zur Geschichte der Pharmazie* 26.2 (1974) 9–14.

Howard D., "I Barbaro come collezionisti rinascimentali", in Hochmann M. – Lauber R. – Mason S. (eds.), *Il collezionismo d'arte a Venezia. Dalle origini al Cinquecento* (Venice: 2008) 193–205.

Huber-Rebenich G., "Marschalk, Nikolaus", in Worstbrock F.J. (ed.), *Deutscher Humanismus 1480–1520. Verfasserlexikon, Vol. 2* (Berlin – New York: 2013) 161–203.

Hünemörder C., "Die Geschichte der Fischbücher von Aristoteles bis zum ende des 17. Jahrhunderts", *Deutsches Schiffahrtsarchiv* 1 (1975) 185–200.

Huth M. – Walter T., "Der Briefwechsel zwischen Conrad Gessner und Johannes Crato von Krafftheim", in Leu U. – Opitz P. (eds.), *Conrad Gessner (1516–1565). Die Renaissance der Wissenschaften / The Renaissance of Learning* (Berlin: 2019) 377–404.

Irblich E. (ed.), *Thesaurus Austriacus: Europas Glanz im Spiegel der Buchkunst: Handschriften und Kunstalben von 800 bis 1600* (Vienna: 1996).

Jacquemard C. – Gauvin B. – Lucas-Avenel M. (eds.), *Hortus Sanitatis. Livre IV: Les Poissons* (Caen: 2013).

Kolb K., *Graveurs artistes et hommes de science: Essai sur les traités de poissons de la Renaissance* (Paris: 1996).

Koning J. De – Uffelen G. Van – Zemanek A. – Zemanek B. (eds.), *Drawn after nature. The complete botanical watercolours of the 16th-century Libri Picturati* (Zeist: 2008).

Kris E., "Georg Hoefnagel und der wissenschaftliche Naturalismus", in Weixlgärtner A. – Planiscig L. (eds.), *Festschrift für Julius Schlosser zum 60. Geburtstag* (Zurich: 1927) 243–253.

Kusukawa S., "Image, Text and Observation: The Codex Kentmanus", *Early Science & Medicine* 14.4 (2009) 445–475.

Kusukawa S., "The sources of Gessner's pictures for the *Historia animalium*", *Annals of Science* 67.3 (2010) 303–328.

Kusukawa S., *Picturing the Book of Nature. Image, Text and Argument in Sixteenth-Century Human Anatomy and Medical Botany* (Chicago – London: 2012).

Laven P.J., *Daniele Barbaro, Patriarch-Elect of Aquileia with special reference to his circle of scholars and to his literary achievement* (Ph.D. dissertation, University of London: 1957).

Leonhard K., "Shell collecting. On 17th-century conchology, curiosity cabinets, and still life painting", in Enenkel K. – Smith P.J. (eds.), *Early Modern Zoology. The construction of animals in science, literature and the visual arts*, Intersections 7 (Leiden – Boston: 2007) 177–214.

Leu U., *Conrad Gessner (1516–1565). Universalgelehrter und Naturforscher der Renaissance* (Zurich: 2016).

Lemerle, F. – Zara V. – Caye P. – Moretti L. (eds.), *Daniele Barbaro 1514–1570: Vénitien, patricien, humaniste* (Turnhout: 2017).

Lewis G., "The debt of John Ray and Martin Lister to Guillaume Rondelet of Montpellier", *Notes and Records of the Royal Society* 66 (2012) 323–339.

Luca M.E. De – Faietti M. (eds.), *Jacopo Ligozzi "Altro Apelle"* (Florence: 2014).

Ludwig G., "Archivalische Beiträge zur Geschichte der venezianischen Malerei", *Jahrbuch der Königlich Preussischen Kunstsammlungen* 24 Beiheft (1903) 1–109.

Marcon S. – Moretti L. (eds.), *Daniele Barbaro 1514–70. Letteratura, scienza e arti nella Venezia del Rinascimento*, exh. cat. Venice (Venice: 2015).

Maselis M.C. – Balis A. – Marijnissen R.H., *The albums of Anselmus de Boodt (1550–1632): natural history painting at the court of Rudolph II in Prague* (Ramsen: 1999).

Mauriello A., "La 'Siracusa' di Paolo Regio e la tradizione letteraria napoletana tra primo e secondo Cinquecento", *Studi rinascimentali* 6 (2008) 91–97.

Mazal O., "Tierbilder aus der Ambraser Kunst und Wunderkammer Erzherzorg Ferdinands von Tirol (Cod. Ser. N. 2669 der Österreichische Nationalbibliothek)", *Codices Manuscripti* 8 (1982) 12–38.

McBurney H. – Rolfe I. – Napoleone C. – Findlen P. et al., *The Paper Museum of Cassiano dal Pozzo, Series B (parts 4 and 5). Natural History: Birds, Other Animals and Natural Curiosities*, 2 vols. (Turnhout – London: 2017).

Minonzio F., "Diffrazioni pliniane prima di Belon (1553): descrizione e classificazione di pesci in Paolo Giovio, Francesco Massari e Simone Porzio", in Maraglino V. (ed.), *Scienza antica in età moderna* (Bari: 2012) 401–439.

Morel P., *Les Grotesques. Les figures de l'imaginaire dans la peinture italienne de la fin de la Renaissance* (Paris: 1997).

Morelli J., *Notizia d'opere di disegno, pubblicata e illustrata da D. Jacopo Morelli, Seconda edizione, riveduta ed aumentata per cura di Gustavo Frizzoni* (Bologna: 1884).

Nardo D., "Brevi cenni storici sui progressi dell'Adriatica Fauna da Oppiano fino a' di' nostri", *Commentario della fauna, flora e gea del Veneto e del Trentino* 3.1 (1868) 125–126.

Nellen W. – Dulčić J., "Evolutionary steps in ichthyology and new challenges", *Acta Adriatica* 49 (2008) 201–232.

Nelles P., "Conrad Gessner and the Mobility of the Book: Zurich, Frankfurt, Venice (1543)", in Bellingradt D. – Nelles P. – Salman J. (eds.), *Books in Motion in Early Modern Europe. Beyond Production, Circulation and Consumption* (Cham: 2017) 39–66.

Nicolò M.L. De, *Il Mediterraneo nel Cinquecento tra antiche e nuove maniere di pescare* (Pesaro: 2011).

Nissen C., *Schöne Fischbücher. Kurze Geschichte der ichthyologischen Illustration* (Stuttgart: 1951).

Nissen C., *Die zoologische Buchillustration. Ihre Bibliographie und Geschichte.* Band II (Stuttgart: 1978).

Olmi G., *L'Inventario del mondo. Catalogazione della natura e luoghi del sapere nella prima età moderna* (Bologna: 1992).

Olmi G., "Bologna nel secolo XVI: una capitale europea della ricerca naturalistica", in Frommel S. (ed.), *Crocevia e capitale della migrazione artistica. Forestieri a Bologna e bolognesi nel mondo (secoli XV–XVI)* (Bologna: 2010) 61–80.

Olmi G. – Simoni F. (eds.), *Ulisse Aldrovandi. Libri e immagini di Storia naturale nella prima Età moderna* (Bologna: 2018).

Olmi G. – Tongiorgi Tomasi L., *De piscibus. La bottega artistica di Ulisse Aldrovandi e l'immagine naturalistica* (Rome: 1993).

Peers G., "Thinking with Animals: Byzantine Natural History in Sixteenth-Century France", *Bibliothèque d'Humanisme et Renaissance* 68 (2006) 457–484.

Peers G., "Forging Byzantine animals: Manuel Philes in Renaissance France", *Rivista di Studi Bizantini e Neoellenici* N.S. 49 (2012–2013) 79–103.

Perfetti S., *Aristotle's Zoology and its Renaissance Commentators (1521–1601)* (Leuven: 2000).

Pinon L., "Clématite bleue contre poissons séchés: Sept lettres inédites d'Ippolito Salviani à Ulisse Aldrovandi", *Mélanges de l'École Française de Rome* 114 (2002) 477–492.

Pyle C., *Das Tierbuch des Petrus Candidus. Codex Urbinas Latinus 276. Eine Einführung* (Zurich: 1984).

Pyle C., "The Art and Science of Renaissance Natural History: Thomas of Cantimpré, Pier Candido Decembrio, Conrad Gessner and Teodoro Ghisi in Vatican Library MS Urb. lat. 276", *Viator* 27 (1996) 265–321.

Rava A., "Il 'Camerino delli antigaglie' di Gabriele Vendramin", *Nuovo Archivio Veneto* 39 (1920) 155–181.

Rijks M., "A painter, a collector, and a horseshoe crab. Connoisseurs of art and nature in early modern Antwerp", *Journal of the History of Collections* 31.2 (2019) 343–361.

Rijks M., "'Unusual Excrescences of Nature': Collected Coral and the Study of Petrified Luxury in Early Modern Antwerp", *Dutch Crossing* 43.2 (2019) 127–156.

Rijks M., "Scales, Skins, and Carapaces in Antwerp Collections", in Bol M. – Spary E. (eds.), *The Matter of Mimesis: Studies on Mimesis and Materials in Nature, Art and Science* (Leiden – Boston: 2023) 295–320.

Rikken M., *Dieren verbeeld. Diervoorstellingen in tekeningen, prenten en schilderijen door kunstenaars uit de Zuidelijke Nederlanden tussen 1550 en 1630* (Leiden: 2016).

Salazar A., "Arias Montano y Pedro de Valencia", *Revista de estudios extremeños* 15 (1959), 475–493.

Sallent del Colombo E., "Natural History Illustration between Bologna and Valencia: The Aldrovandi-Pomar Case", *Early Science and Medicine* 21 (2016) 182–213.

Sallent del Colombo E. – Pardo Tomás J., "Materiali aldrovandiani in Spagna: l'enigmatico caso del *Códice Pomar*", in Olmi G. – Simoni F. (eds.), *Ulisse Aldrovandi. Libri e immagini di Storia naturale nella prima Età moderna* (Bologna: 2018) 37–48.

Schaffrath U., "Läuse, Muscheln und Tabak – Das Herbar Ratzenberger", *Philippia. Abhandlungen und Berichte aus dem Naturkundemuseum im Ottoneum zu Kassel*, 15.3 (2012) 191–214.

Scheichl A., "Wer war(en) Jorg Kolderer, Innsbrucker Hofmaler und Tiroler Baumeister", in Sternath M.L. – Michel E. (eds.), *Emperor Maximilian and the age of Dürer*, exh. cat. Vienna (Munich: 2012) 81–90, 181–182, 315–319.

Schmitter M., "'Virtuous Riches': The Bricolage of Cittadini Identities in Early-Sixteenth-Century Venice", *Renaissance Quarterly* 57.3 (2004) 908–969.

Sestini D. (ed.), *Tractatus per epistolam Francisci Maurolyci ad Petrum Gillium de piscibus siculis*, Messina, 1543, in Sestini D., *Viaggi e Opuscoli diversi* (Berlin: 1807).

Smith P.J., "Rabelais ichtyologue", in Garnier I. – La Charité C. – Menini R. – Pouey-Mounou A.-P. – Réach-Ngô A. – Tran T. – Viet N. (eds.), *Narrations fabuleuses. Mélanges en l'honneur de Mireille Huchon* (Paris: 2022) 439–452.

Smith P.J., *Tussen taal en natuur: de poëzie van Du Bartas* (Leiden: 2020).

Smith P.J., "Ronsard and Du Bartas in the Low Countries: Evidence from Early Modern Dutch Private Libraries and a *Vanitas* Still-Life by Edwaert Collier (ca. 1664)", in Pouey-Mounou A.P. – Smith P.J. (eds.), *Ronsard and Du Bartas in Early Modern Europe* (Leiden – Boston: 2021) 207–228.

Smith P.J., "Ichthyological Topics of the European Reception of Du Bartas" in Pouey-Mounou A.P. – Smith P.J. (eds.), *Ronsard and Du Bartas in Early Modern Europe* (Leiden – Boston: 2021) 255–279.

Smith P.J. – Trijp D. van, "Dynamiques européennes de l'humanisme érudit dans l'histoire naturelle. Le cas de l'ichtyologie, de Belon, Rondelet et Gessner à Willughby et Ray", in Crouzet D. – Crouzet-Pavan E. – Desan P. – Revest C. (eds.), *L'humanisme à l'épreuve de l'Europe (xv*[e]*–xvi*[e] *siècle). Histoire d'une transmutation Culturelle* (Ceyzérieu: 2019) 167–181.

Soldato E. Del, "Un catalogo naufragato: il *De piscibus* di Simone Porzio", in Bellis E. de (ed.), *Aristotele e la tradizione aristotelica. Nuove tematiche per il turismo culturale* (Soveria Mannelli: 2008) 149–176.

Soldato E. Del, *Simone Porzio. Un aristotelico tra natura e grazia* (Rome: 2010).

Thijsse G., "'Tusschen Pampier Geleyt.' Ontstaan, verspreiding en gebruik van de vroegste herbaria", in IJpelaar L. – Chavannes-Mazel C.A. (eds.), *De Groene Middeleeuwen. Duizend jaar gebruik van planten (600–1600)* (Eindhoven: 2016) 64–93.

Timm W., "Die Holzschnitte zu Nikolaus Marschalks *Historia aquatilium latine ac grece cum figuris*. Rostock 1517–20", *Wisschenschaftliche Zeitschrift der Universität Rostock – Gesellschafts- und Sprachwissenschaftliche Reihe* 17.9/10 (1968) 799–802.

Tice L.N., "Collecting in the garden. Inventories of *casini* in late sixteenth- and early seventeenth-century Rome", *Journal of the History of Collections* 23.2 (2011) 315–331.

Tongiorgi Tomasi L., "Il problema delle immagini nei 'Commentarii'", in Ferri S. (ed.), *Pietro Andrea Mattioli: Siena, 1501–Trento, 1578: la vita e le opere* (Perugia: 1997) 369–376.

Tongiorgi P., "Dalle profondità dei mari del xvi secolo", in Alessandrini A. – Ceregato A. (eds.), *Natura Picta. Ulisse Aldrovandi* (Bologna: 2007) 89–93.

Tosi A., "Contrivances of art. The power of imagery in the early modern culture of curiosity", in Beretta M. – Conforti M. (eds.), *Fakes. Hoaxes, Counterfeits and deception in early modern Science* (Sagamore Beach: 2014) 153–175.

Tosi A., "Acconciare, seccare, dipingere: pratiche di rappresentazione della natura tra le "spigolature" aldrovandiane", in Olmi G. – Simoni F. (eds.), *Ulisse Aldrovandi. Libri e immagini di Storia naturale nella prima Età moderna* (Bologna: 2018) 49–58.

Tosi A. (ed.), *Ulisse Aldrovandi e la Toscana. Carteggio e testimonianze documentarie* (Florence: 1989).

Vignau-Wilberg T., *Joris and Jacob Hoefnagel. Art and Science around 1600* (Berlin: 2017).

Violand G., "Historische Fischökologie des Bodensees. Das Fischbuch des Gregor Mangolt, 1557", *Österreichs Fischerei* 59 (2006) 169–171.

Vossilla F., "Cosimo I, lo scrittoio del Bachiacca, una carcassa di capodoglio e la filosofia naturale", *Mitteilungen des Kunsthistorischen Institutes in Florenz* 37.2/3 (1993) 381–395.

Walter H., "An illustrated incunable of Pliny's Natural History in the Biblioteca Palatina, Parma", *Journal of the Warburg and Courtauld Institutes* 53 (1990) 208–216.

Weiler C. (ed.), *Von Fischen, Vögeln und Reptilen. Meisterwerke aus den kaiserlichen Sammlungen* (Vienna: 2011).

Wheeler A. – Davis P.S. – Lazenby E., "William Turner's (c 1508–1568) notes on fishes in his letter to Conrad Gessner", *Archives of natural history* 13 (1986) 291–305.

Zahn G., "Das Herbar des Dr. Caspar Ratzenberger (1598) in der Herzoglichen Bibliothek zu Gotha", *Mitteilungen des Thüringischen Botanischen Vereins* N.F. 16 (1902) 50–121.

Zaunick R., "Fragmente der ältesten sächsischen Fischfauna des Dr. Johannes Kentmann (1518–1574)", *Sitzungsberichte und Abhandlungen der Naturwissenschaftlichen Gesellschaft Isis* (1915) 15–36.

Zaunick R., *Das älteste deutsche Fischbüchlein vom Jahre 1498 und dessen bedeutung für die spätere Literatur* (Berlin: 1916).

Zimmermann T.C.P., *Paolo Giovio* (Princeton: 1995)

Zucker A., "Zoologie et philologie dans les grands traités ichtyologiques renaissants", *Kentron. Revue pluridisciplinaire du monde antique* 29 (2013) 135–174.

CHAPTER 7

Ichthyology and Related Topics in MS Urb. lat. 276 (13th–17th Centuries)

Cynthia M. Pyle

The purpose of this short chapter is to call attention to the relevance of the remarkable Vatican manuscript Urbinas latinus 276 to the history of natural history, and specifically here to the history of ichthyology.[1] Very briefly, this manuscript bears witness to three centuries of natural history, from the mid-13th-century text, *De natura rerum*, by Thomas de Cantimpré (1201–1272), found, anonymous, two centuries later in some old manuscripts in Naples, and portions revised as *De animantium naturis* by the Lombard humanist Pier Candido Decembrio (1399–1477). Decembrio had a fair copy made at the request of Marchese Ludovico Gonzaga of Mantua in the mid-15th century. This copy was then illustrated, over 130 years later in the 1580s and 1590s, probably, as I concluded in 1984 and 1996, by Teodoro Ghisi of Mantua (1536–1601), basing himself largely on the woodcuts of Conrad Gessner of Zurich's (1516–1565) multi-volume *Historiae Animalium* and *Icones* volumes (1551–1560, and the posthumous volume on serpents of 1587).

More recently, a further connection has been made with the natural history illustrations of the Antwerpian father and son artists who ended their careers in the Hapsburg court, Joris and Jacob Hoefnagel (1542–1600 and 1573–1632), who also relied often on Conrad Gessner's works for the outlines of many

1 The manuscript was published in facsimile in 1984, with an introductory volume, first published, in German translation only, and sporting an earlier scholar's title (it is not a moralizing bestiary, but rather a natural history, based on the encyclopedic tradition), in 1984: Pyle C.M., *Das Tierbuch des Petrus Candidus. Codex Urbinas Latinus 276. Eine Einführung*, tr. T. Honref, J. Schlechta (Zurich: 1984; Codices e Vaticanis Selecti, LX). Two further translations of the introductory volume, in Italian and Spanish, followed, both in 1985. There then appeared an updated article on the MS in 1996: Pyle C.M., "The Art and Science of Renaissance Natural History: Thomas of Cantimpré, Pier Candido Decembrio, Conrad Gessner and Teodoro Ghisi in Vatican Library MS Urb. lat. 276", *Viator* 27 (1996) 265–321. Some specific entries are addressed in Eadem, "Insects in [Renaissance] Art", in Smith S. (ed.), *A Cultural History of Insects in the Renaissance* (London: Forthcoming). The editors of the present volume are kind enough to offer space here to bring this remarkable document to scholars' attention. A second revised edition, in English and with a new title, of the 1984 volume (never published in the language it was written in) is in progress.

© CYNTHIA M. PYLE, 2024 | DOI:10.1163/9789004681187_008

of the creatures they depicted.[2] There are thus three traditions behind the Hoefnagels' work, and to which they contribute in a major fashion: the painted natural history tradition, the moralizing emblematic tradition, and, emphatically, the printed graphic tradition in the arts and what we now call the natural sciences. In preparing this chapter, further exciting possibilities, many beyond the scope of the present context, have arisen, and will be dealt with in a second edition of the 1984 book.

Pier Candido Decembrio's work, *De animantium naturis* (ca. 1460), which appears in three other manuscript drafts,[3] is in six books: *On Four-footed Animals*; *On Birds*; *On Marine Monsters and Fish* (our principal focus here); *On Serpents*; *On Lowly Creatures* (*Vermes*); and *On Things Worthy of Memory and Note*. Cantimpré's text was found in Naples by Pier Candido, who resided there from 1456 to 1459, first in the court of King Alfonso d'Aragona (Alfonso il Magnanimo, 1396–1458), then briefly at that of his successor, Ferrante d'Aragona (1424–1494). While in Naples, as Decembrio recounts in his opening paragraph, he found a number of anonymous old books, whose texts he studied before rewriting them, in part for stylistic reasons and in part for content, and presenting them in 1460 to Ludovico Gonzaga, who asked that a fair copy be made and – most significantly – that room be left at the bases of the folios for a fine painter to depict the creatures described, 'so that I may better understand the text'. This request, while not fulfilled during the potential patron's lifetime, very clearly moves the more than five hundred animal images we do find in the manuscript from the realm of illumination (or decoration) to that of illustration. And the illustrations, through good fortune being created by a

2 Pyle C.M., "Art as Science in Joris Hoefnagel and Teodoro Ghisi", *Center 21. National Gallery of Art. Center for Advanced Study in the Visual Arts, Record of Activities and Research Reports, June 2000–May 2001* (Washington: 2001) 112–114; Eadem, "Insects", in Smith (ed.), *A Cultural History of Insects in the Renaissance*. Cf. Joris Hoefnagel, *Animalia Aquatilia et Co[n]chiliata (Aqua)*, III, fig. LVI, National Gallery of Art, Washington; Jacob Hoefnagel, *Archetypa studiaque patris Georgii Hoefnagelii 1592. Natur, Dichtung und Wissenschaft in der Kunst um 1600. Nature, Poetry and Science in Art around 1600*, ed. T. Vignau-Wilberg (Munich: 1994), esp. Pars Prima, Plate 10. On the Hoefnagels, see the thesis of Wilberg Vignau-Schuurman T.A.G., *Die emblematischen Elemente im Werke Joris Hoefnagels*, 2 vols. (Leiden: 1969); Vignau-Wilberg T., *Joris and Jacob Hoefnagel. Art and Science around 1600* (Berlin: 2017) and her other publications in between; Hendrix M.L., *Joris Hoefnagel and the Four Elements: A Study in Sixteenth-Century Nature Painting* (Ph.D. Dissertation, Princeton University 1984); Bass M.A., *Insect Artifice. Nature and Art in the Dutch Revolt* (Princeton: 2019). These sources together illustrate the combined emblematic and scientific-naturalistic interests of the Hoefnagels.

3 Bologna, Biblioteca Universitaria, MS 593 (which belonged to the naturalist Ulisse Aldrovandi); Munich, Staatsbibliothek, clm 508 (which bears the words "emptus 1561" and annotations in a hand very like, if not identical to, Conrad Gessner's); Florence, Accademia Colombaria, MS 231; for descriptions, see Pyle, *Das Tierbuch* 73–76.

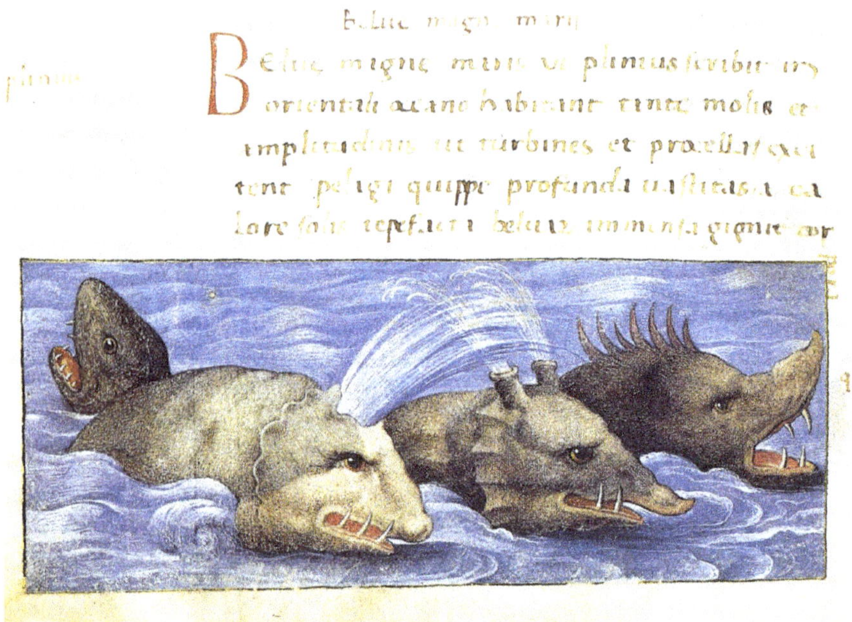

FIGURE 7.1 Teodoro Ghisi (attrib.), Vatican MS Urb. lat. 276, fol. 126v (detail): *Beluae*
BIBLIOTECA APOSTOLICA VATICANA, VATICAN CITY

16th-century naturalist-artist (one who could not have been found with this expertise when the manuscript was prepared, in spite of the scientific quality of practices like humanistic philology and history in the fifteenth century), do indeed follow and illustrate the text closely. The codex was transcribed and, yes, decorated for Ludovico in the fifteenth century just after his request was made, the Gonzaga arms on fol. 1 being painted, and the book and chapter initials being rubricked, in the atelier of the Maestro d'Ippolita Sforza, who was active in Milan 1450–1465.[4]

4 Zanichelli G., "Il Maestro di Ippolita Sforza e il suo atelier: ipotesi di un percorso", *Artes* 14 (2008) 5–31; the Gonzaga arms on fol. 1 of MS Urb. lat. 276 were attributed to this master in Pyle C.M., "Harvard MS Richardson 23: A'Pendant' to Vatican MS Urb. lat. 276 and a Significant Exemplar for P.C. Decembrio's *Opuscula historica*", *Scriptorium* 42, 2 (1988) 191–198. Ippolita herself (1445/6–1488) was betrothed to Ferrante's son, Alfonso II d'Aragona, in 1455 and resident primarily in Naples from their wedding in 1465 until her death in 1488. On her, now see: *Ippolita Maria Sforza Duchess and Hostage in Renaissance Naples: Letters and Orations*, ed. and tr. D. Robin – L.L. Westwater (Toronto and Tempe, Arizona: 2017). There is also Wood J.M., *Ippolita Maria Sforza. The Renaissance Princess Who Linked Milan and Naples* (Jefferson, North Carolina: 2020).

A number of the over 500 tempera illustrations are diagnostic for the sources consulted by the manuscript artist, and several of these crucial illustrations occur among the animals described in Book III, *De beluarum maritimar[um] et piscium naturis omnium* (fols. 121v–168). The third image in this third book, at fol. 126v, is a unique illustration in the manuscript, consisting of a gold-framed rectangular patch of blue water, with the heads of four whale-like *beluae* emergent from the ocean waves [Fig. 7.1]. This image can be explained by the artist's source, Conrad Gessner's *Historiae animalium*,[5] but was probably modeled more on Gessner's *Icones* volumes [Fig. 7.2].[6]

This particular iconography is odd, even in Gessner's volumes, and can be explained in turn by Gessner's artists' use of the *Carta Marina* of Olaus Magnus as *their* model[7] [Fig. 7.3] Gessner's engravers simply copied onto their woodblocks portions of the *Carta Marina* containing "beluae" or monsters of the sea (often whale-like creatures) which are indeed reversed in Gessner's volumes, copying rectangles of sea along with the (elaborate) marine animals, and our manuscript artist used Gessner's copies as his models for the tempera illustration. If my 1984 attribution to Teodoro Ghisi (younger brother of the great engraver, Giorgio Ghisi, and therefore highly familiar with the printed tradition) is correct – and it has not been substantively or successfully disputed to my knowledge, though I am now in this chapter suggesting a nuanced approach to the question – then his (and possibly his collaborators', as below) use of primarily printed sources throughout the manuscript and even in his oil paintings, like *Symbolum Apostolorum*, is explained.[8]

One of the many examples is the "*Orcha*" (fol. 137v), which some now term unsympathetically the killer whale. In our manuscript it does not sport its striking black and white coat, but other characteristics, such as its conical teeth and its blow-spout are present.[9] These tamer whale-like illustrations, with their

5 Gessner Conrad, *Historiae animalium: Liber IIII qui est de piscium et aquatilium animantium natura* (Zurich, Christopher Froschauer: 1558) 137–139 and 249.

6 Gessner Conrad, *Nomenclator aquatilium animantium. Icones animalium aquatilium in mari et dulcibus aquis degentium* (Zurich, Christopher Froschauer: 1560) 176–177: *Cete* (Whales).

7 Olaus Magnus, *Carta marina et descriptio septentrionalium terrarum ac mirabilium rerum in eis contentarum diligentissime elaborata anno dni 1539* (Venice: 1539).

8 Pyle, *Das Tierbuch* 98–102, 104–105, nn. 20, 26–33; Eadem "The Art and Science", 268, and passim. It should be noted that Jacob Hoefnagel was also an engraver, and thus himself undoubtedly deeply familiar with the printed tradition.

9 Species are identified, where possible (and under revision in some cases) in Pyle, *Das Tierbuch* 166–195. There are nowadays videos available on line of orcas following ships like dolphins, and blowing through their spouts as they "gyre and gimble in the wabe" (Lewis Carroll, *The Jabberwocky*, available at: https://www.poetryfoundation.org/poems/42916/jabberwocky).

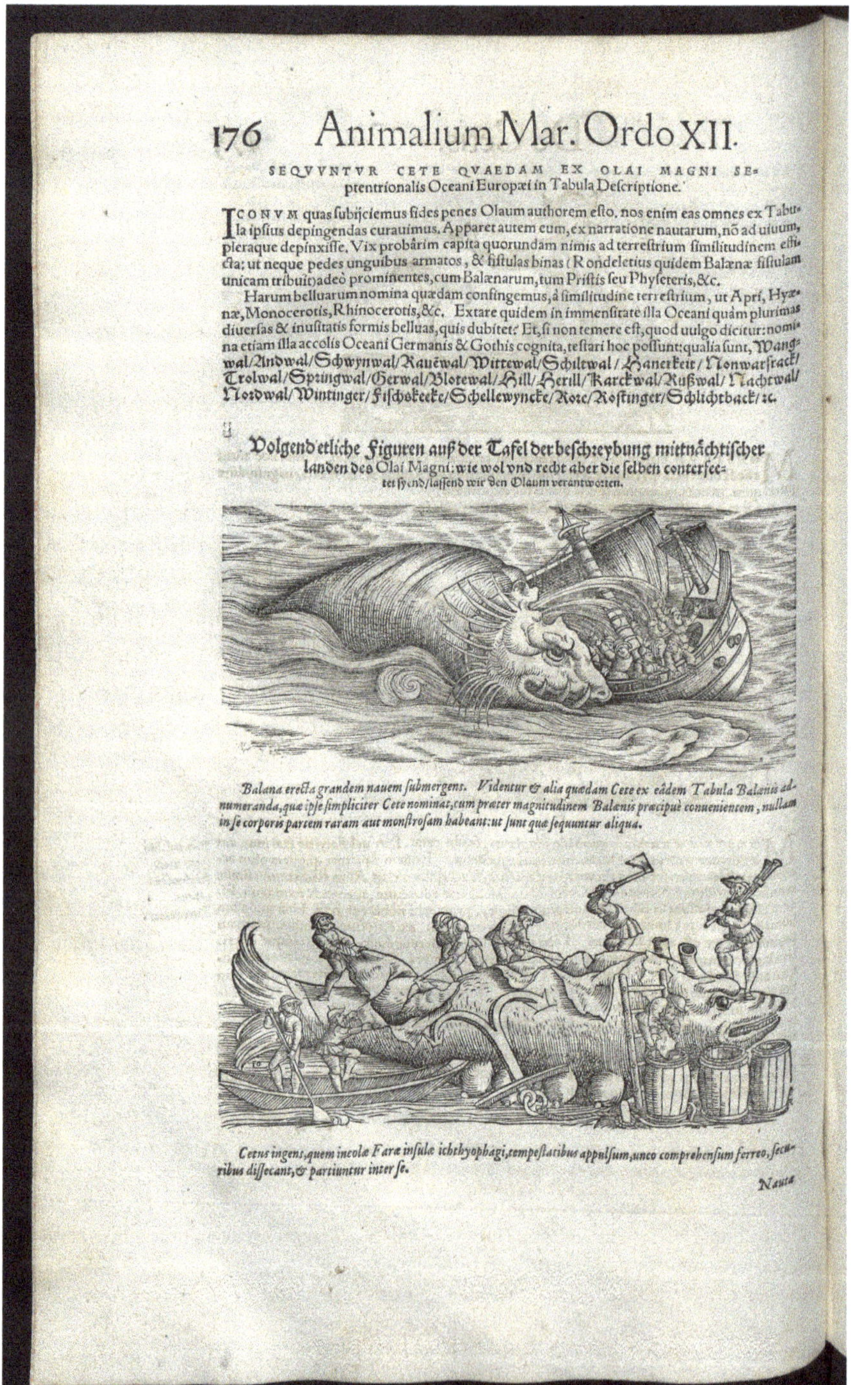

176 Animalium Mar. Ordo XII.

SEQVVNTVR CETE QVAEDAM EX OLAI MAGNI SEptentrionalis Oceani Europæi in Tabula Descriptione.

ICONVM quas subijciemus fides penes Olaum authorem esto, nos enim eas omnes ex Tabula ipsius depingendas curauimus. Apparet autem eum, ex narratione nautarum, nõ ad uiuum, pleraque depinxisse. Vix probârim capita quorundam nimis ad terrestrium similitudinem efficta; ut neque pedes unguibus armatos, & fistulas binas (Rondeletius quidem Balænæ fistulam unicam tribuit) adeò prominentes, cum Balænarum, tum Pristis seu Physeteris, &c.

Harum belluarum nomina quædam confingemus, à similitudine terrestrium, ut Apri, Hyænæ, Monocerotis, Rhinocerotis, &c. Extare quidem in immensitate illa Oceani quàm plurimas diuersas & inusitatis formis belluas, quis dubitet? Et, si non temere est, quod uulgo dicitur: nomina etiam illa accolis Oceani Germanis & Gothis cognita, testari hoc possunt: qualia sunt, Wangwal/Andwal/Schwynwal/Rauẽwal/Wittewal/Schiltwal/Hanerkeit/Nonwarsrack/Trolwal/Springwal/Gerwal/Blotewal/Hill/Herill/Karckwal/Rußwal/Nachtwal/Nordwal/Wintinger/Fischokeeke/Schellewyncke/Rore/Rostinger/Schlichtback/etc.

Volgend etliche Figuren auß der Tafel der beschreybung mittnächtischer landen des Olai Magni: wie wol vnd recht aber die selben conterfeetet syend/lassend wir den Olaum verantworten.

Balana erecta grandem nauem submergens. Videntur & alia quædam Cete ex eâdem Tabula Balænis adnumeranda, quæ ipse simpliciter Cete nominat, cum præter magnitudinem Balænis præcipuè conuenientem, nullam in se corporis partem raram aut monstrosam habeant: ut sunt quæ sequuntur aliqua.

Cetus ingens, quem incolæ Faræ insulæ ichthyophagi, tempestatibus appulsum, unco comprehensum ferreo, securibus dissecant, & partiuntur inter se.

Nauta

FIGURE 7.2A Gessner Conrad, *Nomenclator aquatilium animantium. Icones animalium aquatilium* (Zurich, Christopher Froschauer: 1560) 176: *Cete* (Whales)
BIBLIOTHÈQUE DU MUSÉUM NATIONAL D'HISTOIRE NATURELLE, PARIS

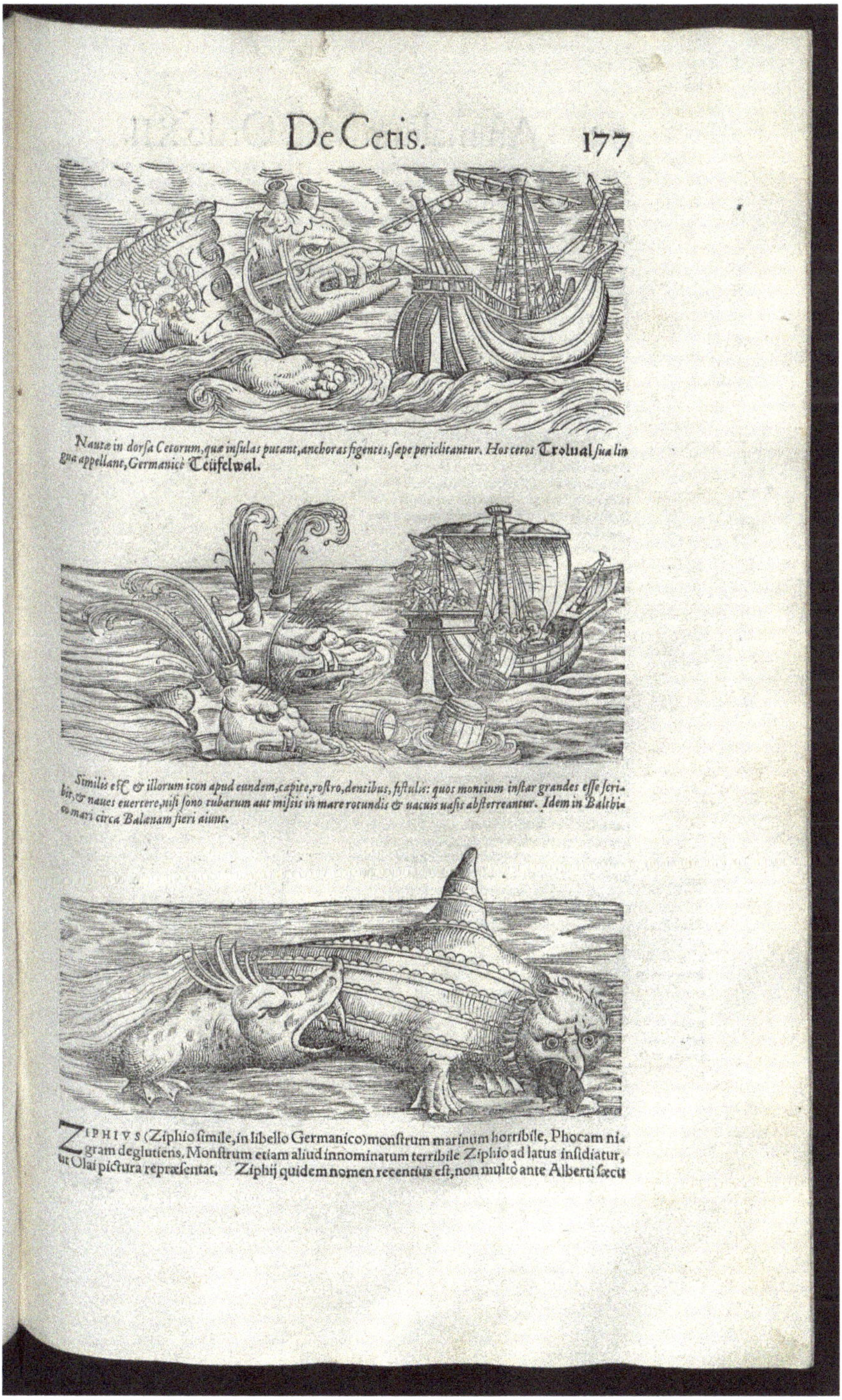

De Cetis. 177

Nautæ in dorsa Cetorum, quæ insulas putant, anchoras figentes, sæpe periclitantur. Hos cetos Trolual *sua lingua appellant, Germanicè* Teüfelwal.

Similis est & illorum icon apud eundem, capite, rostro, dentibus, fistulis: quos montium instar grandes esse scribit, & naues euertere, nisi sono tubarum aut missis in mare rotundis & uacuis uasis absterreantur. Idem in Balthico mari circa Balænam fieri aiunt.

ZIPHIVS (Ziphio simile, in libello Germanico) monstrum marinum horribile, Phocam nigram deglutiens. Monstrum etiam aliud innominatum terribile Ziphio ad latus insidiatur, ut Olai pictura repræsentat. Ziphij quidem nomen recentius est, non multò ante Alberti sæcu

FIGURE 7.2B Gessner Conrad, *Nomenclator aquatilium animantium. Icones animalium aquatilium* (Zurich, Christopher Froschauer: 1560) 177: *Cete* (Whales)
BIBLIOTHÈQUE DU MUSÉUM NATIONAL D'HISTOIRE NATURELLE, PARIS

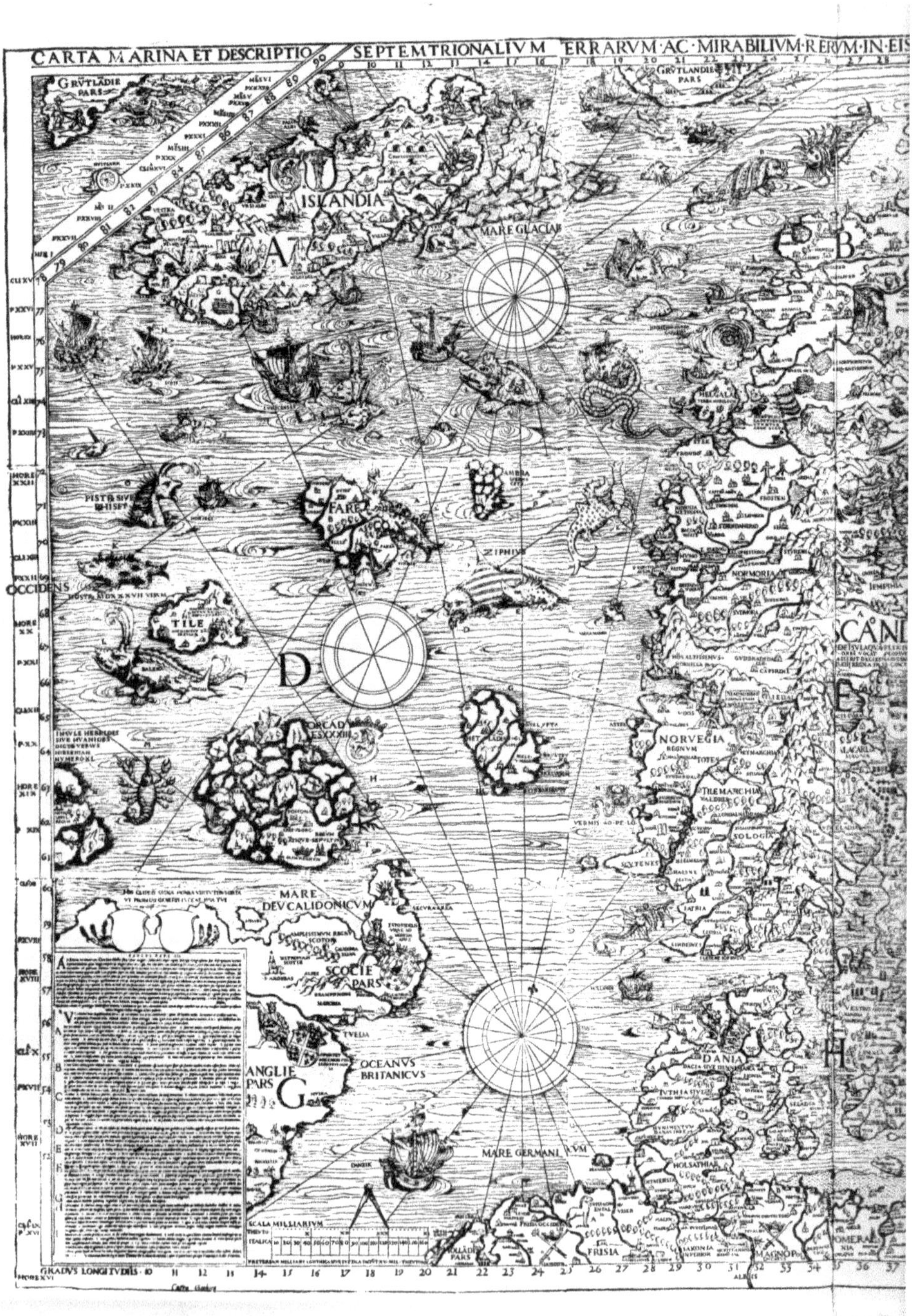

FIGURE 7.3 Olaus Magnus, *Carta Marina*, 1539 (detail)
BAYERISCHE STAATSBIBLIOTHEK, MÜNCHEN. SHELFMARK: MAPP. VII,1

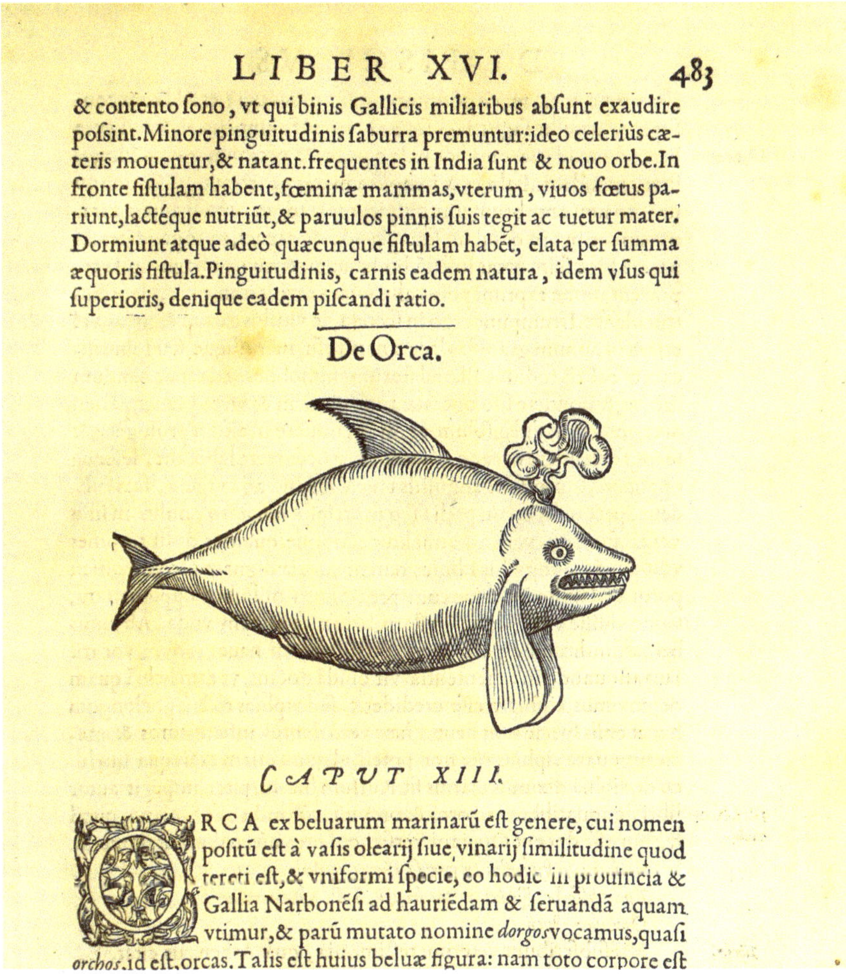
LIBER XVI. 483

& contento ſono, vt qui binis Gallicis miliaribus abſunt exaudire poſsint.Minore pinguitudinis ſaburra premuntur:ideo celeriùs cæteris mouentur,& natant.frequentes in India ſunt & nouo orbe.In fronte fiſtulam habent,fœminæ mammas,vterum, viuos fœtus pariunt,lactéque nutriũt,& paruulos pinnis ſuis tegit ac tuetur mater. Dormiunt atque adeò quæcunque fiſtulam habẽt, elata per ſumma æquoris fiſtula.Pinguitudinis, carnis eadem natura, idem vſus qui ſuperioris, denique eadem piſcandi ratio.

De Orca.

CAPVT XIII.

ORCA ex beluarum marinarũ eſt genere, cui nomen poſitũ eſt à vaſis olearij ſiue vinarij ſimilitudine quod tereti eſt,& vniformi ſpecie, eo hodie in prouincia & Gallia Narbonẽſi ad hauriẽdam & ſeruandã aquam vtimur,& parũ mutato nomine *dorgos* vocamus,quaſi *orchos*.id eſt.orcas.Talis eſt huius beluæ figura: nam toto corpore eſt

FIGURE 7.4 Guillaume Rondelet, *De piscibus marinis* (Paris, 1554), 483 (detail): *Orca*
SMITHSONIAN INSTITUTION LIBRARIES

typically fanciful cloud-like spouts are unmistakable, and probably familiar to all readers of this book as originating in Guillaume Rondelet's (1507–1566) *De piscibus marinis* (1554) [Fig. 7.4].[10] The illustration in MS Urb. lat. 276, however,

10 Rondelet Guillaume, *Libri de piscibus marinis in quibus verae piscium effigies expressae sunt*, 2 vol., t. II: *Universae aquatilium historiae pars altera, cum veris ipsorum imaginibus* (Lyon, Macé Bonhomme: 1554–1555) 483.

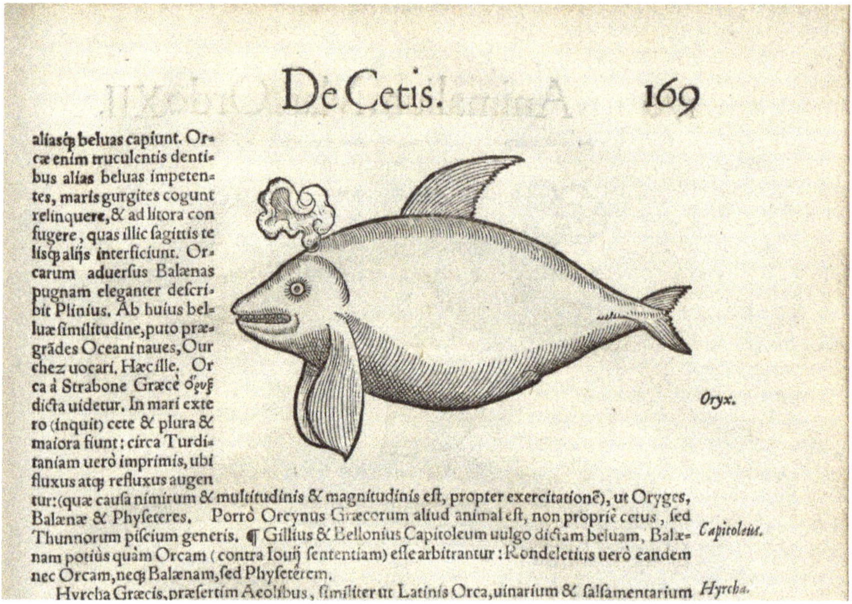
De Cetis. 169

aliasq; beluas capiunt. Orcæ enim truculentis dentibus alias beluas impetentes, maris gurgites cogunt relinquere, & ad litora confugere, quas illic sagittis telisq; aliis interficiunt. Orcarum aduersus Balænas pugnam eleganter describit Plinius. Ab huius beluæ similitudine, puto prægrãdes Oceani naues, Ourchez uocari. Hæc ille. Orca à Strabone Græcè ὄρυξ dicta uidetur. In mari extero (inquit) cete & plura & maiora fiunt: circa Turditaniam uerò imprimis, ubi fluxus atq; refluxus augentur: (quæ causa nimirum & multitudinis & magnitudinis est, propter exercitationẽ), ut Oryges, Balænæ & Physeteres. Oryx. Porrò Orcynus Græcorum aliud animal est, non propriè cetus, sed Thunnorum piscium generis. ¶ Gillius & Bellonius Capitoleum uulgo dictam beluam, Balænam potius quàm Orcam (contra Iouij sententiam) esse arbitrantur: Rondeletius uerò eandem nec Orcam, neq; Balænam, sed Physeterem. Capitoleus.

Hyrcha Græcis, præsertim Aeolibus, similiter ut Latinis Orca, uinarium & salsamentarium Hyrcha.

FIGURE 7.5 Gessner Conrad, *Nomenclator aquatilium animantium. Icones animalium aquatilium* ... (Zurich: 1560), 169 (detail): *Orca*
ERNST MAYR LIBRARY, HARVARD UNIVERSITY

is headed left, since our artist usually resorts to Conrad Gessner's woodcut images, where the reversal occurs through the engraving process [Fig. 7.5].[11]

In only a few instances (notably the electric ray, or "Torpedo", at fol. 165v, shown both dorsally and ventrally), does the manuscript artist consult both Gessner's *Nomenclator .... Icones* III and Gessner's source, in this case Pierre Belon's *De aquatilibus*, resulting in a confusion in the twist of the tail in the two views.[12]

Among the Monsters of the Sea there figures, too, a *Maris homo* (fol. 136v), whose model's physique and beard closely resemble those of a lean human male figure in a lower right-hand panel of Teodoro Ghisi's *Symbolum Apostolorum*, created while he was in the court at Graz. If Teodoro Ghisi is the artist, this

11 Gessner, *Nomenclator aquatilium animantium* 169; for the color illustration at fol. 137v of MS Urb. lat. 276, see: https://digi.vatlib.it/view/MSS_Urb.lat.276#.

12 Gessner, *Nomenclator* 125; Belon Pierre, *De aquatilibus* (Paris, Charles Estienne: 1553) 90–91. Cf. Pyle, *Das Tierbuch* 91–92; Eadem, "The Art and Science" 301–304. For the color illustration at fol. 165v of MS Urb. lat. 276, see: https://digi.vatlib.it/view/MSS_Urb.lat.276#.

would reinforce our dating of at least some of the manuscript illustrations during his three years in Graz (1587–1590).[13]

The second half of Book III (fols. 142v–158) is devoted to marine and river fish closer to reality, including swordfish, mullet, scorpion fish, sturgeon, carp, etc. Many of these are modelled on the woodcut illustrations of Pierre Belon and Guillaume Rondelet, again through their use and reversal by the woodcut engravers of Conrad Gessner of Zurich.

Other illustrations in the manuscript appear to require dates in the 1590s. When one reaches Book V, *Vermes*, including insects, frogs, arachnids, and another aquatic animal, a leech, many of these can also be traced to the printed version of the wash and gouache painted drawings of Joris Hoefnagel published by his son, Jacob Hoefnagel in 1592.[14] And one critical folio (fol. 201v) in the manuscript, where two aquatic worms (the smooth ascarid horsehair worm and the segmented annelid worm, the medicinal leech) are juxtaposed, is particularly important for the tracing of the relationships of these illustrations to others [Fig. 7.6].[15] It is through this folio that one is able to link the work of the Vatican manuscript artist with that of the Hoefnagels, for on this folio, the unsegmented horsehair worm, *Gordius sp.* ("*Seta*"), appears to the left of the segmented leech, *Hirudo medicinalis* ("*Sanguisuga*"), just as in the watercolor by Joris Hoefnagel in his *Four Elements* (dated internally at 1575–1582, and probably worked on for much of his life[16]) [Fig. 7.7]. However, the manuscript artist's illustration of a leech, while clearly inspired by this image, and even copying the juxtaposition of the two aquatic worms, is reversed and far less finely drawn, which tells us that it was taken from the son Jacob Hoefnagel's engraved image in his 1592 *Archetypa studiaque patris Georgii Hoefnagelii* [Fig. 7.8].[17] This of course also provides us with evidence

13 Pyle, *Das Tierbuch* 100–102; Eadem, "The Art and Science" 291–296; for the color illustration at fol. 136v of MS Urb. lat. 276, see: https://digi.vatlib.it/view/MSS_Urb.lat.276#. The *Symbolum Apostolorum* now hangs in the Alte Galerie Schloss Eggenberg of the Universalmuseum Joanneum, Graz.

14 *Archetypa* (above, n. 2) 116–117; the leech is understandably termed 'Nicht bestimmbares Objekt' (unidentifiable object) by Vignau-Wilberg. It was partly through the headings, "Sanguisuga," in MS Urb. lat. 276 and in *Archetypa*, I, 10, that I was able to identify it as *Hirudo*. The fact that Jacob Hoefnagel's (bapt. December 1573) age is given on the title page as 17 rather than 19 could indicate that the engravings were already in existence by 1590 or 1591, though published only in 1592.

15 For the color illustrations at fol. 201v of MS Urb. lat. 276, see: https://digi.vatlib.it/view/MSS_Urb.lat.276#.

16 Hendrix, *Joris Hoefnagel* 39 and 85, n. 32; Bass, *Insect Artifice* 3.

17 Jacob Hoefnagel after Joris Hoefnagel, in *Archetypa studiaque patris Georgii Hoefnagelii* (Frankfurt: 1592), I, 10; cf. Vignau-Wilberg (ed.), in Hoefnagel, *Archetypa* 116–117.

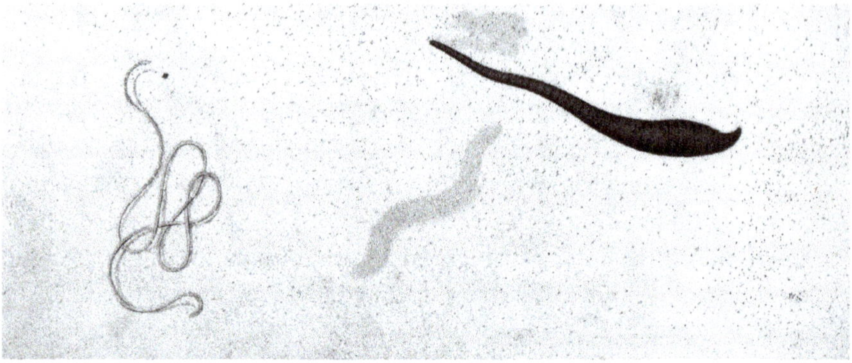

FIGURE 7.6 Teodoro Ghisi (attrib.), MS Urb. lat. 276, fol. 201v (detail: horsehair worm, leech): *Seta* and *Sanguisuga*
BIBLIOTECA APOSTOLICA VATICANA, VATICAN CITY

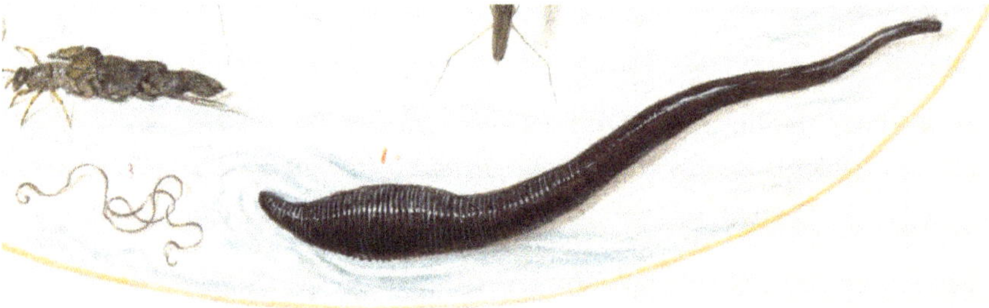

FIGURE 7.7 Joris Hoefnagel, MS *The Four Elements, Animalia aquatilia et co(n)chiliata (Aqua)*, Plate LVI (detail: horsehair worm, leech)
THE NATIONAL GALLERY OF ART, WASHINGTON, D.C. GIFT OF MRS. LESSING J. ROSENWALD, ACCESSION NUMBER: 1987.20.7

for dating at least the later manuscript illustrations in the codex as around or after 1592 (while granting that the manuscript artist may have been shown engravings before they were actually published).[18]

18 *The Four Elements*, III, LI, includes another leech, this one probably the model for Ulisse Aldrovandi's image (reversed from Hoefnagel's) in *De animalibus insectis* (1602) 765. Teodoro Ghisi also worked for Aldrovandi (Pyle, *Das Tierbuch* 91–92; Eadem, "The Art and Science" 291; Tellini Perina C., "Teodoro Ghisi: l'immagine fra Maniera e Controriforma", in Franchini D.A. et al. [eds.], *La scienza a corte, collezionismo eclettico natura e immagine a Mantova fra Rinascimento e Manierismo* [Rome: 1979] 239–268: 248), and could thus have steered him to Hoefnagel's image before Ghisi died in 1601; nor is it impossible for Aldrovandi and the Hoefnagels to have met. I am further investigating these complex interactions. This second leech does not appear in the Vatican manuscript.

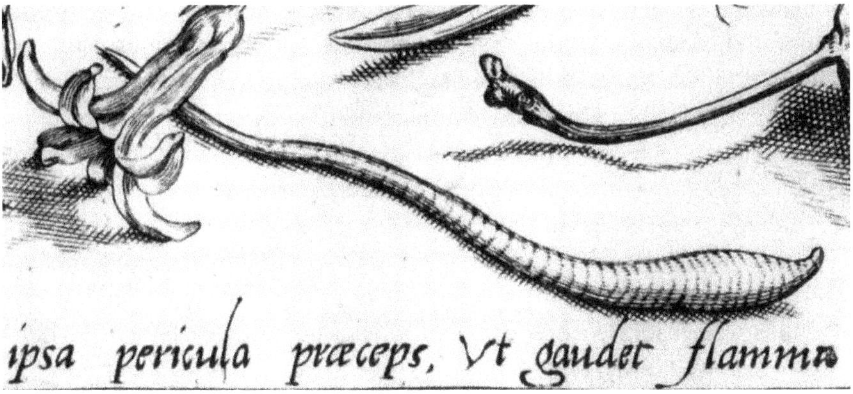

FIGURE 7.8 Jacob Hoefnagel after Joris Hoefnagel, in *Archetypa studiaque patris Georgii Hoefnagelii* (Frankfurt: 1592), I, 10 (detail: leech)
STAATLICHE SAMMLUNG MÜNCHEN

As we have seen, leeches are actually primarily aquatic segmented worms, and so we have a remarkable juxtaposition of two aquatic worms in Hoefnagel's and the manuscript artist's depictions. Was this due to the artists' and naturalists' deep study of the worms' characteristics? Or was it a completely coincidental juxtaposition? The latter seems unlikely. Attempts were clearly being made throughout the sixteenth century to order or classify the plants and animals in the world around us; Conrad Gessner, in keeping with his bibliographical interests, was among the most perceptive of these classifiers.[19]

The probable illustrator of the Urbinas manuscript, Teodoro Ghisi, was in the Graz court of Archduke Charles II Hapsburg of Styria (1540–1590, the brother of Eleonora Hapsburg Gonzaga, Guglielmo Gonzaga of Mantua's wife) for three years (1587-1590), when the Hoefnagels were in Munich, Frankfurt (where *Archetypa* was published in 1592) and in Austria, where they were in residence after 1600 at the court of Rudolf II Hapsburg. If they had not already met in Antwerp when Giorgio was there (1549/50–1555),[20] or during the Hoefnagels' lengthy European peregrinations as Calvinist refugees from Antwerp since 1576, they may well have met and exchanged ideas on their common love of nature while Teodoro Ghisi was residing primarily in Graz. Joris Hoefnagel is known to have studied nature deeply and reverently; his

19 Gessner Conrad, *Bibliotheca universalis* (Zurich, Christopher Froschauer: 1545); cf. Pyle, *Das Tierbuch* 29–31, 34; Eadem, "The Art and Science" 272–273.

20 Boorsch S. – Lewis M. – Lewis R.E., *The Engravings of Giorgio Ghisi* (New York: 1985) 17; Tellini Perina C., "Teodoro Ghisi". See also: Mozzetti F., "Teodoro Ghisi", *Dizionario Biografico degli Italiani* 54 (2000) https://www.treccani.it/enciclopedia/teodoro-ghisi_(Dizionario-Biografico).

son Jacob was clearly devoted to his father, judging from their travels together and from his creation of the *Archetypa* volume, and became court painter to Rudolf II, in whose courts he lived and worked until at least 1612, the year of Rudolf's death. Ghisi too was a naturalist-artist, and certainly painted in oils many religious subjects replete with animal images, including the *Symbolum Apostolorum* (now at the *Alte Galerie des Landesmuseum Joanneum*, Schloss Eggenberg) while he was in Graz.

All this brings to mind yet another possibility, to wit, that Teodoro Ghisi and the Hoefnagels may have collaborated – at least intellectually and verbally – on the illustrations for Decembrio's text in what is now MS Urb. lat. 276. It should be noted that the *beluae* mentioned and shown above, besides being immersed in a rectangle of blue sea, are edged or framed with a fine line of gold (a device of manuscript *illumination*). While gold highlights occur throughout the manuscript for particular animals, such as some birds (including the peacock and the phoenix), some marine creatures and certain insects, the other illustrations are never framed in gold (though the 15th-century Gonzaga arms at fol. 1 are so framed). On the other hand, the illustrations of Joris Hoefnagel's *Four Elements* in the volumes of the National Gallery of Art in Washington, D.C. are all framed in oval lines of gold, tooled fast (possibly with heat and egg white glair, as used in leather book tooling?) onto the parchment in the same way, and of about the same width as the rectangular frame enclosing the *beluae* at folio 126v of MS Urb. lat. 276. This may bespeak a very close working relationship between the Hoefnagels and the manuscript artist, if not an intellectual collaboration, as suggested, on the project. It remains to be seen whether a close comparison and study, with recent or future techniques, of the tempera illustrations of the Urbinas manuscript and the watercolor and gouache drawings of the *Four Elements* and other Hoefnagel works reveal any instances of the same hands at work in both. In other words, did either or both Hoefnagels contribute to the illustration of MS Urb. lat. 276? Not necessarily, but it is not impossible. It may also be significant that Vincenzo Gonzaga, the son of Duke Guglielmo and Eleonora (Hapsburg) Gonzaga, is recorded as having acquired "diversi dipinti di Georg Hoefnagel tra i quali Quattro tavolette miniate" in 1591, and as having purchased "Le Quattro Stagioni" in 1594.[21] Furthermore, the Gonzaga manuscript is absent from archival records from 1460 until 1632 (the

21 Vignau-Wilberg, *Joris and Jacob Hoefnagel* 459; cf. *Gonzaga. La Celeste Galeria. L'esercizio del collezionismo* (Milan: 2002) 290, 344–345; on Jacob, see also Kaufmann T. DaC., *The School of Prague. Painting at the Court of Rudolf II* (Chicago and London: 1988) 211–214.

year of Jacob Hoefnagel's death, but also two years after the Sack of Mantua), when it reappears in the ducal library of Urbino.[22]

Further investigations, archival and technical may, with luck, help to clarify the exact relationship between the artists as well as definitively confirm the authorship of the tempera illustrations of MS Urb. lat. 276. It will be of interest to examine archival records of the various towns and courts where the Hoefnagels and the Ghisis travelled or resided, to attempt a more complete understanding of the relationships among these artists and naturalists.

Bibliography

Aldrovandi Ulisse, *De animalibus insectis* (Bologna, Bapt. Bellagamba: 1602).

Bass M.A., *Insect Artifice. Nature and Art in the Dutch Revolt* (Princeton: 2019).

Belon Pierre, *De aquatilibus Libri duo, cum ειconibus* [sic] *ad vivam ipsorum effigiem, quoad ejus fieri potuit, expressis* (Paris, Charles Estienne: 1553).

Boorsch S. – Lewis M. – Lewis R.E., *The Engravings of Giorgio Ghisi* (New York: 1985).

Carroll L., *The Jabberwocky* (available at: https://www.poetryfoundation.org/poems/42916/jabberwocky).

Gessner Conrad, *Bibliotheca universalis* (Zurich, Christopher Froschauer: 1545).

Gessner Conrad, *Historiae animalium: Liber IIII qui est de piscium et aquatilium animantium natura* (Zurich, Christopher Froschauer: 1558).

Gessner Conrad, *Nomenclator aquatilium animantium. Icones animalium aquatilium in mari et dulcibus aquis degentium* (Zurich, Christopher Froschauer: 1560).

Gonzaga. La Celeste Galeria. L'esercizio del collezionismo, ed. R. Morselli (Milan: 2002).

Hendrix M.L., *Joris Hoefnagel and the Four Elements: A Study in Sixteenth-Century Nature Painting* (Ph.D. Dissertation, Princeton University 1984).

Hoefnagel Jacob, *Archetypa studiaque patris Georgii Hoefnagelii 1592. Natur, Dichtung und Wissenschaft in der Kunst um 1600. Nature, Poetry and Science in Art around 1600*, ed. T. Vignau-Wilberg (Munich: 1994).

Hoefnagel Joris, *Animalia Aquatilia et Co[n]chiliata (Aqua)*, National Gallery of Art, Washington. Gift of Mrs. Lessing J. Rosenwald. Plate LI (Accession number 1987.20.7.52) available at: https://www.nga.gov/collection/art-object-page.69871.html and Plate LVI (Accession number 1987.20.7.57) available at: https://www.nga.gov/collection/art-object-page.69876.html.

Kaufmann T. DaC., *The School of Prague. Painting at the Court of Rudolf II* (Chicago and London: 1988).

22 Pyle, *Das Tierbuch* 68, 72; Eadem, "The Art and Science" 299.

Mozzetti F., "Teodoro Ghisi", *Dizionario Biografico degli Italiani* 54 (2000) https://www.treccani.it/enciclopedia/teodoro-ghisi-(Dizionario-Biografico).

Olaus Magnus, *Carta marina et descriptio septentrionalium terrarum ac mirabilium rerum in eis contentarum diligentissime elaborata anno dni 1539* (Venice: 1539).

Pyle C.M., *Das Tierbuch des Petrus Candidus. Codex Urbinas Latinus 276. Eine Einführung*, tr. T. Honref, J. Schlechta (Zurich: 1984).

Pyle C.M., "Harvard MS Richardson 23: A "Pendant" to Vatican MS Urb. lat. 276 and a Significant Exemplar for P.C. Decembrio's *Opuscula historica*", *Scriptorium* 42, 2 (1988) 191–198.

Pyle C.M., "The Art and Science of Renaissance Natural History: Thomas of Cantimpré, Pier Candido Decembrio, Conrad Gessner and Teodoro Ghisi in Vatican Library MS Urb. lat. 276", *Viator* 27 (1996) 265–321.

Pyle C.M., "Art as Science in Joris Hoefnagel and Teodoro Ghisi", *Center 21. National Gallery of Art. Center for Advanced Study in the Visual Arts, Record of Activities and Research Reports, June 2000–May 2001* (Washington: 2001) 112–114.

Pyle C.M., "Insects in [Renaissance] Art", in Smith S. (ed.), *A Cultural History of Insects in the Renaissance* (London: Forthcoming).

Robin D. – Westwater L.L., *Ippolita Maria Sforza Duchess and Hostage in Renaissance Naples: Letters and Orations* (Toronto and Tempe, Arizona: 2017).

Rondelet Guillaume, *Libri de piscibus marinis in quibus verae piscium effigies expressae sunt*, 2 vol., t. II: *Universae aquatilium historiae pars altera, cum veris ipsorum imaginibus* (Lyon, Macé Bonhomme: 1554–1555).

Tellini Perina C., "Teodoro Ghisi: l'immagine fra Maniera e Controriforma", in Franchini D.A. et al. (eds.), *La scienza a corte, collezionismo eclettico natura e immagine a Mantova fra Rinascimento e Manierismo* (Rome: 1979) 239–268.

Vignau-Wilberg T., *Joris and Jacob Hoefnagel. Art and Science around 1600* (Berlin: 2017).

Wilberg Vignau-Schuurman T.A.G., *Die Emblematischen Elemente im Werke Joris Hoefnagels*, 2 vols. (Leiden: 1969).

Wood J.M., *Ippolita Maria Sforza. The Renaissance Princess Who Linked Milan and Naples* (Jefferson, North Carolina: 2020).

Zanichelli G., "Il Maestro di Ippolita Sforza e il suo atelier: ipotesi di un percorso", *Artes* 14 (2008) 5–31.

CHAPTER 8

A Taste for Fish: Paintings of Aquatic Animals in the Low Countries (1560–1729)

Marlise Rijks

1 Introduction

The fish scales are glittering in the morning sun.[1] Women are strolling by the stalls of Antwerp's fish market, inspecting and buying fish, thinking about lunch or dinner. For sale are herring, cod, ray, plaice, salmon, sturgeon, carp, and eel, but also shrimp, crab, and lobster. Fishmongers chop up salmon filets with their sharp knives, while live carp and pike swim in water-filled barrels and buckets. Behind the Viskoperstoren (Fish merchants' tower) the water of the Scheldt sloshes against the quay. In the sheltered city, the piles of fish are a sign of prosperity – a horn of cornucopia provided by the rivers and seas. Such market scenes with an abundance of products were famously depicted by Pieter Aertsen (*c.*1508–1575) and Joachim Beuckelaer (*c.*1533–*c.*1574) in a new genre of painting in sixteenth-century Antwerp [Fig. 8.1].

Fish was a common type of food in the early modern Low Countries, and in sixteenth-century Antwerp alone there were 75 fishermen selling sea fish and 16 or 17 selling 'river fish', according to a contemporary estimation.[2] Fish was fundamental to the wealth of many cities in the Low Countries. In the words of the chronicler Ludovico Guicciardini (1521–1589), the 'groote wilde Zee' (great wild sea) was not only a constant force of danger threatening the provinces of the Low Countries, the author also thought it to be 'betamelijck hier te verhalen de profyten ende gerieffelijckheden die zy der gantscher Provincien doet' (proper to narrate here of the benefits and conveniences that

1 A generous grant from the Netherlands Organisation for Scientific Research (NWO) for the project *A New History of Fishes. A Long-term Approach to Fishes in Science and Culture, 1550–1800* enabled me to carry out the research upon which this chapter is based. I would also like to thank Florike Egmond and Paul J. Smith for their constructive comments, and Meredith McGroarty for the English editing.

2 Guicciardini Ludovico, *Beschrijvinghe van alle de Nederlanden; anderssins ghenoemt Neder-Duytslandt* (Amsterdam, Willem Jansz: 1612 [first published 1567]) 19 (translation by author).

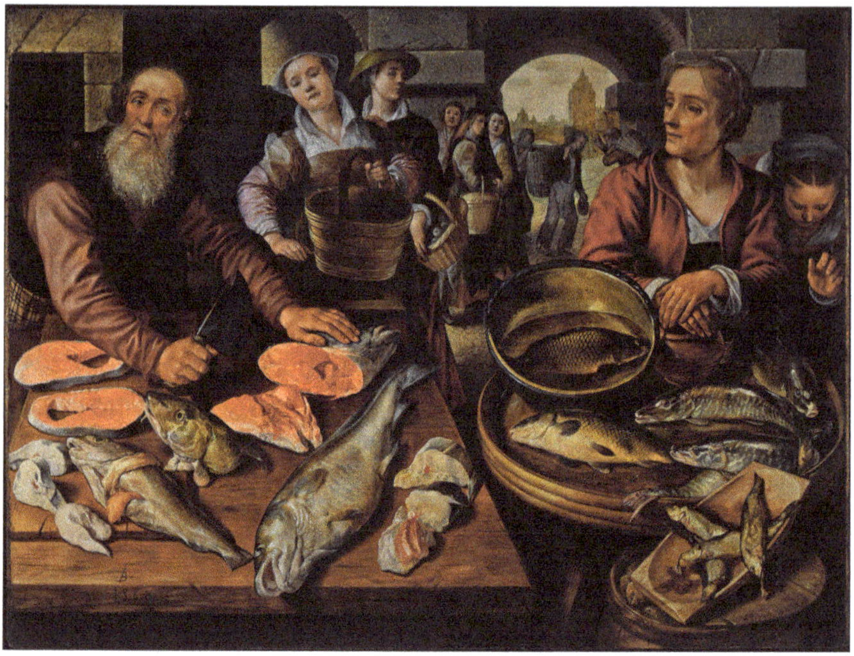

FIGURE 8.1 Joachim Beuckelaer, *Fish market*. 1568. Oil on panel, 128 × 174 cm. New York, Metropolitan Museum

she delivers to all the Provinces).[3] Guicciardini's originally Italian book was a best-seller that was translated into multiple languages, including a Dutch edition for the local market.

In the sixteenth and seventeenth centuries, people started looking differently at the everyday economic buzz of the city. Aware of the economic success of their cities, merchants and artist-artisans became interested in texts and images in which their own activities and their city took centre stage. Plays about the role of merchants and artist-artisans were written and carried out by Chambers of Rhetoric,[4] engravings of cities and local landscapes were printed in successful series, and paintings of markets were made on spec (for the open market).[5] It was as if cities and markets were seen with bright new eyes. The market scenes by Pieter Aertsen and Joachim Beuckelaer immediately come to

3 Guicciardini, *Beschrijvinghe van alle de Nederlanden* 19.

4 For instance, in 1561 at the Antwerp Landjuweel. Honig E.A., *Painting & the Market in Early Modern Antwerp* (New Haven – London: 1998) 7–9.

5 A vivid description of the Amsterdam fish market was given by Gerbrand Adriaensz. Bredero in the play *Moortje* (1615). Thanks to Kornee van der Haven for pointing me to this source.

mind. Close up and in detail, they depicted the goods sold by butchers, vegetable sellers, and fishmongers. Around the turn of the century, the genre of still life painting matured, and fish still life painting became a specialty in the Low Countries. Collectors' cabinets were filled with such new genres of paintings, as well as all sorts of *naturalia*, including aquatic ones, such as shells, corals, and dried fish.[6]

Fish markets were also a source of inspiration for sixteenth-century naturalists *throughout* Europe, such as Conrad Gessner, Pierre Belon, and Ulisse Aldrovandi.[7] Sometimes fishermen became naturalists themselves. Adriaen Coenen (1514–1587) was a fisherman's son who became a fish merchant and fish auctioneer at Scheveningen and The Hague. The experience gleaned from his trade formed the starting point for his natural investigations, which he recorded in his *Visboeck* and *Walvisboeck*: richly illustrated manuscript albums from the 1570s and 1580s.[8] The famous microscopist Antony van Leeuwenhoek (1632–1723) mentions in his letters how he got the help of local fishermen to acquire fish and shellfish for his observations and dissections, but also that he used them as eye-witnesses to learn more about fish when they were just caught and still alive.[9]

---

6 On shell collecting, see: Bass M.A. – Goldgar A. – Grootenboer H. – Swan C. (eds.), *Conchophilia. Shells, Art, and Curiosity in Early Modern Europe* (Princeton – Oxford: 2021). For coral, see: Rijks M., "'Unusual Excrescences of Nature'. Collected Coral and the Study of Petrified Luxury in Seventeenth-Century Antwerp", *Dutch Crossing. Journal of Low Countries Studies* 43 (2019) 127–156.

7 See the Introduction of this volume. In his autobiography, the famous Italian naturalist Ulisse Aldrovandi (1522–1605) writes that it was in the years 1549–1550 that 'I began to be interested in the sensory knowledge of plants, and also of dried animals, particularly the fish that I saw often in the fish markets', see: Findlen P., *Possessing Nature. Museums, Collecting, and Scientific Culture in Early Modern Italy* (Berkeley – Los Angeles – London: 1994) 175–177 (with the quote at 175). Aldrovandi's story of his meeting with Rondelet in 1549, which he describes as a turnaround, was a mystification. See: Smith P.J., "Aldrovandi (Ulisse) (1522–1605)", in Nativel C. (ed.), *Centuriae latinae. Cent une figures de la Renaissance aux Lumières offertes à Jacques Chomarat* (Geneva: 1997) 57–63 (here 57).

8 Egmond F., "On Northern Shores: Sixteenth-Century Observations of Fish and Seabirds (North Sea and North Atlantic)", in MacGregor A. (ed.), *Naturalists in the Field. Collecting, Recording, and Preserving the Natural World in the Fifteenth to the Twenty-first Century* (Leiden – Boston: 2018) 129–148.

9 For the importance of fishermen and fishmongers in the creation of natural historical knowledge, see: Trijp D. van, *Captured on Paper. Fish Books, Natural History and Questions of Demarcation in Eighteenth-Century Europe (ca. 1680–1820)* (Ph.D. dissertation, Leiden University: 2021) 72–107. I will discuss this in more detail below. See, for instance: letter by Antony van Leeuwenhoek to Pieter Rabus, 21–05–1695 (leeu027/0145); letter by Antony van Leeuwenhoek to Frederik Adriaan van Reede van Renswoude, 15–08–1695 (leeu027/0151). Letters available on the website *Circulation of Knowledge and Learned Practices in the 17th-century Dutch Republic*: http://ckcc.huygens.knaw.nl/epistolarium/ (accessed July 2020).

In the last decades, several exhibitions on fish paintings were organized in the Netherlands. In 2004, the Centraal Museum in Utrecht organized an exhibition on fish still lifes from the Northern and Southern Netherlands (*Vis. Stillevens van Hollandse en Vlaamse meesters 1550–1700*).[10] Before that time, fish still lifes had received relatively little attention from museums, scholars, and collectors (compared to, for instance, floral still lifes). In 2019, then, three exhibitions devoted to fish paintings were organized in the Netherlands (Dordrechts Museum, Stedelijk Museum Alkmaar, and Museum Vlaardingen).[11] Notable also were the exhibition on the work of Clara Peeters (*c.*1580–*c.*1621), the first known artist to have created a fish still life (see below), in Antwerp and Madrid in 2016, as well as the exhibition *Slow Food* in The Hague in 2017.[12]

This chapter discusses 'fish paintings' in the early modern Low Countries in the context of the local importance of fishery, the culture of collecting, and the (up and coming) natural knowledge about fish. Through the example of fish paintings, it reflects upon early modern 'genres' of images (and, by extension, genres of texts). In the early modern period, the category 'fish' (or *aquatilia*) contained virtually all aquatic fauna, including sea mammals, crocodiles, turtles, shrimps, and crustaceans.[13] I will use this broad early modern definition. Fish motifs were depicted in paintings representative of some important pictorial innovations in the Low Countries. One thinks here of the new genres of market paintings, still lifes, allegories of the elements (with an abundant number of objects, plants, and animals), gallery pictures (painted collectors' cabinets), and genre painting.

The chapter starts with a section on the local fish trade and fish consumption, and how this was intertwined with natural historical knowledge of fishes and representations of fishes in different media (actual *naturalia*, drawings, prints, and paintings). The development and popularity of fish paintings may be understood in this context of trade, consumption, and knowledge. The local fish trade in the Low Countries is seen through the eyes of Guicciardini, which is combined with historical research on fish consumption. It is argued that sometimes socio-economic realities were implemented in fish paintings. Also,

10 Helmus L.M. (ed.), *Vis. Stillevens van Hollandse en Vlaamse meesters 1550–1700* (Utrecht: 2004).

11 Respectively titled: *Beet! Vissen naar verborgen betekenissen* (Dordrecht: 2019), *Pieter van Schaeyenborgh, Meester van het visstilleven* (Alkmaar: 2019), and *Schoon aan de haak* (Vlaardingen: 2019).

12 *The Art of Clara Peeters* (Antwerp – Madrid: 2016); *Slow Food: Still Lifes of the Golden Age* (The Hague: 2017). See below for the paintings.

13 Egmond F., *Eye for Detail. Images of Plants and Animals in Art and Science, 1500–1630* (London: 2017) 60.

it is argued that fish motifs in paintings could function at the same time as 'descriptive' natural knowledge and as symbols, for instance symbols of fertility and procreation. One particular motif in Beuckelaer's fish markets (a ray on its back) is explained by information from a letter by the seventeenth-century microscopist Antony van Leeuwenhoek. The next section gives a concise overview of the production of fish paintings in the Low Countries, with some of its most successful and innovative painters. Finally, the last sections analyse the ownership of (genres of) fish paintings listed in inventories from Antwerp, Amsterdam, and Haarlem from the seventeenth century (more precisely: 1560–1729, but the dates differ slightly per city).[14] Inventories provide information about ownership, the popularity of genres, and about genre terminology. They are valuable sources to understand how painterly subjects were described at the time – and thus problematize the use of our contemporary genres for the early modern period.

2 Fishery, Natural History, Collections

The Florence-born Ludovico Guicciardini was one of the many foreign merchants who settled in the city of Antwerp in the sixteenth century. In this thriving international city of commerce on the River Scheldt, merchants from all over Europe established trading houses to look after their interests. Guicciardini was unique, however, in that he wrote a thick book about his adopted city and the other cities and provinces of the Low Countries. His chronicles *Descrittione di tutti i Paesi Bassi* from 1567 was reprinted dozens of times and translated into German, French, and Dutch. In the general introduction to his work, Guicciardini devotes numerous pages to the importance of the rivers and sea for the Low Countries.[15] He emphasizes that Antwerp has risen to wealth and importance because of its fortunate position on the Scheldt. Moreover, the Low Countries profit from the 'ontallijcke menichte vande allerleye visschen die daer ghevanghen worden' (countless amounts of all sorts of fish that are caught there), which feed both the rich and the poor. Fishery provided great

14 I have used the great work on inventories by Erik Duverger (Antwerp), John M. Montias (Amsterdam), and Pieter Biesboer (Haarlem). See: Duverger E., *Antwerpse kunstinventarissen uit de zeventiende eeuw*, vols. I–XIV (Brussels: 1984–2009), Montias database: https://research.frick.org/montias (accessed September 2019), Biesboer data on website Getty Provenacnce Index (GPI): https://piprod.getty.edu/starweb/pi/servlet.starweb?path=pi/pi.web (accessed September 2019). For Biesboer, see also the book publication: Biesboer P., *Collections of Paintings in Haarlem, 1572–1745* (Los Angeles: 2002).

15 Guicciardini, *Beschrijvinghe van alle de Nederlanden* 9–21.

wealth because of the export of fish – in particular herring, cod, and salmon – to England, Germany, France, and even as far as Italy.[16]

Guicciardini wrote a chronicle about the history, geography, economy, and customs of the Low Countries, but his work also contains elements of natural history. For instance, Guicciardini pays a lot of attention to the 'nature' of the herring. He describes the appearance of the herring in large schools and how the animal travels from more northern seas in the autumn to the North Sea near the coast of the Low Countries around Christmas.[17] Guicciardini also mentions the 'King' of a school of herrings, which apparently has a red sign on its head, like a crown. The idea of the King of herrings was well known in Guicciardini's time and is also mentioned by the aforementioned Adriaen Coenen in his manuscript. Perhaps the King of the herrings was a myth among fishermen, but perhaps it concerned a real fish with red scales.[18] Guicciardini continues with the taste of herring and the methods for preservation (salting, pickling, and smoking). Later in his work, he also mentions the 14th-century figure of Willem Beukelszoon from Biervliet, who allegedly invented the 'gibbing' (kaken) of herring.[19] Further, he claims to have investigated the number of boats that set sail from Friesland, Holland, Zeeland, and Flanders for the herring catch: he counted at least 700. A calculation of the number of trips with the amount of herring caught per trip per boat comes to a yearly amount of herring worth 490,000 *ponden* (or 1,470,000 *kronen*).[20] Combined with the income gained from fishery of the second two most important fish – cod and salmon (which was probably mostly caught as river fish) – the total value amounts to more than two million *kronen*. In the words of Guicciardini: 'een onspreckelijck ende oneyndelijck wonderwerck' (an unspeakable and infinite

16 Guicciardini, *Beschrijvinghe van alle de Nederlanden* 19.

17 Just as birds travel and change places in different seasons, so do all sorts of fish, according to the chronicler. Guicciardini, *Beschrijvinghe van alle de Nederlanden* 20.

18 Based upon textual descriptions and images by Coenen, it was most likely the red mullet (*Mullus surmuletus*). Earlier suggestions included the Giant oarfish (*Regalecus glesne*) or the John Dory (*Zeus faber*). See Bennema F., "De Haringkoning", *Het Zeepaard* 70.1 (2010) 15–18; Richter C., "Hollandse vissenboeken. De onweerstaanbare kracht van de afbeelding", *Holland. Historisch Tijdschrift* 38 (2006) 161–176 (here 166).

19 This is a myth: gibbing was a practice that came from Scandinavia to the Low Countries. Lauwerier R.C.G.M. – J. Laarman F.J., "Hollandse Nieuwe en de mythe van Willem Beukelszoon", *Holland. Historisch Tijdschrift* 38 (2006) 150–160 (here 154); Guicciardini, *Beschrijvinghe van alle de Nederlanden* 311.

20 Guicciardini, *Beschrijvinghe van alle de Nederlanden* 21.

work of wonder). Like Guicciardini, other writers also praised the importance of fishery in the Low Countries, including Joost van den Vondel (1587–1679).[21]

The consumption of fish had traditionally been related to the Catholic calendar. Throughout Lent, the period between Ash Wednesday and Easter Sunday, people were not allowed to eat products of land animals, but aquatic animals were permitted. During other periods of fasting the same rules applied. The rest of the year, Wednesdays and Fridays (or, in some regions, Fridays and Saturdays) were habitually days that meat was forbidden, and people ate fish. On these days, dairy was allowed, so fish could be prepared with butter. The consumption of fish could also be related to the ancient humoral theory. Fish was seen as moist and cold and considered healthy or unhealthy depending on the humours of the consumer and the manner of preparation.[22] A remarkable proponent of fish was the Antwerp physician Ludovicus Nonnius (1553–1645), whose learned and practical dietary book *Ichtyophagia sive de piscium esu* (Ichtyophagia or of the Eating of Fish) from 1616 was dedicated to the consumption of fish.[23]

In the early modern Low Countries, both freshwater fish and sea fish were consumed. Whereas sea fishery is still happening on a large scale today, the fishery of freshwater fish is no longer a big industry. In the 16th and 17th centuries, migratory fish, such as salmon, sturgeon, and allis shad, were still caught in the Low Countries. On the *Bergse Veld* (nowadays Biesbosch) there were around 150 active fishermen, who caught around 16,000 salmon, 60,000 allis shads, and 200 to 800 sturgeons a year.[24] From Geertruidenberg, many of these fish were transported to cities in Brabant and Flanders, including Antwerp, Brussels, Malines, and Ghent. Fish is prone to decay, so to preserve fish it was

---

21 In Vondel's *Hymnus, ofte Lof-Gesangh over de wijdberoemde scheepsvaert der Vereenighde Nederlanden*. See: De Jongh E., "De symboliek van vis, visser, visgerei en vangst" in Helmus (ed.), *Vis* 75–120 (here 88).

22 Winter J.M. van, "Visrecepten in laat-middeleeuwse en vroeg-moderne kookboeken", in Helmus (ed.), *Vis* 139–153 (here 142–143).

23 Wyssenbach S., "Riches of the Sea: Collecting and Consuming Frans Snijders's Marine Market Paintings in the Southern Netherlands", in Burghartz S. – Burkart L. – Göttler C. (eds.), *Sites of Mediation. Connected Histories of Places, Processes, and Objects in Europe and beyond, 1450–1650* (Leiden – Boston: 2016) 328–352 (here 346). Egmond, 'Books on Natural History' and 'Ludovicus Nonnius's Book on the Consumption of Fish (Antwerp, 1616)', on the website: http://ximenez.unibe.ch/library/natural/fish/ (accessed March 2021). Full title: *Ichtyophagia; sive, De piscium esu commentarius*. Nonnius's book was owned by the wealthy collector Emmanuel Ximenez. Nonnius was the physician of Rubens and portrayed by the painter (portrait now in the National Gallery, London).

24 Martens P., "Visserij en vishandel. De zalm van het Bergse veld", in Helmus (ed.), *Vis* 121–138 (here 138).

often smoked, salted, or pickled. Freshwater fish could also be transported and sold alive. Foreign visitors to Antwerp were amazed by the variety of freshwater fish that was sold alive from buckets.[25] In many paintings of fish markets, such buckets or barrels with living river fish are depicted.

Evidence from various bills and accounts confirms that herring and cod were the most consumed fishes in the early modern Low Countries.[26] This is validated by archaeological research.[27] Salmon was usually reserved for festive days. More extraordinary were also eel, pike, and other river fish. In general, a large percentage of households' and institutions' budgets was spent on food. Most of this money went to grain products, and only a relatively small percentage went to fish.[28] But of course differences in wealth determined how much people spent on food and what they ate. There was a great difference between the meagre and monotonous diet of the poor and the generous and varied diet of the middle and upper classes. The dichotomy between the 'thin kitchen' and the 'fat kitchen' was a popular pictorial theme in the Low Countries. Sometimes these pictures included fish or shellfish. Well known are the prints of the *Thin Kitchen* and the *Fat Kitchen* designed by Pieter Bruegel and published by Hieronymus Cock in Antwerp in 1563 (in the *Thin Kitchen* the poor eat mussels, in the *Fat Kitchen* the figures eat meat). The oldest presently known meal still lifes from the Low Countries are two paintings of *A Rich Man's Meal* and *A Poor Man's Meal* by Hieronymus II Francken from around 1600.[29] In these paintings, Francken combined the still life of the foreground

25 For instance, Vicente Alvarez, who accompanied Prince Philip (later King Philip II) in this tour of the Low Countries in 1549. The writer Alonso Vasquez was also impressed by the variety of freshwater fish sold alive. See Vergara A., "Reflections of Art and Culture in the Paintings of Clara Peeters", in Vergara A. (ed.), *The Art of Clara Peeters* (Antwerp – Madrid: 2016) 13–47 (here 36). The German traveler Aulus Apronius or Adam Ebert also wrote about the variety of fish, mussels, oysters, crabs, and pearl-slugs in Antwerp. Wyssenbach, "Riches of the Sea" 328.

26 Accounts from religious orders, churches, hospitals, orphanages, armies, trading companies, and private households. Van der Wee H., "Voeding en dieet in het Ancien Régime", *Spiegel Historiael. Maandblad voor geschiedenis en archeologie* 1.2 (1966) 94–101.

27 Cod had been common for ages, but herring only became common in the early modern period. Since herring was not present near the coast of the Low Countries in the Middle Ages, fishermen only started fishing for herring (near the English and Scottish coast) when larger ships became common. From the 14th century, the Flemish started fishing for herring (not coincidently the time of Willem Beukelszoon, alleged inventor of gibbing) while the herring business in Holland expanded from the 16th century. Lauwerier – Laarman, 'Hollandse Nieuwe en de mythe van Willem Beukelszoon' 151.

28 The percentage of the budget spent on food by Antwerp workers was around 73%. Van der Wee, "Voeding en dieet" 100.

29 Buvelot Q., "Slow Food: on the Rise and Early Development of Dutch and Flemish Meal Still Lifes", in Buvelot Q. (ed.), *Slow food. Dutch and Flemish meal still lifes 1600–1640*, exh. Cat. Mauritshuis (Zwolle: 2017) 13–32 (here 17).

with a vista to a landscape in the background (so not yet a 'pure' still life and somewhat similar to some of Beuckelaer's market compositions).[30] Prominent in *A poor man's meal* are three herring on the left side, and on the right side a herring cut in pieces on a plate. In *A rich man's meal* there is an oyster next to other delicacies.

Next to socio-economic realities, the development of fish paintings fit with collecting trends in the Low Countries. In both the southern and northern Netherlands, people amassed collections that included aquatic animals. In Antwerp, these *naturalia* were part of collections that also contained many other objects, often with an emphasis on (local) painting. In the northern Netherlands there were also collectors who started to specialize more in natural history.[31] In collections in the Low Countries we find many aquatic *naturalia*: shells, coral, blowfish, lyra, seahorses, turtle shells, whole turtles, crocodiles, horseshoe crabs, snakes, 'tongues of fishes' (which were, in fact, fossilized shark teeth), and generically described 'dried fish'. Notably, Rembrandt's collection contained 'soo see- als lantgedierte' (both sea- and land animals); 'seegewassen' (sea crops, probably coral); 'een groote witte coraelberch' (a large white piece of coral); and 'een groot seegewas' (a large sea crop).[32] Also, there were many luxury items made with or from shells, coral, pearl, tortoiseshell, and whalebone (baleen).[33]

An alternative for actual *naturalia* were drawings of plants and animals. In the sixteenth century, collectors throughout Europe started assembling drawings of plants and animals in albums.[34] In the southern Netherlands, several

30 Several versions of *A poor man's meal* are known today, including one in the Koninklijk Museum voor Schone Kunsten Antwerpen (KMSKA) and Museum Boijmans van Beuningen Rotterdam. Of *A rich man's meal* only one version is known, in a private collection.

31 The first one was probably the Enkhuizen physician and collector Bernardus Paludanus in the late sixteenth century (although his collection also contained exotic weapons, clothing, and artefacts). Jorink E., *Het 'Boeck der Natuere'. Nederlandse geleerden en de wonderen van Gods schepping 1575–1715* (Leiden: 2006) 276–287.

32 Montias inv. 1262, lots 134, 136, 173, 264. See also: Van den Boogert B. (ed.), *Rembrandts schatkamer*, exh. Cat., Het Rembrandthuis (Zwolle 1999) 149–151; Chapman H.P., "Rembrandt on Display. The Rembrandthuis as Portrait of an Artist", *Nederlands Kunsthistorisch Jaarboek* 65 (2015), 202–239.

33 Rijks M., "Scales, Skins, and Carapaces in Antwerp collections", in Bol M. – Spary E. (eds.), *The Matter of Mimesis: Studies on Mimesis and Materials in Nature, Art and Science* (Leiden – Boston: 2023) 295–320.

34 Egmond, *Eye for Detail*; Egmond F. – Rijks M., "Depicting Fish in Early-Modern Venice and Antwerp", in Haar A. van der – Schulte Nordholt A. (eds.), *Figurations animalières à travers les textes et l'image en Europe. Du Moyen âge à nos jours. Essais en hommage à Paul J. Smith* (Leiden – Boston: 2022) 63–77.

animal albums were made in the second half of the sixteenth century, some of them specifically devoted to fish, including one by Hans Bol and another by Joris Hoefnagel, both of whom started composing them in the 1570s.[35] Quickly thereafter, specialized animal series in print appeared in Antwerp. These print series had no equivalent in other European cities at the time and were reprinted for decades to come. The first print series specifically devoted to fish was probably Abraham de Bruyn's *Libellus varia genera piscium complectens*, which appeared *c*.1594, a few years before Adriaen Collaert's fish series *Piscium vivae icons*.[36]

Fish albums, print series, and paintings form the sixteenth century onwards visualized the variety of aquatic animals. For naturalists who wanted to differentiate between species, it was crucial to have detailed and precise images.[37] So, images also played a crucial part in what may be considered the earliest works of ichthyology. As is mentioned in the Introduction to this volume, in the 1550s heavily illustrated fish books were published by Pierre Belon, Guillaume Rondelet, Ippolito Salviani, and Conrad Gessner within a single decade. The above-mentioned Adriaen Coenen formed a link between the worlds of fishery and natural knowledge. His manuscript was a combination of written text and hand-made images. The works of these early 'ichthyologists' often included references to popular belief, long textual traditions, and symbolic meanings of fishes. But empirical information as described in texts or visualized in imagery gained ground. Early modern fish paintings also show this duality: they can be interpreted as carriers of meaning, referring to popular beliefs and older texts, but they can also be read as precise and detailed descriptions of actual species. Often, they are both. But whereas the generic depiction of fish in paintings was age-old (for instance in depictions of the Miraculous catch of fish), the detailed and precise depiction of a much larger number of species was new. This duality is clearly seen in the fish markets of Joachim Beuckelaer, to which we turn next.

Fish motifs could have a symbolic meaning related to sexuality, fertility, procreation, and eroticism. Perhaps the *pekelharing* depicted by Joseph de Bray in

35 Hendrix M.L., *Joris Hoefnagel and the "Four Elements": A Study in Sixteenth-Century Nature Painting* (Ph.D. dissertation, Princeton University: 1984) 15; Rikken M., *Dieren verbeeld. Diervoorstellingen in tekeningen, prenten en schilderijen door kunstenaars uit de Zuidelijke Nederlanden tussen 1550 en 1630* (Ph.D. dissertation, Leiden University: 2016) 37–44; Bass M.A., *Insect Artifice: Nature and Art in the Dutch Revolt* (Princeton – Oxford: 2019) 7.

36 Rikken, *Dieren verbeeld* 69; Egmond – Rijks, "Depicting Fish" 71.

37 For the transition of the comprehensive and philologically oriented Renaissance natural histories to more descriptive and empirical work, see Margócsy D., *Commercial Visions. Science, Trade, and Visual Culture in the Dutch Golden Age* (Chicago: 2014).

his well-known painting *Ode to the Herring* was seen as an aphrodisiac, since salty food in general was considered good for fertility and potency.[38] Various motifs in Aertsen's and Beuckelaer's markets have been ascribed sexual meaning. In Beuckelaer's fish markets, the recurring vibrant red salmon steaks may be related to female genitals. The obscene gazes and hand gestures of some of the figures add to the supposed erotic meaning of these paintings. But more than just erotic meaning, the abundance of products from land and water in market scenes was related to the general themes of fertility and procreation.[39]

One noticeable and recurring motif in several fish markets by Beuckelaar (and others) is a ray positioned on its back.[40] This may have been just a way for the painter to show both sides of the fish (and demonstrate knowledge of the ray's anatomy), but it may also relate to a popular belief that is mentioned by Antony van Leeuwenhoek in one of his letters. Van Leeuwenhoek writes that, according to popular folklore, rays had a monthly period during which they had to be avoided (and not eaten). In the letter from 21 May 1695, he writes that 'dit onnoozel gevoelen' (this silly belief) can be explained by the fact that the intestines of rays often hang out of the body and have a vibrant red colour.[41] This is precisely how Beuckelaer depicted rays in the some of his paintings, presumably referring to this belief – and in line with the references to fertility and procreation in his market paintings. Van Leeuwenhoek denounces this popular belief and argues that the vibrant red colour only becomes visible at the moment the ray dies. This has to do with the way the blood solidifies after death. His proposition also undermines the idea that rays have 'water vessels' next to their blood vessels, an idea that was suggested to Leeuwenhoek by a 'certain gentleman'. Van Leeuwenhoek's observation can be seen in the fierce seventeenth-century debates about the circulation of blood and the question of whether all animals have blood vessels. Notably, Van Leeuwenhoek notes how he got information from local fishermen, who confirmed that certain

38 Weber G.J.M., "'*t Lof van den Pekelharingh.* Von altäglichen und absonderlichen Heringstilleben", *Oud Holland* 101.2 (1987) 126–140.

39 Kaveler E.M., "Erotische elementen in de markttaferelen van Beuckelaer, Aertsen en hun tijdgenoten", *Joachim Beuckelaer. Het markt- en keukenstuk in de Nederlanden 1550–1650*, exh. Cat. Museum voor Schone Kunsten Gent (1986) 18–26. See also: De Jongh, "De symboliek van vis" 103–116; Honig E.A., "Desire and Domestic Economy", *The Art Bulletin* 83.2 (2001), 294–315.

40 For instance, in the paintings now in the Koninklijk Museum voor Schone Kunsten Antwerpen and National Gallery London.

41 Letter by Antony van Leeuwenhoek to Pieter Rabus, 21-05-1695 (leeu027/0145). http://ckcc.huygens.knaw.nl/epistolarium (accessed August 2020).

parts of the ray only become red after the animal dies.[42] As is recently argued by historians of science, fieldwork and eye-witness reports played a crucial part in the development of natural history in the early modern period.[43]

The letter in which Van Leeuwenhoek discusses the ray was addressed to Pieter Rabus (1660–1702), a poet and writer from Rotterdam. Rabus had visited Van Leeuwenhoek in Delft, where he looked through microscopes and witnessed the circulation of blood. Rabus greatly admired the microscopist and wrote him a laudatory poem in 1693.[44] The 1695 letter on rays is a response to an earlier letter by Rabus, in which he had asked Van Leeuwenhoek to comment on this fish. Next to the issue of the blood vessels, Van Leeuwenhoek also writes that rays have different type of eggs compared to other fish. He observed that most fish lay large amounts of small eggs, but that rays in comparison lay large eggs – like the eggs of birds. Notably, Coenen had already observed the eggs of rays a century earlier.[45] Both Coenen's and Van Leeuwenhoek's observations must be seen in the context in which naturalists were debating about the procreation of animals. The interest by someone like Rabus shows how widespread the interest in such issues was.[46]

Van Leeuwenhoek himself was fascinated by the procreation of small animals. He investigated the procreation of oysters and described his observations in a letter to Frederik Adriaan van Reede van Renswoude (1659–1738), a baron and diplomat who served as a deputy at the Peace of Utrecht.[47] Again, the correspondence between Leeuwenhoek and Reede van Renswoude (who also sent Van Leeuwenhoek specimens, such as a large caterpillar[48]) points

42 Letter by Antony van Leeuwenhoek to Pieter Rabus, 21-05-1695 (leeu027/0145), http://ckcc.huygens.knaw.nl/epistolarium (accessed August 2020).

43 MacGregor, *Naturalists in the Field*; Van Trijp, *Captured on Paper* 72–107.

44 Letter by Pieter Rabus to Antony van Leeuwenhoek, 18-08-1693 (leeu027/0125a), http://ckcc.huygens.knaw.nl/epistolarium (accessed August 2020).

45 Coenen also gave several examples of animals that procreated, contrary to the opinion of spontaneous generation (which he does not reject in general terms). Egmond, *Eye for Detail* 206–207.

46 In the aforementioned letter from 1693, Rabus praises Van Leeuwenhoek on his work on the procreation of fleas, proving that they lay eggs (and are not the result of spontaneous generation). Letter by Pieter Rabus to Antony van Leeuwenhoek, 18-08-1693 (leeu027/0125a), http://ckcc.huygens.knaw.nl/epistolarium (accessed August 2020).

47 Letter by Antony van Leeuwenhoek to Frederik Adriaan van Reede van Renswoude, 15-08-1695 (leeu027/0151), http://ckcc.huygens.knaw.nl/epistolarium (accessed August 2020).

48 Letter by Antony van Leeuwenhoek to Frederik Adriaan van Reede van Renswoude, 22-04-1695 (leeu027/0142), http://ckcc.huygens.knaw.nl/epistolarium (accessed August 2020).

to the widespread (elite) interest in natural history.[49] In this letter on oysters, Van Leeuwenhoek describes how he acquired some oysters from a local fisherman in Delft.[50] He observes and describes the offspring of the oysters. When 'vyf voorname Heeren' (five distinguished gentlemen) came to visit Van Leeuwenhoek, they were amazed about the 'volmaaktheyt' (perfection) of the young oysters. Van Leeuwenhoek's observations of the oysters proved for him – again – that the idea of spontaneous generation was false. A point of disagreement among naturalists was whether even the smallest of animals procreated, or if they could (also) come into existence by spontaneous generation. The idea that animals could spontaneously come into existence (in particular in the Element of Water) had been common since antiquity. Throughout the seventeenth century, many scholars did not rule out the possibility of spontaneous generation, in particular that of insects.[51] Muddy pools, swamps, or dead animals were deemed fertile grounds for spontaneous generation, which does not seem all that illogical for anyone who has ever seen mosquitoes at a pool of water or maggots in an animal corpse. Someone like Van Leeuwenhoek disproved the theory of spontaneous generation by showing how even the smallest animals, such as oysters and fleas, procreated.[52]

Natural history was fascinating for relatively large numbers of people among the middle and elite strata of society, which is one factor that explains the exponential growth of natural historical knowledge in the early modern period. In books, old and new knowledge was registered in text and image. In letters, people informed each other of natural phenomena, such as the anatomical features and procreation of rays, or remarkable events, such as beached whales

---

49 As Leeuwenhoek writes in 1695: 'Het genoegen dat zyne Hoog Ed: Geboore Heere schept in myn geringe arbeyt in 't onderzoeken der verborgentheden, als in desselfs aangename Missive gemelt werd, doet my weder de vryheyt nemen dese myne nasporinge toe te zenden'. Letter by Antony van Leeuwenhoek to Frederik Adriaan van Reede van Renswoude, 10-07-1695 (leeu027/0147), http://ckcc.huygens.knaw.nl/epistolarium (accessed August 2020).

50 Letter by Antony van Leeuwenhoek to Frederik Adriaan van Reede van Renswoude, 15-08-1695 (leeu027/0151), http://ckcc.huygens.knaw.nl/epistolarium (accessed August 2020).

51 Jorink, *Boeck der Natuere*, chapter 4 on insects, 187–256. Johannes Swammerdam famously opposed spontaneous generation by demonstrating how insects procreated.

52 In the letter to Reede van Renswoude, Leeuwenhoek writes that 'voortteelingen van de Schulpvissen op zoo een gereguleerde wijse toegaat, en dat die niet uyt slijk of van zelfs voortkomen, gelijk hedendaags nog eenige willen dwars-dryven, en wel meest die geene die de dwalinge van hare Oude leermeesters tragten staande te houden, of niet verder zien als haar neus lang is'. Letter by Antony van Leeuwenhoek to Frederik Adriaan van Reede van Renswoude, 15-08-1695 (leeu027/0151), http://ckcc.huygens.knaw.nl/epistolarium (accessed August 2020).

or a particular large fish caught somewhere.[53] And in collections, *naturalia*, including (parts of) aquatic animals, were on display next to images of nature. In this context, painters invented new genres in which the variety of animal species was depicted with a precision that was thus far unseen.

3 Fish Paintings: Production

The Amsterdam-born painter Pieter Aertsen moved to Antwerp around the same time as Guicciardini. In this thriving commercial city, Aertsen and his nephew and pupil Joachim Beuckelaer came on the market with newly invented paintings of food stalls and market scenes.

These paintings marked the beginning of a new type of imagery and are generally seen as precursors to still life painting.[54] Aertsen's large *Market*, now in the Wallraf-Richartz-Museum in Cologne, is probably the only one in which he depicted fish.[55] However, he was associated with paintings of fish markets by contemporaries, as is revealed in an inventory from 1652 that contains a listing of a 'vischmerckt naer Lange Peer' (a fish market after Lange Peer [the nickname of Pieter Aertsen]).[56] Beuckelaer, on the other hand, painted numerous fish markets – and these compositions were eagerly copied by other painters. Whereas depictions of the biblical story of the miraculous catch of fish had long been popular, Beuckelaer depicted recognizable fish species in incredible detail. Fish was elevated from *bijwerk* (secondary motifs) to main subject. In some cases, the biblical story was moved to the background, while in other cases it was omitted altogether. Compare Beuckelaer's *Miraculous Draught of Fishes* (1563), now in the Getty, with *The Element of Water* (1569), now in the National Gallery in London. The latter painting is a new type of market scene, in which fish take up at least half of the canvas, while in the background (in

53 For example, in an otherwise political letter by Hugo Grotius to Nicolaas van Reigersberch, Grotius informs Van Reigersberch about a large fish that is caught ('Bij de eilanden van Jeres is een groote visch gevangen, een tiger zeer gelijck, briesschende als een stier.'). Letter by Hugo Grotius to Nicolaas van Reigersberch, 19-03-1644 (groo001/6770), http://ckcc.huygens.knaw.nl/epistolarium (accessed August 2020). For beached whales, see also: Jorink, *Boeck der Natuere*.

54 See also Sighem B. van, "Vis op het droge. Visstillevens in Holland en Utrecht in de 17de eeuw", *Holland. Historisch Tijdschrift* 38 (2006) 177–186 (here 179–180).

55 Pieter Aertsen, *Market*, 1550–1575, oil on panel, 127 × 85 cm, Cologne, Wallraf-Richartz-Museum.

56 Inventory of Jan van Meurs (publisher) from 1652. Duverger, *Antwerpse kunstinventarissen* vol. 6, 267.

a vista through a gate), very small, Beuckelaer has depicted the scene of the miraculous draught of fish.

Shortly after the establishment of still life as a mature genre around the turn of the century, Netherlandish painters started to produce specialized fish still lifes. The oldest example presently known is the *Still life with Fish, a Candle, Artichokes, Crab and Prawns* from 1611 by the Antwerp painter Clara Peeters (*c.*1580–*c.*1621), now in the Prado [Fig. 8.2].[57] In another fish still life, now in the Rijksmuseum, Peeters added exotic shells, such as the *conus marmoreus*, which were highly desirable and fashionable collectables. Other early examples from Antwerp are by Osias Beert (*c.*1580–1623), who is known for his tables with oysters and confectionary goods, and which sometimes included fish.[58] Based upon evidence from inventories and works presently known, the most productive fish still life painter from Antwerp was Alexander Adriaenssen (1587–1661).

Frans Snyders's (1579–1657) large canvases of fish market scenes were eagerly collected by Antwerp's elites.[59] He included not only common fish but also seals, otters, porpoises, and even more exotic species. His *Fishmarket* (*c.*1620) in the Kunsthistorisches Museum in Vienna, for instance, shows two horseshoe crabs.[60] Snyders also added a range of exotic and expensive shells (including a nautilus shell and a *conus marmoreus*), which points to the fact that such paintings were a kind of collection in their own right rather than a realist documentation of a fish stall.[61] Moreover, the genre of the 'gallery picture' or '*constcamer* painting' was invented in Antwerp in the 1610s: paintings that depicted (idealized) collections. Its most important representative, Frans Francken the Younger, often included *aquatilia* in his paintings, such as shells,

57 Clara Peeters, *Still life with fish, a candle, artichokes, crab and prawns*, 1611, oil on panel, 50 × 71.6 cm, Madrid, Museo del Prado, inv. no. 1621. For Clara Peeters, see Vergara, "Reflections of Art and Culture in the Paintings of Clara Peeters".

58 Belkin K.L., "Osias Beert", in Brickstock H. (ed.), *The Oxford Companian to Western Art* (online version 2003). DOI: 10.1093/acref/9780198662037.001.0001.

59 Wyssenbach, "Riches of the Sea" 336.

60 Frans Snyders and possibly Anthony van Dyck, *Fishmarket*, *c.*1620, oil on canvas, 253 × 375 cm, Kunsthistorisches Museum, Vienna, inv. no. 383. Wyssenbach, "Riches of the Sea" 336. See also Rijks M., "A Painter, a Collector, and a Horseshoe Crab. Connoisseurs of Art and Nature in Early Modern Antwerp", *Journal of the History of Collections* 31 (2019) 343–361.

61 Göttler C., "The Place of the 'Exotic' in Early Seventeenth-Century Antwerp", in Schrader S. (ed.), *Looking East: Rubens's Encounter with Asia* (Los Angeles: 2013) 93–94; Koslow S., *Frans Snyders. The Noble Estate: Seventeenth-Century Still-life and Animal Painting in the Southern Netherlands* (Brussels: 2006) 140–141. The horseshoe crab and other exotic *aquatilia* – as well as shells – became popular collectables in the course of the sixteenth and seventeenth centuries. See Rijks, "Scales, Skins, and Carapaces", 2023.

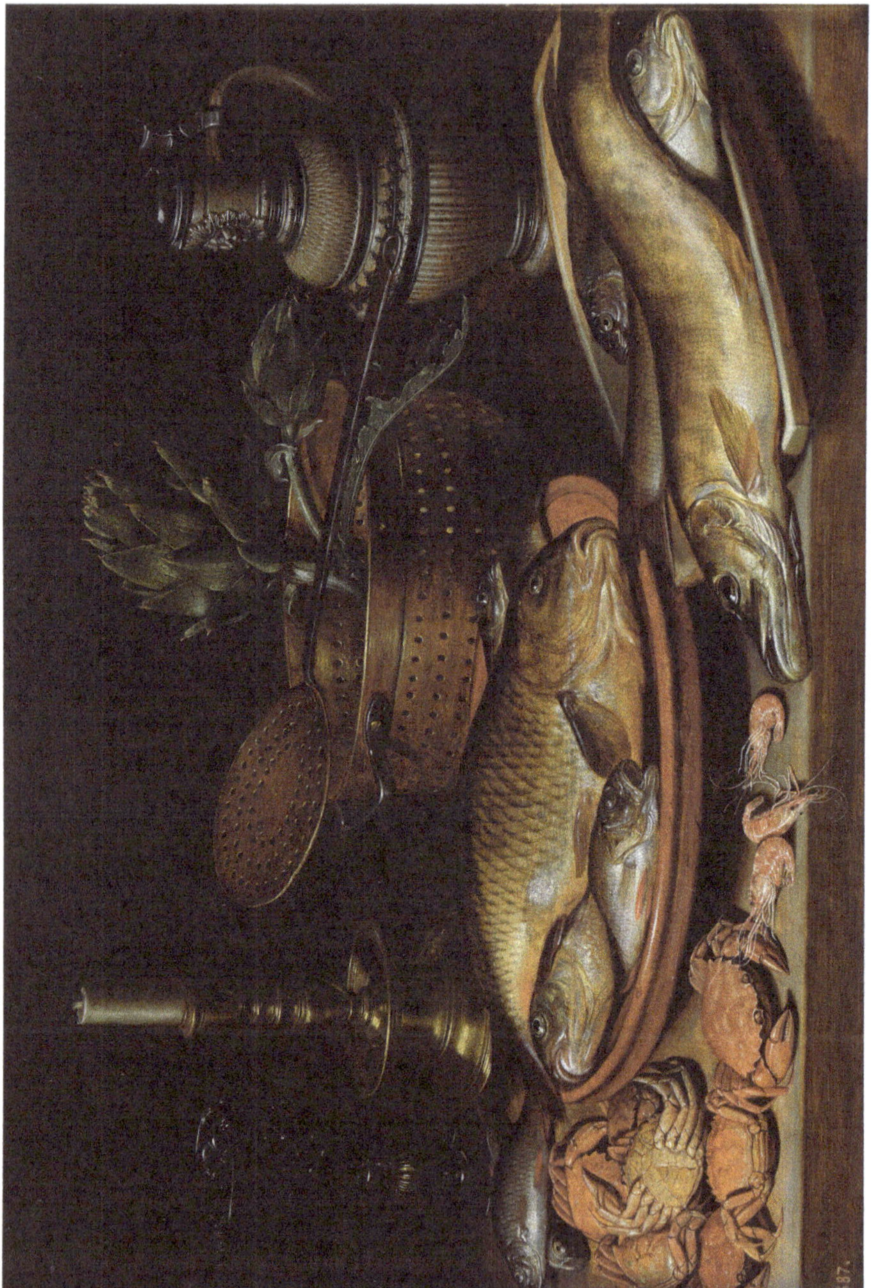

FIGURE 8.2 Clara Peeters, *Still life with Fish, a Candle, Artichokes, Crab and Prawns*. 1611. Oil on panel, 50 × 71.6 cm. Madrid, Museo del Prado
IMAGE © MUSEO DEL PRADO

coral, seahorses, blowfish, sawfish, horseshoe crabs, and dolphin skulls.[62] Finally, traditional depictions of the miraculous draught of fishes were also still made and collected in Antwerp throughout the 17th century. Famous examples presently known are those by Peter Paul Rubens (1577–1640) and Jacob Jordaens (1593–1678).

Painters in the northern Netherlands also made different types of fish paintings. There, Jacob van Nieulandt (1593/94–1634) and Floris van Schooten (*c.*1586–1656) were among the first to paint markets and kitchens with fish. Only a few fish still lifes (thus, without active figures) were painted before 1630. Important centres of fish still life painting in the Dutch Republic were The Hague and Utrecht, but they were also produced in other cities. Pieter de Putter from The Hague was presumably the first specialized fish still life painter in the Dutch Republic.[63] Fish still life painters from Utrecht include Willem Ormea, Jan de Bont, and Jacob Gillig. The most productive Dutch painter in this genre was probably Abraham van Beijeren (1620/21–1690), whose earliest dated fish still lifes are from 1654 or 1655. A herring cut in pieces or oysters were often included in 'breakfast' or 'banquet' still lifes, for instance those by Pieter Claesz (1596/97–1661) and Willem Claesz Heda (1594–1680). Pieter Claesz was born near Antwerp and moved to Haarlem in 1620, where he became a successful still life painter. The still life painter Harmen Steenwijck made several still lifes in which he combined fish with other types of food.[64] Herring and other fishes were also included in genre paintings. Paintings of herring sellers by Leiden *fijnschilder* Gerrit Dou (1613–1675), for instance, could be sold for gigantic sums (see below). Paintings of fishery and fishermen were made, for instance, by the enigmatic Amsterdam painter Arent Arentz Cabel (1585/86–1631).

4 Fish Paintings: Ownership

To investigate the ownership of fish paintings in Antwerp, Amsterdam, and Haarlem, the research on inventories by Erik Duverger, John Michael Montias, and Pieter Biesboer is used.[65] I searched the inventories they collected for spe-

62 Rijks, "Scales, Skins, and Carapaces", 2023.

63 Meijer G.M., "Visstillevens in Holland en Vlaanderen", in Helmus (ed.), *Vis* 13–74 (here 37).

64 Such as the one with fish and peaches currently in the Rijksmuseum Amsterdam: Harmen Steenwijck, *Still Life with Fish and Fruit*, 1652, oil on panel, 23 × 27 cm, SK-A-1529.

65 They are chosen because of the availability of the sources and because these three cities were important centres for the production of paintings. Duverger, *Antwerpse kunstinventarissen*; Montias database, https://research.frick.org/montias (accessed September 2019), Biesboer data on website Getty Provenance Index (GPI): https://piprod.getty.edu

cific records of fish paintings: where the words 'fish', 'fish market', or 'fishing' were used in the description of a painting; or particular species, such as 'herring', 'cod', 'crab', 'lobster', 'oyster', or 'salmon'.[66] The inventories are from the period 1600–1699 (Antwerp), 1597–1681 (Amsterdam), and 1572–1745 (Haarlem). The total number of 'fish paintings' I have counted per city in these periods were: Antwerp 381, Amsterdam 150, and Haarlem 52. This is only a tiny percentage of the total number of paintings in these inventories. In Montias's database of Amsterdam, for instance, a total of 35,839 paintings are listed. Because of the small percentages and the relatively small absolute numbers, one must be careful with drawing general conclusions about the popularity of certain genres or comparative conclusions of these three cities. Other issues are the different dates of the inventories and the fact that one never knows what is missing (which objects or possessions were not listed in inventories). And perhaps most importantly, the terminology used to describe paintings may have differed from notary to notary and city to city. Contemporary descriptions may lead to a distorted idea about the number of certain genres as compared to other genres. To give just one example: Antwerp allegories, which often contained fish motifs, are not described in inventories as having fish motifs (so, they are missing in my numbers). What inventories do provide, however, is a sort of objectivity in that they list actual objects: they give us a unique insight into the ownership of paintings by a wide variety of people. From 17th-century descriptions in inventories from Antwerp, Amsterdam, and Haarlem, we can distinguish four common types of fish paintings: first, fish still lifes; second, fish markets, fish kitchens, and fishmongers; third, scenes of fishing and fishery (including whale hunts); and fourth, biblical scenes with fish or fishing (plus category 5, other – unidentifiable fish paintings), see Table 1.

These categories are based upon the contemporary descriptions I have found in the inventories. For most paintings, there is no painter listed, as was common in this period. Where the painter was mentioned, this usually indicated a relatively higher value.[67] But a majority of the fish paintings in inventories were not valued (in Antwerp even less so than in Amsterdam and Haarlem). Moreover, there were no collectors who specialized in fish paintings: most inventories in which fish paintings were listed contained only one

/starweb/pi/servlet.starweb?path=pi/pi.web (accessed September 2019). For Biesboer, see also the book publication: Biesboer, *Collections of Paintings in Haarlem*.

66 I have excluded seascapes, since they often do not depict fish. Only paintings with 'fish' or a particular species in the description are included. This, of course, may have led to the exclusion of paintings that were in reality 'fish paintings'.

67 As was calculated for the city of Delft by Montias. Montias J.M., *Artists and Artisans in Delft. A Socioeconomic Study of the Seventeenth Century* (Princeton: 1982) 227.

TABLE 1 Fish paintings per genre per city

	Antwerp	Amsterdam	Haarlem
1. Fish still lifes	193 – 51%	66 – 43%	33 – 63%
2. Fishmarkets, fish-kitchens, and fishmongers	86 – 23%	20 – 13%	5 – 10%
3. Fishing and fishery	47 – 12%	53 – 35%	13 – 25%
4. Biblical scenes with fish or fishing	47 – 12%	7 – 5%	0 – 0%
5. Other (unidentified)	8 – 2%	6 – 4%	1 – 2%
Total	381	152	52

or a few such paintings.[68] The only exceptional case was the collection of the Antwerp resident Joanna van Schayenborch, the widow of a fish merchant. She was probably related to the fish still life painter Pieter van Schayenborgh (see below). In Antwerp, there were some other collections with a large number of fish paintings.[69] For instance, the inventory of Suzanna Willemsen from 1657 contained 17 fish paintings (9 still lifes, 5 markets, and 3 fisheries).[70] This is an unusually high number, but it was also an unusual collection: it was one of the most magnificent in Antwerp, containing around 1,500 paintings (so the percentage of fish paintings was still very low).[71]

Notaries and their clerks went from room to room in a house to list the valuables, sometimes accompanied by experts to help them identify paintings and other objects. As mentioned above, their descriptions do not necessarily match our contemporary descriptions or genres. For instance, biblical scenes are often very descriptive ('a St. Peeter fishing'), while scenes of fishery are also quite literal, noted as 'a whale hunt' or 'people fishing'. The described fish markets and 'fish kitchens' probably contained large still-life parts, but also one or more active figures. It is notable that in two versions of an Antwerp will, the

68 Meijer, "Visstillevens in Holland en Vlaanderen" 13.

69 Not in Amsterdam and Haarlem, where there are some inventories with four and three fish paintings, but no more than that.

70 Other examples with a relatively large number of fish paintings are: the inventory of Victor Wolfvoet (painter) with 9 fish paintings; the inventory of Benedictus I van de Walle (surgeon) with 6 fish paintings; and the inventory of Jeremias Wildens (painter and art dealer) with 8 fish paintings.

71 Three of the fish paintings included the name of the painter: two still lifes by Alexander Adriaenssen and a fish market by a certain Christiaen van Dom. Duverger, *Antwerpse kunstinventarissen* vol. 7, 351–401. Suzanna Willemsen was the widow of Jan van Borm, merchant of silk and second-hand clothing.

same painting was once called a 'fish market' and the other time a 'kitchen'.[72] Paintings of fishmongers also include the typical Dutch genre pieces, which often contained only one or a few fish (one herring, for instance). But because these paintings are also a combination of one or more active human figures and a fish motif, I have combined them in category 2 (fish markets, fish kitchens, and fishmongers). Also, there are paintings listed as being or containing a 'pekelharing' (salted herring). In most cases these paintings were probably still lifes, but 'pekelharing' was also the name of a comical figure of a happy drinker.[73] Moreover, the term 'fish market' is often used in inventories, but the term 'still life' is hardly employed. Paintings that are listed summarily as 'a fish' or 'a herring' are categorized here as still lifes (as is common practice). The term 'still life' in connection with fish only appears in a few inventories of my data set – and only in Haarlem. The earliest example is a Haarlem inventory from 1657, with a painting described as '1 stuck van eenige stilstaende kreeften', literally translated as '1 piece of stagnant/still lobsters'.[74] In another one from 1673, there is a painting listed as 'een schilderij vant stille leven met schelle vis' (a painting being a still life with cod).[75] Finally, in an inventory from 1713 we find the description of a 'een creefje stil leeven' (a lobster still life).[76]

5 Fish Still Lifes

Although they were not usually described as such, still life was by far the most popular subgenre of fish paintings in the three cities that were investigated. Still life in general, and in particular fish still lifes, were not very much appreciated by art theorists.[77] The percentages of still lifes among the fish paintings were: Antwerp 51%, Amsterdam 43%, and Haarlem 63%. The number of fish still lifes in Antwerp steadily rose to a peak in the years 1650–1659, only declining in the last decades of the century.[78] In Amsterdam, the peak was reached in the decade 1660–1669. In Haarlem the first fish still lifes are found

72 In the will of Pieter Ryckaert and Barbara Bourelle from 1675 and 1676. Duverger, *Antwerpse kunstinventarissen* vol. 10, 33.

73 See, for instance, the 'Pekelharing' van Judith Leyster in the Frans Hals Museum in Haarlem. https://www.franshalsmuseum.nl/nl/art/pekelharing/ (accessed April 2021).

74 Inventory of Lodewijck van Alteren, 1657, item 18. See: Biesboer (GPI) N-2467.

75 Inventory of Maritje Jaring (widow of Hendrick Simonsz van der Poorten), 8 July 1673, item 4. Biesboer (GPI) N-3160.

76 Inventory of Mattheus Andriesz Stilte, 1 August 1713, item 20. Biesboer (GPI) N-5675.

77 Exemplary is Gerard de Lairesse. Van Sighem, "Vis op het droge" 185–186.

78 The numbers I found in Antwerp inventories are contrary to the claim made in the exhibition catalogue *Vis*, that in the Southern Netherlands the popularity of fish still lifes started

TABLE 2 Fish species as noted in contemporary descriptions of still lifes

	Antwerp	Amsterdam	Haarlem
(Salted) herring	12	11	7
Lobster	36	21	5
Crab	11	8	5
Oysters	7		6
Salmon	3		
Haddock	4		1
Mussels	1		
Cod	1	1	2
Pike	2	1	
Turtle	3		
Shrimp		1	
Plaice			1
Flying fish	1		
'Strange fish'	1		
'Sea fish'		1	2

in 1640–1649, but the numbers are too small to draw general conclusions about the popularity of fish still lifes. Whereas descriptions of paintings in the other categories usually only contain the term 'fish', in the category of fish still lifes there is more variety. Some are more precisely described in terms of species, see Table 2. The most listed species are lobster, herring, and crab. Sometimes more than one species is mentioned, often in the combination of lobster and oysters (or lobster and fruit).

The first still life with aquatic creatures found in Antwerp inventories is listed as 'een lanck paneel met een bancket van eenen creft' (a long panel with a banquet with a lobster) in the estate of the painter Francoi Mirou in 1617.[79] In total, 193 fish still lifes are listed in Antwerp inventories, of which 25 are by Alexander Adriaenssen (while most are listed without the name of the painter). Adriaenssen was the Antwerp specialist in fish paintings, which is also confirmed by the fact that he was sometimes listed with merely his first name, such as 'een schilderye van visschen van Alexander' (a painting of fishes

decreasing from around the middle of the century. Meijer, "Visstellevens in Holland en Vlaanderen" 16.

79 Duverger, *Antwerpse kunstinventarissen* vol. 1, 486.

by Alexander).[80] Only one of Adriaenssen's fish still lifes in Antwerp inventories is appraised, at the relatively large sum of 56 *gulden* and 4 *stuivers*.[81]

Even though Clara Peeters was the first known artist to paint a fish still life, her name occurs only once in Antwerp inventories.[82] A 'stoffasiken van visch van Peeters' (a staffage of fish by Peeters) was owned by the Antwerp surgeon Benedictus I van den Walle in 1652.[83] The Dutch term 'stoffasiken' probably refers to 'stoffage' (staffage), here in the meaning of the 'decorative' fish motifs. In this case, it probably refers not to decorative motifs in a larger scene, but to the main subject of the still life. It is not a very common term in inventories, but in this particular inventory is it used several times, presumably with regard to still life painting.[84] Peeters's name also appears in 1627 in a collection in Rotterdam, which includes a 'fish after Clara Pieters'.[85] This demonstrates that she was already a copied artist early in her career. Paintings by Peeters were further found in seventeenth-century collections in Amsterdam, Haarlem, and Madrid.[86]

Notable was the aforementioned Antwerp collection of Joanna van Schayenborch (d. 1655), widow of Anselmus van den Steen, a fish merchant. In her otherwise relatively modest collection (a total of 17 paintings), there were 6 fish paintings. It seems reasonable to assume that this preference was prompted by the occupation of her late husband and other relatives. Five of the fish paintings were still lifes: two 'schelviskens' (haddocks) by Alexander Adriaenssen and three 'viskens' (fishes) by Peeter van Schaeyenborch.[87] There is some debate about whether the Alkmaar painter Pieter van Schaeyenborgh (a member of the Guild of Saint Luke in Alkmaar since 1635) is the same person as an apprentice of this name in the Antwerp Guild of Saint Luke in

80 In the inventory of Maria de Bodt (widow of Simon Jordaens) from 1659.

81 Statement of Jacques le Roij, about paintings he sold to Joris van Woelput, 11 July 1636. Duverger, *Antwerpse kunstinventarissen* vol. 4, 28.

82 Contrary to the assumption made in the exhibition catalogue *The Art of Clara Peeters*, where it was claimed that there is no evidence of paintings by Peeters in seventeenth- or eighteenth-century Antwerp collections. See Vergara, "Reflections of Art and Culture in the Paintings of Clara Peeters" 17.

83 Inventory Van den Walle from 7 December 1652. Duverger, *Antwerpse kunstinventarissen* vol. 6, 392.

84 In most cases where 'stoffasie' is used, it is in the traditional manner as 'staffage' in a landscape. The term 'gestoffeerd' is also used regarding picture frames.

85 Owned by Lucretia de Beauvois, wife of the painter Herman Saftleven. See Vergara, "Reflections of Art and Culture in the Paintings of Clara Peeters" 17.

86 In 1635 in Amsterdam, in 1685 in Haarlem, in 1637 (and 1655) in Madrid. Vergara, "Reflections of Art and Culture in the Paintings of Clara Peeters" 17.

87 'Dry stuckxkens zynde viskens met swerte leysten geschildert by Peeter van Schaeyenborch'; 'Twee schelviskens geschildert by Alexander Adriaenssens'; 'Een vischerery in grauwen leyste'. Duverger, *Antwerpse kunstinventarissen* vol. 7, 125–126.

1610.[88] It seems likely that it is the same person, who after his apprenticeship in Antwerp moved to Alkmaar. Although there is no conclusive evidence, the painting in Joanna's inventory is a strong indication that it was indeed the same painter, since Antwerp collectors had a strong preference for local Antwerp painters. Paintings from the northern Netherlands were exceptional in Antwerp inventories, and then usually only by the most famous masters. Further, it seems likely that the fish painter Pieter van Schayenborgh (not a very common name) was related to Joanna van Schaeynborch, who had links to the Antwerp fish trade. As stated above, her deceased husband was the fish merchant Anselmus van den Steen. Moreover, the guardian of Joanna's children was François van Schayenborch, presumably the same François van Schayenborch who was the dean of Antwerp's fish merchants' guild. François's own collection, inventoried in 1659, also contained a painting of a haddock, as well as one of a salted herring.[89]

Already in the years 1620–1629 several fish still lifes were listed in Amsterdam inventories. One early example from 1627 is the type associated particularly with Dutch still life painting: 'een ontbijtje van een pekelhering' (breakfast with a salted herring), found in the inventory of Louijs Rocourt.[90] There are three more fish still lifes listed as 'breakfasts' in Amsterdam inventories (one with lobster instead of herring). There are also several paintings listed as 'banquets'. Prices for fish still lifes in Amsterdam inventories ranged between 0.25 and 50 *gulden*. The cheapest one was in the inventory of Grietje Tijmans, a fishmonger living in Moddermolensteeg.[91] The most expensive one was a 'stuck van fruyten ende creeft' (piece of fruit and lobster), valued at 50 *gulden* and 10 *stuivers* in the collection of Jacob Claesen van Hoorn, who also owned a relatively expensive 'schilderij van visscherij' (painting of fishery) of 29 *gulden* and 5 *stuivers*.[92] Another expensive still life was the 'crabbebancket, Pieter Claesz' (crab banquet, Pieter Claesz), worth 42 *gulden*, in the collection of Catharina Thijs.[93] There are many paintings by Claesz presently known that comply with this description. Finally, Amsterdam inventories include still lifes by a certain

88 Meijer mentions that there is no conclusive evidence. Meijer, "Visstillevens in Holland en Vlaanderen" 58.

89 'twee stucken van een grootte: d'een eenen Peeckelharinck ende d'ander eenen Schelvis in swerten leijst'. Inventory Francois van Schaeijenborch, 4 August 1659. Duverger, *Antwerpse kunstinventarissen* vol. 8, 75.

90 Inventory of Louijs Rocourt, 22 May 1627. Montias inv. 6240, lot. 0059.

91 'een schilderytje van een vis met 3 kleijne bortjes' worth 1 *gulden*. Montias takes this as 4 pieces, which would imply a value of 0.25 gulden. Perhaps it was 1 painting worth 1 *gulden*. Inventory Grietje Tijmans, 18 August 1645. Montias inv. 431 lot 0007a–d.

92 Inventory of Jacob Claesen van Hoorn, 20 March 1625. Montias inv. 813 lot 0027 and lot 0018 (Montias records 41938 and 32160).

93 Inventory Catharina Thijs, 15 April 1639. Montias inv. 408 lot 0036 (Montias record 14332).

Juriaen Vlegel, who was perhaps Georg Flegel (*c.*1566–1630), one of the earliest European still life painters.[94] One example by Vlegel is the valuable still life described as 'een nacht stuckken kreeft ende kaers van Jeuriaen Vlegel' (a night scene of a lobster and a candle by Juriaen Vlegel), worth 30 *gulden.*[95] In another inventory, there is a 'peeckelharingh' (salted herring) worth 10 *gulden* by Vlegel.

As Pieter Claesz worked in Haarlem most of his life, it is no surprise to find four fish still lifes by him in Haarlem inventories. Two of them were owned by Hester Cluysaenes, who had them on display in her house on the Lange Begijnenstraat.[96] Other named painters of fish still lifes in Haarlem inventories were Cornelis Cornelisz van Haarlem (1562–1638), Gerrit Adriaensz Berckheyde (1638–1698), and Jan Jansz Treck (1605/06–1652). The 'schilderije met een pekelharingh' (painting with a salted herring) by the Amsterdam still life painter Treck was estimated at 24 *gulden* and the most expensive fish still life in Haarlem inventories.[97] This painting was on display in the kitchen of the wealthy merchant Eduard van Cralen and his wife Agneta van den Heuvel. Their collection of around 70 paintings contained other expensive pieces, such as a landscape by Cornelis Vroom (175 *gulden*), a painting with the story of Tobias by Rembrandt (90 *gulden*), and a piece by Philip Wouwerman (85 *gulden*).[98]

Cornelis Cornelisz van Haarlem is not known for fish still lifes, but in the inventory of Agatha van Stuyvesant there is 'een stuck met vissen van Mr. Cornelis' (a piece with fishes by Mr. Cornelis).[99] The painting was on display

94 Inventory of Hendrick Hoeffslager, 19 March 1625. Montias inv. 579 lot 0047 (Montias record 20619). Hoeffslager owed two other fish paintings: a 'fishery' by Aert Anthonisz worth 18 *gulden* and 10 *stuivers* and an anonymous fish still life worth 4 *gulden* and 15 *stuivers*. Georg Flegel worked in Linz and Frankfurt am Main and was probably trained by Lucas I van Valckenborch (after 1535–1597).

95 Inventory of Samuel Godijn, 26 November 1633. Montias inv. 1123 lot 0061 (Montias record 41323).

96 Cluysaenes was the widow of Symon Wassenaer and lived on the 'Lange Bagijnestraet aende eene zijde de Franse kerck' (so near the Walloon Church that is still in Haarlem). The still lifes were: 'een bancketie van Pieter Claesz. met een crab' (worth 12 *gulden*) and 'een ditto [bancketie van Pieter Claesz.] met een peckelharingh' (worth 6 *gulden*). Inventory Hester Cluysaenes, 28 December 1666. Biesboer N-5057, items 6 and 8.

97 Prices range between 1 and 24 *gulden*. Inventory Eduard van Cralen, 28 April 1675. Biesboer N-4379, item 22.

98 Biesboer, *Collections of Paintings in Haarlem* 239–240.

99 Inventory Agatha van Stuyvesant, 1 October 1646. Biesboer (GPI) N-3710, item 26. Perhaps it was similar to the painting: Cornelis Cornelisz van Haarlem, *A Kitchen Piece with a couple of lovers; in the background a merry company*, 1596, private collection (formerly Stuker Bern Switzerland, 2014), see: https://rkd.nl/nl/explore/images/754 (accessed February 2022).

in the kitchen, together with 11 other pieces by the same painter.[100] Cornelisz van Haarlem was Van Stuyvesant's uncle, which explains the relatively large number of his paintings in her collection of approximately 60 paintings.[101] The painter Gerrit Berckheyde was not known for his fish still lifes either, and perhaps the 'stuck van Bercheyden met een peeckelharingh' (piece by Berckheyde with a salted herring) was not a still life, but in fact one of his known cityscapes with a fish market.[102]

6 Fish Markets, Fish Kitchens, and Fishmongers

In Antwerp, markets, kitchens, and fishmongers make up the second largest category of fish paintings, at 23%. In Amsterdam and Haarlem the category is less prominent, with 13% and 10%, respectively. Contrary to Antwerp, where most of these pieces were described as fish markets, in Amsterdam around half of the paintings in this category were kitchen scenes or genre pieces, such as 'a girl with a herring', 'a women buying herring from a boy', or 'a woman cleaning fish'.

In Antwerp, this category included paintings by Frans Snyders, Joannes Fijt (1609–1661), Adriaen van Utrecht (1599–1652), Alexander Adriaenssen (as well as a copy of Adriaenssens), and a certain Christiaen van Dom. Although most paintings in this category are simply listed as 'a fish market', there is also one specifically described as a 'herring market', as well as an 'Antwerp fish market', 'Venetian fish market', and a 'Scheveningen fish market'. As mentioned above, I have included in this category paintings described as 'a fishmonger' or 'a fish kitchen', since they refer to paintings with active figures, similar to fish markets (and unlike still lifes). The most remarkable of these 'fish genre paintings' was the one in Antwerp described as an 'een vrouken met pekelharinck' (a woman with a salted herring), by Gerrit Dou, appraised at 1,000 *gulden* in 1691.[103] There are several paintings known today by the Leiden *fijnschilder* Dou that match this description, for instance the one in the Hermitage in St. Petersburg or the one in the Leiden Collection in New York [Fig. 8.3].[104]

100 'Elff stucks schilderijtgens van Mr. Cornelis'. Inventory Agatha van Stuyvesant, 1 October 1646. Biesboer (GPI) N-3710, item 28.

101 Biesboer, *Collections of Paintings in Haarlem* 96.

102 But because of the description, the painting is labeled as still life in my data. Inventory of Reyer Willemsz Heus, 18 October 1663. Biesboek N-5248, item 6.

103 Duverger, *Antwerpse kunstinventarissen* vol. 12, 93.

104 Gerrit Dou, *Herring Seller and Boy*, *c.*1670–1675, oil on panel, 41 × 30 cm, Saint Petersburg, Hermitage; Gerrit Dou, *A Boy and an Old Woman with a Herring in the Window of a Shop*, oil on panel, 42.2 × 34.2 cm, *c.*1664, New York City, The Leiden Collection. Further, there

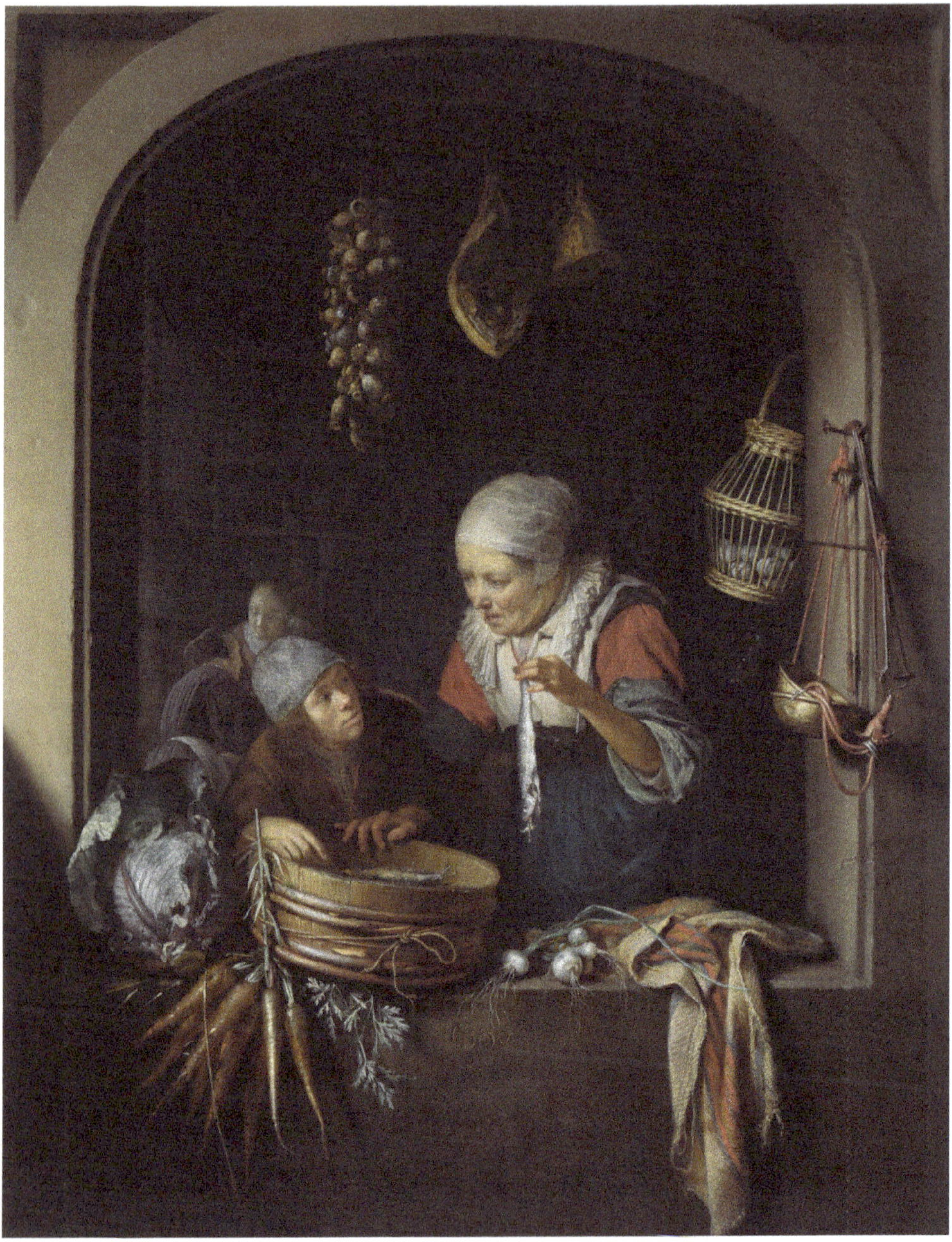

FIGURE 8.3 Gerrit Dou, *Herring Seller and Boy*. *c*.1664. Oil on panel, 43.5 × 34.5 cm. New York, The Leiden Collection
IMAGE COURTESY OF THE LEIDEN COLLECTION, NEW YORK

Dou's incredibly valuable painting was inherited in 1691 by Theresia Maria Anthoine from her father, Antwerp's late postmaster Jan Baptist I Anthoine (d. 1691). The Anthoine collection was one of the most grandiose in Antwerp at this time.[105] In 1664, the Anthoine family was portrayed by Gonzales Coques, which is probably the family portrait by 'Gonsael' mentioned in the inventory.[106] The Anthoine collection contained statuettes, *naturalia* (such as coral, 'see-theeth', and coconuts), reliquaries, objects from India and China, and some books. But his collection of over 300 paintings estimated at a total value of 46,130 *gulden* was particularly impressive.[107] Yet the amount of 1,000 *gulden* for the Dou painting was exceptionally high, even for this collection. Notably, among the other top-priced paintings was another one with fish: 'een cleijn stucxken Moijses vissinge van Paulo Cavillarij Veronese' (a small piece, Moses fishing, by Paulo Caliari Veronese), estimated at 800 *gulden*.[108]

In another of Antwerp's most grandiose collections, there was 'een stucxken van Mostart wesende een Vischmerct op paneel in ebbenhoutte lyste' (a piece by [Gillis] Mostaert being a fish market on panel in ebony frame).[109] This painting was on display in the front room of the house *De Witte Pluym* of Gillis de Kimpe. De Kimpe owned an incredible collection of paintings, drawings, prints, books, instruments, and *naturalia*.[110] As far as we know, he was the only Antwerp collector at this time who owned a horseshoe crab, and it was probably

are similar paintings in Moscow (https://rkd.nl/explore/images/250148), and one in a private collection (https://rkd.nl/explore/images/258801) (accessed March 2021). For the Dou painting in the Leiden Collection, see: Baer, Ronni. "Herring Seller and Boy" (2017). In The Leiden Collection Catalogue, 2nd ed. Edited by Arthur K. Wheelock Jr. New York, 2017–2020. https://theleidencollection.com/artwork/the-herring-seller-and-boy/ (archived May 2020).

105 Timmermans B., *Patronen van patronage in het zeventiende-eeuwse Antwerpen: een elite als actor binnen de kunstwereld* (Amsterdam: 2008) 232.

106 'no. 300 het portrait van de familie van Gonsael', valued at 300 guilders (or perhaps: 'no. 151 een familie van contrefeijtsel van Gonsael', valued at 150 *gulden*). Duverger, *Antwerpse kunstinventarissen* vol. 12, 94, 98.

107 The majority was of local masters, such as Antony van Dyck (36 works), several generations of Brueghels (33), and Peter Paul Rubens (11). Among the northern Netherlandish masters were Adriaen Brouwer (20), Gerrit Dou, Jan Lievens, and Hendrick Cornelisz. van Vliet; among the Italian masters we find work by Tintoretto and Paolo Veronese. Anthoine also owned a 1,000-*gulden* piece of 'two melon-eaters' by the Spanish painter Bartolomé Esteban Murillo – as well as several copies of this work. Timmermans, *Patronen van patronage* 232; Duverger, *Antwerpse kunstinventarissen* vol. 12, 84–99.

108 Duverger, *Antwerpse kunstinventarissen* vol. 12, 93.

109 Next to the fish market by Gillis Mostaert (1528–1598), De Kimpe owned five more paintings and two drawings by the same artist.

110 Inventory Gillis de Kimpe, 23 July 1625. Duverger, *Antwerpse kunstinventarissen* vol. 2, 399–415.

this specimen that was depicted in the gallery pictures by Frans II Francken, who knew De Kimpe.[111] De Kimpe further owned several other aquatic *naturalia* as well as a 'constboeck van alderleye visschen' (art book of all sorts of fishes).[112]

Among the paintings in the category of 'fish markets, fish kitchens, and fishmongers' in Amsterdam inventories, a relatively large share is valued and the values are relatively high.[113] Notable is 'een schilderytge van een vrouw die haring coopt van een jonge' (a painting of a woman buying herring from a boy), probably a genre piece, worth 100 *gulden*.[114] The most expensive piece in Amsterdam was a fish market, by Jan I Brueghel, described as 'een binnewater ofte vismarckt van den jongen Breugel' (an internal water or fish market by the young Brueghel).[115] This valuable painting, worth 240 *gulden*, was on display in Samuel Godijn's house on the Keizersgracht. The wealthy merchant Godijn (1561–1633) was originally from Antwerp and settled in Amsterdam, where he became one of the administrators of the Northern Company and the West Indian Company. Brueghel made several paintings with fish markets located in a riverscape with a large number of figures, like the three paintings now in the Alte Pinakothek in Munich [Fig. 8.4].[116] Like Godijn, the painter Abraham Vinck (1574/75–1619) also emigrated from Antwerp to Amsterdam, where he died in 1619. His inventory lists two fish markets, one described as 'een groote vismarck principael van Vincx' (a large fish market, an original by Vincx) worth 80 *gulden*.[117] In another Amsterdam inventory there is a fish

111 Rijks, "A Painter, a Collector, and a Horseshoe Crab".

112 'een Zeeduyff' (a blowfish), 'een zeespinnecop' (horseshoe crab), 'twee zeepeerden tanden', 'eenen schilt van een schiltpadde', 'een schiltpaddeken'. Duverger, *Antwerpse kunstinventarissen* vol. 2, 399–415.

113 13 out of 20 are valued with an average of 46.5 *gulden*.

114 This painting was 'in een ebbehout cas', so probably in an ebony box instead of a common frame, which may also have added to the value. Inventory Odelia van Arras, Montias inv. 460 lot 0009 (Montias record 13328).

115 Jan I Brueghel was the son of Pieter Bruegel and called 'the young' (confusingly, nowadays he is called 'the Elder' and his son Jan II Brueghel described as 'the young'). Inventory of Samuel Godijn, 26 November 1633. Montias inv. 1123 lot 0021 (Montias record 41269).

116 For instance: *Large Fishmarket* (inv. 1889); *Fishmarket by a River* (inv. 1883); *Harbour with Preaching Christ* (inv. 187). Alte Pinakothek Munich. https://www.sammlung.pinakothek.de/en/artwork/wq4jEKEGWo and https://www.sammlung.pinakothek.de/en/artwork/bwxoBzOGm8 and https://www.sammlung.pinakothek.de/en/artwork/Qlx2dKVGXq (accessed March 2021).

117 Inventory Abraham Vinck, 24 August 1621. Montias inv. 560, lot 0018 (Montias record 20015).

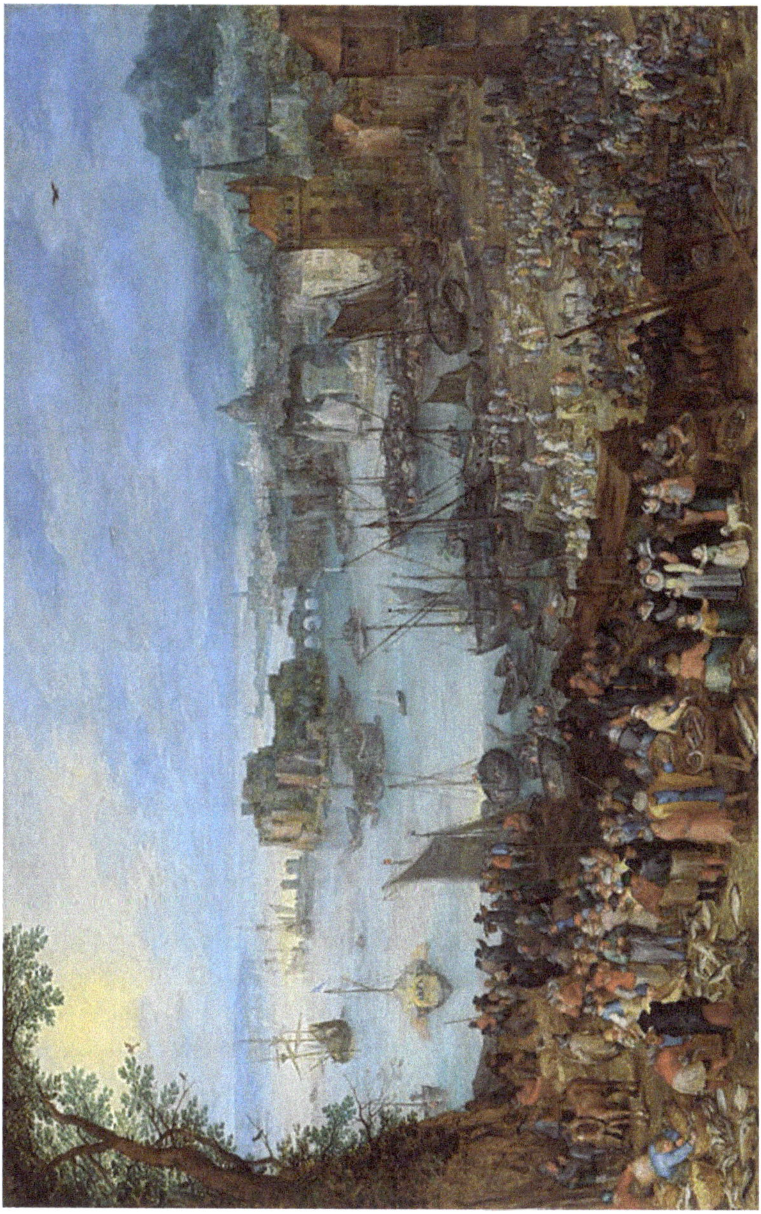

FIGURE 8.4 Jan Brueghel the Elder, *Large Fish Market*. 1603. Oil on panel, 58.5 × 91.5 cm. Munich, Alte Pinakothek

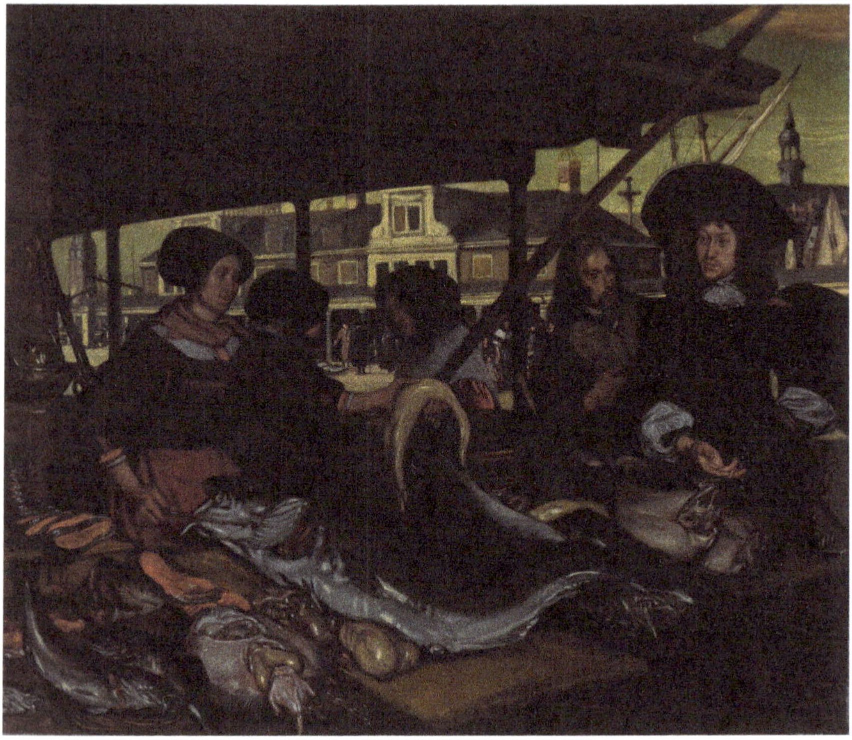

FIGURE 8.5 Emanuel de Witte, *The Nieuwe Vismarkt (New Fish Market) in Amsterdam*. 1655–1692. Oil on canvas, 52 × 62 cm. Amsterdam, Rijksmuseum

market by Emmanuel de Witte (1617–1692), of whom several fish markets are known today, such as the one in the Rijksmuseum Amsterdam [Fig. 8.5].[118]

In Haarlem inventories, there are only five paintings in this category, of which four include the name of the painter. As in Amsterdam, there is a fish market by Emanuel de Witte, which was in the collection of the Haarlem

118 'een vismarckt van Emanuel de Wit met een vergulde lijst'. Inventory of Herman Becker, 19 October 1678. Montias inv. 254 lot 0096 (Montias record 6141). De Witte made several fish markets, such as *The New Fish Market in Amsterdam* now in the Rijksmuseum Amsterdam; the *The New Fish Market in Amsterdam* in the Pushkin Museum in Moscow; *The Fish Market at Evening* in Boijmans van Beuningen in Rotterdam; *Adriana van Heusden and Daughter at the Fishmarket* in the National Gallery in London; and *The Old Fish Market on the Dam, Amsterdam* in the Museo Nacional Thyssen-Bornemisza in Madrid.

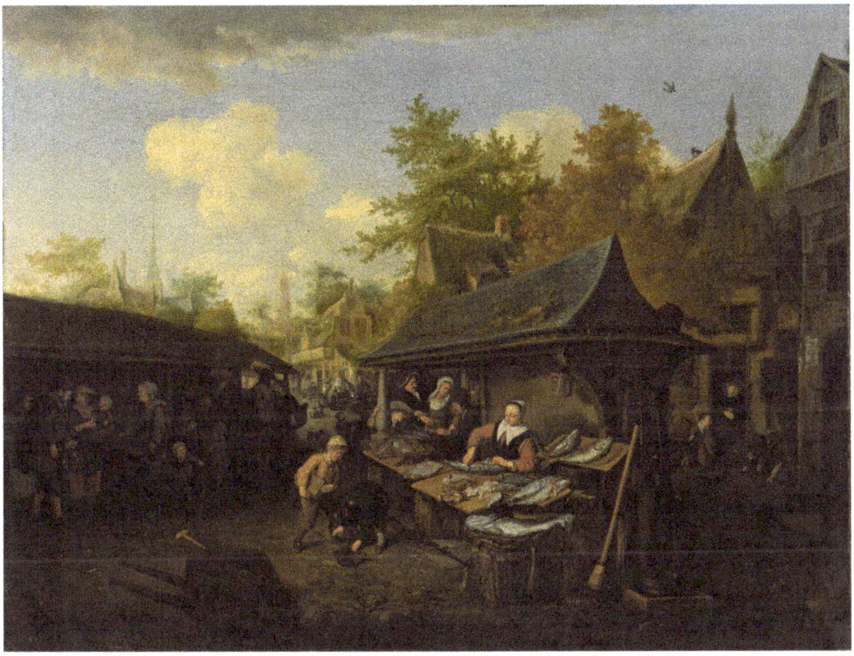

FIGURE 8.6 Cornelis Dusart, *Fish Market*. 1683. Oil on canvas, 67.8 × 90.1 cm. Amsterdam, Rijksmuseum

painter Cornelis Dusart (1660–1704).[119] Dusart also made fish markets himself, such as the one now in the Rijksmuseum [Fig. 8.6].[120] His inventory lists one unfinished fish market by him: 'een gedootverwde vismarckt van Corn. Dusart' (a fish market imprimatura by Cornelis Dusart).[121] The inventory of another Haarlem painter, Jan Miense Molenaer (1609–1668), contained a fish market by a certain 'de oude de vriest' (perhaps the Antwerp painter Guilliam de Vries) and a kitchen scene by Molenaer's wife, Judith Leyster (1609–1660), described as 'een boere keuckentje daer de meyt vis breeckt van Juff. Molenaer' (a famer's kitchen where a maid breaks fish by Miss Molenaer).[122]

119 'een vismartie van Emanuel de Wit'. Inventory Cornelis Dusart, Biesboer (GPI) inv. N-5636, item 63.
120 Cornelis Dusart, *Fish Market*, 1683, oil on canvas, 67.8 × 90.1 cm, Rijksmuseum Amsterdam.
121 Inventory Cornelis Dusart, Biesboer (GPI) inv. N-5636, item 226.
122 Inventory Jan Miense Molenaer, 10 October 1668. Biesboer (GPI) inv. N-5314, items 11 and 106.

7 Fishing and Fishery

The most notable difference between fish paintings in Antwerp, Amsterdam, and Haarlem was the large percentage of fishery paintings in the latter two cities. In Amsterdam and Haarlem fishery was the second largest category, with 35% and 25%, respectively. In Antwerp, this category represents only 12%.[123]

In Antwerp there was a 'visscherije' by Rubens listed in the inventory of Jeremias Wildens in 1653. Father and son Jan (1585/86–1653) and Jeremias (1621–1653) Wildens were painters and dealers of paintings. Jan collaborated with Rubens on several occasions, and he had several paintings by Rubens for sale in his large townhouse.[124] In other Antwerp inventories, there were two whale hunts by Andries van Eertvelt (1590–1652), who was known for his marine pieces.[125]An 'oeverken met visch' (shore with fish) by Jan van Kessel (1626–1679) was in the collection of Knight Joan van Weerden.[126] Van Kessel's painting is included in this category because of the landscape element, but the large and detailed depiction of fish species in many of his paintings is different from most paintings in this category. Notable were also the many descriptions of a 'landscape fishery' or 'landscape (with) fishing' in Antwerp inventories.

The Amsterdam cloth merchant Arent Pietersz Brugman and his wife, Neeltgen Cornelis, owned three fishery paintings, part of a collection of over 60 paintings. One was by Hendrick van Avercamp (1585–1634) and two were by Arent Arentz Cabel (1585/86–1631). Brugman also owed a 'St Pieters visschuyt' (St. Peeter's fishing boat) by Cabel, who was the brother-in-law of Brugman's mother.[127] Cabel was the son of an Amsterdam sailmaker and painted numerous fishing boats and fishermen in summer landscapes. Sometimes the foreground of his paintings contains details of fish just caught. Although Avercamp is primarily known for his winter landscapes, there are similarities with Cabel's work. In the Amsterdam inventories there are two paintings of a 'fisherman' and a 'fishery' by Avercamp, the first appraised at 6 *gulden*. Values in this category range between 1 and 50 gulden in Amsterdam. The most expensive fishery

123 A total of 40 paintings, which is still a lot if one compares it to the total number of 'fish paintings' found in Haarlem inventories (52).

124 Rijks M., *Artists' and Artists' Collections in Early Modern Antwerp. Catalysts of Innovation* (Turnhout – London: 2022).

125 Inventory of Arnout de Bruijne, 19 November 1632. Duverger, *Antwerpse kunstinventarissen* vol. 3, 316; inventory of Anna de Smidt, Duverger, *Antwerpse kunstinventarissen* vol. 7, 110.

126 Inventory of Knight Joan van Weerden, 30 April 1686. Duverger, *Antwerpse kunstinventarissen* vol. 11, 396.

127 Inventory Arent Pietersz Brugman, 1 January 1635. Montias inv. 226 lots. 0010; 0013; 0019; 0034 (Montias records 14627; 5100; 5105; 5131).

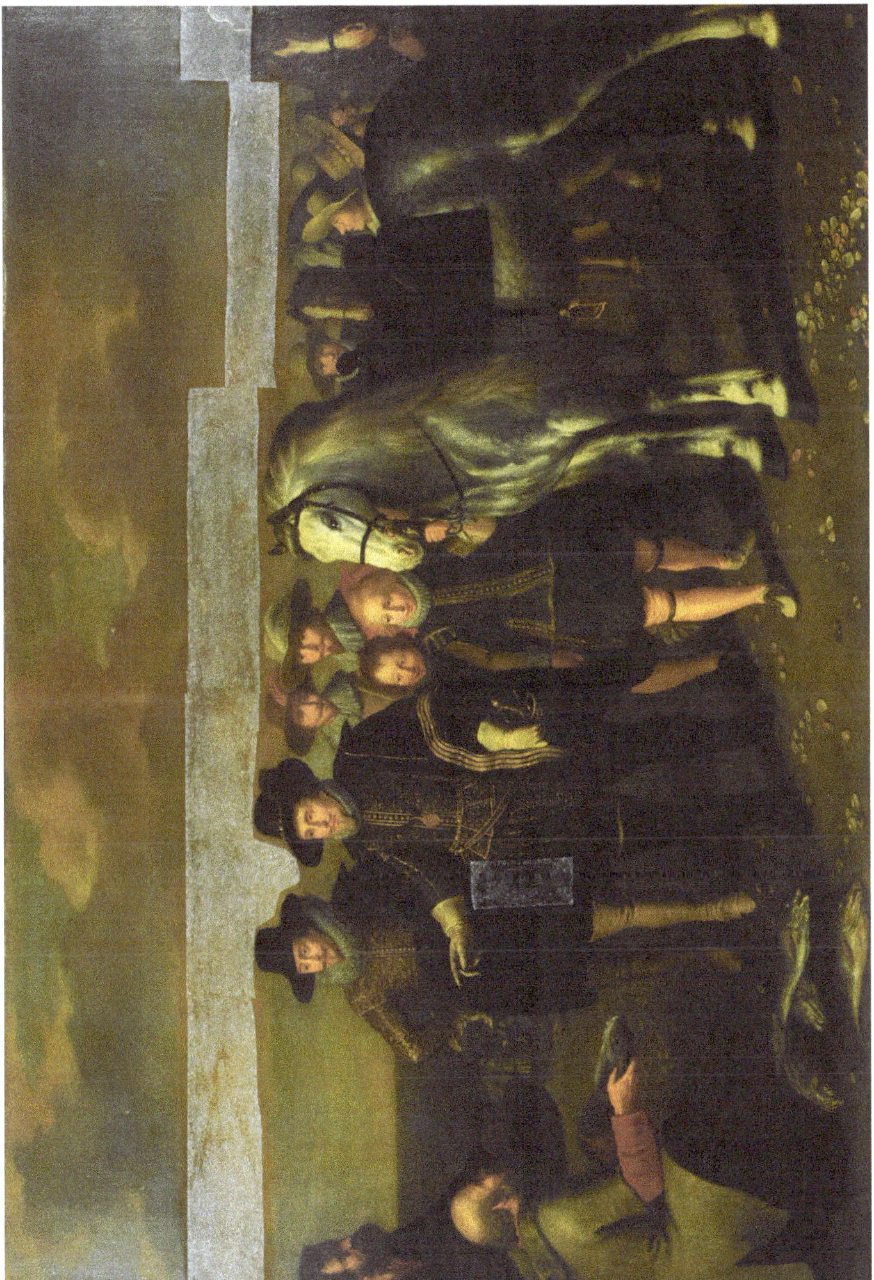

FIGURE 8.7 Adriaen van Nieulandt, *Maurits (1567–1625) and Frederik Hendrik (1584–1647), Princes of Orange, on the Beach at Scheveningen*. 136.3 × 199.3 cm, oil on panel. Amersfoort, Rijksdienst voor het Cultureel Erfgoed (RCE) (formerly Mauritshuis)

at 50 *gulden* was a 'schilderije vande Prins Mauris ende van Prins Hendrick opt strant bijde vissers' (painting of Prince Maurits and Prince Hendrik on the beach with fishermen).[128] This must be the large painting by Adriaen van Nieulandt (1586/87–1658) that still exists today [Fig. 8.7].[129] This scene of the Princes of Orange meeting fishermen on the beach may have been similar to the meetings, decades earlier, that Adriaen Coenen had with William of Orange on the beach of Scheveningen in 1574 and 1581, where he showed the prince his *Visboeck* and told him about fishes and beached whales.[130]

In Haarlem, too, there are a few paintings described as a 'beach with fishermen', but also landscapes and seascapes with fishermen and one of the 'haring vangst' (herring catch). In an inventory of 1717, there is a riverscape by Karel du Jardin (1626–1678).[131] There is also one painting described as a 'schilderij sijnde een henglaertie' (painting being an angler), a more exceptional description.[132]

8 Biblical Fish Paintings

Paintings in the last category – biblical scenes in which fish are particularly mentioned – are mostly found in Antwerp (47 paintings). Only seven of the paintings in this category are found in Amsterdam inventories, and zero are in Haarlem. Both in Antwerp and Amsterdam, most are described as either 'Saint Peter fishing' or 'Moses fishing'. In Antwerp some are described as the 'Apostles fishing' or 'Christ fishing' or 'Christ eating fish'. The literal description 'miraculous draught of fish' (as we would use today) does not occur in Antwerp or Amsterdam inventories from the seventeenth century, but the description of 'Saint Peter fishing' referred to one of the two biblical stories of the miraculous draught of fish (in Luke 5:1–11 and John 21:1–14). Both stories tell of a miraculous catch of fish in which Peter plays a prominent role. Peter was a fisherman when one day Christ came preaching on his boat. Afterwards, Christ told Peter to put out the nets, whereupon he caught an incredible number of fishes. According

128 Inventory Dirck Glaude (embroiderer, lacemaker), 7 January 1644. Montias inv. 508 lot 0009 (Montias record 16572).

129 Adriaen van Nieulandt, *Maurits (1567–1625) and Frederik Hendrik (1584–1647), Princes of Orange, on the Beach at Scheveningen*, 136.3 × 199.3 cm, oil on panel, Rijksdienst voor het Cultureel Erfgoed (RCE), Amersfoort (formerly Mauristhuis).

130 Egmond F., *Het Visboeck. De wereld volgens Adriaen Coenen 1514–1587* (Zutphen: 2005) 30–36.

131 'Vissers aan een beek van Carel du Jardijn'. Inventory Cornelis van der Laan, 11 August 1717. Biesboer (GPI) inv. N-5014, item 6.

132 Inventory Guertie Griecken, 22 January 1682. Biesboer (GPI) inv. N-2586, item 7.

to the Gospel of Luke, Peter fell at Jesus's knees, who told the fisherman: 'from now on you will catch man'. The second story tells of a very similar miracle, but now after Christ's death and resurrection. After Peter's realization that the miracle is caused by the resurrected Christ, Peter jumps into the water to meet him. Finally, 'Saint Peter fishing' can also refer to the miracle of Peter finding the coin in the fish's mouth (Matthew 17:24–27).

The description of 'de Visschinge van Moeijses' or 'Moeyses Vissinge' probably refers to the story of Moses, who as a baby is 'fished' out of the water in his wicker crib by the daughter of the Pharaoh. This is confirmed by one particular description of 'een schilderije schouwstuck affbeldende Moyses Visschinge in de Wiegh' (a painting being a mantelpiece depicting Moses fishing in the crib).[133]

In Antwerp inventories, three paintings in this category are listed with the name of the painter: one by 'sotten Cleve' (probably Joos van Cleve, who mistakenly acquired the nickname 'mad') and 'Breugel' (not indicated which member of the family). Then there is 'een schilderye schouwstuck wesende Vischerye van Sinte-Peeter origineel van Jordaens' (a painting mantelpiece being a fishing of Saint Peter original by Jordaens), which was on display in the *salet* (salon) of the tapestry merchant Michiel Wauters.[134] Several paintings of Saint Peter fishing by Jacob Jordaens (1593–1678) are known today.[135] The way it was described as an 'original by Jordaens' is an indication that it was among Wauters's most precious paintings. Wauters probably admired the work of this painter, as he bought several cartoons (designs for tapestry) from Jordaens's estate.[136]

Rembrandt owned a painting 'een scheepie Petri van Aertie van Leijden' (a ship Peter by Aertgen Claesz. van Leyden).[137] Next to the other paintings with Saint Peter and Moses in Amsterdam, there is also a painting of a 'visgen Jonas' (fish Jona), worth only 16 *stuivers* (in another inventory we find a wooden sculpture of 'Jonasie uytte vis', Jona out of the fish).[138] The most expensive

133 Inventory of Marie Francoise Manaert, 3–15 April 1692. Duverger, *Antwerpse kunstinventarissen* vol. 12, 177.

134 Wauters owed one other fish painting ('een schilderye fruyt met creft'). Inventory of Michiel Wauters, 16 October 1679. Duverger, *Antwerpse kunstinventarissen* vol. 10, 492.

135 For instance, the one in the Musée des Beaux-Arts Strasbourg or the one in the Rijksmuseum Amsterdam.

136 Duverger, *Antwerpse kunstinventarissen* vol. 10, 497.

137 'een scheepie Petri van Aertie van Leijden'. Inventory Rembrandt van Rijn, 26 July 1656. Montias inv. 1262, lot 0095 (Montias record 47698).

138 Inventory Elb[ert?] de Metselaer, 14 February 1609. Montias inv. 726, item 0008 (Montias record 30021). The 'Jonasie uytte vis' was in the inventory of Cornelis Pietersz Kroeger, 27 April 1649. Montias inv. 159, item 0007 (Montias record 1703).

appraised painting in this category in Amsterdam was 'een groot landschap schilderye in vergulde lyste synde vising van Moses' (a large landscape painting in gilded frame being a fishing of Moses), appraised at 20 *gulden* and on display in the 'groote camer' (large room) of the house of Toussain Blanche on the Herengracht.[139]

9 Conclusion

Based upon evidence from inventories from Antwerp, Amsterdam, and Haarlem, paintings with fish motifs made up a small portion of the total number of paintings. Among these fish paintings, fish still lifes were most popular, although the way paintings were described may lead to some distortions about different categories of fish paintings. The most notable difference between the three cities was the larger percentage of biblical fish paintings in Antwerp and the larger percentage of fishery paintings in Amsterdam and Haarlem. This relatively stronger emphasis on biblical history painting is in accordance with earlier research on collecting trends in Antwerp compared to the Dutch Republic.

Changing conceptions of the natural world went hand in hand with the emergence of new pictorial genres, such as market scenes and, around the turn of the century, still life painting. The interest in natural history among relatively large parts of the population may have been one factor that explains the popularity of animal motifs in painting. It probably also worked the other way around: animal motifs in painting may have further spurred the interest in nature. Moreover, the importance of fish in the daily life and economy of the Low Countries, combined with the increasing specialization of painters working for an open market, resulted in specialized fish genres.

In our modern view of painting as a 'fine art', we tend to think of paintings as self-contained works of art with a fixed title. Of course, in the early modern period there were no fixed titles, only descriptions of objects. Only in exceptional cases was the name of the painter included, which points to the fact that ideas about the 'author' of a 'work of art' were only slowly changing. The norm was still to list paintings anonymously, while subject, size, and material (and often picture frame) were much more important. Inventories reveal how the language to describe paintings changed over time and how more or less fixed genres only slowly developed.

---

139 Inventory Tousain Blanche, 11 March 1643. Montias inv. 1191, lot 0017 (Montias record 44572).

In early modern fish paintings, fish prints and fish books, descriptive facts and symbolic meaning were often combined. There was a great overlap between what we would consider the separate domains of 'art' and 'science'. Painters were interested in documenting facts about nature, but also in the meaning of nature. The same is true for natural historians, who were documenting facts about nature, but for a long time also included references to the (symbolic) meaning of nature in their works (inherited from a long and respectable tradition). The same logic underpinned collections. Through the ownership of fish paintings – as well as other types of animal imagery – it may argued that these ideas about nature also ended up in the houses of early modern collectors.

Bibliography

Bass M.A. – Goldgar A. – Grootenboer H. – Swan C. (eds.), *Conchophilia. Shells, Art, and Curiosity in Early Modern Europe* (Princeton – Oxford: 2021).

Bass M.A., *Insect Artifice: Nature and Art in the Dutch Revolt* (Princeton – Oxford: 2019).

Belkin K.L., "Osias Beert", in Brickstock H. (ed.): *The Oxford Companian to Western Art* (online version 2003).

Bennema F., "De Haringkoning", *Het Zeepaard* 70.1 (2010) 15–18.

Biesboer P., *Collections of Paintings in Haarlem*, 1572–1745 (Los Angeles: 2002).

Boogert B. van den (ed.), *Rembrandts schatkamer*, exh. Cat., Het Rembrandthuis (Zwolle 1999).

Buvelot Q., "Slow Food: on the Rise and Early Development of Dutch and Flemish Meal Still Lifes", in Buvelot Q. (ed.), *Slow food. Dutch and Flemish meal still lifes 1600–1640*, exh. Cat. Mauritshuis (Zwolle: 2017) 13–32.

Duverger E., *Antwerpse kunstinventarissen uit de zeventiende eeuw*, vols. I–XIV (Brussels: 1984–2009).

Egmond F., *Het Visboeck. De wereld volgens Adriaen Coenen 1514–1587* (Zutphen: 2005).

Egmond F., *Eye for Detail. Images of Plants and Animals in Art and Science*, 1500–1630 (London: 2017).

Egmond F., "On Northern Shores: Sixteenth-Century Observations of Fish and Seabirds (North Sea and North Atlantic)", in MacGregor A. (ed.), *Naturalists in the Field. Collecting, Recording, and Preserving the Natural World in the Fifteenth to the Twenty-first Century* (Leiden – Boston: 2018) 129–148.

Egmond F. – Rijks M., "Depicting Fish in Early-Modern Venice and Antwerp", in Haar A. van der – Schulte Nordholt A. (eds.), *Figurations animalières à travers les textes et l'image en Europe. Du Moyen âge à nos jours. Essais en hommage à Paul J. Smith* (Leiden – Boston: 2022) 63–77.

Findlen P., *Possessing Nature. Museums, Collecting, and Scientific Culture in Early Modern Italy* (Berkeley – Los Angeles – London: 1994).

Göttler C., "The Place of the 'Exotic' in early seventeenth-century Antwerp", in Schrader S. (ed.), *Looking East: Rubens's Encounter with Asia* (Los Angeles: 2013) 93–94.

Guicciardini Ludovico, *Beschrijvinghe van alle de Nederlanden; anderssins ghenoemt Neder-Duytslandt* (Amsterdam, Willem Jansz: 1612 [first published 1567]).

Helmus L.M. (ed.), *Vis. Stillevens van Hollandse en Vlaamse meesters 1550–1700* (Utrecht: 2004).

Hendrix M.L., *Joris Hoefnagel and the "Four Elements": a Study in Sixteenth-Century Nature Painting* (Ph.D. dissertation, Princeton University: 1984).

Honig E.A., *Painting & the Market in Early Modern Antwerp* (New Haven – London: 1998).

Honig E.A., "Desire and Domestic Economy", *The Art Bulletin* 83.2 (2001), 294–315.

Jongh E. de, "De symboliek van vis, visser, visgerei en vangst" in Helmus L.M. (ed.), *Vis. Stillevens van Hollandse en Vlaamse meesters 1550–1700* (Utrecht: 2004) 75–120.

Jorink E., *Het 'Boeck der Natuere'. Nederlandse geleerden en de wonderen van Gods schepping 1575–1715* (Leiden: 2006).

Kaveler E.M., "Erotische elementen in de markttaferelen van Beuckelaer, Aertsen en hun tijdgenoten", *Joachim Beuckelaer. Het markt- en keukenstuk in de Nederlanden 1550–1650*, exh. Cat. Museum voor Schone Kunsten Gent (1986) 18–26.

Koslow S., *Frans Snyders. The Noble Estate: Seventeenth-Century Still-life and Animal Painting in the Southern Netherlands* (Brussels: 2006).

Lauwerier R.C.G.M. – J. Laarman F.J., "Hollandse Nieuwe en de mythe van Willem Beukelszoon", *Holland. Historisch Tijdschrift* 38 (2006) 150–160.

Margócsy D., *Commercial Visions. Science, Trade, and Visual Culture in the Dutch Golden Age* (Chicago: 2014).

Martens P., "Visserij en vishandel. De zalm van het Bergse veld", in Helmus L.M. (ed.), *Vis. Stillevens van Hollandse en Vlaamse meesters 1550–1700* (Utrecht: 2004) 121–138.

Meijer G.M., "Visstillevens in Holland en Vlaanderen", in Helmus L.M. (ed.), *Vis. Stillevens van Hollandse en Vlaamse meesters* 1550–1700 (Utrecht: 2004) 13–74.

Montias J.M., *Artists and Artisans in Delft. A Socioeconomic Study of the Seventeenth Century* (Princeton: 1982).

Richter C., "Hollandse vissenboeken. De onweerstaanbare kracht van de afbeelding", *Holland. Historisch Tijdschrift* 38 (2006) 161–176.

Rikken M., *Dieren verbeeld. Diervoorstellingen in tekeningen, prenten en schilderijen door kunstenaars uit de Zuidelijke Nederlanden tussen 1550 en 1630* (Ph.D. dissertation, Leiden University: 2016).

Rijks M., '"Unusual Excrescences of Nature". Collected Coral and the Study of Petrified Luxury in seventeenth-century Antwerp', *Dutch Crossing. Journal of Low Countries studies* 43 (2019) 127–156.

Rijks M., "A Painter, a Collector, and a Horseshoe Crab. Connoisseurs of Art and Nature in Early Modern Antwerp", *Journal of the History of Collections* 31 (2019) 343–361.

Rijks M., "Scales, Skins, and Carapaces in Antwerp collections", in Bol M – Spary E. (eds.), *The Matter of Mimesis: Studies on Mimesis and Materials in Nature, Art and Science* (Leiden – Boston: 2023) 295–320.

Sighem B. van, "Vis op het droge. Visstillevens in Holland en Utrecht in de 17de eeuw", *Holland. Historisch Tijdschrift* 38 (2006) 177–186.

Smith P.J., "Aldrovandi (Ulisse) (1522–1605)", in Nativel C. (ed.), *Centuriae latinae. Cent une figures de la Renaissance aux Lumières offertes à Jacques Chomarat* (Geneva: 1997) 57–63.

Timmermans B., *Patronen van patronage in het zeventiende-eeuwse Antwerpen: een elite als actor binnen de kunstwereld* (Amsterdam: 2008).

Trijp D. van, *Captured on Paper. Fish Books, Natural History and Questions of Demarcation in Eighteenth-Century Europe (ca. 1680–1820)* (Ph.D. dissertation, Leiden University: 2021).

Vergara A., "Reflections of Art and Culture in the Paintings of Clara Peeters", in Vergara A. (ed.), *The Art of Clara Peeters* (Antwerp – Madrid 2016) 13–47.

Weber G.J.M., "*'t Lof van den Pekelharingh*. Von altäglichen und absonderlichen Heringstilleben", *Oud Holland* 101.2 (1987) 126–140.

Wee H. van der, "Voeding en dieet in het Ancien Régime", *Spiegel Historiael. Maandblad voor geschiedenis en archeologie* 1.2 (1966) 94–101.

Winter J.M. van, "Visrecepten in laat-middeleeuwse en vroeg-moderne kookboeken", in Helmus L.M. (ed.), *Vis. Stillevens van Hollandse en Vlaamse meesters 1550–1700* (Utrecht: 2004) 139–153.

Wyssenbach S., "Riches of the Sea: Collecting and Consuming Frans Snijders's Marine Market Paintings in the Southern Netherlands", in Burghartz S. – Burkart L. – Göttler C. (eds.), *Sites of Mediation. Connected Histories of Places, Processes, and Objects in Europe and beyond, 1450–1650* (Leiden – Boston: 2016) 328–352.

Websites:

Baer, Ronni. "Herring Seller and Boy" (2017). In The Leiden Collection Catalogue, 2nd ed. Edited by Arthur K.

Wheelock Jr. New York, 2017–20. https://theleidencollection.com/artwork/the-herring-seller-and-boy/ (archived May 2020).

Circulation of Knowledge and Learned Practices in the 17th-century Dutch Republic: http://ckcc.huygens.knaw.nl/epistolarium/.

Egmond F., 'Books on Natural History' and 'Ludovicus Nonnius's Book on the Consumption of Fish (Antwerp, 1616)', on the website: http://ximenez.unibe.ch/library/natural/fish/ (accessed March 2021).

Montias database: https://research.frick.org/montias (accessed September 2019).

Biesboer data on website Getty Provenacnce Index (GPI): https://piprod.getty.edu/starweb/pi/servlet.starweb?path=pi/pi.web (accessed September 2019).

CHAPTER 9

Fishing in the Past: Biodiversity, Art History, and Citizen Science – Preliminary Results

Anne M. Overduin-de Vries and Paul J. Smith

1 Introduction

Fish are important; on the one hand, they are for human society a low-calorie, high-protein food source that also has medicinal value, and also are a source of income and employment, and on the other hand they are a key element for aquatic ecosystems. The importance of fish for human culture is reflected in art and literature. In the early modern period a considerable number of illustrated ichthyological books were published. They often comprised elaborate descriptions and illustrations of fish and other *aquatilia*. These were published not only in Latin but also in the vernacular (Italian, French, English, and German), stressing the importance of these books to a broad readership. Also, artists had access to this information and used these books as a source of information and inspiration for their works. Right after the first ichthyological books appeared, artists, mainly from Italy and the Southern Netherlands, gave a prominent role to fish in their drawings and etchings, often collected and issued in albums devoted entirely to fish (e.g. Giorgio Liberale, Adriaen Collaert, Nicolaes de Bruyn, and Joris and Jacob Hoefnagel).[1] In the early 17th century, detailed realistic oil paintings emerged in Antwerp by artists such as Frans Snyders and Alexander Adriaenssen and were further developed in the Northern Netherlands, for example by Abraham van Beyeren and Jacob Gillig. These oil paintings concerned representations of various topics, such as the element water, kitchen still lives, and market scenes, which became very popular.[2]

---

1 Rikken M., "Abraham Ortelius as Intermediary for the Antwerp Animal Trailblazers", *Jahrbuch für Europäische Wissenschaftskultur* 6 (2011) 95–128.

2 See Helmus L.M. (ed.), *Vis: Stillevens van Hollandse en Vlaamse meesters 1550–1700* (Utrecht: 2004); English translation: *Fish: Still Lifes by Dutch and Flemish Masters 1550–1700* (Utrecht: 2004). See also the article of Marlise Rijks in the present volume.

DOI:10.1163/9789004681187_010

Early modern artists inspired each other and copied certain motifs or scenes from each other. They also based their work on the descriptions or illustrations from the ichthyological books. Thus, Joris Hoefnagel was inspired by both the texts and illustrations of Guillaume Rondelet and Conrad Gessner,[3] whereas Jan Brueghel the Elder had Rondelet's work before him.[4] Apart from drawing inspiration from text and illustration, artists used their daily life experience as an inspiration source. Some painters may have had access to a real specimen, whether from a market, a collection, or a local fisherman.

If we want to know more about the practice of these painters and where they got their inspiration from, it is important that the species in the paintings be identified. When a large body of artwork with labelled species is studied in detail, it is possible to look at several depictions of the same species in books and artwork in order to investigate which works are inspired by second-hand information, who inspired whom and which artworks are based on real specimens. Moreover, it will give insight into the practice of these early modern artists like whether they prefer common species, rare or foreign species, freshwater or marine or combinations of both.

Although the setting of a painting is no guarantee that the depicted objects give a realistic view of the situation, they can provide some information. Thus, fish placed next to kitchenware, such as knives, plates, and bowls, implies consumption.[5] Therefore, paintings including these objects suggest the artist was using fish species meant for consumption as a source for his artwork.

That fish consumption was an important source of inspiration for the painters is confirmed by the exhibition *Vis* (Fish) that was held in 2004 in the Central Museum of Utrecht. This exhibition showed the paintings with fish images by Northern Netherlandish and Southern Netherlandish artists from 1550 to 1700.[6] For the selection of 63 paintings within this exhibition the pictured fish species were identified by Lex Raat.[7] From the 50 species that were represented in these paintings, fish species that were easy to catch with the equipment of that time were pictured most often and rare species did not occur in the paintings. Therefore, the authors of the exhibition catalogue

---

3 See Hendrikx S. – Smith P.J., "Connaissances ichtyologiques au format emblématique: le cas du *sargus*", *RursuSpicae. Transmission, réception et réécriture des textes, de l'Antiquité au Moyen Âge* (2022). http://journals.openedition.org/rursuspicae/2258 ; DOI: https://doi.org/10.4000/rursuspicae.2258.

4 See the Introduction of the present volume.

5 Stupples P., *Art and Food* (Cambridge: 2014).

6 Helmus, *Vis*.

7 Raat L., "Determinatie van de vissen op schilderijen", in Helmus, *Vis* 375–391.

conclude that the Dutch artists usually depicted fish that were available in the Dutch markets of that time.

This availability of fish fluctuates over time and may be noticeable in art when looking at trends of depicting certain fish species. Recent studies have concluded that fluctuations in marine ecosystems are due to overfishing, pollution, filling of estuaries and other shoreline modifications, the introduction of invasive species, global warming, ocean acidification, and other ecological impacts.[8] Therefore, before discussing the occurrence of fish in the visual arts, it is useful to briefly discuss the most important factors that have influenced the availability of fish, namely climate, human fish consumption, and human alterations of the landscape.

2 The Influence of Climate on Fish Biodiversity

Within the last millennium, two cooler periods occurred in the aquatic environment. The first occurred ca. 1400 AD and the second, more drastic one occurred around 1700 AD.[9] Climatic events not only change the water temperature, they also cause severe changes in ocean circulation and currents.[10] Although most adult fish can resist these changes, young individuals are more affected because their survival depends on the timing of algae blooms and zooplankton availability.[11] Indeed, genetic analysis of remains of the Icelandic cod population revealed that the population declined due to the change in climate around 1400.[12] Around 1700 the diversity of aquatic *invertebrae* and the amount of organic matter in the oceans were at minimal levels.[13] The drop in

8 Southward A.J. – Langmead O. – Hardman-Mountford N.J. – Aiken J. a.o., "Long-Term Oceanographic and Ecological Research in the Western English Channel", in Southward A.J. – Tyler P.A. – Young C.M. – Fuiman L.A. (eds.), *Advances in Marine Biology* 47 (2005) 1–105.

9 Luoto T.P. – Nevalainen L. – Sarmaja-Korjonen K., "Multiproxy Evidence for the 'Little Ice Age' from Lake Hamptrask, Southern Finland", *Journal of Paleolimnology* 40 (2008) 1097–1113.

10 Bianchi G.G. – McCave I.N., "Holocene Periodicity in North Atlantic Climate and Deep-Ocean Flow South of Iceland", *Nature* 397 (1999) 515–517.

11 Pepin P., "Effect of Temperature and Size on Development, Mortality, and Survival Rates of the Pelagic Early-Life Stages of Marine Fish", *Canadian Journal of Fisheries and Aquatic Sciences* 48 (1991) 503–518.

12 Olafsdottir G.A. – Westfall K.M. – Edvardsson R. – Palsson S., "Historical DNA Reveals the Demographic History of Atlantic Cod (*Gadus morhua*) in Medieval and Early Modern Iceland", *Proceedings of the Royal Society B-Biological Sciences* (2014) 281.

13 Luoto a.o., "Multiproxy Evidence for the 'Little Ice Age'".

sea temperature had a negative impact on the Atlantic cod population in the 16th century.[14] Another example of the effect of climate change on the extinction of a fish species comes from the greater weever (*Trachinus draco*), which was abundant in the southern North Sea at the end of the 19th and beginning of the 20th century, but has disappeared almost completely[15] since the strong winter of 1963.[16] A difference in effect of climate change is expected between fish with different reproductive tactics, *i.e.* fish that lay many small eggs or fish that lay fewer large eggs. In theory, fish that spawn larger numbers of eggs (mostly pelagic spawners) are better capable of dealing with environmental change than species with low numbers of eggs (demersal spawners).[17] Therefore, it is expected that after 1400 and after 1700 there was a drop of demersal spawning species. It would be interesting to see if this drop and the drop of cod and greater weever are also visible in the occurrence of these species in paintings.

3 Fisheries, Fish Trade, and Consumption through Human History

Fishing is an ancient activity in human culture that already started from 300,000 BC, after which its intensity increased until 5700 BC, when humans started to have an impact on marine ecosystems.[18] The relative importance of marine and freshwater fish has fluctuated during human history. This predicts changes in human pressures that are different for both ecosystems.

The importance of fish in the human diet fluctuates over time, as does the relative importance of different species. In early medieval Europe (5th–9th centuries AD) mostly freshwater fish were consumed, but there was a rise in marine

14 Geffen A.J. – Hoie H. – Folkvord A. a.o., "High-Latitude Climate Variability and Its Effect on Fisheries Resources as Revealed by Fossil Cod Otoliths", *ICES Journal of Marine Science* 68 (2011) 1081–1089.

15 Daan N. – Bromley P.J. – Hislop J.R.G. – Nielsen N.A., "Ecology of North-Sea Fish", *Journal of Sea Research* 26 (1990) 343–386.

16 Bennema F.P. – Rijnsdorp A.D., "Fish Abundance, Fisheries, Fish Trade and Consumption in Sixteenth-Century Netherlands as Described by Adriaen Coenen", *Fisheries Research* 161 (2015) 384–399.

17 Duarte C.M. – Alcaraz M., "To Produce Many Small or Few Large Eggs – A Size-Independent Reproductive Tactis of Fish", *Oecologia* 80 (1989) 401–404.

18 Erlandson J.M. – Rick T.C., "Archaeology Meets Marine Ecology: The Antiquity of Maritime Cultures and Human Impacts on Marine Fisheries and Ecosystems", *Annual Review of Marine Science* 2 (2010) 231–251.

fish consumption from the 10th century.[19] In the early modern (from 1500) period, the importance of fish in human diets increased due to the Christian calendar which prohibited eating meat during the fasting period of 40 days and weekly on Fridays, but fish consumption was allowed. Mainly dried fish was consumed and freshwater fish. In this period also ponds with fresh water fish appeared mostly attached to monasteries. Only rich people could afford marine fish.[20] Freshwater species: perch (*Perca fluviatilis*), carp (*Cyprinus carpio*), bream (*Abramis brama*), and pike (*Esox lucius*) were most popular. Marine and diadromous species: sole (*Solea solea*), flounder (*Platichthys flesus*), turbot (*Scophthalmus maximus*), halibut (*Hippoglossus hippoglossus*), tench (*Tinca tinca*), bleak (*Alburnus alburnus*), eel (*Anguilla anguilla*), sea lamprey (*Petromyzon marinus*), salmon (*Salmo salar*), trout (*Salmo trutta*), tuna spp., mackerel (*Scomber scombrus*), sturgeon (*Acipenser sturio*) (eggs), cod (*Gadus morhua*), herring (*Clupea harengus*), sardine (*Sardina pilchardus*), anchovy (*Engraulis encrasicolus*), shark spp., and ray spp. were consumed in smaller amounts.[21]

Around 1500 a major change in fisheries occurred. From 1300 to 1500 cod was mainly consumed fresh from the sea locally on the coast. After John Cabot discovered the immense cod stock at Newfoundland around 1500, fisheries from Norway, Iceland, Spain, Portugal and France crossed the Atlantic Ocean, which probably led to a 15-fold catch volume of cod.[22] Although the Low Countries took part in the trade of Newfoundland fish,[23] it is unknown how much they contributed to the Newfoundland fisheries. At the same time, the cod population in Iceland declined due to the change in climate during the Little Ice Age.[24] In 1991 the cod stock at Newfoundland collapsed and it is unknown how much both anthropogenic harvesting and climate change contributed to this collapse. In fact, this accounts for most of the changes in the abundance of fish species. Quantitative records of European fisheries before 1750 are few.[25]

19 Ervynck A. – Boudin M. – van den Brande T. – Van Strydonck M., "Dating Human Remains from the Historical Period in Belgium: Diet Changes and the Impact of Marine of Marine and Freshwater Reservoir Effects", *Radiocarbon* 56 (2014) 779–788.

20 Ibidem.

21 Albala K. – Allen R.W., *Food in Early Modern Europe* (Westport, Connecticut – London: 2003).

22 Holm P. – Ludlow F. – Scherer C. – Travis C. et al., "The North Atlantic Fish Revolution (ca. AD 1500)", *Quaternary Research* (2019) 1–15.

23 Glerum-Laurentius D., *A History of Dutch Activity in the Newfoundland Fish Trade from about 1590 till about 1680* (Master's thesis, Memorial University of Newfoundland: 1960).

24 Olafsdottir a.o., "Historical DNA".

25 Michell A.R., "The European Fisheries in Early Modern History", in C.H. Wilson C.H – Rich E.E. (eds.), *The Economic Organization of Early Modern Europe*. [*The Cambridge Economic History of Europe: Volume 5*] (Cambridge: 1977) 133–184.

Most fisheries data are country based and it is difficult to calculate total output rates of certain fishing grounds.[26] Therefore, it is important that there be more data available from this time period about the occurrence of other species and fluctuations in the consumption of fish species, including cod, across countries.

Nowadays, marine fish consumption has surpassed freshwater fish consumption. The European marine fish consumption in 2013 was 1.8 million tonnes, approximately seven times higher than the 0.25 million tonnes of freshwater fish consumption.[27]

Typically human fisheries first target relatively large and long-lived carnivores (whales, cod, tuna, etc.). These large-sized species are characterized by late maturation and slow growth rates that make them particularly sensitive to human impact.[28] After these fisheries lead to large species decline or collapse, they switch to smaller species (herring, lobster, shrimp, etc.). This change represents a switch from higher trophic level species to the species at lower levels.[29] It is obvious that the human impact on commercial species is considerable, but there are also side effects, since predators at high trophic levels have a disproportionate influence on the occurrence of organisms at lower trophic levels.[30] Moreover, non-commercial species are affected too when taken incidentally as by-catches, by poaching or ghost fishing by lost or abandoned gear.[31] In the Gulf of Maine repetitive shifts in targeted species eventually have led to a trophic dysfunctional ecosystem and an accelerated decline in average trophic level[32] known as the trophic cascade.[33] There is debate on

26 Holm a.o., "The North Atlantic Fish Revolution (ca. AD 1500)".

27 Food and Agriculture Organization of the United Nations (FAO) FOA statistics. Accessed January 2021 from https://www.fao.org/faostat/en/#data/FBS.

28 Brander K., "Disappearance of Common Skate Raia-Batis from Irish Sea", *Nature* 290 (1981) 48–49; Jennings S. – Reynolds J.D. – Mills S.C., "Life History Correlates of Responses to Fisheries Exploitation", *Proceedings of the Royal Society B-Biological Sciences* 265 (1998) 333–339.

29 Pauly D. – Trites A.W. – Capuli E. – Christensen V., "Diet Composition and Trophic Levels of Marine Mammals", *ICES Journal of Marine Science* 55 (1998) 467–481.

30 Worm B. – Barbier E.B. – Beaumont N. – Duffy J.E. a.o., "Impacts of Biodiversity Loss on Ocean Ecosystem Services", *Science* 314 (2006) 787–790; Hairston N.G. – Smith F.E. – Slobodkin L.B., "Community Structure, Population Control, and Competition", *American Naturalist* 94 (1960) 421–425; Paine R.T., "Food Web Complexity and Species Diversity", *American Naturalist* 100 (1966) 65–75; idem, "Food Webs – Linkage, Interaction Strength and Community Infrastructure – the 3rd Tansley Lecture", *Journal of Animal Ecology* 49 (1980) 667–685.

31 Dayton P.K., "Ecology – Reversal of the Burden of Proof in Fisheries Management", *Science* 279 (1998) 821–822.

32 Steneck R.S. – Vavrinec J. – Leland A.V., "Accelerating Trophic-Level Dysfunction in Kelp Forest Ecosystems of the Western North Atlantic", *Ecosystem* 7 (2004) 323–332.

33 Jensen O.P. – Branch T.A. – Hilborn R., "Marine Fisheries as Ecological Experiments", *Theoretical Ecology* 5 (2012) 3–22.

whether fishing down the trophic level is a global phenomenon or whether it is restricted to certain geographical regions or periods in time.[34] It would be interesting to see if these effects are seen in other regions and with other species, but since ecosystems are often more complex than that of the Gulf of Maine, more longitudinal studies are necessary to confirm this. On a global scale, there is a drop in the average trophic level of fisheries catch visible from 1950 until 2000, especially in the North Atlantic Ocean (Millennium Ecosystem Assessment 2005).[35] Some indication of a decline of high trophic species comes from a comparison of data from the Dutch fisheries in the 16th and 19th centuries.[36] This study concluded that mainly large species (common smooth hound (*Mustelus mustelus*), common skate (*Dipturus batis*), common stingray (*Dasyatis pastinaca*), blonde ray (*Raja bracyura*), sturgeon (*Acipenser sturio*), pollack (*Pollachius pollachius*), saithe (*Pollachius virens*), and ling (*Molva molva*)) started to disappear in the 19th century. Although the trophic levels of these species are rather high and they are all piscivorous, the trophic level varies between the species (TL on a 1–5 range: starry smooth hound, 3.6 ± 0.3; common smooth hound, 3.8 ± 0.3; skate, 3.5 ± 0.6; common stingray, 4.1 ± 0.63; blonde ray, 3.8 ± 0.61; sturgeon, 3.5 ± 0.51; pollack, 4.3 ± 0.3; saithe, 4.3 ± 0.4; ling, 4.4 ± 0.2).[37] If the artists from the Low Countries were using the fish that were available at the markets it is expected that the mean trophic level of fish species would depict decreases over time.

Which fish species were caught, traded, and consumed in Holland in the early modern period (starting from 1500) is nicely described by Coenen.[38] Characterizing the importance of the herring and plaice trade for Holland is Coenen's description of it: 'the golden mountain of Holland'. Dutch fisheries were active along the Dutch coast and stretched far beyond the Shetland Isles, while in the south, along the French coast, mackerel was targeted. Fresh fish was sold in all Dutch cities and Antwerp, Brussels, Leuven, and Mechelen. Dried fish was exported to Germany. Species which were often consumed in the early modern period were herring, sole, flounder, and cod. Cod was caught in large volumes. To a lesser extent, sturgeon, lesser-weever (*Echiichthys vipera*), and small ray species were consumed by certain parts of the population, although ray was

34 Branch T.A. – Watson R. – Fulton E.A. a.o., "The Trophic Fingerprint of Marine Fisheries", *Nature* 468 (2010) 431–435.

35 Millennium Ecosystem Assessment, 2005. Ecosystems and Human Well-being: Synthesis. Island Press, Washington, DC.

36 Bennema – Rijnsdorp, "Fish Abundance".

37 Froese R. – Pauly D. (eds.), FishBase. World Wide Web electronic publication. 2022. www.fishbase.org.

38 Egmond F., *Het Visboek: de wereld volgens Adriaen Coenen (1514–1587)* (Zutphen: 2005).

more often exported to Germany. Herring and European plaice (*Pleuronectes platessa*) were also exported in large quantities. In contrast with the substantial consumption of freshwater fish and the exclusivity of marine fish in early modern Europe,[39] in Holland (and Flanders) the consumption of freshwater fish seems to have been relatively inferior to that of marine fish if one relies on Coenen. He does mention that freshwater fishery was a valuable source of employment,[40] but few species are mentioned as consumed species. Bream is mentioned as food for the common people, the rich, and the wealthy; eel was consumed by everyone; and the anadromous viviparous eelpout (*Zoarces viviparus*) was consumed by the poor and common people. Other freshwater or anadromous fish were exclusively consumed by the rich and the wealthy (salmon, sturgeon, catfish (*Silurus glanis*), and lampern), or by the poor (smelt (*Osmerus eperlanus*) and ruffe (*Gymnocephalus cernua*)). Although perch, carp, bream, and pike were consumed at high rates in early modern Europe,[41] these species are not mentioned as consumed species in Coenen's fish book.[42] This difference may just be a bias caused by Coenen's interest in spectacular marine species, or it may reflect a real difference in fish consumption between the Low Countries and the rest of Europe. Possibly, Dutch fisheries facilitated access to fresh marine fish for the common people, creating a difference between Dutch fish consumption and that of the rest of Europe. However, other sources indicate that until the late 16th century the Dutch consumption of freshwater fish was more important than that of herring and other marine fish.[43] Clearly there was a shift in popularity towards less freshwater and more marine fish consumption at some point after 1500, but more research is needed to confirm the details about the timing of this shift, differences between the Low Countries and the rest of Europe, and whether in the Low Countries there were different freshwater species consumed than in the rest of Europe. Labelling fish species in early modern paintings may help in studying these trends in fish consumption. It is expected that at some point after 1500 the relative proportion of artwork with freshwater fish in a food context would drop compared with that of marine fish.

39 Ervynck a.o., "Dating Human Remains".
40 Bennema – Rijnsdorp, "Fish Abundance".
41 Ervynck a.o., "Dating Human Remains".
42 Bennema – Rijnsdorp, "Fish Abundance".
43 van der Woude A.M. – de Vries J., *Nederland 1500–1815. De eerste ronde van moderne economische groei* (Amsterdam: 1995).

4 How Crowdsourcing Can Be Used to Label Artwork

Paintings with fish are quite numerous. If one only considers Dutch and Flemish artists, there are already more than 2200 pieces of art involving fish from the collection of the Rijksmuseum, Amsterdam, and the online image database of the RKD (Rijksbureau voor Kunsthistorische Documentatie), The Hague. Several of these works include multiple species. It would take too much time for a single scientist to identify all the species on all these pieces of art. Moreover, identifying species from pictures of varying quality is trivial even for specialists and scientists, who do not always agree on the identification of a pictured fish.[44] Therefore, citizen science is a helpful tool, not only to speed up the process of labelling species but also to improve the accuracy. Citizen science is a proven effective instrument in studies investigating the species composition of wildlife, by labelling species from camera traps.[45] Although the answer from a single untrained, non-expert volunteer may be incorrect, aggregated answers of multiple volunteers give reliable data. In a large-scale study with 1.51 million African wildlife images, the aggregated answers of 28,000 volunteers were correct in 98% of the images when compared to the consensus answers of experts.[46] In fact, aggregated answers of volunteers were more reliable than that of a single expert.[47] Although labelling African wildlife may be easier than labelling fish species, this difficult task is suitable for citizen science as well. When untrained volunteers were asked to label fish species from video footage, the agreement between aggregated volunteer answers and the expert ratings was equal to the agreement between expert ratings (He et al. 2016).[48]

If species in artwork are labelled and the data are shared in open access databases or publications, scientists from various disciplines may use the information for their research.

---

44 He J. – Spampinato C. – Boom B.J. – Kavasidi I., "Data Groundtruthing and Crowdsourcing", in Fisher R.B. – Chen-Burger Y.-H. – Giordano D. a.o. (eds.), *Fish4Knowledge: Collecting and Analyzing Massive Coral Reef Fish Video Data* (n.p.: 2016) 207–227.

45 Swanson A. – Kosmala M. – Lintott C. – Packer C., "A Generalized Approach for Producing, Quantifying, and Validating Citizen Science Data from Wildlife Images", *Conservation Biology* 30 (2016) 520–531.

46 Swanson a.o., "A Generalized Approach".

47 Swanson a.o., "A Generalized Approach".

48 He a.o., "Data Groundtruthing and Crowdsourcing".

5 Results and Discussion

From the original set of 2272 subjects (RKD and Rijksmuseum combined), 1676 subjects (74%) were positively judged to contain identifiable fish and entered into the fish identification task. The subjects that were retired concerned, according to the volunteers, artwork with fish that lacked the amount of detail necessary for identification (16%), fantasy fish (7%), or artwork in which no fish could be found (4%).

The reliability of the identifications made by the crowd in the current data set does not allow for labelling individual artwork on the species level (see ANNEX). Nevertheless, when compared with expert identifications, the fish species identified by the crowd were correct in more than half of instances. Therefore, if the Zooniverse project is continued for a longer period, increasing the number of classifications per object it is possible to make a selection of artworks with a certain level of reliability based on Pielou's index and the fraction of support (see ANNEX). With the current data set only a fraction of the artwork is reliably labelled. For the majority of artwork, the volunteers' answers are not reliable enough to say whether a pictured fish was cod or pollock, but if the consensus species is that it is a cod, it is very likely a species from the Gadidae family and almost certainly a marine fish. Therefore, it is still possible to look at larger trends, such as differences between centuries, or comparing groups of fish species, such as freshwater versus marine fish.

6 Which Species Are Depicted in Early Modern Art?

Consensus species per artwork are reported in the online available data set https://doi.org/10.6084/m9.figshare.19501324. There were 66 different species identified as consensus species from the artworks. The species that was most often identified was cod (*Gadus morhua*) (186 works), followed by pike (*Esox lucius*) (175 works), carp (*Cyprinus carpio*) (149 works), European perch (*Perca fluviatilis*) (123 works), and European flounder (*Platichthys flesus*) (110 works).

From 1485 until 1900 the proportion of artwork with only freshwater or only marine fish was almost equal, with a slightly higher number of marine species, and stayed constant over time [Fig. 9.1]. After 1900 the proportion of purely freshwater fish paintings increased, while that of purely marine species decreased. The proportion of artwork that depicted both marine and freshwater species in one work decreased throughout the early modern period [Fig. 9.2].

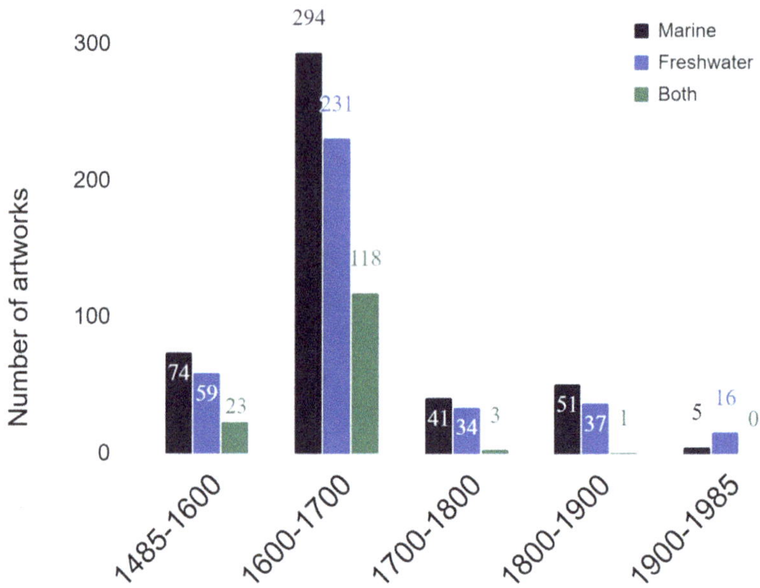

FIGURE 9.1 The number of pieces of artwork that pictured freshwater fish species only, marine fish species only, or both types of fish in one work. Numbers above bars indicate the exact number of paintings in each category

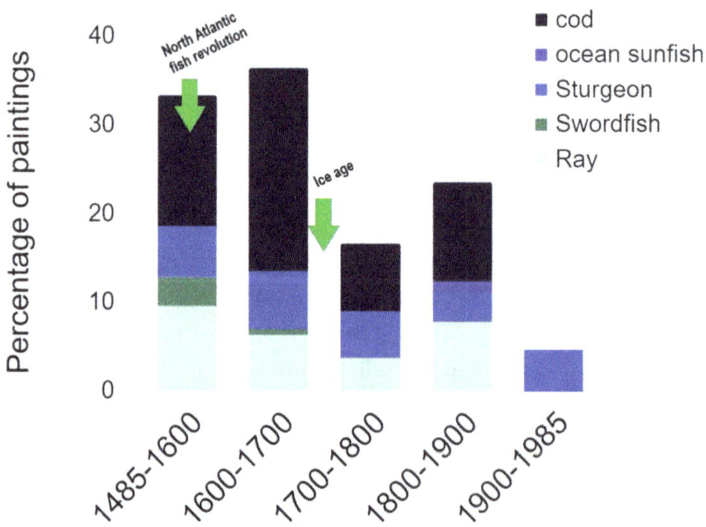

FIGURE 9.2 The proportion of artwork (calculated as the number of paintings depicting a certain species divided by the total number of artworks with fish in that period) that pictured large higher trophic level species

After the discovery of the Newfoundland cod stock in 1500, the proportion of cod in artwork increased from the 15th to the 16th century (see Fig. 9.2). After the Little Ice Age around 1700, the proportion of large-sized fish decreased from 36% in 17th-century art to 17% of 18th-century art (see Fig. 9.2). After the 19th century large higher trophic level species were seldomly pictured except for the sturgeon. The decline of larger fish species in 20th-century art coincides with the global drop in the average trophic level of fishery catches visible from 1950,[49] and it comes after the drop in Dutch fisheries in the 19th century.[50] This indicates that the depiction of these species in art was related to the availability in the markets of that time. Surprisingly, the proportion of sturgeons stays relatively constant throughout the centuries compared with the other large fish species, varying from 3.4% in the 18th century to 6.5% of the works in the 16th century. This was not expected because European sturgeon (*Acipenser sturio*) populations declined from the mid-19th century and disappeared from the Netherlands starting in 1952.[51] Also the morphologically identical *A. oxyrinchu* became extinct in the North Sea from the 19th century.[52] This indicates that artists from the Low Countries were less restricted by what was available at local markets than was expected when judging from the conclusions from the Utrecht exhibition. Various explanations are possible. First, artists may have used specimens from overseas markets or collections that they visited or that were sent to them. Second, they may have been inspired by sources other than real specimens, such as fish books, older works, etc. Third, there may be a bias for appealing species in art. Sturgeons are remarkable for the osseous plates on their skin, which is an interesting study object for artists.

When looking at artwork in a food-related context there is a clear visible shift in the proportion of freshwater and marine species [Fig. 9.3]. In the 15th and 16th centuries, freshwater fish are pictured in a food-related context in 40% of the artworks involving freshwater fish, compared to 20% of the marine fish artworks. This is in contrast with the low importance of freshwater fish in the Low Countries reported by Coenen.[53] But in line with other sources[54] and

49 Millennium Ecosystem Assessment, 2005. Ecosystems and Human Well-being: Synthesis. Island Press, Washington, DC.

50 Bennema – Rijnsdorp, "Fish Abundance".

51 Williot P. – Rochard E. – Castelnaud G. – Rouault T. a.o., "Biological Characteristics of European Atlantic Sturgeon, *Acipenser sturio*, as the Basis for a Restoration Program in France", *Environ Biol Fish* 48 (1997) 359–372.

52 Spikmans F. – Kranenbarg J. – Veenvliet P. – van Emmerik W. a.o., *Standaardlijst namen zoetwatervissen van Nederland en Vlaanderen anno 2019. Achtergronddocument. Stichting RAVON* (Nijmegen: 2019).

53 Bennema – Rijnsdorp, "Fish Abundance".

54 van der Woude – de Vries, *Nederland 1500–1815*.

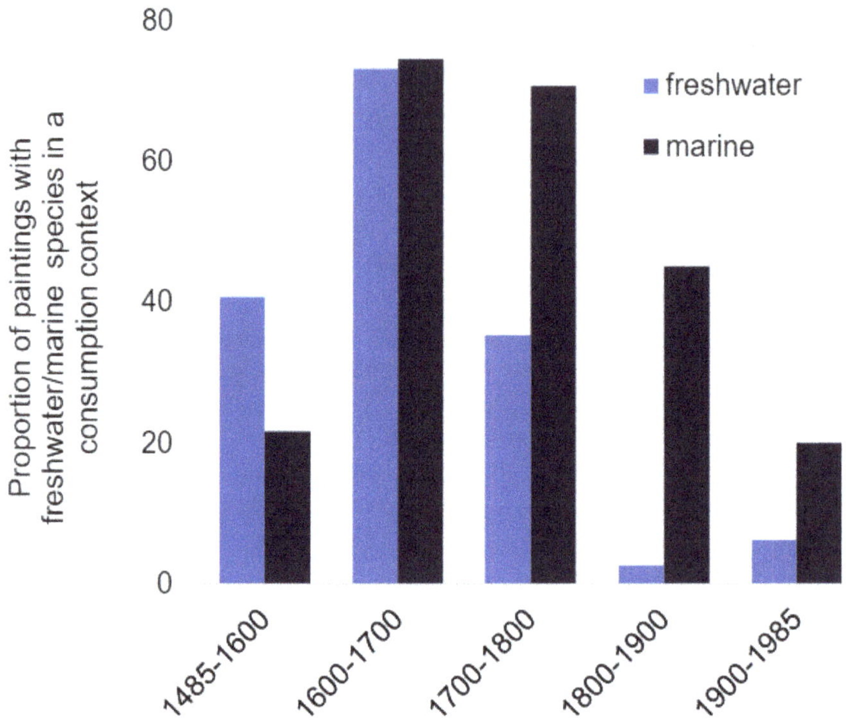

FIGURE 9.3 Proportion of freshwater/marine fish in a consumption-related context

the trends in other parts of Europe where freshwater fish formed an important part of the diet in the 15th and 16th centuries.[55] This suggests that the fish consumption trends in Holland were more similar to the rest of Europe than it appears from Coenen's fish book. The proportions of artwork in a food context increased for both freshwater and marine fish after 1600. This rise may be caused by a growing number of consumption-related artworks in general or a rise in the popularity of fish in the human diet. Not much is known about the fish consumption in the Low Countries in the 17th century. Comparisons with other food items in artwork, such as meat or cheese, may reveal whether there was a change in diet or a general increased interest in producing food-related art. In the 18th to the 20th century marine fish is more often depicted in a food context than freshwater fish. This is in line with the shift in the human diet from freshwater to marine fish. Remarkably, in the 20th century fresh fish were more often seen in artworks than marine fish were, but when looking at the proportion of paintings in a food context, that of marine fish is larger.

55 van der Woude – de Vries, *Nederland 1500–1815*.

Therefore, we can conclude that the overall proportion of fish species in artwork is not merely a representation of the species that were consumed at that time. Only when taking the context into account do the trends in artwork reflect the trends in the human diet.

7 Conclusion

Identification of fish species in artworks by the general population is not as reliable as the identification of, for example, African wildlife from wild traps. There is not much agreement about fish species between volunteers who scored the same artwork. With the current data set, only a fraction of the subjects is reliably labelled.

Nevertheless, when looking at the larger picture, such as the presence of marine, freshwater, or larger higher trophic level species in the artwork, some interesting trends are noticeable. These trends in fish art correspond with major environmental changes and changes in the human diet. Freshwater species in paintings increased after the 19th century. Large fish species are less often depicted after the second Little Ice Age; most of these species are no longer seen in art after 1900. Sturgeons remain constant in paintings throughout the centuries. When looking at artwork involving a consumption-related context, the proportion of freshwater/marine fish changes over time from mostly freshwater in the 15th and 16th centuries to mostly marine in the 18th to the 20th century.

This Zooniverse project not only shows the dependence of the visual arts on the fluctuating ichthyological biodiversity through the centuries – thanks to the online database,[56] this project also provides a tool for motif research in art history. The database makes it possible to localize fish species in a large body of artwork. How important species identification – both botanical and zoological – can be for interpretating works of art is apparent from recent studies.[57] Moreover, crowdsourcing projects like this, where volunteers identify species in art, have a general recruiting function in the context of interdisciplinarity. It makes art historians aware of the importance of flora and fauna

56 Overduin Anne, Results for the Zooniverse Fishing in the past project. figshare. Data set. 2022. 10.6084/m9.figshare.19501324.

57 See, for instance, Segal S. – Alen K., *Dutch and Flemish Flower Pieces. Paintings, Drawings and Prints up to the Nineteenth Century* (Leiden – Boston: 2020), and Rikken M. – Smith P.J., "Jan Brueghel's *Allegory of Air* (1621) from a Natural Historical Perspective", *Netherlands Yearbook for History of Art* 61 (2011) 87–115.

and biologists aware of the importance of visual arts. Lastly, but importantly, it brings the non-specialized crowd in contact with both art and nature.

ANNEX: Methods and Reliability of the Data[58]

Methods

Within this study we selected artwork depicting fish from Dutch and Flemish artists and labelled the species by means of an online Zooniverse[59] citizen science project.

Selection of Artwork

The two major sources of information on Dutch and Flemish art are the collection of the Rijksmuseum, Amsterdam and the online database of the RKD. From the RKD database all the works with the standardized keyword ("onderwerpstrefwoord") labelled "vis" (Dutch for "fish") were selected by Reinier van 't Zelfde, connected as information architect at the RKD. This resulted in 1961 hits. After a quick visual scan of the images by Anne Overduin, most irrelevant pictures were removed, and 1895 subjects from RKD were added to the list of subjects in the Zooniverse project "fishing in the past". Rijksmuseum subjects were selected by Anne Overduin using the Application Programming Interface (API) from the Rijksmuseum, which enables users to make a selection of artwork based on certain criteria. The following selection criteria were used: time frame, 1500–1880; type of work, painting, drawing, etching, or engraving; origin, Dutch; keywords, fishes, bony fishes, other fishes, deep sea fishes, eels, cartilaginous fishes, fishes (with NAME), and the Dutch keyword "vissen" (plural for fish). This resulted in 606 records from the Rijksmuseum database. These hits included some photographic duplicates from the same artwork as well as some artwork not from the Low Countries (since selection on origin was not possible for paintings and drawings), and irrelevant pictures with, for example, fantasy creatures or market scenes without clearly distinguishable fish. After a visual scan of the images, these irrelevant records were removed and 377 unique pieces of artwork were entered as subject in the Zooniverse project. In total there were 2272 subjects added to the project, including 1594 paintings, 301 drawings, 294 etchings or engravings, and 83 other types of work. Each piece of art was represented by 1–5 (mean: 1.1 ± SD 0.4) pictures. All pictures

58 The innovative nature of our Zooniverse project makes it necessary to provide in this Annex an extensive report on the project's method and execution.

59 https://www.zooniverse.org/about.

that belonged to a single artwork were presented to the volunteers in a single subject, which received a single classification per volunteer.

Selection of Common Species

The main task for the volunteers was to identify species from a predefined species list. Facultatively, they could also manually add other species in a second task (see next paragraph). We expected that most of the fish painted by Dutch and Flemish painters were species that have commercial value and that occur in European waters. Therefore, we composed a list of fish species based on the list of European commercial fish species from the EUMOFA (European Market Observatory for fisheries and aquaculture).[60] A total of 59 species from the EUMOFA list were selected.

The number of freshwater species on the EUMOFA list is minimal, probably because currently these fish are less interesting with regard to commercial use. Therefore, 12 freshwater species that are common in European waters were added. In order to include species that were present in the early modern time, we added species that were abundant in the Netherlands in the 16th century. We added the 14 species that were not yet in our list and that were reported as "common" or "plentiful" around 1600.[61] After a first trial period (21-11-2019 until 31-03-2020), 549 pieces of art were successfully classified by 155 volunteers, with 1–5 (mean 1.28 ± SD 0.59) classifications per subject. The 20 species that were only chosen for one painting or less in this trial period were removed from the choice list for consequent classifications.[62] Moreover, the 8 species that were added manually during this trial period and occurred on more than one painting were added to the list.[63]

In order to make identification of the species easier for the volunteers, pictures of the species were added to the citizen science project. For every species at least one picture depicted the entire lateral (or in case of angler and ray sp., dorsal) view of each fish. These pictures were also used as a thumbnail within the species list. The background of these pictures was removed such that even in the small-scaled thumbnails the outline of the fish was obvious. Additional pictures were provided for most species (N = 53), from other angles, with details or with another appearance of the species (e.g. juvenile). Apart from a short description of each species, easily confused species were listed.

60 EUMOFA. 2018. European market observatory for fisheries and aquatic products. Metadata 2 – Data management. ANNEX 1 List of Commodity groups and Main commercial species.

61 Bennema – Rijnsdorp, "Fish Abundance".

62 See the species table at https://doi.org/10.6084/m9.figshare.19497548.

63 See the species table at https://doi.org/10.6084/m9.figshare.19497548.

These included mostly species with the same body shape. Details were given on how to distinguish between each pair of often confused species. In order to enable untrained, non-expert volunteers to make a deliberate choice, all the species from the list could be filtered according to a number of characteristics. In the trial period these included overall body shape, colour, pattern, caudal fin shape, mouth shape, scale size, first dorsal fin shape, and number of dorsal fins. After the first trial period the characteristics "first dorsal fin shape" and "scale size" were removed, because scale size is not clearly defined, and fish are often pictured in their typical out-of-water appearance with clamped fins obscuring fin shape. A characteristic of a given species was not restricted to one option. For example, if a species is brown on the back and silver on the flanks, both colours applied to that species.

Workflow on Zooniverse

Selection of Subjects with Identifiable Fish

In the first trial period, volunteers first had to indicate whether the presented pictures included fish that could be identified. If they answered "yes", they could immediately identify the species; otherwise, they were taken to the next image.

After the first trial period, this first question was split off as a separate "fish or no fish" task. This allowed for the quick retirement of artworks. This task was completed by 392 volunteers from 13-3-2020 until 04-05-2020, when all subjects had been classified by at least two independent volunteers as picturing identifiable fish, or they received three negative classifications and were retired from the project.

Identification of Fish

The identification of fish species was done in two steps. In the first step, volunteers had to select which species of the preselected species list[64] were present on a particular subject. They could click on the species name from an alphabetically sorted list, or they could filter the species list based on their characteristics (body shape, colour, number of dorsal fins, colour pattern, shape of the mouth, shape of the caudal fin). Traditionally, identification of species is done according to a dichotomous key, where a series of questions in a predefined order leads to the correct species. The disadvantage of this system is that the order of questions is fixed and if one does not know the answer to one

64 See the species table at https://doi.org/10.6084/m9.figshare.19497548.

of the first questions, for example, because a crucial part of the animal is not visible, the determination is impossible. Therefore, we used a multi-entry key where users could decide for themselves which characteristic to start with and how many choices they made, before they started comparing the images of the species.

Once a species was chosen, volunteers could click on the name of an often confused species, which displayed the picture of the often confused species with details on how to discriminate between the two. For each of the selected species volunteers had to indicate how many individuals were present and whether they saw any indication of commercial use (whether the fish were cut, cooked, dried, or otherwise prepared; in a market; consumed (on a plate/ dining table); or none of the above). In the second step, volunteers could identify species that were not on the list, by marking a fish and entering a species, genus, or family name manually.

Selection of Volunteers

The higher the number of volunteers that rate a single picture, the higher the accuracy. With 28 volunteers per image 98% of correct labelling was achieved in the *Snapshot Serengeti* project.[65] With 5 volunteers per image already 90% of correct labelling was found in the same data set.[66] Common or easily recognizable species can be reliably labelled with only 2 or 3 volunteers, but rare, difficult, or undetailed pictured species may need 10 or more volunteers to achieve acceptable reliability values.[67] Similarly to automatically taken photos from camera traps, the paintings from our selection often depict only a piece of the fish, and there are also a lot of bad-quality images and roughly painted fish, which made identification difficult. Therefore, we expected the difficulty of labelling species in our project to be comparable to that of the *Snapshot Serengeti* project. We set the number of volunteers that rated each artwork to 14, and we expected the aggregated labels to be correct at values between 90% and 98%. Moreover, we calculated an evenness score for every single piece of art. That way we could select subjects with sufficient evenness scores from our data set to obtain a set of reliably labelled artworks. In the *Snapshot Serengeti* project, when discarding images with evenness higher than 0.5, 97% of images were correctly labelled when considering the answers of 5 volunteers per subject.

65 Swanson a.o., "A Generalized Approach".
66 Swanson a.o., "A Generalized Approach".
67 Swanson a.o., "A Generalized Approach".

Volunteers consisted of the general group of Zooniverse users that were already present at the Zooniverse community. Additionally, groups of people with interest in fish were attracted by using Instagram, Facebook, and articles in Dutch general journals (*Trouw* and *Nederlands Dagblad*)[68] and journals of specialized Dutch communities (RAVON, a conservation organization for Dutch reptiles, fish, and amphibians) and Sportvisserij Nederland (the Dutch sport fishing organization), *Vroege Vogels* (a popular Dutch public radio program),[69] and a presentation at Naturalis Biodiversity Centre, Leiden.

Aggregation of the Data

For subjects where at least 10 volunteers had selected at least one species from the list (N = 1029) the answers of the volunteers were combined in an aggregated answer. First, the number of consensus species N was calculated, as the median number of chosen species. For each of the species that was mentioned by the volunteers, the proportion of volunteers that recognized it was calculated. For each subject the species with the highest proportion of volunteers that recognized it were selected as a consensus answer from the top answer until the Nth species (rounded up in cases of a tie).

Calculation of Reliability

For individual subjects, two values are calculated: evenness and fraction support.[70] Evenness was calculated for all the subjects where at least 10 volunteers had selected at least one species from the list (N = 1029). For the calculation of evenness we used Pielou's evenness index:[71] $-(\sum_{i}^{S}=p_i \ln p_i)/\ln S$, where S is the number of different species reported by all volunteers, and p_i is the proportion of classifications received by species i. When all classifications were in agreement, we assigned a value of zero. The maximum value for this index is 1.0, indicating high disagreement among classifications.

For practical reasons, fraction support was calculated only for the pictures where the median of the number of species was equal to 1 (N = 565). It was calculated as the fraction of classifications that supported the aggregated answer.

Additionally, we have a selection of subjects that have been identified by specialists. We have a selection of subjects that were already identified by Lex

68 K. Moons in *Trouw*, 12 January 2020, and *Nederlands Dagblad*, 20 January 2020.

69 https://www.nporadio1.nl/nieuws/binnenland/08f664b8-e982-4c30-85d4-026b74b7a3d8/vissen-van-de-geschiedenis-herken-jij-de-vissen-op-het-schilderij.

70 Conform with Swanson a.o., "A Generalized Approach".

71 Pielou E.C., "Species-Diversity and Pattern-Diversity in the Study of Ecological Succession", *Journal of Theoretical Biology* 10 (1966) 370–383.

Raat for the 2004 Utrecht *Fish* exhibition.[72] Additionally, specialists from Ravon and Sportvisserij Nederland were asked to judge a selection of paintings within our Zooniverse project. For a total of 56 pieces of artwork we compared the identification of the specialists with the aggregated answer of the volunteers.

Reliability

Pielou's evenness index ranged from 0 (full agreement) to 1 (Fig. 9.1, mean, 0.90 ± 0.11; N = 1029), indicating that there was little agreement between volunteers about the identification of fish species [Fig. 9.4].

The fraction of support for the artwork with only one consensus species (N = 565) ranged from 0.1 to 1 (all supported the consensus species) (see Fig. 9.4) (mean, 0.43 ± 0.21), indicating that there was little support for the consensus answer.

When comparing the consensus answers from the volunteers with the identifications made by the experts (N = 56), the consensus answers agreed in 50% of the cases with the experts' answers. In 27% of the cases only some of the species from the volunteers corresponded with the expert identifications, and in 23% of the cases none of the species from the volunteers corresponded with the experts' annotations.

Therefore, we conclude that the classification of fish species from paintings by the general crowd is more complicated than we thought. Compared to the *Snapshot Serengeti* project, one would need more volunteers per subject in order to get an acceptable reliability (with an evenness <0.5) of the consensus

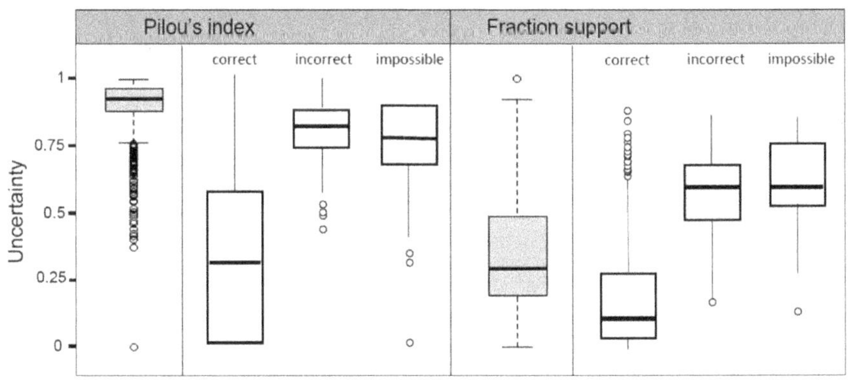

FIGURE 9.4 Boxplots of Pielou's evenness and Fraction Support in the current study (grey) compared to those in the *Snapshot Serengeti* project (white). Note that the inverse of fraction support is plotted in order to correspond with the direction of Pielou's evenness, i.e., a low number corresponds with low uncertainty

72 Raat, "Determinatie van de vissen op schilderijen".

answers. Therefore, with the current data set, identifications on the species level are only reliable for a fraction of the paintings (N = 20). For the other paintings with an evenness of >0.5 the identifications by the crowd cannot be used to label individual artworks. Nevertheless, when comparing the consensus answers of volunteers with those of the experts, the majority of species identifications made by the volunteers are correct. Probably, the volunteers recognized some of the fish, but it was often difficult to discriminate between several similar-looking species. However, if we look at the larger picture, focussing on averages across centuries and differences between particular groups of fish species, such as freshwater and marine fish, a certain margin of error is acceptable, and trends could be interesting. For example, the volunteers' answers are not reliable enough to say whether a pictured fish was cod or pollock, but if the consensus species is a cod, it is very likely a species from the Gadidae family and almost certainly a marine fish.

Acknowledgements

We are thankful to the NWO (Dutch Research Council) for financially making this project possible within the larger NWO project titled "A New History of Fishes. A long-term approach to fishes in science and culture, 1550–1880". We also thank the Naturalis Biodiversity Centre for hosting this project.

Bibliography

Albala K. – Allen R.W., *Food in Early Modern Europe* (Westport, Connecticut – London: 2003).

Bennema F.P. – Rijnsdorp A.D., "Fish Abundance, Fisheries, Fish Trade and Consumption in Sixteenth-Century Netherlands as Described by Adriaen Coenen", *Fisheries Research* 161 (2015) 384–399.

Bianchi G.G. – McCave I.N., "Holocene Periodicity in North Atlantic Climate and Deep-Ocean Flow South of Iceland", *Nature* 397 (1999) 515–517.

Branch T.A. – Watson R. – Fulton E.A. et al.,"The Trophic Fingerprint of Marine Fisheries", *Nature* 468 (2010) 431–435.

Brander K., "Disappearance of Common Skate Raia-Batis from Irish Sea", *Nature* 290 (1981) 48–49.

Daan N. – Bromley P.J. – Hislop J.R.G. – Nielsen N.A., "Ecology of North-Sea Fish", *Journal of Sea Research* 26 (1990) 343–386.

Dayton P.K., "Ecology – Reversal of the Burden of Proof in Fisheries Management", *Science* 279 (1998) 821–822.

De Vos J.M. – Joppa L.N. – Gittleman J.L. a.o., "Estimating the Normal Background Rate of Species Extinction", *Conservation Biology* 29 (2015) 452–462.

Duarte C.M. – Alcaraz M., "To Produce Many Small or Few Large Eggs – a Size-Independent Reproductive Tactis of Fish", *Oecologia* 80 (1989) 401–404.

Dudgeon D. – Arthington A.H. – Gessner M.O. a.o., "Freshwater Biodiversity: Importance, Threats, Status and Conservation Challenges", *Biological Reviews* 81 (2006) 163–182.

Duffy J.E. – Cardinale B.J. – France K.E. a.o., "The Functional Role of Biodiversity in Ecosystems: Incorporating Trophic Complexity", *Ecology Letters* 10 (2007) 522–538.

Egmond F., *Het Visboek: de wereld volgens Adriaen Coenen (1514–1587)* (Zutphen: 2005).

Erlandson J.M. – Rick T.C., "Archaeology Meets Marine Ecology: The Antiquity of Maritime Cultures and Human Impacts on Marine Fisheries and Ecosystems", *Annual Review of Marine Science* 2 (2010) 231–251.

Ervynck A. – Boudin M. – van den Brande T. – Van Strydonck M., "Dating Human Remains from the Historical Period in Belgium: Diet Changes and the Impact of Marine of Marine and Freshwater Reservoir Effects", *Radiocarbon* 56 (2014) 779–788.

EUMOFA. 2018. European market observatory for fisheries and aquatic products. Metadata 2 – Data management. ANNEX 1 List of Commodity groups and Main commercial species.

Fernandes P.G. – Cook R.M., "Reversal of Fish Stock Decline in the Northeast Atlantic", *Current Biology* 23 (2013) 1432–1437.

Food and Agriculture Organization of the United Nations (FAO) FOA statistics. Accessed January 2021 from https://www.fao.org/faostat/en/#data/FBS.

Geffen A.J. – Hoie H. – Folkvord A. a.o., "High-Latitude Climate Variability and Its Effect on Fisheries Resources as Revealed by Fossil Cod Otoliths", *ICES Journal of Marine Science* 68 (2011) 1081–1089.

Glerum-Laurentius D., *A History of Dutch Activity in the Newfoundland Fish Trade from about 1590 till about 1680* (Master's thesis, Memorial University of Newfoundland: 1960).

Gusick A.E. – Joslin T.L. – Brasket K.S. – Tennesen K.E., "Bias in Fish Bone Recovery from Differential Collection Techniques and Its Impact on Cultural interpretation in California", *Journal of Field Archaeology* 43 (2018) 507–521.

Hairston N.G. – Smith F.E. – Slobodkin L.B., "Community Structure, Population Control, and Competition", *American Naturalist* 94 (1960) 421–425.

Harmand S. – Lewis J.E. – Feibel C.S. – Lepre C.J. a.o., "3.3-million-year-old stone tools from Lomekwi 3, West Turkana, Kenya", *Nature* 521 (2015) 310–315.

He J. – Spampinato C. – Boom B.J. – Kavasidi I., "Data Groundtruthing and Crowdsourcing", in Fisher R.B. – Chen-Burger Y.-H. – Giordano D. a.o. (eds.), *Fish4Knowledge: Collecting and Analyzing Massive Coral Reef Fish Video Data* (n.p.: 2016) 207–227.

Helmus L.M. (ed.), *Vis: Stillevens van Hollandse en Vlaamse meesters 1550–1700* (Utrecht: 2004); English translation: *Fish: Still Lifes by Dutch and Flemish Masters 1550–1700* (Utrecht: 2004).

Hendrikx S. – Smith P.J., "Connaissances ichtyologiques au format emblématique: le cas du *sargus*", *RursuSpicae. Transmission, réception et réécriture des textes, de l'Antiquité au Moyen Âge* (2022) http://journals.openedition.org/rursuspicae/2258; DOI: https://doi.org/10.4000/rursuspicae.2258.

Holm P. – Ludlow F. – Scherer C. – Travis C. a.o., "The North Atlantic Fish Revolution (ca. AD 1500)", *Quaternary Research* (2019) 1–15.

Iwamoto E.M. – Myers J.M. – Gustafson R.G., "Resurrecting an Extinct Salmon Evolutionarily Significant Unit: Archived Scales, Historical DNA and Implications for Restoration", *Molecular Ecology* 21 (2012) 1567–1582.

Jakobsdottir K.B. – Pardoe H. – Magnusson A. – Bjornsson H. a.o., "Historical Changes in Genotypic Frequencies at the Pantophysin Locus in Atlantic Cod (*Gadus morhua*) in Icelandic Waters: Evidence of Fisheries-Induced Selection?" *Evolutionary Applications* 4 (2011) 562–573.

Jennings S. – Reynolds J.D. – Mills S.C., "Life History Correlates of Responses to Fisheries Exploitation", *Proceedings of the Royal Society B-Biological Sciences* 265 (1998) 333–339.

Jensen O.P. – Branch T.A. – Hilborn R., "Marine Fisheries as Ecological Experiments", *Theoretical Ecology* 5 (2012) 3–22.

Luoto T.P. – Nevalainen L. – Sarmaja-Korjonen K., "Multiproxy Evidence for the 'Little Ice Age' from Lake Hamptrask, Southern Finland," *Journal of Paleolimnology* 40 (2008) 1097–1113.

Michell A.R., "The European Fisheries in Early Modern History", in Wilson C.H – Rich E.E. (eds.), *The Economic Organization of Early Modern Europe* [*The Cambridge Economic History of Europe: Volume 5*] (Cambridge: 1977) 133–184.

Mihoub J.B. – Henle K. – Titeux N. – Brotons L. a.o., "Setting Temporal Baselines for Biodiversity: The Limits of Available Monitoring Data for Capturing the Full Impact of Anthropogenic Pressures", *Scientific Reports* 7 (2017) 11.

Moyle P.B. – Leidy R.A., "Loss of Biodiversity in Aquatic Ecosystems: Evidence from Fish Faunas", in Fiedler P.L. – Jain S.K. (eds.), *Conservation Biology* (New York: 1992) 128–169.

Olafsdottir G.A. – Westfall K.M. – Edvardsson R. – Palsson S., "Historical DNA Reveals the Demographic History of Atlantic Cod (*Gadus morhua*) in Medieval and Early Modern Iceland", *Proceedings of the Royal Society B-Biological Sciences* (2014) 281.

Paine R.T., "Food Web Complexity and Species Diversity", *American Naturalist* 100 (1966) 65–75.

Paine R.T., "Food Webs – Linkage, Interaction Strength and Community Infrastructure – the 3rd Tansley Lecture", *Journal of Animal Ecology* 49 (1980) 667–685.

Pauly D. – Trites A.W. – Capuli E. – Christensen V., "Diet Composition and Trophic Levels of Marine Mammals", *ICES Journal of Marine Science* 55 (1998) 467–481.

Pepin P., "Effect of Temperature and Size on Development, Mortality, and Survival Rates of the Pelagic Early-Life Stages of Marine Fish", *Canadian Journal of Fisheries and Aquatic Sciences* 48 (1991) 503–518.

Pielou E.C., "Species-Diversity and Pattern-Diversity in the Study of Ecological Succession", *Journal of Theoretical Biology* 10 (1966) 370–383.

Pimm S.L. – Jenkins C.N. – Abell R. – Brooks T.M. – Gittleman J.L. a.o., "The Biodiversity of Species and Their Rates of Extinction, Distribution, and Protection", *Science* 344 (2014) 1246752.

Raat L., "Determinatie van de vissen op schilderijen", in Helmus L.M. (ed.), *Vis: Stillevens van Hollandse en Vlaamse meesters 1550–1700* (Utrecht: 2004) 375–391.

Reid V.R. – Mooney H.A. a.o., *Ecosystems and Human Well-being: Synthesis* (Washington, DC: 2005).

Rikken M., "Abraham Ortelius as Intermediary for the Antwerp Animal Trailblazers", *Jahrbuch für Europäische Wissenschaftskultur* 6 (2011) 95–128.

Southward A.J. – Langmead O. – Hardman-Mountford N.J. – Aiken J. a.o., "Long-Term Oceanographic and Ecological Research in the Western English Channel", in Southward A.J. – Tyler P.A. – Young C.M. – Fuiman L.A. (eds.), *Advances in Marine Biology* 47 (2005) 1–105.

Spikmans F. – Kranenbarg J. – Veenvliet P. – van Emmerik W. a.o., *Standaardlijst namen zoetwatervissen van Nederland en Vlaanderen anno 2019. Achtergronddocument. Stichting RAVON* (Nijmegen: 2019).

Steneck R.S. – Vavrinec J. – Leland A.V., "Accelerating Trophic-Level Dysfunction in Kelp Forest Ecosystems of the Western North Atlantic", *Ecosystem* 7 (2004) 323–332.

Stupples P., *Art and Food* (Cambridge: 2014).

Swanson A. – Kosmala M. – Lintott C. – Packer C., "A Generalized Approach for Producing, Quantifying, and Validating Citizen Science Data from Wildlife Images", *Conservation Biology* 30 (2016) 520–531.

van der Woude A.M. – de Vries J., *Nederland 1500–1815. De eerste ronde van moderne economische groei* (Amsterdam: 1995).

Williot P. – Rochard E. – Castelnaud G. – Rouault T. a.o., "Biological Characteristics of European Atlantic Sturgeon, *Acipenser sturio*, as the Basis for a Restoration Program in France", *Environ Biol Fish* 48 (1997) 359–372.

Worm B. – Barbier E.B. – Beaumont N. – Duffy J.E. a.o., "Impacts of Biodiversity Loss on Ocean Ecosystem Services", *Science* 314 (2006) 787–790.

Worm B. – Branch T.A., "The Future of Fish", *Trends in Ecology & Evolution* 27 (2012) 594–599.

PART 3

Fish and Society in Europe

CHAPTER 10

Piscatorial Elements in 16th-Century Literature in Bruges: Fantasy Scenes and Compassionate Eulogies

Dirk Geirnaert

Being an important seaport town in the Middle Ages, it's not exactly surprising that Bruges, more than once and occurring still in the 16th century, was a place where literary texts with sometimes remarkable maritime and piscine elements were written. These texts may not immediately be of great importance for our insight into the specific ichthyological knowledge of that time, but from a cultural-historical point of view this use of marine themes provides an interesting phenomenon. In this contribution I will discuss four literary Brugean texts of which the piscatorial elements they hold are found in two different settings: (1) in a fairy-like, wondrous, or pseudo-mythological environment, and (2) in two eulogies singing the praises of fishermen. A nice by-catch of this research is that it also reveals some intertextual relationships, leading to a better understanding of the texts in question.

1 An Imaginary Marine World

The first Brugean author to present here is the rhetorician Cornelis Everaert (1480/85–1556). By profession he was a dyer, fuller, and clerk of the controlling authority in the cloth industry, of old one of the most important trades in Bruges. Everaert was a talented and prolific playwright who collected 35 of his plays in an autograph manuscript, now kept in the Royal Library in Brussels.[1]

In his *Esbatement vanden Visscher* (Farce of the Fisherman)[2] a fisherman and his wife share some thoughts on the tough but honourable task they have in society: the work is heavy and dangerous, but they find comfort in the fact that their job links them to the disciples of Christ who also earned their daily

1 Call number of this manuscript: Brussels, Royal Library, 19036. This autograph was edited twice: Muller J.W. – Scharpé L. (eds.), *Spelen van Cornelis Everaert* (Leiden: 1920) and Hüsken W.N.M. (ed.), *De spelen van Cornelis Everaert*, 2 vols. (Hilversum: 2005).

2 Hüsken, *De spelen*, 628–644; see also Muller, *Spelen*, 317–326 en 615–616.

bread by fishing. After this morale-boosting discussion (vs. 1–31), they put out to sea. Having arrived at their fishing grounds, the wife sees wondrous things beneath the waves (vs. 48–66): a shrimp is hunting a codfish, a crab is carrying a ray on his back, immobilizing the fish by holding it by its tail, an eel is hunting a halibut and a plaice, an oyster is trying to poke out the eyes of a haddock, and a sole is holding a speech.[3] This remarkable underwater scene, undoubtedly inserted here to amuse the spectators of the performance and to captivate their attention for what follows, is a nice example in Dutch popular literature of the so-called stringing together of impossibilities or *adynata*, according to E.R. Curtius a preliminary stage of the *mundus inversus*-topos, picturing a reality turned upside down.[4]

A second Brugean poet showing us an imaginary marine world is Everaert's fellow rhetorician Eduard de Dene (1505–ca. 1578). Contrary to Everaert, whom we may consider to be merely a local literary celebrity, De Dene surely was one of the important authors in the Low Countries in the 16th century. In 1561 he collected most of the poems he had written and penned them in a large manuscript of 466 folios, calling this collection *Myn Testament Rhetoricael* (My Last Will in Rhetorician Verse).[5] In the varied literary output found in this autograph and elsewhere, Eduard de Dene proves to be an interesting example of the 16th-century author, who more than once skilfully combines traditional ideas and methods with new insights and products of humanism and the Renaissance. Some striking cases in point here are the borrowings and imitations of François Rabelais in his own work and the fact that he wrote the texts for *De Warachtighe Fabulen der Dieren* (The Truthful Fables of the Animals – published in Bruges in 1567), a book that is often considered to be the first homegrown emblematic work in Dutch literature; in this collection of emblematic fables De Dene offers us also the first Dutch translations or adaptations of Alciato.[6]

---

3 The author introduces here a play on words, as the Dutch word for *sole* is *tong*, a homonym of *tong*, 'tongue'.

4 See Robert-Nicoud V., *The World Upside Down in 16th-Century French Literature and Visual Culture* (Leiden – Boston: 2018) 2. The reference is to Curtius E.R., *European Literature and the Latin Middle Ages*, transl. R. Trask (Princeton: 1973) 94–98.

5 For the edition of this work, see Coigneau D. – Waterschoot W. et al. (eds.), *Eduard de Dene, "Testament Rhetoricael"*, in *Jaarboek "De Fonteine"* 26, 28, 30 (Gent – Oudenaarde: 1976–1980); online: https://www.dbnl.org/tekst/dene001test01_01/index.php.

6 For further information on his importance, see Pleij H., *Het gevleugelde woord. Geschiedenis van de Nederlandse literatuur – 1400–1560* (Amsterdam: 2007) 385–393; Geirnaert D., "A Gift for Hanno. The Fictitious Booklist of Eduard de Dene", in Pouey-Mounou A.-P. – Smith P.J. (eds.), *Early Modern Catalogues of Imaginary Books. A Scholarly Anthology* (Leiden – Boston: 2020) 112–134 (here 112–114).

De Dene had an active social life. He was the *factor*, i.e. artistic leader, of the Brugean Chamber of Rhetoric *De Drie Sanctinnen* (the Three Female Saints), and he was a frequent visitor of the local archers' guild *Sint Sebastiaen*. Meant as a gift to this fraternity, he wrote a long ballad (7 stanzas, 140 vs.) titled *Tcaeckebeen* (The Jawbone),[7] explaining the presence of a peculiar piece of furniture in the meeting room of the fraternity: a large table made of the bones of a whale's head. In this ballad De Dene describes a fictitious battle from long ago ('nearly twenty thousand years') that took place in the vicinity of Bruges, on the waterway connecting the town with the sea. It's in fact a highly personal, very free elaboration by De Dene of the antique *gigantomachia*, the (account of the) war between the giants and the gods, as the fight De Dene reports on, was waged between Neptune and his allies on the one side and some mythological giants, centaurs, cyclops, and Olympian gods on the other. Neptune, helped by the swordfish Gladius Marinus,[8] whales, the Nereids or sea nymphs, and his sons Aeolus and Tryton, leads his troops while riding a huge whale. But Neptune and his army face fearful odds: they must taste defeat and turn back to the sea, leaving behind Neptune's whale, killed by an abundance of arrows. In the refrain line at the end of each stanza of this ballad De Dene reveals to the archers of Saint Sebastian the origin of their table: 'to remember all this, Neptune himself left us here this jawbone'.

2 *The Generous Seaman*[9]

In April 1560 Eduard de Dene wrote *Myn Langhen Adieu* (My Long Farewell), one of the closing poems in his *Testament Rhetoricael*.[10] In this ballad he bids farewell to everyone and everything. He also says goodbye to Flanders and its towns and villages, always incorporating the nicknames given to the inhabitants of the location in question. These nicknames often focus on the assumed food preferences of the citizens, and so we make the acquaintance of the Pike-Eaters (*snouck-eters*) of Axel, the Codfish-Eaters (*cabeliau-eters*) of Nieuwpoort, the Bream-Eaters (*mackel-eters*) of Dendermonde, the Ray-Eaters (*roch-eters*) of Monnikerede, and the Mussel-Eaters (*mossel-eters*) of

7 *Testament Rhetoricael*, fols. 118r–119v.

8 *Gladius maris* or *marinus* is one of the medieval scientific names for the swordfish; in this way it gives an ichthyological tinge to the information given here by De Dene.

9 For the following paragraphs, I rely largely on the findings in Geirnaert D., "Visser versus boer, of de lof der mildheid", in Porteman K. – Verbeke W. – Willaert F. (eds.), *Tegendraads genot. Opstellen over de kwaliteit van middeleeuwse teksten* (Leuven: 1996) 173–186.

10 *Testament Rhetoricael*, fols. 440r–444v.

Boekhoute.[11] This is in fact merely anecdotal information on the piscatorial culinary habits in De Dene's time, but in one of his songs, while fully singing the fisherman's praises, he also gives us a more profound view on the work and life of the fisherman and on his function in society as an important and indispensable provider of food.

The song involved is entitled *Den Milde Zeeman* (The Generous Seaman).[12] In this song De Dene gives us an animated and realistic glimpse into fishing life, and in particular into the different types of greedy customers a fisherman has to serve. De Dene accomplishes this by describing a series of tragicomic situations the seaman must deal with after having put into the harbour with his catch. The song opens with the statement that the fisherman, being the maritime colleague of the peasant, risks his life day and night at sea to ensure the food supply for many; although his profits are low, he does the utmost to kindly and generously distribute the fish among everyone approaching him (stanza 1). The first to take advantage of the wet and weary fisherman bringing his catch ashore are police officers, bailiffs, and other officials abusing their power and authority to pursue self-interest (stanza 2). In stanzas 3–7 De Dene obviously enjoys himself in depicting a motley crew of customers, crowding round the fisherman and his boat: innkeepers and fishmongers try to get the merchandise they need as cheap as possible; mendicant friars, beguines, and sextons try to get their share in exchange for some fancy talk and devout prayers; travelling singers and musicians try to pay for their fish with their music; barflies promise him free drinks; prostitutes hold out the prospect of free admission to the brothel; whimpering charlatans and swindlers try to elicit the fisherman's pity to get fish for free. To put it briefly: even before he gets to the fish market with his goods, a great part of the merchandise and the possible earnings have already disappeared, because people trespass upon his kindness and generosity. And finally (stanza 8), high taxes are levied on the caught plaice, halibut, herring, codfish, lobster, crab, and shrimp; in short, however important the fisherman may be for the society, he's hardly able to make a decent living by his fishing. Notwithstanding this, he always remains generous and kind-hearted. De Dene concludes (stanza 9) that someone like him really deserves our never-ending praise and gratitude, and he prays to God to always help and protect the fisherman and his family because of his great generosity.

11 *Testament Rhetoricael*, fols. 440v–441r.

12 *Testament Rhetoricael*, fols. 232v–233v. For the text and its translation, see the appendix at the end of this contribution.

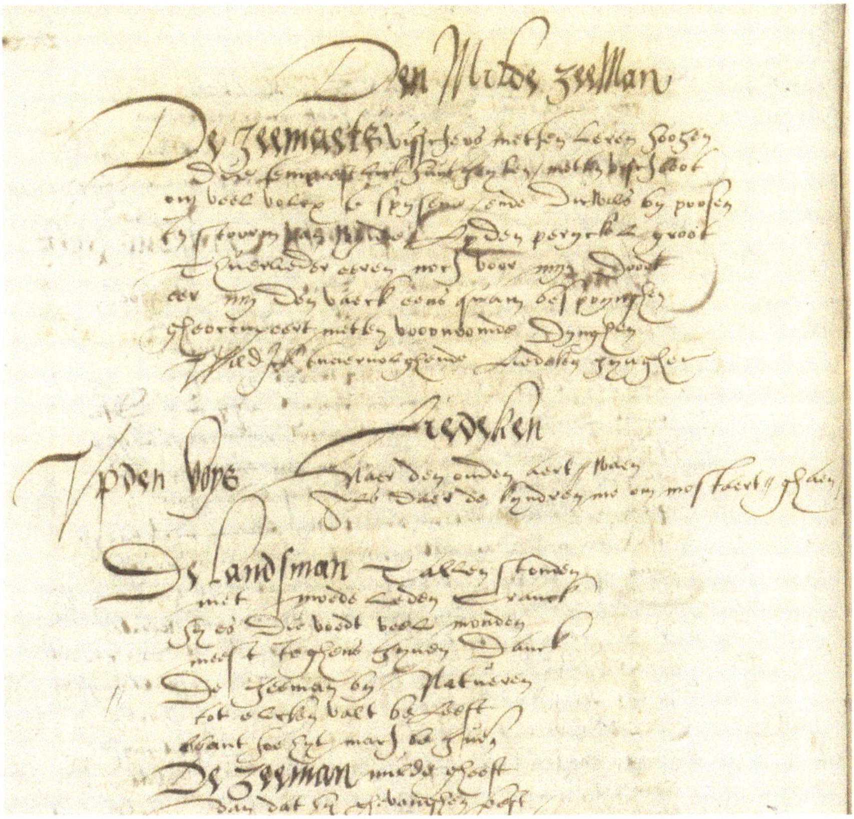

FIGURE 10.1 Dene Eduard de, *Testament Rhetoricael*, fol. 232v. Introduction to and beginning of *Den Milde Zeeman* (University Library Gent, ms. 3330)

Having read this song, one might ask: Is this really the right interpretation of that poem? Enjoying this highly tragicomic *Den Milde Zeeman*, one could viciously be inclined to think that the author wants us to consider the attitude of the fisherman towards all those opportunistic freeloaders and profiteers as silly naïveté or even stupidity, rather than as generosity. Someone who allows others to take advantage of him in such a way deserves to bleed for it. De Dene, however, is not that ironic or sarcastic, and the high degree of tragicomedy in the song does not mean that he makes the fisherman a laughingstock, as the text is clearly written with great sympathy and empathy for the hard-working seaman. Concrete proof of his positive view toward the generous seaman can be found in the fact that the poem is incorporated in De Dene's *Testament Rhetoricael* in a large section devoted to the second of the seven deadly sins (greed) and its counterpart (generosity).[13] If De Dene had intended to label the fisherman's approach as naïveté or stupidity, he certainly would have put the song elsewhere in his *Testament*. Furthermore, in the introductory verses preceding the song, De Dene states explicitly that he wants to give praise to the seamen who will not be stopped by storm or danger to provide the people with food. It's clear that we safely may conclude that there is no satire or mockery involved, and we can read the text as a genuine song of praise with a touch of social commitment: the fisherman is honoured, his customers are criticized.

3 *Ode to the Fisherman*

A second text paying tribute to the fisherman is a poem written down as a contemporary filler on a blank page in a register of the Castellany of Ypres, after a charter dated 13 July 1549.[14] The poem, entitled *Lof vanden Visschere* ('Ode to the Fisherman'), is incomplete: judging by the rhyme scheme, the second to last verse of the first stanza is missing, the fourth stanza only has four instead of eight verses, and moreover, one obviously gets the impression that after this fourth stanza the end of the poem was not yet in sight.[15]

The poem opens with the author pondering on the fact that he recently found some texts, once more speaking of the peasants in the highest terms, for they are the providers of food for everyone, rich and poor, the high lords as well as the most modest folk. This is getting a bit on the writer's nerves, because according to him, the kind fisherman also deserves praise, as he also works hard to feed all the people (stanza 1). Next, just like in De Dene's song, we get a vivid and colourful description of the people taking advantage of the weary fisherman, who had entered the harbour after a wakeful and cold night at sea (stanzas 2–4).

The resemblances between *The Generous Seaman* and this *Ode to the Fisherman* are most striking. The introductory stanzas of the song and the poem run analogously, as do the two storylines; the core points of both texts (i.e. the indispensability of the fisherman, his hard but poorly rewarded work, and the fact that everyone imposes on his kindness) are identical; even more eye-catching are the several similarities in the vocabulary used.[16] It's not immediately clear how to explain the close relationship between a song in De Dene's

13 *Testament Rhetoricael*, fols. 225v–240r.

14 It was discovered and published by E.I. Strubbe in 1940 (Strubbe E.I., "'Lof van den Visschere'. Een vers uit Eduard de Dene's tijd", *Biekorf* 46 (1940–1945) 27–28).

15 For the original text and its translation, see the appendix at the end of this contribution.

16 For these lexical similarities, see the words in bold in the original text of the song and the poem in the appendix.

Testament Rhetoricael and an incomplete poem written down as a filler in a register with official documents. Perhaps we may assume that both texts are products of one and the same author (De Dene), writing two variations on one theme. Such a hypothesis would account for the many resemblances. But these similarities are also accounted for when two different writers (De Dene and a fellow member of his Brugean chamber of rhetoric) created the song and the poem after the same model. This was not an unusual practice, as we can read in the foundation charter of the Ghent chamber *De Fonteine*: one of the requirements is that at each meeting of the chamber, which convened every three weeks, one of the rhetoricians is chosen to write a poem that is to be used by the other members as guiding example for their own literary contribution in the next gathering.[17] As a slight variation on this second explanation, it certainly is also very well conceivable that the song of De Dene must have been the example to follow, as he was, after all, the artistic leader of the Brugean chamber.

4 Eulogies for the Peasant

Be that as it may, in the research on *The Generous Seaman* and the *Ode to the Fisherman* we must introduce here an important new element. Independently of each other, both the song and the poem show the influences of two other texts that must have played an essential role in their creation.

To discover this common ground, we should return to the opening stanza of the *Ode to the Fisherman*, where the author obviously hints to poems singing the peasant's praises. Most probably, this singing can be taken literally, as the opening lines unmistakably seem to refer to two songs, entitled, respectively, *Vanden Edelen Landtman* (On the Noble Peasant) and *Vanden Landtman* (On the Peasant). The songs were published a couple of years before the writing down of *Ode to the Fisherman*, in a small booklet, printed in 1544 and now known as the *Antwerp Songbook*. This songbook is one of the highlights of 16th-century literature in the Low Countries, containing a collection of no fewer than 220 songs in the vernacular. The two songs in question are nr. 176 and nr. 201 of this *Antwerp Songbook*. The high degree of similarity in structure, content, and vocabulary of these two peasant eulogies is even more striking than the resemblances we saw in *The Generous Seaman* and *Ode to the Fisherman*. The explanation for this likeness has already been given by several

17 Coigneau D., "9 december 1448. Het Gentse stadsbestuur keurt de statuten van de rederijkerskamer De fonteine goed. Literaire bedrijvigheid in stads- en gildeverband", in *Nederlandse literatuur, een geschiedenis* (Groningen: 1993) 106.

scholars: the songs must have been written by two members of the same chamber of rhetoric, after a guiding example written in advance by the factor of the chamber.[18] This explanation ties neatly in with the second and third of the hypotheses accounting for the resemblances between *The Generous Seaman* and *Ode to the Fisherman*.

Without any doubt, both peasant songs can be considered as the agrarian counterparts of *The Generous Seaman* and *Ode to the Fisherman*. Their subject matter is, *mutatis mutandis*, identical: peasants deserve our praise and respect because everyone, from the highest to the lowest, can be fed thanks to their hard work; however, a peasant is hardly able to make a decent living, as young couples destroy his crop while making love; mendicant friars, nuns, sextons, and beguines approach him to get food for free with sweet talk, flattery, or in exchange for some prayers; he is swindled by impostors and treacherous millers, and suffers extortion by soldiers and hunters; wolves and foxes kill his cattle and poultry, weasels steal his eggs, and moles churn up his fields. Just like the two fisherman's eulogies, the peasant songs too get a religious twist in the epilogue stanza: the authors link the peasant to Jesus Christ, as he provides the essentials for Christ's transubstantiation in the Holy Mass (wheat for the sacred host and grapes for the holy wine); besides, Christ also honoured the peasant by the fact that, after His resurrection, He appeared unto Mary Magdalene in the shape of a gardener;[19] finally, the songs end by expressing the wish that Christ might always be at the peasant's side. Next to these parallelisms in structure and content, we also see some similarities in vocabulary and literary techniques: De Dene's *The Generous Seaman* has the same rhyme scheme and iambic metre as song nr. 201 of the *Antwerp Songbook*, and in both eulogies for the fisherman we can discover words and phrases also found in the two peasant's songs.

5 A Nice By-Catch

Connecting the peasant's songs to *The Generous Seaman* via 'the missing link' *Ode to the Fisherman* gives us – to speak in fishing terms – some nice by-catch as far as De Dene's song is concerned. His song is no longer just one of the very many poems in his *Testament Rhetoricael*, it's also a song we can place in the right context now, as it illustrates, together with *Ode to the Fisherman*, how

18 See Kalff G., *Het lied in de Middeleeuwen* (Leiden: 1884) 402; and also the most recent edition of the *Antwerp Songbook* (Poel D. van der et al. (eds.), *Het Antwerps Liedboek* (Tielt: 2004), II, 399 and 453).

19 See John 20:15.

a particular theme (the praise of the peasant) and two specific texts (*On the (noble) Peasant*) find their way in rhetorician circles. We are also able to determine a more precise year of origin for De Dene's text: it is no longer 1561 (date of his *Testament Rhetoricael*), because we now know that he must have written it somewhere between 1544 (year of publication of the *Antwerp Songbook*) and July 1549 (date of the only known copy of *Ode to the Fisherman*). But the nicest gain is probably the fact that, by linking De Dene's song to song nr. 201 of the *Antwerp Songbook*, we discovered how his *The Generous Seaman* had to be sung, as the melody used for *On the Peasant* was one of the melodies that could be traced back in 16th-century songbooks where song texts were given together with their music notation.[20]

6 Final Remarks

Apart from the text-inherent results of this research, already put into words in the previous paragraph, we can also draw a more general conclusion on the songs and poems with which we dealt.

From the 17th century onwards, poems on the country life and its inhabitants (peasants and shepherds) often belong to the genre of arcadian or bucolic literature, idealizing the rural and pastoral life in an idyllic, rustic setting. A maritime subvariant of this type of literature was introduced by Jacob Sannazaro's *Eclogae piscatoriae* (Naples, 1526). In this contribution we presented a small corpus of 16th-century Brugean texts using maritime and piscatorial elements as well. These elements were used in fairy-like or pseudo-mythological scenes, and particularly in two eulogies singing the praises of the fisherman. These eulogies, however, differ in a fundamental way from the Renaissance piscatorial eclogues, as there is no idealization of a fisherman's work or of the life on the seashore. Instead of this, the texts emphasize the importance and indispensability of the fisherman as a supplier of food, linking this positive message (1) to a realistic description of the hard, far from enviable life of the fisherman, and (2) to a tragicomic description of the many ways in which people try to take advantage of his generosity. At the end of his song, De Dene praises and thanks the fisherman once again, hoping that God will always stand by him.

How well-intentioned and full of empathy these eulogies – including the two related peasant songs of the *Antwerp Songbook* – may be, they also mean nothing but a rather scant consolation for the fisherman (and peasant). Although

20 See the information on *Vanden Landtman* in the Dutch Song Database (www.liederenbank.nl), a database containing more than 180,000 songs in the Dutch and Flemish language, from the Middle Ages through the 20th century.

the authors state explicitly that they are fully aware of the difficulties of the fisherman (and peasant) and that they sincerely admire him for how he deals with these hardships, this awareness and admiration are not accompanied by a call on the people involved to change the way in which they misbehave, nor do they entail a demand to improve the fisherman's (or the peasant's) life because of his great merits. Especially after having read the religious clinchers in the final stanzas it looks as if the songs and the poem also can be considered as an incentive, whether or not intentional, to patiently accept this fate as an inescapable fact of life in the late medieval class-ridden society. This patient, long-suffering attitude is also revealed in Cornelis Everaert's *Farce of the Fisherman*, in the short talk between the fisherman and his wife before they witness the wondrous underwater scene with which we opened this contribution. In brief, the texts dealt with not only express respect and sympathy for the fisherman (and peasant), but they also illustrate a society where everyone has his fixed function and immutable position, a society where everyone has his cross to bear – hopefully with the help of God.[21]

APPENDIX: Text and Translation of the Two Eulogies for the Fisherman

1) *Den milde zeeman* (E. de Dene, *Testament Rhetoricael*, fols. 232v–233v)[22]

Original text	**Translation**
Den milde zeeman	*The Generous Seaman*
De zeemaets, visschers metten leren hoozen, die tempeestlick huutzeylen metten vischboot om veel volcx te spysene, ende dicwils by poosen in stoorm van wynde lyden peryckel groot. Thuerlieder eeren, noch voor myn doodt, eer my den vaeck eens quam bespryghen, gheoccupeert metten voornoomde dynghen, wild'ick tnaervolghende liedeken zynghen.	*The seamen, fishermen with their leather trousers, leave the harbour with their fishing boats even in tempestuous circumstances, to provide food for many; in stormy weather they often fall into great danger. In their honour, and before I die, I wanted to sing the following song, that I wrote one night, just before falling asleep, as I was brooding on them and the dangers they had to face.*

21 See also Pleij H., *Het gilde van de Blauwe Schuit. Literatuur, volksfeest en burgermoraal in de middeleeuwen* (Amsterdam: 1979) 142–143.

22 Coigneau, *Testament Rhetoricael*, 28, 141–143.

Liedeken, upden voys naer den ouden aert waen / als daer de kyndren me om mostaert ghaen.	*A little song, to be sung on that old melody children use when they go to fetch mustard*
1. De **landsman**, tallen stonden met moede **leden** cranck, hy es die **voedt** veel monden, meest ieghens zynen danck. De zeeman by natueren tot elcken valt beleeft, want hoe hyt mach **bezueren,** De zeeman milde gheeft van dat hy ghevanghen heeft.	1. *A peasant never abandons his task to feed the mouths of many, weary or not, but also often reluctantly. The seaman by nature treats everyone benevolently and, no matter how much effort it takes, the fisherman generously distributes his catch.*
2. De zeeman mat en moede ghecommen nat an tlandt, man metter langhe roede cryght eerst een visch in dhandt. Die noch om hebben gaepen ghy niet van daer en dreeft: baetzouckers / en **tsheeren cnaepen**, de zeeman etc.	2. *When the seaman, soaked and dead tired, steps ashore, the bailiff is the first to claim a fish. Others, such as shameless profiteers and police officers, can't be chased away either, and so, the fisherman generously etc.*
3. Broer lollaerts ende **costers**, in tblanden zeere frisch, met lesen **pater nosters** ghecryghen zy tzeemans visch. Zy cryghense zonder dynghen, hoe ghy daer ieghens keeft, pypers ende die zynghen: den zeeman etc.	3. *Mendicant friars and sextons, experts in flattery and fancy talk, succeed in wheedling fish out of the fisherman in exchange for praying some Our Fathers.* *Musicians and singers too get their fish, they just take it for granted, no matter what, and so, the fisherman etc.*

4.
De weerden zeere ghierich
die commen oock an tboordt,
de mynders oock zeer vierich
die volghen rechtevoort;
om hebben end' om cryghen
elck anden vischboot cleeft.
Tot dat zylieden zwyghen,
de zeeman etc.

4.
Greedy innkeepers come aboard the fisherman's ship, immediately followed by eager fishmongers; everyone is crowding around the boat to get fish. And so, to silence them all, the fisherman etc.

5.
Bancknechten in taveerne
cryghen een visch zeer schoon,
met schuuftueghen gheerne
wordt des zeemans loon.
Die liefver naeyen dan spinnen
fraey up tghetauwe weeft:
die zoo huer broyken winnen,
de zeeman etc.

5.
Barflies get a fish from the seaman, in return for the promise of a free drink. He's allowed to weave on the loom[23] *of those women who prefer sewing*[24] *to spinning for their livelihood, and so the fisherman etc.*

6.
Quaczalven en **zusterluten**
die roupen: 'liefve **maet,**
gheeft ons wat lecx om fruten,
kendt onsen aermen staet!'
Hem blyft een magher coopken,
die styf van couden beeft …
Hoe cleen dat es zyn hoopken,
de zeeman etc.

6.
Quacks and beguines call out to him: 'Dear mate, please, look at our poor situation, give us something tasty to fry!' And so, there is only a meagre gain left for him, who's still frozen to the bone … How small the remaining amount of his merchandise may be, the fisherman etc.

23 *To weave on a woman's loom*: metaphorical expression meaning 'to have sexual intercourse with a woman'.

24 *To sew*: metaphorical expression meaning 'to have sexual intercourse'.

7.
De clerck met zyn practycken
cryght **eilbut** of **pladys;**
den zeeman zietmen strycken
een magher **grand merchys**.
Elck maectter of een rente,
hoe hy tempeestich sneeft.
Eer zyn visch comt ter vente,
de zeeman etc.

7.
Clerics, with their cunning and guile, get a portion of halibut or plaice; the seaman hardly receives a small thanks in return. Everyone profits, whereas the seaman dearly suffers. Before he gets his catch to the fish market, the fisherman etc.

8.
Hy es die vischt om haeryngh,
duer haeghel ende dau.
Hy en es oock gheen spaerlynck
van zynen cabeliau;
al waer hy hem maer voorziende
van mussel, gheernaert, creeft,
het moet al gheuen thiende.
De zeeman etc.

8.
He is the one, fishing for herring in hailstorm and in fog. He never withholds the codfish caught; and, even when he only gets mussels, shrimps, or lobsters aboard, he is obliged to pay his taxes. The fisherman etc.

9.
Van des zeemans spysen
heeft menich man te bet,
dies moet ick eeuwich prysen
schip visscher ende net.
God wacht hem dies van griefve,
al dat hem nicht of neeft
en brynghen thuus met liefve,
want hy zo milde gheeft
van dat hy ghevanghen heeft.

9.
Many a man benefits from the food provided by the seaman, and that's why I always will sing the praises of ship, fisherman, and net. God save him and his kin from grief, and may He always bring him safely back home, as he always hands out his catch so generously.

2. [*Lof vanden visschere*][25]

Original text

[*Lof vanden visschere*]

1.

Onlancx vandt ic ghephantazeert,
vanden **landtsman** vandt ic bescreven
want **hij tal voedt** ende sustineert:
paeus, keyser, coninghen moeten bij hem leven.
Doens docht ick, den visschere ware oock goet verheven
diet zoo **zuerlicke** met zijnen **leden** beslaeft:
want hij zoo minnelicke den mensche met spijse laeft.

2.

Den visscher snachs niet en heeft gheslapen,
compt smoorghens thuus, coudt als een ijs.
Dan compt den bailliu ende **sheeren cnapen**:
deen grijpt eenen **but** ende dander eenen **pladijs**,
dus es deerste handtghifte **gremmescijs**.
Die **ruuters van den bancke** namen oock **gheerne**
eenen visch omme den zelven prijs
ende betalen die **met schuuftijeghen in taveerne.**

3.

Dan commen dobservanten, papen ende **costers**,
precaers, broerkins ende **zusterluuten;**
die betalen den visch **met pater nosters**,
zij bidden voor de zeeman, scip ende scuute.
Danne commen dese vraukins die qualick sluuten,
vriendelick spreken zij den visschere toe:
'Heer, **maet, gheeft ons wat lecx om fruuten**';
zij betaelent met waire, men weet wel hoe.

Translation

Ode to the Fisherman

1.

Not so long ago, I was lost in gloomy thought, because I found a text on the peasant, explaining how he supports everyone by providing food: it's thanks to him that pope, emperor, and kings may live. Then I thought, also a fisherman, always moiling and toiling so hard, deserves praise as he too kind-heartedly feeds many people.

2.

After a sleepless night, the fisherman returns home in the morning, frozen to the bone. Then arrives the bailiff with his police officers: one grabs a halibut, others a plaice, and so his first reward is nothing else but a quick 'thank you'. Barflies too expect a fish at that same cheap price and pay with the promise of a free drink in the pub.

3.

Then come the mendicant friars, priests and sextons, Dominicans, other monks, and beguines; they all pay for the fish just with some Our Fathers, praying for seaman, ship, and boat. Next, we get those women, who find it hard to keep their legs closed, and friendly they address the fisherman: 'Please sir, dear mate, give us something tasty to fry'; and they pay it in kind, you know how.

25 Strubbe, *Lof*, 27–28.

4.
Dan commen de ledeganghers, deze groote trawanten,
die in alle canten scuwen pijne;
zij zegghen dat zij besmit zijn van quanten,
Deene vanden cleene graveene, ende dander vander fledecine.

4.
Then arrive the loafers, these big scoundrels, always avoiding any kind of exertion; they say that they have been infected by their comrades, one with kidney stones, and the other with gout.

Bibliography

Coigneau D., "9 december 1448. Het Gentse stadsbestuur keurt de statuten van de rederijkerskamer De fonteine goed. Literaire bedrijvigheid in stads- en gildeverband", in *Nederlandse literatuur, een geschiedenis* (Groningen: 1993) 102–108.

Coigneau D. – Waterschoot W. et al. (eds.), *Eduard de Dene, "Testament Rhetoricael"*, in *Jaarboek "De Fonteine"* 26, 28, 30 (Gent – Oudenaarde: 1976–1980).

Curtius E.R., *European Literature and the Latin Middle Ages*, transl. R. Trask (Princeton: 1973).

Geirnaert D., "Visser versus boer, of de lof der mildheid", in Porteman K. – Verbeke W. – Willaert F. (eds.), *Tegendraads genot. Opstellen over de kwaliteit van middeleeuwse teksten* (Leuven: 1996) 173–186.

Geirnaert D., "A Gift for Hanno. The Fictitious Booklist of Eduard de Dene", in Pouey-Mounou A.-P. – Smith P.J. (eds.), *Early Modern Catalogues of Imaginary Books. A Scholarly Anthology* (Leiden – Boston: 2020) 112–134.

Hüsken W.N.M. (ed.), *De spelen van Cornelis Everaert*, 2 vols. (Hilversum: 2005).

Kalff G., *Het lied in de Middeleeuwen* (Leiden: 1884).

Muller J.W. – Scharpé L. (eds.), *Spelen van Cornelis Everaert* (Leiden: 1920).

Pleij H., *Het gilde van de Blauwe Schuit. Literatuur, volksfeest en burgermoraal in de middeleeuwen* (Amsterdam: 1979).

Pleij H., *Het gevleugelde woord. Geschiedenis van de Nederlandse literatuur – 1400–1560* (Amsterdam: 2007).

Poel D. van der et al. (eds.), *Het Antwerps Liedboek*, 2 vols. (Tielt: 2004).

Robert-Nicoud V., *The World Upside Down in 16th-Century French Literature and Visual Culture* (Leiden – Boston: 2018).

Strubbe E.I., "'Lof van den Visschere'. Een vers uit Eduard de Dene's tijd", *Biekorf* 46 (1940–1945) 27–28.

CHAPTER 11

What Are the Fish Silent about? Selected Historical Facts on the Use of Fish in Medieval Medicine

A Qualitative Study Based on Sources from The Middle Low German Dictionary *Archive*

Sabina Tsapaeva

Historical pharmacopoeias and recipe collections are a wonderful source of research into the Middle Low German language, as well as the Northern German and Northern European culture of the Middle Ages.[1] After significant prose texts from the domain of historiography as well as legal codification, the 14th–16th centuries produced an abundance of Middle Low German pharmacopoeias, herbal and medical recipe books.[2] These were of great importance

1 Borchling C., "Zur mittelniederdeutschen Medicin", *Korrespondenzblatt des Vereins für niederdeutsche Sprachforschung* 22.5 (1901) 69–71; Idem, "Die mittelniederdeutschen Arzneibücher", *Janus* 7 (1902) 131–134, 175–179, 245–251; Idem, "Zur Handschriftenkunde der mittelniederdeutschen Medizin und Naturwissenschaft", *Mitteilungen zur Geschichte der Medizin und der Naturwissenschaften* 1 (1902) 66–70; Keil G., "Randnotizen zum 'Stockholmer Arzneibuch'", *Studia Neophilologica* 44 (1972) 238–262; Lindgren A., *Ein Stockholmer mittelniederdeutsches Arzneibuch aus der zweiten Hälfte des 15. Jahrhunderts* (Stockholm – Gothenburg – Uppsala: 1967); Idem, "Ein Kopenhagener mittelniederdeutsches Arzneibuch aus dem Ende des 15. Jahrhunderts", *Würzburger medizinhistorische Mitteilungen* 4 (1986) 135–178; Oefele F. von, "Zur mittelniederdeutschen Medizin", *Korrespondenzblatt des Vereins für niederdeutsche Sprachforschung* 22.4 (1901) 49–50; Temmen M., *Die medizinische Rezepthandschrift Burgsteinfurt Hs. 15. Edition und Untersuchung einer Handschrift aus dem 16. Jahrhundert* (Bielefeld: 1998); Idem, *Das 'Abdinghofer Arzneibuch'. Edition und Untersuchung einer Handschrift mittelniederdeutscher Fachprosa* (Cologne – Weimar – Vienna: 2006).

2 Albrecht [van Borgunnien], *Albrecht van Borgunnien's Treatise on Medicine* (*Sloane Ms. 3002, British Museum*), ed. W.L. Wardale (London – Edinburgh – Glasgow: 1936); Alstermark H., *Das Arzneibuch des Johan van Segen* (Stockholm: 1977); Bartholomaeus [Salernitanus], *Angebliche practica des Bartholomaeus von Salerno, Schüler des Constantinus Salernitanus. Introductiones et experimenta magistri Bartholomaei in practicam Hippocratis Galieni Constantini graecorum medicorum. Papier-Handschrift der herzogl. Sachs.-Coburg-Gothaischen Bibliothek, No. 920, fol. 85a bis 104b*, ed. F. von Oefele (Bad Neuenahr: 1894); Brunschwig Hieronymus, *Dat Boek der Wundenartzstedye und der niederdeutsche chirurgische Fachwortschatz*, ed. Ch. Benati (Göppingen: 2012); *Das Breslauer Arzneibuch* [*R 291 der Stadtbibliothek*], ed. C. Külz – E. Külz-Trosse (Dresden: 1908); *Das Promptuarium medicinae.* [*Eyn schone Arstedyge boeck van allerleye ghebreck unnde kranckheyden der mynschen*], ed. P. Seidensticker nach d. Ausg. Magdeburg: 1483 (Lahr: 1990); *Das Utrechter Arzneibuch* (*Ms. 1355,*

 | DOI:10.1163/9789004681187_012

for both Northern German and Scandinavian regions. The majority of these medical texts originated independently of High German sources. That is to say they do not teach entirely new methods of healing. On the contrary, they can be seen as a logical continuation of the medical literature of venerable antiquity and the Middle Ages.

If one looks at the Middle Low German pharmacopoeias, recipe collections and medical instruction manuals in general, a hierarchy of ingredient importance becomes apparent. One is immediately struck by the abundance of herb- and mineral-based recipes; land-based animals and their components are also frequently used. In comparison, fish-based recipes are a rarity, which doesn't mean that they are of less interest.

1 Introduction

The aim of this chapter is to shed light on the role of fish and its concrete use in medieval medicine.[3] The chapter's contents are sourced from the holdings of *The Middle Low German Dictionary* archive[4] (University of Hamburg), including

16°, Bibliotheek der Rijksuniversiteit Utrecht), ed. A. Lindgren (Stockholm: 1977); Doneldey Arnoldus, *Das Bremer mittelniederdeutsche Arzneibuch des Arnoldus Doneldey. Mit Einleitung und Glossar*, ed. E. Windler (Neumünster: 1932); Gallée J.H., "Mittelniederdeutsches Arzneibuch", *Jahrbuch des Vereins für niederdeutsche Sprachforschung* 15 (1889) 105–149; Graeter Chr., *Ein Leipziger deutscher Bartholomaeus* (Borna-Leipzig: 1918); Lindgren, *Ein Stockholmer mittelniederdeutsches Arzneibuch*; Idem, "*Ein Kopenhagener mittelniederdeutsches Arzneibuch*"; Ljungqvist I., *Das mittelniederdeutsche Arzneibuch des Codex Guelferbytanus 1213 Helmstadiensis* (Stockholm: 1971); Norrbom S.E., *Das Gothaer mittelniederdeutsche Arzneibuch und seine Sippe* (Hamburg: 1921); Regel K., "Das mittelniederdeutsche Gothaer Arzneibuch und seine Pflanzennamen", *Programm des Herzoglichen Gymnasium Ernestinum zu Gotha* 1 (1872) [1]–16, 2 (1873) 1–26; Idem, "Zwei mnd. Arzneibücher, Cod. Chart. Goth. 980 und Cod. Wolfenb. 23,3.", *Jahrbuch des Vereins für niederdeutsche Sprachforschung* 4 (1878) 5–26; Röder Sebastian, *Nütte lere vnd vnderricht Doct. Sebastiani Rőder/ Jdtzundes alhier tho Hamborch/ We men sick in dűssen gefarlicken Steruendes lűfften holden/ vnde vor der giff=tigen Pestilentzischen sűcke bewaren schall/ dat men dar vör seker wesen/ vnd nicht darmit beflecket warden mach/ edder sick jo dar van erredden mőge* (Hamburg, Johann Wickradt: 1565); Temmen, *Die medizinische Rezepthandschrift Burgsteinfurt Hs. 15*; Idem, *Das 'Abdinghofer Arzneibuch'*.

3 This chapter is a slightly modified version of the talk given on September 16, 2022 at the International Congress of the Société Internationale Renardienne held at the University of Antwerp (Belgium) and entitled "Of Foxes and Fish: Interdisciplinary Approaches to Medieval Animal Lore and its Afterlife".

4 Cf. *Mittelniederdeutsches Wörterbuch*, online at https://www.slm.uni-hamburg.de/niederdeutsch/ueber-die-abteilung/mittelniederdeutsches-wb.htm, accessed on November 27, 2022.

The Pharmacopoeia by Johan van Segen,[5] *The Bremen Pharmacopoeia* by Arnoldus Doneldey,[6] *The Middle Low German Stockholm Pharmacopoeia*,[7] *The Middle Low German Gotha Pharmacopoeia*,[8] *The Abdinghof Pharmacopoeia*,[9] the Magdeburgian *Promptuarium medicinae* edited by Peter Seidensticker[10] and many others. The following is to be defined as a kind of sub-goal: In addition to the actual recording of the fish examples in the medical recipes that have not been in the focus of research so far, the principles underlying the specific use are to be analysed. Only proper fish are to be considered as fish, i.e., other marine and freshwater animals such as dolphins, whales, cuttlefish, crabs, mussels, frogs, snails, otters and beavers which have more frequently been treated in parallel to proper fish and have most likely been partly perceived or at least declared as such – especially during fasting periods –, are excluded.[11]

The structure of the present chapter is organised as follows. First, a brief overview of the current state of research is given before the methodical approach of the analysis is described. Then, *The Middle Low German Dictionary* research section is to be briefly introduced; thereafter the reader will have the opportunity to have a look into the research text corpus. Subsequently, the research results that could be obtained with the chosen approach are presented, interpreted and compared with the state of the research. Finally, the results obtained are briefly summarised, the limitations of the study are explained and the chosen approach is critically reflected upon in retrospect.

2 Current State of Research

The medicinal use of fish and its components is the subject of a fairly manageable number of papers, whereas the role of fish or aquatic animals in general is mentioned and studied frequently in the context of medieval and early modern dietetics.[12] For context, a brief overview of the current state of research with

5 Alstermark, *Das Arzneibuch des Johan van Segen*.
6 Doneldey, *Das Bremer mittelniederdeutsche Arzneibuch des Arnoldus Doneldey*.
7 Lindgren, *Ein Stockholmer mittelniederdeutsches Arzneibuch*.
8 Norrbom, *Das Gothaer mittelniederdeutsche Arzneibuch*.
9 Temmen, *Das 'Abdinghofer Arzneibuch'*.
10 *Das Promptuarium medicinae*.
11 It is important here to note that since the Council of Constance (1414–1418), everything that lives in water has been classified as fish. Cf. Hoffmann R.C., "A brief history of aquatic resource use in medieval Europe", *Helgoland Marine Research* 59 (2005) 22–30, esp. 23.
12 Cf. further footnote 33.

secondary sources that deal with the medicinal use of fish in order to create a basis of comparison for the main analysis of this paper follows. First, Kamal Sabri Kolta's study focuses, among other things, on the properties and significance of fish in selected medieval Arabic texts that had not been translated at the time of the paper's publication.[13] He points out that the Arabs' reports sometimes mentioned the medicinal use of fish. According to ad-Damiri fish would help against eyewash – presumably the lens opacity or cataract is meant here –, fish bile would be useful against heart palpitations and smelling fish would sober drunks.[14] Sabri Kolta also mentions in passing that the ancient Egyptians are said to have used a certain *ꜣbḏw*-fish for eye conditions, headaches and even poisoning.[15] In ancient Egyptian spells, this *ꜣbḏw*-fish is also recommended against children's diseases.[16] Sabri Kolta also mentions a report by Herodotus that discusses the use of fish together with milk to induce nausea and vomiting for healing purposes.[17] According to Sabri Kolta, another fish (*ꜥꜣḏw*-fish, Flathead grey mullet, *Mugil cephalus*) was recommended as a remedy to solve stiffness of any kind and against greying of the hair.[18] Another use as an impotency remedy together with onions is recorded by most of the authors analysed by Sabri Kolta.[19] In his paper there are many more fish recipes mentioned for medicinal purposes, including for epilepsy attacks, insanity, fever, headaches, etc.[20]

In his short contribution, von Soden looks at fish bile as a remedy for the eyes, namely on the basis of examples from the *Bible* (especially the *Book of Tobit*) and in addition some Assyrian documents that were handed down on clay tablets at the time.[21] He states that fish bile was not known as a remedy for the eyes, but that it should be drunk for diseases of the urinary tract.[22]

In his short article on medieval Russian medicine, Anikin also deals with fish.[23] He notes that in addition to plants, land-based animals and products of

13 Sabri Kolta K., "Der Fisch im Volksglauben und in der Medizin der Araber", *Deutsches Schiffahrtsarchiv* 1 (1975) 211–215.
14 Ibidem 212.
15 Ibidem 212.
16 Ibidem 212.
17 Ibidem 212.
18 Ibidem 212.
19 Ibidem 212.
20 Ibidem 214.
21 Soden W. von, "Fischgalle als Heilmittel für Augen", in Soden W. von – Müller H.-P., *Bibel und Alter Orient* (Berlin – Boston: 1985) 76–77.
22 Ibidem 81.
23 Аникин И.Л., "Некоторые методы лечения, применявшиеся в средневековой русской медицине", *AMHR* 1.20 (1992) 134–143.

animal origin, fish was also used in the treatment of diseases; for example, fish liver from sturgeon, catfish and cod was administered for night blindness.[24]

A particularly significant study[25] for this chapter is the ethnomedical analysis by José Ramón Vallejo and José Antonio González, which looks at the historical and diachronic use of 54 fish species for medicinal purposes using the example of Spanish medical literature (52 sources in total) from *Naturalis historia* by Pliny the Elder[26] (ca. 77 AD) to the most recent treatises by Manuel Pedro Cobo López and Raúl Elías Tijera Jiménez[27] and Montse Rigat et al.[28] Of the sources examined, five belong to the Middle Ages, while three others were published in the 15th–17th centuries. Vallejo and González not only try to identify each fish very specifically, which is particularly problematic with ancient and medieval sources, but also describe the internal as well as external use of the fish or its components. In addition, they try to determine and name a specific disease, i.e., diagnosis, for which fish was used as a remedy or component of a remedy, and additionally comment on the regional assignment of the respective fish evidence.[29]

Cécile Le Cornec deals with the (positive) properties of marine fish from the perspective of dietetics, however her work is useful for this chapter.[30] Although she does not deal with the actual medicinal use of fish, she makes reference to significant medical sources from the 13th–14th centuries. In his paper, Grégory Clesse deals with aquatic animals in Arabic and Latin medical sources, attempts to classify them and also examines their use from a diachronic perspective.[31] His study focuses on five translations from the 12th–13th centuries. Surely the author considers fish aquatic animals, yet they are only given little attention; and their effects are not analysed in depth. However, a general negative effect on the human body is mentioned.[32]

24 Аникин И.Л., "Некоторые методы лечения, применявшиеся в средневековой русской медицине" 136.

25 Vallejo J.R. – González J.A., "Fish-based remedies in Spanish ethnomedicine: a review from a historical perspective", *Journal of Ethnobiology and Ethnomedicine* 10.37 (2014) 1–31.

26 *Naturalis historia* by Pliny the Elder is considered part of ancient Spanish ethnomedical literature along with three other ancient sources. Cf. ibidem 3.

27 Cobo López M.P. – Tijera Jiménez R.E., *Etnozoología de Doñana* (Sevilla: 2013).

28 Rigat M. – Vallès J. – Iglésias J. – Garnatje T., "Traditional and alternative natural therapeutic products used in the treatment of respiratory tract infectious diseases in the eastern Catalan Pyrenees (Iberian Peninsula)", *Journal of Ethnopharmacology* 148.2 (2013) 411–422.

29 Vallejo – González, "Fish-based remedies in Spanish ethnomedicine" 6–22.

30 Le Cornec C., "Les vertus diététiques attribuées aux poissons de mer", in Connochie-Bourgne Ch. (ed.), *Mondes marins du Moyen Âge* (Aix-en-Provence: 2006) 273–284.

31 Clesse G., "Animaux aquatiques dans les sources médicales arabo-latines: continuités et discontinuités d'un discours", *Médiévales* 80.1 (2021) 81–98.

32 Ibidem 95.

A number of other contributions focus more on medieval and early modern fish dietetics; medicinal use in the actual sense is not dealt with.[33] In summary, a lot more research would be needed to make advances in this field of research.

Based on the above, no prominent presence of fish-based medical recipes is expected in the selected research corpus. It is more likely that they were used the same way they were used in older sources, such as eye diseases or problems of the urinary tract.

3 Methodical Approach

Due to the fact that in the selected research corpus a relatively low density of evidence for fish and their components is expected, a qualitative approach is the first choice. The first step is to verify which of the sources have fish mentioned. In the second step, the evidence found will be filtered by distinguishing between genuine use in a medical recipe and, for example, a general dietary recommendation (depending on the sign of the zodiac, month or phase of the moon). Subsequently, the fish mentioned in real recipes will be examined in their context of use. Thus, for each mention it will be recorded whether it is a whole fish or a specific component of a fish, and in which form it is used (fresh, dry, salted, boiled, fried, burnt to powder, etc.). Furthermore, the human organ affected and the disease or diagnosis to be treated will be noted. Finally, the underlying treatment principle (*similia similibus curantur* – with help from the same, *contraria contrariis curantur* – with help from the opposite)[34] will be determined.

However, before demonstrating the research results obtained, a few words about the research section *The Middle Low German Dictionary* should be said; afterwards the research corpus will be presented.

33 Cf. Wickersheimer E., "Zur spätmittelalterlichen Fischdiätetik: Deutsche Texte aus dem 15. Jahrhundert", *Sudhoffs Archiv für Geschichte der Medizin und der Naturwissenschaften* 47.3 (1963) 411–416; Albala K., "Fish in Renaissance Dietary Theory", in Walker H. (ed.), *Fish. Food from the Waters. Proceedings of the Oxford symposium on food and cookery 1997* (Oxford: 1998) 9–19; Buxton M., "Fish-Eating in Medieval England", in Walker H. (ed.), *Fish. Food from the Waters. Proceedings of the Oxford symposium on food and cookery 1997* (Oxford: 1998) 51–59.

34 Cf. Maycock P.P. Jr., "Introduction to the Second Edition", in Buchanan S. (ed.), *The Doctrine of Signatures: A Defence of Theory in Medicine* (Urbana, IL: 1991) xv–xxix; Ullmann M., *Islamic Medicine* (Edinburgh: 1997) 99; Vallejo – González, "Fish-based remedies in Spanish ethnomedicine" 26.

4 Research Section *The Middle Low German Dictionary*

The Middle Low German Dictionary research section at the University of Hamburg[35] has been systematically recording Middle Low German sources and documenting their vocabulary since 1923. In this way, a unique collection of documents has been created with more than one million records. This makes the Low German vocabulary from the 13th to the 17th century in the entire Hanseatic region more accessible for temporal, regional, social and functional research perspectives. In the meantime, more than 80,000 lexemes have been identified and documented. And so, the Hamburgian research section acts as a national and international information centre for enquiries about Middle Low German language and culture and publishes a dictionary of the Middle Low German language, which serves as a foundational work and research tool for numerous needs and is intended to replace the outdated dictionaries of the 19th century.

The dictionary, which was conceived as a hand-dictionary and originally planned for only three volumes, serves the following academic fields in particular:

(1) Linguistics and literary studies, especially for questions of historical linguistics and historical dialectology;

(2) The historical sciences, which are provided with a fundamental aid for working on medieval and early modern sources. A core area here is Hanseatic research, but aspects of regional history and legal history are also becoming increasingly important;

(3) Cultural history, sociology and ethnology.

In addition, the dictionary can also be used as a source of information by laymen interested in regional history and historical linguistics. It is also worth mentioning that *The Middle Low German Dictionary* additionally offers the possibility to consult the material on site and to evaluate it specifically for scientific purposes.[36]

35 Cf. *Mittelniederdeutsches Wörterbuch*, online at https://www.slm.uni-hamburg.de/niederdeutsch/ueber-die-abteilung/mittelniederdeutsches-wb.htm, accessed on November 27, 2022.

36 Cf. Möhn D., "Vorwort zu Bd. II/1", in *Mittelniederdeutsches Handwörterbuch*, ed. A. Lasch – C. Borchling – G. Cordes – D. Möhn – I. Schröder (Kiel – Hamburg: 1928 ff.) vol. II/1, I–XIII, esp. VI; Schröder I. – Tsapaeva S., "Komplexe Semantik im mittelniederdeutschen Wörterbuch", in Harm V. – Lobenstein-Reichmann A. – Diehl G. (eds.), *Wortwelten. Lexikographie, historische Semantik und Kulturwissenschaft* (Berlin – Boston: 2019) 177–202, esp. 178–180.

It has to be mentioned that a strong presence of pragmatic texts can be found in the text corpus, which can be explained as follows: Middle Low German texts are evidence of a differentiating (Hanseatic) urban culture, so that the conditions of medieval and early modern communication are also reflected in the lexis. The main areas of communication are trade, urban institutions, political and legal networks, crafts and *artes*, Church and spiritual life, instruction – edification – and entertainment as well as individual and collective self-assurance, and of course the medical field is also included.[37]

5 Research Corpus

There are 19 pharmacopoeias and promptuaries that were used as medical sources for *The Middle Low German Dictionary*, that also form the basis for this chapter. They are listed in alphabetical order according to their project internal short form[38] which is given in bold below. The information provided is limited to the locations of the respective manuscripts, incunabula and old prints, which can be found next to the reference to the year or the century they were finished or printed in. Further background detail on the editions, partial editions and studies used in the project and the current chapter can be found in the corresponding footnotes.

A. v. B. = Albrecht van Borgunnien's *Treatise on Medicine*, London, British Museum, Sloane Ms. 3002, 15th century.[39]

Abdingh. Arzneib. – *The Abdinghof Pharmacopoeia*, Paderborn, Archiepiscopal Academic Library, Theodoriana VVa 3, mid till end of the 15th century.[40]

---

37 Cf. Schröder I., "Niederdeutsch, niederdeutsche Sprache", in *HanseLexikon* (*HansLex*), ed. Hansischer Geschichtsverband, online at https://www.hansischergeschichtsverein.de/lexikon, accessed on November 27, 2022; Möhn D. – Schröder I., "Lexikologie und Lexikographie des Mittelniederdeutschen", in W. Besch – O. Reichmann – S. Sonderegger, *Sprachgeschichte. Ein Handbuch zur Geschichte der deutschen Sprache und ihrer Erforschung*, vol. 2.2 (Berlin – New York: 2000) 1435–1456; Schröder – Tsapaeva, "Komplexe Semantik im mittelniederdeutschen Wörterbuch" 179.

38 The project intern short form is relevant for the users of *The Middle Low German Dictionary*. One can find this short form completed and complemented by the page or folio reference in the respective dictionary articles beside the corresponding evidences for the lexemes.

39 Albrecht [van Borgunnien], *Albrecht van Borgunnien's Treatise on Medicine*.

40 Temmen, *Das 'Abdinghofer Arzneibuch'*.

Aquae med. ed. Lindgren = *The Aquae Medicinales* from *The Middle Low German Gotha Pharmacopoeia*, Gotha, University and Research Library Erfurt – Gotha, Cod. Chart. A 980, shortly before 1400.[41]

Bartholomaeus Practica = Bartholomaeus' *Practica*, Gotha, University and Research Library Erfurt – Gotha, Cod. Chart. A 980, end of the 14th century.[42]

Boek d. Wundenartzstedye ed. Benati = *Boek der Wundenartzstedye*, the Low German translation of Hieronymus Brunschwig's *Buch der Cirurgia*, Berlin, Berlin State Library – Prussian Cultural Heritage Foundation, Jg 3484; Schwerin, Mecklenburg-Vorpommern State Library Günther Uecker, HSt VII 745, anno 1518.[43]

Brem. Arzneib. = *The Bremen Pharmacopoeia* by Arnoldus Doneldey, Hannover, State Archive, Ms. AA 16, anno 1382.[44]

Bresl. Arzneibuch = *The Wrocław Pharmacopoeia*, Wrocław, University Library, Cod. R. 291, 1st quarter of the 14th century.[45]

Goth. Arzneib. = *The Middle Low German Gotha Pharmacopoeia*,[46] Gotha, University and Research Library Erfurt – Gotha, Cod. Chart. A 980, shortly before 1400.[47]

J. v. Segen = *The Pharmacopoeia* by Johan van Segen, Stockholm, Royal Library of Sweden, MS medic. XII 114, anno 1487.[48]

Kopenh. Arzneib. ed. Lindgren = *The Copenhagen Pharmacopoeia*, Copenhagen, Royal Danish Library, GKS Cod. 1663.4, end of the 15th century.[49]

Leipzig. Bartholomaeus = *The German Bartholomaeus* from Leipzig, Leipzig, University Library, MS 0816, beginning of the 15th century.[50]

---

41 Lindgren A., *Die 'aquae medicinales' des mittelniederdeutschen Gothaer Arzneibuches* (Stockholm: 1979).

42 Bartholomaeus [Salernitanus], *Angebliche practica des Bartholomaeus von Salerno*.

43 Brunschwig, *Dat Boek der Wundenartzstedye*.

44 Doneldey, *Das Bremer mittelniederdeutsche Arzneibuch des Arnoldus Doneldey*.

45 *Das Breslauer Arzneibuch*.

46 It must be said at this point that several medical works are included here, including *Düdesche Arstedie* and the Middle Low German Bartolomaeus' *Practica*.

47 Norrbom, *Das Gothaer mittelniederdeutsche Arzneibuch*.

48 Alstermark, *Das Arzneibuch des Johan van Segen*.

49 Lindgren, "*Ein Kopenhagener mittelniederdeutsches Arzneibuch*".

50 Graeter, *Ein Leipziger deutscher Bartholomaeus*.

Ljungqvist Arzn. = *The Middle Low German Pharmacopoeia* from *Codex Guelferbytanus 1213 Helmstediensis*, Wolfenbüttel, Herzog August Library, Cod. Guelf. 1213 Helmst., mid till end of the 15th century.[51]

Nd. Jb. 15: Utr. Arzneib. = *The Middle Low German Utrecht Pharmacopoeia*, Utrecht, University Library, Ms. 1355 (6 H 23), end of the 14th century – beginning of the 15th century.[52]

Prompt. med. ed. Seidensticker = *Promptuarium medicinae*, Berlin, Berlin State Library – Prussian Cultural Heritage Foundation, 4° Inc 1488.20; Bethesda (MD), National Library of Medicine, WZ 230 P989 1483; Copenhagen, Royal Danish Library, Inc. Haun. 3383; Göttingen, State and University Library of Lower Saxony, 4 MAT MED 34/76 INC; Hamburg, State and University Library Carl von Ossietzky, KB 503; London, British Library, IB.10909; Moscow, Russian State Library, MK Inc.2.135; San Marino (CA), Huntington Library, 90921; St. Petersburg, Russian National Library, 9.11/12.1.23; Stockholm, Royal Library of Sweden, MFBL 98:7; Wolfenbüttel, Herzog August Library, A: 51.3 Med. 2°, anno 1483.[53]

Regel = *The Middle Low German Gotha Pharmacopoeia* and its plant names, Gotha, University and Research Library Erfurt – Gotha, Cod. Chart. A 980, end of the 14th century.[54]

Röder = *Nütte lere vnd vnderricht* [...] *We men sick in důssen gefarlicken Steruendes lůfften holden / vnde vor der giff=tigen Pestilentzischen sůcke bewaren schall* [...], Hamburg, State and University Library Carl von Ossietzky, Scrin A/258; Kiel, University Library, 2 an MK 4499, anno 1565.[55]

Stockh. Arzneib. = *The Middle Low German Stockholm Pharmacopoeia*, Stockholm, Royal Library of Sweden, Ms. X 113, 2nd half of the 15th century.[56]

---

51 Ljungqvist, *Das mittelniederdeutsche Arzneibuch des Codex Guelferbytanus 1213 Helmstadiensis*.

52 Gallée, "Mittelniederdeutsches Arzneibuch".

53 *Das Promptuarium medicinae*. According to the Union Catalogue of Incunabula, or Gesamtkatalog der Wiederdrucke the total number of copies or fragments in public institutions is 14. Cf. 'Promptuarium medicinae' at *Gesamtkatalog der Wiegendrucke*, online at gesamtkatalogderwiegendrucke.de/docs/M35662.htm, accessed on November 27, 2022. Unfortunately the copy at the University and State Library Darmstadt seems to be lost; this copy cannot be traced via the library catalogue.

54 Regel, "Das mittelniederdeutsche Gothaer Arzneibuch und seine Pflanzennamen".

55 Röder, *Nütte lere vnd vnderricht Doct. Sebastiani Rőder*.

56 Lindgren, *Ein Stockholmer mittelniederdeutsches Arzneibuch*.

Utr. Arzneib. ed. Lindgren = *The Utrecht Pharmacopoeia*, Utrecht, University Library, Ms. 1355, 16°, around 1400.[57]

Wolfenb. Arzneib. = *The Middle Low German Wolfenbüttel Pharmacopoeia*, Wolfenbüttel, Herzog August Library, Cod. [Guelf.] 23.3 Aug. 4°, around 1450.[58]

The reader has probably already noticed in the presentation of the sources that certain pharmacopoeias are listed several times. This is mainly due to the fact that the corresponding manuscripts have come into the focus of medievalist research several times – and from different perspectives – and have accordingly also entered the corpus of *The Middle Low German Dictionary* as sources in the updated version. Some of the works mentioned contain editions or partial editions, while others concentrate exclusively on individual aspects of the respective manuscripts or provide word lists or glossaries, which were very helpful at the excerption phase of the project.

After a first thorough review of the sources, however, it was found that not all manuscripts, incunabula and old prints in the corpus contain fish mentions for medicinal purposes. Thus, the 15th-century pharmacopoeia from the Abdinghof monastery,[59] *The Aquae Medicinales* analysed and edited by Lindgren,[60] Bartholomaeus' *Practica*[61] and *The German Bartholomaeus* from Leipzig,[62] *Boek der Wundenartzstedye* by Hieronymus Brunschwig,[63] pharmacopoeias from Bremen,[64] Copenhagen,[65] Utrecht,[66] Wolfenbüttel[67] and Wrocław[68] make no mention of fish or its constituents in the recipes, so the final corpus of investigation is considerably smaller. In a number of the remaining sources examined, fish is mentioned in dietary recommendations, e.g., in *The Middle Low German Pharmacopoeia* from the *Codex Guelferbytanus 1213*

57 *Das Utrechter Arzneibuch.*
58 Regel, "Zwei mnd. Arzneibücher, Cod. Chart. Goth. 980 und Cod. Wolfenb. 23, 3."
59 Temmen, *Das 'Abdinghofer Arzneibuch'.*
60 Lindgren, *Die 'aquae medicinales' des mittelniederdeutschen Gothaer Arzneibuches.*
61 Bartholomaeus [Salernitanus], *Angebliche practica des Bartholomaeus von Salerno.*
62 Graeter, *Ein Leipziger deutscher Bartholomaeus.*
63 Brunschwig, *Dat Boek der Wundenartzstedye.*
64 Doneldey, *Das Bremer mittelniederdeutsche Arzneibuch des Arnoldus Doneldey.*
65 Lindgren, "*Ein Kopenhagener mittelniederdeutsches Arzneibuch*".
66 *Das Utrechter Arzneibuch.*
67 Regel, "Zwei mnd. Arzneibücher, Cod. Chart. Goth. 980 und Cod. Wolfenb. 23,3".
68 *Das Breslauer Arzneibuch.*

Helmstadiensis,[69] *Promptuarium medicinae*,[70] the 16th-century pest prevention guide by Sebastian Röder[71] and partly in *The Middle Low German Stockholm Pharmacopoeia*,[72] but these will not be discussed in this chapter.

6 Research Results

In the selected research corpus, 23 medical recipes were found in which fish or its ingredients were used. Although a qualitative approach is taken in the study, it is nevertheless quite obvious that fresh eel bile occurs most frequently (7 times). In *The Middle Low German Gotha Pharmacopoeia* fresh eel bile is recommended for "watery eyes" (ocular mucous membrane inflammation?), but also for corneal staining after an inflammatory eye disease:

> Weme de oghen vele tranen, de neme enes versschen ales gallen, yserhart vnde vennekelswortelen, make desse krude schone, stot se vnde wrink denne dat sap darvth vnde menghe de gallen darto vnde laet yd tohope wol warm werden vnde syget denne dor enen dok in ene bussen vnde besmere de ogen darmede.[73]

> If you have frequent watering eyes, take the gall of a fresh eel, verbena and fennel root[s], wash these herbs, crush them and then squeeze the juice out of them and mix it with the gall; and let it get warm together and strain it through a cloth into a vessel and smear the eyes with it.[74]

> Jtem nym enes hazen gallen vnde enes ales gallen vnde enes hanen gallen vnde menghe de dree gallen tohope vnde menge dar so vele honnyges zo also van den dren gallen wert vnde laet dat tohope wol warm werden vp deme vure vnde drope des wat in de ogen: dat vordrifft dat mael.[75]

> Take the gall of a hare and the gall of an eel and the gall of a cock and mix the three galls together and add as much honey as you have of the galls

69 Ljungqvist, *Das mittelniederdeutsche Arzneibuch des Codex Guelferbytanus 1213 Helmstadiensis* 49.
70 *Das Promptuarium medicinae* 86.
71 Röder, *Nütte lere vnd vnderricht Doct. Sebastiani Rǒder* B 1r–1v.
72 Lindgren, *Ein Stockholmer mittelniederdeutsches Arzneibuch* 97–98.
73 Norrbom, *Das Gothaer mittelniederdeutsche Arzneibuch* 73.
74 The translations were done by the author of the present chapter.
75 Norrbom, *Das Gothaer mittelniederdeutsche Arzneibuch* 76.

and let it become well heated together upon fire and drop some of it into the eyes; it will drive away the staining.

Furthermore, in the same pharmacopoeia, eel bile is mentioned as a remedy for watery or bleeding eyes and for swellings[76] – a specific body part is not mentioned here:

> Weme de oghen rynnen van watere efte van blode, [...] so nym ales gallen vnde sap van ysenharte vnde sap van vennekoleswortelen vnde menghe dat tosamende by deme vure vnde syge yt denne dor enen lynen dok vnde do et in eyn blyen vaet vnde strik yt denne buten vmme de oghen, so werden se droghe.[77]

> To whom the eyes run with water or blood; [...] take the gall of an eel and the juice of verbena and the juice of fennel root and mix them together by fire and then strain it through a linen cloth and put it into a leaden vessel and then spread it round the outside of the eyes; and they will become dry.

Johan van Segen also records the use of eel bile for aphasia after a blow to the head or for concussion of the brain, as well as for sty:

> Item, der op den cop geslagen ist, dat he nycht sprechen en kan, dem gŭit [sic!] jn de oren von j ale de galle, getempert myt frauwenmylch.[78]

> Furthermore, he who has been struck on the head so that he cannot speak, give into his ears the gall of an eel tempered with woman's milk.

> Nym eynnes hannen gal vnd de gal van eym ale vnd make se warm myt der butter, de vngesalten sij. Do darto huinch [sic!]. Menge dat tohaufe jn ein copperen vat. Darmyt bestrich de awen, de dar ser sint den awent. Sige et dur eyn duch.[79]

76 Norrbom, *Das Gothaer mittelniederdeutsche Arzneibuch* 189. Cf. also 7 Interpretation below.

77 Ibidem 182.

78 Alstermark, *Das Arzneibuch des Johan van Segen* 82.

79 Ibidem 107.

> Take the gall of a cock and the gall of an eel and make them warm with butter that is unsalted. Add honey. Mix this together in a copper vessel. In the evening, spread it on the eyes that are affected. Strain it through a cloth.

Finally, eel bile mixed with chicken bile, hare bile, water and honey is used in *The Middle Low German Stockholm Pharmacopoeia* as a remedy for an unspecified eye disease:

> […], so nym de ghallen van eynem hanen vnde de gallen van eynem ale vnde de ghallen van eynem hasen vnde menge se myt renem warmen watere vnde myt honnighe vnde do dat in eyn kopperuat, vnde alse do to bedde geist, so bestryk dine oghen dar mede eyn lutting, so warden se di alto hant schon vnde klar.[80]

> […], so, take the gall of a cock and the gall of an eel and the gall of a hare and mix them with pure warm water and with honey and put that into a copper vessel; and when you go to bed coat your eyes with it a little; and they will become sharp [in sight; comment. S.T.] and clear to you immediately.

As one can easily see, eel bile is used especially for eye ailments. One can certainly be sceptical about the pharmaceutical effect of bile, as Matiegkova[81] does, for example, but fish bile as a remedy for eyes has been known for longer (there are some ancient Greek and ancient Egyptian sources as well as clay tablets in cuneiform writing, etc.).

Now let us look at the different uses of fish oil and especially fish oil from eel. This is suggested as a remedy in the Albrecht van Borgunnien's *Treatise on Medicine* in the case of eye opacity taken together with honey and aloe juice and in the case of middle ear inflammation together with houseleek (*Sempervivum tectorum* or *Sempervivum globiferum*).

> Nym dat vette van enem verschen vische, make dat warm vnd do dar to en cleyne honniges vnde aloes, lat id tosamende seden vnde schume id wol, drope dat clare in de ogen, dat helpet sere.[82]

---

80 Lindgren, *Ein Stockholmer mittelniederdeutsches Arzneibuch* 138–139.

81 Matiegkova L., "Tierbestandteile in den altägyptischen Arzneien", *Archiv Orientální* 26 (1958) 529–560, esp. 555.

82 Albrecht [van Borgunnien], *Albrecht van Borgunnien's Treatise on Medicine* 23.

> Take the fat from a fresh eel, make it warm and add some honey and [the juice of the] aloe [plant], let it boil together and skim it [the scum; comment. S.T.] off properly, drip this clear into the eyes; this helps a lot.
>
> De nicht wol høren kan. De neme huslok vnde drucke dat sap vth vnde neme dat vette van enem witten verschen ale, do dyt to hope vnde make it warm vnde drope des enen dropen in dat ore, dat helpet.[83]
>
> To whom who cannot hear well. Take houseleek and squeeze out the juice and take the fat from a fresh white eel, put it together and make it warm and drop a drop of it into the ear; that helps.

The Middle Low German Gotha Pharmacopoeia mentions fish oil from eel in a recipe dealing with the treatment of otitis media (used again together with houseleek):

> Stot huslok vnde wringk dat sap vth, vnde do darto dat vette van deme ale vnde make id warm. To deme ersten male | drope dat in dat beste ore vnde kere dat sulue ore vpwart, dat yd wol indrincke; des anderen dages edder des nachtes so do yd in dat andere ore vnde kere id ok vp; des drudden dages so do yt in dat erste ore, darna do yt in dat andere ore. Dyt do so lange bet du gesunt werst.[84]
>
> Grind houseleek and wring out the juice and add the fat from an eel and make it warm. For the first time, drop it into the best [probably: the healthier; comment. S.T.] ear and turn it upwards so that it absorbs the liquid; the next day or night, put it into the other ear and turn it upwards; on the third day, put it into the first ear, then into the other ear. And do this until you get well.

Eel as a whole is attested in a prescription for inflammation of the middle ear due to water in the ears in *The Pharmacopoeia* by Johan van Segen. The following is recommended:

> [...] neme eyn fetten alle; vnd bestrich den myt salt vnd brait en, als men etten sulde. Wan dat water daraf getrofen is, so do dat fette darvan; vnd nemen [sic!] dan cipollen; vnd machen [sic!] eyn lach darjn vnd do dat

83 Albrecht [van Borgunnien], *Albrecht van Borgunnien's Treatise on Medicine* 25.

84 Norrbom, *Das Gothaer mittelniederdeutsche Arzneibuch* 85.

> fette darjn. Brade de cipolle, dat se weich werde, vnd brinck et dan dur eyn reyn duch vnd do dat warm jn den oren.[85]

> […] take a fat eel; and coat it with salt and fry it as it should be eaten. When water has formed on it, take the fat from it; and then take onions; and make a bed in it and add the fat to it. Fry the onion until it becomes soft; and then put it through a clean cloth and put it warm in the ears.

Other fish species are also attested in the selected research corpus. For example, ruffe is mentioned in *The Middle Low German Gotha Pharmacopoeia* and is recommended alongside pike for the disease with four-day recurring fevers and cold sweats (*quartana*), although this evidence borders very closely on a dietary recommendation:

> Wen desse suke warende is, de holde syn ethent aldus: et grone swynenvleisch, junghe hunre, hennen de wol gesaden synt myt speke, mer dat speck eneth nicht. Jtem eth kamen, zafferan, negelken, annys, wetenbroet, culebarse vnde hekede.[86]

> If the disease persists for a long time, the affected person should maintain his eating habits as follows: Eat green [i.e., fresh; comment. S.T.] pork, young chickens, hens that are well boiled with lard, but do not eat the lard. Furthermore, eat cumin, saffron, cloves, star anise, wheat bread, ruffe and pike.

However, since no dependence on the lunar cycle, zodiac sign or season was recognised here, these two examples can still be categorized as medicinal uses.

Fish that are not specified are mentioned in the Magdeburgian *Promptuarium medicinae* for jaundice[87] and for melancholy and so-called loin weakness (in German: *Lendenschwäche*) – presumably meaning the English disease, i.e., potency problems. The second instruction in particular is described in great detail:

> Bystu uele swarmodich mydt vele dancken iffte swaren melancoli=en ock swarten kolre edder heffstu de lenden sucke van vele vuchticheit des flecma so sut polipodium mit annysze vnde sede dat in sode dar erst

85 Alstermark, *Das Arzneibuch des Johan van Segen* 114.

86 Norrbom, *Das Gothaer mittelniederdeutsche Arzneibuch* 140.

87 *Das Promptuarium medicinae* 86.

vissche edder honre ynne ghesoden synt do dar sucker tho icht du one heffst vnde drinck dat auent vnde morghen warm.[88]

If you are melancholic with [too] many thoughts, or heavy melancholy, or black bile, or if you have loin weakness from the excess of moisture of the phlegm, boil polipodium [a type of fern; comment. S.T.] with star anise; and boil that in the stock where fish or chicken were first boiled, add sugar to it; and drink that while it's still warm in the evening and in the morning.

Not only fish oil is used as an ingredient in the recipes, but also a number of other fish parts. For example, fresh fish guts are prepared in *The Middle Low German Gotha Pharmacopoeia* to battle the clouding of the lens in the eye, i.e., scotoma (in German: *Augenschimmel*), prepared as follows:

Jtem nim dat versche butte van eneme vissche warm gemaket vnde do darto eyn cleyne honnyges vnde olyes vnde lat dat tosamende smelten vnde schumret wol vnde drope des wat in de ogen.[89]

Furthermore, take fresh fish guts made warm and add a little honey and oil and let it all melt together and skim off [any scum] properly and drip some [of the mixture] into the eyes.

In contrast, the skin of stockfish (cod) is used in *The Pharmacopoeia* by Johan van Segen as a cooling plaster for inflammatory wounds:

So saltu neme j stackfuschhut [sic!] vnd seden de jn water vnd schum dat rein. Nym dan de hut vnd lege se of de wonde vnd ker de witte sitte negest.[90]

So, you shall take a stockfish skin and boil it in water and remove the scum that has formed. Then take the [stockfish] skin and place it on the wound and turn the white side closer [to the wound].

Pike gills are included in the same pharmacopoeia as a *corsif* ingredient:

88 *Das Promptuarium medicinae* 120.

89 Norrbom, *Das Gothaer mittelniederdeutsche Arzneibuch* 74.

90 Alstermark, *Das Arzneibuch des Johan van Segen* 88.

> Eyn ander corsif. Nym gebranten allun vnd run halwort. Stot dat tosamen. Do darto de kynbacken van dem hechte; de lege jn win /xvj/ xiiij stunden. Dan so bern se to pulwer. Dat menge tosamen. Do darto spangrŭn.[91]

> Another *corsif*. Take burnt alum and round-leaved birthwort. Pound that together. Add pike gills; soak them in wine for /16/ 14 hours. Then burn them to powder. Mix all this together. Add blue vitriol.

Bones of salted fish burnt to powder and fish-bone powder, among others from pike, are mentioned in *The Middle Low German Gotha Pharmacopoeia* in a recipe against intestinal prolapse and umbilical hernia:

> Js de nauel vthgheghaen edder de pagderme, vnde synt se noch warm, | js yt de pagdarm, so schere er dat haer af by der schemede vnde legghe se to bedde vnde laet se myt deme sterte hoger ligghen wen myt deme houede, vnde make denne eyn dingk het vp deme vure dat het apostolicon vnde bestrick de hant darmede edder myt boemolye vnde steck ene wedder yn, vnde stot wedewinden vnde drucke dat sap vth vnde nym dat sap vnde allun vnde hart vnde sede dyt myt honnighe edder myt lynolye vnde myt weken peke vnde laet dyt tohope seden vnde stot sweuel kleyne vnde do den darto vnde legghe dat so warm darvp vnde bestrick dat hol myt blode vnde berne graden | van enen solten vissche to asschen vnde strouwe de darvp.[92]

> If the navel is open, or the rectum, and they are still warm, if it is the rectum, shear off the hair by her [the patient's; comment. S.T.] pubic region and lay her on the bed; and let her lie with her rump higher than with her head; and then heat a thing, that is called apostolion, and coat a hand with it or with tree oil [probably: olive oil; comment. S.T.] and put it [the rectum; comment. S.T.] inside again; and pound bindweed and squeeze out the juice from it; and take the juice and alum and resin and boil that together with honey or with linseed oil and with soft pitch and boil that together and pound sulphur made small and add that to it and put that on it while it's still warm and coat the hole with blood and burn bones from a salted fish to ashes and sprinkle those on it.

91 Alstermark, *Das Arzneibuch des Johan van Segen* 97.

92 Norrbom, *Das Gothaer mittelniederdeutsche Arzneibuch* 130.

According to Johan van Segen they are helpful against fistulas, i.e., deep tubular ulcers:

> De vnnaturlichen meister haint [sic!] de gewonheit, dat se snydent de fistel vit vnd maken se wit myt der /vijr/ kunst, de se dŭn, vordriwen den menschen vnd hain [sic!] ein gewonheit dat to dun myt sterkem pulwer, als herna geschriwen steit. Dit pulwer saltu machen also: Nym funderatum vnd blee gebrant to pulwer vnd fischebeinpulwer vnd gledepulwer vnd hechtesbeinpulwer, bligwitpulwer. Dat mache allet tosamen – vnd nym eyns so vil als des anderen – vnd streuwe dat pulwer jn de wonde vnd jn schaden, de da flissent.[93]
>
> The unnatural masters are in the habit of cutting open the fistula and making it wide with the help of the art of fire, which they do, expel the man, and have a habit of doing this with a strong powder, as is described hereafter. This powder you shall make as follows: take funderatum and lead burnt to powder and fish-bone powder and limb powder and pike-bone powder, white lead powder. Put them all together and take of one as much as of the other and sprinkle the powder on the wound and on the noxious substance that flows there.

In the case of cystitis and uncontrolled urination, Albrecht van Borgunnien recommends fresh unpreserved fish bladders burnt to powder and added to wine, beer or, better, vinegar:

> Nym versche vischeblasen, berne de to puluere vnde dat puluer do in wine efte in bere vnde drinke dat. Noch were it beter in etike gedrunken. Vnde drinke dat ene weken vmme.[94]
>
> Take fresh fish bladders, burn them to powder and put this powder into wine or into beer and drink it. It would be even better drunken in vinegar. And drink it for a week.

The last example to be mentioned isn't really a reference to fish itself, because it is about the water in which fish was washed. According to *The Middle Low German Gotha Pharmacopoeia*, this water was used for the treatment of fistulas:

93 Alstermark, *Das Arzneibuch des Johan van Segen* 91.

94 Albrecht [van Borgunnien], *Albrecht van Borgunnien's Treatise on Medicine* 35.

> Wedder de fistelen, dat ys eyn swer | mit enen engen munde vnde myt wyden grunden. De bekenne hirby: dar geit wytte vulnisse vth alzo water dat visch ynne gewasschen sy vnde breket dor de aderen, vnde dar geit ok swarte vulnisse vth.[95]
>
> Against the fistulas; that is an ulcer with a narrow mouth and with wide grounds. The person recognises here: white putrefaction comes out, that is, water in which fish has been washed; and break it through the veins, so [that] black putrefaction also comes out.

7 Interpretation

A closer look revealed that in all but one recipe, the fish and its components were thermally processed before they could be administered. In most recipes, fresh fish or fresh fish bile, fish guts, etc. are first mixed with other ingredients (honey, water, butter, woman's milk, other bile types, herbs, herbal juices, etc.) and then heated or even boiled. In a number of recipes, fish bones – whether from fresh or salted fish – or even fish bladders are first burnt to powder or ash and only used afterwards. Pike gills are first soaked in wine for several hours (14) and only then burnt to powder and used further. Only in one recipe the eel bile isn't heated. This is an ointment recipe to help with swellings. Here, the eel bile is soaked in a mixture of tartar and vinegar for several days (four or more) and then not further processed:

> Aldus make ene salue jeghen alle swelle: nym eyn punt wynstenes de reyne sy, vnde beneye den in eme | doke vnde legghe dat denne in guden etick vnde laet dat stan ouer ene nacht vnde legghe ed denne in gloiende ameren ij stunde van enen daghe vnde nym ed denne vt vnde legghe yt denne an anderen etick. Dyt do drie ouer vnde do yt denne in eyn kupperen vat vnde do darto gallen van alen vnde laet ed so stan iiij dage efte mer, dat wert gude | salue to allen swellen vorwar.[96]
>
> In this way, make an ointment against all types of swellings: take a pound of tartar that is pure and sew it up in a cloth and then put it into a good vinegar and let it stand about a night; and then put it into the burning ashes for two hours during the day and then take it out again and put it

95 Norrbom, *Das Gothaer mittelniederdeutsche Arzneibuch* 97.
96 Ibidem 189.

> into another vinegar again. Repeat this three times and then put it in a copper vessel and add eel bile and let it stand for four days or more; this will be a good ointment for all swellings indeed.

If we look at the modes of use and action of the remedies in the recipes studied and the relevant fish ingredients, we primarily find examples of the *contraria contrariis curantur* principle of traditional humoral pathology; which was the leading principle from Galen until the development of modern medicine and homeopathy. In part, some uses can be seen more as examples of the *similia similibus curantur* principle, but neither in the homeopathic nor in the classical humoral pathological sense (here I refer in particular to the prescription of fish bladder for cystitis and uncontrolled urination). According to humoral pathology, fish is considered "cold" and "moist", so it is obvious that it is used for inflammatory – "fiery" – diseases, but also in the form of cooling plasters. The perceived excessive coldness could be reduced by the heating process or boiling, if necessary, and the moist attribute would still remain. Finally, one should not forget that most of the remedies described were applied externally, directly to the affected area (eyes, ears, wounds, swollen body parts) and not internally, e.g., orally. Thus, no direct comparison is possible between the mode of action of fish as a food (in dietetics) and of fish as a component of a remedy.

8 Conclusion

In conclusion, as hypothesized, fish-based medical recipes are marginally represented in the selected research corpus. However, the uses of fish and its components are not limited to the already known and more frequently recorded eye diseases and diseases of the urinary tract, but we find a much broader spectrum of uses, such as for aphasia, middle-ear inflammation, inflammatory wounds, fistulas, intestinal prolapse, loin weakness (impotency), etc. As expected, the main underlying treatment principle for the uses of fish and its various components has emerged as the *contraria contrariis curantur* principle of traditional humoral pathology; which was the leading principle in the Middle Ages and the early modern period.

For further research, it would be undoubtedly useful to take a closer look at all available Middle Low German pharmacopoeias, herbal books and recipe books in order to be able to make generalised statements and possibly discover further uses of the fish and its components that were previously unknown to science. Due to the exclusive restriction of the research corpus to the sources of *The Middle Low German Dictionary*, and thus to a clear amount of evidence, it is unfortunately not possible to make any generally valid statements. Nevertheless, even after such an exemplary analysis, it can be stated that this kind of preliminary research may be fruitful. For example, some uses – especially of fish bones or pike gills – were found that were previously unknown to scholars.

Bibliography

Primary Sources

Albrecht [van Borgunnien], *Albrecht van Borgunnien's Treatise on Medicine (Sloane Ms. 3002, British Museum)*, ed. W.L. Wardale (London – Edinburgh – Glasgow: 1936).

Alstermark H., *Das Arzneibuch des Johan van Segen* (Stockholm: 1977).

Bartholomaeus [Salernitanus], *Angebliche practica des Bartholomaeus von Salerno, Schüler des Constantinus Salernitanus. Introductiones et experimenta magistri Bartholomaei in practicam Hippocratis Galieni Constantini graecorum medicorum. Papier-Handschrift der herzogl. Sachs.-Coburg-Gothaischen Bibliothek, No. 920, fol. 85a bis 104b*, ed. F. von Oefele (Bad Neuenahr: 1894).

Brunschwig Hieronymus, *Dat Boek der Wundenartzstedye und der niederdeutsche chirurgische Fachwortschatz*, ed. Ch. Benati (Göppingen: 2012).

Das Breslauer Arzneibuch [R 291 der Stadtbibliothek], ed. C. Külz – E. Külz-Trosse (Dresden: 1908).

Das Promptuarium medicinae. [Eyn schone Arstedyge boeck van allerleye ghebreck unnde kranckheyden der mynschen], ed. P. Seidensticker nach d. Ausg. Magdeburg: 1483 (Lahr: 1990).

Das Utrechter Arzneibuch (Ms. 1355, 16°, Bibliotheek der Rijksuniversiteit Utrecht), ed. A. Lindgren (Stockholm: 1977).

Doneldey Arnoldus, *Das Bremer mittelniederdeutsche Arzneibuch des Arnoldus Doneldey. Mit Einleitung und Glossar*, ed. E. Windler (Neumünster: 1932).

Gallée J.H., "Mittelniederdeutsches Arzneibuch", *Jahrbuch des Vereins für niederdeutsche Sprachforschung* 15 (1889) 105–149.

Graeter Chr., *Ein Leipziger deutscher Bartholomaeus* (Borna-Leipzig: 1918).

Lindgren A., *Ein Stockholmer mittelniederdeutsches Arzneibuch aus der zweiten Hälfte des 15. Jahrhunderts* (Stockholm – Gothenburg – Uppsala: 1967).

Lindgren A., *Die 'aquae medicinales' des mittelniederdeutschen Gothaer Arzneibuches* (Stockholm: 1979).

Lindgren A., "Ein Kopenhagener mittelniederdeutsches Arzneibuch aus dem Ende des 15. Jahrhunderts", *Würzburger medizinhistorische Mitteilungen* 4 (1986) 135–178.

Ljungqvist I., *Das mittelniederdeutsche Arzneibuch des Codex Guelferbytanus 1213 Helmstadiensis* (Stockholm: 1971).

Norrbom S.E., *Das Gothaer mittelniederdeutsche Arzneibuch und seine Sippe* (Hamburg: 1921).

Regel K., "Das mittelniederdeutsche Gothaer Arzneibuch und seine Pflanzennamen", *Programm des Herzoglichen Gymnasium Ernestinum zu Gotha* 1 (1872) [1]–16, 2 (1873) 1–26.

Regel K., "Zwei mnd. Arzneibücher, Cod. Chart. Goth. 980 und Cod. Wolfenb. 23,3.", *Jahrbuch des Vereins für niederdeutsche Sprachforschung* 4 (1878) 5–26.

Röder Sebastian, *Nütte lere vnd vnderricht Doct. Sebastiani Rǒder/ Jdtzundes alhier tho Hamborch/ We men sick in dǔssen gefarlicken Steruendes lǔfften holden/ vnde vor der giff=tigen Pestilentzischen sǔcke bewaren schall/ dat men dar vör seker wesen/ vnd nicht darmit beflecket warden mach/ edder sick jo dar van erredden mǒge* (Hamburg, Johann Wickradt: 1565).

Temmen M., *Das 'Abdinghofer Arzneibuch'. Edition und Untersuchung einer Handschrift mittelniederdeutscher Fachprosa* (Cologne – Weimar – Vienna: 2006).

Secondary Sources

Albala K., "Fish in Renaissance Dietary Theory", in Walker H. (ed.), *Fish. Food from the Waters. Proceedings of the Oxford symposium on food and cookery 1997* (Oxford: 1998) 9–19.

Borchling C., "Zur mittelniederdeutschen Medicin", *Korrespondenzblatt des Vereins für niederdeutsche Sprachforschung* 22.5 (1901) 69–71.

Borchling C., "Die mittelniederdeutschen Arzneibücher", *Janus* 7 (1902) 131–134, 175–179, 245–251.

Borchling C., "Zur Handschriftenkunde der mittelniederdeutschen Medizin und Naturwissenschaft", *Mitteilungen zur Geschichte der Medizin und der Naturwissenschaften* 1 (1902) 66–70.

Buxton M., "Fish-Eating in Medieval England", in Walker H. (ed.), *Fish. Food from the Waters. Proceedings of the Oxford symposium on food and cookery 1997* (Oxford: 1998) 51–59.

Clesse G., "Animaux aquatiques dans les sources médicales arabo-latines: continuités et discontinuités d'un discours", *Médiévales* 80.1 (2021) 81–98.

Cobo López M.P. – Tijera Jiménez R.E., *Etnozoología de Doñana* (Sevilla: 2013).

Hoffmann R.C., "A brief history of aquatic resource use in medieval Europe", *Helgoland Marine Research* 59 (2005) 22–30.

Keil G., "Randnotizen zum 'Stockholmer Arzneibuch'", *Studia Neophilologica* 44 (1972) 238–262.

Le Cornec C., "Les vertus diététiques attribuées aux poissons de mer", in Connochie-Bourgne Ch. (ed.), *Mondes marins du Moyen Âge* (Aix-en-Provence: 2006) 273–284.

Matiegkova L., "Tierbestandteile in den altägyptischen Arzneien", *Archiv Orientální* 26 (1958) 529–560.

Maycock P.P. Jr., "Introduction to the Second Edition", in Buchanan S. (ed.), *The Doctrine of Signatures: A Defence of Theory in Medicine* (Urbana, IL: 1991) xv–xxix.

Möhn D., "Vorwort zu Bd. II/1", in *Mittelniederdeutsches Handwörterbuch*, ed. A. Lasch – C. Borchling – G. Cordes – D. Möhn – I. Schröder (Kiel – Hamburg: 1928 ff.) vol. II/1, I–XIII.

Möhn D. – Schröder I., "Lexikologie und Lexikographie des Mittelniederdeutschen", in Besch W. – Reichmann O. – Sonderegger S. (eds.), *Sprachgeschichte. Ein Handbuch zur Geschichte der deutschen Sprache und ihrer Erforschung*, vol. 2.2 (Berlin – New York: 2000) 1435–1456.

Oefele F. von, "Zur mittelniederdeutschen Medizin", *Korrespondenzblatt des Vereins für niederdeutsche Sprachforschung* 22.4 (1901) 49–50.

Rigat M. – Vallès J. – Iglésias J. – Garnatje T., "Traditional and alternative natural therapeutic products used in the treatment of respiratory tract infectious diseases in the eastern Catalan Pyrenees (Iberian Peninsula)", *Journal of Ethnopharmacology* 148.2 (2013) 411–422.

Sabri Kolta K., "Der Fisch im Volksglauben und in der Medizin der Araber", *Deutsches Schiffahrtsarchiv* 1 (1975) 211–215.

Schröder I., "Niederdeutsch, niederdeutsche Sprache", in *HanseLexikon* (*HansLex*), ed. Hansischer Geschichtsverband, online at https://www.hansischergeschichtsverein.de/lexikon, accessed on November 27, 2022.

Schröder I. – Tsapaeva S., "Komplexe Semantik im mittelniederdeutschen Wörterbuch", in Harm V. – Lobenstein-Reichmann A. – Diehl G. (eds.), *Wortwelten. Lexikographie, historische Semantik und Kulturwissenschaft* (Berlin – Boston: 2019) 177–202.

Soden W. von, "Fischgalle als Heilmittel für Augen", in Soden W. von – Müller H.-P. (eds.), *Bibel und Alter Orient* (Berlin – Boston: 1985) 76–77.

Temmen M., *Die medizinische Rezepthandschrift Burgsteinfurt Hs. 15. Edition und Untersuchung einer Handschrift aus dem 16. Jahrhundert* (Bielefeld: 1998).

Ullmann M., *Islamic Medicine* (Edinburgh: 1997).

Vallejo J.R. – González J.A., "Fish-based remedies in Spanish ethnomedicine: a review from a historical perspective", *Journal of Ethnobiology and Ethnomedicine* 10.37 (2014) 1–31.

Wickersheimer E., "Zur spätmittelalterlichen Fischdiätetik: Deutsche Texte aus dem 15. Jahrhundert", *Sudhoffs Archiv für Geschichte der Medizin und der Naturwissenschaften* 47.3 (1963) 411–416.

Аникин И.Л., "Некоторые методы лечения, применявшиеся в средневековой русской медицине", *AMHR* 1.20 (1992) 134–143.

CHAPTER 12

The Invisible Fisherman: The Economy of Water Knowledge in Early Modern Venice

Pietro Daniel Omodeo

Over three decades ago, Steven Shapin, one of the main scholars responsible for the "culturalist turn" in the history of science, directed the attention of historians of early modern science toward the importance of those practitioners who had materially accomplished the most famous experiments in the time of Robert Boyle and the Royal Society.[1] As Shapin argued, the skillful activities of these practitioners must be seen as a substantial contribution to the construction of modern science, although they have disappeared from sight. This has happened not only because of the negligence or ideological preconceptions of historians of the Scientific Revolution, but also because the historical sources themselves tend to be silent about these early modern practitioners. Yet, some traces of their presence remain, for instance in images depicting them in the sublimated form of assistant putti or cherubs. Much progress has been made towards the reevaluation of these "practical bearers" of knowledge since Shapin's remarks on the importance of submerged micro-sociological contexts. They were people from the working classes such as craftsmen, miners and midwives.[2] Fishermen also belong to this world of

1 Shapin S., "The Invisible Technician", *American Scientist* 77, 6 (1989) 554–563. Shapin coauthored with Simon Schaffer the work that is most commonly regarded as a watershed between the earlier contextualist sociology of science and the new culturalist sociology of scientific knowledge, *Leviathan and the Air-Pump* (1985). Cf. the authors' "Introduction to the 2011 Edition: Up for Air", in idem (eds.), *Leviathan and the Air-Pump: Hobbes, Boyle and Experimental Life* (Princeton: 2011, first edition 1985) xi–l. On the historiographical problem of changing perspectives on the Scientific Revolution, see Cohen F., *The Scientific Revolution: A Historiographical Inquiry* (Chicago: 1994) and Omodeo P.D., "Scientific Revolution, Ideologies of the", in Jalobeanu D. – Wolfe C.T. (eds.), *Encyclopedia of Early Modern Philosophy and the Sciences* (n.p.: 2020).

2 This is a sort of Marxist triad of knowledge-bearers from below. Back in the 1930s and 1940s, Boris Hessen and Edgar Zilsel already pointed to the importance of craftsmen for the construction of modern science, in their works on the socio-economic roots of early modern science. Substantial contributions to the understanding of the micro-sociology of practical knowledge have been made in more recent years. See, among others, Long P.O., *Artisan/*

expertise from below. Florike Egmond has made us aware of their relevance through several case studies, the most prominent of which is that of the Dutch merchant and student of marine life, Adriaen Coenen (1514–1587), himself the son of a fisherman. Through his mixed expertise, Coenen acted as a bridge between various social and epistemological communities. Indeed, he established his fame by collecting mirabilia, and interviewing fishermen and travelers in order to gather marine information. Throughout his life, he exchanged specimens, drawings and knowledge with common people, university scholars and prominent figures.[3] Following Egmond's invitation to investigate the roots of ichthyology at the point of encounter between different groups, I will here turn my attention to the question of the link between erudition and fishermen's practical knowledge in the Renaissance, with Italy and Venice as my main areas of inquiry. I will begin with some remarks on the importance of practice in connection to nomenclature, on the basis of evidence drawn from the most reputed authors who wrote on ichthyology in the 16th and early 17th centuries. Secondly, I will point to some important aspects of fishermen's water expertise in Venice by resorting to historical documents on fishing legislation and water management. My attempt is to shift the focus from the culture of the elites to that of water laborers. Eventually, I will look at one specific fish species, the gobius (which was and is still called 'go' in Venice), as an object of inquiry from a variety of epistemological and social perspectives. This fish, as the sources tell us, constituted an important ingredient in people's diet in the area of Venice area from antiquity to early modernity.

1 Fish Nomenclature and Practical Knowledge

The problem of nomenclature is ubiquitous in early-modern books on plants and animals. An understanding of the classics in this field presupposed that a correct correspondence be established not only between Greek and Latin sources – a problem for *literati* – but also between ancient names and those in

Practitioners and the Rise of the New Sciences, 1400–1600 (Corvallis: 2011); Smith P., *The Body of the Artisan: Art and Experience in the Scientific Revolution* (Chicago: 2004); Valleriani M., *The Structures of Practical Knowledge* (Cham: 2017); and Klein U., *Nützliches Wissen: die Erfindung der Technikwissenschaften* (Göttingen: 2016).

3 Egmond F., "On Northern Shores: Sixteenth-Century Observations of Fish and Seabirds (North Sea and North Atlantic)", in MacGregor A. (ed.), *Collecting, Recording and Preserving the Natural World from the Fifteenth to the Twenty-First Century* (Leiden – Boston: 2016) 129–148.

use among ordinary people. Only on the basis of such translation efforts across different languages could knowledge of the past be reactivated, assessed and, if necessary, corrected and expanded. Indeed, the practical experience of the lower classes was a repository of empirical knowledge from which science and philology could greatly benefit.

In one of the most elegant Renaissance works on fish, *Aquatilium animalium historiae liber* (History of Aquatic Animals) (Rome 1554), the physician Ippolito Salviani of Rome (1514–1572) pitted observation against bookish knowledge and labeled the former as "historical", in accordance with Pliny the Elder. Although Salviani, a learned humanist, held the ancients in high esteem (especially Aristotle, Pliny, Solinus, Oppian, Aelian, Athenaeus and Eustathius), he argued that "reality", rather than opinions, counts. He opposed *res* – "things", which lie at the basis of historical truth (*historiae veritas*) – to men's authority (*hominum auctoritas*):

> Our purpose has been to affirm nothing else than that which we learned and observed in reality itself [*re ipsa*]. Thus, we have often been forced to correct the writings of others, not in order to contradict them (as might seem) but for the sake of truth, which should be a greater friend to us than Plato and Socrates.[4]

In order to reassess and advance ichthyology, Salviani started from names. The first, extensive part of his *Aquatilium animalium historiae liber* deals with *nomina*. It is an alphabetical list of fish from 'Ablennis' to 'Zygaena'. A column to the left lists the Latin names, while a series of additional columns on two consecutive pages present the following information (in this order): the corresponding Greek names; the Italian ones (occasionally with the specification of local variants used in Rome and Venice); the main characteristics of the different species; and the relevant passages in Aristotle, Oppian, Pliny, Athenaeus, Aelian and *varii auctores* (various authors) [Fig. 12.1A–B]. This was by no means an erudite exercise performed for its own sake, because its scientific purpose was clear. It was a necessary premise for any critical engagement with ichthyology and the connection between ancient knowledge and the empirical knowledge that could be gathered from the present.

4 Salviani Ippolito, *Aquatilium animalium historiae* (Rome, Ippolito Salviani: 1554) fols. 8r–v: 'Nobis propositum fuit nihil affirmare, nisi quod ita se habere re ipsa didicimus, ac perspeximus, unde saepe coacti fuimus aliorum scripta reprehendere, non sane contradicendi studio (ut videmur) sed veritatis gratia, quae nobis amicior Platone et Socrate, esse debet [...]'.

These synoptic tables are followed by the "empirical" part. This features wonderful engravings of various fish species accompanied, whenever possible, by their Greek, Latin and Italian names. The text is presented according to a precise scheme that Salviani explains to his readers:

> First, we offer the name of each individual fish in Greek, Latin and the vernacular (as far as this is possible). Then, we describe the appearance of its entire body and, after that, we address its nature and habits. Additionally, we abundantly teach what its specific juices are, as well as its nutritional properties and the diseases they cure. In this manner, in my opinion, nothing is missing for a complete account [*universa historia*].[5]

The visual element has a fundamental epistemological function. Indeed, Salviani's engravings serves as a substitute for direct observation [Fig. 12.2].[6] The origin of the drawings is not specified, but I consider them to be a witness of exchanges with the world of practice, or of the "fieldwork activity" of this author.[7] Yet, one can only speculate about the exchanges that took place between Salviani and fishermen, because he conceals his encounters with working-class people behind the ostentatious dialogue between himself and classical authors (such as Aristotle, Strabo and Galen) and the high clergy in Rome.

One can find more references to practical contexts in other works from the same period, for instance Paolo Giovio's *De Romanis piscibus libellus* (Booklet on Roman Fish) (Rome 1524). This book, which predates Salviani's, deals with the problem of nomenclature in connection with both philology and nature itself. Giovio (1483–1552) claims that a major difficulty in gaining knowledge

5 Salviani, *Aquatilium animalium historiae* fol. +8r: 'Primum exposuimus, quo singuli pisces nomine, tum Graece, tum Latine, tum etiam vulgari gentium lingua (quantum consequi potuimus) appellentur, dein totius corporis figuram descripsimus, denique naturam, moresque illorum persecuti sumus, *ad haec quaque arte et capi*, et condiri debeant, qualis singulorum succus sit, quale nutrimentum, quibusque morbis medeantur copiose docuimus, ut universam historiam (iudicio meo) plane nihil deesse videatur'.

6 On "epistemic images" see Egmond F., "Aldrovandi, Truthfully Drawing Naturalia, and Local Context", *Acta ad archaeologiam et artium historiam pertinentia* 32, 18 (2020) 81–95.

7 Cf. Egmond, "On Northern Shores" 131: 'Some sixteenth-century naturalists who described fishes of the Mediterranean (e.g., Guillaume Rondelet, Pierre Belon, Ippolito Salviani) are known to have personally gathered at least part of their information out at sea, on the coast, in the ports, at fish markets, and via conversation with fishermen. This is as close as we can get to "field" work in early modern marine research, and although the term is perhaps not perfectly suited to marine research, it does help to distinguish personal observation and fieldwork once removed (via fishermen) from book learning'.

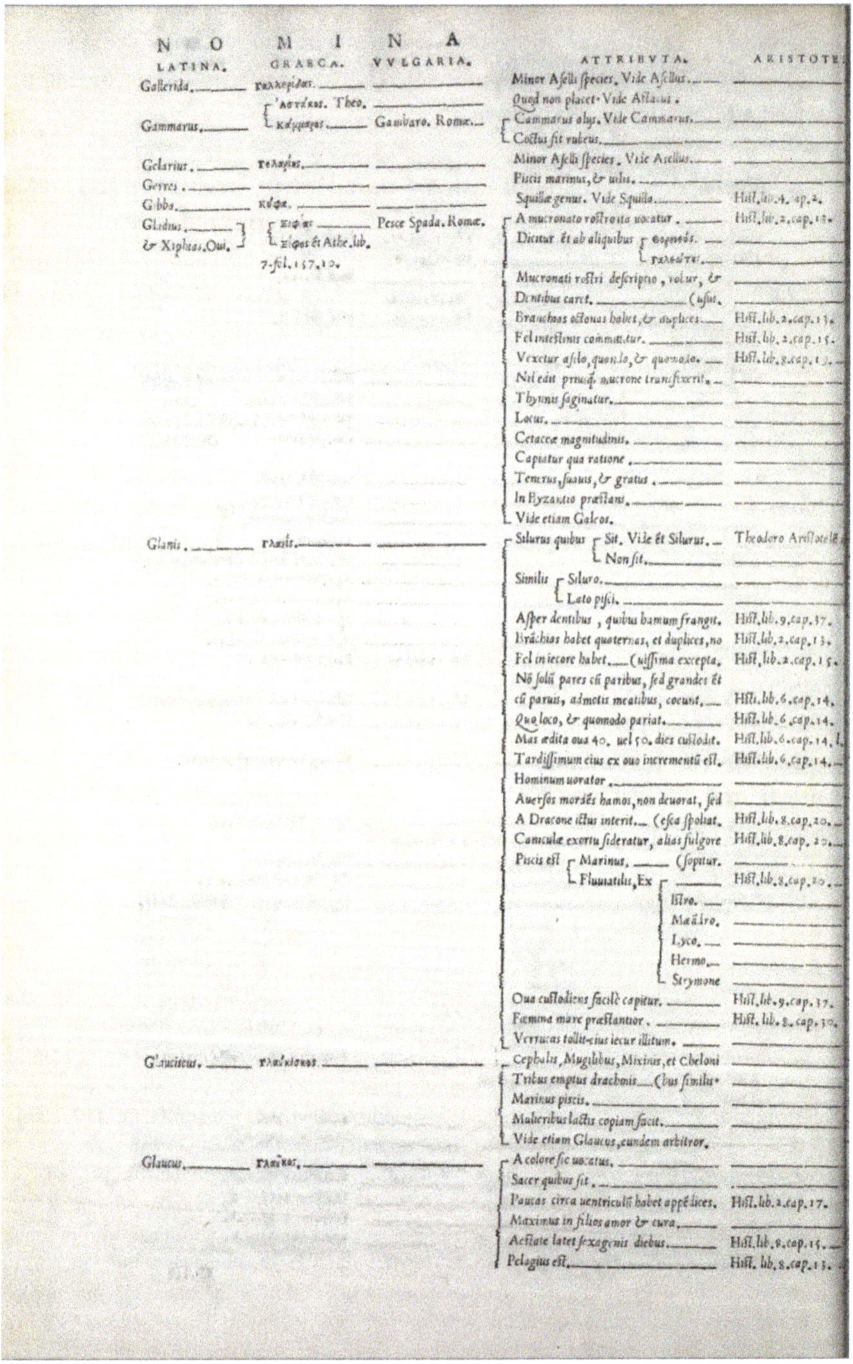

NOMINA LATINA.	GRAECA.	VVLGARIA.	ATTRIBVTA.	ARISTOTE[LES]
Gallerida.	γαλλερίδας.		Minor Aselli species. Vide Asellus.	
			Quod non placet. Vide Asellus.	
Gammarus.	ἀστακός. Theo.		Cammarus alijs. Vide Cammarus.	
	κάμμαρος	Gambaro. Romae.	Coctus fit rubeus.	
Gelarius.	γελαρίας.		Minor Aselli species. Vide Asellus.	
Gerres.			Piscis marinus, & uilis.	
Gibba.	κύφα.		Squillae genus. Vide Squilla.	Hist. lib. 4. cap. 2.
Gladius, & Xiphias. Oui.	ξιφίας	Pesce Spada. Romae.	A mucronato rostro ita uocatur.	Hist. lib. 2. cap. 13.
	ξίφος et Athe. lib. 7. fol. 157. 10.		Dicitur et ab aliquibus θρανίας.	
			γαλεώτης.	
			Mucronati rostri descriptio, uolur, & (usus.	
			Dentibus caret.	
			Branchias octonas habet, & duplices.	Hist. lib. 2. cap. 13.
			Fel intestinis committitur.	Hist. lib. 2. cap. 15.
			Vexetur asilo, quando, & quomodo.	Hist. lib. 8. cap. 19.
			Nil edit priusquam mucrone transfixerit.	
			Thynnus saginatur.	
			Locus.	
			Cetaceae magnitudinis.	
			Capiatur qua ratione.	
			Tenerus, suauis, & gratus.	
			In Byzantio praestans.	
			Vide etiam Galeos.	
Glanis.	γλάνις.		Silurus quibus Sit. Vide et Silurus.	Theodoro Aristotele
			Non sit.	
			Similis Siluro.	
			Lato pisci.	
			Asper dentibus, quibus hamum frangit.	Hist. lib. 9. cap. 37.
			Brãchias habet quaternas, et duplices, no (uissima excepta.	Hist. lib. 2. cap. 13.
			Fel in iecore habet.	Hist. lib. 2. cap. 15.
			Nõ solũ pares cũ paribus, sed grandes et cũ paruis, admotis meatibus, coeunt.	Hist. lib. 6. cap. 14.
			Quo loco, & quomodo pariat.	Hist. lib. 6. cap. 14.
			Mas aedita oua 40. uel 50. dies custodit.	Hist. lib. 6. cap. 14.
			Tardissimum eius ex ouo incrementũ est.	Hist. lib. 6. cap. 14.
			Hominum uorator.	
			Auersos mordẽs hamos, non deuorat, sed (esca spoliat.	
			A Dracone ictus interit.	Hist. lib. 8. cap. 20.
			Caniculae exortu sideratur, alias fulgore (sopitur.	Hist. lib. 8. cap. 20.
			Piscis est Marinus.	
			Fluuiatilis, Ex	Hist. lib. 8. cap. 20.
			Istro.	
			Maeãdro.	
			Lyco.	
			Hermo.	
			Strymone	
			Oua custodiens facilè capitur.	Hist. lib. 9. cap. 37.
			Foemina mare praestantior.	Hist. lib. 8. cap. 30.
			Verrucas tollit eius iecur illitum.	
Glaucicus.	γλαυκίσκος.		Cephalis, Mugilibus, Mixinis, et Cheloni (bus similis.	
			Tribus emptus drachmis.	
			Marinus piscis.	
			Mulieribus lactis copiam facit.	
			Vide etiam Glaucus, eundem arbitror.	
Glaucus.	γλαῦκος.		A colore sic uocatus.	
			Sacer quibus sit.	
			Paucas circa uentriculũ habet appẽdices.	Hist. lib. 2. cap. 17.
			Maximus in filios amor & cura.	
			Aestate latet sexagenis diebus.	Hist. lib. 8. cap. 15.
			Pelagius est.	Hist. lib. 8. cap. 13.

FIGURE 12.1A Salviani Ippolito, *Aquatilium animalium historiae* (Rome, Ippolito Salviani: 1554). Multicolumn tables of fish names concordances and fish information, fol. 19v

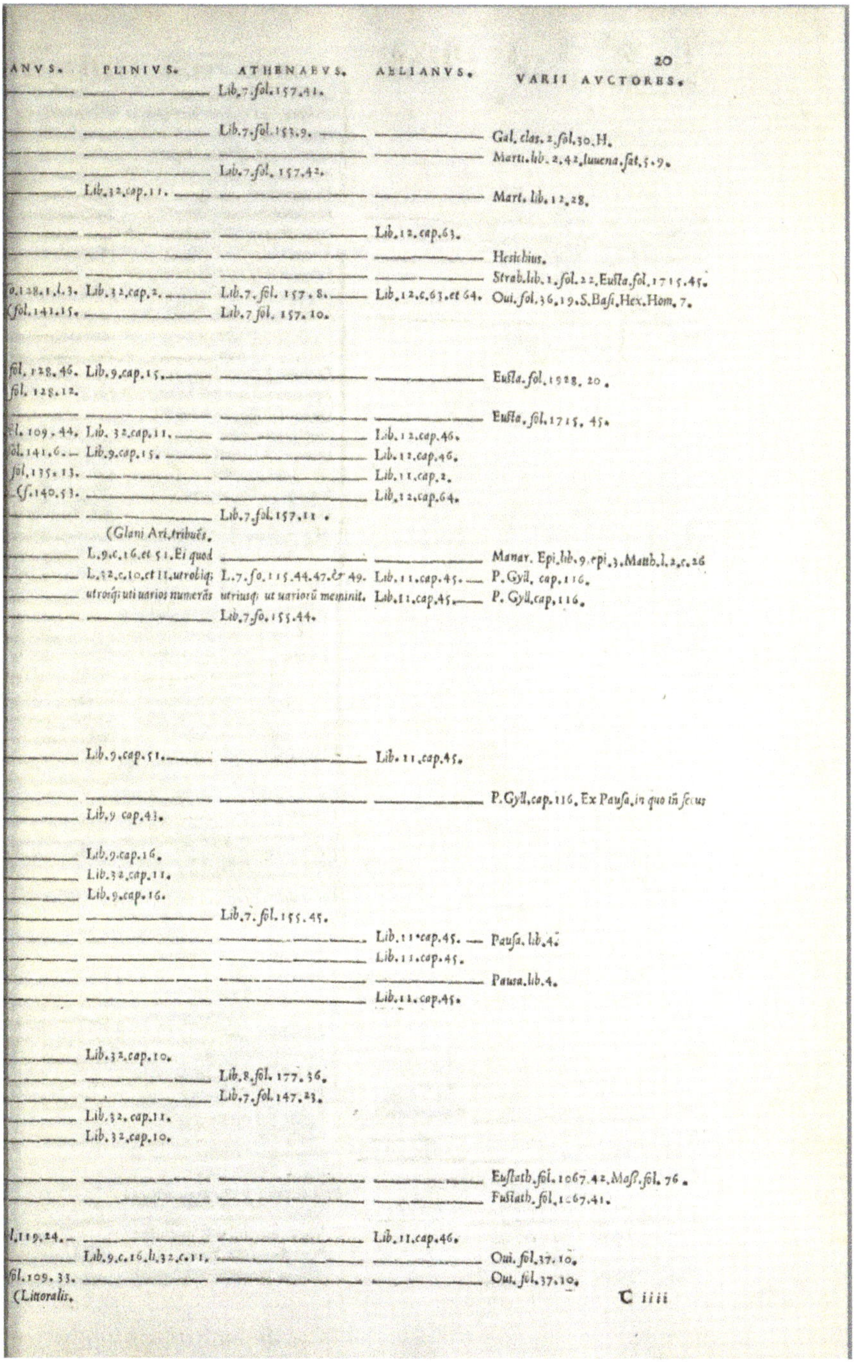

20

ANVS.	PLINIVS.	ATHENAEVS.	AELIANVS.	VARII AVCTORES.
		Lib.7.fol.157.41.		
		Lib.7.fol.153.9.		Gal. clas. 2.fol.30.H.
				Martı.lib. 2.42.Iuuena.sat.5.9.
		Lib.7.fol. 157.42.		
	Lib.32.cap.11.			Mart. lib. 12.28.
			Lib.12.cap.63.	
				Hesichius.
				Strab.lib.1.fol.22.Eusta.fol.1715.45.
o.128.1.l.3.	Lib.32.cap.2.	Lib.7.fol. 157.8.	Lib.12.c.63.et 64.	Oui.fol.36.19.S.Basi.Hex.Hom. 7.
fol.141.15.		Lib.7.fol. 157.10.		
fol. 128.46.	Lib.9.cap.15.			Eusta.fol.1928. 20.
fol. 128.12.				
				Eusta. fol.1715. 45.
l. 109.44.	Lib. 32.cap.11.		Lib.12.cap.46.	
ol.141.6.	Lib.9.cap.15.		Lib.12.cap.46.	
fol.135.13.			Lib.11.cap.2.	
f.140.53.			Lib.12.cap.64.	
		Lib.7.fol.157.11.		
	(Glani Ari.tribuēs,			
	L.9.c.16.et 51.Ei quod			Manar. Epi.lib.9.epi.3.Matth.l.2.c.26
	L.32.c.10.et 11.utrobiq;	L.7.fo.115.44.47.& 49.	Lib.11.cap.45.	P.Gyll. cap.116.
	utrosq; uti uarios numerās	utriusq; ut uariorū meminit.	Lib.11.cap.45.	P. Gyll.cap.116.
		Lib.7.fo.155.44.		
	Lib.9.cap.51.		Lib.11.cap.45.	
				P.Gyll.cap.116.Ex Pausa.in quo in fecus
	Lib.9 cap.43.			
	Lib.9.cap.16.			
	Lib.32.cap.11.			
	Lib.9.cap.16.			
		Lib.7.fol.155.45.		
			Lib.11.cap.45.	Pausa. lib.4.
			Lib.11.cap.45.	
				Pausa.lib.4.
			Lib.11.cap.45.	
	Lib.32.cap.10.			
		Lib.8.fol. 177.36.		
		Lib.7.fol.147.23.		
	Lib.32.cap.11.			
	Lib.32.cap.10.			
				Eustath.fol.1067.42.Mass.fol. 76.
				Eustath.fol.1067.41.
l.119.24.			Lib.11.cap.46.	
	Lib.9.c.16.li.32.c.11.			Oui. fol.37.10.
fol.109.33.				Oui. fol.37.10.
(Littoralis,				C iiii

FIGURE 12.1B Salviani Ippolito, *Aquatilium animalium historiae* (Rome, Ippolito Salviani: 1554). Multicolumn tables of fish names concordances and fish information, fol. 20r

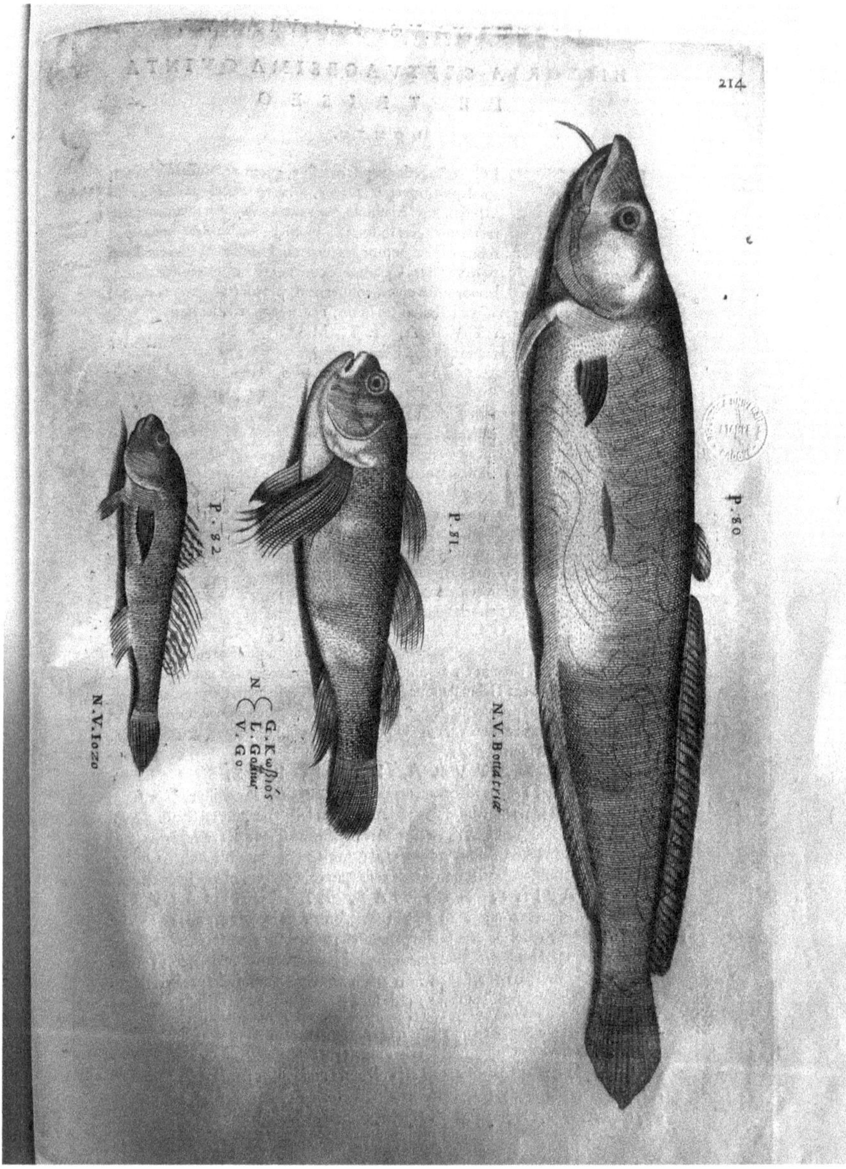

FIGURE 12.2 A fish engraving in Salviani Ippolito, *Aquatilium animalium historiae* (Rome, Ippolito Salviani: 1554)

about fish lies in the 'almost infinite variety' of aquatic species. According to him, problems of languages and natural variability concur to make ichthyology an arduous field of investigation. This complexity partially explains the divergences that, in his view, mark the distance between the classics and the living experience of contemporary fishing:

> It should come as no surprise for curious people that the works of the ancients largely conflict with the fishing of our times. Even if the animal species of land and sea may have always remained the same according to the perpetual order of nature, there is no doubt that they seem to have varied and changed, owing to many causes.[8]

Giovio remarks that fish migrate, just like birds. This fact accounts for the relevant changes that often occur in marine animal populations.

> Fish spontaneously migrate [...]. Fishermen are [at times] surprised at what fish they have caught with their nets. One ought to believe that they come from the extreme recesses of the seas or the depths of the profoundest eddies, or that they have arrived in the Mediterranean Sea [*Mare nostrum*] through the strait of Gibraltar [*Gaditanas fauces*], in the same manner as, in certain years, unusual birds will fly into Italy from very distant regions, as has been reported.[9]

Giovio also mentions the fish market as a place of encounters. However, he does not present it as a place in which knowledge about fish is gathered, but rather as one in which it is important to already have adequate knowledge, in order not to be cheated by the fishmongers. Moreover, he considers some of the difficulties that scholars face in trying to understand ancient sources on fish to be the result of discontinuities in fishing practices. Hence, he acknowledges the importance of these practices, at least indirectly: 'As fishing was neglected, the old names for fish have been radically forgotten'.[10]

Two French authors stand out alongside these Italians for their exploration of aquatic regions and linguistic aspects. In his systematic *Libri de piscibus marinis, in quibus verae piscium effigies expressae sunt* (Books on Marine Fish, in Which the True Images of Fish Are Presented) (Lyon 1554), the Montpellier professor of medicine, Guillaume Rondelet (1507–1566), lists as many

8 Giovio Paolo, *De Romanis piscibus libellus* (Rome, F. Minitius Calvus: 1524) fol. A3r: 'Neque mirum curiosis esse debuerit, si quae veteres prodidere, nostrae tempestatis piscationi, omni ex parte, minime consenserint. Nam, et si easdem semper animantium species perpetuo natuae ordine terra marique fuisse, nequaquam sit dubitandum, multa tamen variis de causis variata, atque immutata videntur'.

9 Giovio, *De Romanis piscibus libellus* fol. A3v: 'Peregrinantur enim sua sponte pisces [...]. Novos etiam pisces retibus exceptos piscatores admirati sunt, quos de extremis maris recessibus, imisque profundi gurgitibus provenire, vel per Gaditanas fauces in mare nostrum a vastissimo oceano irrumpere credendum est, sicuti etiam certis annorum curriculis inusitati generis aves e diversa terrarum regione in Italiam advolasse traduntur'.

10 Giovio, *De Romanis piscibus libellus* fol. A4v: 'Neglecta piscatione, vetera piscium nomina penitus exolverint'.

vernacular names as possible in addition to the Latin and Greek ones, including French and Provençal words, but also very local expressions used in harbors such as those of Marseille and Italian seaside towns. The encyclopedic naturalist, Conrad Gessner (1516–1565), who only saw the sea twice and sojourned in Venice for one month, compiled one of the most monumental works on water animals of his time by drawing information about Mediterranean fish from secondary sources, including Rondelet, whose linguistic explanations he highly praised:

> Rondelet displayed uncommon diligence in his highly accurate inquiry into the true and ancient names of fish, their descriptions (for which he personally traveled to Belgium and Italy), learned references, and explanations of dubious and obscure passages in the reference authors.[11]

Gessner also integrated his knowledge through the collection of images, which he received and widely exchanged. They included images of fish, some of them of Venetian provenance [Fig. 12.3].[12] Another prominent French author, Pierre Belon (1518–1564), published a booklet of fish images and descriptions, *De aquatilibus, libri duo* (Two Books on Water Animals) (Paris 1554), which began with a proud celebration of his own explorations:

> I dare to say only this: there is nothing imaginary or conjectural [in this work], but only the expression of that which I observed at various moments in Pontus and Hellespont, the Tyrrhenian, Eritrean and Adriatic seas, and our ocean.[13]

11 Gessner Conrad, *Historia animalium liber IIII, qui est de piscium et aquatilium animantium natura, […] continentur in hoc volumine Guilelmi Rondeletii […] Petri Bellonii […] de aquatilium singula scripta* (Zurich, Christopher Froschauer: 1558) fols. b1r: 'Rondeletius diligentiae summae circa indaganda vera ac vetera piscium nomina, eorumque descriptiones (peregrinatus etiam ipse ad Belgas et Italos), variam eruditionem, et in explicandis dubiis obscurisque authorum locis haud volgarem solertiam adiunxit'.

12 I am very thankful to Florike Egmond for informing me about these images. For their context, see her essay, "A Collection within a Collection: Rediscovered Animal Drawings from the Collections of Conrad Gessner and Felix Platter", *Journal of the History of Collections* 25, 2 (2013) 149–170. See also Egmond F. – Kusukawa S., "Gessner's Fish: Images as Objects", in Leu U. – Opiz P. (eds.) *Conrad Gessner (1516–1565): Die Renaissance der Wissenschaften / The Renaissance of Learning* (Berlin – Boston: 2019) 581–606.

13 Belon Pierre, *De aquatilibus Libri duo, cum ειconibus* [*sic*] *ad vivam ipsorum effigiem, quoad ejus fieri potuit, expressis* (Paris, Charles Estienne: 1553) fol. aiiir: 'Hoc unum affirmare ausim, nihil hic esse confictum, aut supposititium, sed ita expressum, quemadmodum nos aliquando in Ponto, Hellesponto, Tyrrheno, Eithraeo, Adriatico, nostroque Oceano conspeximus'.

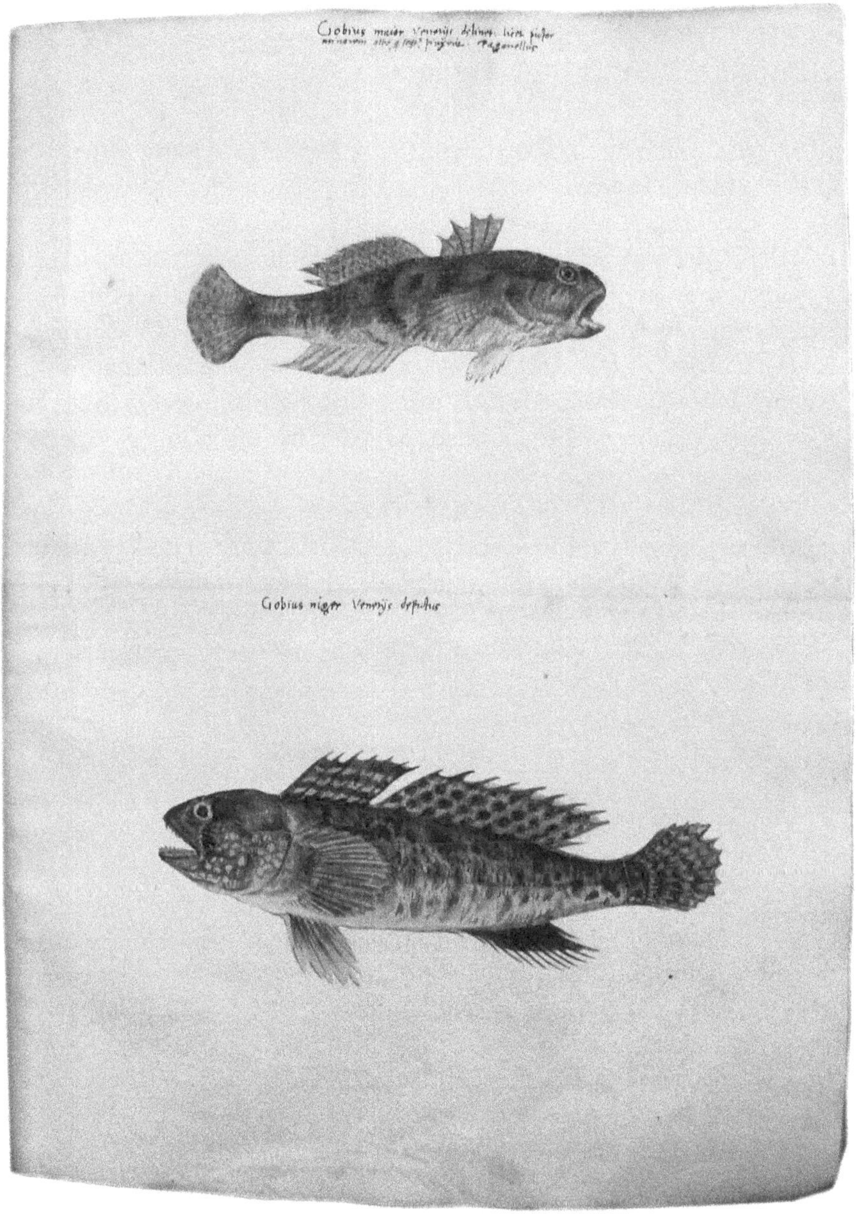

FIGURE 12.3 Two species of gobius from Gessner's collection of watercolours, which served as a basis for his *Historia animalium*. Provenance: MS University Library Amsterdam C III 22–23

Still, in spite of Belon's self-celebratory observational attitude, many legendary animals find their place in his *De aquatilibus, libri duo*.[14]

2 The Sociology of Fishermen's Water Expertise in Early Modern Venice

In light of the texts just discussed, the question arises of the origin of the empirical knowledge possessed by the many Renaissance authors who wrote lengthy works on ichthyology. To what extent did scholars like Giovio, Salviani, Belon, Rondelet and Gessner interact with seafaring people? What was fishermen's epistemic status, for instance? Do we have any access to it? From what sources can we gather information about water practitioners' experience and their contribution to science? I propose to investigate Venetian sources on water policies and fishing legislation as a vantage point to newly address these questions and contribute to a reconstruction of early modern knowledge about marine animals in its socio-epistemological multidimensionality.

Knowledge about aquatic environments and water policies were areas of common concern across the classes that constituted the republican body politic of early modern Venice. Rather than taking the form of top-down decisions and technical solutions, water policies – as a 'public domain' – had a more circular character than in Italian provinces ruled by princes.[15] A technical institution, the 'Magistrate for the Waters' (according to its later name), was specifically entrusted with overseeing all water-related matters, including the creation and maintenance of channels, river diversions, the defense of the lagoon and the protection of the coastline against alterations of anthropic or natural origin. The *savi alle acque* (water magistrates) and *proti* (practitioners with engineering expertise) working within this institution also relied on the knowledge of fishermen, whose epistemological status as experts on the

14 For instance, the legendary monkfish. For a brief yet exact overview of Renaissance ichthyology, see my grandfather's book Omodeo P., *Alle origini delle scienze naturali (1492–1632)* (Soveria Mannelli, Catanzaro: 2001), chap. IV, "Studi sui pesci".

15 In archival documents, one can find expressions such as *publico dominio* and *publiche aque*. Cf. Archivio di Stato di Venezia [State Archives of Venice, henceforth ASV], *Compilazione leggi Pesca, Pescaria, Pescatori, Pesce (1314–1786)* fol. 1065r. See my paper on the political epistemology of early modern Venetian hydrogeology, 'Hydrogeological Knowledge from Below: Water Expertise as a Republican Common in Early Modern Venice'. Much has been written about science in court society. See esp. Biagioli M., *Galileo, Courtier: The Practice of Science in the Culture of Absolutism* (Chicago-London: 1993). See also Omodeo P.D. – Renn J., *Science in Court Society: Giovanni Battista Benedetti's Diversarum speculationum mathematicarum et physicarum liber (Turin, 1585)* (Berlin: 2019).

lagoon is witnessed by the documentary evidence. As one reads in a decree of 1536, the water officers had to take into account fishermen's advice whenever they endeavored to carry out engineering work in the lagoon:

> Since no one understands the flow and movements of the waters of our lagoon better than the fishermen who travel through it by day and by night, the Gastaldo [the chief of the fishermen's community] and the fishermen's guild of San Nicolò shall choose two of the most sensible and practical elderly fishermen, or retired fishermen [...] the choice of another person must be made by the Gastaldo and the fishermen of Sant'Agnese; another will be chosen by the fishermen of Murano, two by those of Burano and two by those of Chioggia. All eight of them, when we deal with matters related to this lagoon, should be summoned to this Office to share their opinions and recollections about the matter proposed, for the benefit of our lagoon.[16]

Many archival sources bear witness to existing forms of collaboration between the water officers and the local fishing community.[17] In some cases, the fishermen carried out ambitious technical projects such as the mapping of the lagoon to define public and private waters. The Venetian Senate backed the fishermen's request to define these boundaries. For them, it was a means to ascertain where they could fish freely. Through a proclamation of 7 June 1684, the political authorities specifically requested the water officers to conduct such a survey. Their explicit goal was the safeguarding of public waters.[18] The connection between fishing and the preservation of the morphology of

---

16 Scarpa G., "Premessa", in *Mariegola della Scuola di Sant'Andrea de' Pescadori 1569–1791* (Sottomarina: 1996) 9–44, here 28 (own translation): 'Perché niuno meglio intende il corso et andamenti de le acque de queste nostre lagune de quello farà li pescadori che il zorno et nocte le practicano però sia preso che per el gastaldo et scuola dei pescadori de San Nicolò sia facta electione de duo dei più sensati vechi e pratici pescadori, o che siano stati pescadori che potranno trovar. Et questo sotto debito de Sacramento; et simile electione far debi de uno altro il gastaldo et pescadori di Sant'Agnese, et di uno altro li pescadori de Muran, et de duo altri quelli de Buran, et duo quelli de Chioza, li quali tuti otto quando se tracterano materia tantum de questa lacuna se debino far intervenir in questo Collegio per haver da loro, le loro opinion et aricordi circa dicta materia fosse proposta per benefitio de questa nostra laguna'.

17 On the fishing community of San Niccolò, see above all Zago R., *I Nicolotti: Storia di una comunità di pescatori a Venezia nell'età moderna* (Padova: 1982) and Rivoal S., "Agir en être collectif: L'État, la communauté des Nicolotti et l'approvisionnement de Venise à l'époque moderne", *Tracés: Revue de Sciences humaines* 29 (2015) 65–84.

18 ASV, *Compilazione leggi Pesca, Pescaria, Pescatori, Pesce (1314–1786)*, fol. 1050r.

the lagoon also emerges from the statutes of the fishing community of San Niccolò in Venice. In the so-called *Mariegola vecchia della comunità di San Nicolò* (the old guild statutes of the community, today preserved in the Library of the Correr Museum in Venice), privileges, regulations and decisions of relevance for the fishermen were brought together. Among other documents, the *mariegola* includes the transcription of an order that the water officers issued in 1549 and which, anticipating the aforementioned plan to map the waters of the lagoon, called for an assessment of the effects on the lagoon of barriers delimiting fish farms, the so-called 'valli da pesca' (literally, "fishing valleys", or fishing enclosures inside the lagoon).[19] The officers argued that barriers erected for this purpose could damage the lagoon bed. As these structures also hindered the free movement of fishermen, the Magistrate for the Waters and the fishermen had a common interest to contrast the "privatization" of the public waters. For the sake of fishing and the preservation of the lagoon, artificial structures and anything that might restrict movement in the waters of the lagoon were to be reduced to a minimum. There are many other documents that bear witness to this alliance.[20] These sources also preserve indirect information about the fauna of the lagoon in the past. For instance, one reads in another sixteenth-century document that the area of San Raffaele in Venice, today part of the stony city center, was full of reeds and populated by ducks. The document was indeed a ban on hunting them.[21] The great respect for the

19 *Mariegola vecchia della comunità di San Nicolò al Angelo Rafael de Mendicoli*, in Biblioteca del Museo Correr, coll. Cicogna, no. 2789, Ms. IV, no. 100, fol. 42v.

20 For instance, an order was issued in 1549 to eliminate all barriers constructed in a large area of the lagoon, because they altered its flows (ibidem, fol. 43r). Another document from the sixteenth century forbade the canons of Torcello to charge fishermen wishing to fish in their area. The document made it clear that that the area in question was not private: 'We tell them that it is our intention that they pay no rent because the Lagoon is public' (ibidem, fol. 45v: 'Gli dicemo che intention nostra è che non gli paghino fitto per esser la Laguna publica'). Another similar document, issued by the water officers in 1565, guaranteed fishermen's right to freely exercise their trade in the public waters, in accordance with the orders of the most important political magistracy, the Council of Ten (ibidem, fol. 85v).

21 Ibidem, fol. 32r. The problem of the extraction of proxy data for the environmental sciences and geomorphological history needs to be taken very seriously and constitutes a challenge for geoanthropological studies. See Camuffo D. et al., "A Novel Proxy and the Sea Level Rise in Venice, Italy, from 1350 to 2014", *Climatic Change* 143, 1–2 (2017) 73–86, and Camuffo D. et al., "When the Lagoon was Frozen over in Venice from A.D. 604 to 2012: Evidence from Written Documentary Sources, Visual Arts and Instrumental Readings," *Méditerranée: Revue géographique des pays méditerranéens, Varia* (2017) 1–68 (https://journals.openedition.org/mediterranee/7983) (accessed 23 November 2020). "When the Lagoon Was Frozen over in Venice from A.D. 604 to 2012"; Trevisani and Omodeo, "Earth Scientists and the Sustainable Development Goals".

fishermen's community and their knowledge in Venice is further witnessed by the existence of questionnaires about their trade and interviews with them conducted by water officials. Together with some colleagues, I have already investigated a specific set of interviews in the context of a study of the geomorphological history of the Venice lagoon.[22] I should now like to turn my attention to these and similar documents as a source of information about fishermen's practical knowledge and link them to ichthyology.

3 Archival Evidence of the Social Circulation of Knowledge about Fish in Early Modern Venice

In a set of questionnaires from 1623 [Fig. 12.4], designed to assess the consequences of the diversion of the river Brenta, fishermen first were asked about their work and what areas of the lagoon they usually fished in:

> What is his profession?
> Where do they usually fish?
> What are their observations about the state of the lagoon after the diversion of the Brenta?[23]

The interviewers expected the fishermen to report any observations they might have made in the lagoon and also to express their opinions about the causes of the changes they witnessed, for instance concerning water flows. The officers were particularly interested in the state of harbors. They wanted to know whether they had been damaged by the diversion of the river Brenta. Fishermen could share their opinions quite freely and often offered advice on how to solve specific problems, for instance ones related to navigation. One of the questions directly concerned fishing: 'Has fishing remained the same [as before the river diversion]?'[24] The documents show that the fishermen were quite critical about the recent diversion. They were especially concerned about the alteration of the quality of the waters, in particular stagnation,

---

22 For a preliminary study, focused on hydrogeological issues, see Omodeo P.D. – Trevisani S. – Senthil B., "Benedetto Castelli's Considerations on the Lagoon of Venice: Mathematical Expertise and Hydro-Geomorphological Transformations in Seventeenth-Century Venice", *Earth Science History* 39, 2 (2020) 420–446.

23 ASV, *Savi ed esecutori alle acque*, Atti, pezzo 123, fol. 7r: 'Qual sia la sua professione. / Dove son soliti a pescar. / Che osservatione han fatto del stato della laguna doppo levata la Brenta'.

24 Ibidem: 'Se le pesche sono le medesime'.

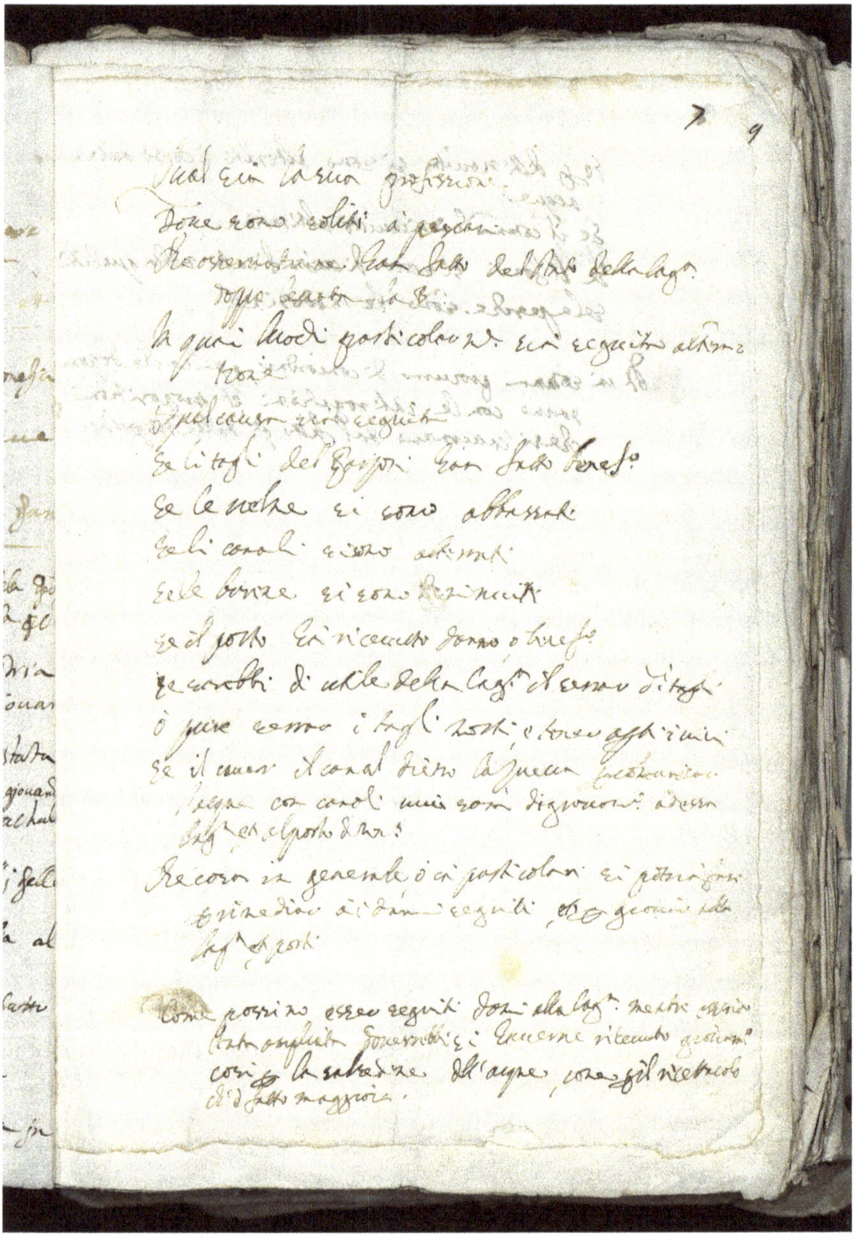

FIGURE 12.4 A questionary devised by the water officers of Venice in 1623 to interview fishermen on the consequences of a river diversion. Provenance: Archivio di Stato di Venezia, *Savi ed esecutori alle acque*, Atti, pezzo 123, fol. 7r

which was deadly for fish. They expressed a series of concrete criticisms of the consequences of the Brenta diversion of 1610. They claimed that waters had become shallower in the areas in which the river used to enter the lagoon. The water was not flowing as it used to, more algae were growing and fish were becoming rarer. Additionally, some navigable water channels in the lagoon had become difficult to navigate. Fishermen also reported about the erosion of saltmarshes and changes relative to the semi-submerged islands, so-called "velme". Additionally, new sandbanks had been formed at the lagoon inlets.

On 28 June 1623, Nadalin Gritti, a 64-year-old fisherman, had the following exchange with the water officers:

> Question: 'Has fishing changed?'
> Answer: 'Yes, Sir, because where this hair grows nothing is born.'
> Question: 'Where does this hair come from?'
> Answer: 'From the dead waters, which have no motion, and from that filth, which comes in from the channels'.[25]

Several other fishermen had noted the same effects: water stagnation, algae and a drop in the number of fish. For instance, Andrea Spinazzi of San Niccolò, a 42-year-old man, declared:

> I repeat that this is what I have observed. Now that the Brenta is no longer there, one does not fish as before. In fact [the river branch] Melison was the best possible place and one could fish everything there. But now it is full of hair.[26]

Another interviewee, Giacomo Ingiostro of the Giudecca, added that stagnating waters infected the air and that this constituted a public health hazard.[27] The water officers recorded all remarks, suggestions and criticisms. On some

25 Ibidem, fols. 14r–v: 'E domandato: "Le pesche si sono elle mutate?" Rispose: "Signor sì, perché dove vi è questo pelo non vi nasce niente." / E domandato: "Da che nasce questo pelo?" / Rispose: "Dalle acque morte che non hanno moto et da quel sporchezzo che vien zo dei tagli."'

26 Ibidem, fol. 19r: 'Io torno a dire che trovo così et dopo che non vi è la Brenta non vi si pesca più come si faceva perché Melison era il miglior luoco che fosse et vi si prendeva d'ogni cosa *et adesso è tutto pelo*'. Giacomo Ferro, a 72 year-old, gave the same witness on 3 July 1623. Cf. ibidem, fols. 21r–v: 'Dove capitava la Brenta si prendeva del pesce che non si prende adesso perché all'hora el si nutriva con quell'acqua et anco li fondi si sono mutati'.

27 Ibidem, fol. 23r: 'Et però io credo che a serrar questi tagli si faria bene, li quali tagli con il sporchezzo che conducono causano anco danno all'aria et faranno ammorbar Venetia'.

occasions, the officers rebuked fishermen for their skepticism about the human possibility of governing the elements and for their excessive reliance on Providence. The officers sometimes corrected the individual perception of the alterations on the basis of the historical records preserved in the archives of their institution. Moreover, instead of appealing to God or nature as the force responsible for the pristine conditions of the lagoon, whenever the discussion touched upon such theological and philosophical topics, the officers would tell the fishermen that the lagoon landscape was not immutable but the result of a long-term process, which depended on human decisions and interventions. The officers clearly pointed out to the leader of the fishing community, their 'Gastaldo' Domenico Papacica, that 'the river Brenta did not use to naturally end where we had it, for it had been brought there artificially [*condotta con arte*]'.[28] In this manner they countered his observation that the lagoon should be left alone, as God had created it.

In spite of these disagreements between the fishermen and the water officers, such exchanges bear witness to the fact that a process of cultural and scientific negotiation took place between the two sides. I would call these exchanges forms of knowledge circulation from below. Fishermen were conscious of the relevance of their work and their experience for the city. Menego Balbi of Sant'Agnese, a 74-year-old fisherman, directly criticized his interviewers on the 1st of August 1623 for previously disregarding the advice of those who were most familiar with the lagoon: 'It was a big mistake not to collect information from us when the Brenta was diverted, as we are so familiar [*havemo tanta pratica*] with these places'.[29]

The socially shared character of water knowledge in early modern Venice is also witnessed by fishing regulations. They incorporate knowledge about the reproduction times of the various aquatic species and techniques that can endanger their well-being. An excerpt from a medieval document shows that Venice imposed restrictions on the times of the year in which fish could be caught as early as 1314:

> It is ordered that no fisherman should dare to catch young fish with a net until the feast of Saint Peter [the 29th of June] […]
>
> If someone catches any [such fish], he should throw it back into the water and should not dare sell them or let anyone else sell them […]

28 Ibidem, fol. 11r: 'La Brenta non capitava dove voi havete detto naturalmente ma ne vi era stata condotta con arte'.

29 Ibidem, fol. 28r: 'L'è sta un grand'error quando si ha levato detta Brenta non prender anco information da noi altri che havemo tanta pratica dei luochi'.

> If someone contravenes this [order], he will forfeit his equipment and pay a fine, which will be more or less [steep] depending on the judges' decision.[30]

These regulations established what fishing techniques were lawful. In particular, nets' mesh should not be too thick, for otherwise young fish would be killed before reaching maturity.

Some Renaissance *Orders Issued by the Most Excellent College of Fishmongers* in 1595 (*Ordini presi nell'Eccellentissimo Collegio delle Pescarie*) offer some insight into many details related to fishing laws and policies that the authorities adopted in order to mitigate any decline in fish populations. Some of these regulations concerned the protection of fish during reproduction periods:

> No one should dare to take the fish gobius with his hands when it lays its eggs. The punishment for those who transgress [this law] is to row in a galley for two years with chains on his feet, and to pay 25 ducats, half of which will go to the complainant, while the other half [will be assigned] at the discretion of the Lords of [the Magistracy of] the Rason Vecchie. Whoever is not suitable for a galley will be locked up in prison for five years.[31]
>
> It is totally forbidden to take *menole* [*spicara maena*] in Istria in the season when they breed and lay eggs, especially in the months of March, June, July and August [...].[32]

30 ASV, *Compilazione leggi Pesca*, fol. 516r: 'Ordinatum fuit quod nullus piscator a modo usque ad festum Sancti Petri [29 giugno] sit ausus capere pisces vaninos cum tractis ... Et si quis ceperit eos debeat eos proiicere in aquam, et non sit ausus eos vendere, nec vendi facere. Item si quis fecerit, vel facere fieri voluerit tracturos de nocte debeat accipere ... Item quod nullus audeat ire ad tratturos pedem per palludos. Item, si quis vult ponere cucullos, vel sorborarcios in aqua, debeat ispos ponere de vero in prima campana et in mane ante tercias ipsos debet elevare: et si quis contra haec omnia [...] fuerit debeat perdere ipsas artes, et insuper solvat bannus intergum, et plus et minus ad voluntatem Dominorum Iustis'.

31 Collegio delle Pescarie, *Ordini presi* ([Venice], Antonio Pinelli: 15 December 1595), fol. A2r: 'Che niuno ardisca preder gò à brazzo al tempo, che hanno gettato l'ove predette fatto pena a chi contrafacesse d'esser condannato in Galea per anni doi a vogar il remo con li ferri alli piedi, et di pagar Ducati 25 applicati la mità al denontiante, et l'altra mità ad arbitrio delli Signori delle Rason vecchie, et non essendo buono da Galea di stare per anni 5 in pregion serrato'.

32 Ibidem, fols. A2r–v: 'Che sia totalmente prohibito il prender le menole [*Spicara maena*] nell'Istria al tempo che vanno in frega, et sono da ove, et particolarmente nelli mesi di Marzo, Zugno, Lugio et Agosto [...].'

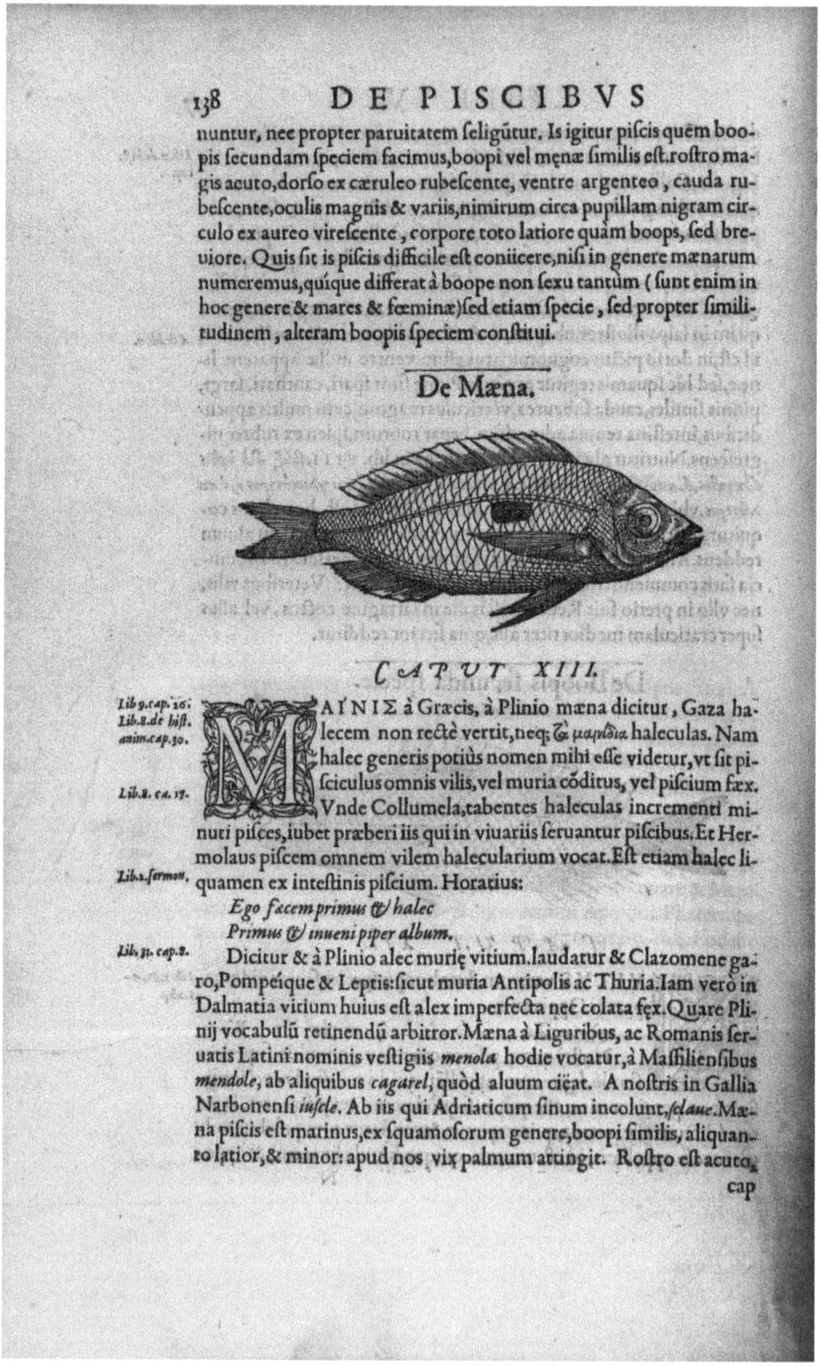

138 DE PISCIBVS

nuntur, nec propter paruitatem ſeligũtur. Is igitur piſcis quẽm boopis ſecundam ſpeciem facimus,boopi vel mænæ ſimilis eſt.roſtro magis acuto,dorſo ex cæruleo rubeſcente, ventre argenteo, cauda rubeſcente,oculis magnis & variis,nimirum circa pupillam nigram circulo ex aureo vireſcente, corpore toto latiore quàm boops, ſed breuiore. Quis ſit is piſcis difficile eſt coniicere,niſi in genere mænarum numeremus,quíque differat à boope non ſexu tantùm (ſunt enim in hoc genere & mares & fœminæ)ſed etiam ſpecie, ſed propter ſimilitudinem, alteram boopis ſpeciem conſtitui.

De Mæna.

CAPVT XIII.

Lib 9.cap.26. Lib.8.de hiſt. anim.cap.30. ΜΑΙΝΙΣ à Græcis, à Plinio mæna dicitur, Gaza halecem non rectè vertit,neq; τὰ μαινίδια haleculas. Nam halec generis potiùs nomen mihi eſſe videtur,vt ſit piſciculus omnis vilis,vel muria cõditus, vel piſcium fæx. Lib.8. ca.17. Vnde Collumela,tabentes haleculas incrementi minuti piſces,iubet præberi iis qui in viuariis ſeruantur piſcibus. Et Hermolaus piſcem omnem vilem halecularium vocat. Eſt etiam halec liquamen ex inteſtinis piſcium. Lib.2.ſermon. Horatius:

Ego fæcem primus & halec
Primus & inueni piper album.

Lib.31.cap.8. Dicitur & à Plinio alec muriæ vitium.laudatur & Clazomene garo,Pompeique & Leptis:ſicut muria Antipolis ac Thuria.Iam verò in Dalmatia vitium huius eſt alex imperfecta nec colata fęx.Quare Plinij vocabulũ retinendũ arbitror.Mæna à Liguribus, ac Romanis ſeruatis Latini nominis veſtigiis *menola* hodie vocatur,à Maſſilienſibus *mendole*, ab aliquibus *cagarel*, quòd aluum ciẽat. A noſtris in Gallia Narbonenſi *iuſcle*. Ab iis qui Adriaticum ſinum incolunt, *claue*. Mæna piſcis eſt marinus,ex ſquamoſorum genere,boopi ſimilis, aliquanto latior,& minor: apud nos vix palmum attingit. Roſtro eſt acuto,

cap

FIGURE 12.5 Engraving of a *maena* in Rondelet Guillaume, *De piscibus marinis* (Lyon, Macé Bonhomme: 1554–1555) 138. Provenance: Munich, Bayerische Staatsbibliothek

With the same goal of protecting fish, their habitat and reproduction, a regulation of 2 May 1726 prohibited walking on the shallow marshes and semi-submerged *velme* to harvest oysters, because this practice caused the devastation of the seabed and the death of black fish (*pesce negro*), presumably the *gobius niger*.[33]

4 The Gobius and People's Diet in Early Modern Venice

Practical knowledge about fishing and fish thus entered Venetian legislation. Did this also lead to its introduction into erudite scientific literature? To what extent did the social and epistemic status of a community like that of the Venetian fishermen contribute to the advancement of science? At present, I do not have enough evidence to answer this question, which ultimately concerns the codification practical knowledge on fish in erudite scientific literature. For the time being, I will only examine the socio-epistemological multidimensionality of a specific species, the *gobius*, typical of the Lagoon of Venice. This fish is repeatedly mentioned in Venetian fishing legislation and is also described in the most relevant early modern works on aquatic animals. Most of them refer to Venice in connection to the specific variant of this species, the *gobius marinus niger*, which was typical of its lagoon. Belon, for one, begins his section on the gobius by mentioning its Venetian name, before adding a series of other vernacular names:

> The Venetian call the sea gobius 'goi', the Genoese 'guigiones', the Romans 'missori', although the name 'missor' is also applied to many other fish. The inhabitants of the towns of La Spezia, Porto Venere and Genoa call them 'zoseros'.[34]

Gessner quotes this passage literally in his illustrated *Historia animalium liber IIII, qui est de piscium et aquatilium animantium natura* (History of Animals, Book IV, Dealing with the Nature of Fish and Other Aquatic Animals) (1558).[35]

The old statutes of the fishing community of San Niccolò comprise, among the earliest documents, the transcription of a deliberation of the justice

33 ASV, *Compilazione leggi Pesca*, fol. A2r.

34 Belon, *De aquatilibus* 233: 'Gobiones marini Venetis Goi, Genuensibus Guigiones, Romanis Missori vacantur, quamquam Missoris vox ad plerosque alios pisces transferatur. Incolae Urbis de le Specie et qui Portum Veneris ac Genuam inhabitant, Zoseros nominant'.

35 Gessner, *Historiae animalium liber IV* 466.

magistrates (the *provveditori comuni della vecchia giustizia*) of 14 November 1503 which prohibited "unsustainable" fishing techniques. It includes the prohibition to fish the gobius by hand in the lagoon, because this practice damaged the seabed and fish eggs.[36] The regulations concerning the fishing of the gobius were then revised. The prohibition later only concerned specific times of the year: the reproductive season for this fish.

It should be remarked that a constant concern about the fishing of gobius emerges from the extant archival legislation. This fish must have been an important ingredient in the local diet since antiquity. The first-century poet Martial wrote the following verses:

> Although in the region of Venetia people have lavish banquets
> Usually, gobius is eaten at the beginning of the dinner.[37]

The Bologna naturalist Ulisse Aldrovandi (1522–1605) quotes these verses in the section devoted to the gobius in his treatise *De piscibus et de cetis* (On Fish and Cetacea). Aldrovandi begins his presentation of this fish with a description (*descriptio*) accompanied by some illustrations. For this purpose, he includes some images taken from Rondelet and Gessner [Fig. 12.6]. He then discusses the gobius' habitat, its reproduction (*locus, partus*), its nutritional properties (in a longish section entitled '*Usus in cibis*') and finally its medical properties (in a very brief section entitled '*Usus in medicina*'). The culinary part stresses the gobius' poor taste, at least in comparison to other kinds of fish:

> Marcello Virgilio indicated that the gobius is a very bad fish by quoting Juvenal's verses 'Do not desire a goatfish [*mullus*] if you only have a gobius'.[38]

36 *Mariegola vecchia della comunità di San Nicolò*, fol. 10r: 'Item per avanti el fu proibido, che non se podesse per algun modo pigliar go a brazo in algun tempo per esser sta robba el libro dove era nota tal ordene, et el sia venuto in consuetudine che ogni uno pesca, et piglia i detti go a brazo guasta e rompe le ove de tal pesci con gran detrimento nostro. Per tanto ordenemo, et volemo che de cetero el non sia alguno sia chi esser si voglia che ardisca piar né far piar go a brazo sotto penna de lire cento de pizoli e star mesi tre in preson per ogni volta. Lequal tutte penne sia divise per mittà tor la mittà all'accusador et l'altra mittà dell'officio preditto, et sia publicada nelle pescarie a Rialto et S. Marco a notitia'.

37 'In Venetis sint lauta licet convivia terris / Principium caenae Gobius esse solet'. Quoted from Aldrovandi Ulisse, *De piscibus et de cetis* (Bologna, Giovanni Baptista Bellagambia: 1613) 99. The same verses can be found in other ichthyological sources, as well.

38 Ibidem: 'Marcellis Virgilius ex hoc Iuvenalis versu "Nec Mullum cupias cum sit tibi Gobio tantum" Gobium vilissimum piscem esse indicavit'.

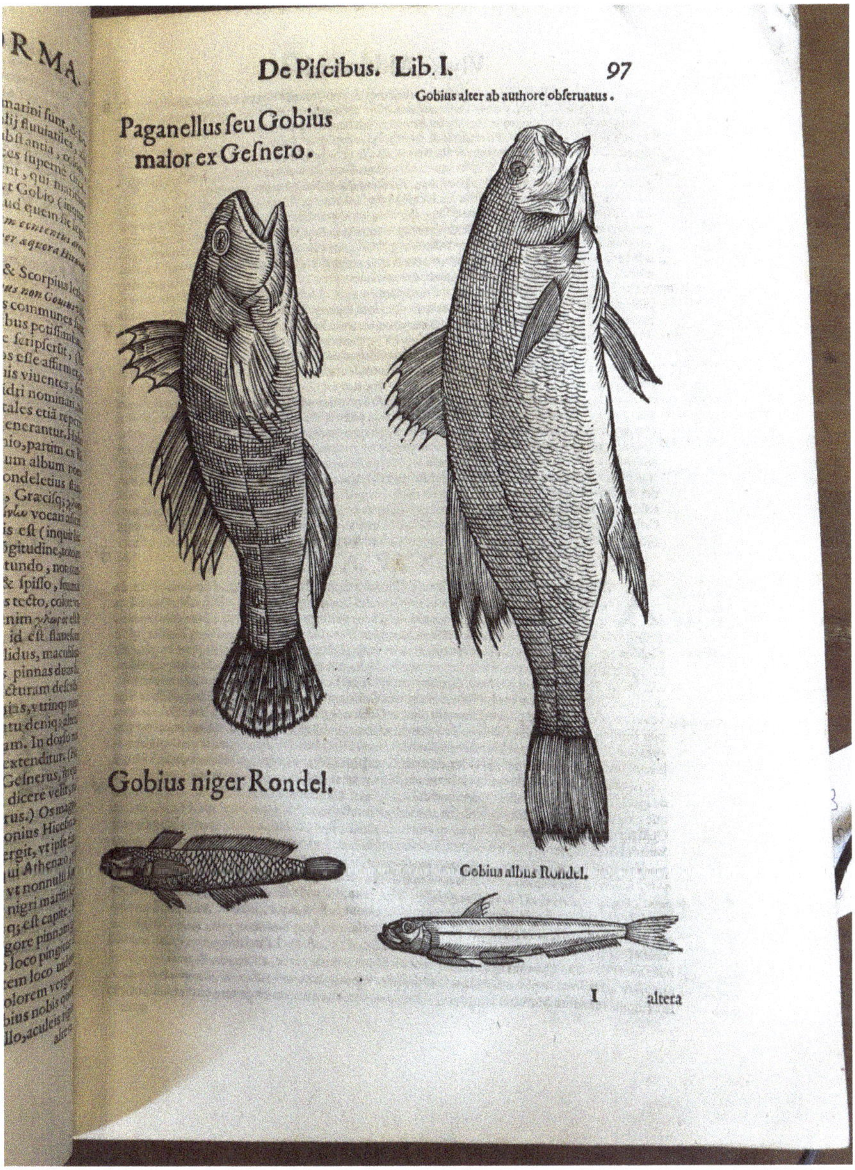

FIGURE 12.6 Images of gobius, taken from Gessner and Rondelet, in Aldrovandi Ulisse, *De piscibus et de cetis* (Bologna, Giovanni Baptista Bellagambia: 1613). Provenance: Biblioteca Universitaria di Padova

In his answer to the question whether the gobius is a nutritious food (*An nutriat*), Salviani relies on ancient authorities, who argued that it is only in quantitative terms that this poor animal can contribute to human nutrition.

> As Athenaeus reports on the basis of Hicesius, 'the gobius has a lot of juice, which can easily be extracted; they do not nourish much, and produce no good juice. When it turns golden, its meat has little substance to it; it is less fat, the juice lighter and less abundant; but owing to its bigger size, it is more nourishing'.[39]

According to these sources, the gobius is a juicy fish but does not have much flesh. In the Venice area, it is still used for a typical risotto, a rice dish in which the gobius essentially adds flavor but not much substance. Concerning the reputation of this fish (*quantae sit aestimationis*), Salviani remarks that 'although the gobius cannot be compared to the most noble fish, as Juvenal witnesses [...], nonetheless it should not be despised, because it has a fairly good taste. For this reason, the ancients held it in some esteem'.[40] Galen, for example, considered its taste to be very pleasant (*praestantissimum ad voluptatem*).[41]

Judging from the constant references to the fishing of the gobius in the lagoon and the constant efforts to regulate it in Venetian legislation, it must have been a favorite ingredient across the centuries. Independently of the high or, actually, rather poor esteem it enjoyed among scientists and physicians, the gobius continued to be present in the local diet as a protein source. Fish was seen as an important source of nourishment in general. Indeed, the sale of fish was strictly regulated by the Venetian authorities. Official lists of fish were issued for the markets of San Marco and Rialto, to set maximum prices. From such *tariffe* (prices), one discovers that the gobius was one of the cheapest fish. In 1760, one gobius cost 8 or 6 soldi, depending on its size. This was almost the same price as a frog (8 and 5 soldi, again depending on its size) [Fig. 12.7].[42]

---

39 Salviani, *Aquatilium animalium historiae liber*, fol. 215r: 'Ut refert Athenaeus authore Hicesio, "Gobiones multi succi sunt, facile excernuntur, *non multum* nutriunt, nec bonum succum gignunt. Flavescentium autem caro substantiae est rarioris, minus pinguis, succumque tenuiorem, neque ita copiosum praebet; ob magnitudinem tamen magis *nutriunt*"'.

40 Ibidem: 'Etsi Gobius cum nibilissimis piscibus conferri non debeat, ut testari videtur Iuvenalis [...]; haud tamen spernendum est, cum iucunde satis sapiat; et propterea apud etiam veteres in aestimatione fuit'.

41 Ibidem.

42 Provveditori sopra la giustizia vecchia, *Nuova tariffa per la vendita del pesce* (17 May 1760) 9.

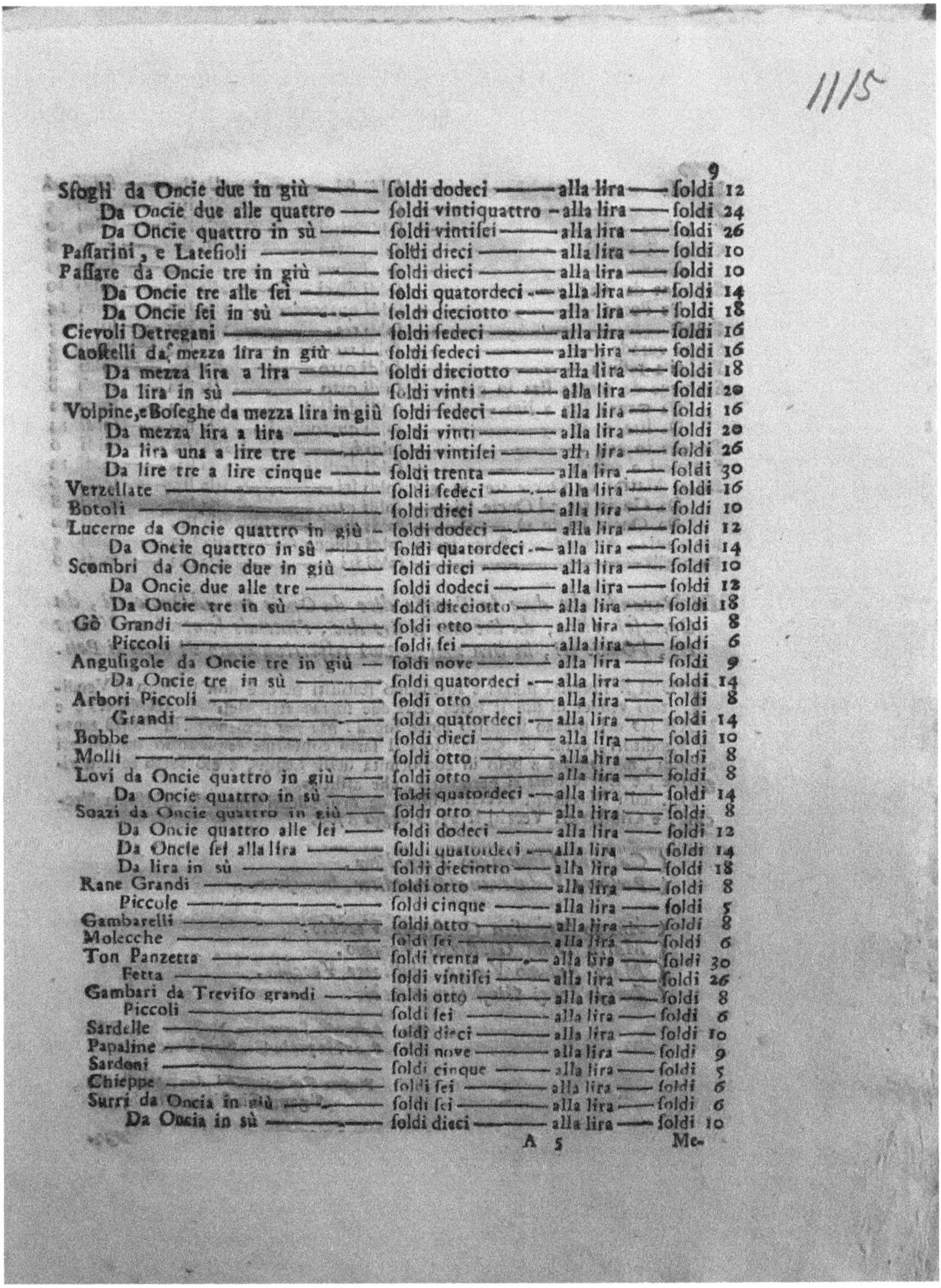

1115

9

Sfogli da Oncie due in giù	ſoldi dodeci	alla lira	ſoldi	12
Da Oncie due alle quattro	ſoldi vintiquattro	alla lira	ſoldi	24
Da Oncie quattro in sù	ſoldi vintiſei	alla lira	ſoldi	26
Paſſarini, e Lateſioli	ſoldi dieci	alla lira	ſoldi	10
Paſſare da Oncie tre in giù	ſoldi dieci	alla lira	ſoldi	10
Da Oncie tre alle ſei	ſoldi quatordeci	alla lira	ſoldi	14
Da Oncie ſei in sù	ſoldi dieciotto	alla lira	ſoldi	18
Cievoli Detregani	ſoldi ſedeci	alla lira	ſoldi	16
Caoſtelli da mezza lira in giù	ſoldi ſedeci	alla lira	ſoldi	16
Da mezza lira a lira	ſoldi dieciotto	alla lira	ſoldi	18
Da lira in sù	ſoldi vinti	alla lira	ſoldi	20
Volpine, e Boſeghe da mezza lira in giù	ſoldi ſedeci	alla lira	ſoldi	16
Da mezza lira a lira	ſoldi vinti	alla lira	ſoldi	20
Da lira una a lire tre	ſoldi vintiſei	alla lira	ſoldi	26
Da lire tre a lire cinque	ſoldi trenta	alla lira	ſoldi	30
Verzellate	ſoldi ſedeci	alla lira	ſoldi	16
Botoli	ſoldi dieci	alla lira	ſoldi	10
Lucerne da Oncie quattro in giù	ſoldi dodeci	alla lira	ſoldi	12
Da Oncie quattro in sù	ſoldi quatordeci	alla lira	ſoldi	14
Scombri da Oncie due in giù	ſoldi dieci	alla lira	ſoldi	10
Da Oncie due alle tre	ſoldi dodeci	alla lira	ſoldi	12
Da Oncie tre in sù	ſoldi dieciotto	alla lira	ſoldi	18
Gò Grandi	ſoldi otto	alla lira	ſoldi	8
Piccoli	ſoldi ſei	alla lira	ſoldi	6
Anguſigole da Oncie tre in giù	ſoldi nove	alla lira	ſoldi	9
Da Oncie tre in sù	ſoldi quatordeci	alla lira	ſoldi	14
Arbori Piccoli	ſoldi otto	alla lira	ſoldi	8
Grandi	ſoldi quatordeci	alla lira	ſoldi	14
Bobbe	ſoldi dieci	alla lira	ſoldi	10
Molli	ſoldi otto	alla lira	ſoldi	8
Lovi da Oncie quattro in giù	ſoldi otto	alla lira	ſoldi	8
Da Oncie quattro in sù	ſoldi quatordeci	alla lira	ſoldi	14
Soazi da Oncie quattro in giù	ſoldi otto	alla lira	ſoldi	8
Da Oncie quattro alle ſei	ſoldi dodeci	alla lira	ſoldi	12
Da Oncie ſei alla lira	ſoldi quatordeci	alla lira	ſoldi	14
Da lira in sù	ſoldi dieciotto	alla lira	ſoldi	18
Rane Grandi	ſoldi otto	alla lira	ſoldi	8
Piccole	ſoldi cinque	alla lira	ſoldi	5
Gambarelli	ſoldi otto	alla lira	ſoldi	8
Molecche	ſoldi ſei	alla lira	ſoldi	6
Ton Panzetta	ſoldi trenta	alla lira	ſoldi	30
Fetta	ſoldi vintiſei	alla lira	ſoldi	26
Gambari da Treviſo grandi	ſoldi otto	alla lira	ſoldi	8
Piccoli	ſoldi ſei	alla lira	ſoldi	6
Sardelle	ſoldi dieci	alla lira	ſoldi	10
Papaline	ſoldi nove	alla lira	ſoldi	9
Sardoni	ſoldi cinque	alla lira	ſoldi	5
Chieppe	ſoldi ſei	alla lira	ſoldi	6
Surri da Oncia in giù	ſoldi ſei	alla lira	ſoldi	6
Da Oncia in sù	ſoldi dieci	alla lira	ſoldi	10

A 5 Me-

FIGURE 12.7 List of fish prices for the fish market in 1760 for the months of June and July. Provenance: Archivio di Stato di Venezia, *Compilazione leggi Pesca, Pescaria, Pescatori, Pesce* (1314–1786). A big gobius (*go*) costs 8 soldi, just like a big frog (*rana grande*)

As one of the most affordable fishes on the market, gobius must have been important for the sustenance of less affluent people. Indeed, the political reasons for the publication of *fish prices* by the Venetian authorities was to protect their citizens from overpricing, as is clearly stated in a ducal proclamation of 30 November 1765:

> The excessive price at which fish are sold, creating universal discontent and difficulties for poor families, who are damaged in their livelihood, aroused the zeal of your Excellencies. Given their role, it was their duty to identify the guilty reasons [for this]. They clearly determined that this [problem] stems from the fact that so many Public Laws are not respected, fish is not brought into the public market, nor is it delivered to fishmongers. Instead, the food passes to traffickers. In this manner, an artificial famine is always created, even in the most plentiful fishing times and seasons. And by passing from hand to hand, the product becomes more expensive.[43]

Similarly, fishing techniques were regulated so as to ensure the future prosperity of the city by protecting fish as a source of food. As one reads in the *Ordini in proposito della pesca del pesce novello* (Orders Concerning the Fishing of Young Fish) of 1760 (reissued in 1774–1775):

> The wisdom of our ancestors always aimed to protect the fishing of young fish through excellent and beneficial laws, since the abundance of an indispensable food largely depends on this. They therefore distinguished the times, places and different uses of nets and techniques [...] because if fish are allowed to grow and are caught only in the appropriate and permitted seasons, this brings a happy abundance to everyone's benefit, thanks to copious fishing.[44]

43 Il Serenissimo Prencipe, *Fa sapere* ([Venice]: stampato per li figliuoli di Z. Antonio Pinelli, 30 Novembre 1765): 'L'eccessivo strabocchevole prezzo, a cui viene venduta la Vittuaria del Pesce con universale reclamo, e con sbilanzio delle povere Famiglie, altamente pregiudicate nella loro economia, chiamò il zelo di Sue Eccellenze per dover della propria Deputazione a rintracciarne i rei motivi, ed hanno chiaramente conosciuto che ciò deriva perché in sprezzo di tante Pubbliche Leggi non viene condotto il Pesce nelle Pubbliche Pescarie, né s'esequisse la messa de' Compravendi, ma passa la Vittuaria in potere de Sbazzegari; cosicché comparisse sempre una procurata carestia, anche ne' tempi, e staggioni più fertili della Pesca, e col passaggio da mano a mano s'incarisse il prodotto'.

44 Ibidem, fol. 538r: 'La Sapienza de' Maggiori con ottime, e salutari avvertenze ebbe sempre in vista di custodire la pesca del pesce novello, come quello da cui dipende in gran parte l'abbondanza d'una Vittuaria indispensabile, distinguendo i tempi, i luoghi il diverso uso

5 Concluding Remarks

Renaissance knowledge about fish is an interesting field to focus on in order to investigate the multilayered nature of early modern science at the encounter between people and groups belonging to different epistemological and social circles. In this essay I have explored this multidimentionality starting from the detection of traces of practical knowledge in the empirical sections of works by the most erudite early modern authors dealing with ichthyology: Paolo Giovio, Ippolito Salviani, Pierre Belon, Guillaume Rondelet, Conrad Gessner and Ulisse Aldrovandi. I then shifted my focus from the top-down cultural attitude of these learned elites to a bottom-up perspective by considering archival material pertaining to the fishing community of Venice. These documents range from water officers' interviews with fishermen for the assessment of the state of the lagoon and its engineering to the community statutes of the fishermen of San Niccolò and fishing regulations. All these documents provide information about knowledge about fish, showing that it lay at the basis of early modern environmental policies – for instance, the protection of fish during the reproductive season and the prohibition of overfishing through non-standard nets. After suggesting a multiple way of accessing early modern knowledge about fish, I chose to focus on a species, the gobius, that is typical of Venice and which constituted an important component of the local diet. As I have pointed out, information about this fish is found across a range of fields and disciplines. It was protected by Venetian laws and was described and engraved in most sixteenth-century scientific publications. This is a brief case study of the multidimensional social epistemology of knowledge in early modernity. By zooming in on the gobius, which today is considered of little gastronomic relevance, I have stressed the connection between knowledge about fish, people's diet and food policies. As I have also pointed out, economic was an important factor in the production and circulation of knowledge about fish. The Venetian authorities were very concerned about securing food and regulating the market for the benefit of everybody, but especially the less affluent classes who could buy fish only if overpricing was avoided. This essay is an attempt to highlight the importance of fishermen's practical knowledge, as well as the broader societal paradigms of early modern science.

---

delle Reti, ed Arti [...] poiché se il pesce si lasciasse crescere, ed alle sole opportune permesse stagioni si pescasse apporterebbe una gioconda ubertà a benefizio universale, e nella felicità delle Pesche copiose [...]'.

Acknowledgments

This essay is part of a project that has received funding from the European Union's Horizon 2020 Research and Innovation Programme (GA 725883 ERC EarlyModernCosmology). It also contributes to the following research projects: FARE *EarlyGeoPraxis* (Grant of the Italian Ministry of University and Research, cod. R184WNSTWH) and Max Planck Partner Group *The Water City* (Max Planck Institute for the History of Science, Berlin, in collaboration with Ca' Foscari University of Venice). I also wish to thank THE NEW INSTITUTE Center for Environmental Humanities (NICHE), Venice and the Max Planck Institute of Geoanthropology, Jena.

Bibliography

Aldrovandi Ulisse, *De piscibus et de cetis* (Bologna, Giovanni Baptista Bellagambia: 1613).

Belon Pierre, *De aquatilibus Libri duo, cum εiconibus* [*sic*] *ad vivam ipsorum effigiem, quoad ejus fieri potuit, expressis* (Paris, Charles Estienne: 1553).

Biagioli M., *Galileo, Courtier: The Practice of Science in the Culture of Absolutism* (Chicago-London: 1993).

Camuffo D. et al., "A Novel Proxy and the Sea Level Rise in Venice, Italy, from 1350 to 2014", *Climatic Change* 143, 1–2 (2017) 73–86.

Camuffo D. et al., "When the Lagoon was Frozen over in Venice from A.D. 604 to 2012: Evidence from Written Documentary Sources, Visual Arts and Instrumental Readings," *Méditerranée: Revue géographique des pays méditerranéens, Varia* (2017) 1–68 (https://journals.openedition.org/mediterranee/7983) (accessed 23 November 2020).

Cohen F., *The Scientific Revolution: A Historiographical Inquiry* (Chicago: 1994).

Collegio delle Pescarie, *Ordini presi* ([Venice], Antonio Pinelli, 15 December 1595).

Egmond F. - Kusukawa S., "Gessner's Fish: Images as Objects", in Leu U. – Opiz P. (eds.) *Conrad Gessner (1516–1565): Die Renaissance der Wissenschaften / The Renaissance of Learning* (Berlin – Boston: 2019) 581–606.

Egmond F., "A Collection within a Collection: Rediscovered Animal Drawings from the Collections of Conrad Gessner and Felix Platter", *Journal of the History of Collections* 25, 2 (2013) 149–170.

Egmond F., "On Northern Shores: Sixteenth-Century Observations of Fish and Seabirds (North Sea and North Atlantic)", in MacGregor A. (ed.), *Collecting, Recording and Preserving the Natural World from the Fifteenth to the Twenty-First Century* (Leiden – Boston: 2016) 129–148.

Egmond F., "Aldrovandi, Truthfully Drawing Naturalia, and Local Context", *Acta ad archaeologiam et artium historiam pertinentia* 32, 18 (2020) 81–95.

Gessner Conrad, *Historia animalium liber IIII, qui est de piscium et aquatilium animantium natura, [...] continentur in hoc volumine Guilelmi Rondeletii [...] Petri Bellonii [...] de aquatilium singula scripta* (Zurich, Christopher Froschauer: 1558).

Giovio Paolo, *De Romanis piscibus libellus* (Rome, F. Minitius Calvus: 1524).

Il Serenissimo Prencipe, *Fa sapere* ([Venice]: li figliuoli di Z. Antonio Pinelli, 30 Novembre 1765).

Klein U., *Nützliches Wissen: die Erfindung der Technikwissenschaften* (Göttingen: 2016).

Long P.O., *Artisan/Practitioners and the Rise of the New Sciences, 1400–1600* (Corvallis: 2011).

Omodeo P., *Alle origini delle scienze naturali (1492–1632)* (Soveria Mannelli, Catanzaro: 2001).

Omodeo P. D. – Renn J., *Science in Court Society: Giovanni Battista Benedetti's Diversarum speculationum mathematicarum et physicarum liber (Turin, 1585)* (Berlin: 2019).

Omodeo P.D., "Scientific Revolution, Ideologies of the", in Jalobeanu D. – Wolfe C.T. (eds.), *Encyclopedia of Early Modern Philosophy and the Sciences* (n.p.: 2020).

Omodeo P.D., "Hydrogeological Knowledge from Below: Water Expertise as a Republican Common in Early Modern Venice," *Berichte zur Wissenschaftsgeschichte: History of Science and Humanities* 45, 4 (2022) 538–560.

Omodeo P.D. – Trevisani S. – Senthil B., "Benedetto Castelli's Considerations on the Lagoon of Venice: Mathematical Expertise and Hydro-Geomorphological Transformations in Seventeenth-Century Venice", *Earth Science History* 39, 2 (2020) 420–446.

Provveditori sopra la giustizia vecchia, *Nuova tariffa per la vendita del pesce* ([Venice], figliuoli di Z. Antonio Pinelli: 17 May 1760).

Rivoal S., "Agir en être collectif: L'État, la communauté des Nicolotti et l'approvisionnement de Venise à l'époque moderne", *Tracés: Revue de Sciences humaines* 29 (2015) 65–84.

Salviani Ippolito, *Aquatilium animalium historiae* (Rome, Ippolito Salviani: 1554).

Scarpa G., "Premessa", in *Mariegola della Scuola di Sant'Andrea de' Pescadori 1569–1791* (Sottomarina: 1996) 9–44.

Shapin S. – Schaffer S., "Introduction to the 2011 Edition: Up for Air", in idem (eds.), *Leviathan and the Air-Pump: Hobbes, Boyle and Experimental Life* (Princeton: 2011 (first edition 1985) xi–l.

Shapin S., "The Invisible Technician", *American Scientist* 77, 6 (1989) 554–563.

Smith P., *The Body of the Artisan: Art and Experience in the Scientific Revolution* (Chicago: 2004).

Trevisani S. – Omodeo P.D., "Earth Scientists and the Sustainable Development Goals: Geocomputing, New Technologies, and the Humanities", *Land* 10 (2021) no. 294 (https://doi.org/10.3390/land10030294).

Valleriani M., *The Structures of Practical Knowledge* (Cham: 2017).

Zago R., *I Nicolotti: Storia di una comunità di pescatori a Venezia nell'età moderna* (Padova: 1982).

CHAPTER 13

'Um Grande Peixe, Dona Baleia da Costa': The Whale in Portuguese Early Modern Natural History

Cristina Brito

1 Introduction: Setting the Stage for Whales

Today we could sit anywhere along the coast of Peniche – in Papoa, within sight of the island of Berlenga and Cape Carvoeiro, on the small "island" of Baleal or on the isthmus that connects it to the mainland, or even further inland in the village of Atouguia de Baleia – and see before our eyes cetaceans that today populate the coastal waters and those that once populated the waves of this region's environmental history. In the past, when different species of whales migrated along the western Portuguese coasts, specimens would have run aground or been captured by local fishermen; only a few lost bones, local stories, and a couple of written records that survived in historical sources remain.

Beyond what we see today, we can imagine a naturalist, an enthusiast of the natural world, sitting in his home fishing village sometime in the early 18th century, who must also have observed his present and his past in search of these giants of the sea and many other creatures. Doctor Domingos Franco Quaresma must have seen and read, studied, and written about the marine animals of Portugal, and he left us a testimony of the regional biodiversity of his time and clear evidence of the importance of the whale to Portuguese society, as we will describe.

History, natural sciences, and art, including more recently visual and digital media, have placed whales as relevant elements of oceanic and natural realities and different socio-cultural realities. Whales are now known and seen as the largest living animal, an archetypal sea monster, a deep-swimming marine mammal with highly developed cognitive and behavioural characteristics, the aquatic species that connects to land. Whales have a long history of interactions with human coastal communities. As whales have continued to approach the same regions on the coasts during their long-distance oceanic peregrinations, they have become sources of food and fuel, mythological and literary

DOI:10.1163/9789004681187_014

figures, and ultimately symbols of human ruthlessness and ecological danger.[1] Whales served – and still serve today – as agents and subjects of history and science and became a place of cultural meaning.[2] They are marine mammals that represent the interface between water and air, and between water and land. They are somehow hybrids and dual; and in all their paradoxes they can be perceived and analyzed as a hyper-object. In their own kaleidoscopic and multi-existing forms of cognitive beings and conceptualized entities, whales shape humans in their quest for knowledge and understanding, which influence their distribution and movements in the oceans, as well as their practices and choices on land.[3]

Considering that people have been using aquatic systems – inland, coastal, and ocean – for thousands of years, with strong impacts in the last few hundred years, and have been modifying these environments, the human component is also essential to understanding the long-term dynamics in these ecosystems. We must begin to consider the existence and influence of humans in ecological systems and niches that coexist with other natural elements. Likewise, it is essential to include non-human agents in the construction of historical and cultural narratives. As elements of the same biomes, humans and other species are interdependent and interrelated. We must come to assume, in understanding the human past and in constructing historical narratives, the "multi-species" systems that populate the Earth.

Whales have never been (and are not) just one animal, one place, or one geography, but rather a complex atlas of multiple ecologies, states, and emotions. Studying the multiple meanings of the whale allows us to understand (or come close to better understanding) the importance of marine ecosystems to different human societies or maritime and coastal communities. The whale allows us to recreate concepts, complicate and disorganize categories

1 Ritcher V., "'Where things meet in the world between sea and land': Human-Whale Encounters in Littoral Space", in Kluwick U. – Ritcher V. (eds.), *The Beach in Anglophone Literatures and Cultures: Reading the Littoral Space* (Farnham: 2015) 155–173.

2 Colby J., "Change in Black and White: Killer Whale Bodies and the New Pacific Northwest", in Nance S. (ed.), *The Historical Animal* (New York: 2015) 19–37; Brito C., "Beauties and Beasts: Whales in Portugal, from Early-Modern Monsters to Today's Flagship Species", *Arcadia* 21 (2018) doi.org/10.5282/rcc/8449; Giggs R., *Fathoms. The World inside the Whale* (New York: 2020); Brito C. – Vieira N., "Uma construção cultural de ser baleia: A história ambiental de dois arrojamentos na Lisboa ribeirinha e das pessoas que os observaram e descreveram", SCAENA – *Revista do Museu de Lisboa – Teatro Romano. O rio como horizonte: o outro palco do teatro romano*, no. 3. EGEAC, EM / Museu de Lisboa – Teatro Romano (Lisbon: 2022).

3 Brito – Vieira, "Uma construção cultural de ser baleia".

of the natural, illuminate stories and feed beliefs, and become constituents of socio-economic and eco-cultural systems.[4] This is particularly important for Portugal (or Iberia), where much is still unknown about the evolution of natural history and natural philosophy in relation to the ocean and the coast. However, an interesting amount of written and iconographic sources produced in Portugal on the marine life of Portugal have come to light in recent years and may shed new light on this topic.[5]

We hope to address the development of early modern maritime knowledge about the use and perceptions of marine mammals and the development of Iberian perceptions and practices about the peoples-animals-oceans relationship. Supported by, but not limited to, two unpublished and mostly unknown manuscripts of natural history or fish and fisheries history in Portugal, we will reveal not only the background of the documents and their authors, but also a narrative of the perception of nature from the point of view of naturalist-humanists and Portuguese society in general. Other categories of sources that are not formal scientific productions also offer good insights into the occurrence and importance of whales and local perceptions and knowledge about them.

2 The Portuguese Natural History of Aquatic Animals

Portuguese historiography on fish and fisheries goes back mostly to the middle or late 18th and early 19th century,[6] and this period can be referred to as the landmark for the start of ichthyology – the branch of zoology dedicated to the study of fish. In Portugal, first studies of natural history came around at the same time, together with the arrival of Domingos Vandelli[7] to Portugal and the establishment of the country's main scientific-museological complex, the Botanical Garden and Royal Cabinet of Natural History of Ajuda, in 1768 (which would later turn into the current institution, that is the National Museum of Natural

4 Brito – Vieira, "Uma construção cultural de ser baleia".

5 See the full review of the documentary sources that served as basis for the narrative presented in this chapter, in an open access document produced by the author and colleagues: Lacerda T. – Vieira N. – Brito C., *Fontes documentais para uma história natural das baleias em Portugal*. Zenodo. https://doi.org/10.5281/zenodo.6417799. Online document (2022).

6 Ceríaco L., *O 'Arquivo Histórico Museu Bocage' e a história natural em Portugal. Professor Carlos Almaça (1934–2010) – Estado da Arte em Áreas Científicas do Seu Interesse* (Lisbon: 2014).

7 Domingos Vandelli (Pádua, 1735–Lisboa, 1816). Italian naturalist who came to Portugal, where he developed studies on natural history and chemistry. He directed the first works for the creation of the Ajuda Botanical Garden and was its director between 1787 and 1788.

History and Science) in Lisbon[8] and the foundation of the Royal Academy of Sciences. The emergence of ichthyology in Portugal draws on the *memórias* written by Constantino Lacerda Lobo and published by the Lisbon Academy of Sciences in 1812 and 1840,[9] even though it only really took off on a national level from the mid-19th century onwards with the investigations and publications of Baldaque da Silva, Barbosa du Bocage and Britto Capello, Balthazar Osorio, and King D. Carlos, while the first national aquarium, marine stations, and exhibitions were being established.[10] Since the mid-18th century, and within the enlightenment spirit of the period related to the scientific knowledge about nature in Portugal and in other regions of the Portuguese overseas empire,[11] authors such as the above-mentioned Domingos Vandelli and Constantino Lacerda Lobo dedicated some time to the study of marine fauna, revealing a renewed interest in this subject. The unparalleled work of Baldaque da Silva's *Estado actual das pescas em Portugal*, dating from 1891, is truly a significant mark in the study of fish and fisheries in Portugal, but in fact, more than a century before its publication Doctor Domingos Franco Quaresma, a poorly renowned naturalist and physician, and a native of Peniche in Portugal, wrote a natural history of Portuguese fishes. He was doctor of the king's party, of the Peniche place and its royal hospital. Domingos' father, Francisco Franco Quaresma, graduated with a bachelor's degree from the University of Coimbra, and also a native of Peniche, which means that the Quaresmas were members of the educated small nobility.[12]

8 Ceríaco L., *O 'Arquivo Histórico Museu Bocage'* (Lisbon: 2014).

9 Amorim I., "A decadência das pescarias portuguesas e o constrangimento fiscal – entre a Ilustração e o Liberalismo", in Ribeiro da Silva F. – Cruz M.A. – Martins Ribeiro J. – Osswald H. (eds.), *Estudos em Homenagem a Luís António de Oliveira Ramos* (Porto: 2004) 153–164.

10 Gamito-Marques D., "A Space of One's Own: Barbosa du Bocage, the Foundation of the National Museum of Lisbon, and the Construction of a Career in Zoology (1851–1907)", *Journal of the History of Biology* 5 (2017) 223–257; Amorim I. – Pinto B., "Portugal in the European Network of Marine Science Heritage and Outreach (19th–20th Centuries)", *Humanities* 8.1 (2019) 14. https://doi.org/10.3390/h8010014; Pinto B. – Amorim I., "A Biodiversidade Marinha nos Museus de Portugal continental: Uma introdução", *Museologia & Interdisciplinaridade* 7.14 (2019) 107–127.

11 Brigola J., *Domingos Vandelli e a circulação de conhecimentos na rede de naturalistas europeus* (Évora: 2016); Roque A.C., "Towards a Scientific Approach of Natures: Looking at the Southern Africa Biodiversity throughout the 16th-Century Portuguese Records on Marine Fauna", in Polónia A. – Bracht F. – Conceição G.C. – Palma M. (eds.), *Cross-cultural Exchange and the Circulation of Knowledge in the First Global Age* (Porto: 2018) 75–100.

12 Francisco Franco Quaresma qualified for the position of familiar of the Holy Office. Cf. IAN/TT, Tribunal do Santo Ofício, Conselho Geral, Habilitações, Francisco, mç. 16, doc. 475.

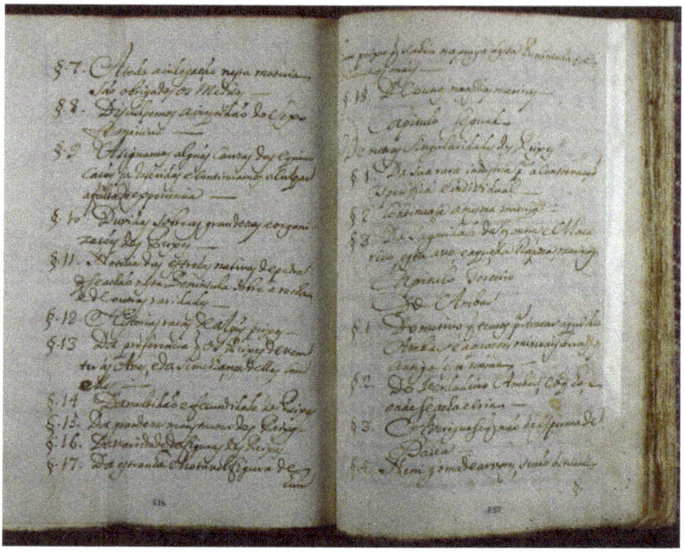

FIGURE 13.1 Quaresma Domingos Franco, Index of *Piscilegio lusitano, Nova, Exacta, Natural, e Medicinal Noticia dos nomes e qualidades dos Peixes que se pescão nos mares e rios da Costa de Portugal damnos e proveitos que do seu bom, ou mao uso no comer podem resultar aos sãos, e enfermos*, unpublished manuscript (*c.*1750). Reproduced with the permission of Samuel Iglesias

In mid-18th century Domingos Franco Quaresma wrote what seems to be the first comprehensive treatise on the Portuguese aquatic fauna, the methods used to catch the fish. The manuscript is titled *Piscilegio lusitano, Nova, Exacta, Natural, e Medicinal Noticia Dos nomes e qualidades dos Peixes que se pescão nos mares e rios da Costa de Portugal*,[13] hereafter *Piscilegio lusitano* [Fig. 13.1]. The 650-page-long manuscript, possibly dated *c.*1750, is devoted primarily to the study of ichthyology, and marine and freshwater fisheries; the author lists and describes 135 aquatic species (including bony fish, elasmobranchs, aquatic mammals, and invertebrates), highlights topics such as whaling, the quality

13 *Piscilegio lusitano, Nova, Exacta, Natural, e Medicinal Noticia Dos nomes e qualidades dos Peixes que se pescão nos mares e rios da Costa de Portugal damnos e proveitos que do seu bom, ou mao uso no comer podem resultar aos sãos, e enfermos; Com outras raridades e curiosas advertências importantes ao bem comum da Saude* (…) *Exposto pela curiosidade do Sr. Dr. Franco Quaresma natural da Praça de Peniche, médico de S. Mag.*, (*Piscilegio lusitano, New, Exact, Natural, and Medicinal News of the names and qualities of Fishes that are caught in the seas and rivers of the Coast of Portugal* …). Unpublished manuscript by Domingos Franco Quaresma (*c.*1750).

and use of fishery products or ambergris, and the virtues of hot springs. He includes a wide variety of fish, using the scholarly knowledge of classical and Renaissance authors, but also the experience of "seafaring men". The author does not follow Carl Linnaeus' biological classification of fish. This is a manuscript with only one known copy. The work was certainly composed in the first half of the 18th century, as Barbosa Machado[14] wrote that "it was in the licenses of printing", but why it never came to know the printed form is unknown.

Piscilegio lusitano significantly precedes the pioneering work of the 19th-century naturalist Baldaque da Silva. The erudition and completeness of this document should have marked the coeval Portuguese scientific panorama, possibly like the *Traité général des pesches* by Duhamel du Monceau and La Marre (1769–1782) in France, or the *Ensayo de una historia de los peces* by D. José Cornide (1788) in Spain. But the devastating Lisbon earthquake in 1755 and its consequences for economic activity[15] certainly thwarted the publication of this work, which has remained unpublished and not unknown to the scientific community.

The first readings of this text reveal valuable and original information on the exploitation and use of marine animals and products in 18th-century Portugal. *Piscilegio lusitano* deals with marine species and their exploitation on a national scale. At first glance, it appears quite different from the Renaissance European ichthyologist treatises, such as the ones by Pierre Belon, Guillaume Rondelet, Conrad Gessner, Adriaen Coenen, Ulisses Aldrovandi, and Ippolito Salviani,[16] both in structure and purpose. No documents alike are found in the Portuguese production of natural history for the 16th to the 18th century. The one we might be able to compare with it is the 16th-century Spanish natural history treatise titled *Bestiario de Don Juan de Austria*.[17] The memoirs on the fisheries of Lacerda Lobo from the late 18th century onwards and the lists

14 Machado, Diogo Barbosa, *Bibliotheca Lusitana, historica, critica, e chronologica, na qual se comprehende a noticia dos authores portuguezes, e das obras, que compozeraó desde o tempo de promulgaçaó da ley da graça até o tempo presente*. Vol. IV. (Lisbon, [Lisboa Occidental: Na officina de Antonio Isidoro da Fonseca]: 1759) 159.

15 Pereira A., "The Opportunity of a Disaster: The Economic Impact of the 1755 Lisbon Earthquake", *The Journal of Economic History* 69.2 (2009) 466–499.

16 Gudger E., "The Five Great Naturalists of the Sixteenth Century: Belon, Rondelet, Salviani, Gesner and Aldrovandi: A Chapter in the History of Ichthyology", *Isis* 22.1 (1934) 21–40; Kraemer F. – Zedelmaier H., "Instruments of Invention in Renaissance Europe: The Cases of Conrad Gesner and Ulisse Aldrovandi", *Intellectual History Review* 24.3 (2014) 321–341; Egmond F., *Eye for Detail: Images of Plants and Animals in Art and Science 1500–1630* (London: 2017).

17 García Gil J.J. – Molinero Hernando P. (eds.), *Bestiario de D. Juan de Austria. S. XVI. Estudios y transcripción de la edición facsimilar* (Burgos: 2000).

of fishes by Balthazar Osorio in late 19th century, by the Lisbon Academy of Sciences, the *Apontamentos para a ichthyologia de Portugal* by Barbosa Du Bocage and Britto Capello (1866), the *Catalogo dos peixes* de Portugal by Britto Capello (1880), and the late-19th-century treatise by Baldaque da Silva also deal with this topic. Preliminary research reveals that this manuscript is cited only in publications in 1759, 1878, and 2018.[18]

The rediscovery of this "ready to be printed" manuscript by the French biologist Samuel Iglesias[19] is a unique opportunity to complete a history of Portuguese natural history, notably concerning ichthyology, sea and freshwater fishing, and whaling. We do not intend to go deep into this and other historical sources but rather to briefly review them to address knowledge about whales in Portugal prior to modern times. Future endeavours regarding *Piscilegio lusitano* will include its full transcription, and translation of relevant parts to English, with comments and notes from biologists and historians.

Before *Piscilegio lusitano*, we can find sparse indications of the value of knowledge about marine fauna as well as true compilations of the Portuguese marine fauna and biodiversity, as shown recently by Herold, Horst, and Leitão[20] (see also Bernardo Herold's and João Paulo S. Cabral's contribution to the present volume) in their study of a mid-16th-century manuscript, which includes a list of aquatic animals. The manuscript, untitled but dubbed 'The Natural History of Portugal' (1555–1556) by the research team responsible for its study, is divided into different parts revealing the author's interests. In the German text, Leonhard Thurneysser often refers to plants, animals, etc. by the Portuguese names he collected in loco. In addition to the obvious botanical, pharmacological, zoological, geographic, palaeoecological, and anthropological interest of his observations, certain commercial and industrial aspects that he intersperses with the afore-mentioned main themes are also noteworthy. The Second Part has a full title by the author but regarding our main interest can be reduced to 'Small Aquatic Animals from Lisbon' (*Animaizinhos*

18 Machado, *Bibliotheca Lusitana*; Anonymous, *Catalogo dos preciosos manuscriptos da bibliotheca da casa dos marquezes de Castello Melhor. Documentos officiales, grande numero de autographos obras originaes e ineditas* (Lisbon: 1878); Iglésias S.P. – Mollen F.H., "Cold Case: The Early Disappearance of the Bramble Shark (*Echinorhinus brucus*) in European and Adjacent Waters", *Oceans Past News* 10 (2018) 1–2.

19 The manuscript was found and purchased at the Salon International du Livre Rare et de l'Objet d'Art 2017 in Paris (Grand Palais, 7–9 April 2017), and is currently part of a private collection of antique books and manuscripts dedicated to ichthyology, in Concarneau (Finistère, France).

20 Herold B. – Horst T. – Leitão H., "A 'História Natural de Portugal' de Leonhard Thurneysser zum Thurn, ca. 1555–1556", *Ágora. Estudos Clássicos em Debate* 19 (2017) 305–334.

Aquáticos que se encontram em lisboa) (fols. 111r–127v).[21] And here are listed several species from Portugal, including marine and estuarine fish (namely from the Tagus River), mollusks, and crustaceans, and among the fish, dolphins and whales are included. The manuscript, written in German by Thurneysser, an academic friend of Damião de Góis, shows that the author was aware of the Portuguese scientific scene at the time, providing relevant information for a better perception of Portuguese scientific thought of that period.[22] Unfortunately, a translation of the document transcript for this part of the fish list is not yet accessible, and so far, the index is the only information available for analysis and comparison.

Possibly coeval to this work another one of relevance for the understanding of local studies and interest in marine fauna must have existed. What has reached our present day are mere fragments of a document that we are calling here a 'Portuguese Latin dictionary of marine animal names',[23] but it is enough to enlighten us about a "scientific" interest in marine fauna. The translation of fish names from Portuguese into Latin indicates a concern on the part of the author or humanists of the time to understand, categorize, and eventually transfer information about the natural world and particularly animals of the seas.

Whether with complete and rich documents, or with pamphlets, single pages, or even poems, we can begin to draw a history of the natural and cultural history of whales in early modern Portuguese society. The first evidence of such a conceptualization and categorization of species, in an attempt to understand the natural order of the ocean and its animals, seems to emerge in Portugal as early as the 16th century. At this point, as it needs to be systematized, I will start with one of the "first" animals, the whale, the mighty Leviathan.

21 *Índice e Descrição de vários Animais e especialmente de Animaizinhos Aquáticos que seencontram na Lusitânia, mas que na nossa Terra não se costumam ver. Iniciadoem Lisboa. Ano de Cristo 1555 e 1556. As viagens e caminhadas de uma pessoapensadora profunda e hábil são superiores a quase todos os AcademiarumStudijs ou dedicação à filosofia. Plutarco: Peregrinatio alit sapientiam. Quemcaminha por país estrangeiro, fica a conhecer muitas coisas estranhas; alcançaassim sabedoria, compreensão e muitas vezes grande fortuna.* According to Herold – Horst – Leitão, "A 'História Natural de Portugal'" 318.

22 Herold – Horst – Leitão, "A 'História Natural de Portugal'". Herold B. – Horst T. – Leitão H. (eds.), "A *História Natural de Portugal* de Leonard Thurneysser zum Thurn, ca. 1555–1556" (Lisbon: Academia das Ciências de Lisboa, 2019).

23 [Dicionário português-latino de nomes de animais marinhos], Biblioteca Pública de Évora, Cód. CLXIX, 1–26, no. 42.

3 Mrs. Whale, '*Um Grande Peixe, Dona Baleia da Costa*', in Natural History

According to Domingos Franco Quaresma, fish exceed land animals both in number and in beauty. The erudite author supports his opinion with Pliny, Aristotle, and the Bible. Among the fish, he highlights the whale, writing: 'and it seems that this one was enough for no terrestrial animal to dispute majorities'.[24] But he also goes about describing local events of sightings or strandings of large whales, according to those who have witnessed them in prior times. He states, on 22 April 1575, in Peniche, a 'monstrous fish' appeared on the beach that 'no one knew about'. This news was noted by Father Luís de Granada[25] in his work *Introduction to the Symbol of Faith*, and a drawing of the monster was sent to D. Henrique of Portugal [Fig. 13.2]. Again, on 10 January 1723, another sea monster, already dead, was shipwrecked on the Cacilhas pontoon and nicknamed 'sombreiro'.[26] As we will see further down in the text, such an astonishing event inspired several poets, made headlines in the press, and was disseminated across and beyond Portugal [Fig. 13.3].

In *Piscilegio lusitano*, Domingos Franco Quaresma copies a piece by an anonymous poet who wrote about the event using both an informative poetic way and a fable kind of wording. Observing it in real space and time, the poet portrayed the great interest and expectations of the people of Lisbon and Almada.[27] They crowded around to observe its grandeur and strangeness, as well as to witness the human effort to rescue it from the waters of the Tagus. Using fable as an approach to the world of humans, the poet used marine fauna to comment on the inequality between the strong and the weak. 'You have been eating the little ones / with great tyranny / that in the belly of the whale / no eel escapes you'.[28] Whether the result of direct observation or the exercise of imagination, the poet described the agony of the animal after it ran aground in Cacilhas. The rocks destroyed its body and the 'fresh water' from the 'Ocean Tagus' made it vomit. Doctors and medics were called to relieve the animal, but nothing more could be done. The author of the poem, in addition to his knowledge of marine fauna, was also comfortable with the medical issues of the time. Interestingly, the hypothetical treatments applied to the sea

24 Quaresma, *Piscilegio* 40–41.

25 Granada Luís de, *Obras de Fray Luis de Granada* (Madrid, La Publicidad, Imprenta de M. Rivadeneyra, 1848) tomo I, p. 238.

26 Quaresma, *Piscilegio* 44.

27 Quaresma, *Piscilegio* 46.

28 Quaresma, *Piscilegio* 49.

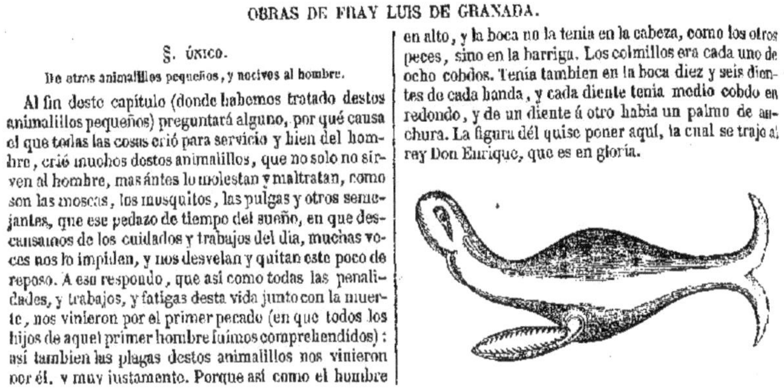

OBRAS DE FRAY LUIS DE GRANADA.

§. ÚNICO.

De otros animalillos pequeños, y nocivos al hombre.

Al fin desto capítulo (donde habemos tratado destos animalillos pequeños) preguntará alguno, por qué causa el que todas las cosas crió para servicio y bien del hombre, crió muchos destos animalillos, que no solo no sirven al hombre, mas ántes lo molestan y maltratan, como son las moscas, los mosquitos, las pulgas y otros semejantes, que ese pedazo de tiempo del sueño, en que descansamos de los cuidados y trabajos del dia, muchas veces nos lo impiden, y nos desvelan y quitan este poco de reposo. A esto respondo, que así como todas las penalidades, y trabajos, y fatigas desta vida junto con la muerte, nos vinieron por el primer pecado (en que todos los hijos de aquel primer hombre fuimos comprehendidos): así tambien las plagas destos animalillos nos vinieron por él. Y muy justamente. Porque así como el hombre

en alto, y la boca no la tenia en la cabeza, como los otros peces, sino en la barriga. Los colmillos era cada uno de ocho cobdos. Tenía tambien en la boca diez y seis dientes de cada banda, y cada diente tenia medio cobdo en redondo, y de un diente á otro habia un palmo de anchura. La figura dél quise poner aquí, la cual se trajo al rey Don Enrique, que es en gloria.

FIGURE 13.2 Granada Luís de, *Obras de Fray Luis de Granada* (Madrid, La Publicidad, Imprenta de M. Rivadeneyra: 1848), part I, 238 (detail). Biblioteca Digital Hispanica with open access license CC-BY: http://bdh-rd.bne.es/viewer.vm?id=0000052692&page=1

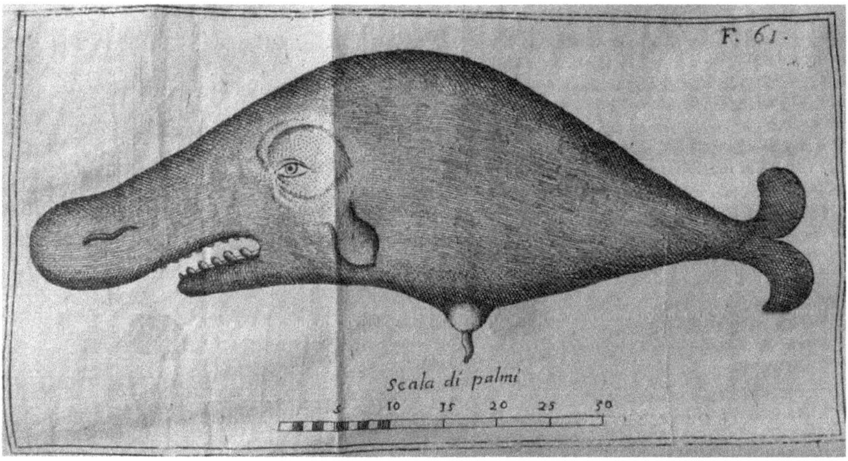

FIGURE 13.3 Montigore Antonino, *Della Sicilia ricercata nelle cose più memorabili* (Palermo, Francesco Valenza: 1742–1743), vol. 2, 61. Illustration of a sperm whale that was stranded along the coast near Mazzara, Sicily, on 20 November 1734. This illustration was used by Antonio Mongitore as the basis for *Physeter urganantus*, Rafinesque-Schmaltz C.S., *Précis des découvertes et travaux somiologiques de Mr C.S. Rafinesque-Schmaltz entre 1800 et 1814 ou Choix raisonné de ses principales découvertes en zoologie et en botanique* (Palermo: 1840)

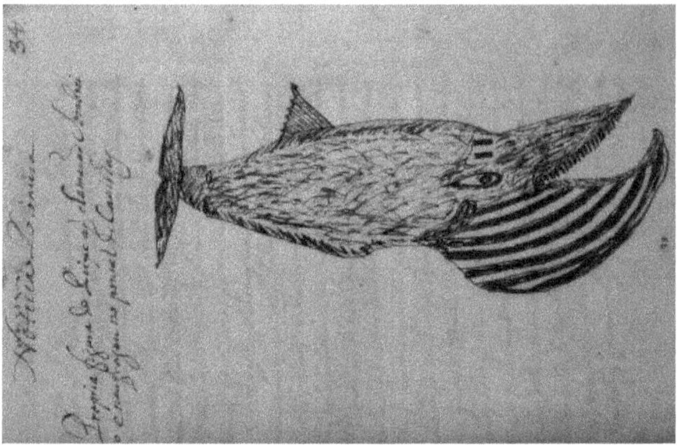

FIGURE 13.4 Stranded whale in Quaresma Domingos Franco, *Piscilegio lusitano, Nova, Exacta, Natural, e Medicinal Noticia dos nomes e qualidades dos Peixes que se pescão nos mares e rios da Costa de Portugal damnos e proveitos que do seu bom, ou mao uso no comer podem resultar aos sãos, e enfermos*, unpublished manuscript (*c.*1750). Reproduced with the permission of Samuel Iglesias. This illustration of a stranded whale in Cacilhas (Lisbon) bears the caption *"Própria figura do peixe a que chamão Sombreiro e naufragou no pontal de Cassilhas"*; it is the only illustration included in the manuscript *Piscilegio Lusitano*. Reproduced with the written permission of Samuel Iglesias

monster were the same as those used on humans. Among the possible causes of the animal's death, the author pointed out dropsy. After being removed from the Tagus, the monster was taken to Ribeira das Naus, where it was exposed for three days, arousing the curiosity of the people of Lisbon, who could not stand the stench, 'that one could feel from afar' [Fig. 13.4].[29]

While Domingos Franco Quaresma was writing his *Piscilegio lusitano*, other news reached him that he included in his manuscript – in the port of Viana do Castelo an already dead sea monster appeared that was also not identified with any species. The animal was already in a state of putrefaction, which is why it was burned to avoid an epidemic.[30] After being pulled from the sea, the 'monstrous fish' were displayed in the Ribeira da Naus, a large and airy space where they could stretch out the creatures' bodies. After a few days, when

29 Quaresma, *Piscilegio* 53–54.
30 Quaresma, *Piscilegio* 56.

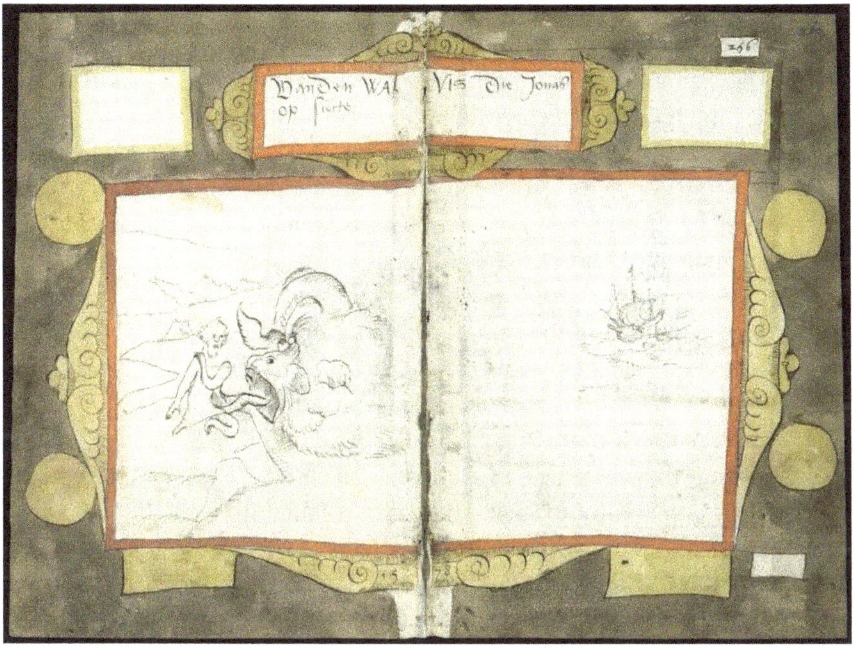

FIGURE 13.5 Jonah and the whale in Adriaen Coenen's *Visboeck*, 1577–81, Ms 78 E 54, fols. 259v–260r

the animal was already showing signs of rotting, the corpse was burned. The news also shows aspects and characteristics of the whales' lives and behaviors, or at least attempts to offer such a view of the eco-ethology of the animals. For instance, Quaresma refers to the whale's instinct to protect its young. In case of a storm or the proximity of a possible predator, the whale swallows its young to protect them, throwing them back into the sea as soon as the danger passes. The author compares this behaviour to the that in the biblical story of Jonah [Fig. 13.5]. This news about the harrowing of sea monsters testifies to the impact that these events had on the populations, arousing the interest of various social groups, from the humblest to the most educated, and even the aristocracy. As we have seen, the drawing of one of these animals reached the eyes of King D. Henrique.

Domingos Franco Quaresma, dealing not only with the biology of the species he addresses but also the products and activities of the extraction and utilization of the animals, concluded in his work that the ambergris was not whale sperm. This would be easily verified, he wrote, because in Bahia, where whale fishing was organized in a contract, which yielded many hunters, there were no reports of ambergris being found inside these animals.[31] To support his claim, Quaresma resorted, beyond experience, to the opinion of the Spanish physician and botanist Nicolás Monardes (*c.*1493–1588). According to him, the cause of the confusion was due to the fact that ambergris is light and floats and fish eat it. However, few whales could hold amber in their stomachs for long, as the smell of amber nauseated them. The arguments presented in the previous chapters made credible the hypothesis that amber was marine bitumen. Against this opinion was the argument of Dr. Jacob de Castro Sarmento that amber was generated in the sacs in the genital region of male whales called *sperma ceti*. Quaresma contested this position because ambergris was found on all the coasts of the world, and whales are not found in all seas.

He also mentions that Father Benedict Jerónimo Feijó helped spread the false idea that the whale had an esophagus so narrow that it could only swallow one sardine at a time. Such a difficulty resulted from the divine punishment meted out to the whale that swallowed Jonah. Father Benedict's theory was discredited by the testimony of the Florentine scholar John Fabri, who claimed that a man on horseback could fit down the throat of a whale that sank off the Italian coast in 1624. Fishermen in Portugal and Brazil also said the same thing. In addition, he claims that some authors have classified the whale as a cartilaginous animal, basing this claim on Aelian's work. However, this is a misinterpretation of the Roman author, whose intention was only to affirm that the whale was not scaly. According to Quaresma, the whale and other "monster fish" were not scaly or cartilaginous, because their spines are bony, as is notorious to anyone who knows the whale's beard and back. The taste of the meat of this animal was bad and in Brazil only slaves ate it, he writes.

There is also a very vivid description of a fight between a swordfish and a whale: 'the whale rises from the water with his tail in the air, and strikes him with it and comes upon the swordfish with his whole body and height tries to kill him; he at the same time tries to hurt it and take its life with blows, always looking for it on the sides, where he knows the skin is softer.' Quaresma is said to have witnessed these battles in the seas of Peniche, and this empirical knowledge, whether the piece just described or another about different marine mammals and fish, is an indication of his motivation to approach the

31 See the PhD monograph by Nina Vieira for a review of the Basque-style shore-based whaling developed by the Portuguese in colonial America: Vieira N., *A taxonomia da baleação portuguesa entre os séculos XV e XVIII: Uma história atlântica do mar, das baleias e das pessoas*. Faculdade de Ciências Sociais e Humanas da Universidade NOVA de Lisboa. Tese de Doutoramento (2020).

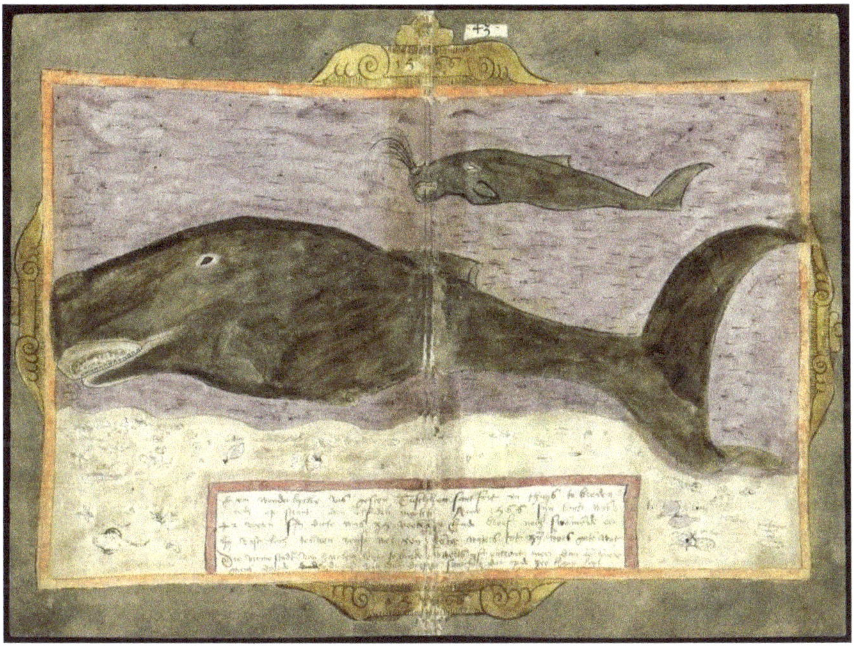

FIGURE 13.6 Stranding of sperm whales in Adriaen Coenen's *Visboeck*, 1577–1581, Ms 78 E 54, fols. 51v–52r

natural history of animals while compiling a long and exhaustive list of existing marine species. He was a renowned physician, certainly a well-read scholarly author who paid attention to the source of information he shared with his readers while critiquing it according to his own interpretation, but he was also an on-the-spot observer. Possibly living, or knowing very well, the local peculiarities of a lively and developing coastal fishing village, he would have taken advantage of this fact and accumulated experience of his own. Whale catches, strandings, and coastal sightings could have been quite common in the region, as in many other coastal areas throughout Europe [Fig. 13.6] and, of course, in Portugal. In the long run of this story of Portuguese natural history (or of marine animals) he was not alone in these kinds of observations.

In the above-mentioned manuscript by Leonhard Thurneysser, a list of the species approached is available to our current analysis. His work includes, among many different fishes of the Portuguese shores, and some marine mammals, sea monsters in what may be seen as a reminiscence of the medieval bestiaries. From folio 64 until almost the end of the first part of this book, there is a new narrative and description of the fish caught in Lisbon. There we can find the '*Balena*' and the '*Balena altera species balenae*' as the first two entries under "Pisces in Portugallia".[32]

The 'Dictionary of marine animal names' is a manuscript with four unnumbered pages, which contains an armillary sphere under a six-pointed star as a watermark that can be dated to the 16th century.[33] No information about the document itself or the author is available. Although the references to classical authors are many, an excerpt from the Dictionary mentions the French naturalist Pierre Belon (1517–1564), which allows us to say that the document was written in the second half of the 16th century. We simply have an index or a list, from 'M' to 'T', where we find: dolphin (*toninha – plinio aeste Tursiogrece focena*); and manatee/the ox-fish of Brazil (*pexe boi do brazil – latine vitulus maris oceani*). With only two references to marine mammals, and whales not among them, we are left with our imagination to believe that whales (under the 'B' for the Portuguese *baleia*) might have been included in it.

4 Mrs. Whale, '*Um Grande Peixe, Dona Baleia da Costa*', in Poetry

As to imagination, there is also plenty of room for the admiration of these large and impressive animals in the literature – mostly poetry – found in Portuguese documentary sources.[34] An unpublished manuscript from the Lisbon Academy of Sciences is an example of such; with no specific title, the poem is included in the hand-written book by Father Manoel de Santa Maria (1723).[35] Here, he describes in amusing satirical verses the royal critter, Mrs. Whale of the shores, 'a bixa real, Dona Balea da Costa'.[36] The extraordinary animal fulfilled the people's taste for novelty.

32 Herold – Horst – Leitão, "A 'História Natural de Portugal'" fols. 124v to 127v.

33 The document is found among writings from the 16th to the 18th century, some of which came from convents in the district of Évora and was part of the estate of Joaquim Heliodoro da Cunha Rivara, director of the Évora Public Library from 1838 to 1855 (Silveira L., *Manuscritos de filologia latina da Biblioteca Pública e Arquivo Distrital de Évora* (Évora: 1941) 37–41).

34 Freitas J.G. – Brito C., "A Bixa Baleia. Ou a história de um manuscrito sobre o maravilhoso do mar", in Clamote Carreto C.F. – Moreira Sousa L.M. (eds.), *Imaginários do mar: uma antologia crítica* (Lisbon: 2021) 203–209. Lacerda T. – Vieira N. – Brito C., *Fontes documentais para uma história natural das baleias em Portugal* (2022).

35 *Este Livro he uso do P.M. Fr. Manoel de Sta Maria Leytor de vespera neste conv.^{to} de N.S.^{ra} de Jesus na era de 1723 tem noventa e oito folhas. neste anno appareceo na ribeira de L^{xa} huã balea p^{a} cuja vista concorreo toda a gente da cidade* (Santa Maria: 1723).

36 See Academia das Ciências de Lisboa (1986) – *Catálogo de Manuscritos, Série Vermelha*, II (no. 500–980). Lisbon: Publicações do II Centenário da Academia das Ciências de Lisboa.

The poem "Testament in verse of a whale to the *baleato*"[37] is undated, but the lyrics appear to be from the 18th century.[38] The verses seem to be inspired by real facts, possibly in the 1723 stranding in Cacilhas, similarly to the above-mentioned poem. In this one, the whale character expressed the desire to be embalmed, probably this would be the end that the author of the poem would like a real whale to have known, since its magnificence and strangeness impressed the curious and knowledge-lovers. In his will, the whale determined that his body should be accompanied by the people of Cacilhas to the Ribeira das Naus on the other bank of the Tagus. This is yet another poetic evocation of what would happen when a whale was beached. The capture of the animal and its transport created a commotion among the people, and it is easy to imagine the curious jumping onto boats and battalions to watch the pulling maneuvers up close. In Ribeira, those attending the funeral ceremonies would be treated to 'beer, oranges, wine, and brandy', evoking the possible festive atmosphere that the rushing would cause. The whale burial mimicked in many ways the human one and should be 'English style' because 'it made more nobility'. To demonstrate the dead man's importance, a procession was essential, as such, the whale ordered that 'on the wharf some boys follow, / And the Dutch, who are most able'. Like royal wills, the whale's fictional will also named the executors. Later in the poem, there is the confession 'that all that was said, was lied about', going on to relate what happened to a bold whale. The dead animal was butchered, and its flesh divided. The poem mentions the parts that were used: the guts, the lungs, the heart, the bacon, the neck, the throats, and the large loins.

The poem 'To the Most Reverend Father Br. Vicente for not having fulfilled the Testament of *Balca*'[39] is a satire to expose the bad behavior of the executor. The choice of the 'swimmer, brave and strong monster' who came to die on the sands of the Tagus as a motto for satire is due to the possible ocular observation of a boldness or simply to the fame that these phenomena garnered. It also served the purpose of ridiculing the executor, since he was compared to a whale: 'for with him [the sea monster], there you only resemble / That figure of yours, so tremendous / That another one but seen, even more hideous'. The executor was accused of not doing his job properly, wanting to

37 *Baleato* is the Portuguese name for a small whale, or a different species of whale.

38 It is part of a miscellany of various bound handwritten documents that includes sonnets, verses, and satires. *Testamento em verso de uma baleia ao baleote*, Biblioteca Geral de Coimbra, Ms. 512.

39 *Ao Reverendíssimo Padre Fr. Vicente por não haver dado cumprimento ao Testamento da Balea*. Biblioteca Geral de Coimbra, Ms. 512. This is an undated manuscript, also included in the former miscellany, and it possibly dates from the late 18th century as it refers to an observed occurrence, most certainly the Cacilhas stranding.

misappropriate the testamentary property – 'that you want, to put everything, into the Belly.' In the poem there was even an implicit threat 'you're bound to get it in the snouts.' The choice of the whale as a monstrous animal served as an allegory for the criminal act: 'but I believe that you, who ate it / And forgot the Testament'; still, it is a monstrous case.

It should be noted that the use of animals, especially sea animals, by thinkers and poets as a way of creating allegorical moral lessons was recurrent, and the three poems commented on are examples of this. This tradition went back to Classical times, at least from Aesop, through Aelian. The existence of poems about whales and their boldness shows how these animals were not only of interest to scholars but had an impact on the common person. Their rarity generated popular curiosity, the circulation of pamphlets, of news, and was the motto for the creativity of poets, particularly in the creation of satires. Even if directed toward a different audience – the commoners, and, through the use of illustrations, also the illiterate – all these poems seem to have been inspired by the real events, as coeval printed news also testifies.

5 Mrs. Whale, *'Um Grande Peixe, Dona Baleia da Costa'*, in the News

Hand leaflets, letters, written news, and newspapers articles, at least from the middle 18th century onwards, included descriptions of stranded whales mainly in central Portugal shores. These events are worth printing, disseminating, and talking about; they reveal the interest of different audiences in the natural history of these mammals and shed light onto the general curiosity that involved the presence at the shore of a large marine animal.

Just before, during, or after the massive 1531 Lisbon earthquake, whales were stranded on the shores of the Tagus River. As much as the catastrophic event, whales made the news that was produced in Portugal and conveyed elsewhere in Europe. We are familiar with the German leaflet referring to it as well as the Book of Miracles,[40] but the source of the information must have been the 'piece' by Gaspar Correia.[41] Here, the chronicler describes on 26 January 1531 a great earthquake felt in Portugal, Castile, Flanders, Rome, and France. With

40 Brito C., "The Voice of Skogula in 'Beasts Royal' and a Story of the Tagus Estuary (Lisbon, Portugal) as Seen through a Whale's-Eye View", *Humanities* 8 (1) 47 (2019) 1–16; Brito – Vieira, "Uma construção cultural de ser baleia".

41 Correia Gaspar, *Crónicas de D. Manuel e de D. João III* (Lisbon: 1992) 300–301. Gaspar Correia (1495–1565) belonged to the Royal Chamber. In 1512, he left for India, where he served as clerk to Afonso de Albuquerque. He held several other posts until his death in 1565. Between 1532 and 1534, he wrote the *Chronicas dos Reys de Portugal*, containing

it, 30 or so dead whales washed up on the Portuguese coast. In Sesimbra there came to die a fish, which came through the air to land, which had 33 wings and whose appearance was unknown to all. For this reason, it was brought before D. João III (king of Portugal).

Another whale stranded in the Tagus River in Lisbon 1723 became famous. The poems I showed before were inspired directly by it and the observation of all involving such an extraordinary event, or eventually based on the news posted on coeval press. According to the *Gazeta de Lisboa Occidental* the large fish, a species of *balea* or *sombrero*, was massive and never seen[42] opening space to more detailed and illustrated news a week later in the same journal [Fig. 13.7].[43] This news was so significant that crossed national borders[44] and was presented at the Royal Society of London: "Lisbon, 21 January 1723. The great fish, that came into this harbor last week nobody can say to have a certain knowledge of his species".

Another bit of news refers to another 'Memory of a large fish that died in the Tagus'.[45] On 11 January 1724, small boats in the service of the Ribeira das Naus brought to the said Ribeira a large fish that was 85 palms long and 14 palms high, and the mouth measured 19 palms. The English called this animal *Baleato*, and the Dutch cut it up to make oil. D. João VI (king of Portugal) ordered the corpse to be shown to the nuns at the Odivelas Convent. The choice of this convent was not random, since Mother Paula of Odivelas, the king's mistress, lived here.

According to 'News of the fish that arrived in the Beach of Moita Village',[46] the naturalist Domingos Vandelli, in charge of the Ajuda Botanical Garden,[47] bought a large fish that was in Ribeira das Naus. With this acquisition Vandelli intended to enrich the garden's collection of natural history, a subject that was part of the education of princes. The animal was skinned for about 13 coins.

---

summaries of the lives of kings Afonso Henrique to João III, as well as the main events of the Portuguese kingdom.

42 Mascarenhas, José Freire de Monterroio (ed.), *Gazeta de Lisboa Occidental*, 14 January 1723. Mascarenhas was the editor of the newspaper (that ran from 1670 to 1760) for that period.

43 Mascarenhas José Freire de Monterroio (ed.), *Gazeta de Lisboa Occidental*, 21 January 1723.

44 Vieira C.C., "Observing the Skies of Lisbon. Isaac de Sequeira Samuda, an *estrangeirado* in the Royal Society", *Notes and Records of the Royal Society* 68.2 (2014) 135–149.

45 "*Memória de hum grande Peyxe que morreo em o Tejo*", Biblioteca do Museu Nacional de Arqueologia, MS/Pp DIV, cx. 12, no. 767, fol. 62v.

46 "*Noticia do peixe que apareceo na Praya da villa da Moita*", Biblioteca do Museu Nacional de Arqueologia, [manuscrito], MS/Pp, DIV, cx. 12, no. 767.

47 The Royal Botanical Garden of Ajuda (Lisbon) was founded in 1768. The garden was to maintain and study the largest nu.mber of plant species. In addition to botany, it was to instruct princes in the natural sciences.

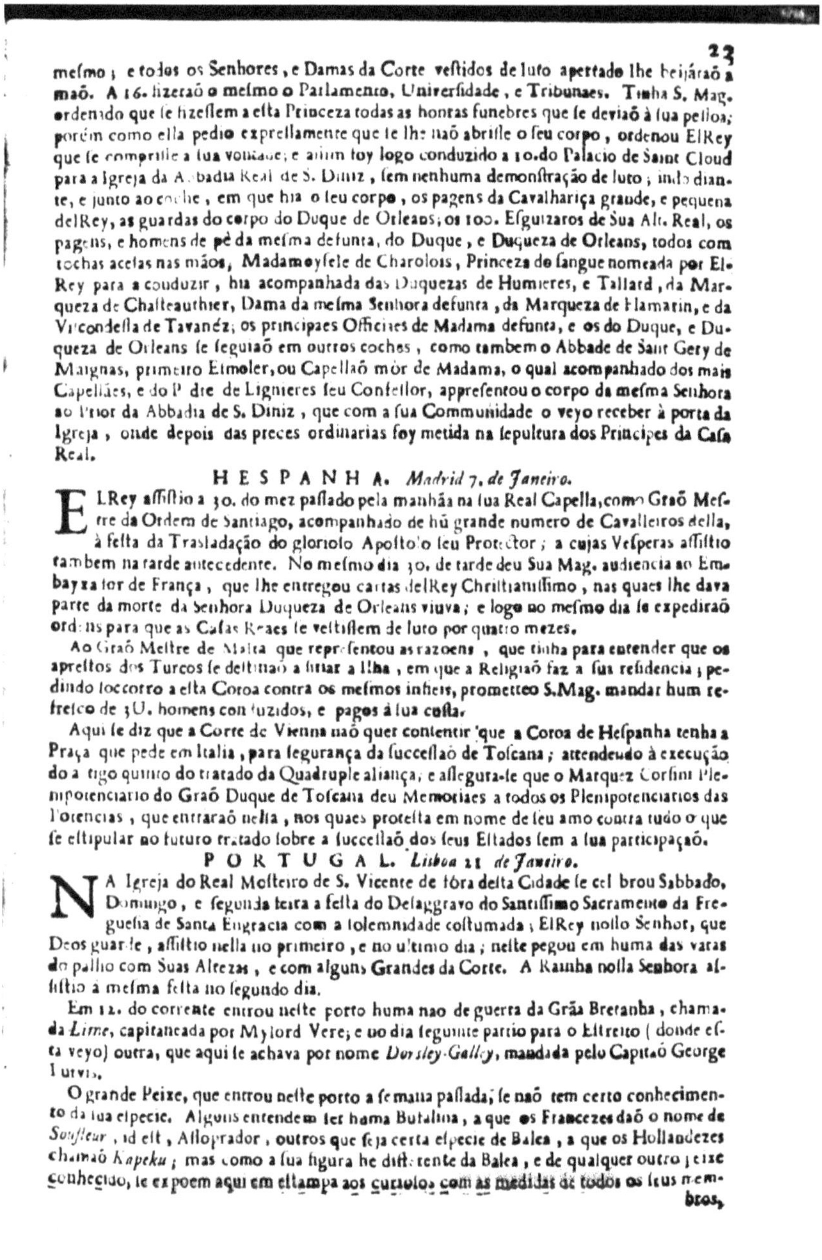

23

meſmo; e todos os Senhores, e Damas da Corte veſtidos de luto apertado lhe beijáraõ a maõ. A 16. fizeraõ o meſmo o Parlamento, Univerſidade, e Tribunaes. Tinha S. Mag. ordenado que ſe fizeſſem a eſta Princeza todas as honras funebres que ſe deviaõ à ſua peſſoa; porém como ella pedio expreſſamente que ſe lhe naõ abriſſe o ſeu corpo, ordenou ElRey que ſe compriſſe a ſua vontade; e aſſim foy logo conduzido a 10. do Palacio de Saint Cloud para a Igreja da Abbadia Real de S. Diniz, ſem nenhuma demonſtração de luto; indo diante, e junto ao coche, em que hia o ſeu corpo, os pagens da Cavalhariça grande, e pequena delRey, as guardas do corpo do Duque de Orleans; os 100. Eſguizaros de Sua Alt. Real, os pagens, e homens de pè da meſma defunta, do Duque, e Duqueza de Orleans, todos com tochas aceſas nas mãos, Madamoyſele de Charolois, Princeza do ſangue nomeada por ElRey para a conduzir, hia acompanhada das Duquezas de Humieres, e Tallard, da Marqueza de Chaſteauthier, Dama da meſma Senhora defunta, da Marqueza de Flamarin, e da Viſcondeſſa de Tavanéz; os principaes Officiaes de Madama defunta, e os do Duque, e Duqueza de Orleans ſe ſeguiaõ em outros coches, como tambem o Abbade de Saint Gery de Maignas, primeiro Eſmoler, ou Capellaõ mór de Madama, o qual acompanhado dos mais Capellães, e do Padre de Lignieres ſeu Confeſſor, appreſentou o corpo da meſma Senhora ao Prior da Abbadia de S. Diniz, que com a ſua Communidade o veyo receber à porta da Igreja, onde depois das preces ordinarias foy metida na ſepultura dos Principes da Caſa Real.

HESPANHA. *Madrid 7. de Janeiro.*

ElRey aſſiſtio a 30. do mez paſſado pela manhãa na ſua Real Capella, como Graõ Meſtre da Ordem de Santiago, acompanhado de hũ grande numero de Cavalleiros della, à feſta da Trasladação do glorioſo Apoſtolo ſeu Protector; a cujas Veſperas aſſiſtio tambem na tarde antecedente. No meſmo dia 30. de tarde deu Sua Mag. audiencia ao Embayxador de França, que lhe entregou cartas delRey Chriſtianiſſimo, nas quaes lhe dava parte da morte da Senhora Duqueza de Orleans viuva; e logo no meſmo dia ſe expediraõ ordens para que as Caſas Reaes ſe veſtiſſem de luto por quatro mezes.

Ao Graõ Meſtre de Malta que repreſentou as razoens, que tinha para entender que os apreſtos dos Turcos ſe deſtinaõ a ſitiar a Ilha, em que a Religiaõ faz a ſua reſidencia; pedindo ſoccorro a eſta Coroa contra os meſmos infieis, prometteo S. Mag. mandar hum ſocorro de 3U. homens conduzidos, e pagos à ſua coſta.

Aqui ſe diz que a Corte de Vienna naõ quer contentir que a Coroa de Heſpanha tenha a Praça que pede em Italia, para ſegurança da ſucceſſaõ de Toſcana; attendendo à execução do artigo quinto do tratado da Quadruple aliança; e aſſegura-ſe que o Marquez Corſini Plenipotenciario do Graõ Duque de Toſcana deu Memoriaes a todos os Plenipotenciarios das Potencias, que entraraõ nella, nos quaes proteſta em nome de ſeu amo contra tudo o que ſe eſtipular no futuro tratado ſobre a ſucceſſaõ dos ſeus Eſtados ſem a ſua participaçaõ.

PORTUGAL. *Lisboa 21 de Janeiro.*

Na Igreja do Real Moſteiro de S. Vicente de fóra deſta Cidade ſe celebrou Sabbado, Domingo, e ſegunda feira a feſta do Deſaggravo do Santiſſimo Sacramento da Freguezia de Santa Engracia com a ſolemnidade coſtumada; ElRey noſſo Senhor, que Deos guarde, aſſiſtio nella no primeiro, e no ultimo dia; neſte pegou em huma das varas do pallio com Suas Altezas, e com alguns Grandes da Corte. A Rainha noſſa Senhora aſſiſtio à meſma feſta no ſegundo dia.

Em 12. do corrente entrou neſte porto huma nao de guerra da Grãa Bretanha, chamada *Lime*, capitaneada por Mylord Vere; e no dia ſeguinte partio para o Eſtreito (donde eſta veyo) outra, que aqui ſe achava por nome *Dursley-Galley*, mandada pelo Capitaõ George Purvis.

O grande Peixe, que entrou neſte porto a ſemana paſſada, ſe naõ tem certo conhecimento da ſua eſpecie. Alguns entendem ſer huma Butalina, a que os Francezes daõ o nome de *Souffleur*, id eſt, Aſſoprador, outros que ſeja certa eſpecie de Balea, a que os Hollandezes chamaõ *Kapeku*; mas como a ſua figura he differente da Balea, e de qualquer outro peixe conhecido, ſe expoem aqui em eſtampa aos curioſos com as medidas de todos os ſeus membros,

FIGURE 13.7 Stranded whale in Mascarenhas José Freire de Monterroio (ed.), *Gazeta de Lisboa Occidental* (21 January 1723) 23–24. Open access in Hemeroteca Nacional de Lisboa (http://hemerotecadigital.cm-lisboa.pt)

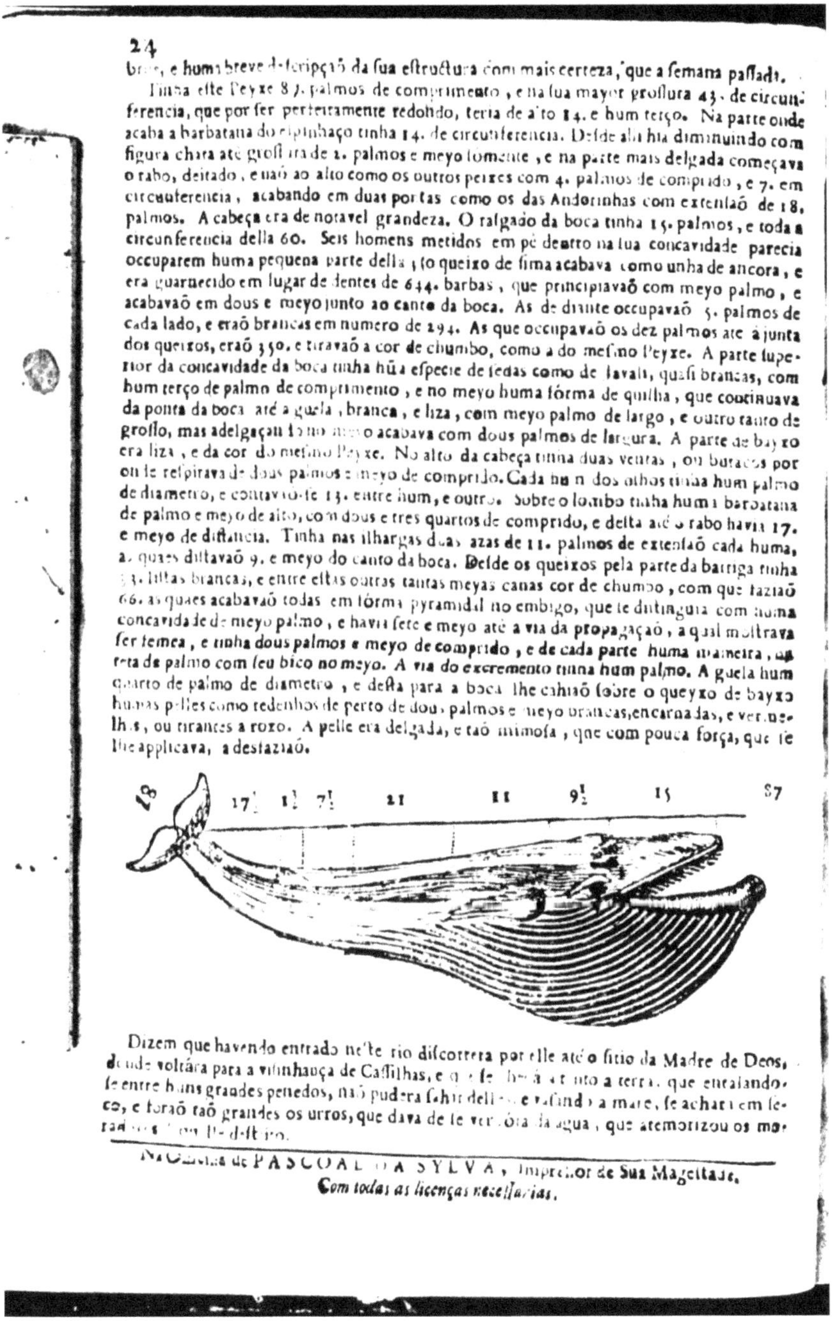

24

[illegible], e huma breve descripçaõ da sua estructura com mais certeza, que a semana passada.

Tinha este Peyxe 87. palmos de comprimento, e na sua mayor grossura 43. de circunferencia, que por ser perfeitamente redondo, teria de alto 14. e hum terço. Na parte onde acaba a barbatana do espinhaço tinha 14. de circunferencia. Desde alli hia diminuindo com figura chata até grossura de 2. palmos e meyo somente, e na parte mais delgada começava o rabo, deitado, e naõ ao alto como os outros peixes com 4. palmos de comprido, e 7. em circunferencia, acabando em duas pontas como os das Andorinhas com extensaõ de 18. palmos. A cabeça era de notavel grandeza. O rasgado da boca tinha 15. palmos, e toda a circunferencia della 60. Seis homens metidos em pé dentro na sua concavidade parecia occuparem huma pequena parte della; o queixo de sima acabava como unha de ancora, e era guarnecido em lugar de dentes de 644. barbas, que principiavaõ com meyo palmo, e acabavaõ em dous e meyo junto ao canto da boca. As de diante occupavaõ 5. palmos de cada lado, e eraõ brancas em numero de 294. As que occupavaõ os dez palmos ate á junta dos queixos, eraõ 350. e tiravaõ a cor de chumbo, como a do mesmo Peyxe. A parte superior da concavidade da boca tinha hũa especie de sedas como de javali, quasi brancas, com hum terço de palmo de comprimento, e no meyo huma fórma de quilha, que continuava da ponta da boca até a guela, branca, e liza, com meyo palmo de largo, e outro tanto de grosso, mas adelgaçando [illegible] acabava com dous palmos de largura. A parte de bayxo era liza, e da cor do mesmo Peyxe. No alto da cabeça tinha duas ventas, ou buracos por onde respirava de dous palmos e meyo de comprido. Cada hum dos olhos tinha hum palmo de diametro, e contavaõ-se 13. entre hum, e outro. Sobre o lombo tinha huma barbatana de palmo e meyo de alto, com dous e tres quartos de comprido, e desta até o rabo havia 17. e meyo de distancia. Tinha nas ilhargas duas azas de 11. palmos de extensaõ cada huma, as quaes distavaõ 9. e meyo do canto da boca. Desde os queixos pela parte da barriga tinha 33. listas brancas, e entre estas outras tantas meyas canas cor de chumbo, com que faziaõ 66. as quaes acabavaõ todas em fórma pyramidal no embigo, que se distinguia com huma concavidade de meyo palmo, e havia sete e meyo até a via da propagaçaõ, a qual mostrava ser femea, e tinha dous palmos e meyo de comprido, e de cada parte huma mameira, [illegible] de palmo com seu bico no meyo. A via do excremento tinha hum palmo. A guela hum quarto de palmo de diametro, e desta para a boca lhe cahiaõ sobre o queyxo de bayxo humas pelles como redenhos de perto de dous palmos e meyo brancas, encarnadas, e vermelhas, ou tirantes a roxo. A pelle era delgada, e taõ mimosa, que com pouca força, que se lhe applicava, a desfaziaõ.

Dizem que havendo entrado neste rio discorrera por elle até o sitio da Madre de Deos, donde voltára para a visinhança de Cassilhas, e [illegible] a terra, que entalando-se entre huns grandes penedos, naõ pudera sahir delle [illegible] e vasando a maré, se achara em seco, e foraõ taõ grandes os urros, que dava de se ver [illegible] da agua, que atemorizou os moradores [illegible].

N'Officina de PASCOAL DA SYLVA, Impressor de Sua Magestade. Com todas as licenças necessarias.

After the skin was cleaned of the fat, it was embalmed, and the skeleton was cleaned of all flesh. The animal was set up to be observed by D. José I (king of Portugal). The first skin could not be used, but the second was useful because it had a good hide. The animal's meat yielded four barrels of oil. The fish was 31 palms long and had 2 sets of teeth, the bottom ones smaller and the top ones larger, like those of horses. The animal was gray, with white spots, 'and had by its head two large wings the size of two cartwheels'.

King, nobles, scholars, the people – all wanted to see the whale. Definitively, Lisbon was amazed by the whale.

6 Building Knowledge about Whales in Early Modern Portugal

Across time and regions, whales have been hunted, used, venerated and worshipped, kept, hated, and remembered. Whales have been appropriated by societies and social groups in many ways, as they were food for the body but also food for the mind. Both in more recent science and in the realms of natural philosophy and secular natural history, these animals have been described as abundant, large, monstrous, valuable, impressive, unknown, amazing, beautiful. The presence of whales in a particular place or seascape – even if particularly muted in historical sources or in the archaeological and material records, as well as in the science of the last 200 years – can trace, shape, or alter human actions and choices.[48]

The whale – real or conceptualized – has continually been an element of human fascination: an animal that still attracts crowds of people when it strands itself on nearby shores or when it is spotted on the horizon. The whale allows for a close connection of people with the strange, enormous, ambivalent, still largely unknown, ocean.[49] Whales have a power of attraction over humans, and such a fascination is apprehended in scientific and cultural productions that encompass whales and their presence. Surprise and awe have been feelings commonly associated with a sighting of whales in the open ocean, a whaling event, or even strandings and whaling practices. In the words of Father Manoel de Santa Maria and Doctor Domingos Franco Quaresma, as well as many other authors, we can find clear signs of the importance of the two strandings to Portuguese society. The stories of both whales echoed in the

---

48 Colby J., "Change in Black and White"; Colby J., *Orca: How We Came to Know and Love the Ocean's Greatest Predator* (New York: 2018).

49 Brito C. – Vieira N. – Freitas J.G., "The Wonder Whale: A Commodity, a Monster, a Show and an Icon", *Anthropozoologica* 54.3 (2019) 13–27.

national scientific production as in many different cultural milieus, reaching different levels of Portuguese (and European) society. The whales were brought before the eyes of the Portuguese kings, humanists, and naturalists seeking information, as people were curious about them. Those big and strange "fish" moved individuals, ideas, and mental views about nature and animals.

Up to the 18th century, cetaceans were categorized as fishes, as every 'animal that is born and lives in the water, covered with skin, or scales, with gill, fins'.[50] In fact, it was only in the 10th edition of *Systema Naturae* of Carolus Linnaeus (1758–1759) that whales were taxonomically classified as mammals (Class Mammalia; Order Cetacea).[51] But, for instance, following the *Gazeta de Lisboa Occidental* news, the whale is characterized as a different fish, one that breathes air and shows detailed features that differentiate it from other marine animals. In *Piscilegio lusitano* cetaceans are present as elements of the Portuguese marine fauna and in the case of the whales they also seem to be treated in a singular way, with the only depiction of the work being precisely that of a whale, which the author describes as a monstrous fish. Of extraordinary greatness, giving birth to live young and breathing air, the whale was never truly just a fish, but frequently consecrated as the crowning creature of that group.[52] Whales, dolphins, and other marine mammals seem to be an important part of all the elements of natural history produced in Portugal.

But a consistent narrative about the Portuguese natural history that goes back to the beginning of the early modern age, and possibly medieval times, and that informs us about the production and dissemination of knowledge about natural life and biodiversity at that time, remains to be produced. Much of Portuguese science in the early modern period – mostly in the domains of nautical science and technology, medicine, and natural history – depended on the construction of the colonial empire[53] but is not limited to it. During the 15th and 16th centuries local practitioners might have been those more informed about coastal waters and the animals living in it, as were the mariners, explorers,

50 Bluteau Raphael, *Vocabulario portuguez e latino, aulico, anatomico, architectonico, bellico, botanico, brasilico, comico, critico, chimico, dogmatico, dialectico, dendrologico, ecclesiastico, etymologico, economico, florifero, forense, fructifero* [...] *autorizado com exemplos dos melhores escritores portugueses, e latinos* [...] (Coimbra, [no Collegio das Artes da Companhia de Jesu]: 1712–1728) 373.

51 Laist D.W., *North Atlantic Right Whales: from Hunted Leviathan to Conservation Icon* (Baltimore: 2017).

52 Burnett D.G., *Trying Leviathan: The Nineteenth-Century New York Court Case that Put the Whale on Trial and Challenged the Order of Nature* (Princeton – Oxford: 2007).

53 Sánchez A. – Costa, P.F. da – Leitão H., "Introdução ao Volume", in Sánchez A., Costa P.F. da – Leitão H, (eds.), *Ciência, tecnologia e Medicina na Construção de Portugal: Novos Horizontes, Sécs. XV–XVII* (Lisbon: 2021) 17–39.

missionaries, and settlers in the open ocean and the overseas. In fact, most Portuguese natural history is considered that produced in the wake of maritime voyages of expansion to the South Atlantic and Indian Ocean.[54] Renaissance humanists compiled observations about the natural world in order to explain it, but the ocean resisted the kind of systematic inquiry that had been applied to terrestrial environments, animals, and plants.[55] The Wet Globalization[56] that produced and circulated local, transoceanic, and imperial knowledge about the oceans and their animals was built on the backs of these nonhuman actors. The confrontation with unexpected winds, currents, and climates, with never-before-seen animals, plants, and humans, with the biogeography and eco-cultural diversity of the subtropical and tropical regions, demanded from Europeans attention to detail and an attempt at understanding. This exacerbated a systematic form of observation and compilation, categorization and naming, which also reflected what was being done on the Iberian Peninsula. And even if they are not as abundant as early modern records of strandings for the North Sea,[57] there are several records to attest to an effort to understand and document the natural history of whales in Portugal [Fig. 13.8].

Whales arouse curiosity, and if early modern scholars wanted to know what they were, they were equally interested in knowing all the other animals that inhabited the marine environment. If we look at this acquired and accumulated knowledge as significant, even if pre-disciplinary, it will be possible for us to establish a narrative about the value attributed to the animals, their habitats, add information about their uses and associated practices, and provide information about the relationship established between local (or Iberian) societies and the fauna that surrounded them. Here, I take a first step, trying to look at an older chronology of "scientific" events and productions, before the implementation, interpretation, or impact of Linnaeus' *Sistema naturae* in Portugal.[58] The works presented here represent an important corpus of information to

54 Leitão H. – Sánchez A., "Too Much to Tell: Narrative Styles of the First Descriptions of the Natural World of the Indies", *History of Science* 55.2 (2017) 167–186.

55 Pastore C.L., "Knowledges", in Cohen M. (ed.), *A Cultural History of the Sea*. Vol. 3. *A Cultural History of the Sea in the Early Modern Age* (London – New York: 2021) 25–51.

56 Mentz S., *Ocean* (Bloomsbury: 2019); Mentz S., "Introduction", in Cohen, *A Cultural History of the Sea in the Early Modern Age* 1–23.

57 Hoare P., *Albert and the Whale: Albrecht Dürer and How Art Imagines Our World* (New York: 2021).

58 The Linnaean programme for the classification of nature entered the Portuguese teaching system as a result of the reform of the University of Coimbra launched in 1772 by the marquis of Pombal (Costa P.F. da, "The Introduction of the Linnaean Classification of Nature in Portugal", in Gunnarsson B. (ed.), *Languages of Science in the Eighteenth Century* (Berlin: 2011) 227–244).

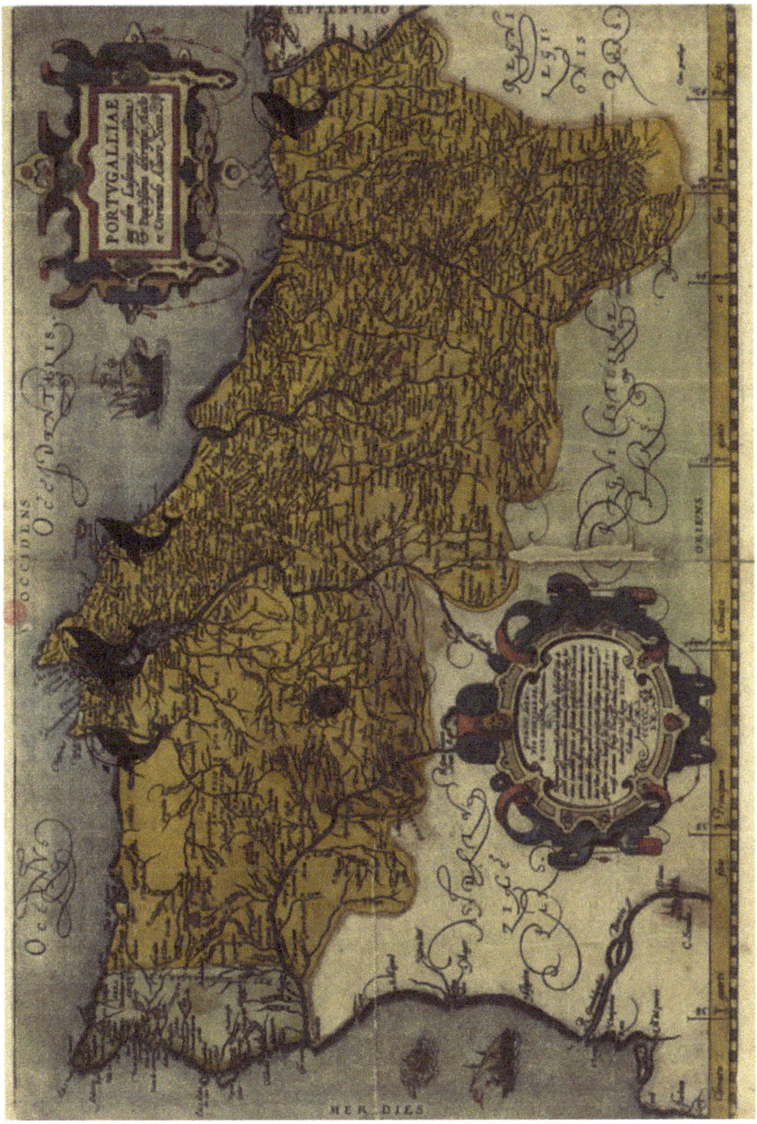

FIGURE 13.8 Whale strandings in Fernando Álvares Seco's map of Portugal (1561). Open access at the Digital National Library of Portugal (permalink: http://id.bnportugal.gov.pt/bib/mod/23506)

be analysed, and their authors should be included in the national panorama of oceanic natural history. The primary task of the natural historian (even if they are not aware that they are) is to name and produce order.[59] We find such concerns in the works of scholars who have produced an organized way of understanding the marine world, marine animals, and people's interactions with them. Whales, if not central, were of great relevance in their efforts. This review reflects exactly on the fact that there was an ongoing and perennial relationship of people with the whale-animal and the whale-symbol, and it was crystallized in multiple formats, including non-formal scientific ones, such as the poems or news reports of early modern Portugal.

Acknowledgements

The writing of this chapter greatly benefited from the multidisciplinary team working in Lisbon, at CHAM – Centre for the Humanities – NOVA FCSH (FCT Strategic project, UIDP/04666/2020), as members of the European research Council (ERC) Synergy Grant 4-OCEANS: Human History of Marine Life, under the European Union's Horizon 2020 research and innovation programme (grant agreement no. 951649).

I would like to acknowledge the incredibly hard work of Teresa Lacerda, who identified and transcribed several of the sources discussed here, while enthusiastically learning about the biology and environmental history of whales. I also sincerely thank Samuel Iglesias for allowing our team to access and work in *Piscilegio lusitano*, Pedro Pinto for his continuous source digging and sharing, and Nina Vieira, Catarina Garcia, and Ana Cristina Roque for their constant insights, comments, and support about the past of marine ecosystems and animals.

Bibliography

Amorim I., "A decadência das pescarias portuguesas e o constrangimento fiscal – entre a Ilustração e o Liberalismo", in Ribeiro da Silva F. – Cruz M.A. – Martins Ribeiro J. – Osswald H. (eds.), *Estudos em Homenagem a Luís António de Oliveira Ramos* (Porto: 2004) 153–164.

59 Knight D., *The Age of Science* (New York: 1986).

Amorim I. – Pinto B., "Portugal in the European Network of Marine Science Heritage and Outreach (19th–20th Centuries)", *Humanities* 8.1 (2019), https://doi.org/10.3390/h8010014.

Anonymous, *Catalogo dos preciosos manuscriptos da bibliotheca da casa dos marquezes de Castello Melhor. Documentos officiales, grande numero de autographos obras originaes e ineditas* (Lisbon: 1878).

Baldaque da Silva A.A, *Estado actual das pescas em Portugal comprehendendo a pesca maritima, fluvial e lacustre em toda o continente do reino, referido ao anno de 1886* (Lisbon: 1891).

Bluteau Raphael, *Vocabulario portuguez e latino, aulico, anatomico, architectonico, bellico, botanico, brasilico, comico, critico, chimico, dogmatico, dialectico, dendrologico, ecclesiastico, etymologico, economico, florifero, forense, fructifero* […] *autorizado com exemplos dos melhores escritores portugueses, e latinos* […] (Coimbra, [no Collegio das Artes da Companhia de Jesu]: 1712–1728).

Brigola J., *Domingos Vandelli e a circulação de conhecimentos na rede de naturalistas europeus* (Évora: 2016).

Brito C., "Beauties and Beasts: Whales in Portugal, from Early-Modern Monsters to Today's Flagship Species", *Arcadia* 21 (Autumn 2018) doi.org/10.5282/rcc/8449.

Brito C., "The Voice of Skogula in 'Beasts Royal' and a Story of the Tagus Estuary (Lisbon, Portugal) as Seen through a Whale's-Eye View", *Humanities* 8.1, 47 (2019) 1–16.

Brito C. – Vieira N., "Uma construção cultural de ser baleia: A história ambiental de dois arrojamentos na Lisboa ribeirinha e das pessoas que os observaram e descreveram". SCAENA – *Revista do Museu de Lisboa – Teatro Romano. O rio como horizonte: o outro palco do teatro romano*, no. 3. EGEAC, EM / Museu de Lisboa – Teatro Romano (Lisbon: 2022).

Brito C. – Vieira N. – Freitas J.G., "The Wonder Whale: A Commodity, a Monster, a Show and an Icon", *Anthropozoologica* 54.3 (2019) 13–27.

Burnett D.G., *Trying Leviathan: The Nineteenth-Century New York Court Case That Put the Whale on Trial and Challenged the Order of Nature* (Princeton – Oxford: 2007).

Ceríaco L., *O 'Arquivo Histórico Museu Bocage' e a história natural em Portugal. Professor Carlos Almaça (1934–2010) – Estado da Arte em Áreas Científicas do Seu Interesse* (Lisbon: 2014).

Colby J., "Change in Black and White: Killer Whale Bodies and the New Pacific Northwest", in Nance S. (ed.), *The Historical Animal* (New York: 2015) 19–37.

Colby J., *Orca: How We Came to Know and Love the Ocean's Greatest Predator* (New York: 2018).

Correia Gaspar, *Crónicas de D. Manuel e de D. João III* (Lisbon: 1992).

Costa P.F. da, "The Introduction of the Linnaean Classification of Nature in Portugal", in Gunnarsson B. (ed.), *Languages of Science in the Eighteenth Century* (Berlin: 2011) 227–244.

Cuvier Georges (ed.), *Historical Portrait of the Progress of Ichthyology, from Its Origins to Our Own Time*, ed. and transl. T.W. Pietsch (Baltimore: 1995).

Egmond F., *Eye for Detail: Images of Plants and Animals in Art and Science 1500–1630* (London: 2017).

Freitas J.G. – Brito C., "A Bixa Baleia. Ou a história de um manuscrito sobre o maravilhoso do mar", in Clamote Carreto C.F. – Moreira Sousa L.M. (eds.), *Imaginários do mar: uma antologia crítica* (Lisbon: 2021) 203–209.

Gamito-Marques D., "A Space of One's Own: Barbosa du Bocage, the Foundation of the National Museum of Lisbon, and the Construction of a Career in Zoology (1851–1907)", *Journal of the History of Biology* 5 (2017) 223–257.

García Gil J.J. – Molinero Hernando P. (eds.), *Bestiario de D. Juan de Austria. S. XVI. Estudios y transcripción de la edición facsimilar* (Burgos: 2000).

Giggs R., *Fathoms. The World inside the Whale* (New York: 2020).

Granada Luís de, *Obras de Fray Luis de Granada* (Madrid, La Publicidad, Imprenta de M. Rivadeneyra: 1848).

Gudger E., "The Five Great Naturalists of the Sixteenth Century: Belon, Rondelet, Salviani, Gesner and Aldrovandi: A Chapter in the History of Ichthyology", *Isis* 22.1 (1934) 21–40.

Herold B. – Horst T. – Leitão H., "A 'História Natural de Portugal' de Leonhard Thurneysser zum Thurn, ca. 1555–1556", *Ágora. Estudos Clássicos em Debate* 19 (2017) 305–334.

Herold B. – Horst T. – Leitão H. (eds.), *A* História Natural de Portugal *de Leonard Thurneysser zum Thurn, ca. 1555–1556* (Lisbon: Academia das Ciências de Lisboa, 2019).

Hoare P., *Albert and the Whale: Albrecht Dürer and How Art Imagines Our World* (New York: 2021).

Iglésias S.P. – Mollen F.H., "Cold Case: The Early Disappearance of the Bramble Shark (*Echinorhinus brucus*) in European and Adjacent Waters", *Oceans Past News* 10 (2018) 1–2.

Knight D., *The Age of Science* (New York: 1986).

Kraemer F. – Zedelmaier H., "Instruments of Invention in Renaissance Europe: The Cases of Conrad Gesner and Ulisse Aldrovandi", *Intellectual History Review* 24.3 (2014) 321–341.

Lacerda T. – Vieira N. – Brito C., *Fontes documentais para uma história natural das baleias em Portugal*. Zenodo. https://doi.org/10.5281/zenodo.6417799. Online document (2022).

Laist D.W., *North Atlantic Right Whales: From Hunted Leviathan to Conservation Icon* (Baltimore: 2017).

Leitão H. – Sánchez A., "Too Much to Tell: Narrative Styles of the First Descriptions of the Natural World of the Indies", *History of Science* 55.2 (2017) 167–186.

Machado Barbosa Diogo, *Bibliotheca Lusitana, historica, critica, e chronológica, na qual se comprehende a noticia dos authores portuguezes, e das obras, que compozeraó desde o tempo de promulgaçaó da ley da graça até o tempo presente.* Vol. 4 (Lisbon, [Lisboa Occidental: Na officina de Antonio Isidoro da Fonseca]: 1759).

Mentz S., "Introduction", in Cohen M. (ed.), *A Cultural History of the Sea.* Vol. 3. *A Cultural History of the Sea in the Early Modern Age* (London – New York: 2021) 1–23.

Mentz S., *Ocean* (Bloomsbury: 2019).

Montigore Antonino, *Della Sicilia ricercata nelle cose più memorabili.* 2 vol. (Palermo, Francesco Valenza: 1742–1743).

Pastore C.L., "Knowledges", in Cohen M. (ed.), *A Cultural History of the Sea.* Vol. 3. *A Cultural History of the Sea in the Early Modern Age* (London – New York: 2021) 25–51.

Pereira A., "The Opportunity of a Disaster: The Economic Impact of the 1755 Lisbon Earthquake", *The Journal of Economic History* 69.2 (2009) 466–499.

Pinto B. – Amorim I., "A Biodiversidade Marinha nos Museus de Portugal continental: Uma introdução", *Museologia & Interdisciplinaridade* 7.14 (2019): 107–127.

Rafinesque-Schmaltz C.S., *Précis des découvertes et travaux somiologiques de Mr C.S. Rafinesque-Schmaltz entre 1800 et 1814 ou Choix raisonné de ses principales découvertes en zoologie et en botanique* (Palermo: 1840).

Ritcher V., "'Where things meet in the world between sea and land': Human-Whale Encounters in Littoral Space", in Kluwick U. – Ritcher V. (eds.), *The Beach in Anglophone Literatures and Cultures: Reading the Littoral Space* (Farnham: 2015) 155–173.

Roque A.C., "Towards a Scientific Approach of Natures: Looking at the Southern Africa Biodiversity throughout the 16th-Century Portuguese Records on Marine Fauna", in Polónia A. – Bracht F. – Conceição G.C. – Palma M. (eds). *Cross-cultural Exchange and the Circulation of Knowledge in the First Global Age* (Porto: 2018) 75–100.

Sánchez A. – Costa P.F. da – Leitão H., "Introdução ao Volume", in Sánchez A. – Costa P.F. da – Leitão H. (eds.), *Ciência, tecnologia e Medicina na Construção de Portugal: Novos Horizontes, Sécs. XV–XVII* (Lisbon: 2021) 17–39.

Santa Maria Fr. Manoel, *Este Livro he uso do P.M. Fr. Manoel de Sta Maria Leytor de vespera* […], Cota Vermelho 569 [manuscrito]. Biblioteca da Academia das Ciências de Lisboa (1723).

Silveira L., *Manuscritos de filologia latina da Biblioteca Pública e Arquivo Distrital de Évora* (Évora: 1941).

Vieira C.C., "Observing the Skies of Lisbon. Isaac de Sequeira Samuda, an *estrangeirado* in the Royal Society", *Notes and Records of the Royal Society* 68.2 (2014) 135–149.

Vieira N., *A taxonomia da baleação portuguesa entre os séculos XV e XVIII: Uma história atlântica do mar, das baleias e das pessoas.* Faculdade de Ciências Sociais e Humanas da Universidade NOVA de Lisboa. Tese de Doutoramento (2020).

CHAPTER 14

'My Eyes Have Never Yet Beheld Him.' Demythologising Arctic Sea Monsters in the Poetry of the Norwegian Priest and Fish Merchant Petter Dass (1647–1707)

Ronny Spaans

'Mit navn er Petter Dass,/ som boer mod Verdens Ende' (My name is Petter Dass, who lives at the end of the world) – in this way the Norwegian poet and priest Petter Dass (1647–1707) presented himself in a poem letter to Dorothe Engelbretsdatter in 1680, a poet in Bergen.[1] Here Petter Dass combines an allusion to classical literature, Ovid's *Tristia, orbis ultimus*, with a reference to the remote place his residence, *Nordland* – "Northland" – in Arctic Norway. The quotation nicely illustrates the paradox inherent in the life and work of the subject of this article, namely his ambition to be perceived as a learned and civilised individual and the dangers and perils that early modern Europe associated with the Northern Waters, such as sea monsters and maelstroms.[2]

Petter Dass is the first modern literary voice of Arctic Norway. He was priest of the parish Alstahaug in the bailiwick Helgeland, county Nordland – this parish was one of the richest and most populous parishes in Norway at that time.[3] Dass was a so-called *embetskjøpmann*: next to his work as clergyman, he was also a prosperous trader selling stockfish to the Hanseatic merchants in Bergen.[4] Dass lived during the Dano-Norwegian Realm, with the Danish king as an absolute ruler of the two countries. Denmark-Norway was one of the

1 Dass Petter, *Viser og rim*, ed. D.A. Seip (Oslo: 1980) 72.

2 Many thanks to Paul Smith, Florike Egmond, the editors of this book, Ivar Roger Hansen, Benedicte Briså, Edda Frankot, Øystein Rian and Rune Spaans for tips and help while writing this article.

3 Although Petter Dass lived and worked in Alstahaug (65° North) in the most southern part of the county (*amt*) "Nordland" of 17th-century Northern Norway, Alstahaug is located just beneath the south of the Arctic Circle. The other geographic regions that Dass describes in *Nordlands Trompet*, include however for the most part territories to the North of the Circle (the northernmost area depicted is Troms bailiwick (*fogderi*) at 69° North). I therefore mean to have reason to call Dass a writer from an Arctic region.

4 Rian Ø., 'Det store samrøret – Embetskjøpmennene på 1600-tallet mellom fyrstestat og undersåtter', *Heimen* 44 (Oslo: 2007) 293–310.

DOI:10.1163/9789004681187_015

most stringent absolute monarchies in Europe. This is visible in the economic and political governance of the Twin Realms. A large number of the civil servants and merchants in Norway, the inferior part of the union, consisted of Danes, Germans and other foreigners. Bergen, Norway's largest city at that time, dominated the export of dried cod from northern Norway. In this trade, the Hanseatic League had still a leading role, together with merchants of other origins who had taken citizenship in Bergen. This was the case with the father of Petter Dass, who was a Scotsman and had come to Bergen in order to trade in fish from Nordland. The maternal side of the family, the Falch family, had Dutch origins in a certain Adrian Rockertsen Falkener, who had emigrated to Norway in the early 16th century.[5]

However, in Dass's poems there are no references to Scotland or the Netherlands. Although there are passages that bear witness to a friction between the Copenhagen-educated priest and the local population, it is evident from his poems that he identifies himself with the local ruling class, which consisted in particular of the Falch family, who had over the years held the most important public offices in the region and taken on a local identity. Dass was thus a representative of a self-conscious merchant and civil servant class in northern Norway. Although he wrote in Danish, he called himself a Norwegian poet and included a number of Norwegian words in his poems.[6] The same goes for the information about his Nordland, which he gathers in his poems, especially about fishing, aquatic fauna and nature phenomena. As I will show, Dass mainly bases his writings on first-hand observations of the nature in Nordland.

In this article, I study the natural knowledge that he presents in his book, *Nordlands Trompet*. I argue that this knowledge is guided by a rational method in which he presents both himself and the fishermen with whom he has contact as reliable observers of the nature of the region. This method is presented in the preface of *Nordlands Trompet*, where Dass expresses the purpose of his book. I will then show how the vernacular knowledge of *Nordlands Trompet* collides with the classic concepts and ideas of the humanist culture of the period and how this contributes to the invention of a modern and rational northern Norway. This I will do by first examining Dass's presentation of

5 Evju H., *Ancient Constitutions and Modern Monarchy: Historical Writing and Enlightened Reform in Denmark-Norway 1730–1814* (Leiden – Boston: 2019). For the Dutch origin of the Falch-family, see Volqvartz Marcus C., *Aandelige Sørge-Sange / De Afdøde til Ære* (Copenhagen, Hartvig Godiche: 1750) 66–67.

6 Midbøe H., *Petter Dass* (Oslo: 1947). This does not, of course, mean that Dass did not have an international orientation in his writings. We find quotations in both German and Dutch in the poems – two languages in which he could probably make himself understood.

northern Atlantic fish species and then his view of more fabulous phenomena attributed to the coast of Norway in the early modern period, the sea serpent and the Moskstraumen, the famous whirlpool of Lofoten. Dass, in a way, demythologises his landscape of trolls and monsters. As we will see, his self-presentation as a learned, rational clerk in the *Ultima Thule*, and his presentation of his homeland as a magic-free, welcoming and nutritious landscape, is relevant both for his fish trade and his career as an early modern poet and gentleman. At the end of this chapter, I will show how this regional self-image involved a critique of Bergen merchants and their dominance of the export of Norwegian fish. In the course of the article, I will also refer to other early modern topographers and naturalists, such as the famous Olaus Magnus and Erik Pontoppidan, both Scandinavian bishops whose books were translated into other languages. However, I will mention lesser-known naturalists as well, such as the Dutch writer and fishmonger Adriaen Coenen and the Dane Diderik Brinch. Their observations and reflections on natural history give us an international and historical perspective of *Nordlands Trompet*.

1 Introducing a Scientific Method

Before we study Dass's project of demythologisation, we need to take look at how Petter Dass has been received in Norwegian literary criticism. Throughout the centuries, Dass has been a popular author, thanks especially to his religiously didactic works. He was also regularly mentioned by critics, but then only as a "minor poet" in Danish-Norwegian literature. It was first in 1854, when the poet and critic Johan Sebastian Welhaven wrote the article "Digteren af Alstahoug" (The Poet of Alstahaug), that Dass's status changed. At that time, Norway was in the process of breaking away from the union with another Nordic country, Sweden, to become an independent nation. Dass was assigned a position as a "father" of the new Norwegian literature and also a regional symbol of northern Norway. Welhaven's "rediscovery" concerned Dass as a writer of fiction and *Nordlands Trompet* as a poetic narrative of rural folklore. The following literary critics described Petter Dass as a harmonious clergyman-poet, fatherly to his parishioners, confident with his Nordic surroundings and critical of the baroque literature on the European continent.[7]

7 For Welhaven's article and other research on Dass, see Hansen I.R., *Petter Dass: Bibliografi og resepsjonshistorie 1678–2022* (Trondheim: 2023). For an English presentation of Dass, see Stokker K., "Oral Tradition, Humanism and the Baroque", in Naes H.S. (ed.), *A History of Norwegian Literature* (Lincoln – London: 1993) 39–52.

Only in recent decades has this view begun to be criticised. On the one hand, Dass has been studied in the light of the power relations of the time; on the other, his poetry has been examined with respect to early modern genre theory and rhetoric.[8] There have also been studies of Dass carried out in terms of natural history, but no detailed systematic study has been made in this direction.[9] My chapter is therefore a contribution to a new understanding of Petter Dass, as it involves a comparative, interdisciplinary approach, where I relate *Nordlands Trompet* to other fields of research, such as history of knowledge and science. By history of science, I mean not only the history of ideas, but also the knowledge associated with crafts and lower professions. I am particularly interested in the experiences that Dass relates to his profession as a fish merchant.[10] Dass did not run a fish shop himself, but his priest's salary consisted of tithes and land tax, which he mainly received in the form of fish, especially stockfish. He transported this fish to Bergen where it was sold to Hanseatic merchants. As a fish trader, he had therefore training in identifying, selecting and sorting fish and fish products.[11]

Dass was not a scientist in the strict sense of the word. When Dass in *Nordlands Trompet* ponders over the phenomenon of glaciers, he writes about it as something for 'physici' (physicists) to figure out; nevertheless, Dass launches a new explanation of the phenomenon. Here he follows a long tradition among early modern Danish-Norwegian priests to act as *dandemænd*,

8 Bruland S.H., "Petter Dass og den klassiske litteraturen" in Alenius M. – Bergh B. – Boserup I. – Friis-Jensen K. – Skafte Jensen M. (eds.), *Latin og nationalsprog i Norden 1500–1800* (Copenhagen: 1991) 195–204; Hansen K., *Petter Dass: Guds øyesteen* 2 vols. (Sandnessjøen: 2018); Spaans R., "Hekser, Kopernikus og nordnorske fiskeslag. Eit kritisk blikk på den seinaste forskinga på Petter Dass", *Edda* 108.2 (2021) 84–97; Lauvstad H., *Helicons Bierge og Helgelands Schiær.* Nordlands Trompets *tekst, repertoar og retorikk* (Oslo: 2006); Dass Petter, *Katekismesanger*, ed. J. Haarberg (1715/2012): https://www.bokselskap.no/boker/katekismesanger/tittelside.

9 Sandnes J., "Lokalhistorisk litteratur til omkring 1900", Bjørkvik H. – Fladby R. – Reinton L. – Sandnes J. (eds.), *Lokal historie i forskning og kulturarbeid gjennom 200 år* (Oslo: 1970) 13–32; Foss G., "Skou-Essen. Om skrift og mat i *Nordlands Trompet*", in Andersen B. – Elisassen K.O. (eds.), *Maskepi og maskerade* (Trondheim: 2005) 63–78.

10 My approach to Dass relies on a shift in the historiography of science that has become established in recent decades. Today the history of science is not simply a history of ideas; science is also studied as the sum total of specific practices out of which science emerges. As a result, the assumption that astronomy, mechanics and optics constituted the core of the Scientific Revolution in the seventeenth century has been replaced by a more ecumenical approach that accords at least equal status to natural history, geography, medicine, etc. See, for example, Cook H., *Matters of Exchange: Commerce, Medicine, and Science in the Dutch Golden Age* (New Haven – London: 2007).

11 He also sold fish products that were not his own, such as the income from the churches he leased, see Midbøe, *Petter Dass* 149; Hansen, *Petter Dass: Guds øyesteen*, vol. 1, 121–124.

"gentlemen", with universal knowledge.[12] The view of Petter Dass as a parson-naturalist, however, depends on how we define *Nordlands Trompet*. The book itself, which was probably written between the 1670s and 1690s and first published posthumously in 1739, is called both a praise poem and a versified topographic description of northern Norway. The book consists of fourteen chapters with different topics. The poem first gives a general description of Nordland, including the weather conditions, birdlife, fisheries, population of the region, trade, and Sámi people. This is followed by descriptions of the five bailiwicks in Nordland, such as Helgeland and Lofoten, where the poet uses a spacious writing style, allowing himself anecdotes and digressions. All in all, *Nordlands Trompet* consists of some three thousand verse lines. Even though it is not a natural history work in the strict sense of the word, the oldest reception of *Nordlands Trompet*, is, in fact, in a scientific context. Some of the oldest extant copies of *Nordlands Trompet* are bound together with works of natural history, such as a description of the sea monster "Kraken"; the title itself, *Nordlands Trompet* (*The Trumpet of Nordland*) was not his own but probably a "sale trick" by the publisher. Dass presumably called the book "Description of Nordlands Amt [County]", a title associated more with natural history than praise poetry.[13]

It is also worth noting that the earliest mentions of Petter Dass were in scientific-topographic works of the 18th century. A telling example is *Det Kongerige Norge, fremstillet efter dets naturlige og borgerlige Tilstand* (*The Kingdom of Norway, Presented According to its Natural and Civil Status*) (1763) published under the name of Erik Johan Jessen-Schardebøll, although the book's real author is Hans Steenbuch. In the first section of his book, Steenbuch assesses the quality of previous studies of Norwegian nature. *Nordlands Trompet*, he said, contains 'Paalidelige Underretninger' (reliable information), especially about Helgeland, because there the author writes about things he himself has experienced. He continues by saying that even though, 'Versene og Stilen undertiden [ere] til Hinder i Materien; dog finder man ham ellers forstandig og erfaren i Landets Tilstand' (the verses and style are sometimes an obstacle to the material; yet one finds him [Dass] otherwise sensible and experienced

12 In the Protestant Scandinavian states, the state-church seems to have played a great role in the exercise of scientific work: 'Particularly in the case of Norway, which had few other institutions, it can be seen in this period as the single most important institution for promoting natural history', Brenna B., "Clergymen Abiding in the Fields: The Making of the Naturalist Observer in Eighteenth-Century Norwegian Natural History", *Science in Context* 2 (2011) 143–166, here 145.

13 Haarberg J., "Hvorfor trompet? – Om tittelen på Petter Dass' nordlandsbeskrivelse", *Edda* 99.1 (2012) 3–13. But it is also possible that the original title was *Buccina Polaris*.

in the status of the country).[14] As a contrast, Steenbuch mentions a work from the same period, written in prose by an author with the ambition of being perceived as a naturalist: *Prodromus e Norvegia, Sive Descriptio Loufoudiæ, Omnium Nordlandiæ Præfecturarum longe celeberrimæ, accuratissima* (*Forerunner from Norway, or Extremely Accurate Description of Lofoten, the Most Famous of All the Nordland's Bailiwicks*). Published in Amsterdam in 1676 and reprinted in 1683 by Diderik Brinch, who called himself a 'naturæ admiratore', it impressed readers with its many realistic explanations juxtaposed with stories about the island of 'Udrøest' (*Utrøst*), a 'huldeland' (secret land), accessible only at high tide, with flying dragons and dangerous whales.[15] It is obvious who his most important source of inspiration was, namely Olaus Magnus, the author of *Historia de gentibus septentrionalibus* (*A Description of the Northern Peoples*), which was printed in Rome in 1555 and based on his *Carta marina* (1539) – the first map of the Nordic countries to give details and place names. Brinch partly agrees with and partly opposes Olaus Magnus in his book, while Steenbuch more or less rejects the whole Brinch's book, on account of its many 'urigtige Ting' (wrong things).

In his lexicon article on Diderik Brinch, Jon Haarberg points to a difference between Brinch and Dass, namely that Brinch does not, unlike Dass, stage his own *persona* in his narrative.[16] Dass's self-fashioning as an inhabitant of the Arctic North forms a thread through *Nordlands Trompet*. A key to understanding the intention of this book is expressed in the preface, "Til Læseren" (To the reader). I quote below the beginning of Dass's preface, because here we also learn what method he used when it comes to obtaining information about his region:

> Mærk fromme Læser, giv det agt, at alt, hvis her er bleven,
> Om Nordland og den kolde Trakt i denne Bog indskreven,
> Det har jeg efter andres Mund i Pennen ladet føre,
> Om Avel, Eng, om Mark og Grund, og hvad der er at gjøre.
> Vel er jeg selv en Nordlands Mand, og Nordland har mig klædet,
> Men Norden for mit Helgeland min Fod har aldrig trædet.
> Thi kand jeg dette sige grandt hvad Helgeland tilhører,
> At det er hver en Tødel sandt, hvis Pennen derom rører.
> Men Norden for og lenger hen, jeg ikke veed saa nøye,

14 Jessen-Schardebøll Erik Johan [Steenbuch Hans], *Det Kongerige Norge* (Copenhagen, Gottmann Friedrich Kisel: 1763) 120–121.

15 Brinch Diderik, *Prodromus e Norvegia* (Amsterdam, Christophori Cunradi: 1676) 10–11, 17–19.

16 Haarberg J., "Diderik Brinch", *Norsk biografisk leksikon* (2009): https://nbl.snl.no/Diderik_Brinch.

Jeg har det kun af Hørselen, men ikke seet med Øye.
Hvad andre for mig haver sagt, og mundtlig mig opregnet,
Det har jeg paa Papiret lagt, og særligen optegnet.
[…][17]

My pious reader, man of worth, observe that all here written
Of Nordland and the frosty north, I have from others gotten.
I have it from the mouth of men, my pen has it recorded,
Concerning harvest, field and glen, as it to me reported.
True, I myself am Nordland's son, a Nordland's son of toil;
But farther north I've never gone than Helgeland's dear soil.
Thus I must tell, all, forsooth, what here I have related
Is every whit and part the truth if Helgeland is stated.
But farther north, in distant sphere, I have not ever been at;
So that I only know by ear – my eye has never seen it.
What others did to me relate, and orally made a word of,
That I have on the paper set and specially recorded.
[…][18]

Here Dass reveals a source criticism that is unique for his time. Either Dass bases what he writes on his own sensory experiences, on what he himself has *seen* in Helgeland, or he bases it on what he has *heard* from others – who also have it from sensory experiences. This is, as we saw earlier, what Steenbuch valued about Dass: his place of residence near the nature he describes makes him a reliable scientific source. Moreover, as we will soon see, the well-known parson-naturalist Erik Pontoppidan had the same vision on Dass. However, just as interesting as what Dass says his sources are, is what he indirectly states are *not* his sources: The knowledge of Arctic Norway of classical-humanist origin.

2 "Swimming Animals in the Northern Ocean"

The most interesting chapter in *Nordlands Trompet*, "Svemmende Dyr i det Nordlandske Hav" (Swimming animals in the Northern Ocean), is in this context also the chapter that entails many of the poet's most famous quotations,

---

17 I quote the edition of *Nordlands Trompet* published by Dass's grandson, Albert Christian Dass: *Beskrivelse over Nordlands Amt* (Copenhagen, Hartvig Godiche: 1763) 21.

18 Dass Petter, *The Trumpet of Nordland*, trans. by Theodore Jorgensen (Minnesota: 1954) 4. In this quote, I have taken the liberty to change the graphic layout in Jorgensen's translation in accordance with Albert Dass's edition.

according to Norwegian literary criticism. By using the term "swimming animals", Dass refers to the early modern sense of the term "fish" – as stated by Paul Smith in the introduction of this volume – as *aquatilia*: all aquatic animals. By fish from "det Nordlandske Hav", Dass probably means fish from the sea off the coast of Nordland.[19] Norwegian literary critics have pointed to the great poetic imagination Dass exercises in his descriptions of fish, hence this chapter is considered a highlight in early modern Scandinavian literature. I quote here his description of the halibut, the flatfish that is still today considered a delicacy:

Du smukkeste Qvæite, Du Dronning i Vand,
Hvor flad er din Boelig paa dybeste Sand,
  Hvorpaa du fremskrider saa sagte!

Du farer spagfærdig paa Grunden omkring,
Og hviler naar andre de kiøre i Ring,
  Det kand vore Fisker' vel agte.

Hvor findes din liige blant svæmmende Kræe?
Din Ryg er som Ravnen og Bugen som Snee,
  Ja viider' end Skiæle paa Sanden!

Og vilde man salte din eeniste Krop,
Du fylder et pakkede Tønde-Rum op,
  En Tønde! ja stundom halv-anden.[20]

O beautiful flounder! O queen of the sea!
On flat bottom sand you will swim true and free,
  With dignity and with composure.

You are in no hurry, you take your own time;
You know what is proper in northernmost clime;
  The fishermen honor you, Lady!

You have not an equal 'mong fishes I trow;
Your back is like raven, your breast is like snow;
  Much whiter than shells in the sunlight.

19 This sea is a part of the sea area that is both called the Norwegian Sea, the North Sea and Mare Septentrionale on early modern maps.

20 Dass, *Beskrivelse over Nordlands Amt* 70–71.

And if you should salt but one flounder to boot,
It filled a whole barrel, it packed it forsooth;
Sometimes also nearly another.[21]

We are impressed by the ease of narration and the striking metaphors. However, what is equally important is what Dass really conveys through these verses. First, he gives a description of the fish's behaviour in the water, e.g., that we are dealing with a deep-sea fish. Then he gives an accurate depiction of the fish's appearance, before instructing the reader on how much space the fish takes up in cargo when it is salted and ready to be shipped. Another recurring theme in his descriptions of fish species is the fishing method to be used. That is demonstrated in his portrait of the following fish, the Greenland shark – which today is still a difficult fish to catch [Fig. 14.1]. Here, too, Dass uses words in an imaginative but striking way. This fish 'render til Krogen, som Rytter til Storm' (runs to the bait like the knight to the storm):

Saa snart hun fornemmer, den Angel er fast,
Omtuller hun strax sig den sneedige Gast,
Derover optaves det Snøere:

Det skeer og med saadan en hastendes Iil
At Huuden paa Hende som raspende Fiil,
Skiær Snoeren før man det kand andse;

Men hvilken der tænker den Seyer at faae,
En Favn Jern-Lænker han lave sig maae,
Saa lær han den Kiærling at dandse.[22]

But when it discovers the bait has a line,
It churns and it twists as no word can define,
And thereby the line is in danger.

For sharkey will move with terrible bile,
With scales that can work like the keenest of files;
The line may be cut in a jiffy.

21 Dass, *The Trumpet of Nordland* 41–42.

22 Dass, *Beskrivelse over Nordlands Amt* 71.

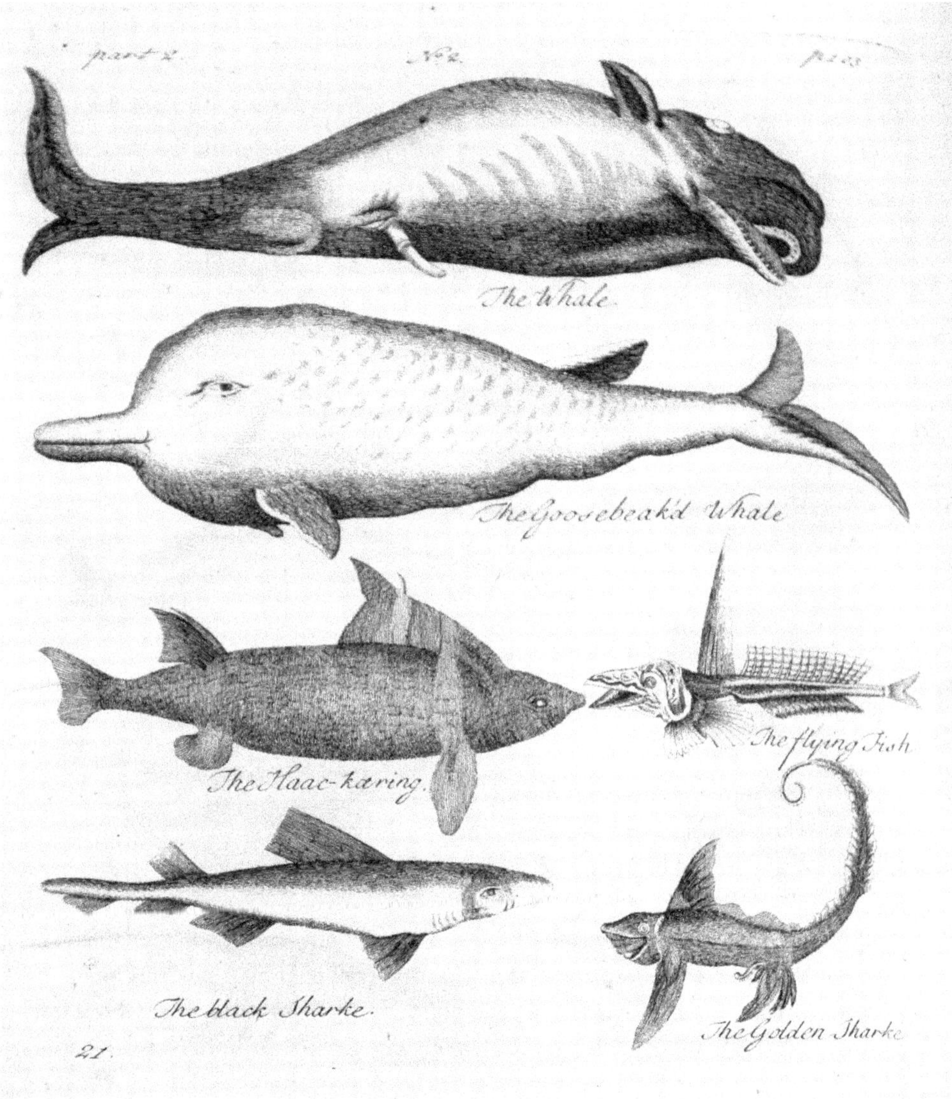

FIGURE 14.1 In the middle is the Greenland shark, here called 'the Haac-kæring'. Engraving, taken from Pontoppidan Erik, *The Natural History of Norway*, vol. 2 (London: A. Linde: 1755) 46

But if you are anxious to win this game,
You must have a line that you never need blame,
A fathom or two of iron.[23]

In *Nordlands Trompet*, other types of information related to fishing are also dealt with: Dass mentions good fishing grounds along the *Nordland* coast and mountains and landscapes to navigate by, such as the mountain range "The Seven Sisters", right by the parsonage that he lived in [Fig. 14.2]. As elsewhere in the poem, Dass writes with a great image-creating ability:

Ved Alsterhougs Præstegaard findes at see,
Syv Søstre, som fletter sin' Lokker i Snee
Syv Damer, og alle saa viide

Hvoraf i Land-Carterne meget gaaer Rye,
Hvis Toppe opstiger anseelig i Skye,
Og sees i Havet saa vide;[24]

By Alstahaug parsonage nature will show
The sisters whose braids are as white as the snow –
The seven who stand in the mountain.

By them also fame has been widely spread.
The maps show them raising their silverly heads;
To far out horizon one sees them.[25]

A closer look at the structure of the whole poem clarifies the inclusion of detailed aquatic information and instructions related to the catching and shipping of fish: It is as much the fish merchant Petter Dass as the poet who is talking. The poem opens with a greeting to all who live in northern Norway, people of low and high birth. This sequence has been interpreted as Dass's dedication of the poem to all the people of northern Norway, as a fatherly gesture to his parishioners. Nevertheless, this greeting is probably primarily rhetorical, as Skirne Helg Bruland has pointed out, for the poet speaks to different groups during the poem.[26] Dass most often addresses one person: a traveller to the

23 Dass, *The Trumpet of Nordland* 42.
24 Dass, *Beskrivelse over Nordlands Amt* 116.
25 Dass, *The Trumpet of Nordland* 79.
26 Bruland, "Petter Dass og den klassiske litteraturen" 203.

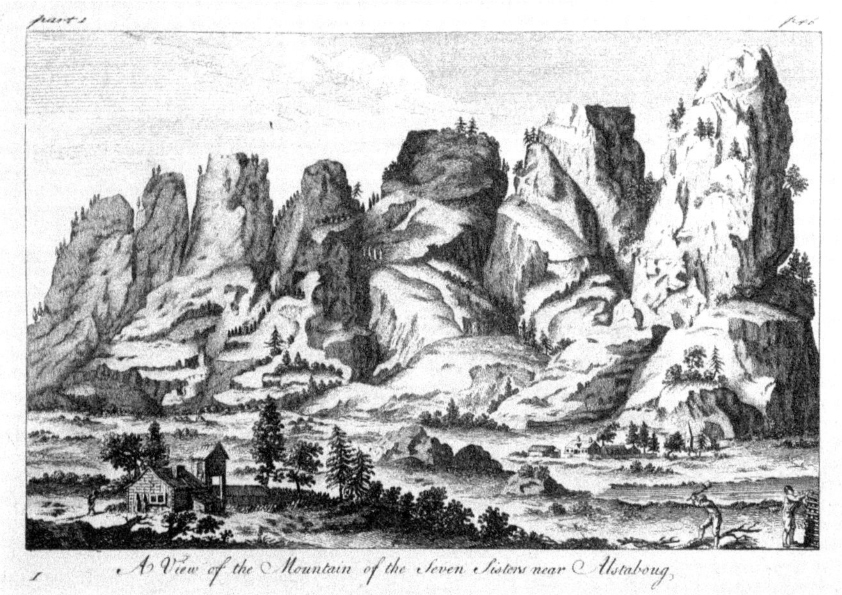

FIGURE 14.2 The mountain range "The Seven Sisters" with Alstahaug's parsonage. Engraving, taken from Pontoppidan Erik, *The Natural History of Norway*, vol. 1 (London, A. Linde: 1755) 102
IMAGE © UNIVERSITY LIBRARY OF OSLO

region, as demonstrated by such verses as 'Paa toe-rømmit Færing om Øerne roe, / Kun jeg og du eene selv anden' (By two-seaters rowboat to islands we row; We two by ourselves and no others.).[27] Dass does not specify whom he is inviting on this fictional boat journey, but he reveals that he is a travelling *dandemand*, "gentleman". A possible candidate is the so-called *uteliggerborger*: a trader in Nordland through citizenship of Bergen or Trondheim, just like the poet's Scottish father.

We already have a candidate for such a reader. In Dass's occasional poetry handed down to us there is an exchange of poem letters between Dass and the *uteliggerborger* Ole Nysted. In his letter to Dass, signed 1678, Nysted actually writes that he has part of *Nordlands Trompet* in his possession – which he received from the county prefect of Nordland.[28] Already in the 1670s there were thus unfinished versions of the poem in circulation – both among traders and higher officials. We can assume that Nysted used this copy actively in his merchant business. In his letter he wrote to Dass, he explained that he had

27 Dass, *Beskrivelse over Nordlands Amt* 49; Dass, *The Trumpet of Nordland* 26.
28 Dass, *Viser og rim* 280–281.

lived in northern Norway for many years but learned several things from reading the poem. However, Nysted's letter is primarily about something else: he is in financial difficulties and is willing to provide Dass with a report on the area where he is a merchant, Ofoten, to the east of Lofoten, in exchange for a loan. As we know, Dass was a member of the powerful Falch-dynasty and therefore possibly able to help Nysted. Moreover, we also know that Dass was also interested in information on the bailiwicks to the north of his Helgeland. Thus Nysted had undoubtedly chosen the right man for his request. Unfortunately, we do not know how this contact ended. In his reply, Dass expresses reservations about lending money. With or without help from Dass, Nysted died some years later and his estate was then insolvent.[29]

There is no doubt that *Nordlands Trompet* is an expression of Petter Dass's literary ambitions – that fact is also evident from the reflections on poetry in the correspondence between Dass and Nysted. However, as we have seen, Petter Dass's descriptions of fish are also closely related to fish trade. From that point of view, it is interesting to investigate whether there exist parallels to Petter Dass in early modern European literature. One name in particular stands out in this regard: the Dutch writer Adriaen Coenen, who lived in Scheveningen, a hundred years earlier, from 1514 to 1587. Like Dass he was trader of fish and a collector of all kinds of aquatic animals and knowledge, which he shared with visitors from outside Scheveningen, even the Prince of Orange. This information was gathered in two large manuscripts handed down to us: *Walvisboeck* (*Whale Book*) and *Visboeck* (*Fish Book*). The text in Coenen's books is mainly prose, but there is some verse in between. In their presentation on Coenen, Florike Egmond and Peter Mason state that these books are among 'the oldest manuscripts in the world to be entirely devoted to whales and other marine mammals, fish and other creatures that in lived in the waters of north-west Europe'.[30]

Although there has not been a thorough study of literary representation in early modern times regarding fish from the northern part of the Atlantic Ocean, I dare say that the Dass's fish descriptions deserve the same label.[31] Coenen and Dass describe many of the same fish species, but it comes as no surprise

29 Hansen, *Petter Dass: Guds øyesteen*, vol. 1 99.

30 Coenen Adriaen, *The Whale Book*, ed. F. Egmond – P. Mason (London: 2003) VIII.

31 A possible exception is Peder Claussøn Friis (1545–1614), from whom Dass obtained some information. He also wrote descriptions of fish and fishing, both in his manuscript *Om Diur, Fiske, Fugle oc Trær udi Norrig* (*About Animals, Fish, Birds and Trees in Norway*), and in *Norrigis Beskriffuelse* (*The Description of Norway*), published in 1632. But he himself was from southern Norway, and his descriptions of northern Norwegian fish is not as detailed as in Dass's book, and the chapter on fish is also filled with warnings about deadly whales:

that where Coenen emphasises herring fishing, for which Scheveningen was famous, cod gets the most attention from Dass. Coenen calls the herring 'den coninck boven alle vischen onsen groten gouden berch in Hollant doer Godts grasij almachtich oock Zeelant ende Vlaenderen' (the king above all fishes, our great golden mountain in Holland by God's grace almighty, also for Zeeland and Flanders).[32] Dass portrays the cod with the same royal metaphors:

Nu maa jeg mig snoe til den Nordlandske Torsk,
Som Fiskerne kalde monn' Skreien paa Norsk,
  Han nævnes maae Nordmandens Krone,

Han kroner vor Gielde han kroner vor Skiaae
O, Sæel est du Bonde, som Torsken kand faae,
  Han føeder baad' dig og din Kone.[33]

Well, now I must come to the king of the fish,
The aim and the object of northerners' wish,
  The cod, called the "skrei" in Norwegian.

It hangs on the rocks, and it fills all the stores;
Praise God that this fish comes each year to our shores;
  It feeds both the wives and the husbands.[34]

The close relationship these two *ichthyologists* had with the local population is visible in their texts. Firstly, their social commitment is expressed in their descriptions of poverty among fishers and the dangers of fishing at sea. Secondly, local fishermen served as their main source of information. Coenen was probably out on fishing boats in the North Sea at a young age, but most of the information about fish and fishing he acquired mainly on land, in conversation with fishermen, claim Florike Egmond and Peter Mason. The same goes for Petter Dass; as the leader of a large parish with many employees under him, he was hardly responsible for practical tasks such as fishing. His knowledge must have been largely based on the exchange of information with local farmers and fishermen. This access to first-hand experiences made both Dass and

---

Friis Peder Claussøn, *Samlede Skrifter*, ed. G. Storm (Kristiania: 1881) 60–120. I will come back to him soon.

32 https://www.kb.nl/visboek, fol. 26.

33 Dass, *Beskrivelse over Nordlands Amt* 57–58.

34 Dass, *The Trumpet of Nordland* 32–33.

Coenen critical sources for many of the wonderful tales of sea monsters that circulated during this period. Nevertheless, there are also differences between them. Unlike Dass, Coenen provided his books with many illustrations. The drawings are mostly realistic, but in a couple of places in his books we also encounter the Renaissance concept of attributing symbolic character to natural phenomena. Nature as a carrier of hidden meaning was in this period particularly expressed through the notion of comets as a supernatural warning of impending danger, but fish could also be equipped with warnings on their skin. Coenen mentions a so-called "Geuzenvis", a fish that foreshadows an incident in the conflict between the "Geuzen", the Dutch rebels in the Eighty Years' War, and the Spanish king. The sober Coenen does not fully accept the idea, but he does not reject it either. Coenen also sometimes follows the humanist practice of interpreting animals and natural phenomena as allegorical *exempla* – as moral narratives with a universal message. For example, he tells of how the hands of a fisherman were paralysed when he tried to catch a poisonous torpedo fish. Coenen interprets it as an image of God's providence. None of this is to be found in Dass's representation on 'swimming animals' in *Nordlands Trompet*.[35]

What is most striking about Coenen is his great interest in marine wonders, sea monsters and serpents, where references to Norway are frequent. Olaus Magnus has apparently made a great impression on the Dutch author. Although he expresses scepticism about some of Magnus's sea monsters – 'Olaus scryft ons zeer vonderlycken van deze vischen dat niet wel te loeven en is' (Olaus writes very extraordinary things about these fishes, but they are not credible) – there are plenty of illustrations and texts about them in his books.[36]

At first glance, marvels do not seem to be a subject that interested Dass, but on closer inspection it turns out that it actually forms a common thread throughout *Nordlands Trompet*. A reader would probably expect descriptions of churches in a book written by a parson-poet, but there is little of that. Nor does *Nordlands Trompet* contain any discussion of major theological issues – apart from a reflection of the ethical dilemma associated with burial at sea. Throughout the chapters we find interest in various marvels, ranging from Arctic wind conditions, menhir's from Old Norse times at Tjøtta, a ghost in Steigen, and various natural wonders such as Svartisen, two glaciers located

35 Coenen, *Het walvisboek* 172; Egmond F., *Visboek: de wereld volgens Adriaen Coenen (1514–1587)* (Zutphen: 2005) 121–123. In Peder Claussøn Friis's writings, we find mentions of fish with such semiotic patterns. We know, however, that Dass wrote a poem on a comet that appeared in 1680, but unfortunately, the text did not survive.

36 Coenen, *Het walvisboek* 86; Coenen, *The Whale Book* 86.

at Rana and Rødøy, and the two maelstroms in Nordland, the Saltstraumen at Bodø, and the Moskstraumen in Lofoten. The existence of sea serpents is also discussed. What characterises all these wonder phenomena is that Dass tries to explain them rationally.[37]

We could argue that this difference is due to the large time gap between these authors, but that is still not a good explanation. The 1676 description of northern Norway by the aforementioned Diderik Brinch is also full of sea monsters from Olaus Magnus's book; Brinck mentions monsters like the 'Tro[l]dqualle' (Troll whale) or 'Kaars-Trold' (Cross-Troll), which is so powerful that it can pull ships down into the sea. Brinch, like Coenen and Dass, was not an outsider. He worked as an inspector of the fish trade in northern Norway. Brinch also relied on testimonies from locals in his representation of magical phenomena, such as the secret *huldeland*. I think the explanation here lies in Petter Dass's programme for *Nordlands Trompet*, which he must have adhered to more strictly than Brinch. Dass's family, moreover, had long roots on Helgeland, unlike the Dane Brinch. This probably explains why he did not accept everything he heard from fishermen. He demanded high standards of perception and experience, and this is also reflected in Dass's description of Arctic wonder phenomena. I will below study two of them in more detail, namely the sea serpent and the Moskstraumen in Lofoten, as they illustrate Dass's demands for rationality in knowledge acquisition.

3 Oceanic Marvels

Olaus Magnus intentionally made his comprehensive map and detailed report to disseminate knowledge about the Nordic region in Europe. Ironically, his works led in part to the opposite, at least with respect to northern Norway. His book contributed like no other work to associating the Norwegian coast with monsters, as we saw in Coenen's books.[38] The previously mentioned Hans

37 Spaans, "Hekser, Kopernikus og nordnorske fiskeslag". In *Nordlands Trompet* there are also references elsewhere to the early modern interest in wonder and rarities. For early modern marvels, see Daston L. – Park K., *Wonders and the Order of Nature, 1150–1750* (New York: 1998).

38 The monsters of Olaus Magnus had different purposes. Some were there mainly for their symbolic significance, and they also had different origins, see Sandmo E, "Dwellers of the waves: Sea monsters, classical history, and religion in Olaus Magnus's Carta Marina", in *Norsk Geografisk Tidsskrift – Norwegian Journal of Geography*, 74.4 (2020) 237–249. Many were taken from classical works, but some actually had their origins in an older travel account of the most northern part of Norway, Finnmark, from 1520, written by Erik Walkendorf, archbishop of Nidaros (the old name for Trondheim). In Walkendorf's

Steenbuch writes that Magnus's map was in his time 'det beste og rigtigste' (the best and truest); however, the map is 'langt fra ikke uden Feil både i henseende til navn og situation' (far from without error both in terms of names and situations). He also states that, although Olaus Magnus filled his book with fables, there are 'en og anden ubekiendt Sandhed' (some unknown truths) in his works.[39] The line between fable and truth is blurred in this period. A scientific work from the 18th century may actually contain more fabulous material than a book from the 17th century. We must therefore go to the individual author in this period to discover what their actual view of sea monsters was. This brings us to Erik Pontoppidan, the bishop of Bergen and the most well-known parson-naturalist in Norwegian history, the author of *Forsøk til Norges naturlige historie* (1752 and 1753), translated into English with the title *The Natural History of Norway* in 1755.[40]

Pontoppidan refers to Dass in several places in his book, regarding him to be a reliable source. In his discussions on sea serpents, he quotes the whole passage, in which Dass discusses observations of the sea monster as a 'kind of testimony to the existence and properties of this extraordinary creature'. I render some lines here, in the English translation from 1755:

> The greate Sea snake's the subject of my verse;
> For tho' my eyes have never yet beheld him,
> Nor ever shall desire the hideous sight;
> Yet many accounts of men of truth unstain'd,
> Whose ev'ry word, I firmly do believe,
> Shew it to be a very frightful monster.
> [...]
> When on the sea he lies, streched at his length,
> He seems a hundred loads; so vast his bulk!
> [...]
> Methinks he seems another Behemoth,
> Or the Leviathan, who doth despise
> All arms, as swords, and guns, and glittering spears.[41]

travelogue we encounter for the first time the concept of the 'trolhwal' (troll whale), which we recognise from topographical works discussed in this article. Walkendorf writes that it is the name the inhabitants use for sea monsters. For the Latin original and a Norwegian translation of the text, see Walkendorf E, *Finmarkens Beskrivelse: brev til Pave Leo den 10de fra erkebiskop Erik Walkendorf*, ed. K.H. Karlsson – G. Storm (Kristiania: 1901), here 8.

39 Jessen-Schardebøll [Steenbuch], *Det Kongerige Norge* 19.

40 Pontoppidan Erik, *The Natural History of Norway*, 2 vols. (London, A. Linde: 1755). Pontoppidan's book was also translated into German and Dutch.

41 Pontoppidan, *The Natural History of Norway*, vol. 2, 204–206.

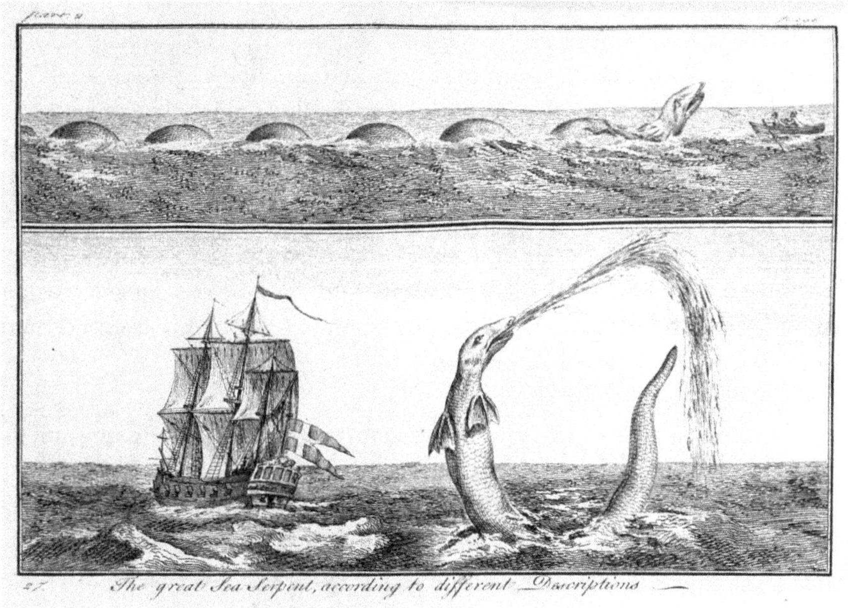

FIGURE 14.3 The sea serpent. Engraving, taken from Pontoppidan Erik, *The Natural History of Norway*, vol. 2 (London, A. Linde: 1755) 196
IMAGE © UNIVERSITY LIBRARY OF OSLO

The last lines are not included in Pontoppidan's book. I quote them here: 'Men hvorfore skal jeg bemøye mig saa, / At grunde de Ting, som jeg ey kand forstaae?' (But why should I put forth an effort to see / The things that forever a mystery must be?).[42] Pontoppidan uses Dass as a starting point for a long discussion on similarities between the whale and the Biblical Leviathan and Behemoth, what he calls the 'pole-serpent' and 'crooked serpent', before he finally comes to a conclusion concerning its existence [Fig. 14.3]. He ends up with the same assessment as his predecessor in *Nordland Trompet*: 'The reason of his [the serpent's] proceedings cannot; nor ought to be comprehended by us'.[43] In my judgment, the passage on the sea serpent forms an exception in *Nordlands Trompet*; this is the only place where Dass still lingers in the world of fables. I think it is possible to detect a greater reservation in Dass than in Pontoppidan on the prospect of the sea serpent, even though they both conclude that it is hubris to brood further on the monster's existence. Dass wrote in his preface that the most reliable information is related to what he himself has seen. As we saw in the quote, 'my eyes have never yet beheld him'.

42 Dass, *Beskrivelse over Nordlands Amt* 75; Dass, *The Trumpet of Nordland* 45.
43 Pontoppidan, *The Natural History of Norway*, vol. 2, 208.

Another aspect that illustrates the difference between Dass and Pontoppidan is the space they devote to this subject. Unlike Dass, Pontoppidan wrote a whole chapter on "Sea-monsters or strange and un-common Sea-animals". He lists several witnesses to the existence of marvellous sea creatures, both oral sources and written authorities, such as Olaus Magnus. After the "sea serpent" he discusses several testimonies of the "Kraken", the legendary monster of enormous size said to appear off the coasts of Norway. Pontoppidan identified the kraken both as a gigantic crab – based on the word's similarity with Norwegian "krabbe" – and a large octopus, and discusses its dangerous character; it has a reputation of pulling down ships. A testimony from northern Norway is interesting in this context:

> In the year 1680 a Krake (perhaps a young and careless one) came into the water that runs between the rocks and cliffs in the parish of Alstahoug. [...] It happened that is extended long arms, or antennæ, which this creature seems to use like the Snail, in turning about, caught hold of some trees standing near the water, which might easily have been torn up by the roots; but beside this, as it was found afterwards, he entangled himself in some openings or clefts in the rock, and therein stuck so fast, and hung so unfortunately, that he could not work himself out.[44]

There the kraken died: 'The carcase, which was a long while decaying, and filled great part of that narrow channel, made it almost impassable by its intolerable stench'. Alstahaug was, as mentioned, the parish of Petter Dass. We must assume that Dass knew of the story of this fantastic sea creature but chose not to include it in the *Nordlands Trompet*. It was contrary to the principle of his book: to rely on first-hand sources.[45] In that respect, Dass differs from both his predecessors and successors. Another parson-naturalist, Peder Claussøn Friis (1545–1614) – who was influenced by Olaus Magnus and whom Dass refers to in his book – wrote *Om Diur, Fiske, Fugle oc Trœr udi Norrig* (*About Animals, Fish, Birds and Trees in Norway*), wherein he states that northern Norway is famous for all 'Throld-huale' (the troll whales) that 'ere gierrig paa Mennischens Kiød at æde' (are eager to eat human flesh). They live in that part of the country,

44 Pontoppidan, *The Natural History of Norway*, vol. 2, 183–218, here 213.

45 Such finds were clearly not isolated cases in Alstahaug. Pontoppidan also refers to what one of his informants, Peter Angel, observed in his youth in the parish, in 1719: Angel 'along with several other inhabitants of Alstahoug in Nordland, saw what is called a Mer-man, lying dead on a point of land near the sea, which had been cast ashore by the waves, along with several Sea-calves, and other dead Fish' (Pontoppidan, *The Natural History of Norway*, vol. 2, 190).

because they are attracted to the large fish deposits there. The whales of southern Norway are, on the other hand, small and harmless.[46] Lucky for him, because Friis himself was from that part of the country!

One of the dangers assigned to the Kraken was the strong whirlpool it could create when submerging. This brings us to the second marine marvel I will discuss here. In the early modern period, the Moskstraumen was one of the most famous natural phenomena of the Nordic countries. The Moskstraumen is a maelstrom, a system of tidal eddies and whirlpools, and is one of the strongest maelstroms in the world. It is located in Lofoten, between the island of Moskenesøya and the island of Mosken. The maelstrom is clearly marked on early modern maps of northern Norway, and again, it is Olaus Magnus who is the originator of its fame. In his book, he attributes to it monstrous forces and compares it with the Greek sea monster Charybdis, known from Homer's *Odyssey*. The epic tells how Charybdis was turned into a maelstrom and tried to get rid of Odysseus by pushing him against the rock demon Scylla. The view of the Moskstraumen as a malicious maelstrom spread to other works of the time, as Magnus was published in new editions throughout the 17th century. Other early modern writers attributed the whirlpool to other fantastic properties. Athanasius Kircher based his version on the idea, which was widespread among the humanists of the Renaissance, that the earth had an underground network of water passages. He argued that the Moskstraumen was connected to a subterranean channel that led to the Gulf of Bothnia. Hence if you were sucked into the whirlpool, the wreckage of your ship could be found 700 kilometres away. Peder Claussøn Friis also included the Moskstraumen in his book on Norwegian natural history. Friis points to high cliffs and caves on the seabed as the source of the Moskstraumen. The water is drawn through these caves at certain times and then cascaded out of the earth through other holes at other times, explained Friis, who spiced his explanation with stories about whales that made the ground shake and tremble as they got caught in the current.

All these ideas about the Moskstraumen are discussed in an influential work from this period, *Dissertationes De admirandis mundi cataractis supra & subterraneis* (1678) (*Dissertations on the wonderful waterfalls of the world above and underground*) by the German Naturalist Johannes Herbinius, who among other places lived in Scandinavia.[47] In his book we find illustrations of the ideas of the stream in Northern Norway [Figs. 14.4 and 14.5]. The top image shows the waves that are created, which are dangerous for ships. The bottom

46 Friis, *Samlede skrifter* 67.

47 Herbinius Johannes, *Dissertationes De admirandis mundi cataractis supra et subterraneis* (Amsterdam, Johannes Janssonius van Waesberge: 1678) 126–135.

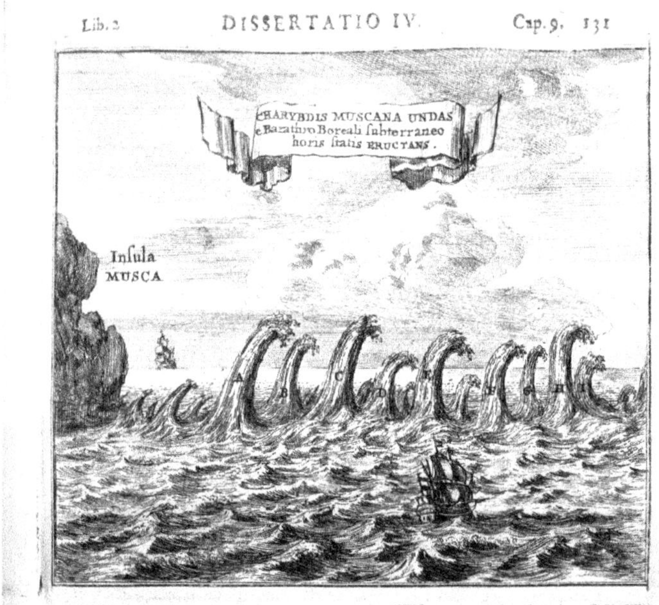

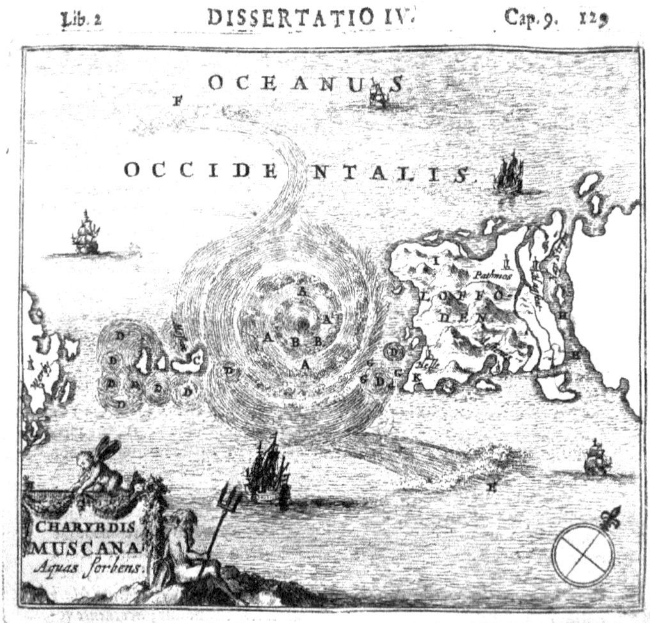

FIGURES 14.4 AND 14.5 Depictions of the Moskstraumen. Engraving, taken from Herbinius Johannes, *Dissertationes De admirandis mundi cataractis supra et subterraneis* (Amsterdam, Johannes Janssonius van Waesberge: 1678) 129 and 131

picture shows how the sea flows between the islands of Lofoten, forming a large maelstrom.

Before we look at *Nordlands Trompet,* I would also like discuss what the Danish poet Anders Arrebo and the aforementioned Diderik Brinch wrote about the Moskstraumen. Anders Arrebo, who in his young years was bishop of Trondheim, also included the maelstrom his *Hexaëmeron* (from 1631–1637, published posthumously in 1661). He also refers to cliffs and caves on the seabed, where the water swirls between the gates of the cliffs, before being sent up to the surface, where it creates waves so big that they obscure the view of the sun. Brinch is more rational in his description. He emphasises the importance of tides for the maelstrom, as others have done before him, but without a proper understanding of the effects of the tides, and also the depth of the sea. We can recognise, as we will see, these factors in *Nordlands Trompet,* but the destructive forces of the phenomenon are still at the centre of Brinch's book. He tells of boats that perish, even in calm weather, and of houses collapsing due to the maelstrom's bouldering.[48]

In his version, Petter Dass polemicises against Anders Arrebo. In fact, Moskstraumen is the natural phenomenon that occupies the largest place in *Nordlands Trompet,* four pages. There we find an explanation in accordance with our modern view of currents. Dass relates the whirlpool with tides by noting that it was the strongest at the full and new Moon and the weakest at half-Moon. The strong currents, Dass writes, must therefore be understood through the interplay between the tides and the topographical conditions of Lofoten. Instead of Arrebo's idea of underwater cliffs and caves, the whirlpools are created by narrow and shallow openings between the islands in Lofoten, through which the water of the Vestfjord must pass on its way out to the Atlantic:

> Er Bundene dybe, som inden for er,
> Og ude ved Gabet er Grundere Skiær,
>   Da vil jo Naturen det lære:
>
> Jo dybere Bunde, jo sagtere Fart,
> I grundeste Vande gaaer Strømmen jo hart,
>   Og veed sig ey selv at regiere.[49]

48 For the different views on the Moskstraumen through the centuries, see Lockert S.S., *Havsvelget i nord. Moskstraumen gjennom årtusener* (Stamsund: 2011); Brinch, *Prodromus e Norvegia* 7–9.

49 Dass, *Beskrivelse over Nordlands Amt* 150–151.

If inlets are deep that run into the land,
And if they are shallow at outermost strand,
Then nature will teach us as follows:

The deeper the bottom, the slower the speed,
On shallower stretches there always is need
To run with more violent motion.[50]

Sigri Skjegstad Lockert, who has studied the different views on the maelstrom through the centuries, calls *Nordlands Trompet* the most descriptive and explanatory of the works on the Moskstraum during this period.[51] Bjørn Gjevik, professor of hydrodynamics, is even more laudatory in his praise: 'Tatt i betraktning av at *Nordlands Trompet* ble skrevet omtrent samtidig med at Isaac Newton ga den første banebrytende forklaring av tidevannets årsak, et arbeid Petter Dass neppe kjente til, er hans beskrivelse fremtidsrettet' (Considering that *Nordlands Trompet* was written around the same time that Isaac Newton gave the first groundbreaking explanation of the cause of the tide, a work Petter Dass hardly knew, his description is forward-looking). Gjevik elaborates the parallels of Dass and Newton as follows: 'Han gav ingen dypere vitenskapelig forklaring av fenomenene som Newton gjorde. Men med sine detaljerte og nøkterne beskrivelser, i blomstrende poesi, tydeligvis basert på faktiske observasjoner, utmerket han seg likevel på en fordelaktig måte i samtiden' (He gave no deeper scientific explanation of the phenomena that Newton did. But with his detailed and sober descriptions, in florid poetry, clearly based on factual observations, he nevertheless distinguished himself in a beneficial way in his time).[52]

Of course, Dass was not a natural philosopher as was Newton, but he still had a scientific method that he followed. We become better acquainted with it in the passage where Dass mentions Arrebo:

Den Gisning jeg lade vil staae ved sit værd,
Den Salig Mands Skrifter heel kyndig og lærd
Som Gulvet jeg agter at giemme;

Men Mandens Beslutning om Moske-strøms Magt,
Hvis ikke Forfarenhed andet har sagt,
Da kunde jeg med den vel stemme.

50 Dass, *The Trumpet of Nordland* 109.
51 Lockert, *Havsvelget i nord* 87.
52 Gjevik B., *Flo og fjære langs kysten av Norge og Svalbard* (Jar: 2009) 12–13, 122.

Jeg vil det kun an med Eenfoldighed gaae,
Naturen os lærer det best at forstaae,
  Hvor Strømme sin Magt monne tage;[53]

May Arreboe's guess fairly stand at its worth!
The blessed man's writings are gold in the north,
  Forever I'll treasure and keep them.

If experience didn't otherways tell,
I might also say he had judged it all well
  Concerning the Moskenes current.

By means of simplicity I want to say
That nature can best give us hint of the way
  That currents gain power in ocean.[54]

Here we acquire direct evidence of how the programme Dass laid out in his preface guides his way of obtaining empirical information. Dass emphasises what he calls 'Forfarenhed' (experience), 'Eenfoldighed' (simplicity) and 'Naturen'. These concepts deserve greater clarification. As we remember from Steenbuch's discussion on the writing of Dass, he is reliable thanks to his residence close to the Artic nature he is presenting. Pontoppidan also discusses different explanations of the Moskstraumen. Here *Nordlands Trompet* is also mentioned, and here Dass's explanation comes out as the best one for exactly the same reason: 'Mr. Peder Dass, *who lives on the very spot* [...]'.[55] Dass may not have lived *on the very spot*, in Lofoten, but lived close to informants who could enlighten him about the maelstrom based on their direct sensory experiences – their 'Forfarenhed'.

By "simplicity", we also must understand a distancing from the authorities of renaissance humanism. Dass's informants have offered knowledge that is more "simple" and therefore more reasonable than those complicated explanations Dass found in the writings of contemporaries, such as Arrebo. Besides Arrebo and Friis, Dass does not actually refer to any authorities in *Nordlands Trompet*

53 Dass, *Beskrivelse over Nordlands Amt* 150.

54 Dass, *The Trumpet of Nordland* 108. Jorgensen has a good translation, which takes into account both the content and rhymes, but there are two lines in this quote, which are an inaccurate rendering of Dass's original. I have taken the liberty of carefully rewriting these lines. Here is Jorgensen's version of the lines: 'If I had no other experience to tell'; 'But I will with all my due humbleness say'.

55 Pontoppidan, *The Natural History of Norway*, vol. 2, 79. My italics.

but occasionally rejects explanations of which he is sceptical by calling them "lies". It is likely that he then alludes to authorities such as Olaus Magnus.

Dass was not unique for his time. His emphasis on rational method is part of a larger European discovery of the indigenous nature and topography.[56] We have already seen how Adriaen Coenen made demands for empirical observation in his description of different fish species. Although the belief in Olaus Magnus's sea monsters did not disappear in the 17th century, there is no doubt that the demand for rationality and source criticism became increasingly strong towards the 18th century. It is worth looking at two men who lived at the same time as Petter Dass: Francis Willughby (1635–1672) and John Ray (1627–1705), who are regarded as the two earliest English parson-naturalists. Both these men were Fellows of the Royal Society, and they are known for their work on *Historia piscium* (1686) (*Description on fishes*). Didi van Trijp has studied the method these researchers used to acquire knowledge. What she writes here reminds us of Petter Dass:

> Indications of direct observation are present in the fish book in various ways. Willughby and Ray, for example, added 'I have seen' (*vidi*) or 'we have seen' (*vidimus*) to certain species descriptions. In other cases, they punctuated statements with appeals to 'experience' (*experientia*), as in the case of the salmon.[57]

As we have noticed, Dass also points to first-hand, sensory observation when explaining the nature and fauna of his region. Next to his programme in his preface, we find sentences like 'Jeg selv det med største Forundring har seet' (I have myself seen, in amazement, this thing), 'Der siges, at Gaasen dra'r Væir som en hund, / Jeg haver den Sagn udaf Skytternes Mund' (For wild geese have noses as keen as a hound's; / I learned that from hunters who much are around) and 'Og mange med det mig kand sande' (Yes, many can prove what I'm saying).[58] The observations he describes apply thus to his own visual experiences, to those of others, or to a visual observation that many share. Didi van Trijp has also examined the kinds of informants the English naturalists have used. Although Ray and Willughby do not name their informants, it appears that they are mostly fishermen and fishmongers, having provided them with

56 *Cf.* Cooper A., *Inventing the Indigenous. Local Knowledge and Natural History in Early Modern Europe* (Cambridge: 2007).

57 Trijp D. van, "Fresh Fish: Observation Up Close in Late Seventeenth Century England", *Royal Society of London: Notes Rec.* 75 (2021) 311–332; here 316.

58 Dass, *Beskrivelse over Nordlands Amt* 127, 49, 137; Dass, *The Trumpet of Nordland* 87, 26, 97.

both information and specimen of fishes. As earlier discussed, the same was probably the case with Dass. In addition to personal experiences as fish merchant, he refers to others, as in the example above, where he relates the knowledge of a goose hunter. Moreover, we have also seen that Dass was in contact with travelling merchants, such as Ole Nysted, who offered him information in return for a money loan. It is possible Dass paid fishermen and farmers who offered him their knowledge. However, as we have also seen, he did not accept everything that he was presented; some information he perceived as fabulous and unreliable. Here he also uses the same method as Willugby and Ray, but Dass is in his turn not as strict as they are, in that *Nordlands Trompet* is ultimately not a scientific work in the strict meaning of the term. Thorough descriptions of fish and natural phenomena are accompanied by anecdotes, episodes, folk life from Nordland.

Another question is whether Dass's source criticism stemmed from a philosophical programme that Dass had adopted, for example, during his studies in Copenhagen. A scientific method based on experience (*experientia*) immediately makes us think of Francis Bacon, his principle of sober science and the Scientific Revolution. Van Trijp asks the same question in relation to the two English naturalists. Bacon was obviously important to the Royal Society, but as highlighted by Van Trijp and other scholars, we should not underestimate an older tradition of vernacular science that existed among artisans, farmers, fishers, etc. The ideals formulated by Bacon thus have methods of observation and knowledge acquisition in common with the fishing population with which Dass was in contact.[59] It is also possible that Dass was inspired to follow a method while collecting local information. Peder Hansen Resen, a legal scholar and professor at the University of Copenhagen, encouraged priests in Denmark-Norway to collect and submit information from their parishes (in 1666, 1682 and 1686). A closer investigation should determine whether Resen gave the priests concrete advice for the method of gathering information.[60]

4 Opposition to Merchants in Bergen

It is important to understand Dass's identity as a rational parson-naturalist in light of the political situation wherein Dass was living. As mentioned above, the merchants of Bergen had since the Middle Ages secured privileges

59 See note 10.

60 Hens H.A., "Peder Hansen Resen", *Dansk biografisk lexicon* (2011): https://biografisklek sikon.lex.dk/Peder_Hansen_Resen.

concerning the lucrative trade of stockfish. This trade had been mainly in the hands of Hanseatic merchants. However, since the 15th century, they had faced competition from not only the Norwegians but also the Dutch and later merchants from other countries who took citizenship in Bergen. Nevertheless, the Hanseatic League was still an important power in Bergen – after they had lost a foothold in other European towns – and it was to Hanseatic merchants that Petter Dass sold his fish. In the early modern period, there did not exist a market town in northern Norway. The northerners had two options: Sell their fish to the so-called *uteliggerborgere* or transport the fish themselves down to Bergen. This meant sailing across dangerous stretches of sea. More and more *uteliggerborgere* were from Trondheim, a city that had begun to compete with Bergen, but the economic centre for stockfish was still in Bergen. Dass himself owned ships, which transported stockfish to Bergen.[61] When one of his boats sank in 1692 at Stadhavet and a valuable cargo was lost, he felt he had to speak out. In the last chapter of *Nordlands Trompet*, on the bailiwick of Troms, Dass is far away from his Helgeland, and the lack of material about Troms gives him the opportunity to make some personal statements. He writes about his loses and expresses his wishes for the future:

> Men dersom Nord-Farernes Troe var saa stoer,
> De kunne faae Bergen henflyttet i Nor
> Omtrent hen ved Hundrede Miile,
>
> Hvor skulle den gandske Nordledingens Tract
> Af inderste Hierte sig fryde med Magt,
> Med lystige Ansigter smile!
>
> Men hvad vil jeg ynske! jeg faaer ikke ja,
> De Bergenske flytte dog aldrig derfra,
> Vi maae vore Dage saa slide,[62]
>
> But if we the northernes our faith could prove,
> The city of Bergen we surely would move
> Some six hundred miles farther northward.

61 Schreiner J., *Hanseatene og Norge i det 16. århundre* (Oslo: 1941); Hansen, *Petter Dass: Guds øyesteen*, vol. 1 163–165.

62 Dass, *Beskrivelse over Nordlands Amt* 182. Dass also wrote a poem on the incident: Dass, *Viser og rim* 119–123.

How would this entire great northernly land
At innermost heart be a jubilant band,
If such a thing happened among them!

But, much as I wish it, I know it is vain.
The merchants of Bergen will such talk disdain;
We have to accept our hard fortune.[63]

The last conciliatory statement we must understand as more than mere politeness, for Dass was not alone in these thoughts. A backdrop for the struggle was the memory of an economic centre in northern Norway in Old Norse time: Vågan in Lofoten. In addition, in that period, the situation was reversed: Then it was the merchants of Bergen themselves who had to travel up to northern Norway and buy the fish. In 1664, Tage Lobitz sent a note about the history of the old Vågan to the Danish Chancellery and asked for the establishment of a new northern Norwegian market town, so that one can put an end to 'all the danger and cost that the northerners have to endure'. The county prefect of Nordland, Preben von Ahnen, also suggested that a northern Norwegian market town should be founded. As an argument, the officials referred to the poverty of the fishing population. Many fishermen had debts to the merchants of Bergen, which were passed down from generation to generation. However, we should not underestimate the fact that the officials also had a self-interest. They wanted to build their own economic and political elite in northern Norway.

This was not about plans of secession from Denmark – there are no such thoughts in Dass's poetry – rather it is about a desire to keep the economic benefits of the fish trade for themselves. During the 16th and 17th centuries, a number of merchants from Denmark and the Dutch Republic were given a so-called "pass", or license, to trade in northern Norway, much to the frustration of the Bergen merchants. The king also granted selected individuals royal privileges to trade in certain goods. In 1661, for example, the famous Dutch admiral Michiel de Ruyter was granted a monopoly on the trade in northern Norwegian cod roe. In addition, both foreign and domestic merchants traded illegally in Nordland. Local officials like Petter Dass benefited economically from all these breaches of the Bergen dominance. Nevertheless, the Bergen merchants were always on their guard and usually prevailed when they complained to the governmental authorities about violations of their privileges.[64]

63 Dass, *The Trumpet of Nordland* 135–136.

64 Kiil A., *Nordlandshandelen i det 17. århundre* (Svorkmo: 1940) 120–131; Ytreberg N.A., *Malangen bygdebok* (Trondheim: 1943) 8–10. This does not mean, however, that Dass's

Petter Dass not only expresses frustration about the ship loss in the last part of *Nordlands Trompet*; in the same chapter he mentions an illegal trade that the Dutch had run in Malangen in Troms, in positive terms. From other sources we also know that Dass had a conflict with the Hanseatic merchant in Bergen with whom he traded, and that the Falch-dynasty, Dass included, collaborated with the *uteliggerborger* Bendix Nebel, who was known for his smuggling activities in northern Norway. Bendix Nebel, like the above-mentioned Ole Nysted, was a person who would probably have been just as anxious to get hold of a copy of the *Nordlands Trompet*.[65]

5 Conclusion

It is in this context that we must read *Nordlands Trompet* and Dass's self-fashioning as a sober and reliable priest and fish merchant. When he presents himself as 'en Nordlands Mand' (a Nordland's son of toil), and 'en Klerk / Iført en Nordlands Messe-Serk, / Og boer blant kolde Tuer' (a clerk / dressed in a Nordland's surplice / and who lives in [a landscape of] cold tussocks), it is not only in accordance with the modesty topos in early modern poetry.[66] By presenting himself as an integral part of Nordland, he builds up a credibility – both as a disseminator of reliable knowledge and as a potential partner for fish trade. Readers of *Nordlands Trompet* would understand that northern Norway is not an area of dangerous monsters and maelstroms, but a welcoming and nutritious landscape. As I have shown, the merchant Ole Nysted had a part of the manuscript in his possession, and was able to acquaint himself better with northern Norway through it. It is likely that other traders have used it as a "guidebook" while traveling to Nordland during Dass's lifetime.

Considering the diverse and rich content of *Nordlands Trompet*, it would, however, be reductive to understand the book from only an economic perspective. Dass as an ambitious "gentleman" and poet has had a further aim with the poem. Since *Nordlands Trompet* was written in the vernacular, it has not had a major impact on the European mapping of the Northern Atlantic, as today's readers may have wished. Moreover, we do not know to what extent contemporary poets and scholars have consulted with the work. However, as I have

relationship with Bergen was simply one of opposition. Dass went to the Latin school in the city, and he wrote poems about Bergen after the city fire of 1702, in which he wished for a new economic and cultural revival for the city.

65 Kiil, *Nordlandshandelen i det 17. århundre* 126; Hansen, *Petter Dass: Guds øyesteen*, vol. 1 258–273; Spaans, "Hekser, Kopernikus og nordnorske fiskeslag" 92–93.

66 Dass, *Beskrivelse over Nordlands Amt* 16.

shown, Dass was read and appreciated by scientists in the 18th century – he was in some respects even more accurate than Erik Pontoppidan. Moreover, as I also have demonstrated, his knowledge of Lofoten was more correct than his contemporary Diderik Brinch's, who actually lived in Lofoten for a period and who even called himself a *naturæ admiratore*, a naturalist. Unfortunately, Brinch's book likely had greater impact than *Nordlands Trompet*. The book was written in Latin, the international language of scholarship, and was published in two editions in Amsterdam, right at the centre of the knowledge production in contemporary Europe.

According to the historian Rasmus Brandt, the Baltic and North Seas had been *terra incognita* for the enlightened European, due to the Hanseatic trade privileges. In maps from the 16th century, the new world's coastlines were therefore far better mapped than the Nordic countries. It was undoubtedly better off at the end of the 17th century, but it is still noticeable that the coast of Nordland in many maps – especially Dutch maps, because it was the Dutch who first drew the coast of Norway – were still filled with inaccuracies. Helgeland was marked as an island off the Norwegian coast well into the 17th century.[67] An example is Claes Jansz. Vooght's map of the Norwegian Sea from 1692, where the Moskstraumen is also marked as a large spiral in Lofoten [Fig. 14.6]. We must assume that it is not only due to the Hanseatic merchants, but also old fabulous narratives that were passed on through writings such as Brinch's book.

According to another researcher on this period, Jørn Sandnes, we do not find a similar source critique in the writings of many of those who come after Petter Dass and who call themselves historians, such as Jonas Ramus. This is evident when we compare these authors with scholars from the latter half of the 18th century, as here we again find articulation of a strict scientific method similar to that of Dass.[68]

Now at the end it is worth dwelling a little on the mentioned author, namely the priest and historian Jonas Ramus (1649–1718). It was his and not Dass's version of the Moskstraumen that gained fame in hindsight. Five years before Petter Dass died, in 1702, Ramus published *Ulysses et Otinus Unus et idem* (*Odysseus and Odin are one and the same*). Here he presents the idea that Odysseus and the Norse god Odin were the same person, and that many ancient stories should actually be set in Norway. The idea of Moskstraumen as

67 Brandt R., "Raritas et curiositas. Amtmann Hans Hansson Lilienskiold, en 1600-talls europeer", in Johannesson H.E. (ed.), *Mimesis förvandlingar: tradition och förnyelse i renässansens och barockens litteratur* (Stockholm: 2002) 359–373, here 370.

68 Sandnes, "Lokalhistorisk litteratur til omkring 1900" 17.

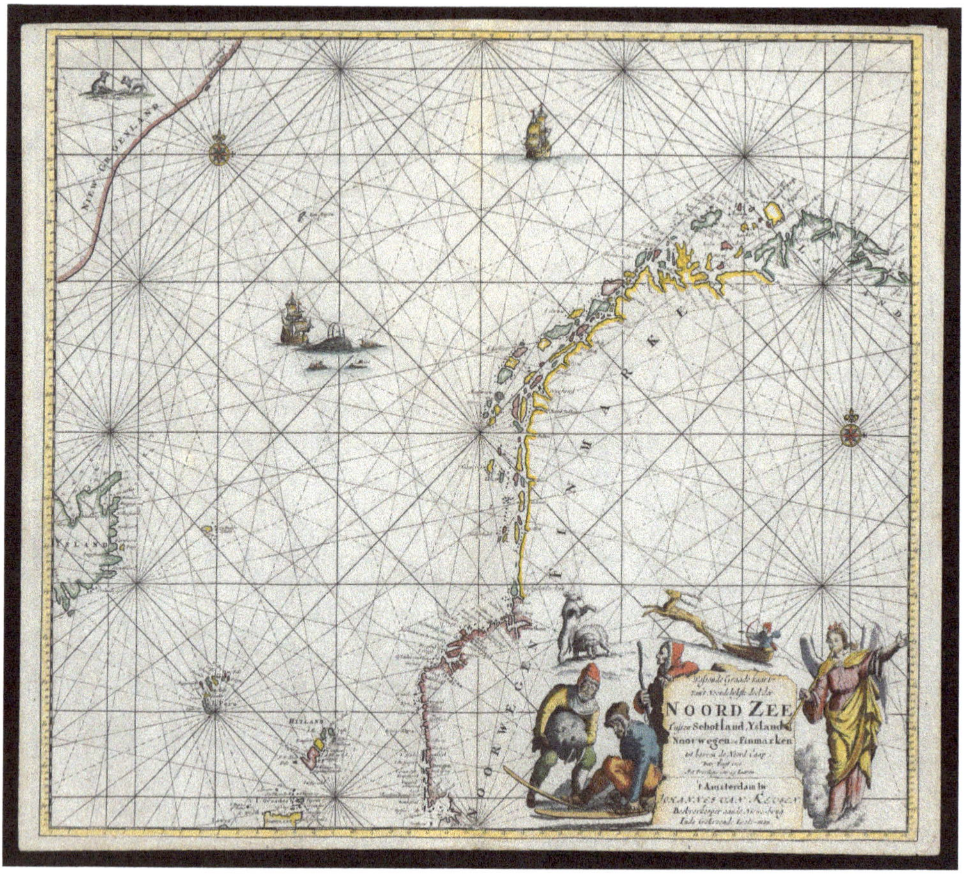

FIGURE 14.6 Claes Jansz. Vooght's map of the Norwegian Sea from 1692. Coloured engraving: "Wassende Graade Kaart van 't Noordelykste Deel der Noord Zee tussen Schotland, Ysland, Noorwegen en Finmarken tot Booven de Noord Caap", from Keulen Johannes van, *De groote nieuwe vermeerderde Zee-Atlas ofte Water-Werelt*. Nr. 8B.9a, in Ginsberg William B., *Sea Charts of Norway*, 1585–1812 (2012) © NATIONAL LIBRARY OF OSLO

Odysseus's Charybdis and Scylla is revitalised, and Ramus discusses whether *Utrøst* in Lofoten – the secret *huldeland* – is actually Odysseus's 'Island of the Sirens', while Træna, not far from Dass's residence, is given the honour of being 'Thrinacia – the Island of the Sun'. His project recalls similar attempts in Renaissance Scandinavia to recapture the splendour of antiquity, notably Olaus Rudbeck's idea of Sweden as Atlantis, the cradle of civilisation. Ramus refers to Peder Claussøn Friis and Diderik Brinch throughout his whole work. The work was republished in 1713 and 1716. Ramus's fantasies formed the basis of some of the most well-known and quoted literary works of modern times: Edgar Allan Poe's short story "A Descent into the Maelström" (1841), Jules Verne's *Vingt mille lieues sous les mers* (1869–1870) (*Twenty Thousand Leagues Under the Sea*), and Herman Melville's *Moby-Dick* (1851), where the Moskstraumen is mentioned by Captain Ahab. There, the famous captain swears that he will pursue the white whale around every corner of the world: 'Aye, aye! and I'll chase him round Good Hope, and round the Horn, and round the Norway Maelstrom, and round perdition's flames before I give him up'.

At this point, however, we have moved far from the history of early modern literature and knowledge and firmly placed ourselves in the world of modern fiction.[69]

Bibliography

Brandt R., "Raritas et curiositas. Amtmann Hans Hansson Lilienskiold, en 1600-talls europeer", in Johannesson H.E. (ed.), *Mimesis förvandlingar: tradition och förnyelse i renässansens och barockens litteratur* (Stockholm: 2002) 359–373.

Brinch Diderik, *Prodromus e Norvegia, Sive Descriptio Loufoudiæ, Omnium Nordlandiæ Præfecturarum longe celeberrimæ, accuratissima* (Amsterdam, Christophori Cunradi: 1676).

Bruland S.H., "Petter Dass og den klassiske litteraturen" in Alenius M. – Bergh B. – Boserup I. – Friis-Jensen K. – Skafte Jensen M. (eds.), *Latin og nationalsprog i Norden 1500–1800* (Copenhagen: 1991) 195–204.

Coenen Adriaen, *Het walvisboek*, ed. F. Egmond – P. Mason (Zutphen: 2003).

Coenen Adriaen, *The Whale Book*, ed. F. Egmond – P. Mason (London: 2003).

Egmond F., *Visboek: de wereld volgens Adriaen Coenen* (1514–1587) (Zutphen: 2005).

Cook H., *Matters of Exchange: Commerce, Medicine, and Science in the Dutch Golden Age* (New Haven – London: 2007).

Cooper A., *Inventing the Indigenous. Local Knowledge and Natural History in Early Modern Europe* (Cambridge: 2007).

Dass Petter, *Katekismesanger*, ed. Jon Haarberg (1715/2012). https://www.bokselskap.no/boker/katekismesanger/tittelside.

Dass Petter, *Beskrivelse over Nordlands Amt*, ed. Albert Christian Dass (Copenhagen, Hartvig Godiche: 1763).

69 Wellendorf J., "Odin of Many Devices: Jonas Ramus (d. 1718) on the Identity of Odin and Odysseus", in Tangherlini T. (ed.), *Nordic Mythologies: Interpretations, Intersections, and Institutions* (Berkeley: 2014) 115–132; Lockert, *Havsvelget i nord* 145–154.

Dass Petter, *The Trumpet of Nordland*, trans. by T. Jorgensen (Minnesota: 1954).

Dass Petter, *Viser og rim*, ed. D.A. Seip (Oslo: 1980).

Daston L. – Park K., *Wonders and the Order of Nature, 1150–1750* (New York: 1998).

Evju H., *Ancient Constitutions and Modern Monarchy: Historical Writing and Enlightened Reform in Denmark-Norway 1730–1814* (Leiden – Boston: 2019).

Foss G. "Skou-Essen. Om skrift og mat i Nordlands Trompet", in Andersen B. – Elisassen K.O. (eds.), *Maskepi og maskerade* (Trondheim: 2005) 63–78.

Friis Peder Claussøn, *Samlede Skrifter*, ed. G. Storm (Kristiania: 1881).

Gjevik B., *Flo og fjære langs kysten av Norge og Svalbard* (Jar: 2009).

Hansen I.R., *Petter Dass: Bibliografi og resepsjonshistorie 1678–2022* (Trondheim: 2023).

Hansen K., *Petter Dass: Guds øyesteen*, 2 vols. (Sandnessjøen: 2018).

Haarberg J., "Diderik Brinch", *Norsk biografisk leksikon* (2009): https://nbl.snl.no/Diderik-Brinch.

Haarberg J., "Hvorfor trompet? – Om tittelen på Petter Dass' nordlandsbeskrivelse", *Edda* 99.1 (2012) 3–13.

Hens H.A., "Peder Hansen Resen", *Dansk biografisk lexicon* (2011): https://biografisk leksikon.lex.dk/Peder-Hansen-Resen.

Herbinius Johannes, *Dissertationes De admirandis mundi cataractis supra & subterraneis* (Amsterdam, Janssonio-Waesbergois: 1678).

Jessen-Schardebøll Erik Johan [Steenbuch Hans], *Det Kongerige Norge, fremstillet efter dets naturlige og borgerlige Tilstand* (Copenhagen, Gottmann Friedrich Kisel: 1763).

Lauvstad H., *Helicons Bierge og Helgelands Schiær. Nordlands Trompets tekst, repertoar og retorikk* (Oslo: 2006).

Lockert S.S., *Havsvelget i nord. Moskstraumen gjennom årtusener* (Stamsund: 2011).

Kiil A., *Nordlandshandelen i det 17. århundre* (Svorkmo: 1940).

Midbøe H., *Petter Dass* (Oslo: 1947).

Pontoppidan Erik, *The Natural History of Norway*, 2. vols (London: A. Linde: 1755).

Rian Ø., 'Det store samrøret – Embetskjøpmennene på 1600-tallet mellom fyrstestat og undersåtter', *Heimen* 44 (Oslo: 2007) 293–310.

Sandmo E, 'Dwellers of the waves: Sea monsters, classical history, and religion in Olaus Magnus's Carta Marina', in *Norsk Geografisk Tidsskrift – Norwegian Journal of Geography*, 74.4 (2020) 237–249.

Sandnes J., "Lokalhistorisk litteratur til omkring 1900", in Bjørkvik H. – Fladby R. – Reinton L. – Sandnes J. (eds.) *Lokal historie i forskning og kulturarbeid gjennom 200 år* (Oslo: 1970) 13–32.

Schreiner J., *Hanseatene og Norge i det 16. århundre* (Oslo: 1941).

Spaans R., "Hekser, Kopernikus og nordnorske fiskeslag. Eit kritisk blikk på den seinaste forskinga på Petter Dass", *Edda* 108.2 (2021) 84–97.

Stokker K., "Oral Tradition, Humanism and the Baroque", in Naes H.S. (ed.), *A History of Norwegian Literature* (Lincoln – London: 1993) 39–52.

Trijp D. van, "Fresh Fish: Observation Up Close in Late Seventeenth Century England", *Royal Society of London: Notes Rec.* 75 (2021) 311–332.

Ytreberg N.A., *Malangen bygdebok* (Trondheim: 1943).

Volqvartz Marcus C., *Aandelige Sørge-Sange / De Afdøde til Ære* (Copenhagen, Hartvig Godiche: 1750).

Walkendorf Erik, *Finmarkens Beskrivelse: brev til Pave Leo den 10de fra erkebiskop Erik Walkendorf*, ed. K.H. Karlsson – G. Storm (Kristiania: 1901).

Wellendorf J., "Odin of Many Devices: Jonas Ramus (d. 1718) on the Identity of Odin and Odysseus", in Tangherlini T. (ed.), *Nordic Mythologies: Interpretations, Intersections, and Institutions* (Berkeley: 2014) 115–132.

CHAPTER 15

The Historical Truth behind the "Salmon-Servant" Myth

Rob Lenders

'Gone are the days when the apprentices of Attercliffe petitioned the Crown that they be not made to eat salmon more than twice a week.' With these words, Charles, Prince of Wales, addressed the attendees of the Second International Atlantic Salmon Symposium in Edinburgh in 1978,[1] a conference devoted entirely to the decline and projected recovery of a single fish species: Atlantic salmon (*Salmo salar*). In his opening address, Prince Charles probably referred to young working-class apprentices employed in the rapidly emerging 18th- and 19th-century crafts and industry of Attercliffe, now an industrial suburb of Sheffield, England. The apprentices' request to the Crown exudes an atmosphere of a then overwhelming abundance of salmon, such that even the lower classes of society became fed off consuming it. However, at the turn of the 20th century, the situation seems to have changed completely and in most British rivers salmon is hardly caught at all anymore. From then, consumption of salmon seems to have been reserved for the upper classes of society only. What had happened in those mere decades since the Attercliffe petition, or had nothing actually changed significantly at all?

Charles' "rescued from oblivion" anecdote is not the only one that has gone or, for that matter, still goes around Europe.[2] It is also not the case that the anecdote, or similar anecdotes, related only to industrial apprentices, or that it is only known from England; neither are stories of historical salmon abundance and its subsequent demise only set in the 19th century, nor was salmon the only subject of tales of profusion and successive demise. In what I have coined "the salmon-servant anecdote" it are mostly domestic servants and, to a lesser degree, farm hands and apprentices that are subject to alleged limitation of the number of times they had to eat salmon each week. In these anecdotes,

1 Went A.E.J., *Atlantic Salmon: Its Future. The Proceedings of the Second International Atlantic Salmon Symposium, Edinburgh 1978* (Farnham: 1980) 1.

2 In the most recent version of the anecdote, which is considered accurate and true by the author, Breuers tells of domestic servants in the Cologne area who, at the beginning of the 20th century, allegedly had their contracts limited to eating salmon three times a week. Breuers D., *Colonia im Mittelalter. Über das Leben in der Stadt* (Cologne: 2013) 279.

 | DOI:10.1163/9789004681187_016

they either complained about the large quantities of salmon they were served and which they abhorred, or they stipulated a maximum number of times of salmon consumption per week in their contracts or in city ordinances. The essential message of the anecdote – which is always set in a distant or less distant past and/or in other countries – is that there had been a time when or place where salmon was so abundant that even the lower classes could eat it as often as they wanted (a proverbial "Land of Cockaigne", so to say), or – on the contrary and more often – had to eat it so frequently that they were disgusted with it. The anecdote was undoubtedly meant to remind one of better times when rivers teemed with salmon, but was perhaps even more reminiscent of human hubris, illustrating that exploitation and environmental destruction had led to the collapse of salmon stocks.

In this contribution, I will take a closer look at the backgrounds of the anecdote and see if there is possibly a historical-ecological truth behind the myth. I will also explore further what the reason and purpose of the anecdote may have been. Although the anecdote is also told about other species of fish (especially sturgeon, allis shad, burbot and tuna) or even features lobsters, partridges and hares in the leading role, I shall limit myself here to those anecdotes which concern salmon, if only because they far outnumber those about other species. Geographically, the search has been limited to (former) German-, English-, French- and Dutch-speaking areas in Western Europe, although I have also found an occasional reference to other countries (such as Norway). The main reason for limiting myself to the above-mentioned areas has been the accessibility of sources and my own inability to read and understand Polish, Scandinavian and Baltic texts. This probably underestimates the geographical spread of the anecdote in these areas, but I am nevertheless convinced that I have been able to cover the geographical core of the anecdote sufficiently.

Mentions of anecdotes were sought in a wide variety of historic and scientific sources, including newspaper and magazine articles, books and even parliamentary records. Search terms were formed by combining 'salmon' with 'per week' or 'weekly' and/or 'servant' in four different languages (English, French, German, Dutch).[3] The information found was stored in a database where the

3 A few special sources that are a compilation of a similar search deserve special mention here because they formed a valuable starting point for my research. It concerns in particular the following publications: Danker-Carstensen P., "Stör oder Lachs – aber auf Keinen Fall mehr als Zweimal in der Woche? Legendenbildung und Erzähltradition in einem Kapitel Deutscher Fischereigeschichte", in Pelc O. (ed.), *Mythen der Vergangenheit. Realität und Fiktion in der Geschichte* (Göttingen: 2012) 265–285; Schwarz K., "Der Weserlachs und die Bremischen Dienstboten. Zur Geschichte des Fischverbrauchs in Norddeutschland", in Elmshäuser K. (ed.), *Bremisches Jahrbuch* 74/75 (Bremen: 1995/1996) 134–173; Schwarz K., "Nochmals: der

records were formed by a specific mention of the anecdote for a specific place and time. If several sources mentioned the same anecdote, anecdotes from younger sources were only included as separate records if they had any surplus value for interpretation of data, mainly because something was added or changed to it in the course of time. If in a certain source several places or times were mentioned in the same anecdote, these have been included as separate records. The number of records is thus not equal to the number of different anecdotes found in literature. The bibliographical data of all sources consulted have been included per record. In addition, per record also the following information was noted (if stated): data concerning the original source (e.g. author and year of publication; especially when this source was not or not easily traceable), the relevant quote in the language of the source and in an English translation, the presence of evidence according to the author of the consulted source, the (maximum) number of times that a certain social group had to eat salmon according to the anecdote, a specification of that social group, information concerning the state/condition of salmon to be consumed (e.g., fresh or conserved, rotting, kelts, parr or smolts) and a specification as precise as possible of the place where and the time when the anecdote was supposed to have taken place. In the end, 229 records were collected in this way.

In total the anecdote was recorded for 13 different countries across Europe. Surprisingly, England was with 5 records (including Prince Charles' quote) only number 5 in the ranking of countries with the most mentions. The top 4 consisted of France (107 records), Germany (48 records), The Netherlands (28 records) and Scotland (18 records). The level of geographical precision of the anecdotes varied considerably: 84 on the level of specific places (towns, villages), 53 on the level of regions (e.g., Normandy, Devon, Saxony), 35 on the level of rivers/river catchments (e.g. Rhine, Severn, Dee, Loire) and 57 on the level of countries only. The number of *unique* geographical locations varied from 20 regions and 20 rivers to 38 specific places [Fig. 15.1]. Temporal precision also varied: only in 7 cases was a precise year mentioned in which the event of the anecdote was supposed to have taken place; other anecdotes gave an approximate century or even spoke only in terms of "former times" and the likes. In case a century could be determined with some accuracy (171 instances), this ranged from the 12th to the 20th century. Of these, only 8 referred to the

Lachs und die Dienstboten. Eine Nachlese", in Elmshäuser K. (ed.), *Bremisches Jahrbuch* 77 (Bremen: 1998) 277–283; Thibault M. – Garçon A.-F., "Un problème d'écohistoire: le saumon dans les contrats de louage, une origine médiévale?", in Benoit P. – Loridant F. – Mattéoni O. (eds.), *Actes des Premières Rencontres Internationales de Liessies 'Pêche et pisciculture en eau douce: la rivière et l'étang au Moyen-Age, 27, 28 et 29 avril 1998'* (Lille : 2004).

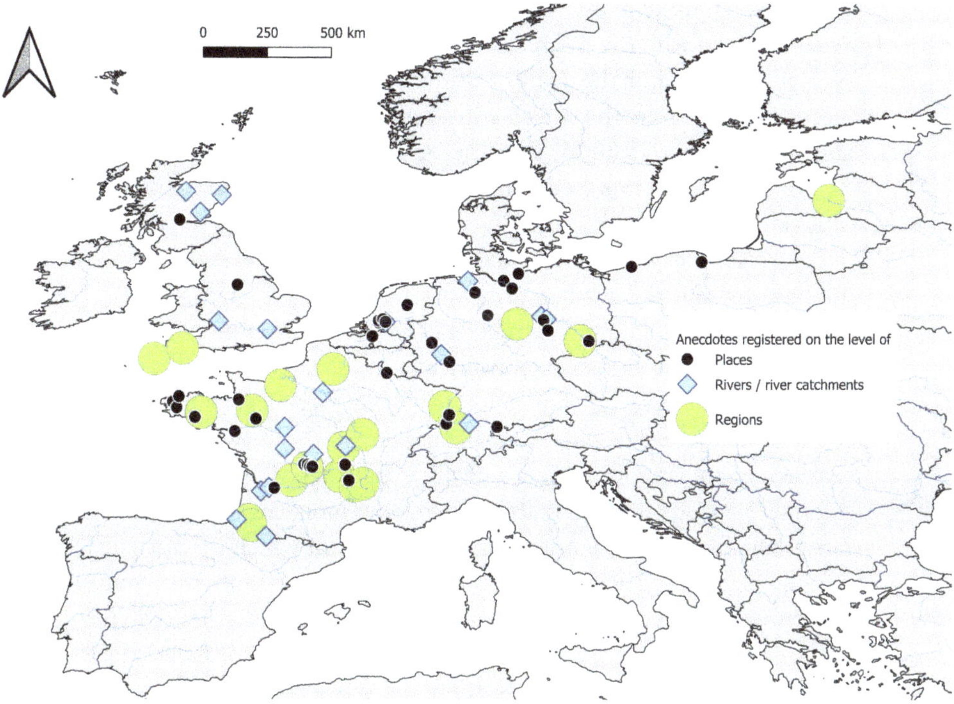

FIGURE 15.1 Distribution of the "Salmon-Servant" anecdote in Europe on several levels of scale, 16th–21st century
MAP MADE WITH FREE AND OPEN SOURCE QGIS

Middle Ages, 79 referred to the Early Modern Period (16th–18th century), while 76 referred to the 19th century and even still 8 to the 20th century.

The wide spatial and temporal spread of the records indicates that it is highly unlikely that the anecdotes were real events in all cases and times. The fact that the anecdote was told from the far north of Europe (Scotland, Norway and the Baltic States) to the south of France combined with the alleged topicality of the events described in a broad temporal spectrum from the Middle Ages to the 19th and even the 20th century makes it inconceivable that we are dealing with actual events in all cases. This implausibility is enforced by the fact that if we have a fairly precise indication of the time in which the anecdote is supposed to have taken place, the spatial precision is rather coarse (on the level of regions, river basins or even countries) or that – from the eye of the beholder – the event took place somewhere abroad; on the other hand, if the geographical indication is fairly precise, the temporal indication is imprecise ('a few hundred years ago', 'long ago', 'in the past'). In many cases, however, both the geographical *and* the temporal indications remain rather vague which made it difficult or even impossible to pinpoint the described events in space and time with the desired accuracy.

Of all the records in the database, almost 70% relate to an anecdote in which domestic servants are the subjects; for farm labourers and miller's servants the figure is about 27%; and for the other social groups it ranges from about 1% each (army officers, monks and prisoners) to about 2.5% for apprentices such as those of Prince Charles' anecdote.[4] The number of times servants or others had to eat salmon varied in the anecdotes from once to even six times a week. The latter is said to have been the case in an alleged (and undated) regulation of Basel, Switzerland.[5] A miller's help living somewhere in France in an unmentioned century had the misfortune of having to eat salmon up to five times a week;[6] however, convincing proof for these large numbers of times servants were forced to eat salmon is missing.

A lack of substantial evidence of the events described in the anecdotes is generally a serious concern in this regard. If we look only at domestic servants, (farm) labourers and apprentices, social groups for which we can expect written agreements to be in place, it appears – according to the sources consulted – that 77% concern individual contracts or arrangements (from the 14th to the 20th century) and just under 10% speak of municipal ordinances (16th–19th century). In the remaining cases, there are only "complaints" or "refusals to work" recorded, or it is not known what form the limitation of the number of times to eat salmon had taken. The sources consulted, however, are usually silent on whether their authors have seen any written evidence themselves or not. Although it is understandable that individual contracts have not or hardly been preserved, one may expect that city regulations are largely well-kept-up in municipal archives. Some authors specifically mention that they searched for evidence in relevant archives or otherwise, but that they failed to find any.[7] The authors of three sources,[8] all concerning the German city of Hamburg,

4 In some cases, the anecdotes referred to more than one social group, for instance servants *and* apprentices. Hence, percentages can add up to more than 100%.

5 Schwarz, "Der Weserlachs und die Bremischen Dienstboten" 144.

6 Bazin 1973, cited in Thibault – Garçon, "Le saumon dans les contrats de louage".

7 E.g., Paulze d'Ivoy de la Poype, 1901; Jenkins, 1825; Seguin, 1938; Chimits, 1963; Wilkins, 1989 (all cited in Thibault – Garçon, "Le saumon dans les contrats de louage"); Storck A., *Ansichten der Freien Hansestadt Bremen und ihrer Umgebungen* (Frankfurt am Main: 1822) 388; Burema L., *De Voeding in Nederland van de Middeleeuwen tot de Twintigste Eeuw* (Assen: 1953).

8 Büsch J.G., "Verhandlungen der Gesellschaft über den Verfall der Fischereien unserer Gegend, besonders in der Elbe, und über die Mittel zur Verbesserung derselben", *Verhandlungen und Schriften der Hamburgischen Gesellschaft zur Beförderung der Künste und nützlichen Gewerbe* 4 (1797) 1–48; Rambach J.J., *Versuch einer physisch-medizinischen Beschreibung von Hamburg* (Hamburg: 1801) 110; Anonymous, No title. *Abendzeitung*, January 29, 1824.

assert that in their time an ordinance was still read in public twice a year, proclaiming that it was forbidden to serve salmon to servants more than twice a week. In these cases too, however, no written evidence was found in the archives despite extensive searches[9] and the "evidence" is thus no more than literally hearsay.

The number of records that mention individual contracts or agreements is much larger: 145 records mention such agreements but are silent on whether there is actual written evidence for them; 16 records hint that such evidence is probably not available or even state very clearly that evidence is completely lacking. Baude for instance states:

> Nous avons tous entendu qu'en Ecosse les domestiques stipulent dans leurs contrats de louages les jours de la semaine où ils seront dispensés de manger du saumon. Je n'ai lu aucun de ces contrats.[10]
>
> We have all heard that in Scotland domestic servants stipulate in their contracts of lease the days of the week on which they will be exempted from eating salmon. I have not read any of these contracts.

Of the nine records of which the authors claim that they *did* see such contracts, most remain rather vague and no details of the agreements are given. Three remaining records relate to one and the same contract allegedly concluded on 17 June 1842 in Saint-Antoine de Breuilh, in Dordogne, France.[11] I will come back to this specific case later, which is in fact the only existing "proof" substantiating the salmon-servant anecdote.

If the essence of the anecdote is true, regardless of whether there is conclusive evidence for it or not, one would expect the price of salmon to be very low during the times the anecdotes refer to. However, across Europe, this does not appear to be the case in any of these times. In 15th-century Basel, for instance, salmon cost as much as 15 sacks of rye.[12] In the 17th century the average price

9 Anonymous, *Hamburgische Burspracken vom Jahre 1594. Nach dem Alten Originale, wornach dieselben Jährlich Abgelesen worden, mit Anmerkungen* (Hamburg: 1810).

10 Baude J.-J., "L'empoissonnement des eaux douces", *Revue des Deux Mondes* 31 (1861) 293–344.

11 The existence of this contract was first reported on by Pustelnik, G., *Les saumons de la rivière Dordogne. 1189–1981* (Paris – Sarlat: 1982); copies are published in Thibault – Garçon, "Le saumon dans les contrats de louage", and Vibert R., "Le saumon atlantique: origine et caractéristiques essentielles", in Gueguen J.C. – Prouzet P. (eds.), *Le saumon atlantique. Biologie et gestion de la ressource* (Brest: 1994) 11–25.

12 Meier E.A., *z'Basel an mym Rhy. Von Fähren und Fischergalgen* (Basel: 2013).

of salmon in the Netherlands was equal to the weekly wage of a labourer.[13] In the period 1777–1781 salmon cost an average of about 120 sols apiece in the Low Countries, seven to twelve times the daily wage of a labourer.[14] On the fish market of Cologne (Germany) in 1550, fresh salmon cost slightly more than the highly appreciated carp and pike. Fifty years later, the price of salmon was even about twice as high as that of other valued fish.[15] A similar development can be seen in Lower Normandy at the end of the Middle Ages: the price of salmon had risen by a factor of 4.5 compared to other protein rich foods (mutton, pork, partridge) over a period of 150 years (1260/1270 to 1410/1420).[16] The (inflation-adjusted) prices paid for salmon in the 15th and 16th centuries were thus evidently already relatively high and rising sharply. Consequently, at least from the 17th and 18th centuries onwards, salmon was far beyond the reach of the lower social classes as an affordable food in large parts of Europe. The implicit assertion in the anecdotes that salmon was an abundant staple food in these centuries is therefore in obvious contradiction to the figures presented by historical economic statistics.

According to several city accounts, Dutch, Belgian and German river towns bought salmon for their own dignitaries (mayors and aldermen) but also for their local lords. In 1511, the city of Dordrecht donated a salmon to its landlady at a banquet.[17] Earlier, the town of Deventer had donated salmon to various of its relations, such as the Bishop of Munster, the Lord of Ghemen at Bredervoerd and the 'richter' (bailiff) of Arnhem.[18] Salmon was also given by the city of Worms (Germany) to King Sigismund in 1414 and to Emperor Maximilian and his consort in 1494; Bern gave salmon to Pope Martin V in 1418 and Basel to Archduke Matthias of Austria in 1596.[19] In 15th century France, freshly caught salmon from Normandy was even sent on horseback

13 Martens P.J.M., *De Zalmvissers van de Biesbosch. Een Onderzoek naar de Visserij op het Bergse Veld 1421–1869* (Tilburg: 1992).

14 Van Buyten L., "Données historiques sur le commerce de poissons à Louvain (Brabant, Belgique) au 18ème siècle et leur apport à l'archéozoologie", in Neer W. van (ed.), *Fish Exploitation in the Past.* Annales du Musée Royal de l'Afrique centrale, Sciences Zoologiques 274 (Tervuren: 1994) 151–161.

15 Kuske B., "Der Kölner Fischhandel vom 14.–17. Jahrhundert", *Zeitschrift für Geschichte und Kunst* 24 (1905) 227–313.

16 Halard X., "La pêche du saumon en Normandie du XIe au XVe siècle", *Journal of Medieval History* 9 (1983) 173–178.

17 Dam P.J.E.M. van, "Feestvissen en Vastenvissen. Culturele, Ecologische en Economische Aspecten van de Visconsumptie in de Nederlanden in de Late Middeleeuwen", *Tijdschrift voor Sociale Geschiedenis* 29 (2003) 467–496.

18 Lobregt P. – Os J. van, *De Laatste Riviervissers* (Heerewaarden: 1977).

19 Kuhn G., *Die Fischerei am Oberrhein* (Stuttgart: 1976).

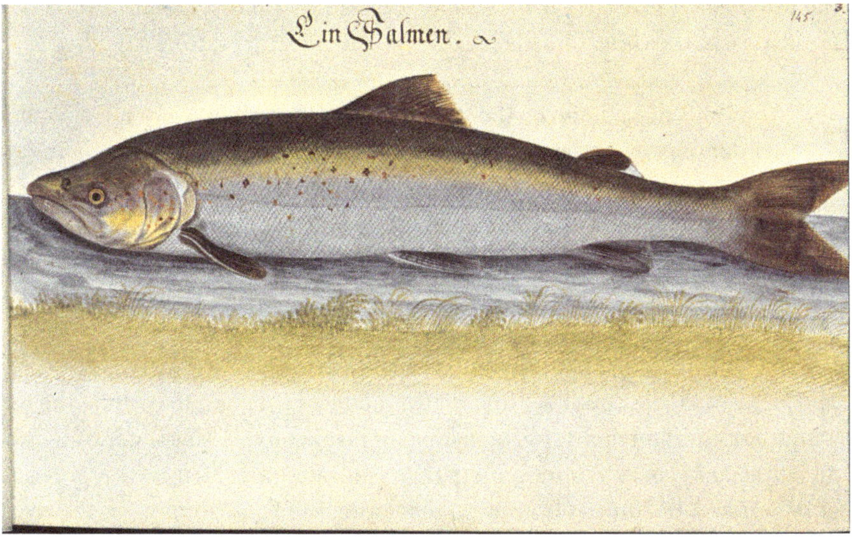

FIGURE 15.2 Baldner Leonard, *Salmon*. Coloured drawing. In Baldner Leonard, *Vogel-, Fisch- und Thierbuch*. (British Library, Add MS 6485)

express to the royal court in Paris.[20] Scarcity and value of salmon in Early Modern Times also becomes apparent from the following story. When in 1586 the town of Doesburg (The Netherlands) wanted to serve salmon during the annual banquet for the local dignitaries, organised since 1389, it appeared to be impossible to get hold of salmon from the nearby IJssel river. A courier sent to neighbouring town of Deventer, instructed to buy salmon there, returned empty-handed.[21] Salmon was therefore not only expensive, but also at times difficult to obtain. At the end of the Middle Ages and the beginning of the Early Modern Era, salmon was – as Leonard Baldner, a fisherman from Strasbourg, aptly put it in 1666 – 'Herrenfisch und köstlich in der speiß' (Fish for the Lords and delicious to eat)[22] [Fig. 15.2] and by no means the "cat food" as salmon was termed according to an anonymous source from 1770 by the inhabitants of Lekkerkerk, the Netherlands.[23]

In other ways, too, it appears that salmon was generally not intended for the lower classes but rather for the higher ones, as is evident, for example, from

20 Halard, "La Pêche du Saumon en Normandie".
21 Deelden C.L., "De Overvloedige Zalmvangsten", *Visserij* 35 (1982) 60–62.
22 Baldner Leonard, *Vogel-, Fisch- und Thierbuch*. British Library, Add MS 6485.
23 Anonymous, *De Koopman of Bydragen ten Opbouw van Neêrlands Koophandel en Zeevaard. Tweede Deel* (Amsterdam, Gerrit Bom: 1770).

the established medieval fishing rights on salmon (and sturgeon) in the Low Countries. In the *Groot Charterboek* of Van Mieris (1753–1756) we read:

> Voirt alle, die zalmen en die stoeren, die ghevanghen worden [...] vinden wi den Grave, het en sy of yemand betoghen mach, dat sine is mitten rechte.[24]
>
> Further, all, salmon and sturgeon, that are caught [...] belong to the count we judge, unless someone can make it plausible that it rightfully belongs to him.

In other words, the count of Holland was allowed (in 1306) to include the revenues from salmon and sturgeon fishing in his income. Also in later times, the higher social classes retained their rights to salmon fishing or the proceeds of fisheries. A document from 1730, regulating the fishing rights on the rivers Rupel, Dijle, Nete and Zenne (Belgium), stated that if salmon was caught on these rivers, it had to be offered at a cheap price to the Court of Rumst.[25]

The high market price of salmon also invited unlawful activities. In the 15th and 16th centuries, Flemish sailors with so-called water vessels (kaerschepen), suitable for transporting live fish, increasingly tried to buy salmon illegally from local fishermen in the Biesbosch before they delivered the catch at the fish auctions, sometimes even before they were moored in the harbour.[26] In this way, the official fish auctions were side-lined. In 1574 King Philip II intervened; he promulgated a decree which stated:

> dat van nu voortaen alle versschen Visch, ende namentlyck Salm, Steur ende Elft, binnen Hollandt gevangen, ghebracht ende verkocht zullen worden ter Plaetse van haren behoorlycken ende ghewoonlycken afslagh, ende nerghens anders.
>
> that from now on all fresh fish, and in particular salmon, sturgeon and shad, caught within Holland, shall be presented and sold at the appropriate and customary fish auction, and nowhere else.

---

24 Mieris Frans van, *Groot Charterboek der Graaven van Holland, van Zeeland, en de Heeren van Vriesland* (Leiden, Pieter van der Eyk: 1753–1756).

25 Vrielynck C. – Belpaire C. – Stabel A. – Breine J. – Quataert P., *De Visbestanden in Vlaanderen 1840–1950* (Groenendael – Hoeilaart: 2002).

26 Martens, *De Zalmvissers van de Biesbosch*.

An explanation sometimes given for the striking discrepancy between what the hard economic figures and legal regulations show and the "salmon-servant" anecdote is that the anecdotes could be about temporary large numbers of salmon being caught and marketed, or about salmon of questionable quality. The latter could be so-called kelts or parr and smolts.[27] The sources are not very conclusive about this. Of the 22 records from which something can be deduced about the condition of the salmon to be consumed, 18 (approximately 8% of the total number of records) refer to "fresh" salmon. However, this could also be the aforementioned kelts, or parr or smolt, which the authors of the sources considered possible for 4 and 2 of these cases of "fresh" salmon, respectively. The texts of the cited anecdotes themselves, however, do not always lead to the conclusion that these would concern inferior salmon; it is obviously the authors of these sources who interpret it as such.

The other cases of fresh salmon mentioned in the anecdotes could involve large numbers of salmon which became temporarily available during the fishing season and which could not all be consumed or processed immediately due to "overwhelming catches". The local population would therefore have been forced to consume the almost rotting salmon themselves at the end of the fishing season. The contracts and ordinances would only have been in force during this part of the year. The fact that most of the anecdotes are set in towns and regions that were renowned for their salmon fishing could support such an explanation. Some authors claim that it was therefore primarily a problem of poor transport facilities. When rail transport became available, the "problem" of salmon abundance would have solved itself. The famous Charles Dickens wrote in 1865:

> Why, in the old days, before railways established a ready and rapid communication with the London markets, the servants of country gentlemen residing on the banks of the Severn, the Tay, the Dee, and the Spey, made a stipulation in their terms of engagement that they were not to be fed upon salmon more than three times a week.[28]

27 Most salmon die after spawning. A small proportion of spawned salmon survive and manage to start making their way back to the sea. These are called 'kelts'. Usually these animals are completely exhausted and covered in fungal infections, and die on their way back to sea. Parr and smolt both refer to young salmon. Parr is juvenile salmon growing up in their natal river; smolt refers to older juveniles ready to out-migrate towards their marine foraging grounds where they reach maturity.

28 Dickens C., "Concerning the Cheapness of Pleasure", *All the Year Round* (May 6, 1865).

Others are of the opinion that it is largely about lack of appropriate preservation methods. The *Saturday Magazine* of May 1, 1841, for instance, writes about late 18th century Scottish salmon fisheries:

> Forty or fifty years ago, the art of packing salmon in ice for the London markets was brought to perfection, and since that period the value of the fisheries has risen incalculably; the salmon have become dear in proportion.[29]

According to the anonymous author of the article, salmon had since the late 1700s become unaffordable for the common people, while previously domestics made agreements with their masters on the maximum number of times a week they would be fed on salmon. However, such explanations ignore the fact that methods of preserving salmon had been known since the Middle Ages and that preserved salmon was a popular product. Fresh as well as preserved salmon (especially salted) was regularly purchased by the kitchens of English manors and colleges (1293–1691).[30] The prices for both did not differ much. At the end of the 16th century, the price of 'geräucherter Salm' (smoked salmon) on the Cologne fish market was as high as or even slightly higher than that of 'Krimpsalm' (fresh salmon).[31] The export of barrelled salmon from Scotland (1311–1541)[32] to, among others, Yerseke in the Netherlands[33] (certainly from 1499 onwards, but probably from 1472 until 1570) shows that conserved salmon was a valuable trading product. Therefore, those fisheries that had specialised in salmon fishing in particular would have quickly mastered available conservation methods and adopted them instead of dumping the salmon on local markets or distributing it to the local population in a half rotten state. In seven cases (3% of the records), the authors state that the quality of preserved salmon did not match that of fresh salmon. In these cases, however, it seems that the authors were seeking an explanation for an anecdote they found difficult to believe. They *assumed* that the anecdote concerned preserved salmon

29 Anonymous, "Fresh-water fish. II The Salmon (concluded)", *The Saturday Magazine* 18, May 1, 1841.

30 Thorold Rogers J.E., *A History of Agriculture and Prices in England from the Year after the Oxford Parliament (1259) to the Commencement of the Continental War (1793)*, volumes 2–4 (London: 1866–1882).

31 Kuske, "Der Kölner Fischhandel".

32 Gemmil E. – Mayhew N., *Changing Values in Medieval Scotland. A Study of Prices, Money, and Weights and Measures* (Cambridge: 1995).

33 Unger W.S., *De Tol van Iersekeroord. Documenten en Rekeningen 1321–1572* (The Hague: 1939).

of bad quality, rather than there being evidence that this was actually the case. They could not interpret the "salmon-servant" anecdote in any other plausible way, knowing that fresh salmon was already in short supply at the time the anecdote was set.

As mentioned earlier, there is only one physical piece of "proof" that the anecdote has some basis in truth [Fig. 15.3]. It concerns an employment contract dated 17 June 1842 between Henri Benoist de Fonroque, landowner in the Laroques-Vigneron municipality of Saint-Antoine-de-Breuilh, situated on the Dordogne (France), and the spouses Vigouroux who were hired as coachman and cook. In the conditions of entry into service, it is stated:

> Il est stipulé que selon les usages locaux et constants, il ne sera pas donné à la cuisine de saumon frais plus de trois fois par semaine de février à l'Assomption d'août.[34]
>
> It is stipulated that according to local and customary practice, fresh salmon will not be given to the kitchen more than three times a week from February to the Assumption of August.

It is clear from the quote that in this case it was indeed *fresh* salmon being served for about six months of the year and that the working conditions imposed, which limited the number of times salmon to be consumed, were according to local custom. Most historical-ecological sources now agree that salmon catches were already declining in large parts of Europe well before the mid-19th century.[35] Whether this was also the case for the Dordogne is unknown, but highly probable. Also, the included labour condition could be a relic from earlier times when there was indeed an abundance of salmon. It is known that some clauses were still included in employment contracts for a long time, even though they were no longer applicable.[36] However, it should also not be entirely excluded that the contract is a 20th-century forgery and

34 Pustelnik, *Les saumons de la rivière Dordogne*; Pustelnik G. – Roguet M. – Tinel C. – Soumastre J. – Roux M. – Simonet F., "Historique, cartographie écologique de la rivière Dordogne et évaluation de son potentiel d'accueil pour le saumon atlantique", in Thibault M. – Billard R. (eds.), *Restauration de rivières à saumons* (Paris: 1987) 53–64.

35 Hoffmann R.C., "Economic Development and Aquatic Ecosystems in Medieval Europe", *American Historical Review* 101 (1996) 631–669; Lenders H.J.R. – Chamuleau T.P.M. – Hendriks A.J. – Lauwerier R.C.G.M. – Leuven R.S.E.W. – Verberk W.C.E.P., "Historical Rise of Waterpower Initiated the Collapse of Salmon Stocks", *Scientific Reports* 6 (2016) no. 29269; Lenders, H.J.R., "Fish and Fisheries in the Lower Rhine 1550–1950: A Historical-ecological Perspective", *Journal of Environmental Management* 202.2 (2017) 403–411.

36 Thibault – Garçon, "Le saumon dans les contrats de louage".

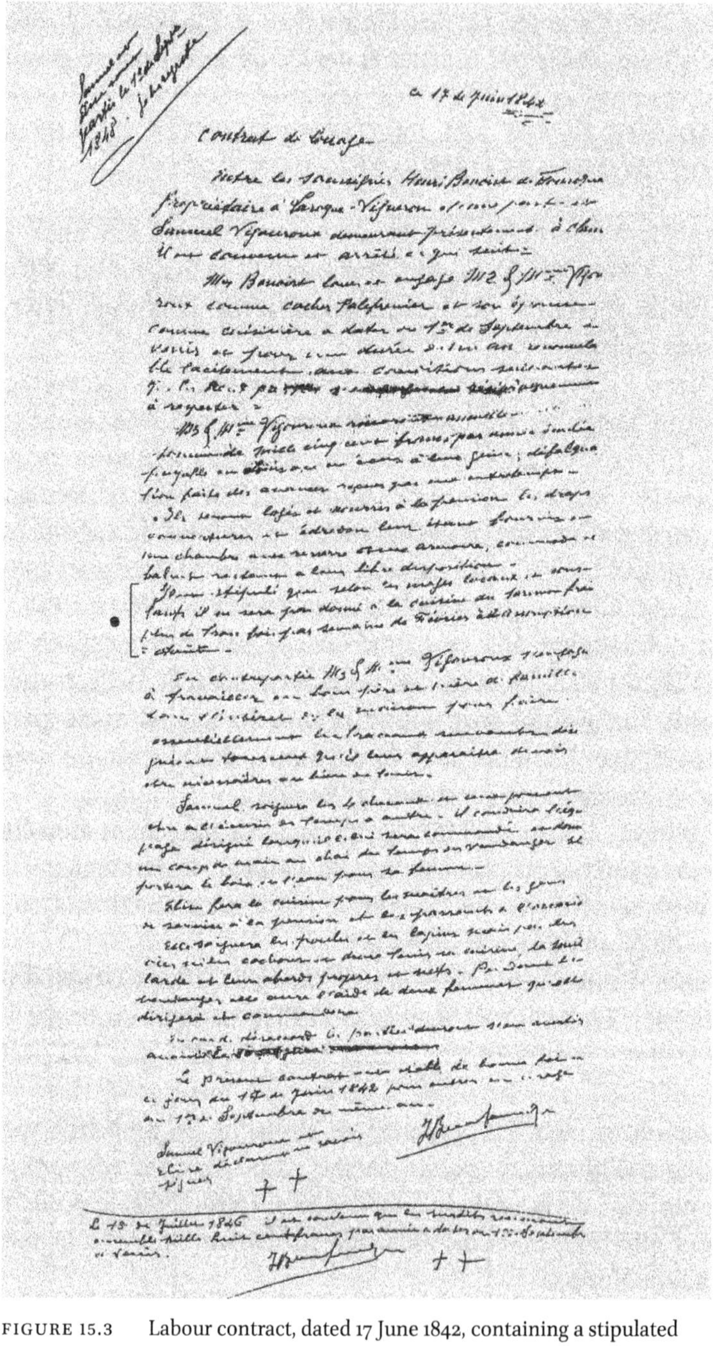

FIGURE 15.3 Labour contract, dated 17 June 1842, containing a stipulated number of times that fresh salmon could be served to the employees (passage indicated by a bar with a dot)
SOURCE: THIBAULT – GARÇON, "LE SAUMON DANS LES CONTRATS DE LOUAGE"

therefore not authentic. In that case, it may have been an instrument to create social and political interest in the recovery of decimated salmon stocks. Pustelnik *et al.* (1987) themselves have doubts about the significance of the contract for the interpretation of salmon abundance in earlier times: 'Son interprétation quant à l'abondance passée du saumon reste toute hypothèse'[37] (Its interpretation of past salmon abundance remains speculative).

In any case, authentic or not, the contract is by no means the oldest mention of the anecdote. When older versions of the same or similar stories exist, it is likely that authors of the younger versions were inspired by or simply copied older versions. The challenge, therefore, is to track down the oldest version, as that may be the "mother of all anecdotes", from which all other anecdotes ultimately stemmed. For the oldest version of our anecdote, dealing unmistakably with salmon, and traced so far, we have to go back to 17th-century Holland, at the time of the Dutch Republic. In his book *Out-Hollandt, Nu Zuyt-Hollandt,* dated 1654, Jacob van Oudenhoven tells about the cities of Dordrecht, Geertruidenberg, Rotterdam, Schoonhoven, Lekkerkerk and Krimpen, then renowned for their fisheries and fish auctions. One of the passages deals with domestic servants and the abundance of salmon and reads:

> Ende wort geseyt, dat de Dienstboden in haer Huyr plachten te bedingen, datse maer tweemael in de Weeck Salm wilden eten.[38]

> And it is said, that the domestic servants used to let take up in their contracts of lease, that they wanted to eat salmon no more than twice a week.

With this quote, Van Oudenhoven wanted to make clear that the salmon catches in the Dutch delta, where Rhine and Meuse confluence, had strongly decreased, already in his time. Again, it is striking that the anecdote is situated in a vague past; the exact time is unclear, but it must have been well before 1654, possibly as early as the 16th century, because Van Oudenhoven did not know it from his own experience, as evidenced by his "hearsay" formulation. This "Van Oudenhoven" version of the anecdote is probably the source of two of the oldest non-Dutch anecdotes, which, remarkably, still concern (domestic) servants in the Netherlands. The French biographer Bouquet (1878) cites from the memoires of Pierre Thomas, sieur du Fossé, written down in approximately 1667 about a visit at the Nantes market which surprised him by the prodigious quantity of fresh salmon:

37 Pustelnik – Roguet – Tinel – Soumastre – Roux – Simonet, "Historique, cartographie écologique de la rivière Dordogne".

38 Oudenhoven Jacob van, *Out-Hollandt, nu Zuyt-Hollandt* (Dordrecht, Andriesz.: 1654).

> je compris alors très facilement ce qui se dit d'un certain canton des Flandres ou d'Hollande, que les servantes, en se louant, mettent à leur marché qu'on ne leur fera manger de saumon frais qu'un certain nombre de jours de chaque semaine.[39]
>
> I then understood very easily what is said of a certain canton of Flanders or Holland, that the maids, in hiring themselves out, put in their contract that they will only be made to eat fresh salmon on a certain number of days of each week.

The oldest known source referring to France itself – apart from the contract mentioned above – dates back to only 1862 and refers to an unknown time and unknown places which seems to emphasise that the author himself was also not very convinced of the truthfulness of the anecdote. If salmon was still abundant in France only twenty years earlier, as the 1842 contract suggests, this author would most probably have been aware of this and he would not have formulated his concern as follows:

> L'abondance était telle que, dans les ferme [sic] et les habitations voisines de quelques-unes de nos rivières, les domestiques, en entrant en gages, stipulaient qu'ils ne mangeraient du saumon qu'une ou deux fois par semaine.[40]
>
> Such was the abundance that, in the farms and dwellings adjacent to some of our rivers, the servants, on taking up employment, stipulated that they would eat salmon only once or twice a week.

The other early-modern source probably derived from the "Van Oudenhoven" anecdote version concerns a quote from the Swiss author König (1682), who mentioned the anecdote while writing about the (natural) scarcity of salmon in Switzerland:[41]

---

39 Bouquet F., *Mémoires de Pierre Thomas, sieur du Fossé, publiés en entier, pour la première fois d'après le manuscrit original avec une introduction et des notes* (Rouen: 1878).

40 De Saint-Prix, 1862, cited in Thibault – Garçon, "Le Saumon dans les Contrats de Louage".

41 Originally in Latin; here in the translated German version by Schwarz, "Der Weserlachs und die Bremischen Dienstboten".

> im entlegenen Mündungsgebiet des Stromes in Holland machten die Knechte mit ihren Herrschaften aus, daß sie nicht zu oft in einer Woche Lachs essen müßten.
>
> in the remote estuary of the river in Holland, the servants agreed with their masters that they would not have to eat salmon too often in a week.

Does this make the "Van Oudenhoven" version the "mother of all anecdotes"? Most probably not. In fact, a slightly older version of a similar anecdote, to which Van Oldenhoven is clearly indebted, was already circulating in Holland. It concerns a passage from the book *'t Begin van Hollant in Dordrecht* by Johan van Beverwijck from 1640, dealing with the history of the city of Dordrecht. In this work, van Beverwijck states:

> Daer van noch vertelt vvert, dat de Dienstmaeghden in ouden tijden, eer sy in haer huyre quamen, plachte te bedingen, datse maer tvveemael 's vveecks roode visch zouden eten.[42]
>
> It was also said that in ancient times, before being hired, the maids used to stipulate that they would only have to eat red fish twice a week.

'Roode visch' (red fish) probably refers to salmon, but that is not entirely certain. According to Burgersdijk (1873), Van Beverwijck referred to 'roode visch' not (only) as salmon but also as sturgeon and shad.[43] It would, however, not change the reasoning about the origin of the anecdote very much, because dating and geographical location hardly differ. Again, even in this oldest of all investigated anecdotes, we see a reference to 'ouden tijden' (ancient times), which most likely places the perceived lost abundance of salmon in the 16th century or even earlier. Geographically, the oldest traceable version of the anecdote seems to be situated in the Netherlands. From there, the anecdote may have spread to the rest of Europe. The oldest references not relating to the Rhine-Meuse delta in the Netherlands are dated about half a century later than the "Van Oudenhoven" anecdote, and are located in the German city of

42 Beverwijck Johan van, *'t Begin van Hollant in Dordrecht. Mitsgaders der Eerster Stede Beschrijvinge, Regeringe, ende Regeerders: als oock de Gedenckvvaerdige Geschiedenissen aldaer Gevallen* (Jasper Gorissz., Dordrecht: 1640).

43 Burgersdijk L.A.J., *De Dieren Afgebeeld, Beschreven en in hun Levenswijze Geschetst. Derde deel: Kruipende Dieren, Visschen, Ongewervelde Dieren* (Leiden: 1873).

Hamburg, situated on the river Elbe.[44] On the way to further dissemination, the anecdote was sometimes exaggerated (especially with regard to the maximum number of salmon consumed per week) or adapted to local conditions.

In 1995 Daniel Pauly published his idea of the shifting baseline syndrome.[45] The essence of this concept is that each generation of fisheries biologists implicitly uses a reference of fish stock composition and volume that goes back to their own youth but no further. Over time, this gives us an increasingly distorted picture of fish stocks in the past, especially of those before human impacts. In this context, there is also talk of so-called "generational amnesia", each generation forgetting what its predecessors' reference was. The "salmon-servant" anecdote fits seamlessly with this, although the time span in the anecdotes often exceeds that of a single generation. The salmon stocks were, according to the anecdotes, much larger "in ancient times", "in the past", "in the last century", regardless of the year in which the anecdote was written down. The advantage over the largely lost references of early fisheries biologists is that many anecdotes, probably because of their appealing story-telling nature, are during their time recorded in written sources dating back to as early as the 17th century. Collectively, the consecutive anecdotes portray a scenario of continuously dwindling salmon populations in large parts of Europe, although the "reference", the "salmon-servant" anecdote, is sometimes grossly exaggerated, probably to convince the reader that there was something seriously wrong with salmon stocks. Thus, for centuries, people have been aware of the decline in salmon stocks, but they did not realise that their "reference" was a reflection of already decimated salmon stocks. The reconstruction of salmon abundance in North-western Europe on the basis of quantitative and qualitative historical data demonstrates this.[46] Presently, it is almost inconceivable that the hundreds of thousands of salmon fished annually in the Rhine at the end of the 19th century represented less than 1% of medieval stocks. Europe's rivers must have been teeming with salmon before water mills, dams and weirs,

44 Janibal, 1695 (an unpublished chronicle, Staatsarchiv Hamburg, 731–1 Handschriftensammlung 80a), cited in Schwarz, "Der Weserlachs und die Bremischen Dienstboten"; Adelungk Wolffgang Henrich, *Kurtze Historische Beschreibung der Uhr-Alten Kayserlichen und des Heiligen Römischen Reichs Freyen-Ansee-Kauff- und Handels-Stadt Hamburg* (Hamburg, Conrad Neumann: 1696).

45 Pauly D., "Anecdotes and the Shifting Baseline Syndrome of Fisheries", *Trends in Ecology & Evolution* 10 (1995) 430.

46 Lenders – Chamuleau – Hendriks – Lauwerier – Leuven – Verberk, "Historical Rise of Waterpower"; Lenders, "Fish and Fisheries in the Lower Rhine"; Lenders H.J.R. – Verberk, W.C.E.P., "'Dar enkan noch Laeis noch Vijsch up gegayn'. De Atlantische Zalm in Noordwest-Europa in de Middeleeuwen en Vroegmoderne Tijd", *RAVON* 69 (2018) 26–29.

river regulation and overfishing took their toll. In the early Middle Ages, one might have witnessed, year after year, the migration of millions and millions of salmon on their way to their spawning grounds in the upper reaches of many European rivers. Our capacity of imagination cannot cope with such numbers of salmon anymore, but a distant memory of it remains in the dozens of anecdotes about the "salmon and the servant" handed down to us from generation to generation.

Bibliography

Adelungk Wolffgang Henrich, *Kurtze Historische Beschreibung der Uhr-Alten Kayserlichen und des Heiligen Römischen Reichs Freyen-Ansee-Kauff- und Handels-Stadt Hamburg* (Hamburg, Conrad Neumann: 1696).

Anonymous, *De Koopman of Bydragen ten Opbouw van Neêrlands Koophandel en Zeevaard. Tweede Deel* (Amsterdam, Gerrit Bom: 1770).

Anonymous, *Hamburgische Burspracken vom Jahre 1594. Nach dem Alten Originale, wornach dieselben Jährlich Abgelesen worden, mit Anmerkungen* (Hamburg: 1810).

Anonymous, "No title", *Abendzeitung*, January 29, 1824.

Baldner Leonard, *Vogel-, Fisch- und Thierbuch*. British Library, Add MS 6485.

Baude J.-J., "L'empoissonnement des eaux douces", *Revue des Deux Mondes* 31 (1861) 293–344.

Beverwijck Johan van, *'t Begin van Hollant in Dordrecht. Mitsgaders der Eerster Stede Beschrijvinge, Regeringe, ende Regeerders: als oock de Gedenckvvaerdige Geschiedenißen aldaer Gevallen* (Jasper Gorissz., Dordrecht: 1640).

Bouquet F., *Mémoires de Pierre Thomas, sieur du Fossé, publiés en entier, pour la première fois d'après le manuscrit original avec une introduction et des notes* (Rouen: 1878).

Breuers D., *Colonia im Mittelalter. Über das Leben in der Stadt* (Cologne: 2013).

Burema L., *De Voeding in Nederland van de Middeleeuwen tot de Twintigste Eeuw* (Assen: 1953).

Burgersdijk L.A.J., *De Dieren Afgebeeld, Beschreven en in hun Levenswijze Geschetst. Derde deel: Kruipende Dieren, Visschen, Ongewervelde Dieren* (Leiden: 1873).

Büsch J.G., "Verhandlungen der Gesellschaft über den Verfall der Fischereien unserer Gegend, besonders in der Elbe, und über die Mittel zur Verbesserung derselben", *Verhandlungen und Schriften der Hamburgischen Gesellschaft zur Beförderung der Künste und nützlichen Gewerbe* 4 (1797) 1–48.

Buyten L. van, "Données historiques sur le commerce de poissons à Louvain (Brabant, Belgique) au 18ème siècle et leur apport à l'archéozoologie", in Van Neer W. (ed.), *Fish Exploitation in the Past*. Annales de Musée Royal de l'Afrique Centrale, Sciences Zoologiques 274 (Tervuren: 1994) 151–161.

Dam P.J.E.M. van, "Feestvissen en Vastenvissen. Culturele, Ecologische en Economische Aspecten van de Visconsumptie in de Nederlanden in de Late Middeleeuwen", *Tijdschrift voor Sociale Geschiedenis* 29 (2003) 467–496.

Danker-Carstensen P., "Stör oder Lachs – aber auf Keinen Fall mehr als Zweimal in der Woche? Legendenbildung und Erzähltradition in einem Kapitel Deutscher Fischereigeschichte", in Pelc O. (ed.), *Mythen der Vergangenheit. Realität und Fiktion in der Geschichte* (Göttingen: 2012) 265–285.

Deelden C.L., "De Overvloedige Zalmvangsten", *Visserij* 35 (1982) 60–62.

Dickens C., "Concerning the Cheapness of Pleasure", *All the Year Round* (May 6, 1865).

Gemmil, E. – Mayhew, N., *Changing Values in Medieval Scotland. A Study of Prices, Money, and Weights and Measures* (Cambridge: 1995).

Halard X., "La pêche du saumon en Normandie de XIe au XVe siècle", *Journal of Medieval History* 9 (1983) 173–178.

Hoffmann R.C., "Economic Development and Aquatic Ecosystems in Medieval Europe", *American Historical Review* 101 (1996) 631–669.

Kuhn G., *Die Fischerei am Oberrhein* (Stuttgart: 1976).

Kuske B., "Der Kölner Fischhandel vom 14.–17. Jahrhundert", *Zeitschrift für Geschichte und Kunst* 24 (1905) 227–313.

Lenders H.J.R., "Fish and Fisheries in the Lower Rhine 1550–1950: A Historical-ecological Perspective", *Journal of Environmental Management* 202.2 (2017) 403–411.

Lenders H.J.R. – Verberk W.C.E.P., "'Dar enkan noch Laeis noch Vijsch up gegayn'. De Atlantische Zalm in Noordwest-Europa in de Middeleeuwen en Vroegmoderne Tijd", RAVON 69 (2018) 26–29.

Lenders H.J.R. – Chamuleau T.P.M. – Hendriks A.J. – Lauwerier R.C.G.M. – Leuven R.S.E.W. – Verberk W.C.E.P., "Historical Rise of Waterpower Initiated the Collapse of Salmon Stocks", *Scientific Reports* 6 (2016) no. 29269.

Lobregt P. – Os J. van, *De Laatste Riviervissers* (Heerewaarden: 1977).

Martens P.J.M., *De Zalmvissers van de Biesbosch. Een Onderzoek naar de Visserij op het Bergse Veld 1421–1869* (Tilburg: 1992).

Meier E.A., *z'Basel an mym Rhy. Von Fähren und Fischergalgen* (Basel: 2013).

Mieris Frans van, *Groot Charterboek der Graaven van Holland, van Zeeland, en de Heeren van Vriesland* (Leiden, Pieter van der Eyk: 1753–1756).

Oudenhoven Jacob van, *Out-Hollandt, nu Zuyt-Hollandt* (Dordrecht, Andriesz.: 1654).

Pauly D., "Anecdotes and the Shifting Baseline Syndrome of Fisheries", *Trends in Ecology & Evolution* 10 (1995) 430.

Pustelnik, G., *Les Saumons de la Rivière Dordogne. 1189–1981* (Paris – Sarlat: 1982).

Pustelnik G. – Roguet M. – Tinel C. – Soumastre J. – Roux M. – Simonet F., "Historique, cartographie écologique de la rivière Dordogne et évaluation de son potentiel d'accueil pour le saumon atlantique", in Thibault M. – Billard R. (eds.), *Restauration de rivières à saumons* (Paris: 1987) 53–64.

Rambach J.J., *Versuch einer Physisch-Medizinischen Beschreibung von Hamburg* (Hamburg: 1801).

Schwarz K., "Der Weserlachs und die Bremischen Dienstboten. Zur Geschichte des Fischverbrauchs in Norddeutschland", in Elmshäuser K. (ed.), *Bremisches Jahrbuch* 74/75 (Bremen: 1995–1996) 134–173.

Schwarz K., "Nochmals: der Lachs und die Dienstboten. Eine Nachlese", in Elmshäuser K. (ed.), *Bremisches Jahrbuch* 77 (Bremen: 1998) 277–283.

Storck A., *Ansichten der Freien Hansestadt Bremen und ihrer Umgebungen* (Frankfurt am Main: 1822).

Thibault M. – Garçon A.-F., "Un problème d'écohistoire: le saumon dans les contrats de louage, une origine médiévale?", in Benoit P. – Loridant F. – Mattéoni O. (eds.), *Actes des Premières Rencontres Internationales de Liessies 'Pêche et pisciculture en eau douce: la rivière et l'étang au Moyen-Age, 27, 28 et 29 avril 1998'* (Lille: 2004).

Thorold Rogers J.E., *A History of Agriculture and Prices in England from the Year after the Oxford Parliament (1259) to the Commencement of the Continental War (1793)*, volumes 2–4 (London: 1866–1882).

Unger W.S., *De Tol van Iersekeroord. Documenten en Rekeningen 1321–1572* (The Hague: 1939).

Vibert R., "Le saumon atlantique: origine et caractéristiques essentielles", in Gueguen J.C. – Prouzet P. (eds.), *Le saumon atlantique. Biologie et gestion de la ressource* (Brest: 1994) 11–25.

Vrielynck C. – Belpaire C. – Stabel A. – Breine J. – Quataert P., *De Visbestanden in Vlaanderen 1840–1950* (Groenendael-Hoeilaart: 2002).

Went A.E.J., *Atlantic Salmon: Its Future. The Proceedings of the Second International Atlantic Salmon Symposium, Edinburgh 1978* (Farnham: 1980).

CHAPTER 16

Public Opinion on Seals in Dutch Newspapers 1725–1900

Paul J. Smith

The relationship between seals and humans in the Netherlands has been studied from different perspectives, mainly in book-length studies addressing inland premium hunting,[1] arctic sealing,[2] and recent Dutch seal populations.[3] For a more direct approach to the Dutch perception of the seals, scrutinizing newspapers proves to be an invaluable instrument, facilitated by the recent availability of the website Delpher. Delpher was created by the Dutch Royal Library, providing full-text Dutch-language digitized historical newspapers.[4] About 1 million newspapers are available from the period between 1618 and 1995 (which represents about 10% of the total published newspapers in this period).

1 Method

The period under discussion in this case study runs from 1725 to 1900. The year 1725 was chosen because in this year the term *zeehonden* (seals) was used for the first time in the digitized Delpher newspapers. The year 1900 was chosen as end date, because in this year a new era started with the introduction of a country-wide premium system on shooting seals. As a search term I opted for the Dutch plural *zeehonden* (seals, literally "sea-dogs"), and not for the singular *zeehond* (seal), not only because of the large number of hits (7,453) but also because an important part of these hits pertains to the proper name *Zeehond*,

1 't Hart P., *Zeehondenjacht in Nederland, 1591–1962*, PhD dissertation, Free University Amsterdam, Amsterdam, The Netherlands, 2007.

2 Dekker P., *De laatste bloeiperiode van de Nederlandse arctische walvis- en robbevangst 1761–1775* (Zaltbommel: 1971); Schokkenbroek J.C., *Harpoeniers en robbenjagers. Nederlandse walvisvaart en zeehondenjacht in de 19de eeuw* (Zutphen: 2008).

3 Brasseur S.M.J.M., *Seals in motion. How movements drive population development of harbour seals and grey seals in the North Sea*. PhD dissertation, Wageningen University, Wageningen, The Netherlands, 2017.

4 https://www.delpher.nl/.

a favourite name for vessels, such as the Dutch warship De Zeehond, which is mentioned weekly from 1785 to 1797, or the marine 'instructiebrik De Zeehond' (instruction brig The Seal), which has been mentioned almost daily in the newspapers' shipping reports since 1850. I therefore only searched incidentally and in a non-systematic manner, using the singular *zeehond* as the search term. Nor were the common synonyms of *zeehond*, such as *rob* or *zeerob*, chosen as search terms, because these words have different meanings: *Rob*, derived from *Robrecht*, is a common proper name for persons, and *rob* and *zeerob* frequently have the derived meaning of "sailor". The search term *zeehonden* yields 1,786 hits in the Delpher corpus of 18th- and 19th-century newspapers. References to newspapers in this article are always to the newspaper's title and publishing day (date indication: dd-mm-yyyy), making it possible to easily find the referenced newspaper on the Delpher website.

Quantitative and qualitative analysis of these hits is complicated by two factors:

(a) terminological ambiguity. At first sight, there seems to be no ambiguity in the use of the term *zeehonden*, because the only species living in the Dutch coastal waters was the harbour seal (*Phoca vitulina*), of which the males can weigh 70 to 150 kg, and the females 60 to 110 kg. The much larger grey seal (*Halichoerus grypus*), weighing 105 to 186 kg for the females and 170 to 310 kg for the males,[5] disappeared from the Dutch waters starting in the early Middle Ages, to return only in the late 20th century. However, in some cases, as with the indicated weight of 200 or 250 kg,[6] this could concern a grey seal. Moreover, the Dutch newspapers show that the term *zeehond* is ambiguous, indicating, incidentally, very different animals, such as the otter (*Lutra lutra*)[7] or the small-spotted catshark or sandy dogfish (*Scyliorhinus canicula*).[8] More common was the confusion between seal (*zeehond*) and sea lion (*zeeleeuw*), fur seal (*zeebeer*), and other *phocidae* and *otariidae* – all these species that do not occur in the Netherlands but are regularly mentioned in the news regarding international sealing, are invariably referred to as *zeehond*. The papers were clearly behind the scientific knowledge of these animals. Only since 1875 have some newspapers regularly distinguished between seals and sea lions.[9] In 1881 one finds the first Dutch newspaper distinction between four

5 Information from http://www.soortenbank.nl/index.php.

6 *Algemeen Dagblad* 12-07-1889; *Rotterdamsch Nieuwsblad* 04-03-1899.

7 At least in the Dutch East Indies: *Bataviaasch handelsblad* 25-08-1886; *Soerabijasch handelsblad* 16-11-1886.

8 *Provinciale Noordbrabantsche en 's Hertogenbossche courant* 15-09-1890.

9 *Provinciale Noordbrabantsche en 's Hertogenbossche courant* 19-04-1877, see also *De standaard* 09-04-1878.

Greenlandic species, with their Dutch and scientific names: *zadelrob* or harp seal (*Phoca groenlandica*, now *Pagophilus groenlandicus*), *klapmuts* or hooded seal (*Cystophora cristata*), *baardzeehond* or bearded seal (*Phoca barbata*), and the *ijsschotsen-zeehond* (no scientific name was given – probably the ringed seal [*Phoca hispida*]).[10] However, these are exceptions: the confusion between the different species persisted well into the 20th century.

(b) absence of instruments for analysing news impact. For newspapers before 1900 media studies did not develop a method of analysis permitting the quantification and qualification of the impact of a newspaper message (assuming its length, place in the newspaper [front page or elsewhere], section [state news, financial news, fishing reports, local news, mixed news, letters, advertisement], target group of intended readers [local, regional, national, or colonial newspapers], publication frequency [daily, every two days, weekly, etc.], and quotation of news from or by other newspapers).

Therefore, as newspaper analysis is difficult to apply, this present article limits itself to giving a general overview of the main newspaper information on seals. The main topics in the news coverage are: (a) international sealing; (b) domestic seal hunting; (c) inland seal spotting; and (d) public perception of seals (empathy and amusement).

2 International Sealing

The coverage in 18th-century media did not pay attention to the Dutch local seal population. One even has the impression that seals were relatively unknown to the 18th-century readership, even though the word *zeehond* in Dutch has been attested since 1293.[11] In the first release on the seal hunt to Greenland, the newspaper felt obliged to give a definition of the animal: 'Het is een dier, 't welck soo wel op 't Land als in Zee left, doorvoed is met Vis, maer aen 't hooft en verscheyde andere gedeeltens seer na een Hond gelijckt' (It is an animal that lives both on land and in sea, it feeds on fish, but, as for the head and other body parts, it is very similar to a dog).[12] The media coverage was exclusively focussed on international arctic sealing, especially in Greenland[13] but also in Iceland and other Nordic areas. The Dutch had traditionally been a people of whaling and sealing, but that period had long since passed. One newspaper

10 *Provinciale Overijsselsche en Zwolsche courant* 04-10-1881.

11 Sijs N. van der, *Chronologisch woordenboek: de ouderdom en herkomst van onze woorden en betekenissen* (Amsterdam: 2002 (2nd edition)) 150 (s.v. *zeehond*).

12 *Oprechte Haerlemsche courant* 07-08-1725. All translations from the Dutch are mine.

13 For example, *Amsterdamse courant* 30-08-1732; *Leeuwarder courant* 25-07-1759; *Oprechte Haerlemsche courant* 11-03-1760; *Oprechte Haerlemsche courant* 22-07-1760.

article looked back with nostalgia to the time when czar Peter the Great sent his seal hunters to the Dutch for instruction – an initiative that was doomed to failure for political reasons but which could open the door for renewed sealing agreements with Russia in 1893. As the newspaper sums it up:

> In 1732 begonnen de Russen onder leiding van deze Hollanders te visschen. De uitslag beantwoordde evenwel niet aan de verwachting. Jaren achtereen werd er zeer weinig gevangen. De Russen wisten wel den walvischen en den zeehonden met harpoenen zware wonden toe te brengen, maar in de meeste gevallen ontsnapten de dieren. In 1731 werd de zaak opgegeven. De Hollandsche visschers, die als leermeesters aangesteld waren geweest, werden ontslagen en gingen naar hun land terug. Sommige personen uit de omgeving van [de] czaar beweerden, dat die Hollandsche leermeesters zich hadden laten omkoopen door de Hollandsche walvisch-maatschappij en dat zij den Russen het vak niet goed leerden, zodat dezen niets vingen.
>
> Czaar Peter was trouwens reeds in 1725 overleden en zijn opvolgsters en opvolgers waren niet in zulke mate met de Hollanders ingenomen, als hij dit was geweest. Zeker is het evenwel dat de Hollanders als visschers in de noordelike zeeën een groote rol gespeeld hebben. Mocht Rusland er nog eenmaal toe besluiten aan buitenlanders concessiën te verleenen betreffende de visscherij in de noordelijke wateren, dan zouden de Nederlanders daartoe meer aangewezen zijn dan de Franschen of Engelschen.[14]

> In 1732 the Russians started fishing under the leadership of these Dutchmen. However, the result did not meet expectations. For years in a row, very few fish were caught. The Russians managed to inflict severe wounds on the whales and seals with harpoons, but in most cases the animals escaped. In 1731 the business was abandoned. The Dutch fishermen, who had been appointed as tutors, were dismissed and returned to their country. Some people in the czar's circle claimed that those Dutch tutors had let themselves be bribed by the Dutch whaling society and that they did not teach the Russians the trade well, so that they caught nothing.
>
> By the way, Czar Peter had already died in 1725 and his successors and successors were not as pleased with the Dutch as he had been. It is certain, however, that the Dutch played a major role as fishermen in the northern seas. Should Russia once more decide to grant concessions to

14 *Algemeen Handelsblad* 29-01-1893.

foreigners regarding fishing in the northern waters, then the Dutch would be better qualified for this than the French or English.

Noteworthy are the precision and frequency with which sealing yields and revenues from abroad are listed by the Dutch newspapers.[15] A good example is the detailed rendering in the form of a table of the revenues of 10 English sealing ships – with a total revenue of 41,000 guilders in two months.[16] In this case, providing these details had an explicit goal: a recommendation to create a Dutch Nordic Company for whaling and sealing. But not every report was positive, for newspapers also paid attention to the risks, both financial and human: the great number of shipwrecks and casualties among the sealers was reported regularly.[17] This is the detailed coverage of such a disaster:

> Op haar zeehondenvangst, terwijl de bemanning zich op de ijsschotsen bevonden, stak een verblindende sneeuwstorm op, vergezeld van strenge koude, waarin 25 man het leven verloren. 23 anderen werden vermist en 66 leden zoo door de koude dat ongeveer 20 hunne handen en voeten bevroren en dezelve afgezet zullen moeten worden.[18]
>
> While the crew [of the sealing ship] was on the ice floes, a blinding blizzard arose, accompanied by severe cold, in which 25 men lost their lives. 23 others were missing and 66 suffered from the cold so much that about 20 of them had their hands and feet frozen, which have to be amputated.

Mainly for financial, economic, and political reasons, Dutch readers are concerned, for example, with the seals on Jan Mayen Island, where the seal population risked extinction by overhunting, in particular by the British. This concern fitted well into more general thoughts on eradication, as can be seen in a review article of one of the publications by the American zoologist Joel Asaph Allen.[19] New to the discussion is that those in favour of stopping

15 *Algemeen Handelsblad* 18-08-1884; idem 26-06-1685; idem 19-11-1886 (e.g., '1,879 big seals for 22 krona, 7,578 lesser seals for 6.5 krona').

16 *De standaard* 08-06-1876.

17 *Nederlandsche staatscourant* 30-05-1872: 'New York, 28 Mei. Vier stoomboten en omstreeks 40 zeilvaartuigen, die op de zeehonden vangst uit waren, zijn op de kust van Labrador vergaan. De bemanning dier vaartuigen is daarbij omgekomen' (New York, May 28. Four steamers and about 40 sailing craft, seeking seals, were lost on the Labrador coast. The crew of those vessels perished). See also *De Volksstem* 16-03-1892; *Nieuwsblad van het Noorden* 23-08-1893.

18 *De Volksstem* 30-03-1998.

19 *De Volksstem* 21-09-1898. It is not specified which work by Allen is meant here.

or regulating the seal hunting also proposed modern-style arguments, such those hinging on as animal cruelty. A newspaper mentioned that when the Jan Mayen adult seals are slaughtered, the young ones start to cry like children, 'hartverscheurend, volgens 't getuigenis der waarlijk niet overgevoelige robbenjagers zelf' (heart-breaking, according to the testimony of the not really hypersensitive sealers themselves).[20]

The Dutch readers were very interested in the international developments around the Bering Sea conflict. In 1867 the US bought Alaska from Russia. From that moment on the US considered the Bering Sea as a *mare clausum*, in which they could freely fish and hunt – which went against the British interests in Canada. Not only were the political and legal aspects of the Bering Sea conflict reported and discussed (who was right in this conflict?),[21] but the economic and ecological aspects of seal hunting gained a lot of attention as well. Important names from abroad were quoted in the Dutch press, particularly Henry Wood Elliott (1846–1930), who would later become famous as a champion of the regulation of seal hunting in the Bering Sea. From his book *An Arctic Province. Alaska and the Seal Islands* (1886) passages were quoted addressing the endangered life of the Northern fur seal (*Callorhinus ursinus*) [Fig. 16.1]. A Dutch reviewer remarked: 'Het slachten dezer dieren is zóó afschuwelijk, dat menige dame, na het lezen der beschrijving in Eliott's boek, geen robbevel meer zal dragen' (The slaughter of these animals is so horrible, that many a lady will wear no seal fur anymore, after reading this description in Elliott's book).[22] Newspaper readers were horrified by the cruel details presented with heavy irony by the Norwegian polar explorer Carsten Egeberg Borchgrevinck (1864–1934):

> Het vangen, dooden en villen der zeehonden is een zeer interessante en aangename bezigheid, die in het bijzonder aan te bevelen is voor menschen met een medelijdend hart. Slechts zelden sterft een zeehond van de twee of drie slagen, welke men hem toebrengt. Al is het dier echter nog niet geheel dood, men stroopt het toch de huid af. De robbenvangers beweren zelfs dat het villen het gemakkelijkst gaat, zoolang het dier nog half in leven is en voelt wat er met hem gebeurt. In zijn verschrikkelijke lijden trekt de zeehond alle spieren samen en vergemakkelijkt daardoor onwillekeurig den wreeden jager het werk.[23]

---

20 *Het nieuws van den dag* 13-03-1875.

21 *Algemeen Handelsblad* 23-12-1890; *De Volksstem* 20-04-1892.

22 *Algemeen Handelsblad* 03-04-1887.

23 *Rotterdamsch nieuwsblad* 27-11-1899.

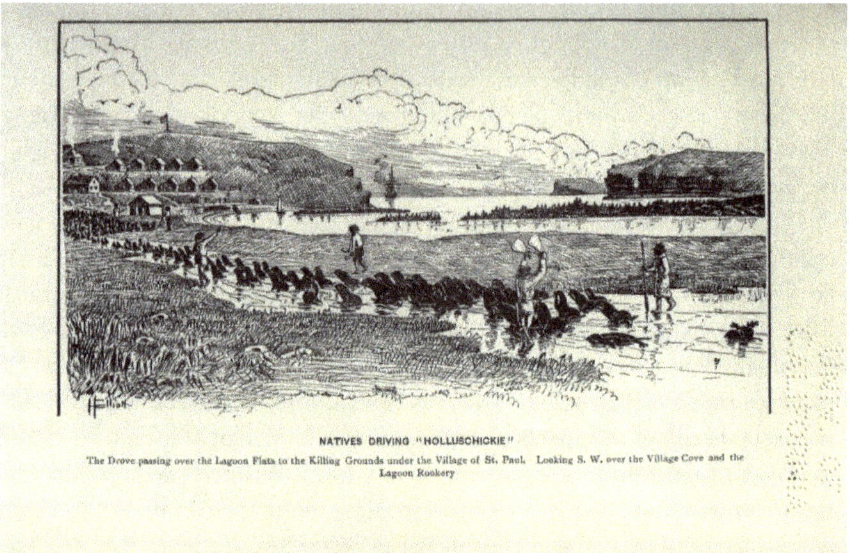

FIGURE 16.1 Henry Wood Elliott, *An Arctic Province. Alaska and the Seal Island* (New York: 1906 [1886]) 336

> The catching, killing, and skinning of the seals is a most interesting and pleasurable activity, which is especially recommended to those of a compassionate heart. Rarely does a seal die of the two or three blows inflicted upon it. Although the animal is not yet completely dead, the skin is still peeled off. The seal-catchers even claim that the skinning is easiest as long as the animal is still half alive and feels what is happening to it. In its terrible suffering the seal contracts all its muscles and thereby involuntarily facilitates the work of the cruel hunter.

Indeed, in the 1890s Dutch readers realized with a shock that seals were often skinned alive.[24] In reaction to an ad in which a manufacturer in glacé-gloves asked for 'ongeboren huidjes' (unborn skins), a long and indignant letter was published about the cruel skinning of unborn baby seals.[25]

The Dutch newspapers also presented detailed analyses of the final report of the International Arbitration Committee, presided over by Elliott, leading to the treaty which finally stopped the conflict.[26] The problematic implementation of

24 *De locomotief: Samarangsch handels- en advertentie-blad* 04-07-1891.

25 *Rotterdamsch nieuwsblad* 06-12-1900.

26 *Algemeen Handelsblad* 20-05-1893. The report, the newspaper emphasizes, is based on counts: 'In 1873 waren er 3,192,000, in 1890 slechts 959,455 zeehonden op de [Pribyloff-]eilanden, zodat hun aantal in 16 jaren met 2,232,545 verminderd was.' (In 1873

the treaty also got attention.[27] The measure taken by the US to mark fur seals with a glowing rod, so that their skin became unsalable for English sealers, was likewise mentioned in the Dutch newspapers.[28] Interesting in the Dutch colonial context is that the Bering Sea issue was connected to pearl hunting by Australians in the Dutch Indies, as stated in an anonymous main article titled "Paarloesters en zeehonden" (Pearl Oysters and Seals), in the colonial journal *Java-bode*:

> Sedert het vorig jaar vertoonen zich in den Indischen archipel Australische schepen, die er naar paarloesters komen visschen. Vroeger deden zij dit uitsluitend langs de kust van Nieuw-Holland, doch daar hebben zij het terrein leeggevischt, en wanneer men hen hun gang laat gaan, zullen zij met de streek, waar zij nu visschen hetzelfde gedaan hebben, daar zij alles weghalen, groot en klein, wat onder hun bereik komt en zoodoende de paarloester geheel uitroeien.[29]

> Since last year, Australian ships have been showing up in the Indian archipelago to fish for pearl oysters. They used to do this exclusively along the coast of New Holland, but there they have fished out the grounds, and if they are allowed to do their thing, they will have done the same in the way they fish now, as they take everything away, large and small, which comes under their reach and thus completely exterminate the pearl oyster.

3 Domestic Seal Hunting

From 1811 the first reports appeared in Dutch newspapers about the harmful influence of the seal on fish stocks, first in the foreign news about Hamburg[30] and then about Copenhagen.[31] Shortly afterwards, the discussion reached the Netherlands: The native harbour seal became the subject of a pricing competition: "Over de schadelijke vermeenigvuldiging der Zeehonden op de Zeeuwsche Kusten; de daardoor vermoedelijke vermindering van Vischvangst, en de middelen daar tegen aen te wenden" (On the harmful multiplication of seals on the Zeeland coasts, the presumed reduction in fish catches as a result,

---

there were 3,192,000, in 1890 only 959,455 seals in the [Pribyloff] Islands, so that in 16 years their numbers were reduced by 2,232,545). See also *Algemeen Handelsblad* 28-05-1893.

27 *Leeuwarder courant* 17-08-1893.

28 *Bataviaasch nieuwsblad* 06-09-1897.

29 *Java-Bode. Nieuws-, Handels- en Advertentieblad voor Nederlandsch-Indië* 21-09-1893.

30 *Advertentieblad van het Departement Wester-Eems* 04-07-1811.

31 *Utrechtsche courant* 12-08-1816.

and the means to be used against it).[32] From about 1820 there was a discussion in the newspapers about related aspects, first of all in the practical context of the regulations on the premiums, particularly in Zeeland.[33] Newspapers from those days inform us that the premium on every dead seal was 3 guilders (3 guilders being the equivalent to a week's salary for a rural worker). The newspapers published the regulations: the premium was paid if, in the presence of an official, the dead animal's 'snuit' (snout) was chopped off. Then, starting in 1856, when the premium was abolished,[34] a vibrant discussion began on seal hunting and the reintroduction of the premium. There were further price competitions published in order to find a solution for the seals.[35] Almost every year there were complaints from fishermen, in almost identical terms, first from the fishermen of Zeeland, and then from those of the Zuiderzee (Urk and Lemmer). For modern and contemporary readers as well it is difficult to evaluate the amount of exaggeration in typical recurrent notifications, such as:

> Zij zwemmen bij honderden achter de schuiten, verscheuren de netten en vernielen de vangst, zoodat met genoodzaakt is reeds tegen den morgen in de haven terug te keeren. [...] Men begroot het getal zeehonden, dat wel eens op het zelfde oogenblik nabij het eiland Urk is waargenomen, op meer dan 1000 stuks. Iedere hond verslindt dagelijks wel 200 haringen.[36]

> they swim at hundreds behind the barges, they tear up the nets and destroy the catch, so that the fishermen are forced to return to the port already before daybreak. [...] The number of seals, spotted at the same time near the island of Urk, is even estimated at more than 1,000 specimens. Each seal devours daily 200 herrings.

In this case, the daily damage caused by the seals is estimated to be 4,500 guilders. Other newspapers reported seals in which 300 herrings were found.[37] The Zuiderzee fishermen were obliged to fish at night without light – which was

32 *Rotterdamsche courant* 23-01-1817.

33 Vooys C.G.N. de – Brasseur S. – Meer J. van der – Reijnders P.J.H., "Zeehondenjacht in Zeeland: het effect van premies", *Lutra* 17.3 (2012) 55–65.

34 *Opregte Haarlemsche Courant* 30-01-1857.

35 For example, *Opregte Haarlemsche Courant* 09-08-1850; *Nederlandsche staatscourant* 08-11-1853.

36 *Provinciale Overijsselsche en Zwolsche courant* 30-01-1872.

37 *Provinciale Overijsselsche en Zwolsche courant* 13-01-1862 (information repeated at least five times in other newspapers); see also *Algemeen Handelsblad* 16-03-1876.

prohibited by law – in order not to draw the attention of the seals.[38] The seals were so frequent that their very exceptional (and inexplicable) absence in the years 1862 and 1878 was also reported in the newspapers.[39] In perspective relativization of the problem was rare, for example in this report, which mentions four different causes for the disappearance of the herring, without saying which is correct: (1) there are too many fishermen; (2) there are too many seals; (3) there is the 'disturbance by steamers'; and (4) fishermen use so-called miracle trawl nets with too fine a mesh.[40] Rarely, indeed, were the arguments of both proponents and opponents of the premium hunt given.[41] From ca. 1880 it was reported that not only did the herring and flounder fishery suffer damage from the seals, but the salmon fishery did as well, especially in Zeeland. More and more salmons were found that were gnawed by seals, and therefore worth less in sale.[42]

All this led to even more actions in the 1890s. The petitions of fishermen from Urk and Zeeland made news,[43] ultimately leading to the reintroduction of the premium system (one dead female seal, 3 guilders; one dead male, 2.50 guilders). The official entry form, to fill in by the premium hunters, was published in the Dutch Government Gazette.[44] From this moment on, the news reports on seals focussed mainly on the practical consequences of the implementation of the premium: the problems with payment[45] and the designation of the municipalities where the premiums could be collected.[46] Also, the first results of the premiums were reported and discussed.[47]

Inland seal hunting was presented as adventurous and exciting, as can be read in the articles of reporters who accompanied the sealers on their hunt.[48] Iconic for the seal's presence in the Dutch press was the news coverage of the Visser family living in Westernieland (Groningen). This family took on almost

38 *Algemeen Handelsblad* 29-12-1898.

39 *Algemeen Handelsblad* 29-03-1878.

40 *Rotterdamsche courant* 22-10-1862. See also *De Gooi- en Eemlander: nieuws- en advertentieblad* 31-01-1874, which proposes the very cold winter as a possible cause for the disappearance of the seals.

41 *De grondwet* 09-06-1885 and *Algemeen Handelsblad* 21-08-1885. Against the arguments of the fishers, opponents of the premium hunt argue that sealing is of primary importance for the professional seal hunters, such as the Visser family (see below).

42 *Algemeen Handelsblad* 19-02-1893; *Rotterdamsch nieuwsblad* 18-01-1898.

43 *Algemeen Handelsblad* 03-03-1898; *Rotterdamsch nieuwsblad* 20-12-1898.

44 *Nederlandsche staatscourant* 29-05-1900.

45 *Middelburgsche courant* 26-06-1900.

46 *Middelburgsche courant* 09-07-1900.

47 E.g., *Middelburgsche courant* 12-07-1900.

48 *Algemeen Handelsblad* 02-10-1887; *De Telegraaf* 22-07-1899.

heroic proportions in the press because of their ingenuity to survive in harsh conditions. Dutch readers seemed to recognize in this family their own idealized heroic past. The poetic epitaphs of two of their ancestors (1779–1871 [Fig. 16.2][49] and 1815–1873[50]) were quoted in order to highlight the respectable history of their profession.[51]

The numbers of seals caught annually were reported[52] – for example, it is reported that, at the introduction of the premium system, the Visser family earned a premium sum of 500 guilders.[53] Their inventiveness was praised: the seals were not only killed for skins and train: 'de gedroogde vellen worden verkocht op Borkum, en te Bremen met groote winst verwerkt tot vloekleedjes, randen m petten, geldbeursjes, sigarenpijpjes enz., welke voorwerpen dan o.a. naar Nederland worden uitgevoerd' (the dried skins are sold at Borkum [an island in north-western Germany], processed for great profit in Bremen into rugs, trims for caps, money pouches, cigar pipe holders etc, objects that are exported to the Netherlands).[54] The Visser family hunted originally with clubs, but from 1898 on they developed a wide mesh net, the exact fabrication of which was jealously kept secret.[55] From ca. 1880 the Visser family organized hunting parties with rifles in the Wadden area. People from across the country, particularly army officers, participated,[56] as did some from Britain.[57] The

49 'Tot aan zijn vijf en vijftig Jaren,
Heeft hij de wadden steeds bevaren,
Zeehonden vangen was zijn werk.
Bij vlijt en zuinigheid zoo sterk,
Om zes en dertig jaar alhier
Te leven als een rentenier'
Up to the age of 55 he always sailed the Wadden. His job was to hunt seals. So that through diligence and strong frugality he could live here (in Westernieland) for 36 years as a rentier.

50 'Ik was schipper op het Wad,
En ontving geen koude of nat.
Daar ging ik menig zeehond slagen.
Daar redde ik menschen van den dood,
Tot dat een ziekte op kwam dagen,
En mij de dood mijne oogen sloot.'
I was skipper on the Wad, and I didn't get cold or wet there. There I went to kill many a seal. I saved people from death there. Until a disease came and death closed my eyes.

51 *De Telegraaf* 04-08-1898.

52 *Provinciale Overijsselsche en Zwolsche courant* 15-08-1870; *Algemeen Handelsblad* 05-10-1898.

53 *Algemeen Handelblad* 02-12-1900.

54 *Provinciale Drentsche en Asser courant* 31-10-1888.

55 *De Telegraaf* 27-07-1898; idem 04-08-1898.

56 *Nieuwsblad van het Noorden* 08-08-1896.

57 *De Maasbode* 15-07-1897.

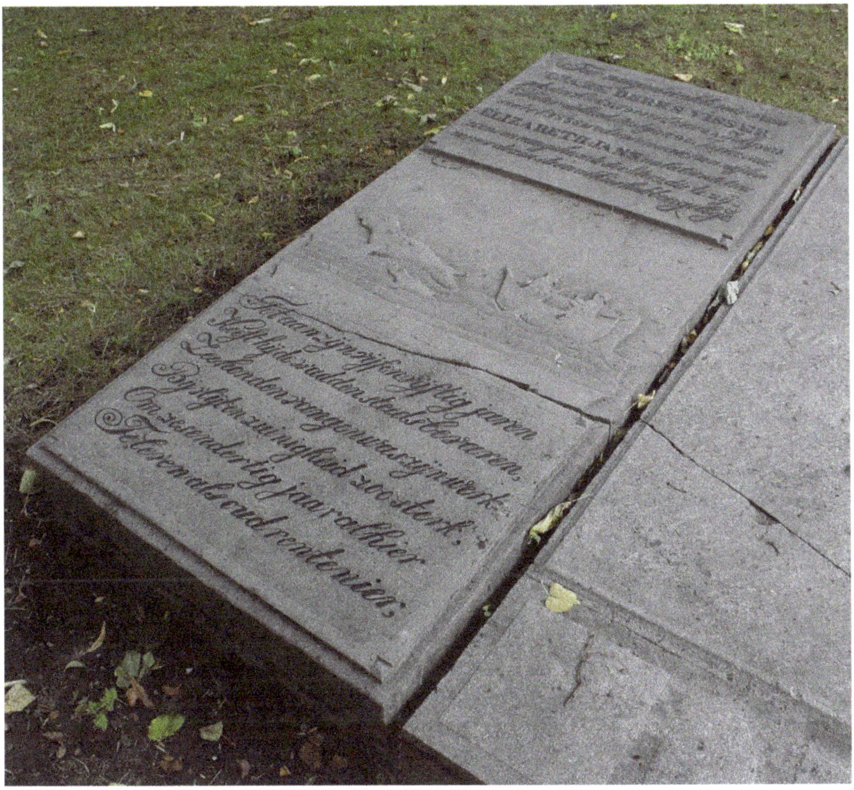

FIGURE 16.2 Tombstone of Tjark Derks Visser
PHOTO: HTTPS://WWW.DEVERHALENVANGRONINGEN.NL/ALLE-VERHALEN/ROBBENJAGERS-EN-MONNIKENWERK-IN-WESTERNIELAND

shot animals were skinned on the spot, and the skins were salted and given to the hunters as a hunting trophy.[58] The Visser family also caught living animals, which were sold as pets.[59] They were also sold or given to zoos (the Artis Zoo in Amsterdam, and the zoos of Cologne and Basel)[60] and to the University of Groningen. At the instigation of a certain G. Steenhuis, who also lived in Westernieland and who published well-argued letters in the national press,[61] a contact was made between the Visser family and G.F. Westerman, the director and co-founder of the Artis Zoo, who was enthusiastic about the idea of

58 *Provinciale Drentsche en Asser courant* 05-06-1898.

59 *Het nieuws van den dag: kleine courant* 16-01-1891 (advertisement).

60 *Het nieuws van den dag: kleine courant* 30-06-1897; *De standaard* 25-06-1885.

61 E.g., *Algemeen Handelsblad* 04-02-1898; *De Tijd: godsdienstig-staatkundig dagblad* 14-12-1898.

processing seal meat for animal nutrition.[62] The plan failed because the distance between Westernieland and Amsterdam was too great to preserve the meat of the seals, killed in the summer.[63] A counterpart of the Visser family was the Van de Klooster family from Burghsluis (Zeeland). This family was much less known, but interestingly, their capture of 118 seals in 1899 was reported in great detail: 41 of the captured seals were sold alive to Belgium, 12 to the Netherlands, and 8 to France; 57 were killed for seal oil.[64]

In some cases the Dutch seal became an object of dispute with neighbouring Germany. The Germans attributed the decline of the salmon to the large numbers of seals in the Dutch coastal waters, while the Dutch blamed the German nurseries of pike-perch (sander) on the great rivers: they believed that these sanders fed upon young salmon.

Another example concerns cases of food-fraud when seal meat was sold as pork;[65] later it became clear that this meat was imported from the Netherlands.[66] However, according to a letter from Steenhuis, published in several newspapers, it is difficult to sell seal meat as pork: this is only possible with very heavy animals, which are non-existent in the Netherlands. He therefore suspected that the German reports were fake news, aiming to discourage the Dutch import of pork into Germany.[67]

4 Inland Seal Spotting

The local press frequently reported inland sightings and catches of seals (especially on the large rivers, the Rhine, the Lek, and the Maas, and in the Zeeland ports). Invariably it was said that those sightings were rare.[68] Once in a while a meteorological explanation was suggested, for instance stormy weather.[69] And often the violent reaction of the local people was described, who hunted down the spotted seal 'out of antipathy towards the animal'.[70] Sometimes, however, the animal was captured alive to be displayed as a curiosity, as happened with

62 Ongoing reports in *Rotterdamsch nieuwsblad* 24-10-1888, *Provinciale Drentsche en Asser courant* 31-10-1888, *De standaard* 17-01-1889.

63 *Algemeen Handelsblad* 15-09-1895.

64 *Rotterdamsch nieuwsblad* 29-09-1899.

65 *Het nieuws van den dag: kleine courant* 23-07-1895.

66 *De Tijd: godsdienstig-staatkundig dagblad* 12-09-1895.

67 *Algemeen Handelsblad* 15-09-1895.

68 *Leeuwarder courant* 02-03-1776; *Groninger courant* 25-02-1825; *De Tijd: godsdienstig--staatkundig dagblad* 06-03-1885; *Provinciale Noordbrabantsche en 's Hertogenbossche courant* 03-0-1892; *Provinciale Overijsselsche en Zwolsche courant* 14-12-1895.

69 *Het nieuws van den dag: kleine courant* 17-11-1881; idem 29-11-1881.

70 *De standaard* 15-03-1878.

a seal caught in the Rhine near Xanten (Germany).[71] During several weeks in 1899 two seals that escaped from a basin in Liège (Belgium) made the news. The animals swam for days in the Meuse. The fishing tenants, who feared for their fish stocks, were roused to action. The hunted animals were driven into a lock, and one was captured alive by way of a net made of bed linen sewn together. The second escaped but was later shot dead. The incident ended tragically: one of the hunters hit the water and drowned. A watching lock keeper died of a stroke.[72]

5 Empathy and Amusement

The image of the seal, as is apparent from the above points, is two-fold: the seal as an object of the hunt and as a competitor of man. In the course of the 19th century other attitudes of the public in relation to seals can be seen. In addition to the above-mentioned reactions against animal cruelty in the case of sealing in the Bering Sea and on Jan Mayen, the reports on seal hunting on the island of Lobos must also be mentioned.[73] All these reactions were consistent with an empathy for seals, and even a certain humanization of these animals. An article noted the similarities between a human head and a seal's.[74] A feuilleton talks about the seals' 'groote, ronde zwarte, vochtig glanzende oogen' (great, round, black, moist, shining eyes).[75] Critical for the positive image regarding the seal are the news reports on the intelligence of these animals. In the first half of the 19th century, this coverage was still incidental. A newspaper article mentioned a monk seal in Italy that was trained and could speak a few words: dad, mom, and thank you.[76] Another report is about seals trained to draught vessels.[77] Occasionally there were reports of tame seals, both abroad (a tame Baikal seal)[78] and locally.[79] Domesticated young seals were sometimes offered for sale per ad.[80] An American ordered young seals from Dutch fishermen in order to teach them tricks.[81] Reports were published repeatedly about an

71 *Algemeen Handelsblad* 19-07-1831.

72 Continuous news coverage: *Middelburgsche courant* 06-05-1899; *Venloosch nieuwsblad* 06-05-1899; *Provinciale Noordbrabantsche en 's Hertogenbossche courant* 17-05-1899; *Middelburgsche courant* 18-05-1899; idem 20-05-1899; idem 26-05-1899.

73 *Rotterdamsch nieuwsblad* 13-03-1897.

74 *De Curaçaosche courant* 03-08-1833.

75 *Rotterdamsch nieuwsblad* 03-04-1884.

76 *De Curaçaosche courant* 03-08-1833.

77 *Vlissingsche courant* 02-02-1844.

78 *De Sheboygan Nieuwsbode* 09-06-1857.

79 *Het nieuws van den dag: kleine courant* 24-08-1883.

80 *Het nieuws van den dag: kleine courant* 27-07-1886.

81 *Provinciale Drentsche en Asser courant* 01-06-1895.

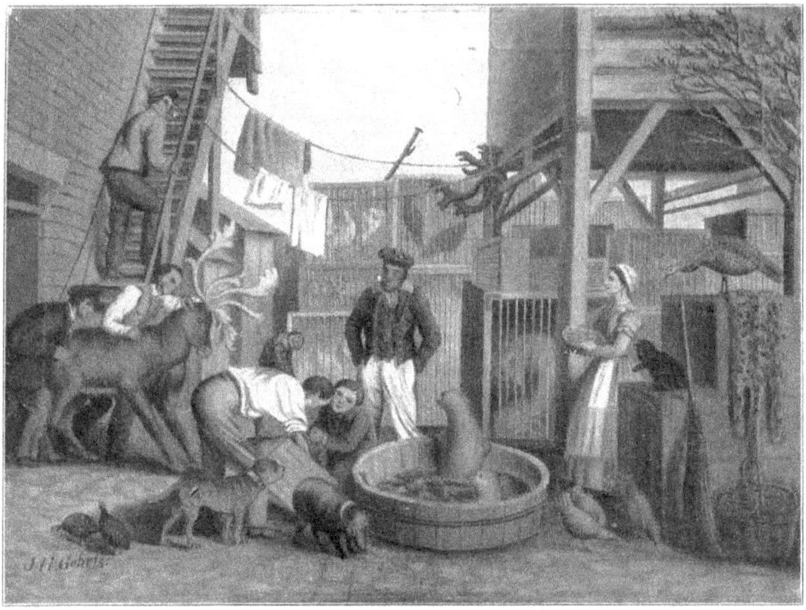

FIGURE 16.3 J.H. Gehrts after Heinrich Leutemann, "Bad der Seehunden", in C. Hagenbeck, *Von Tieren und Menschen* (Berlin: 1908) 51 https://upload.wikimedia.org/wikipedia/commons/f/fc/Hagenbeck4.jpg

Englishman (one Lord Murphy) who hunted in the Dollard with seals, dressed to retrieve the culled fowl.[82]

From ca. 1860 there are recurring news reports about performances with tamed seals,[83] first abroad in the major European cities of London and Vienna.[84] In Germany the animal dealer Carl Hagenbeck (1844–1913) organized variété programmes with dressed seals; other companies followed.[85] In 1887, the first seal performance in the Netherlands was reported.[86] Later circus companies had great success with seals in the Netherlands. In 1892 Hagenbeck's shows received much publicity and attention in the Netherlands.[87] In a long interview Hagenbeck told how he had come to his current profession: thanks to three seals that his father had given to him when he was a boy [Fig. 16.3].[88]

82 *Het nieuws van den dag: kleine courant* 09-09-1884.
83 *Rotterdamsch nieuwsblad* 26-01-1885.
84 *Provinciale Noordbrabantsche en 's Hertogenbossche courant* 11-06-1885; *Bataviaasch handelsblad* 09-04-1887.
85 *Provinciale Overijsselsche en Zwolsche courant* 14-09-1885.
86 *Rotterdamsch nieuwsblad* 12-05-1887.
87 E.g., *Rotterdamsch nieuwsblad* 31-08-1892; *Algemeen Handelsblad* 09-09-1892.
88 *Nieuwsblad van het Noorden* 28-05-1895.

Several reviews reported the tricks performed by the seals: they made music, sang, juggled, smoked pipes, etc. These tricks showed that in almost all cases the animals concerned were in fact sea lions rather than seals. In 1899 for the first time it was clearly stated that a show with Captain Webb was with 'sea lions and seals' [Fig. 16.4].[89] Another successful American seal trainer, Captain Wills, put on many shows in the Netherlands before entering the service of the Barnum and Bailey circus.[90] In 1900 an interview was published in which he explained the differences between seals and sea lions (i.e. the California sea lion (*Zalophus californianus*)).[91]

The positive appreciation of the seal undoubtedly accompanied the 19th-century interest in the peoples of the Arctic. Frequent were the travel reports, often in feuilleton form, to northern areas, in which seals were an obligatory topic.[92] There are detailed descriptions of seal hunting by the Inuit, invariably called 'Eskimos'.[93] The newspapers reported extensively on the great ethnological exhibitions in which seals also were exhibited. These exhibitions were to be seen in Paris, Brussels, and Cologne, and had a huge success. It was stated that the Paris exhibition on the afternoon of All Saints' Day alone drew more than 20,000 visitors.[94] This ethnological interest in the seal is certainly connected with a certain historical interest in the animal: there is coverage in the archives of Middelburg,[95] archaeological finds in Wisby (Gothland, Sweden),[96] and in New Zealand.[97] This interest is undoubtedly the result of a sentiment of superiority on the part of the 'civilized' Dutch, who looked down upon the 'barbarian' eating habits of other people from past eras and the present day. The Dutch did not eat seal meat. The meat of the killed animals was thrown away, even by the above-mentioned Visser family.[98] The Dutch esteemed eating seal meat to be medieval (the eating habits of their medieval ancestors

89 *Het nieuws van den dag: kleine courant* 22-12-1899.

90 *Rotterdamsch nieuwsblad* 09-04-1900.

91 *Rotterdamsch nieuwsblad* 15-03-1900.

92 For example, *Provinciale Overijsselsche en Zwolsche courant* 15-08-1883.

93 *Rotterdamsch nieuwsblad* 14-03-1887.

94 *De Gooi- en Eemlander: nieuws- en advertentieblad* 11-11-1877.

95 *Algemeen Handelschblad* 20-05-1888. One reads in the newspaper: 'Zeehonden vond men in die tijd smakelijk eten. In 1501 ontving mw. Van Ravenstein een zeehond van 110 pond, en gekocht was voor 5.5 gr. het pond' (In those days seals were considered tasty food. In 1501 Mrs. van Ravenstein received a seal of 110 pounds, which was bought for 5.5 gr. the pound).

96 *Delftsche courant* 30-09-1888. The newspaper reports that the excavation first revealed 'beenderen van paarden, ossen, schapen, varkens, vogels en vissen. Hoe verder naar beneden, hoe meer beenderen van zeehonden' (bones of horses, oxen, sheep, pigs, birds, and fish. The further down, the more seal bones), amid 'brokken vuursteen en priemen van been' (chunks of flint and bone awls).

97 *Bataviaasch nieuwsblad* 03-05-1890.

98 *Rotterdamsch nieuwsblad* 24-10-1888.

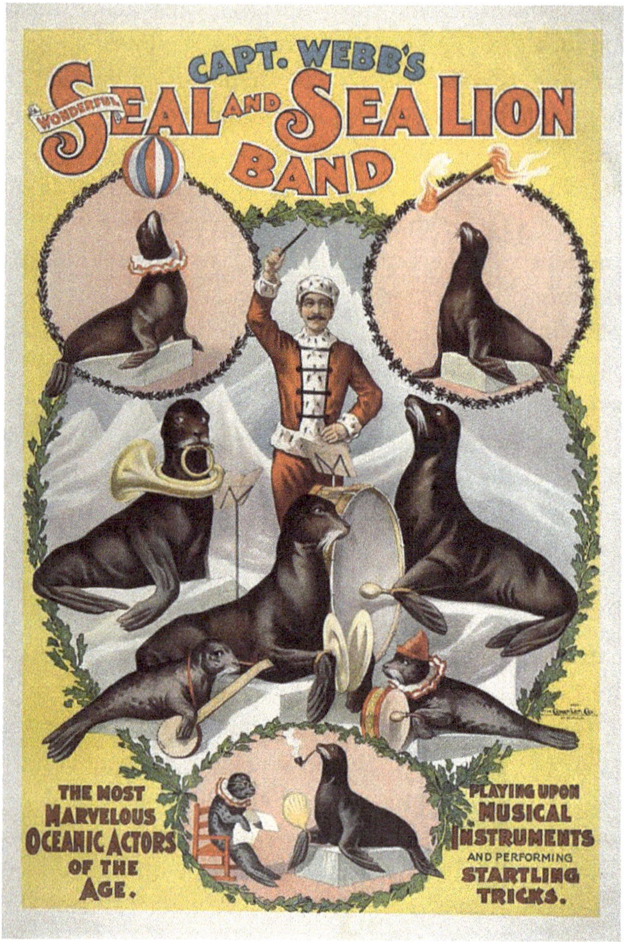

FIGURE 16.4 *Captain Webb's Seal and Sea Lion Band*, Courier Co., no date (American, 19th century) https://emuseum.ringling.org/emuseum/objects/9973/unknown-captain-webbs-seal-and-sea-lion-band?ctx=6452c4c5-6bbc-45fa-8ada-ad2b5f9adf0d&idx=26

were mentioned, as these can be drawn from old accounts)[99] and uncivilized, matching the barbarian peoples from the North, and acceptable only out of dire necessity, e.g. when it is done to survive after a shipwreck.

99 *De Tijd: godsdienstig-staatkundig dagblad* 17-03-1899.

6 Conclusions

The following nuanced picture of public opinion on the subject of seals emerges from the present newspaper analysis. From the 18th century on, seal hunting in the far north was considered economically attractive because of the skin and oil of the animals. The Dutch newspapers kept their readers well informed about the economic and later political consequences of the international seal hunt. The domestic seal hunt, for skin and tears, but also the pleasure hunt received ample attention in the newspapers. From the beginning of the 19th century, seals were mainly hunted domestically because the animals, preying on herring, plaice, and salmon, were considered formidable competitors of the fishermen. Reports of seals in the interior regularly made the headlines. The public's response to these stray seals was mostly negative. But in the second half of the 19th century, seals also aroused empathy among the general public, both internationally and in the Netherlands. The seal became the symbol against unregulated seal hunting and animal cruelty. From the 1880s, trained seals were a source of entertainment in zoos and itinerant shows (although reporting did not always distinguish between seal and sea lion). The newspapers thus show that Dutch public opinion on the seal was multifaceted, contradictory, and changeable – a situation that would continue well into the 20th century. It goes without saying that the results of my newspaper analysis are only a starting point for further research. Verification of my conclusions is required through the consultation of other sources: one can think of ego documents, such as travel reports, and literary works and visual art.

Acknowledgements

This article was written as part of the research programme *A New History of Fishes. A long-term approach to fishes in science and culture, 1550–1880*, based at LUCAS – Leiden University Centre for the Arts in Society, and financed by the NWO (Netherlands Organisation for Scientific Research). I thank Jan Hillgärtner and Meredith McGroarty for their reading of this article.

Bibliography

Brasseur S.M.J.M., *Seals in motion. How movements drive population development of harbour seals and grey seals in the North Sea*. PhD dissertation, Wageningen University, Wageningen, The Netherlands, 2017.

Dekker P., *De laatste bloeiperiode van de Nederlandse arctische walvis- en robbevangst 1761–1775* (Zaltbommel: 1971).

Schokkenbroek J.C., *Harpoeniers en robbenjagers. Nederlandse walvisvaart en zeehondenjacht in de 19[de] eeuw* (Zutphen: 2008).

Sijs N. van der, *Chronologisch woordenboek: de ouderdom en herkomst van onze woorden en betekenissen* (Amsterdam: 2002 (2nd edition)).

’t Hart P., *Zeehondenjacht in Nederland, 1591–1962*, PhD dissertation, Free University Amsterdam, Amsterdam, The Netherlands, 2007.

Vooys C.G.N. de – Brasseur S. – Meer J. van der – Reijnders P.J.H., “Zeehondenjacht in Zeeland: het effect van premies”, *Lutra* 17.3 (2012) 55–65.

PART 4

Ichthyological Knowledge from Afar

CHAPTER 17

The Travelling Nautilus: Spaces of Circulation from the Indian Ocean to Britain

Melinda Susanto

An intricately carved and adorned shell takes pride of place in the Rare Treasures Gallery in the Natural History Museum, London [Fig. 17.1].[1] It was once part of a marine creature, the exterior of a *Nautilus pompilius*, which lived in the depths of the Indo-Pacific region. Its exhibition label describes how Sir Hans Sloane, its collector and founder of the Natural History Museum, would have

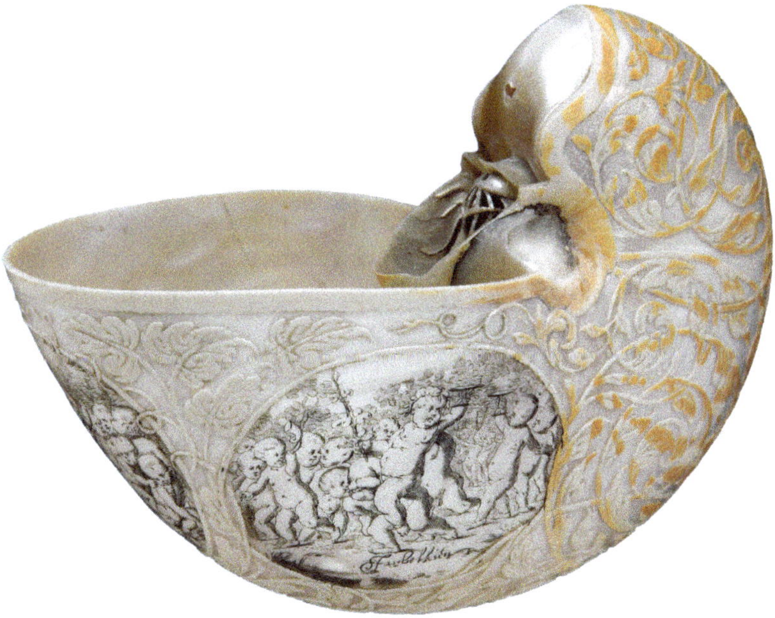

FIGURE 17.1 Jan Bellekin, A 17th-century carved nautilus shell showing three techniques used by Dutch craftsmen: cameo work, engraving and an opening resembling a helmet cut into the wall of the chamber, late 1600s, 15.8 × 11.7 cm. London, Natural History Museum
IMAGE © TRUSTEES OF THE NATURAL HISTORY MUSEUM, LONDON

1 An earlier version of this chapter was presented at the LUCAS Graduate Conference *Animals: Theory, Practice, Representation*, Leiden University (2019). I am grateful to Joanna Woodall

© MELINDA SUSANTO, 2024 | DOI:10.1163/9789004681187_018

marvelled at the beauty of its logarithmic curls and appreciated the added value from its decoration.[2] Little, however, is said about its geographic origins, or the complex global networks of correspondence and exchange that would have delivered such a coveted object into Sloane's collection. Dated to the 17th century, this shell is the only known fully-signed piece by Jan Bellekin. It has a unique history linking the Indo-Pacific region, the early modern Netherlands and Britain. How did the transformation from a marine creature to this object of display occur, and what did such an object signify to its audiences as it traverses space and time?

1 Liminality and Spaces of Circulation

This Nautilus shell, as an object of distant origins oscillating between *naturalia* (natural wonders) and *artificialia* (man-made creations), would have been treasured in a collector's cabinet in early modern Europe. Though there have been many object-based analyses in early modern scholarship, the context of reception in early modern Europe and the history of European collecting have often taken centre stage, even if highlighting global connections and recovering local perspectives from regions of origin have become increasingly important.[3] The connection between commercial networks and the proliferation of such objects has been understood in terms of Harold Cook's *Matters of Exchange*, which demonstrates how the Dutch East India Company facilitated the transfer of goods and ideas leading to developments in natural history and medicine in the Dutch Republic.[4] Subsequent scholarship has continued to expound upon the role of trading companies as facilitators of knowledge.[5]

---

and Eric Jorink for introducing me to this object, and to Andreia Salvador, Hellen Pethers and Kathyn Rooke at the Natural History Museum, London for facilitating access to the object and Hans Sloane's collection catalogues.

2 The object on display with this label was last visited in the Rare Treasures Gallery, Natural History Museum, London in April 2019.

3 Recent scholarship incorporating the complexities of global encounters include Spary E.C., "On the Ironic Specimen of the Unicorn Horn in Enlightened Cabinets", *Journal of Social History* 52.4 (2019) 1033–1060. For the history of collecting, see Mette H., *Der Nautiluspokal: wie Kunst und Natur Miteinander Spielen* (Munich: 1995); Tait H., *Catalogue of the Waddesdon Bequest in the British Museum, III: The Curiosities* (London: 1991). For interpretations of Nautilus shells, see Kehoe M.L. "The Nautilus Cup Between Foreign and Domestic in the Dutch Golden Age", *Dutch Crossings* 35.3 (2011) 275–285; Zuroski E. "Nautilus Cups and Unstill Life", *Journal 18* 3 (2017) https://www.journal18.org/1493.

4 Cook H.J., *Matters of Exchange: Commerce, Medicine and Science in the Dutch Golden Age* (New York: 2007).

5 Various case studies tracing the lives of individual figures associated with trading companies have been undertaken. Of particular relevance to the present analysis is Leuker M-T.,

In line with these recent approaches, this chapter attempts to give more agency to the Jan Bellekin Nautilus shell as an object that *travels* through spaces of circulation, which in turn constituted global connections in the early modern period.[6] To unravel this further, I adopt the description of shells as 'liminal things', as used by Anne Goldgar in her introduction in *Concophilia*, which surveys the myriad ways in which a fascination with shells flourished in early modern Europe, particularly the Low Countries.[7]

What does it mean for this Jan Bellekin Nautilus shell to be a 'liminal thing'? In a recent chapter by Martin Mulsow discussing history of knowledge approaches, with a response from Lorraine Daston, Mulsow uses case studies to reconstruct global connections to either a court space or an individual.[8] This Jan Bellekin shell, however, had neither a fixed space nor a fixed identity. As the Nautilus shell travels, the values accorded to it remains mutable. In this chapter, I trace its travels through space and time, and reconstruct its value in each space of circulation, discussing the knowledge practices and actors involved. Knowledge in this instance is understood as information which has been processed. The act of processing information about the object constitute the knowledge practices in each space. Each space of circulation is embedded within a particular 'knowledge culture', as Mulsow defines it.[9] Space is used to denote both literal and discursive space. This Nautilus shell as an object that *travels* reveals its own capacity to occupy the spaces between particularity and generality.

2 From Primordial Patterns to Contemporary Omens

The first space of circulation which the Nautilus inhabits is the biogeographic space of its origin, the Indo-Pacific region. What was its value to peoples across this vast oceanic region? The early modern Nautilus shell was perceived as an object of nature that reflect patterns of nature across time. Making associations

"Knowledge Transfer and Cultural Appropriation: Georg Everhard Rumphius's *D'amboinsche rariteitkamer* (1705)", in Siegfried H. – Jong J.L. de – Kolfin E. (eds.) *The Dutch Trading Companies as Knowledge Networks* (Leiden – Boston 2010) 145–170.

6 Gerritsen A., – Riello G., "The Global Lives of Things: Material Culture in the First Global Age", in Gerritsen A. – Riello G. (eds.), *The Global Lives of Things: The Material Culture of Connections in the Early Modern World* (London – New York: 2016) 1–29.

7 Goldgar A., "Introduction: For the Love of Shells", in Bass M.A. – Goldgar A. – Grootenboer H. – Swan C. (eds.), *Conchophilia: Art, Curiosity, in Early Modern Europe* (Princeton – Oxford: 2021) 1–17, here 4.

8 Mulsow M. – Daston L., "History of Knowledge", in Tamm M. – Burke P. (eds.), *Debating New Approaches to History* (London: 2019) 159–179, here 160–163, 169.

9 Ibidem.

between the logarithmic curls of its shell and prehistoric fossil counterparts constituted one of the earliest developments for the study of cephalopods (the scientific class to which the species *Nautilus pompilius* belongs today) in the 17th and 18th centuries.[10] The long history of association of molluscs with human life can also be understood through the use of molluscs as a food resource across the region.[11]

In addition to these fundamental concerns, the Nautilus has also long occupied symbolic value in some Indo-Pacific communities. Archaeological evidence suggests that Nautilus shells had been used for ornamentation purposes some 42,000 years ago in the Timor Leste region.[12] Collated evidence across archaeological finds suggest a sort of continuity, that 'the use of Nautilus shell was restricted by social conventions important enough to be upheld and continued over many thousands of years'.[13] It further demonstrates how communities in the Indo-Pacific region held not only the practical knowledge required to prepare molluscs for consumption, but also maintained a social system with material culture, preserving ways of passing down knowledge through generations. On the other side of the spectrum, a recent anthropological study of the Buli community in North Maluku notes that upturned, empty Nautilus shells are often seen as bad omen.[14] This range of evidence from prehistoric times to contemporary accounts, from which practical knowledge, social hierarchies or long-held beliefs could be inferred, serves to demonstrate that the Nautilus has been invested with different values particular to each 'knowledge culture'.[15]

The vastness of the biogeographical space of the Nautilus and the scale of deep time run the risk of falling into generalisation, that each shell is inherently mutable across every space and time.[16] However, two particular instances suggest that shells already inhabited a space of circulation within the Indo-Pacific region and constituted shared material or symbolic values across different 'knowledge cultures', likely even before the arrival of Europeans in the region.

---

10 Etter W., "Early Ideas about Fossil Cephalopods", *Swiss Journal of Palaeontology* 134.2 (2015) 177–186.

11 Szabó K. – Amesbury J.R. "Molluscs in a World of Islands: The Use of Shellfish as a Food Resource in the Tropical Island Asia-Pacific Region", *Quaternary International* 239 (2011) 8–18.

12 Langley M.C. – O'Connor S. – Piotto E., "42,000-year-old Worked and Pigment-stained Nautilus Shell from Jerimalai (Timor-Leste): Evidence for an Early Coastal Adaptation in ISEA", *Journal of Human Evolution* 97 (2016) 1–16.

13 Ibidem 13.

14 Bubandt N., *The Empty Seashell: Witchcraft and Doubt on an Indonesian Island* (New York: 2014) 1–5.

15 Mulsow – Daston, "History of Knowledge" 162–163.

16 Ibidem 176–178.

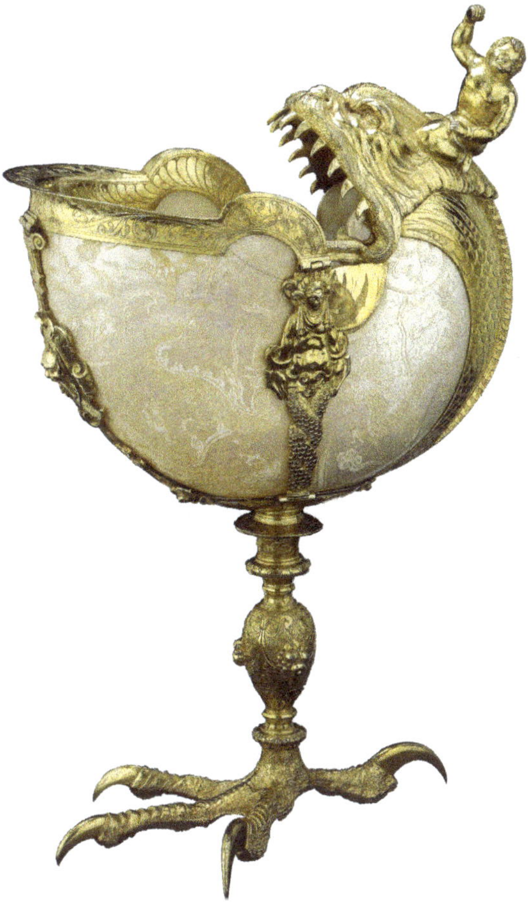

FIGURE 17.2
Nautilus shell mounted in silver, gilt and chased, engraved with dragons among clouds, ca. 1550, 26.1 × 17 × 10.3 cm. London, British Museum
IMAGE © TRUSTEES OF THE BRITISH MUSEUM, LONDON

The first instance of shared circulation across the Indo-Pacific region can be extrapolated from a few carved Nautilus shells in European collections with scales, dragons or floral motifs. These motifs have led scholars to suggest that either the carving was undertaken by Chinese artists, and the shell subsequently mounted in silver in Italy, Germany or the Dutch Republic, or that European artists were imitating Chinese designs to give a foreign flair to these shells.[17] In his analysis of the Nautilus shell with probable 'Chinese' origin in the British Museum collection, Hugh Tait did not find conclusive evidence to confirm the links between Chinese workshops and shells in European collections [Fig. 17.2]. However, Tait's sources highlight a much earlier trade in

17 Tait, *Waddesdon Bequest* 80–87. Grasskamp A., "Shells, Bodies, and the Collector's Cabinet", in Bass M.A. – Goldgar A. – Grootenboer H. – Swan C. (eds.), *Conchophilia: Art, Curiosity, in Early Modern Europe* (Princeton – Oxford: 2021) 49–71, here 62–63.

nacreous shells from Kiau-chï (present-day Northern Vietnam) to be made into cups and small ornaments, as part of the Chinese and Arab trade in the twelfth and thirteenth centuries.[18] Even before the arrival of Europeans in the region, there had already been ornamental trade in shells, with the workshops of Chinese craftsmen possibly situated in coastal regions of Suzhou, Zhangzhou as well as Canton.[19] Tait cites two Chinese historical texts written by Zhang Xie (1574–1640) and Qu Dajun (1630–1696) which describe Vietnam and Hainan island as sources of such shells, suggesting the trade and craftsmanship of shell artefacts in the Chinese workshops were already well-established by the time of their writing.[20]

The second instance of shared circulation can be traced through Malay manuscripts and their regional counterparts which describe shells in literary terms. *Hikayat Indraputra* is a classical Malay text, with the earliest extant written manuscript dated to around 1700.[21] The tale describes the quest of the protagonist, Indraputra, to find medicine to cure the childlessness of a certain king named Syahsyian, incorporating many fantastical elements and miraculous encounters.[22] In one part of his quest, Indraputra comes across a new realm where he becomes overwhelmed by the wealth of God's creations.[23] He is dropped into the sea by his enemy Tamar Boga, whose son he had killed. While looking for a way out of this realm, Indraputra comes upon a river which begins where the water is 'sweet', morphs into saltwater along its middle, and ends with freshwater. In this body of water, he is welcomed by various creatures of God's creations, each welcoming him while at the same time swearing vengeance. The shells are but one of the many *aquatilia* he encounters, including fish and crabs. At each friendly-yet-menacing welcome, Indraputra retaliates by cooking and eating the flesh of the animals and throwing away the shells or carcasses. Each time, whatever animal part he tosses into the waters regenerates thousandfold as living creatures anew, and at the end of each such encounter, Indraputra's inner voice wonders again at the wealth of God's nature.

Some repetitive descriptions of shells were most likely a trope of the literary genre, with bodies of water often described as 'sweet, the shells comprising of pearls and gems'.[24] However, this tale could also be used to extrapolate

18 Tait, *Waddesdon Bequest* 83.
19 Ibidem 84.
20 Ibidem.
21 Mulyadi S.W.R., *Hikayat Indraputra: a Malay Romance* (Dordrecht: 1983) 20.
22 Ibidem 28.
23 Ibidem 93–94.
24 Ibidem 176, 180.

probable shared cultural understandings of *aquatilia* within these interconnected manuscript cultures. The scene of an audience with a mystical being describes how shells can be used as a tribute.[25] In another part of the tale, the hero describes the charm of a man-made space consisting of a garden and a pool, by naming the variety of *aquatilia* that resides in its 'sweet' waters: oysters, snails, algae, shells and crabs.[26]

This tale features in manuscript traditions in multiple languages across Southeast Asia.[27] Mulyadi's dating and regional study of manuscripts suggests that the tale might have been circulating in the region from the 16th century or even earlier.[28] A close reading of the themes and linguistic characteristics leads Mulyadi to conclude that this tale was most likely 'passed down from generation to generation by story-tellers, tinted with Indian elements, before acquiring its Islamic veneer when Islam spread throughout the Malay world'.[29]

Both these spaces of circulation feature sources which do not directly mention the Nautilus. It is contended, however, that these instances still demonstrate how shared material and symbolic values existed in some communities across the Indo-Pacific region, as there likely would have been for the Nautilus shell.

3 'Noteworthy' Shells

One aspect of the Nautilus which has been much discussed is how such a shell was prized for its exoticism and rarity for Europeans. The rarity was not just because it came from the Indo-Pacific region, distant and only accessible through long voyages on ship. It was also because this species of *Nautilus pompilius* lives in the deep sea, and only surfaces when the animal inside was already dying or in trouble, compounding its rarity. It is likely therefore that *Nautilus pompilius* shells were found floating on water surfaces.[30] The process of collecting such shells leads to other spaces of circulation, as the Nautilus moves from being the remnants of an animal into becoming logistical data

25 Ibidem 171.
26 Ibidem 177.
27 This includes Cham language, the peoples who inhabit part of Vietnam and Cambodia today, as well as the Maranao and Mangindanao peoples who live in parts of the Philippines. Malay-language manuscripts circulated in Indonesia, Malaysia and Singapore. In Indonesia, there is also a separate Acehnese tradition. Ibidem 26–27.
28 Ibidem 28.
29 Ibidem 40.
30 Langley – O'Connor – Piotto, "Nautilus shell from Jerimalai" 12–13.

within the Dutch East India Company (VOC) infrastructure. The directors of the VOC produced a list of desirable objects that should be sent back to the Dutch Republic as early as 1623, which included rare animals.[31] While the letters below describe encounters from the 18th century, it can still be illustrative of knowledge practices on the ground within the context of the VOC 'knowledge culture'.

A letter from Ternate dated 13 September 1724 describes how local people could not find sufficient large shells for collecting.[32] Various knowledge practices on the ground could be inferred from this report. Firstly, that the letter writer takes advice from an 'inlander', the term VOC uses to describe indigenous peoples. It was also noted that the governor had sent out different vessels with indigenous sailors to search for shells in various locations including reefs and seabanks within the three kingdoms of Ternate, Tidore and Batchan. This suggests that the VOC needed locals who had familiar geographical knowledge of the region and knew where to look for shells. There is also implicit knowledge of territorial boundaries which people involved in the search must be able to put into practice. The letter also reported that previously the Macassarese had brought some small shells as trading goods, but not in notable quantities. This letter writer uses the term 'naamwaardige' several times, which I have translated as 'noteworthy'.

The nature of what constituted 'noteworthy' shells in quantity and quality is something that might have developed out of incidental encounters, but became codified within the VOC infrastructure as logistical data. The 'knowledge culture' of the VOC meets that of an Asian court in one encounter where the nature of the 'noteworthiness' of the object in discussion comes to the fore. In a letter from Timor dated 22 May 1741, the resident Christiaan Fredrik Brandenburg describes several fortuitous moments over the course of several days. While intending to take a walk along the VOC's garden with the 'burgher' Carel de Clercq, he comes across a broken piece of shell of good white pearl.[33] This first half was found by de Clercq on their walk, while the second half was found the next day by a 'mardijker' named Jonaszoon. The resident was also accompanied by an enslaved person, who came to understand how the resident valued the wholeness of a shell, based on observing the resident's dismay at finding a broken shell. The next day, the enslaved person successfully found

31 Winters R., "The Dutch East India Company and the Transport of Live Exotic Animals in the Seventeenth and Eighteenth Centuries", in Chaiklin M. – Gooding P. – Campbell G., *Animal Trade Histories in the Indian Ocean World* (Cham: 2020) 27–63, here 32.

32 Nationaal Archief Den Haag, *Verenigde Oostindische Compagnie* (VOC) 1.04.02 inv.nr. 8093, fol. 385.

33 NL-HaNA, VOC, 10.04.02, inv.nr. 8334, fol. 171.

another whole shell in the waters of Baban. The day after that, the resident had an audience with Buni, the King of Kupang. In the discussion, the King describes how the animal is used for nutrition, but they typically dispose of the shell. Upon discovering that the resident, and therefore the VOC highly values such a shell, the King promises to set up an exchange with the VOC.

A close reading of this letter reveals other information about the VOC settlement as a space of knowledge circulation. The company had established enough of a presence to have a garden. The language used and the practices of knowledge echoed that of the Dutch Republic. Conversing in gardens, which is a known scholarly activity in early modern Europe, also took place locally in Timor. This letter also shows the range of peoples from different statuses operating within the shared space. This includes the resident, who is the highest VOC authority on the island of Timor, the 'burgher', denoting a free citizen of the colonies, and a 'mardijker', usually a term used in the colonies to describe someone descended from enslaved peoples who had been freed. The motivation given to the 'mardijker' for this collecting process was 'uijt liefhebberij.' 'Liefhebbers', or enthusiasts, in the context of the Dutch Republic refers to someone who pursues knowledge out of love for the subject, here we see one such person operating within a colonial context, as part of the VOC 'knowledge culture'. Aside from the physical spaces associated with VOC power, the interactions between the VOC resident and the King of Kupang took place in a courtly setting, which can be seen as another space of circulation, particularly that of cross-cultural knowledge exchanges.

4 Knowledge in Transit

The establishment of the Dutch East and West India Companies facilitated increased exchange of information and goods, which enabled access to rare natural objects such as Nautilus shells.[34] A steady trade in natural objects began to emerge.[35] Since collecting *naturalia* was an incidental effect and not the official purpose of these voyages, on most ships the *naturalia* trade was limited by space allocations.[36] The shipping and storage of such natural history object therefore entails certain kinds of knowledge practices. These included, for example, knowing how to organise storage space on the ship, or how to maximise the economic value of goods transported. Those of higher rank were

34 Cook, *Matters of Exchange* 416.
35 Ibidem 30.
36 Ibidem.

given more space on a ship and could bring back more found objects from their travels.[37] Thus small and light items, such as the Nautilus shell, which could fetch high value, were deemed most profitable and desirable.[38]

Practical knowledge was also required, including methods to preserve and store inanimate specimens in good condition, or how to take care of live animals in transit. Winters outlines how the lack of knowledge about proper care would have contributed to the estimation that half of all live animals did not survive such long journeys.[39] This included lack of knowledge about the feeding of animals, or knowing which animals could handle enclosed spaces.[40]

Once the ships arrived in the Dutch Republic, natural history objects became eligible for sale or distribution, which saw them enter different spaces of circulation. Some brokers purchased objects dockside and resold them to collectors.[41] Brokers would have had to judge the quality of specimens upon arrival to determine which would fetch the highest price. In auctions of these *naturalia*, the auctioneer would have to determine the grouping of items, the order of the auction, or even which city the auction should take place in.

5 Genealogies of Knowledge

Upon arrival in the Dutch Republic, the Nautilus would have gained capacity to enter different spaces of circulation. First, it entered the spaces of textual knowledge production: as an unadorned object decontextualised from its distant origins, to be identified, named and scrutinised as part of European genealogies of knowledge. Secondly, it entered the spaces of material knowledge and artisanal practices, which will be further discussed below.

The naming of things is the first step in incorporating a new specimen into European intellectual discourses. The *Nautilus pompilius*, as a marine creature, already had associations with the sea at a fundamental level. However, the same could be said for any marine creature. What made this species also distinctive was its etymology which reinforced this oceanic association. The name *Nautilus pompilius* derives from the Greek word 'ναυτίλος' (of a ship), which is also associated with the word ναύτης (a sailor). Aristotle used the name 'ναυτίλος πολύπους' in his *History of Animals* (622 BCE) to describe the animal that

37 Ibidem.

38 Ibidem.

39 Winters, "Transport of Live Exotic Animals" 39.

40 Ibidem 39–40.

41 Cook, *Matters of Exchange* 30.

uses its membranes as a sail, and tentacles as oars.[42] Pliny the Elder's *Natural History* (first century CE) describes the characteristics of the Nautilus echoing that of Aristotle's, only with the analogy to a sailing vessel being made more explicit. Between describing the use of membranes as sail and tentacles as oars, and the subsequent lines description of the animal's reflexive sinking in the event of fear, Pliny adds a line which was not present in Aristotle:

> Afterwards it twists back its two foremost arms and spreads out between them a marvellously thin membrane, and with this serving as a sail in the breeze while it uses its other arms underneath it as oars, it steers itself with its tail between them as a rudder. *So it proceeds across the deep mimicking the likeness of a fast cutter*, if any alarm interrupts its voyage submerging itself by sucking in water.[43]

The identification of species based upon the authority of classical authors such as Aristotle and Pliny was a significant aspect of the humanist tradition which began during the Renaissance. Reconciling new material and textual knowledge coming from Asia that did not fit the boundaries of existing knowledge inherited from Antiquity continued as both a linguistic exercise and scientific practice through to the Enlightenment. The coining of the modern term 'nautilus' for scientific identification can be traced back to Pierre Belon's *L'histoire naturelle des estranges poissons marins* [...] (1551).[44] However, in line with Aristotle and Pliny, Belon's descriptions actually refer to the species which today is known by the scientific name *Argonauta argo*. In Belon's text, the *Nautilus pompilius* is referred as the chambered or pearly nautilus, described as analogous to a large porcelain due to its mother of pearl shell.[45]

Within these genealogies of European knowledge, Nautilus shells also embodied the efforts of learned men like Georg Everhard Rumphius to expand upon existing knowledge by offering more empirical observation. Rumphius was a German who worked for the VOC in Ambon. Though his *D'Amboinsche Rariteitkamer* (Ambonese Curiosity Cabinet) was only published posthumously in 1705, his writings are relevant evidence of the developing trends through the 17th century of acquiring certain knowledge about the expanding

42 Aristotle, *History of Animals, Volume III: Books 7–10*, ed. and trans. D.M. Balme (Cambridge, MA: 1991) IX 37.

43 Emphasis mine. Pliny the Elder, *Natural History, Volume III: Books 8–11*, trans. H. Rackham (Cambridge, MA: 1940) IX 87.

44 Belon Pierre, *L'histoire naturelle des estranges poissons marins* [...] (Paris: Regnaud Chaudiere, 1551).

45 Ibidem 55.

world. Rumphius pioneered the observation of tropical shells in situ, recording accurate locations, as well as local practices related to shells.

While it might be tempting enough to consider Rumphius' contributions as more 'empirical' since it was based on his own observations, the division between earlier Renaissance forms of reasoning and Rumphius' is not as clear-cut. In his description of the paper nautilus, Rumphius takes the analogy of the sailing vessel even further by localising the analogy, saying that 'this little boat is steered like a kind of Javanese sloop, called *Tingang*, commonly known as *Tinan*, which is steered with 2 rudders'. This confusion between the *Nautilius pompilius* and *Argonauta argo* continues with subsequent authors reflecting upon and engaging with previous interpretations. Rumphius also engages with ancient and early modern texts alike to situate his empirical observations, demonstrating how the practices of analogical and linguistic analyses that have been seen as a defining feature of the Renaissance continued well into this period. The scientific naming and classification was eventually resolved by Carl Linnaeus in the 10th edition of *Systema Naturae* in 1758 (first published in 1735). This is evident from the current naming with is appended with L., referring to Linnaeus.

Though Pliny already mentions shellfish like the Nautilus lived in the Indian Ocean, it was not until Rumphius that the local names associated with both *Nautilus pompilius* and *Argonauta argonauta* came to re-associated with the marine creatures again.[46] *Nautilius pompilius* in Rumphius' text is described as 'Nautilus major sive crassus' alongside its Malay names, 'Krang Modang', 'Bia Papeda', 'Bia Coijn', as well as Ambonese 'Kika Lapia'.[47] Rumphius' text also notes how the terms 'Bia' usually only refers to the shell, whereas the term 'Kika' can also include the next type of Nautilus, that is *Argonauta argo*, which is listed as 'Nautilus tenuis' and 'Roema Gorita' in Malay.

Though Rumphius' text did not go further to explain the historic meanings of these names during his time, the translator E.M. Beekman's footnotes provide current information on the etymology of these terms. For *Nautilus pompilius*, the second term 'Papeda' is the name in Moluccan language for a porridge cooked using sago flour.[48] For the *Argonauta*, 'Roema Gorita' can be translated as 'house of the Gorita'.[49] In this case, 'Gorita' is a term generally used to mean octopus or squid.[50] Tracing the reception of Malay words into European

46 Pliny the Elder, *Natural History*, IX 108.

47 Rumphius Georg Everhard, *The Ambonese Curiosity Cabinet*, trans. E.M. Beekman (New Haven: 1999) 89.

48 Ibidem 419 fn. 11.

49 Ibidem 420 fn. 1.

50 Ibidem.

linguistic systems also reveal further spaces in which knowledge circulated. Mahdi's research on the reception and adaptation of Malay words in German publications before 1700 includes analyses of texts by VOC employees.[51] Mahdi concludes that Malay words and their associated knowledge entered German publications through Indo-European or Dutch-associated nomenclature, and not necessarily directly from indigenous Malay speakers themselves. Tracing connections through such linguistic roots further reiterates the many layers of mediations of knowledge between the Indo-Pacific region and Europe in the early modern period.

6 Material Value and Artisanal Skills

Aside from being incorporated into these genealogies of textual knowledge, the Nautilus shell would have also embodied material value as an object of knowledge.[52] There was a fascination with nature's shapes, colours and patterns in the early modern period, which coincided with the drive to acquire more knowledge.[53] Naturalists were curious about the Nautilus shell because of its logarithmic form, which some perceived as representing the ordering principles of the universe.[54] Seen through a cross-section, the Nautilus is made up of spiraling chambers, each preceding chamber proportionate in shape and size to the next [Fig. 17.3].[55] Material objects from far-away places had additional appeal, especially if they were close enough to touch, to handle and to experience their surfaces.[56] For natural, unadorned shells, as Daston and Park note, the value lies in the feeling of unmediated interaction with a distant land.[57] For Jan Bellekin's Nautilus shell, however, I would suggest that the value lies in the very accumulation of mediated contacts through travel and exchange: that the Nautilus shell's journey from the Indo-Pacific region to the early modern Netherlands added to its material value. The Nautilus shell

51 Mahdi W., *Malay Words and Malay Things: Lexical Souvenirs from an Exotic Archipelago in German Publications before 1700* (Wiesbaden: 2007).

52 Daston L. – Park K., *Wonders and the Order of Nature, 1150–1750* (New York – Cambridge, MA: 1998) 67.

53 Allen D., "Tastes and Crazes", in Jardine N. – Secord J.A. – Spary E.C. (eds.), *Cultures of Natural History* (Cambridge: 1995) 394.

54 Kehoe, "The Nautilus Cup" 282.

55 Ibidem.

56 Ibidem.

57 Ibidem 67–68.

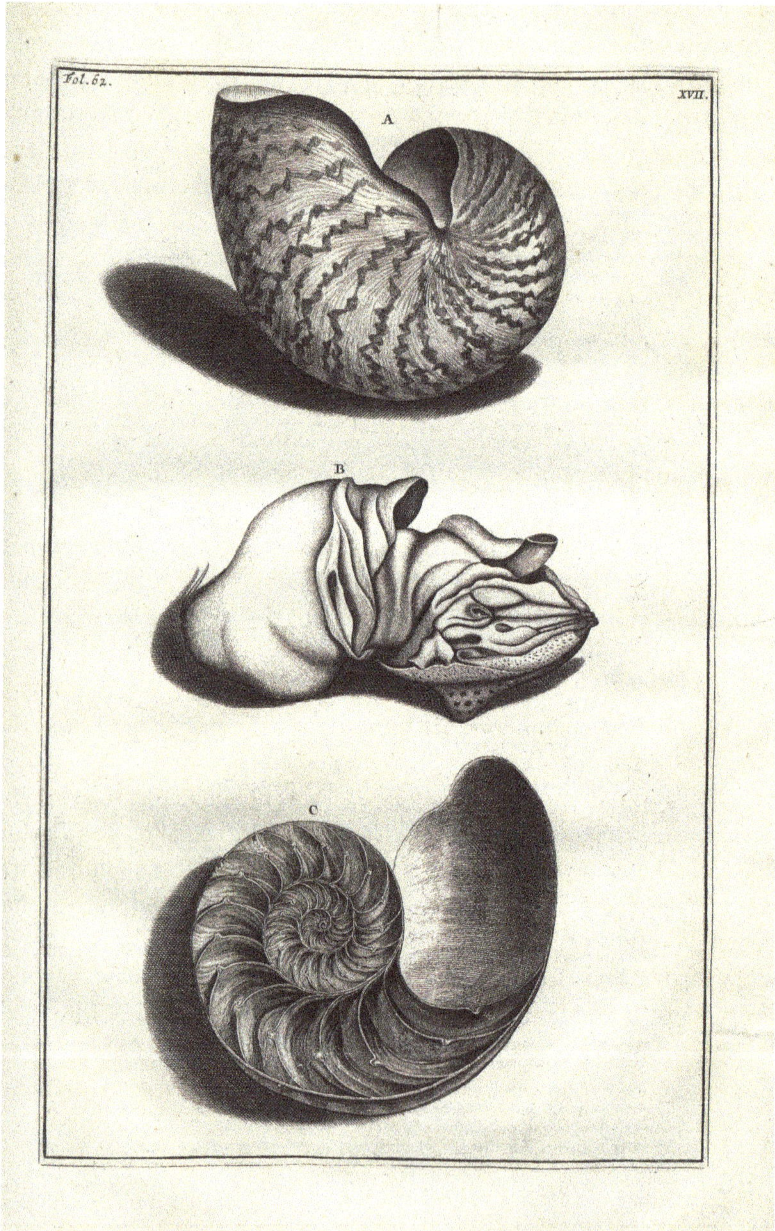

FIGURE 17.3 *Nautilus major sive crassus*, or *Bia Papeda*, depicted with the animal. Engraving. From Rumphius Everhard Georg, *D'Amboinsche rariteitkamer* [...] (Amsterdam, François Halma: 1705) Plate XVII. Leiden, University Library M y 104

would have been even more valued for its intricate decorations, after its transformation from natural object to object of display.[58]

The task of working on this Nautilus shell took skill. Natural objects that displayed technical virtuosity in minute details, such as this shell, were often treasured.[59] It represented a challenge of artisanal skill against the inherent beauty of nature's creations.[60] The surface of this shell is wrapped in delicate relief-carving with a vines-and-leaves pattern. This carving showcases the shell's natural pale brown striations and reveals a mother-of-pearl layer underneath. The patterning winds around the surface of the shell and echoes the spiraling of the shell itself. The minimal use of lines to etch the *putti* figures preserves the delicate surface of the shell while showcasing its natural luminescence.

Objects such as this Nautilus shell would have been admired for the beauty and virtuosity of its creation, and for providing 'fruitful new fields' for artisans.[61] Collectors deemed artistic intervention to add value to the natural object, thus the drive to have shells 'beautified' artificially.[62] 'Shell-doctoring' emerged as a substantial trade in the Netherlands which provided livelihoods for artisans.[63] Rumphius describes the method of preparing the Nautilus shell to become a functional vessel: by rubbing away the outer parts of the shell with something corrosive, followed by washing, then cutting out the chambers.[64] In addition to preparing, carving and adding colour to the engravings using crushed coals, this Nautilus shell is also fitted with a metal helmet in its last chamber, which requires dexterity to insert without breaking the delicate shell.

The Bellekin family were well-regarded in the Netherlands for producing such fine objects of display. This Jan Bellekin shell is the only present-day fully-signed piece by the engraver-artisan. Though the name in the Hans Sloane catalogue is given as 'Jan Belkien', there has been noted spelling variations of the same name in the form of Belkien, Bellekin and Belquin.[65] He was part of

---

58 Ibidem 281.

59 MacGregor A., *Curiosity and Enlightenment: Collectors and Collections from the Sixteenth to the Nineteenth Century* (New Haven: 2007) 47.

60 Daston – Park, *Wonders* 274.

61 Ibidem 213.

62 Allen, "Tastes and Crazes" 395.

63 Ibidem.

64 Rumphius, *The Ambonese Curiosity Cabinet* 90.

65 Seters W.H. van, "Oud-Nederlandse Parelmoerkunst: het Werk van Leden der Familie Belquin, Parelmoergraveurs en Schilders in de 17e eeuw", *Nederlands Kunsthistorische Jaarboek* 9 (1958) 173–238.

the same workshop in Amsterdam established by Jérémie Belquin.[66] The timeline suggests that Jan Bellekin was a contemporary of Cornelis Bellekin, and they were most likely brothers.[67] Cornelis Bellekin had a more prolific output, so his pieces are easily identified today.[68] H. Sander, a traveller to Amsterdam in 1777, wrote about the prized status of a Cornelis Bellekin piece:

> They place a high rarity value on Nautili [Nautilus pompilius L.] which have been carved by Bellekin. He carved heads, coats of arms, flowers & c. on them. Such a piece cost 100 guilders.[69]

7 Nature's Resemblances

In the early modern period, cabinets of curiosity played a role as the site of intersection between art and nature.[70] Daston and Park note that 'objects that displayed the closest resemblances between the two realms' were considered most marvelous.[71] The term 'resemblances' used by Daston and Park requires unpacking. The technique of casting from life is a form of imitating nature and creating a resemblance in another medium based on the form of the natural object.[72] Instead of merely creating a resemblance of the same natural form in a different medium, however, here the artisan had shifted into actively interweaving associations of nature and artifice within one object. The 'resemblances' between objects were no longer about copy and imitation, but about intersection, amalgamation, and the interplay of emulating one another. The visual impact usually came from analogies of form between the natural and artificial materials.[73]

The form of this Nautilus shell mimics a cup for drinking, and it is also adorned with carved patterns echoing its own spiraling form. Pliny the Elder had already established the idea that aberrations of nature reflect the creative

66 Search query "Bellekin" through artist database in: "Ecartico: Linking Cultural Industries in the Early Modern Low Countries, ca. 1475–ca. 1725", *University of Amsterdam* https://www.vondel.humanities.uva.nl/ecartico/.

67 Ibidem.

68 Kehoe, "The Nautilus Cup" 277.

69 Cited in Dance S.P. *A History of Shell Collecting* (Leiden: 1986) 57.

70 Daston – Park, *Wonders*, 296.

71 Ibidem.

72 Silver L. – Smith P.H., "Splendor in the Grass: The Powers of Nature and Art in the Age of Durer", in Smith P.H. – Findlen P. (eds.), *Merchants & Marvels: Commerce, Science and Art in Early Modern Europe* (London: 2002) 29–62, here 47.

73 Daston – Park, *Wonders* 277.

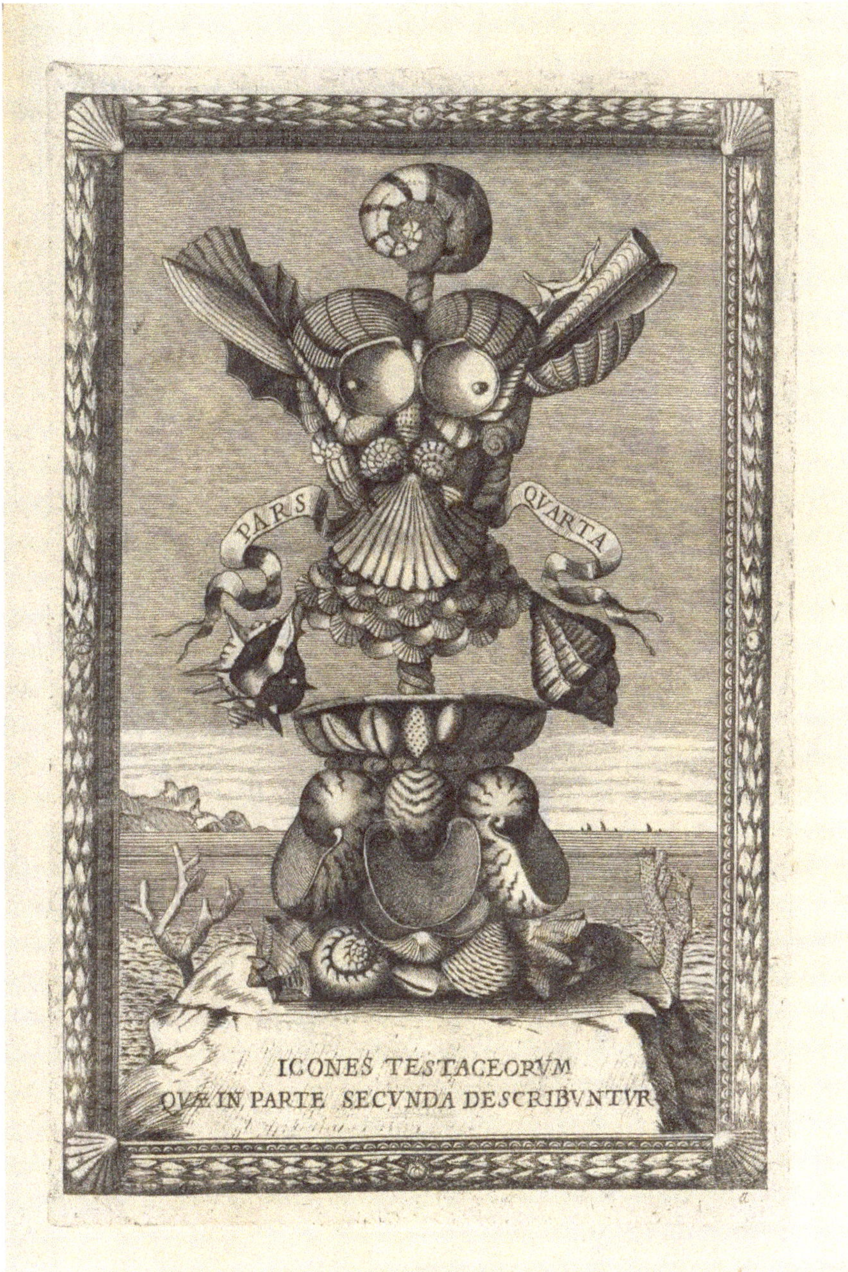

FIGURE 17.4 *Icones testaceorum quae in parte secunda describuntur*. Engraving. From: Bonanni Filippo, *Musaeum Kircherianium* […] (Rome, Giorgio Plancho: 1709). Public Domain. Biodiversity Heritage Library

playfulness of God and nature.[74] Thus the artisan 'played' with form and matter just as nature 'sported' with her flora and fauna from time to time.[75] Artisans who actively created resemblances between the realms of art and nature were themselves converging with the act of nature 'playing'. The playful nature of early modern image-making which relied on resemblances is also evident, for example, in the first plate introducing for the fourth part of Filippo Bonanni's *Musaeum Kircherianum*. In this engraving, the Nautilus shells form a composite assemblage of shells that resembles a human figure [Fig. 17.4]. This may also allude to the associations of the animal bodies of shells with the human reproductive system, as Grasskamp discusses in her article.[76]

8 An Embodiment of Love

Surviving Nautilus shells from the early modern period are engraved with a variety of themes, often biblical or mythological. The question then is, what could the viewer have perceived when confronted with engravings of *putti* on this particular Nautilus shell? In addition to its general association with the sea, the engravings of *putti* evoke ideas of love and desire. In classical mythology, Cupid is the son of Venus, who was born of the sea and known as the goddess of Love and Beauty. The vines on the surface of this shell, as well as the musical instruments played by the *putti*, represent objects which evoke passion.[77] This is compounded by the fact that the Nautilus shell is shaped like a cup for drinking, and the *putti* carry wine glasses. This Bacchic imagery alludes to the potential for being transformed. In parallel to this allusion, the relief-carving and engraving on the surface of the shell itself signifies the human capacity for the love of an object and the possessive act of mark-making left on the object of desire.

74 Bredekamp H., *The Lure of Antiquity and the Cult of the Machine: the Kunstkammer and the Evolution of Nature, Art and Technology*, trans. A. Brown (New Jersey: 1995) 68.

75 Daston – Park, *Wonders* 261.

76 Grasskamp A., "Shells, Bodies, and the Collector's Cabinet", in Bass M.A. – Goldgar A. – Grootenboer H. – Swan C. (eds.), *Conchophilia: Art, Curiosity, in Early Modern Europe* (Princeton – Oxford: 2021) 49–71.

77 Veldman I.M., "Love Emblems by Crispijn de Passe the Elder: Rollenhagen's 'Emblemata', 'Cupid's Bow', 'Youthful Pleasures' and Other 'Charming and Useful' Prints", in Manning J. – Porteman K. – Van Vaeck M. (eds.), *The Emblem Tradition and the Low Countries: Selected Papers of the Leuven International Emblem Conference 18–23 August, 1996* (Turnhout: 1999) 111–156, here 126.

Furthermore, the setting of the *putti* in a garden on this Nautilus shell is also significant, since gardens were seen to reflect a 'third nature', the intersection between 'primal nature' and 'human artifice', a celebration of both usefulness and delight.[78] The Enkhuizen poet Cornelis Biens articulated such a dual concept of usefulness and delight: he wrote in his 1636 treatise on drawing that the visual arts reflect this duality, capable of being aesthetically pleasing while inviting contemplation in the mind.[79]

9 Making Conversation, Making Meaning

Whether adorned or otherwise, the Nautilus shell as a 'liminal thing' further embodies the duality of usefulness and delight as an intricately carved object that draws in conversation. The community of *liefhebbers* or 'enthusiasts' who collected objects such as this Nautilus shell were united through their common interests and the exchange of friendship.[80] The Nautilus shell could thus function as one of the objects around which an intellectual circle revolved, within which learning and the exchange of knowledge could take place.[81] Interested learners would have exchanged observations and specimens among themselves from the 16th century onwards, which would have fostered strong friendship networks by the early 17th century.[82]

The study of shells, or conchology, became more established with publications focusing exclusively on shells. Two earlier contributions came in the form of Filippo Bonanni's *Ricreatione dell'occhio e della mente, nell'osservazion' delle chiocciole* (1681) in Italy and Martin Lister's *Historiae Conchyliorum* (1685–1692) in England.[83] The books by Bonanni, Lister and Rumphius met the shell collectors' demand for precise illustrations, reflecting the popularity of shell collecting which was swept across Europe in the 17th century.[84] Spieß undertook an analysis of inventories which shows that the popularity of Nautilus vessels

78 Ibidem.

79 Ibidem.

80 Swan C., "Collecting *Naturalia* in the Shadow of Early Modern Dutch Trade", in Schiebinger L. – Swan C. (eds.), *Colonial Botany: Science, Commerce and Politics in the Early Modern World* (Philadelphia: 2005) 223–236, here 226.

81 Ibidem.

82 Daston L., "The Empire of Observation, 1600–1800", in Daston K. – Lunbeck E. (eds.), *Histories of Scientific Observation* (Chicago: 2011) 81–114, here 102.

83 Macgregor, *Curiosity and Enlightenment* 136.

84 Dance, *Shell Collecting* 29.

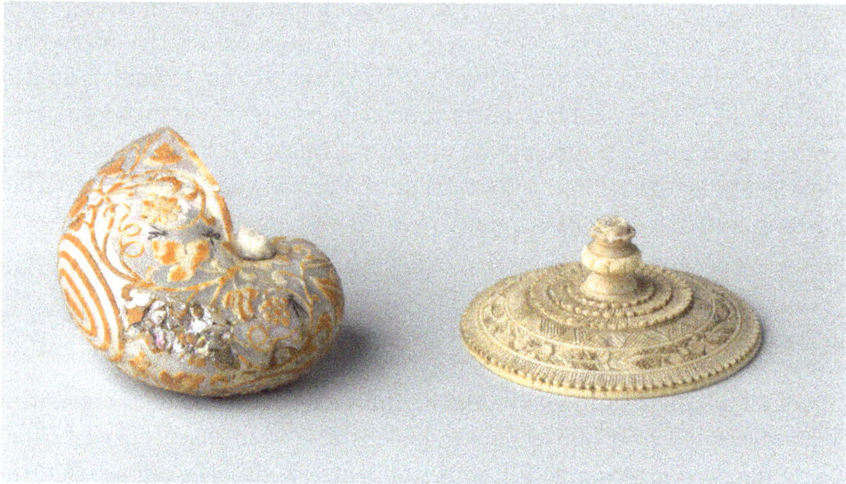

FIGURE 17.5 Cornelis Bellekin, Nautilus shell carved with vines, with a matching holder, 1650–1700, 6.5 × 3.5 × 7 cm. Amsterdam, Rijksmuseum
IMAGE © RIJKSMUSEUM

peaked in the early 17th century.[85] These Nautilus vessels refer to mounted shells with metalwork, unlike Jan Bellekin's shell, which does not appear to be made as a mounted cup. This shell is not accompanied by a mount, nor does the surface indicate any mechanism which connects to a mount. A comparison can be made to Cornelis Bellekin's shell from the same time period, featuring similar carvings, which is accompanied by a mount [Fig. 17.5]. Nevertheless, Spieß's quantitative study is still relevant as evidence of the popularity of objects crafted from Nautilus shells in the 17th century.

An indication of how this particular Jan Bellekin shell might have been intended to be displayed could be extrapolated from the collection catalogues and sale notices from this time, which suggest that carved and decorative shells were also displayed in curiosity cabinets alongside unadorned specimens, laid out in drawers. Two sale notices for 26–27 September 1708 and 28 May 1709 by a certain dealer J.P. Zomer describes the same curiosity cabinet of 78 drawers, which includes various shells, cones and marine creatures, especially highlighting the inclusion of 'extraordinaire groote Paerlemoere Hoornen en Schelpen, door C. Bellekin zeer konstig gesneden' ('extraordinarily large pearly

85 Spieß K., "Asian Objects and Western European Court Culture in the Middle Ages", in North M. (ed.), *Artistic and Cultural Exchanges between Europe and Asia, 1400–1900: Rethinking Markets, Workshops and Collections* (London: 2010) 27–46, here 24.

cones and shells, very finely etched by Cornelis Bellekin').[86] The skills of the collector would have been involved in the organising and ordering of such a shell collection. Therefore, the knowledge and skills require to handle the object would have added to the value to a Nautilus shell such as this.

10 Representations

The Nautilus shell also takes on different meanings once it moves from being a three-dimensional object into becoming a two-dimensional representation in sale catalogues, engravings in natural history books, or still life paintings. A Nautilus shell can oscillate between its particular and general characteristics in representations, as well as threading the fine line between an existing and an imaginary object, which nevertheless fits into the culture of early modern viewers with its capacity for multivalent readings.

An early example of a particular Nautilus shell is depicted in Basil Besler's *Fasciculus rariorum* [...] published in 1616 which depicts his curiosity cabinet list.[87] On the other hand, the Nautilus shell shown in Ulisse Aldrovandi's *De reliquis animalibus* [...] in 1618 might just be decorated in typical designs thought to be 'Asian' by Europeans in order to evoke artisanal skills and highlight its exotic nature, thus possibly representing an imaginary and not necessarily an existing object.[88] On the other hand, another example of an unadorned Nautilus shell which moved from being a particular object to being a representative of a whole is that depicted in the cross-section of Rumphius' plate. It was once owned by Henri D'Acquet and provided the reference for the engraving. Therefore, this engraving refers to a specific shell that had existed in reality, eventually used to represent a general specimen. Given the popularity and widespread translation and circulation of Rumphius' text, it would have taken on a very broad appeal.[89]

This Jan Bellekin Nautilus shell also similarly entered into a wider space of circulation by being represented in the volumes of Albertus Seba's collection, *Locupletissimi rerum naturalium thesauri* [...] (1734–1765) [Fig. 17.6]. The first two volumes appeared in 1734 and 1735 during his lifetime, but upon his death in 1736 the last two volumes were still awaiting publication, which did not take place until 1758 and 1765. The physical collection of Seba had by then been

86 *Amsterdamse courant*, 14 May 1709; *Amsterdamse courant*, 20 September 1708.
87 Leonhard, "Shell Collecting" 191.
88 Grasskamp, "Shells, Bodies" 62.
89 Leonhard, "Shell Collecting" 208.

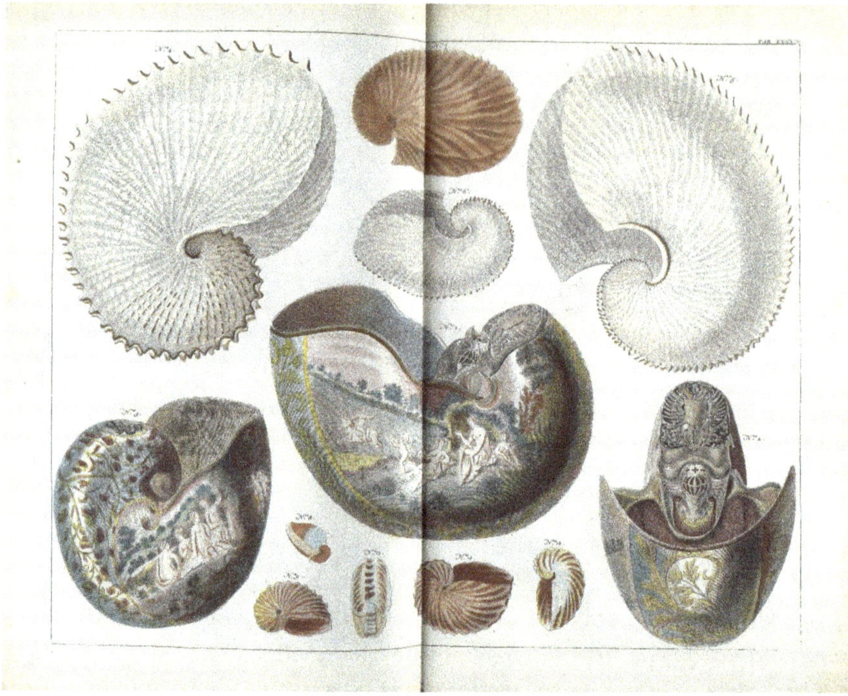

FIGURE 17.6 Jan Bellekin's Nautilus shell is depicted from different sides, labelled with numbers 1 to 3. Hand-coloured engraving. From Seba Albertus, *Locupletissimi rerum naturalium thesauri* [...], 3 vols. (Amsterdam, J. Wetsteen – William Smith – Janssonius van Waesberge: 1734–1765) Plate LXXXIV. Public Domain. Biodiversity Heritage Library

dispersed all over Europe, after being auctioned in 1752.[90] Representations of particular shells could thus be understood as a remembrance of lost objects, once the exact object is no longer within reach. They also take on a generic didactic function, illustrating the knowledge being discussed in the texts.

11 Anglo-Dutch Exchanges in the 18th Century

The story of this particular Nautilus shell continues beyond its transformation from a marine creature originating in the Indo-Pacific region into an object of display or representation in the early modern Dutch Republic. The dispersal

90 Boeseman M., "The Vicissitudes and Dispersal of Albertus Seba's Zoological Specimens", *Zoologische Mededelingen* 44.13 (1970) 177–206.

of some major Dutch collections in the 18th century allows us to consider how knowledge constituted in the Dutch Republic found its way to subsequent spaces of circulation. One endpoint of such collections was Hans Sloane in London.

A recent article utilising digital technologies to analyse Hans Sloane's Miscellanies catalogue, comprising of approximately 2,168 entries, was able to generate a list of the ten most commonly occurring place names, as well as place names alongside a person.[91] The 'East Indies' and 'East India' feature highly on both lists, though China and Japan were the two most prominent place names.[92] Engelbert Kaempfer, who worked as a physician for the VOC in Japan, features in the list of individual names associated with these ten most occurring places.[93] However, this article is based on Sloane's Miscellanies catalogue. The spread of geographies and actors involved might not necessarily be the same across the different catalogues.

Absent the same digital technologies, a qualitative survey of the catalogue inventories of Hans Sloane reveals a few illustrative examples of the global connections which would have enabled rare objects such as this Nautilus shell to come into his collection. The catalogue entry for this particular Nautilus shell did not reveal a clear source.[94] It only describes the specimen numbered 1880 as 'Jan Belkiens [sic] carved & painted nautilus' [Fig. 17.7]. However, other catalogue entries make clear the networks of Anglo-Dutch exchanges in the 18th century. Sloane describes a specimen in the Echinoderm collection as coming from a certain 'Dr. Hermans collection [in] Holland'.[95] This specimen would most likely originate from the auction of Paul Hermann's collection in 1711 where James Petiver bought some items on behalf of Sloane.[96] Paul Hermann was a physician working for the VOC in Ceylon between 1672 and 1677.[97]

---

91 Ortojla-Baird, A. – Nyhan, J., "Encoding the Haunting of an Object Catalogue: on the Potential of Digital Technologies to Perpetuate or Subvert the Silence and Bias of the Early-modern Archive", *Digital Scholarship in the Humanities* (2021) 1–24. https://doi.org/10.1093/llc/fqab065.

92 Ibidem 12.

93 Ibidem 13.

94 *Sloane Manuscript Catalogues: Fossils, Vol. 2: Shells*, fol. 272, specimen 1880.

95 *Sloane Manuscript Catalogues: Echinoderm*, fol. 22, specimen a.30.

96 Macgregor, A., *Sir Hans Sloane: Collector, Scientist, Antiquary, Founding Father of the British Museum* (London: 1994) 107.

97 "About Paul Hermann: The Collection", *Natural History Museum London*, https://doi.org/10.5519/0062484.

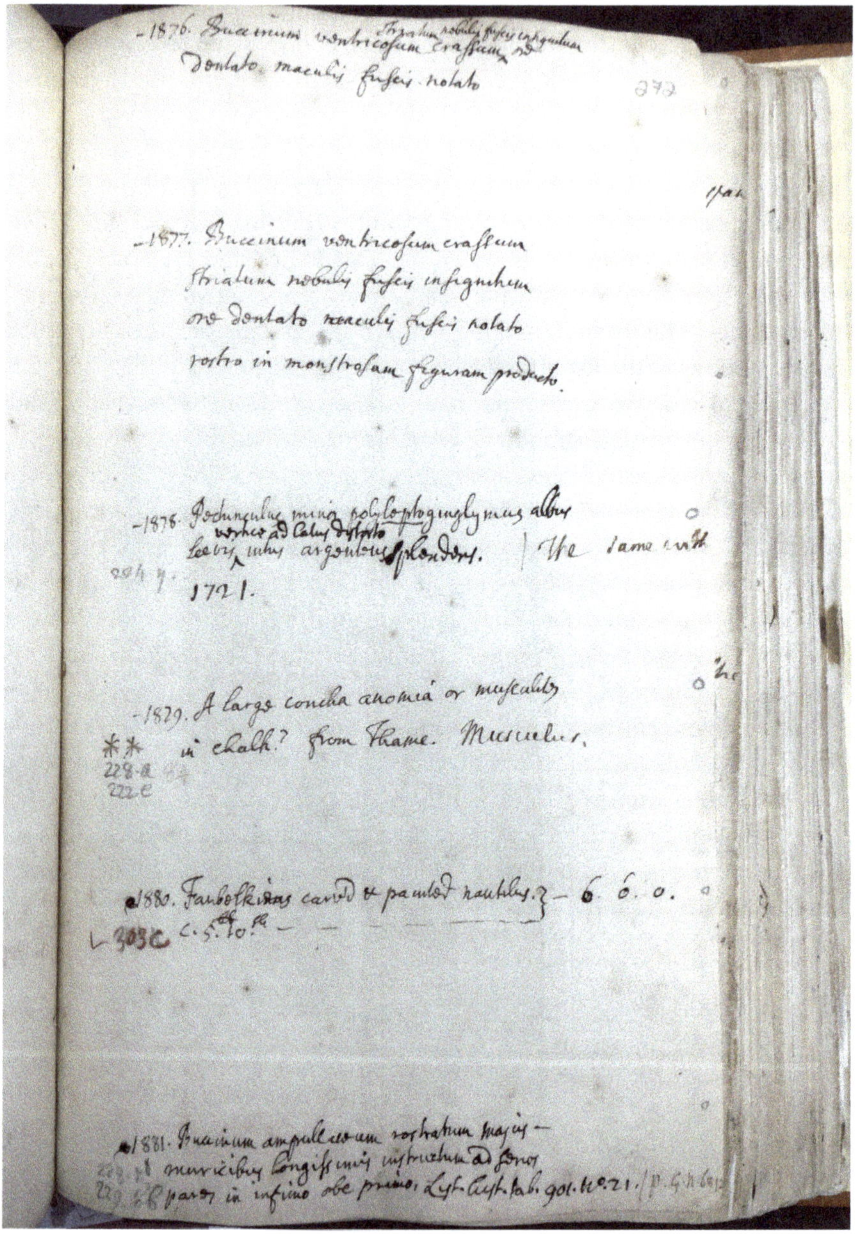

FIGURE 17.7 Hans Sloane, inventory entry for the 'Jan Belkien' shell listed under specimen number 1880. From: *Sloane Manuscript Catalogues: Fossils*, Vol. 2: Shells, fol. 272. London, Natural History Museum

In another instance, a French-language advertisement leaflet of the VOC's upcoming auctions in various Dutch cities is tucked into one of the volumes of Sloane's catalogue.[98] This particular specimen numbered 1716 entered Sloane's collection accompanied with a certificate dated 2 July 1737 from the directors of the VOC, written in Dutch with a corresponding English language translation:

> We the underwritten Directors of the East
> India Company for this Chamber, do certify for truth
> that the Growth out of the Brains of an Elephant.
> Contained in a Golden Case, was Sold at our Sale the
> 13th May 1737, and Sent to the General East India
> Company as a Present by the Very Sultan of Jamby
> in the year 1735.

In the absence of direct description for the source for the Jan Bellekin Nautilus shell, these examples illustrate how other natural history objects entered Sloane's collection from the Dutch Republic: through Dutch intermediaries or VOC auctions.

12 Afterlives of Jan Bellekin's *Nautilus pompilius*

By examining Jan Bellekin's Nautilus shell through its particularities as well as general characteristics, it has been possible to trace the complexities of knowledge practices across spaces and over time. This chapter has also attempted to be reflexive in the use of sources: In trying to unravel the perspectives from Asian 'knowledge cultures', it is necessary to read against the grain of Dutch or other European sources. Conversely, there is insufficient Malay or Indonesian early modern sources which specifically mentioned the Nautilus shell. However, through extrapolations of material culture and literary texts, it has at least been possible to sketch out the spaces of circulation and globalising encounters in the early modern period without privileging European knowledge as the only thread of the 'knowledge culture' under discussion. Having traversed through these spaces of knowledge circulation that can be glimpsed through this one particular Nautilus shell, the multivalent histories contained and left behind reveal how its value as a historic and contemporary object could be better understood, from its journey originating in the Indo-Pacific region, to its acknowledged status as a 'rare treasure' in the Natural History Museum of London today.

98 *Sloane Manuscript Catalogues: Fossils Vol. 5: Fishes, Birds*, fols. 244–245, specimen 1716.

Bibliography

Allen D., "Tastes and Crazes", in Jardine N. – Secord J.A. – Spary E.C. (eds.), *Cultures of Natural History* (Cambridge: 1995) 394–407.

Arens E.H. – Kießling C., "Knowledge and Power: Rumphius' Ambonese Herbal and Ambonese Curiosity Cabinet as Colonial Contact Zones", *European Review* 26.3 (2018) 461–470.

Aristotle, *History of Animals, Volume III: Books 7–10*, ed. and trans. D.M. Balme (Cambridge, MA: 1991).

Belon Pierre, *L'histoire naturelle des estranges poissons marins* […] (Paris: Regnaud Chaudiere, 1551).

Bredekamp H., *The Lure of Antiquity and the Cult of the Machine: the Kunstkammer and the Evolution of Nature, Art and Technology*, trans. A. Brown (New Jersey: 1995).

Boeseman M., "The Vicissitudes and Dispersal of Albertus Seba's Zoological Specimens", *Zoologische Mededelingen* 44.13 (1970) 177–206.

Bubandt N., *The Empty Seashell: Witchcraft and Doubt on an Indonesian Island* (New York: 2014).

Cook H.J., *Matters of Exchange: Commerce, Medicine and Science in the Dutch Golden Age* (New York: 2007).

Dance S.P., *A History of Shell Collecting* (Leiden: 1986).

Daston L. – Park K., *Wonders and the Order of Nature, 1150–1750* (New York – Cambridge, MA: 1998).

Daston L., "The Empire of Observation, 1600–1800", in Daston K. – Lunbeck E. (eds.), *Histories of Scientific Observation* (Chicago: 2011) 81–114.

Delbourgo J., *Collecting the World: Hans Sloane and the Origins of the British Museum* (Cambridge, MA: 2017).

Etter W., "Early Ideas about Fossil Cephalopods", *Swiss Journal of Palaeontology* 134.2 (2015) 177–186.

Gerritsen A. – Riello G., "The Global Lives of Things: Material Culture in the First Global Age", in Gerritsen A. – Riello G. (eds.), *The Global Lives of Things: The Material Culture of Connections in the Early Modern World* (London – New York: 2016) 1–29.

Goldgar A., "Introduction: For the Love of Shells", in Bass M.A. – Goldgar A. – Grootenboer H. – Swan C. (eds.), *Conchophilia: Art, Curiosity, in Early Modern Europe* (Princeton – Oxford: 2021) 1–17.

Grasskamp A., "Shells, Bodies, and the Collector's Cabinet", in Bass M.A. – Goldgar A. – Grootenboer H. – Swan C. (eds.), *Conchophilia: Art, Curiosity, in Early Modern Europe* (Princeton – Oxford: 2021) 49–71.

Kehoe M.L. "The Nautilus Cup Between Foreign and Domestic in the Dutch Golden Age", *Dutch Crossings* 35.3 (2011) 275–285.

Langley M.C. – O'Connor S. – Piotto E., "42,000-year-old Worked and Pigment-stained Nautilus Shell from Jerimalai (Timor-Leste): Evidence for an Early Coastal Adaptation in ISEA", *Journal of Human Evolution* 97 (2016) 1–16.

Leonhard K., "Shell Collecting: On 17th-Century Conchology, Curiosity Cabinets and Still Life Painting", in Enenkel K.A.E. – Smith P.J. (eds.), *Early Modern Zoology: The Construction of Animals in Science, Literature and the Visual Arts* (Leiden – Boston: 2007) 177–214.

Leuker M.-T., "Knowledge Transfer and Cultural Appropriation: Georg Everhard Rumphius's *D'amboinsche rariteitkamer* (1705)", in Siegfried H. – De Jong J.L. – Kolfin E. (eds.) *The Dutch Trading Companies as Knowledge Networks* (Leiden – Boston 2010) 145–170.

MacGregor A., *Curiosity and Enlightenment: Collectors and Collections from the Sixteenth to the Nineteenth Century* (New Haven: 2007).

MacGregor A., *Sir Hans Sloane: Collector, Scientist, Antiquary, Founding Father of the British Museum* (London: 1994).

Mahdi W., *Malay Words and Malay Things: Lexical Souvenirs from an Exotic Archipelago in German Publications before 1700* (Wiesbaden: 2007).

Mette H., *Der Nautiluspokal: wie Kunst und Natur Miteinander Spielen* (Munich: 1995).

Mulyadi S.W.R., *Hikayat Indraputra: a Malay Romance* (Dordrecht: 1983).

Mulsow M. – Daston L., "History of Knowledge", in Tamm M. – Burke P. (eds.), *Debating New Approaches to History* (London: 2019) 159–179.

Ortojla-Baird A. – Nyhan J., "Encoding the Haunting of an Object Catalogue: on the Potential of Digital Technologies to Perpetuate or Subvert the Silence and Bias of the Early-modern Archive", *Digital Scholarship in the Humanities* (2021) 1–24. https://doi.org/10.1093/llc/fqab065.

Pliny the Elder, *Natural History, Volume III: Books 8–11*, trans. H. Rackham (Cambridge, MA: 1940).

Raj, K., "Beyond Postcolonialism … and Postpositivism: Circulation and the Global History of Science." *Isis* 104.2 (2013): 337–347.

Rumphius Georg Everhard, *The Ambonese Curiosity Cabinet*, trans. E.M. Beekman (New Haven: 1999).

Silver L. – Smith P.H., "Splendor in the Grass: The Powers of Nature and Art in the Age of Durer", in Smith P.H. – Findlen P. (eds.), *Merchants & Marvels: Commerce, Science and Art in Early Modern Europe* (London: 2002) 29–62.

Spary E.C., "On the Ironic Specimen of the Unicorn Horn in Enlightened Cabinets", *Journal of Social History* 52.4 (2019) 1033–1060.

Spieß K., "Asian Objects and Western European Court Culture in the Middle Ages", in North M. (ed.), *Artistic and Cultural Exchanges between Europe and Asia, 1400–1900: Rethinking Markets, Workshops and Collections* (London: 2010) 27–46.

Swan C., "Collecting *Naturalia* in the Shadow of Early Modern Dutch Trade", in Schiebinger L. – Swan C. (eds.), *Colonial Botany: Science, Commerce and Politics in the Early Modern World* (Philadelphia: 2005) 223–236.

Swan C., "The Nature of Exotic Shells", in Bass M.A. – Goldgar A. – Grootenboer H. – Swan C. (eds.), *Conchophilia: Art, Curiosity, in Early Modern Europe* (Princeton – Oxford: 2021) 21–47.

Szabó K. – Amesbury J.R. "Molluscs in a World of Islands: The Use of Shellfish as a Food Resource in the Tropical Island Asia-Pacific Region", *Quaternary International* 239 (2011) 8–18.

Tait H., *Catalogue of the Waddesdon Bequest in the British Museum, III: The Curiosities* (London: 1991).

Winters R., "The Dutch East India Company and the Transport of Live Exotic Animals in the Seventeenth and Eighteenth Centuries", in Chaiklin M. – Gooding P. – Campbell G., *Animal Trade Histories in the Indian Ocean World* (Cham: 2020) 27–63.

Van Seters W.H., "Oud-Nederlandse Parelmoerkunst: het Werk van Leden der Familie Belquin, Parelmoergraveurs en Schilders in de 17e eeuw", *Nederlands Kunsthistorische Jaarboek* 9 (1958) 173–238.

Veldman I.M., "Love Emblems by Crispijn de Passe the Elder: Rollenhagen's 'Emblemata', 'Cupid's Bow', 'Youthful Pleasures' and Other 'Charming and Useful' Prints", in Manning J. – Porteman K. – Van Vaeck M. (eds.), *The Emblem Tradition and the Low Countries: Selected Papers of the Leuven International Emblem Conference 18–23 August, 1996* (Turnhout: 1999) 111–156.

Zuroski E., "Nautilus Cups and Unstill Life", *Journal18* 3 (2017) https://www.journal18.org/1493.

CHAPTER 18

François de Meyer's Fish Travelogue (1698)

Paul J. Smith, Didi van Trijp and Alan Moss

1 Introduction

The Dutch National Archives in The Hague houses a special travelogue in manuscript form. It concerns an account written in French of a sea voyage from France to La Martinique from 11 September 1698 to 5 December 1698.[1] Consisting of 24 pages, the manuscript contains 8 coloured drawings of sea creatures. Judging by the consistent handwriting, text written around drawings and only copying errors, this manuscript is most likely a final version, of which we do not have any rough drafts. Nothing is known about the author, a certain 'François d[e] Meyer', except his name, which he himself gives once,[2] and some sparse, circumstantial information he reveals about himself. One can even question his family name, 'de Meyer': its spelling cannot be interpreted unambiguously. We will therefore refer to him in this article by his first name, 'François'. The travelogue is particularly special because of the coloured drawings of marine creatures it contains, and the information that François provides about these as well as about animals that are not depicted. The anonymity of the author, and the fact that it straddles the boundaries between what we tend to see as separate genres – the travel diary and the natural historical account – may be why the manuscript has been scarcely studied by historians. Additionally, the journal is part of the Delprat family archives, which have very little to do with natural history. The travelogue is surrounded by genealogical documents, private letters, copied sermons, and a more haphazard variety of papers of the Delprat family, mostly from the 18th and 19th centuries. Because of this

1 Meyer François de, *Reisjournaal van François de Meyer, opvarende op het schip l'Aigle onder bevel van François Dubois naar Martinique*. 1698–1699. Ms. Nationaal Archief, Den Haag, Collectie 349 Familie Delprat, toegang 2.21.183.16, inv.nr. 106a. The manuscript's pages measure 20 × 15 cm. The cover is probably original and contains draft notes on the inside (calculations and scribbles in pencil by a later hand). The manuscript is part of a miscellaneous section, presumably collected by members of the Delprat family. The relationship with the Delprat family and the Delprat family archive is not clear. Unfortunately, the genealogical papers in the family archive (inv. numbers 1–2, 5 and 7) do not provide an answer. A critical edition of this travelogue, which besides the natural history aspects also discusses the ceremonial aspects of maritime customs, is in preparation.

2 Nationaal Archief (hereafter: NA), Delprat 106a, 1.

archival context, and the fact that the family archive has only been at the Dutch National Archives since 1980, it has not been the focus of scholars of early modern maritime life.[3]

In this article we want to present and analyse François's drawings and descriptions of marine life. His travelogue offers a fascinating insight into how an individual that appears to have received little formal training in the study of nature observed and documented the animals that he encountered during his voyage. In presenting François's drawings and descriptions we will primarily keep the chronological order of the travelogue, because the drawings and descriptions of marine fauna are strongly intertwined with this narrative, showing a progressive argumentative development. Secondarily, we will further interpret the natural historical material, as well as the possible sources and the way the material was presented. Before going into the drawings and descriptions, however, it is important that we briefly present this manuscript from a historical, palaeographic, and linguistic perspective.

2 Context

François's travelogue provides the following information about the ship in its opening lines. The ship is called L'Aigle and the captain's name is François du Bois. In passing, François gives more information further on: the ship has 9 guns, 36 crew members,[4] and 19 passengers (who remain unnamed).[5] More information about this ship can be found at the French Archives Nationales. In an archival record dated 19 May 1702,[6] the ship is referred to as a 'flute', a popular type of sailing vessel with three masts and a broad belly, mostly used for freight transport. In the service of the French king, L'Aigle carried out assignments in the Caribbean around 1700, such as the transport of provisions and materials to the post in Guadeloupe. Guadeloupe and La Martinique had been part of the French colonial empire since 1635. At the beginning of the 18th century, both islands had a number of sugar cane plantations, cultivated by enslaved people. The captain of L'Aigle is mentioned once in this archive document: 'Desbois', corresponding to the name 'Du Bois' given by François.

3 The sole reference that we found in secondary literature is Davids K, *Global Ocean of Knowledge, 1600–1860: Globalization and Maritime Knowledge* (London: 2020) 68.

4 Apart from the captain, one crew member is three times briefly mentioned: a certain 'Des Rivee' [?].

5 NA Delprat 106a, 8. These numbers indicate that this ship is not identical to the warship L'Aigle, which was active at the same time: that ship had 100 guns and 34 crew members in peacetime.

6 19 mai 1702; Cote de communication: COL C8 A 14 fol. 114.

About this Du Bois, François mentions in passing, towards the end of the report, that he is a sailor with 30 years of experience in the Caribbean.

The place of departure is 'Chef debois', according to François: this is Chef de Baie, a small coastal town near La Rochelle, known in the 17th century as 'Chedebois'. The exact location of arrival at La Martinique is not given. The duration of the journey is explicitly given in the closing sentence of the report: 'le 5 a 10 heure du matin nous arivame [...] ala martinique dieu mersij aprais quatre vaingt sinc jours de traversee'[7] (On 5 [December] at 10 o'clock in the morning we arrived [...] at La Martinique – thank God – after a crossing of 58 days). Although thanking God and Holy Providence for returning home unscathed after a long journey is a common trope in travel literature, this almost sounds like a sigh of relief. In fact, it probably is: François twice quotes the captain, who stated that he had never had so much delay in his long career due to the weather conditions (storm, headwind, and calm).

As mentioned, nothing is known with certainty about the identity of François. However, some indirect information about his person and his working methods can be obtained from his spelling, use of language, and areas of interest. First, we are dealing here with a travelogue, in which the journey is tracked from day to day. The travel report is built up from diary notes and sketches that were made on the spot. The voyage report has a logbook-like precision: not only is each day mentioned, but also, for each day, the wind directions and other weather conditions and events the people on board have to deal with are mentioned very precisely, down to the hour. And François furthermore notes all the ships that L'Aigle encounters along the way, which are mentioned sometimes very briefly, if the ship remains far away, sometimes more extensively if there is actual contact or a threat, such as the encounters with an English warship and a Turkish pirate.[8] In this precise recording of the circumstances of the journey, it resembles the general structure of early modern travel diaries.[9]

The manuscript includes eight coloured drawings. While some (Dutch) travel manuscripts often contain doodles of far-away sights or cut-out engravings from printed travel guides, the level of detail and use of colour here are exceedingly rare. It is remarkable that these illustrations only concern sea animals. A number of other interesting things are described but not depicted; the

7 NA Delprat 106a, 24.

8 NA Delprat 106a, 8.

9 Most early modern travelogues were not solely meant as personal souvenirs. While François could have used this paper memento to later reminisce about a past adventure in the far-away Caribbean in the comfort of his own study, family and friends would most likely have perused his journal as well, marvelling at the author's depiction of maritime life.

only exception is a hailstone, drawn in the same ink in which the manuscript is written.[10] It is difficult to precisely reconstruct how the coloured illustrations were made. It seems that a part of them was made as ink-wash, a technique in which the lines of the drawing are made with ink and subsequently coloured in with that same ink in diluted form. The lines of the drawings and their colourization seem similar, executed in brownish hues. Certain drawings, however (notably those of the dolphin fish and the bonito), include brighter shades of green, yellow, and blue. This can indicate that watercolour was also used. That in some cases a rather thick layer of paint or varnish was applied can be inferred from the drawing of the triggerfish, which shows clear signs of what art historians call craquelure. Regardless of the materials used, almost all the drawings are placed prominently across of the page, with the text written around them.

This brings us to the handwriting, spelling, and language of François. The handwriting has indeed been put into the net with obvious care, but was done so in an untrained hand. There are times when writing fatigue seems to have set in, as can be seen in certain spots in the handwriting, in word repetitions, strikethroughs, and insertions of forgotten words (in our quotes we will reflect these corrections to the extent possible). The spelling is phonetic and with scarce and inconsequent interpunction – which makes a precise reading of the text very difficult at times. Sometimes words clump together ('lonnesoroit'[11] should be read as *l'on ne saurait*, or 'saizelle'[12] as *ses ailes*, for example); sometimes they fall apart: *Guadeloupe* is written as 'garde.loupe'[13] and *parfaitement* as 'par faitte mant'.[14] The spelling is also often inconsistent: the spellings 'devand' and 'devent' (for *devant*) can coexist in one sentence.[15] There are also many confusions: 'destentions'[16] or 'dixtaintions'[17] for *distinction*, 'un faittivement'[18] for *effectivement*, or 'en larope'[19] or 'en leuroppe'[20] for *en l'Europe*. For this reason we will add, where necessary, a transcription in modern French in brackets, for a good understanding of the text.

10 'qui estoit comme des zeufe [= œufs] caree comme celle ceij' (egg size, square, as pictured here). NA Delprat 106a, 11.

11 NA Delprat 106a, 4.

12 NA Delprat 106a, 16.

13 NA Delprat 106a, 1.

14 NA Delprat 106a, 4.

15 NA Delprat 106a, 12.

16 NA Delprat 106a, 5.

17 NA Delprat 106a, 6.

18 NA Delprat 106a, 16.

19 NA Delprat 106a, 16.

20 NA Delprat 106a, 6.

The use of language is usually grammatically correct but stylistically unpolished – often with endless paratactical constructions, based on the excessive use of the coordinating conjunction "et" (for the sake of readability, we have not always adopted these paratactic sentence constructions in our English translation). Sometimes, however, we are dealing with fairly complicated hypotactic sentence structures (using the *participe présent*, and concessional clauses). Certain words he uses come from the dialect of the Ile de France. The vocabulary furthermore contains many dialecticisms, especially when it comes to technical subjects, such as tools ('foinne', 'fisson' – types of harpoon), parts of the ship, or names of birds and fish – as in the designation of 'bascouette' (wagtail), 'paille enceus' (tropicbird) and 'touil' (Breton for 'dogfish'). Some of these words we have only found in 19th-century dialect dictionaries, as we shall see. Remarkable is the verb tense used in the travelogue: this is almost consistently the *passé défini* (also called *passé simple*) – a verb tense that, unlike present-day French, in which the *passé défini* has a literary connotation, was commonly used in the 17th century in narrative reporting. The *passé défini* has the function of indicating that it is a completed narration of events that took place in a past that is usually recent.

3 Marine Life: A Chronological Presentation

Let us present François's attention to marine fauna in chronological order. From this chronological perspective, it is remarkable that during the first days of the voyage, no attention is paid to marine fauna. Perhaps François, as a landlubber, had to acclimate to life at sea. Be that as it may, it is not until 16 December, five days after departure, that marine life is mentioned – and then it suddenly explodes. At 6 AM gannets (*Morus bassanus*) are spotted: 'Le 16 a 6 heures du matin nous vimme plusieurs oiseaud qu lon nomme foux.'[21] (On 16 September at 6 AM we saw several birds called gannets).

At 10 AM he sees a wagtail (*Motacilla spec.*):

> sur les 10 heures il viens serepoze un peti oizeaud sur nos vergue qui ne me fu pas unconnut / lon le nome en France Bascouette autre ment pipis[22]

> At 10 o'clock there came on our yards a small bird, which was not unknown to me. In France he is called 'bascouette', also called 'pipis'.

21 NA Delprat 106a, 2.

22 NA Delprat 106a, 3.

At 3 PM something strange happens:

> sur les 3 heure apres midi il luij eut [= il y eut] un de nos pijon qui prit la penne [= peine] de prendre la lavollee pour san nalle [la volée pour s'en aller] mais il ne peutpas fairre sans comparaizon comme la colombe de larche qui sapuya sur le roche pour prendre de louriers [= des lauriers] pour faire son nique [= nid ?] car celuy la sapuija sur les lamme deaux et si notre capitainne navoit pas ueut pitiee de luy qui fit maittre le navire ala cap pour luij donner secours et qui mit sa chaloupe aleaud pour le souvee [= sauver] je crois quil lestoit noiers.[23]

> at 3 o'clock in the afternoon one of our pigeons made an attempt to fly away, but he did not succeed in imitating the pigeon of the Ark [of Noah], which landed upon a rock and plucked laurel to make a nest. This pigeon, however, landed on the waves, and if our captain had not pitied him and let the ship change course to help him and lower the lifeboat into the water to save him, he would have, I think, drowned.

The captain's rescue is remarkable: is it 'pitiee', as François suggests, or is it more superstition (pigeons on a ship bring good luck), based on the biblical story of Noah's Ark? Equally remarkable is François's personal attention given to the pigeon, and especially to the wagtail – although the phenomenon of tired land birds perching on ships is quite common, there are only a few other early modern travelogues in which attention would be paid to such a small and everyday bird.[24]

From this moment on, François focuses on the sea creatures. It begins with a 'marsouin', which is not only described but also depicted, along with the harpoon used to capture the animal [Fig. 18.1]. By the way, the illustration makes it clear that this is not a 'marsouin' (a porpoise) but a dolphin, probably the common dolphin (*Delphinus delphis*), which is pelagic and has an average weight of 75 kilograms.

> Du 17 a 60 lieux enmer a 7 heure du soir un de nos contremaitre nomme des vive monta sur la vergue de sivadiere et prit un dard fait en fleche il darda un marsouin qui pezoit 150 lb [= livres] Et nous le mimme a notre

23 NA Delprat 106a, 3.

24 Sloane Hans, *A Voyage to the Islands Madera, Barbados, Nieves, S. Christophers and Jamaica* (London, B.M. for the author: 1705–1725) 7.

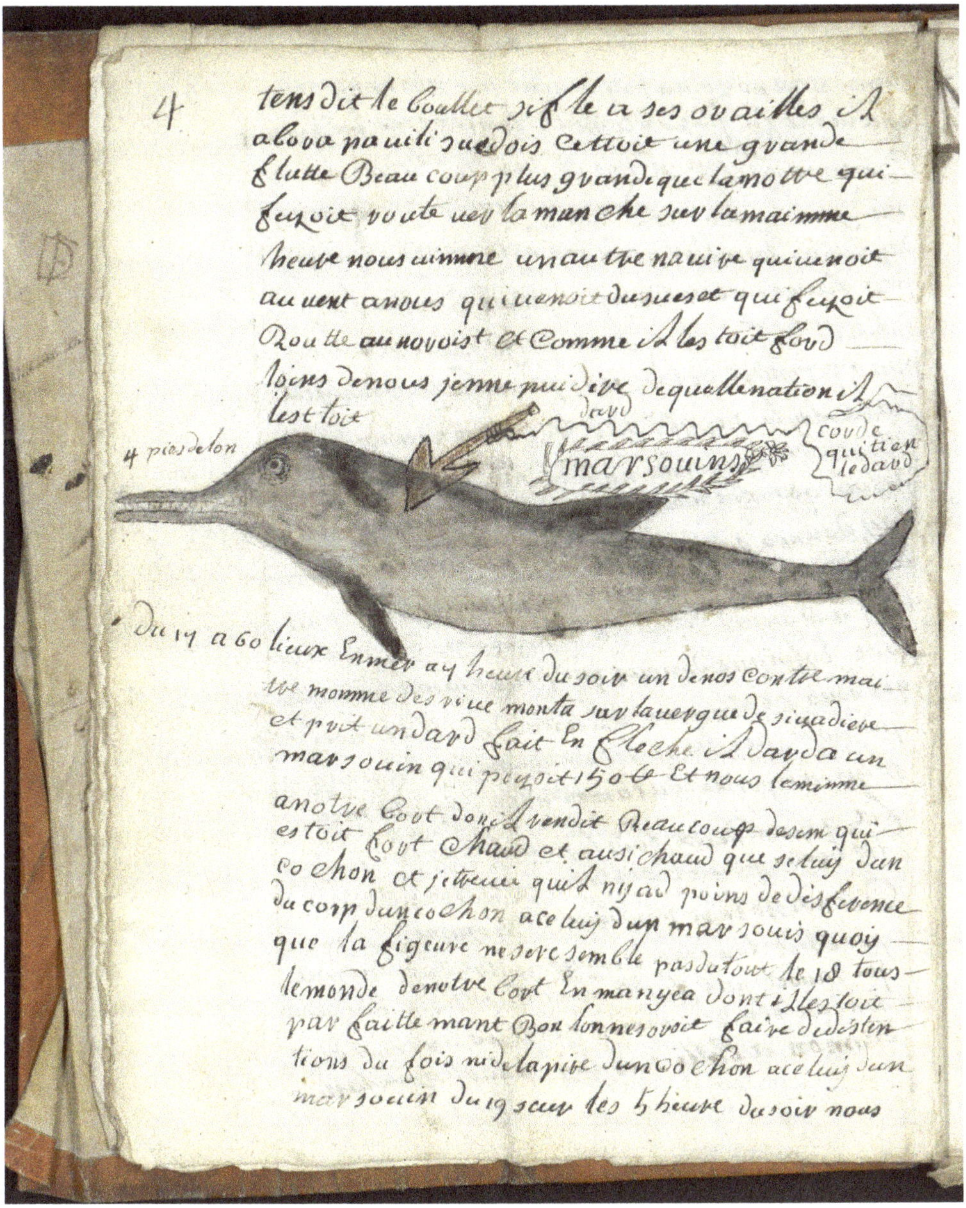

FIGURE 18.1 François de Meyer, Dolphin. Ink-washed drawing. In *Journal de voyage de François de Meyer*, 1698. The Hague, Nationaal Archief, Collection 349, Familie Delprat, 2.21.183.16, Inv. no 106a, p. 4

> bort don il rendit beaucoup de sen qui estoit fort chaud et ausi chaud que se luy dun cochon et je treuvee quil nij ad poins de difference du corp dun cochon a celuy dun marsouin quoi que la figeure ne se resemble pas du tout. Le 18 tous le monde de notre bort en manyea dont il lestoit par faitte mant [= parfaitement] bon. lonnesoroit [= l'on ne saurait] faire de destentions [= distinction] du fois [= foi] ni de la pire[25] dun cochon a celuy dun marsouin.[26]
>
> At 7 o'clock in the evening on 17 September, 60 miles from the coast, one of the commanders named Des Vives mounted the bowsprit, took an arrow-shaped harpoon, and speared a 150-pound dolphin. And we drew it on board, where it gave much blood, which was very warm, as warm as a pig's. And I found that there was little difference between the body of a dolphin and that of a pig, although they look very different in appearance. The 18th, everyone aboard ate the animal, which was particularly tasty. It is not possible to distinguish the liver or lungs of a pig from those of a dolphin.

On 19 September there are a number of large whales (of an indefinable species):

> Du 19 scur les 5 heure du soir nous vimme quantite de poisson que lon nomme soufleurs quipassoit contre notre bord. Ce poisson est une fois plus gros quin beuf que nous ayeons en leuroppe il seroulle sur leaud comme le marsouin il se plonge dans leaud et quand il viens a fleurs deaud il gette la dune pique deaud par un trou quil la sur la tete et fait un soufle beaud coup plus fort que celuy dun beuf.[27]
>
> On 19 September, about 5 o'clock in the evening, we saw a number of fish called "soufleurs" coming to our boat. This fish is twice the size of an ox we know in Europe. It swims in the water like a dolphin. It dives down and when it comes up it sprays a jet of water through a hole he has on its head, and it blows much harder than an ox.

---

25 A dialectical term for 'poumon' (lung). See Puichaud C., "Dictionnaire du patois du Bas-Gatinais (Suite)", *Revue de philologie française et provençale* 7 (1887) 100–137, here 120.

26 NA Delprat 106a, 4.

27 NA Delprat 106a, 5.

On 20 September, a species of dogfish is caught:

> du 20 ilviens contre notre bort quantitte de poisson que lon nomme chins demer don je ne puis pas faire de dixtaintions dun chins demairs a un touil[28] que nous manjon beaucoup a la Rochelle particeulierement notre capitainne voient le carmme[29] mit sa yolle aleau luy et 2 matelot dedens et il prit 2 zin [?] et sans fut a la portee dans jet depierre de son bort et il lans prit 3 en peux de temp.[30]
>
> On 20 September, a quantity of fish called "dogfish" arrived at our ship, which I cannot distinguish from a "touil" (kind of dogfish) that we eat a lot, especially in La Rochelle. When our captain saw this food [?], he launched his sloop, with him and two sailors in it, and he took two nets [?], and at a stone's throw from the ship, he caught three in a short time.

What is striking here, as with the wagtail and other animals, is the attention to naming – we will come back to this in our analysis section. Here, too, the food aspect does not go unmentioned, both prior to the voyage, during François's sojourn in La Rochelle, and the day after, as he pens: 'Le 21 nous en mangimme dont il lestoit admirable.'[31] (We ate it on the 21st: it was delicious).

And the marine life observations keep coming in, that same day:

> le maimme jours sur les 5 heures du soir il viens une quantitte de poisson contre notre bort que lon nome tons un de nos capittainne nomme des rivee prit un arpon fait en fasson dun fisson[32] de serpent. Il monta sur la vergue de sivadiere tc [= et] tiralarpons sur le dos du poisson mais comme se poisson adextremment la peaud durre larpon faussa sur la peaud du poisson / se poisson est dumoins au sigros comme un mouton[33]

28 'toul': Breton word for dogfish. See Danois E., "Les noms de quelques animaux et végétaux marins en dialecte de Léon", *Annales de Bretagne et des pays de l'Ouest* 25.3 (1909) 548–555, here 555.

29 Probably 'carne', 'viande'; cf. Puichaud C., "Dictionnaire du patois du Bas-Gatinais", *Revue de philologie française et provençale* 7 (1887) 18–53, here 31.

30 NA Delprat 106a, 5.

31 NA Delprat 106a, 5.

32 'fisson' is a dialectal term for spear. Cf. Puichaud, "Dictionnaire" 51. 'Fisson de serpent', meaning 'serpent's tongue' or 'serpent's tooth', is mentioned by Puichaud.

33 NA Delprat 106a, 5.

> On the same day at 5 o'clock in the evening a quantity of fish called tuna arrived at our ship. One of our captains, named Des Riccée [?], picked up a harpoon shaped like a "fisson de serpent". He climbed on the bowsprit and threw the harpoon on the fish's back, but because this fish had a very hard back, the harpoon slipped on the fish's skin. This fish is at least as big as a sheep.

Then there is a long period of silence with regard to marine fauna, which is probably due to the weather conditions. François focuses on these conditions (waves as high as mountains, huge hailstorms, St. Elmo's fire), as well as on the impending approach of a Turkish pirate. It was not until 20 October 1698 that attention was again paid to fish, namely the albacore or white tuna (*Thunnus alalunga*), referred to here as 'Bounitte' [Fig. 18.2]:

> Le 20 sur les 9 heure du matin il parut au devend de notre navire quantite de poissons que lon nomme Bounnittes des le moment que nous heumme üeut notre capitainne prit unameson et semit sur lavergue de sivadiere dans le moment il lans pit [?] dix / cepoisson est ~~parfaittement bon~~ [34]ford cour pezent 16 a 17 tt piece danlemoment quil fut pris / notre capitainne donna ordre dan fairre bouilly dons nous en manyamee tous et qui estoit par [inserted] faittement bon.[35]

> At 9 o'clock in the morning on 20 October, a number of fish called "bounittes" appeared in front of our ship. As soon as we saw [these fish], our captain took some fishing bait and stood on the bowsprit. In no time he caught 10 of these fish. This fish ~~tastes excellent~~ is quite short, weighing 16 to 17 pounds each, at the time of catch. Our captain ordered it to be made into soup, which we all ate, and which tasted excellent.

Also, in this case there seems to be a concentration of marine life. On 21 October, it rains again, but after the rain has stopped, the crew sees a huge number of dolphins:

> nous vimme venir soullevend une sigrande quentitte de marsouins que tout lequipage medij quil nans nauoit jamais üeu temps ensemble [= tout l'équipage me dit qu'il n'en avait jamais vu tant ensemble][36]

34 Strikethrough in text.
35 NA Delprat 106a, 9.
36 NA Delprat 106a, 10.

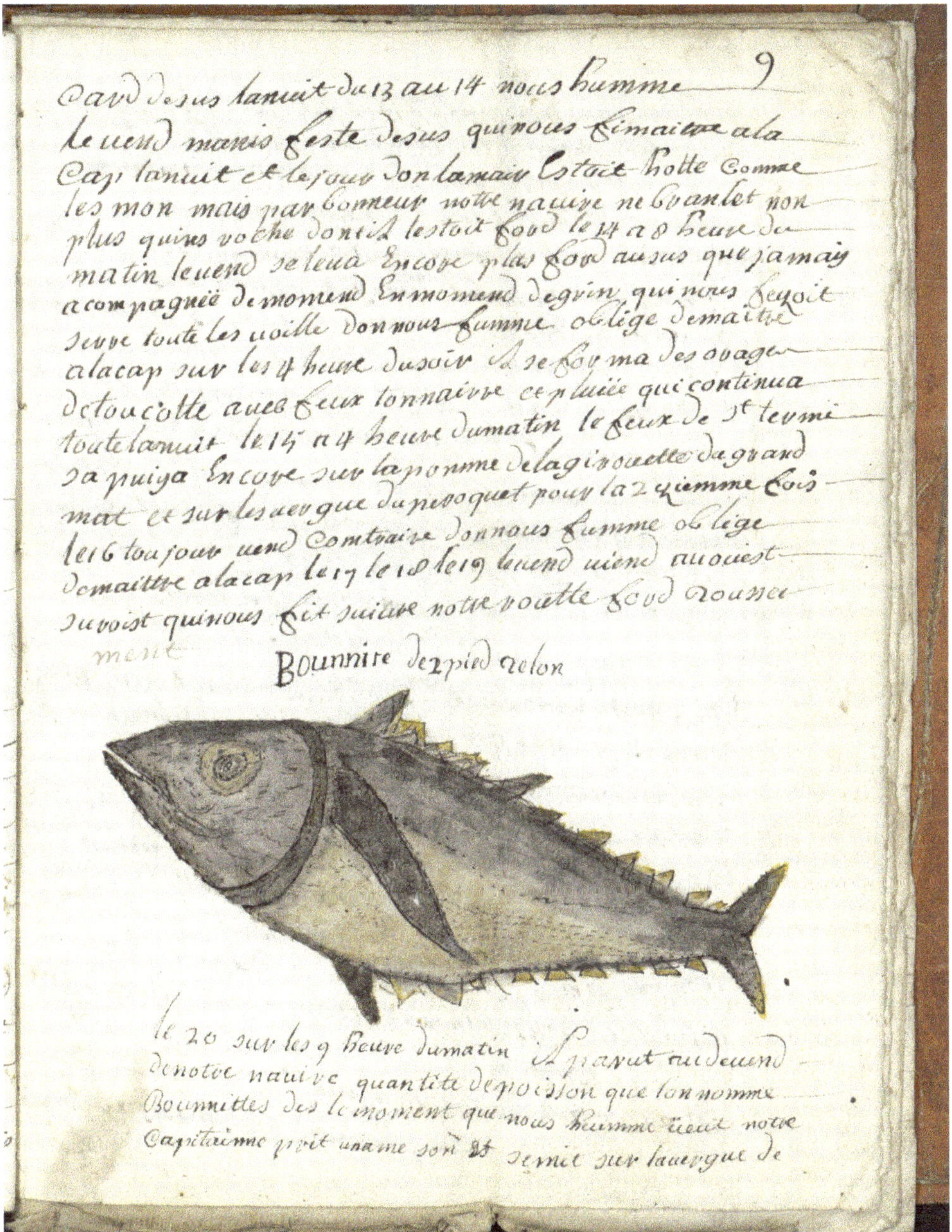

FIGURE 18.2 François de Meyer, Albacore or white tuna (*Thunnus alalunga*). Ink-washed drawing. In *Journal de voyage de François de Meyer*, 1698. The Hague, Nationaal Archief, Collection 349, Familie Delprat, 2.21.183.16, Inv. no 106a, p. 9

> we saw such a large number of dolphins under the wind that all the crew members told me they had never seen so many at once.

It is likely that François sees a causal relationship here with the previously observed albacore tuna. After all, both dolphins and tuna prey on flying fish. At least one piece of this food chain seems clear to François, as can be seen with the next fish caught on 23 October, described and depicted by François, namely the pilot fish (*Naucrates ductor*), which helps the tuna to track down fish [Fig. 18.3]:[37]

> sur les 10 heure du matin notre capitainne prit un poisson contre notre bort que lon nomme pilotte / cest un poisson qui conduitte les bonnitte pour atrape les poisson volland

> at 10 AM, our captain caught a fish called 'pilot fish' near our ship. This is a fish that directs the albacore to the flying fish to catch it.

A few days later, on 28 and 29 October 28, huge numbers of flying fish are observed over two days ('nous vimme quantitte de poisson volland').[38]

Something special is going on with the captured and depicted pilot fish. The size is very small (half a foot, as the legend states). It is not mentioned that several specimens are caught, from which, for example, soup can be cooked, as was the case with the albacore. The mention that it was the captain who caught this fish is also remarkable. After all, the captain and crew were just recovering from a terrible hail storm ('plus ford que jamais'), which ravaged the ship from 3 AM. that same day. The storm was so violent that the captain ordered the carpenters to get the axes ready to cut down the mainmast – which fortunately proved unnecessary. In short, the captain probably had something else on his mind to catch such a small fish. Yet he did it – one has the impression that he did this especially for François.

The fact that the two are on good terms with each other is apparent from the moment the ship crosses the equator, on 30 October. At the traditional baptism of Neptune, the captain acts as godfather (parrain) of François: 'le capitainne me fit lonneur daistre mon parain dont il medonna le nom de mariee gallande'[39] (The captain gave me the honour of being my godfather:

37 In addition to large sea creatures, this species also follows ships. Nowadays it is known that they do not guide the animals to their prey, but only swim along to eat the prey's remains.

38 NA Delprat 106a, 11.

39 NA Delprat 106a, 11.

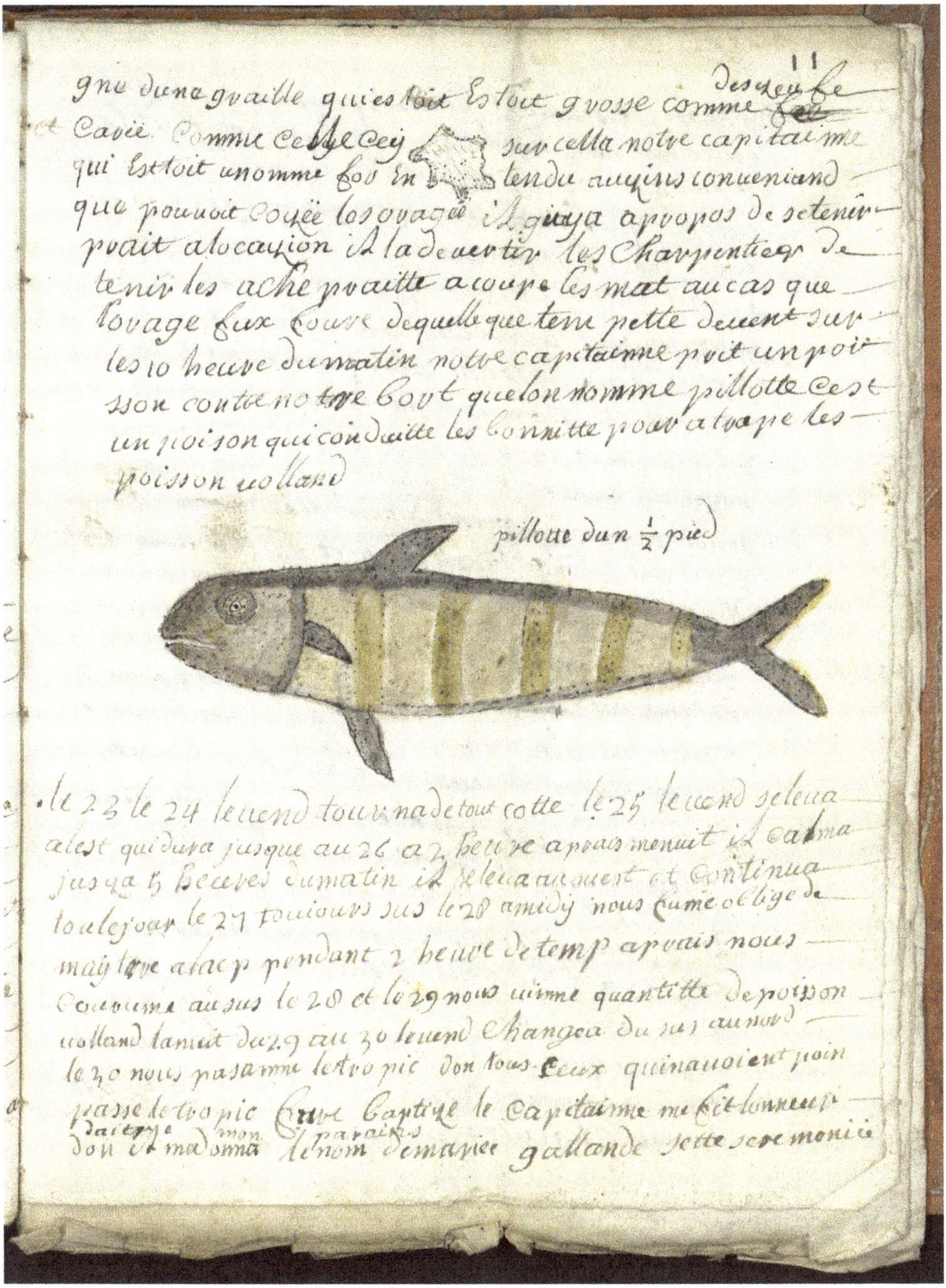

FIGURE 18.3 François de Meyer, Pilot fish (*Naucrates ductor*). Ink-washed drawing. In *Journal de voyage de François de Meyer*, 1698. The Hague, Nationaal Archief, Collection 349, Familie Delprat, 2.21.183.16, Inv. no 106a, p. 11

he baptized me with the name 'mariee gallande')[40] – which seems to be an example of transvestism, quite common in the baptismal ritual. Be that as it may, even at later moments one gets the impression that the fish was caught especially by the captain to be drawn and described by François.

After 29 October, things become a bit quieter in terms of observations. On 5 November at 10 AM a large whale appears: 'a 10 heure du matin il viens contre notre bord un poisson que lon nomme soufleurs qui estoit extraimmement gros'[41] (At 10 o'clock in the morning a huge fish called 'soufleur' arrives at the ship). And on 9 November the captain catches a 'dorade', a dolphin fish or mahi mahi (*Coryphaena hippurus*). This is a large fish, which will no doubt be eaten (this is not mentioned), but of which François will have the time and opportunity to make a drawing and precise notes [Fig. 18.4]:

> a 4 heure du soir notre capitainne prit une dorade avec un extrument de ferd que lonnome foinne [= foène, foëne] comme celle qui est reprezententee sur la tette de la dorade,/ ce poisson est dune beaute en chantee estemp dans leaud particulierement il nait pas malnomme zossi dorade par ce quil est beaud et bon et deplus il porte une fleur delis sur la taitte et il ad ce longeur 4 a 5 pid selon.[42]
>
> At four o'clock in the morning our captain caught a dorado with an iron instrument called "foène", as depicted here at the head of the dorado. This fish is of a mesmerizing beauty, especially in the water. It is not just called "dorado" because it is beautiful and tastes good. In addition, it has a lily flower on his head. It is about 4 to 5 feet tall.

François's observations are indeed precise: he notices that the fish loses its beautiful colours as soon as it is out of the water. He takes the time to trace the harpoon used. Moreover, François seems to have talked to the captain or someone else about the lily on the head of the fish. A modern folklorist writes about this: '[Les marins] voient dans l'anatomie des poissons, des dessins qui nous échappent. Ils donnent au *jol* le nom de 'poisson royal' parce qu'il porte une fleur de lys sur la tête'[43] ('[Sailors] see in the anatomy of fish designs that

40 'Marie-Galante' is today the name of a Caribbean island, which was called 'Maria Galanda' by Columbus in 1493 (corruption of the Amerindian word 'aulinagan'). In the 17th century, the island was still called 'île de Saint-Louis'.

41 NA Delprat 106a, 13.

42 NA Delprat 106a, 14.

43 Herber J., "Le folklore de la mer (dans l'Hérault)", *Folklore* 11.3 (1948) 43–48, here 43.

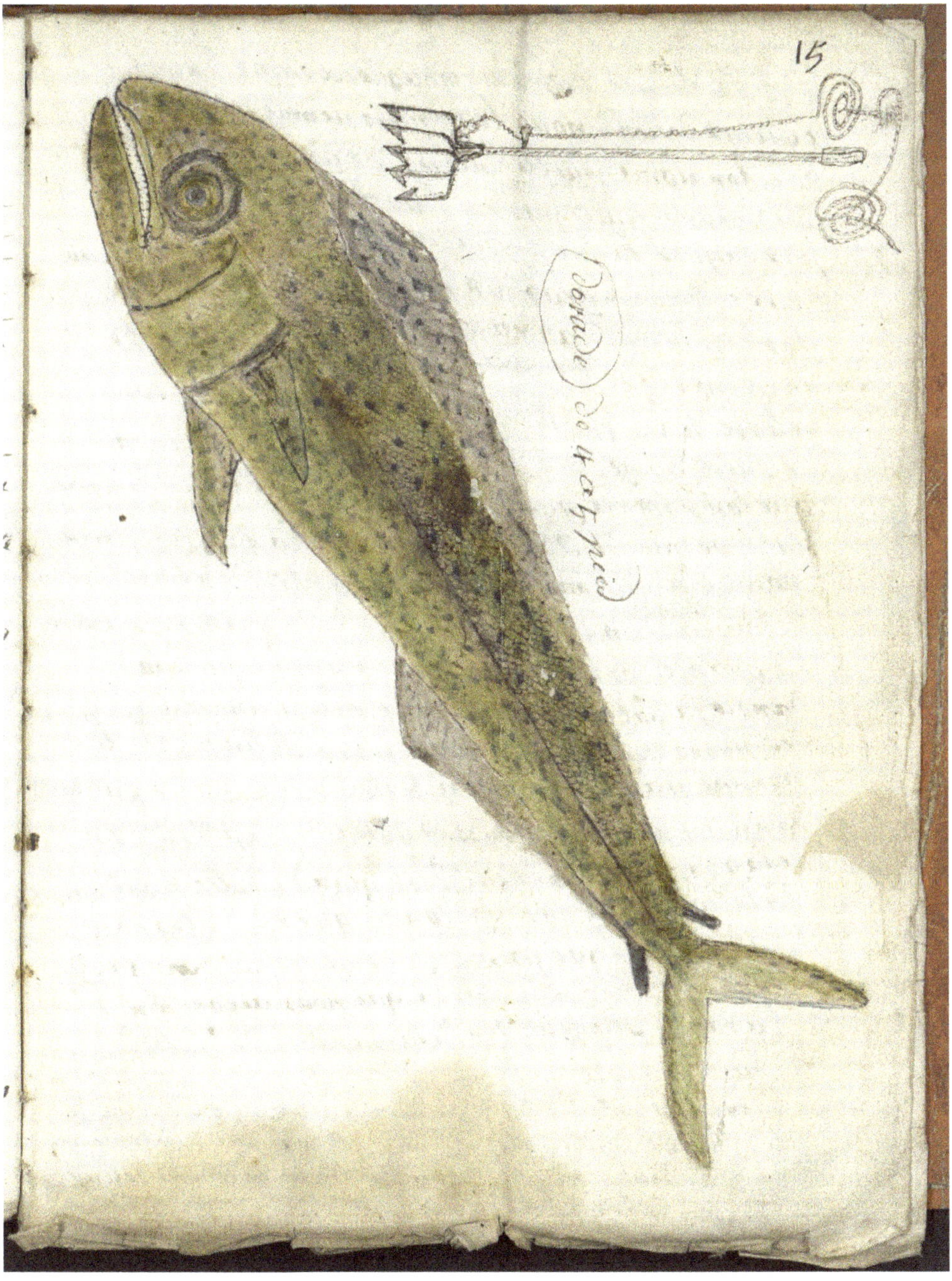

FIGURE 18.4 François de Meyer, Dolphin fish or mahi mahi (*Coryphaena hippurus*). Ink-washed drawing. In *Journal de voyage de François de Meyer*, 1698. The Hague, Nationaal Archief, Collection 349, Familie Delprat, 2.21.183.16, Inv. no 106a, p. 15

escape us. They give the *jol* the name of 'royal fish' because it has a fleur-de-lis on its head).

How much François is focused on the marine fauna is apparent from the following long quote. It describes how on 13 November a tropicbird (*Phaethon spec.*)[44] is observed; for the next three days (14, 15, and 16 November) nothing is said except the wind direction; and on 17 November a frigatebird (*Fregata spec.*) is spotted.[45]

> le 13 nous vimme un oiseau que lon nome paille enceus / et un faittivement [= effectivement] cetoizeaud nait pas mal nome paille enceus par cequil la une plume auderriere qui luij sert de ceüsee de la longeur dun pied et ½ selon / cetoizeau serd resinnal au navigateur pour leszille de lamerique et sinifiee que lon napproche de terre / [...] / le 17 nous vimme un oiseaud que lon nomme gallere / cetoizeaud ne laisse point la terre que de 80 a 100 lieux / cela fit juge ~~notre~~ notre capitainne que nous aprochions deterre / cetoiseaud me re semble a legron que nous zavons en France ormij un paquet de plume quil lad dans le jabot quil luij fait faire une manière desprous degallere et sait pour sellaque lon le nomme galere.[46]

> On 13 November, we saw a bird called 'paille enceus' [straw-in-bottom], and indeed this bird is rightly called 'paille enceus' because it has a feather in its butt that serves as a tail, which is longer than one and a half feet. This bird serves as a sign to the navigator of the islands of America, and signifies that we came near land. [...] On 17 November, we saw a bird called 'galley' [frigate bird]. This bird never goes more than 80 to 100 miles offshore. That made our captain think we were approaching land. This bird resembles the heron we have in France, except that it has in its crop a package of feathers which it can swell like the prow of a galley, hence it is called 'galley'.

These birds are described with the usual care – the information on naming these birds and their appearance (which is a sign of land nearby) comes from the captain or a member of the crew, and is supplemented by personal observation.

---

44 Tropicbirds have long extended tail feathers.

45 The Frigatebird, with an enormous red throat pouch, is indeed mainly found along coasts. It looks like a 'proue de galère' (bow of a galley).

46 NA Delprat 106a, 16.

On 18 November, a flying fish lands on the deck of the ship, which is copied by François [Fig. 18.5] and precisely described:

> / le 18 a 8 heure du soir il saute dans notre navire un poisson volland / ce poisson est de la grosseur dun meuillet qui ad 2 eelle contre la taitte et il sord de leaud quend il les poursui de quelque poisson et il volle dumoins ausij loins comme une caille que nous aieons en larope / saizelle son de la manière de selle dune chauve sourit / laille na point de separation / elle ad le fon dune petitte toille comme la petitte peaud dunuefe[47]
>
> On 18 November at 8 o'clock in the evening a flying fish jumps into our ship. This fish is the size of a mullet, and it has two wings on its head, and it jumps out of the water when chased by a fish, and it flies at least as far as a quail that we have in Europe. Its wings resemble those of a bat. The wing is not articulated, and is made of a thin membrane, like that of an egg.

We will return to this description, which is made up of a series of comparisons, in the analytical part of this article. For the moment, we only draw attention here to the comparison between the flight of the flying fish and that of a quail ('caille'): this could indicate that François lived in the countryside. One notes that this description is more precise than the preceding descriptions. This is a development that continues in the rest of the travelogue. The following description with drawing [Fig. 18.6] concerns a triggerfish, which is caught by the captain on 24 November:

> le matin notre capitainne prit un poisson nomme vielles[48] de la longeur dun pide [inserted] / ce poisson a la peaud extraimmement dure et qui nat aucune escailles il la le fon de la peaud grize avec daistrais desus croizoze qui luij fon fairre une maniere descaille resemblable a une pomme de pin et dans le fons de cette forme des caille il luy ad [= il y a] ad des petitte piceure grosse comme des pointe despaingle qui luij rend la peaud rude comme du chagrin et de plus jay remarque a se poisson quil lat sur

47 NA Delprat 106a, 16.

48 'vieille' is the name for different types of fish. Judging by the illustration, this is a triggerfish (*Balistidae*), probably *Canthidermis sufflamen* (Ocean Triggerfish). Wikipedia gives the following concise description: 'As a protection against predators, triggerfish can erect the first two dorsal spines: The first (anterior) spine is locked in place by erection of the short second spine, and can be unlocked only by depressing the second, "trigger" spine, hence the family name "triggerfish"'.

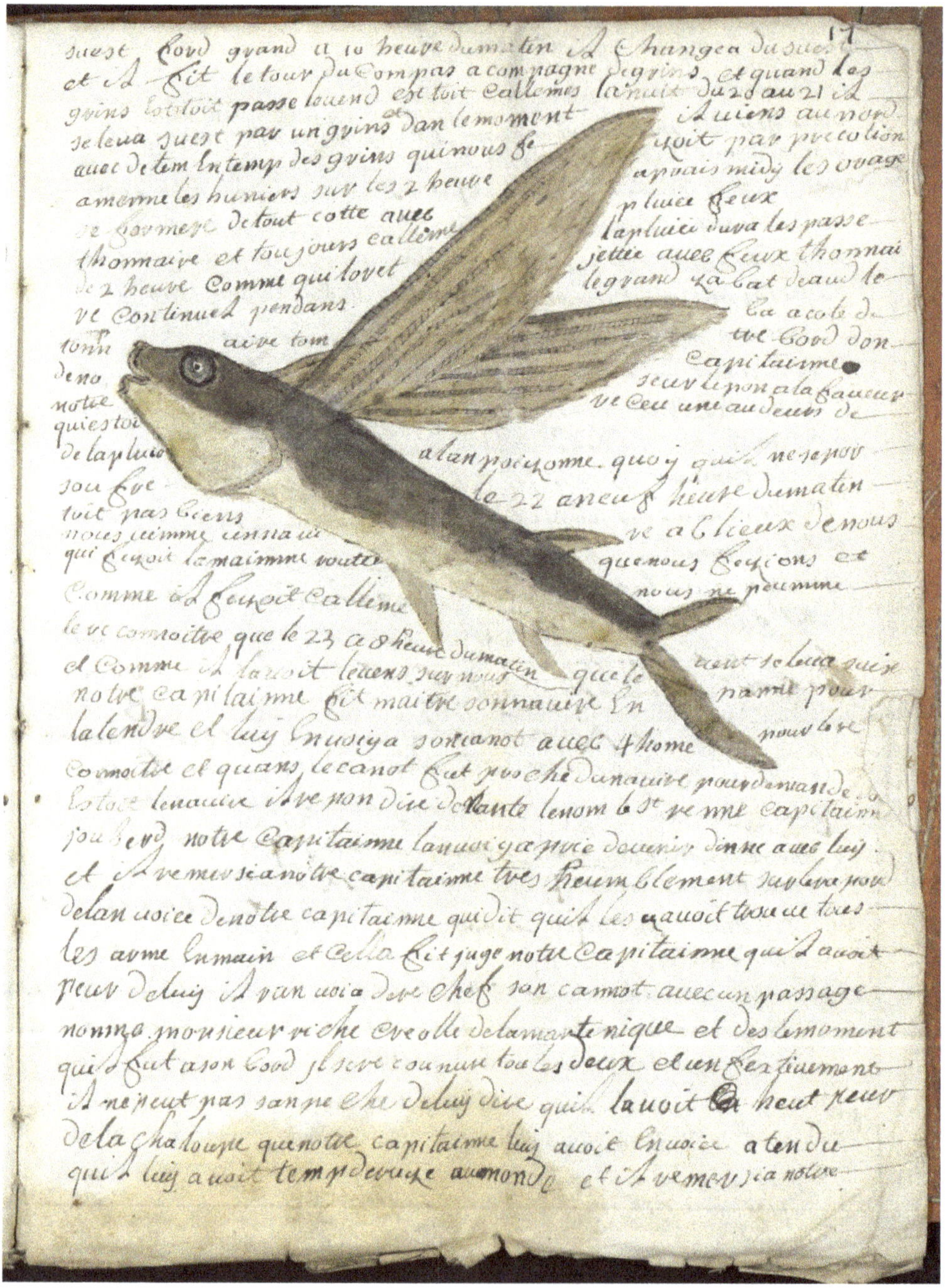

FIGURE 18.5 François de Meyer, Flying fish. Ink-washed drawing. In *Journal de voyage de François de Meyer*, 1698. The Hague, Nationaal Archief, Collection 349, Familie Delprat, 2.21.183.16, Inv. no 106a, p. 17

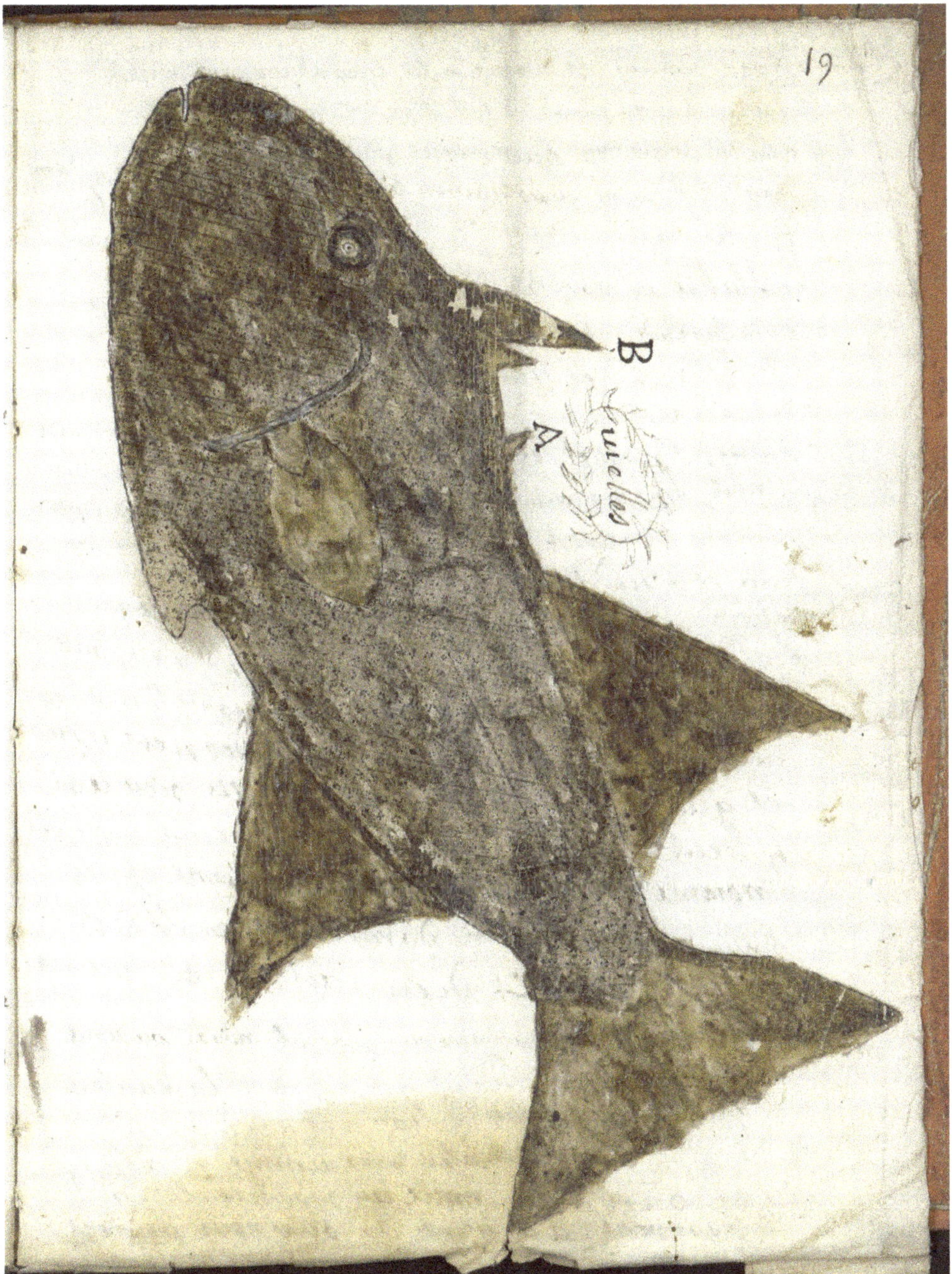

FIGURE 18.6 François de Meyer, Triggerfish. Ink-washed drawing. In *Journal de voyage de François de Meyer*, 1698. The Hague, Nationaal Archief, Collection 349, Familie Delprat, 2.21.183.16, Inv. no 106a, p. 19

> le dos une manière de dard qui ad le dard [repeated word] fait comme la moitiee de la patte dun chancre qui est piquee de petigrins et se dar resemble aune baterie[49] de fuzi et je desfiee alomme defaire baisser **B** sans faire baisser **A** car cest une maniere de resord comme le chiens dun fusir et de le moment que vous touce **A B** se ebuche sur **A** et je nait jamais peux connaitre le resor de son dard dou proveemoit que lon ne pouvoit pas faire baisser **B** sans faire baisser **A** et sepandand touchan **A B** baisse fasillemt tou lelon de son dos[50]
>
> In the morning our captain caught a fish called "vielles" that was about a foot long. This fish has an extremely hard skin without scales. It has grey skin with serrated stripes, which give it a kind of scales like those of a pinecone, and between those scales it has small spines, which make its skin like a donkey's skin. In addition, I noticed about this fish that it has a kind of spine on its back, which resembles a half of a crab claw. This spine is covered with small grains, and it resembles the trigger of a rifle. I challenge anyone to lower **B** without lowering **A**, because it is a mechanism that works like the cock of a gun: as soon as you touch **A**, **B** is activated. I have not been able to find the spring mechanism of the spine that prevents one from lowering **B** without lowering **A**, and that when one touches **A**, **B** goes down easily all the way down the spine.

Compared to the description of the flying fish, François here goes even further into detail for the description. Moreover, François not only describes the fish, but he also tells how he investigates how the 'trigger' of the fish works. The triggerfish is small in size, inedible, and can even be poisonous. That is why it is probable that the captain took this fish out of the water especially for François.

On 26 November, a large group of dolphins appears again: 'et sur les 10 heur du matin il viens une cantite demarsouins qui fire pluzieurs tour de notre navire pendand une demie heur de tem' (and by 10 o'clock in the morning a quantity of 'porpoises' [dolphins] that swam around our ship for half an hour).

On 30 November, a 'becunne' or 'le ceune' (the spelling is uncertain) is caught. It appears from the description and the drawing that it is a barracuda [Fig. 18.7], whose usual abbreviated name "cuda" seems related to the name given by François:

> a 6 heure du soir nous primme entre lille st alouisi et lille de st vainsent un poisson a la ligne nomme le cunne de la longueur de 3 pid de lon don

49 Upright pin of a 17th-century rifle struck by the battle cock.

50 NA Delprat 106a, 18.

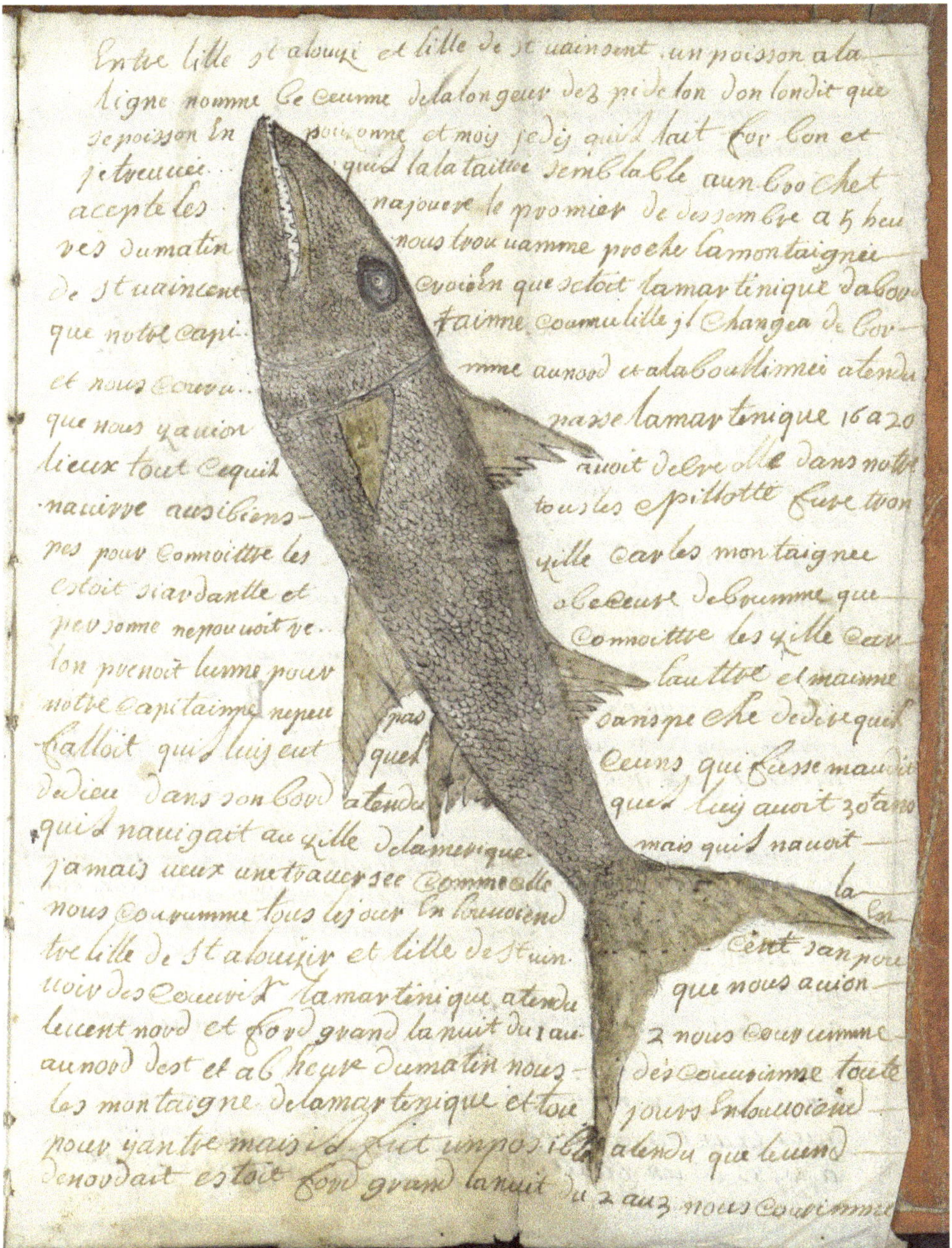

FIGURE 18.7 François de Meyer, Barracuda. Ink-washed drawing. In *Journal de voyage de François de Meyer*, 1698. The Hague, Nationaal Archief, Collection 349, Familie Delprat, 2.21.183.16, Inv. no 106a, p. 21

> lon dit que ce poisson enpoizonne et moij je dis quil lait for bon et je trouvee quil la la taillee semblable a un brochet acepte les najouere[51]
>
> at 6 o'clock in the evening, between Ile de Saint-Louis and Ile de Saint-Vincent, we caught with a fishing line a fish called 'cunne' [barracuda], three feet in length, which is said to be poisonous. But I tell you that this fish tastes really good, and I think it's shaped like a pike, except for the fins.

François's critical, inquisitive attitude, already apparent in the description of the triggerfish, is also apparent here: despite the fact that the fish is said to be poisonous, François eats the fish, and considers it 'for bon'. His keen eye for detail is also evident when he points out both its difference from and its similarity to the pike.

François's development in natural history description and portrayal culminates in the last fish caught and described, a 175-kilogram shark:

> En louvoiend nous primme sur les 2 heure aprois mijdij nous primme[52] un poisson ue lon nomme requin pezent 350 tt / don se poisson est lennemij de lome attendu quan lomme se baigne dans la mair se poisson le coupe en 2 quanstil lattrape et deplus jay remarque ase poisson que la nature luij a fait ne extraordinairre danjandre sais petis en comparaizon des autre poisson / quand celluij la enjandre ses petis comme une chienne enjandre sais petits chiens aulieu que les autres poisson enjandre par les zeufe et jene parlle pas zisij par ouij dire / je parlle pour avoir veut et pour preuvee voilla[53] un des 5 petits comme il les zavoir dans le corp marque **B**[54] tout prais a sortir de son corp / dieu nous benit deleprendre se jour la caraulieux dun que nous croijons prendre nous enprime 6 tous za lafois[55]
>
> Twirling around, we caught a fish called shark at 2 in the afternoon, weighing 350 pounds. This fish is an enemy of man, because when a man goes swimming in the sea, this fish bites him in half, if it catches him. Moreover, what I noticed about this fish is that nature makes it produce its young in a special way compared to other fish. This fish produces her

51 NA Delprat 106a, 21–22.

52 Repeated words.

53 So, the drawing is of a young newborn shark. The drawing serves as proof of the veracity of the viviparous shark story.

54 It is not clear what the letter B refers to, at least not to the accompanying drawing.

55 NA Delprat 106a, 22.

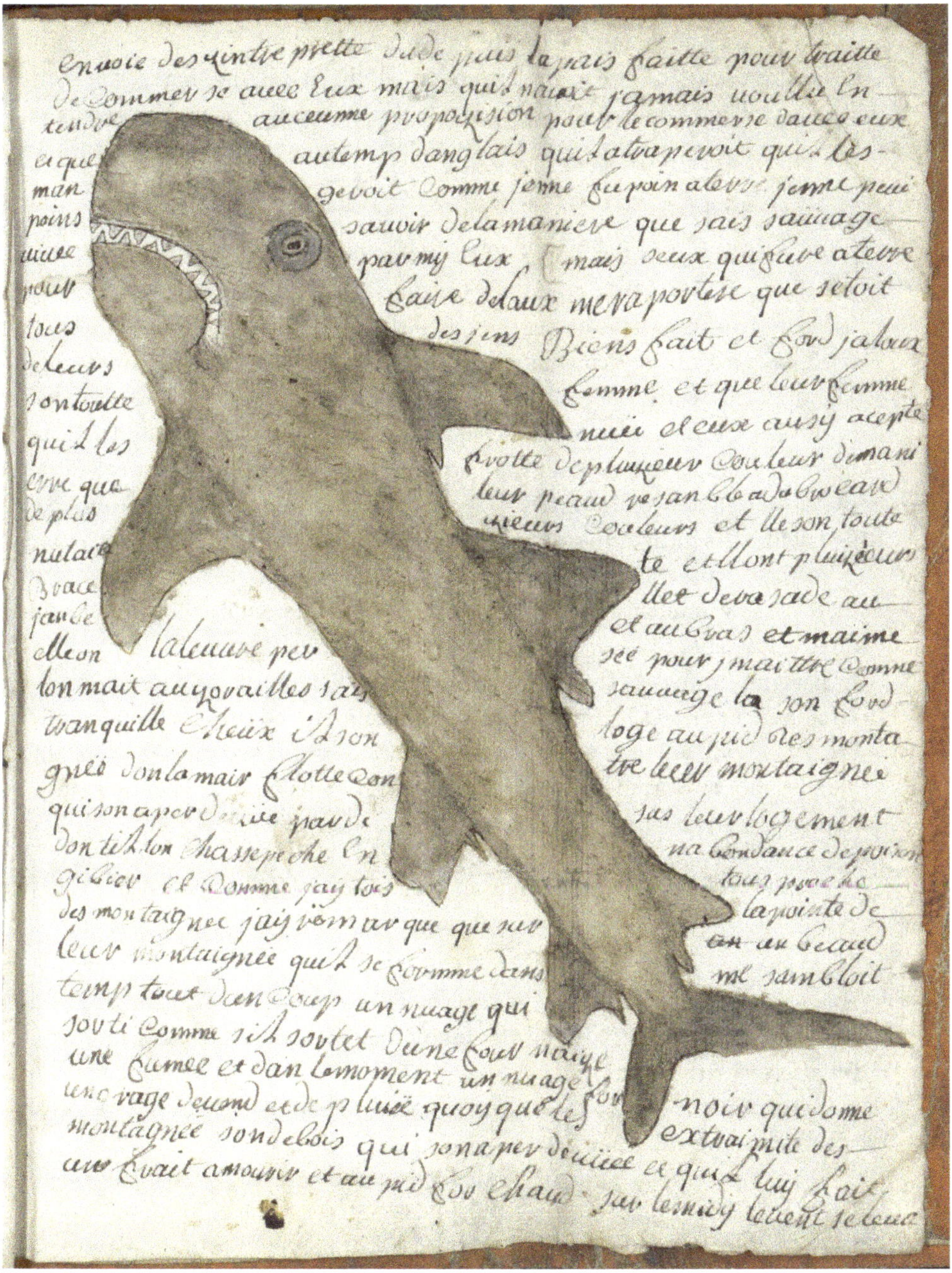

FIGURE 18.8 François de Meyer, Shark. Ink-washed drawing. In *Journal de voyage de François de Meyer*, 1698. The Hague, Nationaal Archief, Collection 349, Familie Delprat, 2.21.183.16, Inv. no 106a, p. 23

> young as a bitch produces her cubs, while other fish reproduce by means of eggs. And I'm not talking from hearsay about this. I speak of it because I have seen it. Here's a drawing of one of the five young sharks, just as the shark had it in her body (marked with a B), ready to leave the body. God had mercy on us to catch the shark just that day, because we thought we were only going to catch a single shark, but we caught six at a time.

It is remarkable that the danger of the animal is only briefly, almost obligatorily, mentioned. Much more interesting for François is the discovery that the fish is viviparous. The drawing [Fig. 18.8] therefore does not depict the shark itself, but one of the five young that were about to leave the mother's body. The drawing, which depicts the unborn shark as large as possible – i.e. half of its size – obliquely across the page, is intended to authenticate what has been described. Also striking is a certain sense of humour that speaks from the closing sentence of this description: it's a good thing we caught this shark just before the birth of the young, so instead of one shark we had six at a time.

As the discussion of these passages indicates, François certainly had a more than passing interest in the birds, fish, and sea mammals that he encountered while aboard l'Aigle. His interest in examining marine life only seems to have intensified as the journey progressed. References to observations of fascinating marine species become more frequent, and they are documented in more detail. We now turn to a closer analysis of his strategies of depicting and describing marine life, placing them in a broader context of early modern natural historical study of the time.

4 Depicting and Describing Marine Life

So far, we have treated François's travelogue as a stand-alone document, without attempting to plot him on the map of early modern people who busied themselves with the study of nature. While the travelogue is a unique document, certain aspects of it conform to more widely shared natural historical practices, as we will show here.

For example, historians have made clear that the study of living nature was by no means the prerogative of university-educated individuals. People with a more practical background, such as fishmongers and fishermen, took to study and even produce manuscripts.[56] We might group François among

56 See, for example, Egmond F., "On Northern Shores: Sixteenth-Century Observations of Fish and Seabirds (North Sea and North Atlantic)", in MacGregor A. (ed.) *Naturalists in*

what Anthony Grafton calls 'practical men', who were better versed in practical techniques than in book learning.[57] That such practical men were valued for their uninhibited look can be seen in Michel de Montaigne's chapter "Des cannibales" in his *Essais* (1580, first edition). In this chapter, Montaigne dwells upon the usefulness of the eyewitnesses of practical men, fulgurating against 'those clever' cosmographers, who always have the tendency to embellish their reports. Montaigne eloquently exemplifies his argumentation by focussing on one particular person from his household, who stayed for a long time in French Brazil:

> I have long had a man with me who stayed some ten or twelve years in that other world which was discovered in our century when Villegaignon made his landfall and named it La France Antartique. [...] That man of mine was a simple, rough fellow – qualities which make for a good witness: those clever chaps notice more things more carefully but are always adding glosses; they cannot help by changing their story a little in order to make their views triumph and be more persuasive; they never show you anything purely as it is: they bend it and disguise it to fit in with their own views. [...][58]

This matter-of-fact perspective of the reliable eyewitness, without frills or exaggeration, is visible everywhere in François's travelogue.

Furthermore, that ships were privileged sites for the study of natural phenomena was evident in the early modern period. Accounts of observations of marine life from aboard a ship en route to faraway islands are not uncommon.[59] For European naturalists journeying overseas, documenting their experiences of nature began upon departure from the harbour, not upon arrival on the dock. Of course, sailors and seamen also made plenty of observations over the course of their employment; we find their accounts, for example, among

---

the Field: Collecting, Recording and Preserving the Natural World from the Fifteenth to the Twenty-First Century (Leiden: 2018) 129–148; Trijp, D. van, "Fresh Fish: Observation Up Close in Late Seventeenth-Century England", *Notes and Records: The Royal Society Journal of the History of Science* 75 (2021) 311–332.

57 Grafton A., *New Worlds, Ancient Texts. The Power of Tradition and the Shock of Discovery* (Cambridge, MA: 2014), 69.

58 Montaigne Michel de, *The Complete Essays*, transl. M.A. Screech (London: 1991) 231.

59 For example, Davids, *Global Ocean of Knowledge* 66–68; Williams G., *Naturalists at Sea: Scientific Travellers from Dampier to Darwin* (New Haven: 2013) 76.

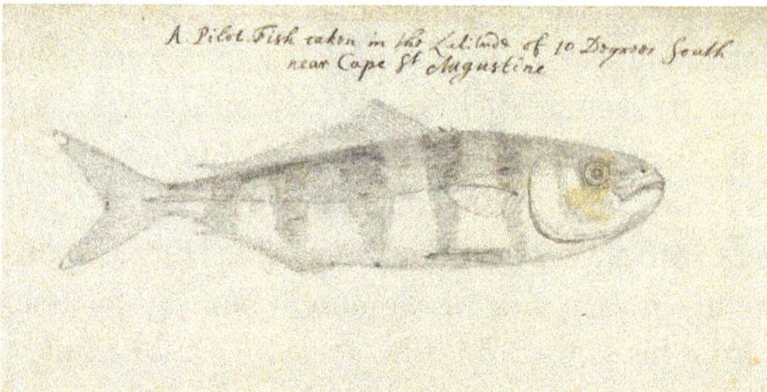

FIGURE 18.9
Edmond Halley, Pilot fish (*Naucrates ductor*). Pencil on paper. 1699–1700. Royal Society Archives MS/131/49. https://pictures.royalsociety.org/image-rs-9364

the works of the Fellows of the Royal Society in London, who drew on their information and interpretation.[60]

We will now look at textual and visual documentation of nature in overseas journeys. In 1698, the same year our travelogue is dated, Edmond Halley (1656–1742) was commissioned to command the voyages of HMS Paramore to the South Atlantic. These voyages, which lasted from September 1699 to September 1700, had as their main aim to take readings of the variations of the magnetic needle in the Atlantic Ocean to improve navigation.[61] Halley also set aside time during this voyage to produce several sketches of the fishes that he came across. Upon return to London, he brought these to a meeting of the Royal Society.[62] Among them were pencil sketches of species that François also depicted: tuna, pilot fish, triggerfish, and flying fish. The inscriptions that Halley added to the sketches indicate that the fish were caught while he was on the ship. The pilot fish, for example, was '[…] taken in the Latitude of 10 Degrees South near Cape St Augustine', whereas the flying fish was '[…] taken in the sight of Palm'[63] [Fig. 18.9].

A little over a decade before François left for Guadeloupe, the physician and collector Hans Sloane (1660–1753) undertook a similar trajectory. He describes his journey on no fewer than 47 pages in the first part of his *Voyage to Jamaica* (London, 1707–1725). From his departure in September 1687 onwards, Sloane reports every few days, often mentioning the directions of the wind and other weather circumstances in the few first lines. Most of his attention, however, is

---

60 Deacon M., *Scientists and the Sea 1650–1900: A Study of Marine Science* (Ashgate: 1997, second edition). 75; Hellawell P., "'The Best and Most Practical Philosophers': Seamen and the Authority of Experience in Early Modern Science", *History of Science* 58 (2019) 28–50.

61 Cook A.H., *Edmond Halley: Charting the Heavens and the Seas* (Oxford: 1998) 256–291.

62 Royal Society Archives, Atlantic tuna: MS/131/48, and unidentified fish, MS/131/1, Flying fish, MS/131/53, Pilot fish, MS/131/49, Doctor fish [a type of triggerfish], MS/131/2.

63 'Palm' likely refers to La Palma, part of the Canary Islands. Royal Society Archives, MS/131/53.

taken up by observations of the natural world. He describes the many animals they pass along the way. Among them are 'many of a kind of *Larus*, or Gull', a lark taking a rest on the ship's rigging, and a giant jellyfish seamen call 'Caravel or Portuguese Man of War' because their impressive size reminds them of a warship, as well as vast amounts of *albacore*.[64]

Just like François, Sloane seems interested in fishing techniques and the taste of different species. Among the tools he mentions is the harping-iron with which seamen catch porpoises for food.[65] One time, a shark was brought aboard. Upon cutting it open, the crew saw several young in its belly, a situation similar to that described by François.[66] When one reads Sloane's and François's reports alongside each other, despite some obvious similarities (the general structure of the travelogue, the trajectory, the species observed) the difference is clear. For Sloane, every phenomenon which he observes sparks off a wealth of comparisons to travel reports, books, and other publications. His description of the common dolphin fish, for example, contains no fewer than 28 references to other authors.[67] That is not the case for François, who nowhere refers to any specific written source.

Yet François is not quite a *tabula rasa*. This is apparent in the first place from his drawings: although these are roughly drawn and are based on autopsy, it requires practice to select and depict precisely those characteristics of a fish that make the fish recognizable as a species. This applies to the shape and proportions of the fish's body as well as the shape and position of its fins. To be able to do this, François could have benefited from the many illustrated books on fish and other aquatic animals that had been widely available in Europe since the 1550s: especially works by Belon, Rondelet, Gessner, Aldrovandi, Jonston, and Willughby and Ray. Not that François had these works in front of him when writing his travelogue, but he could have leafed through these kinds of illustrated fish books.

That he did have illustrated works before his mind's eye is apparent from the way in which some fish are depicted with the harpoon used. For example, in the work of Guillaume Rondelet, a whale is depicted in a similar way with a harpoon [Fig. 18.10].[68] Also, the way the triggerfish mechanism is depicted, with A and B references from the text to the illustration, attests to knowledge of an illustrated technical book – although not necessarily a book on fish.

64 Sloane, *A Voyage* 4, 7, 11.
65 Ibid. 5. More mentions on 11, 20, 22, 26.
66 Ibid. 23.
67 Ibid. 21.
68 Rondelet Guillaume, *L'histoire entière des poissons* (Lyon, Macé Bonhomme: 1558) 351.

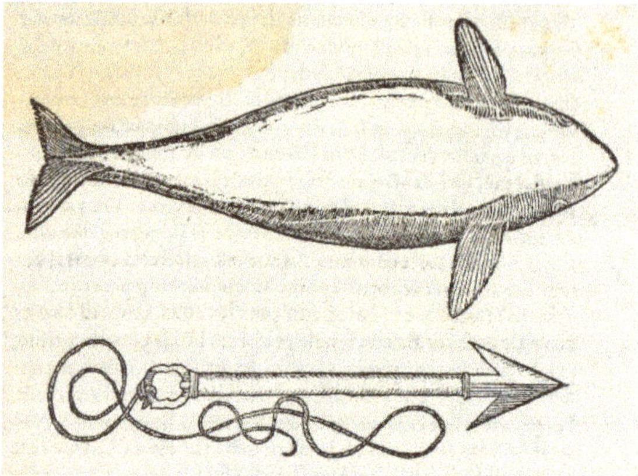

FIGURE 18.10
Whale. Rondelet Guillaume, *L'histoire entière des poissons* (Lyon, Macé Bonhomme: 1558) 351. https://gallica.bnf.fr/ark:/12148/bpt6k1512044f/f367.item

This way of referring was common in medical literature from the 16th century on. Famous is the comparative anatomical image of Pierre Belon, who, using letter references, indicated the similarities between the skeleton of a human and that of a bird.[69] In the later 18th century, this form of illustration would become commonplace; one can think of Diderot's *Encyclopédie*.

The description of the shark is perhaps the most convincing evidence that François gives of a certain bookish knowledge: he appears to be aware of the fact that the natural historical works on fish of the time do not mention that certain shark species are viviparous. This explains why François emphasizes both textually and visually that he has seen this – according to him – unknown fact with his own eyes.

Other descriptions also suggest that François is not completely ignorant of existing literature. For instance, the resemblances between François's travelogue and Jean de Léry's account of his voyage to Brazil (1578)[70] are manifold: Both authors describe dolphins, flying fish, mahi-mahis, white tunas, whales, sharks, frigatebirds in a very similar way.[71] Moreover, like almost all early modern naturalists, François begins each description with the name of the animal described, which is usually given via a fixed formula ('que l'on nomme' …), and sometimes provided with an etymological explanation (wagtail, tropicbird, frigate bird), which testifies to a certain linguistic awareness. The naming of

69 Belon Pierre, *L'histoire de la nature des oyseaux* (Paris, Guillaume Cavellat: 1555) 40–41.

70 Léry Jean de, *Histoire d'un voyage faict en la terre du Bresil* (La Rochelle, Antoine Chuppin: 1578; numerous editions and translations).

71 Smith P.J., "Léry et les poissons: une lecture rapprochée des stratégies descriptives", *Le Verger* 25 (2022) 1–18 (http://cornucopia16.com/blog/2023/01/06/bouquet-xxv-lhistoire-dun-voyage-faict-en-la-terre-du-bresil-de-jean-de-lery/) (last consultation 14 January 2023).

the animals comes from different sources: native dialect (wagtail, dogfish); nautical vocabulary (tropicbird, frigatebird, and various tropical fish species); of the latter category, some are semantically transparent (pilotfish), others (for instance 'becunne' or 'ceune') are not.

What François says about the anatomy of the animal sometimes seems to be a distant echo of what can be read in fish books. For example, the resemblance between the organs of a dolphin and those of a pig is a commonplace, which can be found in the work of Belon, Rondelet, Gessner and Léry.[72] The description of the flying fish is also reminiscent of that of Rondelet, Belon and Léry.[73] François describes the same characteristics as Rondelet, Belon or Léry, with a similar way of comparing. For example, François compares the length of the flying fish with that of a 'meuiller' (a mullet); Rondelet does this with a 'muget' (the words are similar in sound). Belon mentions that the fish fly because they are afraid: 'Quand il ha peur' en la mer, il sort hors' (When he is afraid in the sea, he leaves it); Léry observes: 'ces pauvres poissons volans [...] ne sont jamais en repos: car [...] les Albacores et autres grands poissons les poursuivans pour les manger [...], ils se veulent sauver au vol'[74] (these poor flying fish are never at rest: for as the white tunas and the other large fish pursue them to eat them, they want to escape in flight), and François writes: 'il sord de leaud quend il les poursui de quelque poisson'[75] (He leaves the water when he is chased by some fish). François gives information about the length of the flight (as far as a quail flies): with Belon this is, 'Il vollee quelquesfois iusques a un traict d'arbaleste' (he flies sometimes as far as the arrow of a crossbow); Léry has: 'quelques fois près de cent pas loin' ([they] sometimes [fly] nearly a hundred paces). François's comparison between the wings of a flying fish and those of a bat is furthermore similar to Rondelet's and Léry's.

François makes frequent use of this descriptive strategy, namely the analogy or comparison of the described animal with known animals. The use of analogy, in relating the unknown to the known, can have two different rhetorical effects. The effect can be either confidence-inspiring (there is no difference between the 'chien de mer' and the 'toul' in La Rochelle) or alienating (the 'soufleur' is larger than a European ox). The latter seems to fit into the strategy of the hyperbole: 'huge amount ...', 'the largest ... the crew had ever seen', etc.),

72 For Belon, see Belon Pierre, *La nature et diversité des poissons* (Paris, Charles Estienne: 1555) 4, 11.

73 Belon, *La Nature et diversité des poissons* 191; Rondelet Guillaume, *La Première [seconde] partie de l'Histoire entière des poissons* (Lyon, Macé Bonhomme: 1558) 137.

74 I cite Léry Jean de, *Histoire d'un voyage faict en la terre du Bresil 1578 (2e édition, 1580)*, ed. F. Lestringant (Paris: 1994) 128.

75 NA Delprat 106a.

which is very common in travelogues. But one has the impression that François is never exaggerating. The analogy can apply to the whole animal as well as to the parts (or aspects) of an animal: the flying fish is, as it were, divided into parts, and each part is compared with a corresponding part of another animal. It is noteworthy that within the analogy between the known and the unknown, the geographical context of the known is often made explicit: La Rochelle, France, Europe.

5 Conclusion

As has been stressed in the introduction, little remains known about François de Meyer and his reasons for embarking on this specific journey to Guadeloupe. In this article, we have examined his manuscript in order to shed light on François's unusually lively interest in maritime fauna. Based on an analysis of François's spelling and use of language, he can be characterized as a person with at least a basic education. On the basis of his handwriting and especially his aberrant spelling, we can say that he is not a trained writer. The phonetic spelling and choice of words as well as its references to France suggest that François, despite his (probably) Dutch family name, was French. Judging from his good relationship with the captain, as well as the fact that he has the time, space, and interest to draw up a logbook, he is apparently high in rank.

This article has shown that François is an attentive observer of all that occurs on as well as around the ship. His descriptions are extensive, especially those of maritime fauna, which at one point takes up all his attention. In his travelogue, François shows a developing interest in natural historical study. His accounts of marine fauna differ from those of learned naturalists in that they are not laden with bibliographical references. At the same time, his reports seamlessly adhere to the emphasis on *autopsia* widely shared among naturalists, stressing that he has seen the species which he describes with his very own eyes. The coloured drawings convey and underline these observations. Sources like these offer a wealth of insight into how early modern individuals without a formal training experienced and interpreted the natural world.

Bibliography

Belon Pierre, *L'histoire de la nature des oyseaux* (Paris, Guillaume Cavellat: 1555).
Belon Pierre, *La nature et diversité des poissons* (Paris, Charles Estienne: 1555).
Cook A.H., *Edmond Halley: Charting the Heavens and the Seas* (Oxford: 1998).

Danois E., "Les noms de quelques animaux et végétaux marins en dialecte de Léon", *Annales de Bretagne et des pays de l'Ouest* 25.3 (1909) 548–555.

Davids K., *Global Ocean of Knowledge, 1600–1860: Globalization and Maritime Knowledge* (London: 2020).

Deacon M., *Scientists and the Sea 1650–1900: A Study of Marine Science* (Ashgate: 1997, second edition).

Egmond F., "On Northern Shores: Sixteenth-Century Observations of Fish and Seabirds (North Sea and North Atlantic)", in MacGregor A. (ed.) *Naturalists in the Field: Collecting, Recording and Preserving the Natural World from the Fifteenth to the Twenty-First Century* (Leiden: 2018) 129–148.

Grafton A., *New Worlds, Ancient Texts. The Power of Tradition and the Shock of Discovery* (Cambridge, MA: 2014).

Hellawell P., "'The Best and Most Practical Philosophers': Seamen and the Authority of Experience in Early Modern Science", *History of Science* 58 (2019) 28–50.

Herber J., "Le folklore de la mer (dans l'Hérault)", *Folklore* 11.3 (1948) 43–48.

Léry Jean de, *Histoire d'un voyage faict en la terre du Bresil 1578* (2e *édition, 1580*), ed. F. Lestringant (Paris: 1994).

Meyer François de, *Reisjournaal van François de Meyer, opvarende op het schip l'Aigle onder bevel van François Dubois naar Martinique*. 1698–1699. Ms. Nationaal Archief, Den Haag, Collectie 349 Familie Delprat, toegang 2.21.183.16, inv.nr. 106a.

Montaigne Michel de, *The Complete Essays*, transl. M.A. Screech (London: 1991).

Puichaud C., "Dictionnaire du patois du Bas-Gatinais (Suite)", *Revue de philologie française et provençale* 7 (1887) 100–137.

Rondelet Guillaume, *La Première [seconde] partie de l'Histoire entière des poissons* (Lyon, Macé Bonhomme: 1558).

Rondelet Guillaume, *L'histoire entière des poissons* (Lyon, Macé Bonhomme: 1558).

Sloane Hans, *A Voyage to the Islands Madera, Barbados, Nieves, S. Christophers and Jamaica* (London, B.M. for the author: 1705–1725).

Smith P.J., "Léry et les poissons: une lecture rapprochée des stratégies descriptives", *Le Verger* 25 (2022) 1–18 (http://cornucopia16.com/blog/2023/01/06/bouquet-xxv-lhistoire-dun-voyage-faict-en-la-terre-du-bresil-de-jean-de-lery/).

Trijp D. van, "Fresh Fish: Observation Up Close in Late Seventeenth-Century England", *Notes and Records: The Royal Society Journal of the History of Science* 75 (2021) 311–332.

Williams G., *Naturalists at Sea: Scientific Travellers from Dampier to Darwin* (New Haven: 2013).

CHAPTER 19

The Afterlives of Fish Far from Home: (Mis)Representations in the Iconography of Preserved and Printed Pufferfish in 18th-Century Germany

Dorothee Fischer

1 Pufferfish Far from the Sea

In the period of Enlightenment, research in natural history was less concerned with wondrous singularities as in previous centuries, but primarily endeavoured to generate knowledge of general laws of nature by ordering, classifying, and comparing objects.[1] Hence, a coherent systematisation of natural objects became increasingly important in natural collections of the time. The first most cohesive and widely accepted attempt at such systematisation was made by Carl Linnaeus (1707–1778).[2] In his influential *Systema Naturae* (published in twelve volumes between 1735 and 1768), he differentiates between the Animal, Vegetable, and Mineral Kingdoms and formally introduces binomial nomenclature.[3]

To research and categorise the world around them, naturalists needed exemplar organisms that could permanently demonstrate which characteristics defined the species as a whole. This approach could only be realised with the help of specimens; individual animals that were chosen as representatives of their conspecifics. The value and credibility of information stored in these

1 Heesen A. te – Spary E.C., "Sammeln als Wissen", in Heesen A. te – Spary E.C. (eds.), *Sammeln als Wissen. Das Sammeln und seine wissenschaftsgeschichtliche Bedeutung* (Göttingen: 2001) 7–21, here 14. Rijks M., "Fish out of Water. Collecting Aquatic Animals in the Early Modern Period", in Rijks M. – Smith P.J. – Egmond F. (eds.), *Fish & Fiction. Aquatic Animals between Science and Imagination (1500–1900)* (Leiden: 2020) 48–61, here 49.

2 Siemer S., "Naturkundliches Sammeln im 18. Jahrhundert. Ein Überblick", in Mieth K.M – Museum Waldenburg (eds.), *Das Naturalienkabinett. Sammeln, Forschen, Zeigen* (Chemnitz: 2011) 42–54, here 51. See also: Trijp D. van, "The Murky Waters of Classification. Ordering Fish in Eighteenth-century Europe", in Rijks M. – Smith P.J. – Egmond F. (eds.), *Fish & Fiction. Aquatic Animals between Science and Imagination (1500–1900)* (Leiden: 2020) 76–85.

3 This system is still used today providing scientific names for particular species consisting of two parts referring to the genus and species (e.g., *Homo sapiens*).

 | DOI:10.1163/9789004681187_020

exemplar organisms vary among different animals. From the perspective of the collectors, fish are, in this regard, a complicated case. Until the aquarium became accessible to a broader public in the middle of the 19th century, it was nearly impossible to keep (especially salt) water creatures alive outside of their natural habitat over a longer period of time.[4] Additionally, when fish are taken out of the water, their appearance changes dramatically: As life leaves their bodies, their scales and skins lose their shine and their colours fade. Especially bright colours drastically diminish within a few hours and the contrast of their patterns decreases.[5] In order to study and describe water creatures, it was necessary to find other ways to conserve them, especially for naturalists operating far from the sea. Thus, the characteristics of living fish were transferred into written sources (e.g., notebooks, letters, inventory lists), but also into sketches, drawings and prints. Furthermore, their bodies were physically preserved by transforming them into wet [Fig. 19.1] or dry specimens [Fig. 19.2].[6] These different kinds of (re)presentation of the animals generated, legitimised and further developed knowledge about fish in and beyond the collection rooms, while also posing several problems.[7] How can the relationship between these different media, the animals' preserved bodies, the prints[8] and written descriptions be defined?

Most taxidermic items of the early modern period have not lasted to the present day due to poor preservation techniques that only improved in the late 18th century and, accordingly, documents are frequently the sole witnesses of

4 Further reading: Vennen M., *Das Aquarium. Praktiken, Techniken und Medien der Wissensproduktion (1840–1910)* (Göttingen: 2018). Of course, attempts had been made prior the aquarium; for example, some naturalists contained sweet water fishes temporarily in vessels. Paepke H.-J., "M.E. Bloch's frühe aquatische Versuche. Über einen Pionier der Heimtierhaltung", *T.I. Magazin* 28.129 (1996) 33–36.

5 Schlegel H., "Über das Anfertigen von Amphibien- und Fischbildern", in Nissen C. (ed.), *Die zoologische Buchillustration. Ihre Bibliographie und Geschichte. Band II: Geschichte* (Stuttgart: 1978) 250–252, here 251 and Rijks, "Fish out of Water" 51.

6 It is to be discussed whether all of these "images" are able to represent the living animal properly or, moreover, what exactly can be presented of the animals in the context of collections.

7 Because fish lose characteristics like their colour sometimes within only minutes after leaving the water, they are, until today, an especially difficult animal to conserve. Even with better technology, there barely are attempts to conserve fish taxidermically. It is thus common to use casts, filling the negative moulds and painting on the resulting artificial body. After this process, only a slight trace of the dead fish remains as its organic components are not conserved.

8 Even though drawings chronologically preceded the prints, in this study I focus only on the prints as these synthesise and multiply all the preliminary work and, hence, allow for a spread of knowledge beyond the collections.

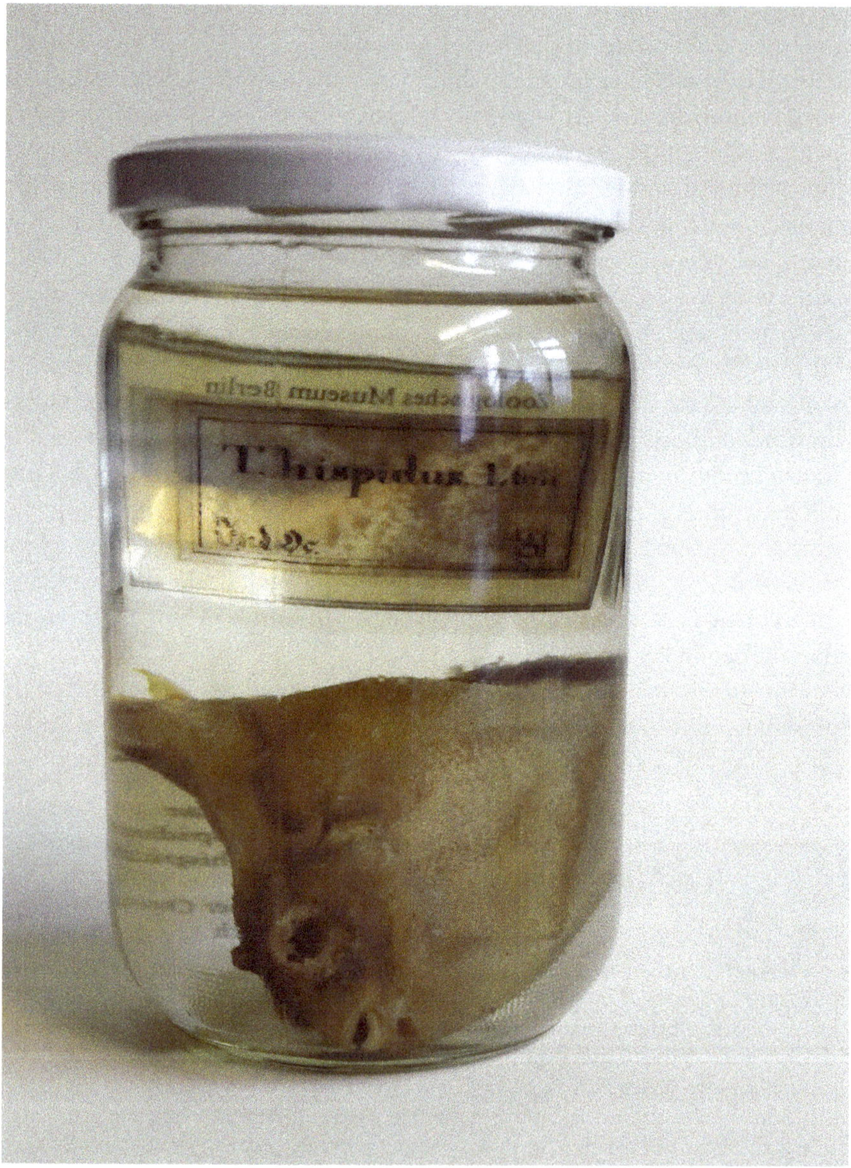

FIGURE 19.1 *Tetrodon hispidus*, wet specimen, 18th century, circa 9 cm × 14 cm (glass vessel), Bloch Collection, ZMB_Pisces_4274, Museum für Naturkunde Berlin (Germany)
IMAGE © DOROTHEE FISCHER 2022

what were once magnificent taxidermic collections.[9] This article focuses on two exemplary collections located in central Europe, far from the sea, both of which are 'among the oldest existing ichthyological collections in the world'[10] and thus make accessible early modern fish taxidermies. The first, owned by the family Linck, was established and continued in Leipzig (Germany) between 1670 and 1807 and, hence, lasted the entire 18th century.[11] The second is the collection of Marcus Elieser Bloch (1723–1799), who started collecting fish only in the late 1770s, but soon composed one of the largest ichthyological collections of his time.[12] Both the Linck and Bloch collections remained unparalleled through centuries in their magnitude and quality.[13]

The universal collection of the pharmacist family Linck comprised, in addition to *scientifica* objects, rarities of botany, mineralogy and zoology. It soon became especially known for its large compilation of snakes, starfish and fish thanks to the family's scholarly output.[14] This collection was established and furthered over the course of three generations: Around 1670 Heinrich Linck (1638–1717) initiated it, afterwards his son, Johann Heinrich Linck the Elder (1674–1734), and his grandson, Johann Heinrich the Younger (1734–1807), continued their ancestor's work by preserving and expanding the collection.

9 Bauernfeind R., "Jona und der Hai. Zu einem frühneuzeitlichen Hai-Präparat zwischen Exegese und Naturgeschichte", *Zeitschrift für Kunstgeschichte* 82 (2019) 166–178, here 167.

10 Paepke H.-J., *Bloch's Fish Collection in the Museum für Naturkunde der Humboldt-Universität zu Berlin: An Illustrated Catalog and Historical Account* (Rugell: 1999) 11 on the Bloch collection. He gives a similar statement about the Linck collection on page 24.

11 Engelmann W.E. – Sterba G.W.H., "Über einige interessante Objekte in der Fischsammlung des Linck'schen Naturalienkabinetts", *Bulletin of Fish Biology* 16.1/2 (2016) 15–32, here 15.

12 Paepke, *Bloch's Fish Collection* 157.

13 Unfortunately, both collections (as so many others of that time) faced an untimely end after their owners' deaths. Since there were no heirs to Johann Heinrich Linck the Younger, the whole collection was auctioned off. The objects were torn from their original architectural context, merged with other collections and were relocated to a newly built museum in Waldenburg, roughly 60 kilometres from Leipzig, where most of the items can still be found today. Due to these rearrangements, however, the provenances of some exhibits remain foggy. Further on the history of the collection: Ross A.S., "Recycling Embryos: Old Animal Specimens in New Museums, 1660–1840", *Journal of Social History* 52.4 (2019) 1087–1109. Bloch's collection was also sold after his death and not spared some changes; moving the collection to a smaller space resulted in selling multiple specimens. Further on this: Paepke, *Bloch's Fish Collection* 20–21.

14 More on the Linck collection generally: Beyrich H., "Das Linck'sche Naturalien- und Kunstkabinett aus Leipzig, jetzt in Waldenburg (Sachsen)", in Grote A. (ed.), *Macrocosmos in Microcosmo. Die Welt in der Stube. Zur Geschichte des Sammelns 1450–1800* (Wiesbaden: 1994) 581–601.

The physician and naturalist Marcus Elieser Bloch was eager to accumulate new insights into natural history, especially where Linnaeus's classification did not suffice.[15] Bloch's effort has often been described as the most influential ichthyological achievement of the 18th century,[16] and entailed collecting a great many fish, ascribing names to new species, and disseminating his knowledge through publications that are still valued today among ichthyologists and laypersons alike. Thanks to Bloch's famous plates of fish illustrations, totalling about 800 in number,[17] many species were presented in vivid colour and impressive detail to a larger audience for the first time.[18] In line with the ideas of the Enlightenment, these universal depictions aimed to synthesise the main characteristics of the described fishes.

Both collections were located far from the sea but in close proximity to each other (with roughly 170 kilometres of distance between them). As both Bloch and the Lincks had the ambition to incorporate the latest research and scientific knowledge into their collecting activities and maintained an international network with other collectors and scholars,[19] it is no surprise that in the Lincks' surviving guest book, there is a record of Bloch visiting their collection in 1789.[20] As has been demonstrated by biologists Wolf-Eberhard Engelmann and Günther H.W. Sterba in 2016 for at least two specimens already,[21] it is entirely plausible that there is a relation between the Linck and Bloch collections to be traced in other fish (depictions) as well.

The current study makes use of a rich material corpus of one particular pufferfish species, then called *Tetrodon hispidus*, in order to understand how the representations of this particular species were generated, influenced the collectors, as well as to shed light on the interplay of the actors and objects

15 The first time Bloch encountered a fish that he could not identify with Linnaeus's system was in 1779, resulting in him describing a moray eel as a new species. Paepke, *Bloch's Fish Collection* 156.

16 Nissen C., *Die zoologische Buchillustration. Ihre Bibliographie und Geschichte. Band II* (Stuttgart: 1978) 153.

17 Paepke, *Bloch's Fish Collection* 18.

18 Ibidem 157.

19 For example, Linck the Elder was in contact with, to name only a few, Jacob Theodor Klein from Gdansk (1685–1759) and the well-known Albertus Seba (1665–1736). Beyrich, "Das Linck'sche Naturalien- und Kunstkabinett" 583.

20 Visitor entry 1000 in the Linck family's guest book: Linck Heinrich, *Rerum naturalium amatoribus et admiratoribus qui huic museo praesentiam commodarunt suam officiosam memoriam spondet musei possessor IO. Heinricus Linckius* (Leipzig, unpublished: 1767–1809), holding institution: Museum Naturalienkabinett Waldenburg.

21 They discuss two wet specimens as original patterns for Bloch's descriptions of *Chaetodon kleinii* and *Premnas biaculeatus*. Engelmann – Sterba, "Über einige interessante Objekte".

involved. While analysing and comparing the preservations of the fish's bodies as well as depictions in the context of the Enlightenment ambition to generate universal knowledge, the focus lies on the 'afterlife' of the animals;[22] that is not on how these animals entered the collections, but the place these fish as *specimens* occupied in the ichthyological discourse. By approaching wet and dry specimens under an art historical lens, I endeavour to demonstrate that examining long-marginalised taxidermy objects as "images" provides valuable insights.[23] This analysis highlights how the knowledge preserved in the taxidermy and other sources relied on each other in a complex process of knowledge production. I attempt to determine whether the knowledge stored in different representations of the pufferfish, and attained by different collectors, remains separated or rather merges. I argue that the various modes of depiction and the knowledge about the species are interdependent and generate (long-lasting) iconographic traditions with a coherent image of the *Tetrodon hispidus*. As will be demonstrated, with specimens showing distinct (and in some cases contradicting) characteristics, especially a publication by Bloch – including a description and print of the pufferfish that claims to be a universal image of the species – lead to long term consequences in establishing an image of this fish in Central Europe.

2 Pufferfish Specimens in 18th-Century Collections

One of the rare fish taxidermies can be found in the *Museum Naturalienkabinett Waldenburg* (Germany) hosting the Lincks' collection today; a dry specimen, staring with round, yellow glass eyes [Fig. 19.2]. From its head to tail fin it is approximately 27 centimetres long and 25 centimetres in height, with an almost triangle-shaped body due to its straight back and saggy abdomen. The object's tough skin is characterised by a dark glaze and wrinkles that merge into drawn-out protrusions towards the bottom. Equally dark, close-fitting spines stand out slightly from the rest of the body, both visually and haptically.

22 Wells A., "History of Animal Collections/Animal Taxonomy", in Roscher M. – Krebber A. – Mizelle B. (eds.), *Handbook of Historical Animal Studies* (Oldenbourg: 2021) 603–618. The animals take on a new existence in that afterlife; from being creatures in their natural habitat outside the human order to existing as the material basis for universal knowledge production.

23 Bauernfeind, "Jona und der Hai" 167–168. Thus, like Robert Bauernfeind, I refer to the discipline's broader understanding as *Bildwissenschaften* (Visual Studies). On this term, see for example: Bredekamp H., "Bildwissenschaft", in Pfisterer U. (ed.), *Metzler Lexikon Kunstwissenschaft. Ideen, Methoden, Begriffe* (Stuttgart – Weimar: 2011) 72–75.

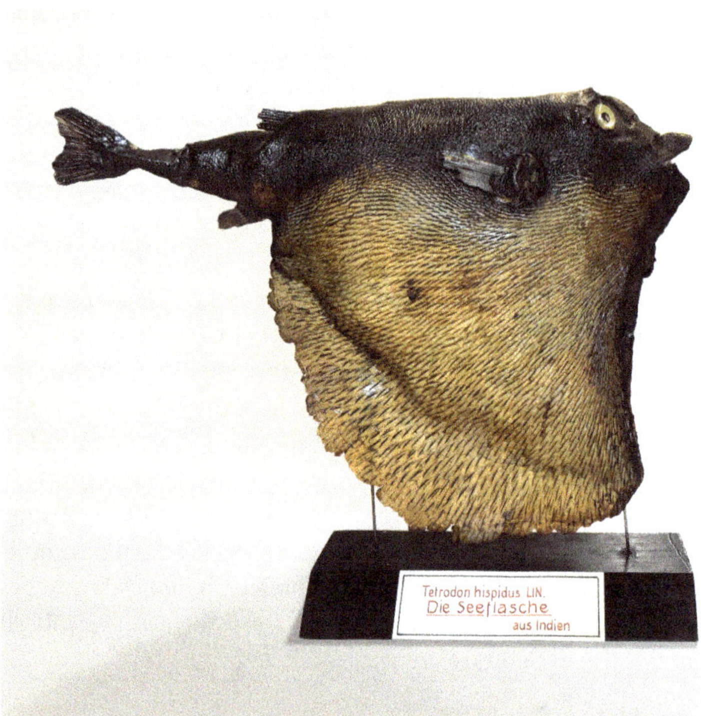

FIGURE 19.2 'Tetrodon hispidus LIN. Die Seeflasche aus Indien', dry specimen, 18th century, circa 27 cm × 25 cm, Linck collection, NAT I 1975 A5, Museum – Naturalienkabinett Waldenburg (Germany)
IMAGE © DOROTHEE FISCHER 2020 | LISA EFFERTZ 2022

The skin structure is distinguishably leathery on top and spikey further down. On the bottom stand, this exhibit is labeled as 'Tetrodon hispidus, LIN. Die Seeflasche aus Indien'. In addition to the species name in binomial nomenclature ('Tetrodon hispidus'), the label lists an abbreviation of Linnaeus ('LIN.'). Also, a vernacular name ('Die Seeflasche') as well as its proposed origin (India) are noted in German.[24] This matches an entry in the three-volume index of the Lincks' collection; the *Index musaei linckiani, oder kurzes schematisches Verzeichnis der vornehmsten Stücke der Linckischen Naturaliensammlung zu Leipzig*, published by Johann Heinrich Linck the Younger in 1783.[25] There

24 More information on the origin and authenticity of the label are yet to be determined.

25 Linck Johann Heinrich, *Index musaei linckiani, oder kurzes schematisches Verzeichnis der vornehmsten Stücke der Linckischen Naturaliensammlung zu Leipzig. Erster Theil* (Leipzig, Buchhandlung der Gelehrten: 1783) 59. Unfortunately, the record of the animal in the book published by Linck the Younger does not disclose when the specimen entered the collection. Since the specimen may have been purchased by a family member living before his time, it will henceforth be attributed to the entire Linck family.

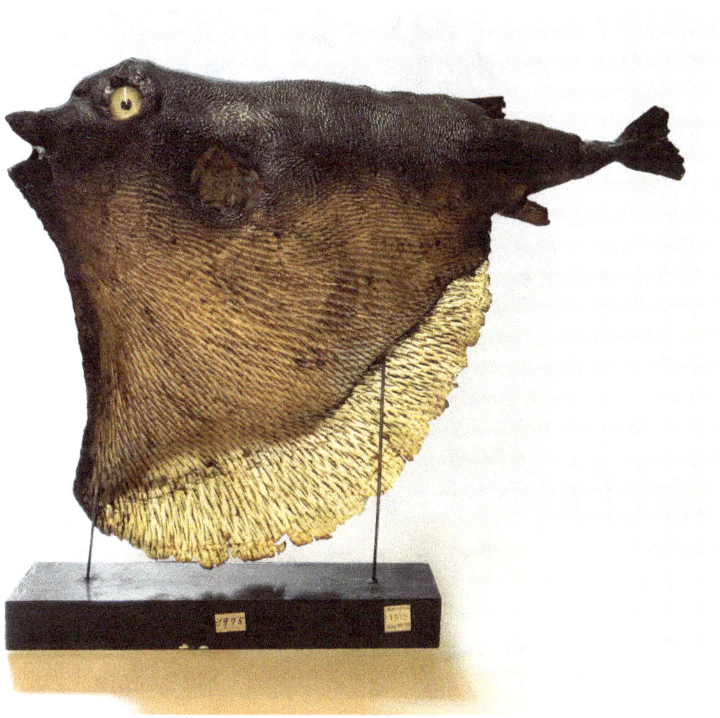

FIGURE 19.3 'Tetrodon hispidus LIN. Die Seeflasche aus Indien', dry specimen, 18th century, circa 27 cm × 25 cm, Linck collection, NAT I 1975 A5, Museum Naturalienkabinett Waldenburg (Germany)
IMAGE © DOROTHEE FISCHER 2020 | LISA EFFERTZ 2022

the pufferfish are classified according to Linnaeus's 12th edition of *Systema Naturae*[26] as 'schwimmende Amphibien' ('swimming amphibians'), supposedly breathing with external organs and lungs.[27]

A prominent seam along its chest reveals the taxidermical process of stuffing, rearranging and, thus, appropriating the animal's body.[28] The current specimen's state of preservation is modest: On the front side (see Fig. 19.2), there is a hole next to the remaining fin. On the other side [Fig. 19.3], the fin is missing completely and instead of it, there is another, large round hole. The position of the skin at the front of the head suggests a mouth, which is absent. Consequently, this literally gives an insight into the specimen's inside; it is filled

26 Ibidem XIV. According to the preliminary report, he refers to Linnaeus's edition of 1766 as well as its German translation by Philipp Ludwig Statius Müller of 1773.

27 Beyrich, "Das Linck'sche Naturalien- und Kunstkabinett" 597.

28 This attests to a violent practice. Regarding the specimen at hand, an in-depth inquiry on the conserving process as well as its embedding in colonial practices is yet to be done.

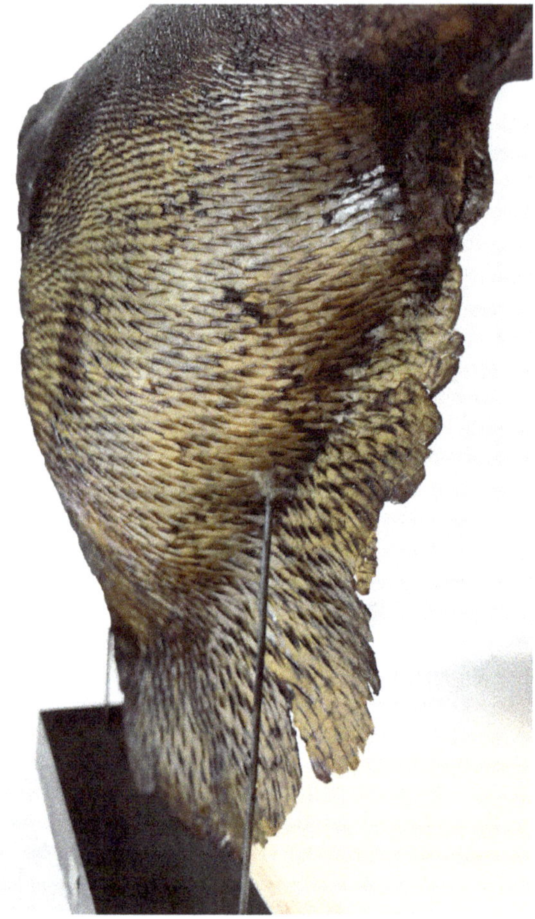

FIGURE 19.4 'Tetrodon hispidus LIN. Die Seeflasche aus Indien', dry specimen (detail), Linck collection, NAT I 1975 A5, Museum Naturalienkabinett Waldenburg (Germany) IMAGE © DOROTHEE FISCHER 2020 | LISA EFFERTZ 2022

with a stuffing of wood wool. The "animal's" filling only extends to a depth of about 20 centimetres, leaving a flat broad rim of two to seven centimetres in width [Fig. 19.4]. This morphological feature almost bears resemblance to the comb of a rooster. The skin on the lower abdomen is paper-thin, appearing to consist of merely one layer. However, viewed against the light, as in Fig. 19.3, a second layer becomes visible because the adjacent spines of the back – which are slightly offset – shine through.

In the collection space of the Linck family, this specific specimen was, according to the published *Index musaei linckiani*, one of 166 wet and 60 dry specimens in total.[29] As it was then presented 'hung up dry', it could be viewed from different perspectives.[30] This presentation evokes thoughts of early modern depictions of so-called *Kunst- und Wunderkammern* where ball-shaped pufferfish were traditionally hung from the ceiling.[31]

Another way of presenting (puffer) fish was (and still is) to preserve them in spirits and display them in glass jars. A great many of these jars can be found in the *Museum für Naturkunde Berlin* (Germany), also housing Marcus Elieser Bloch's collection today. 'There are still about 800 specimens left out of the original 1,400',[32] a major amount (almost three quarters) which 'originally consisted of specimens in alcohol, the rest were dried.'[33] One of these wet specimens can be seen in Fig. 19.1, contained in a glass measuring circa nine centimetres in diameter and circa 14 centimetres in height. In contrast to the Lincks' exhibit, it is difficult to perceive the animal's full size and body proportions due to the refraction of light in the transparent glass filled with liquid, which additionally precludes the possibility of touching the animal directly. This wet specimen is more displaying than hiding the fact that the animal is dead; in contrast to the dry specimen, this body is completely enclosed, entrapped rather than exposed, as if it had become a portable commodity in its afterlife. The lifeless skin is coming off in some areas and its eyes are hollow, disclosing that the inside of the body is empty. The specimen's organs have been removed, as a noticeable seam underneath the anal fin indicates. The base colour of its gently spiked skin is pale, but divided into a dark brown and a lighter brown part. In between those, a chocolate brown stripe can be noticed on each side of the body. On the upper side of the body in particular, closer inspection reveals uniform white spots a few millimetres large [Fig. 19.5] – a detail not observable in the Lincks' exhibit. Nevertheless, this specimen is also classified as *Tetrodon hispidus*.

---

29 This information emerges from an inventory list in the printed index, which was supplemented by handwriting until 1794. Of these objects, 112 wet and 24 dry specimens can still be observed in Waldenburg today. Engelmann – Sterba, "Über einige interessante Objekte" 16. On another pufferfish specimen of the Linck collection: Dreyer N. – Fischer D., "Migration vom Ozean in Wissensordnungen des 18. Jahrhunderts. Ein Kugelfisch-Präparat des Linck'schen Naturalienkabinetts", in Ullrich J. – Middelhoff F. (eds.), *Tierstudien* (*Tiere und Migration*) 19 (2021): 43–54.

30 Translation of 'trocken aufgeh.', Linck, *Index musaei linckiani* 59.

31 Although the animal has been hung, it does not appear as round in shape, but rather as a flat and elongated fish. This effect is caused by the specimen's creator's intentions. More likely it is due to the physiognomy of this particular animal.

32 Paepke, *Bloch's Fish Collection* 11.

33 Ibidem 30.

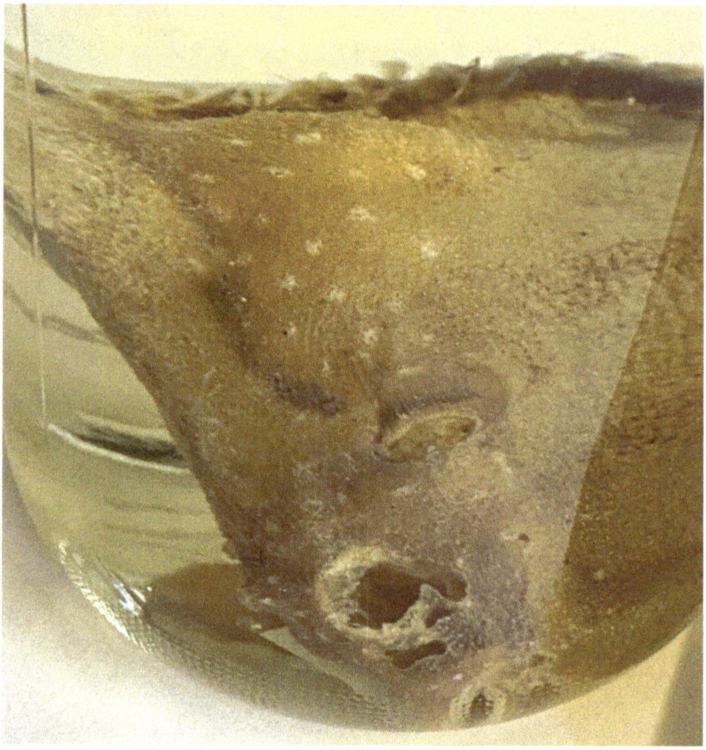

FIGURE 19.5 *Tetrodon hispidus* (detail of Fig. 19.1), wet specimen, 18th century, Bloch Collection, ZMB_Pisces_4274, Museum für Naturkunde Berlin (Germany)

Next to this wet specimen, Bloch's collection contains a second, dry *Tetrodon hispidus* specimen. Fig. 19.6 shows this almost globe-shaped and brown shaded "fish", currently presented on a wooden base. In total, the short and compact body is roughly 17 centimetres in length and ten centimetres in height. The specimen's blown abdomen is considerably expanded, coloured in a uniform caramel brown, becoming slightly lighter (yellowish-brown) towards the bottom. The specimen's almost monochrome skin is studded with small spines but does not show any white spots. On some parts, the skin has cracked apart, providing a glimpse of the stuffing. In place of the eyes, there are only holes. Resembling lips or even a bird's beak, the animal's dental plate conspicuously protrudes from the upper body.

These three fish representations could not be more different, yet they all are classified as the same species, *Tetrodon hispidus*, after Linnaeus. Comparing the three individual specimens demonstrates the variety of differences and issues in presenting the animals' bodies out of water in their afterlives far from home.

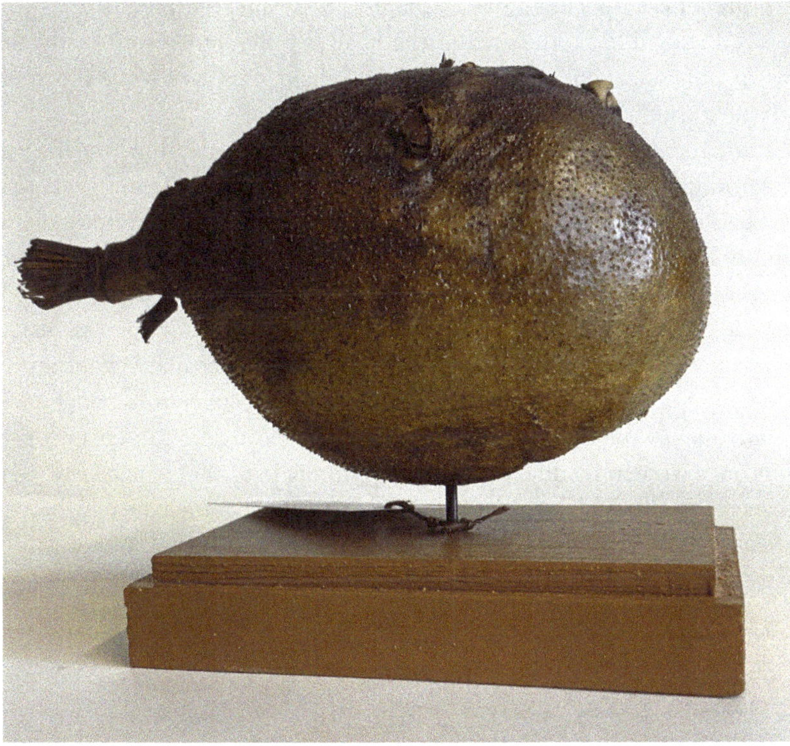

FIGURE 19.6 *Tetrodon hispidus*, dry specimen, 18th century, circa 17 cm × 10 cm, Bloch Collection, ZMB_Pisces_4275, Museum für Naturkunde Berlin (Germany)
IMAGE © DOROTHEE FISCHER 2022

While all three bodies share basic morphological features like the same number of fins and have in common that their biological sex is not recognisable,[34] their appearance and the way they are presented differ significantly.[35] The dry specimens showcase the animals' full size and proportions. They emphasise body features like the huge abdomen and heterogeneous skin texture covered with spines. Nevertheless, the mere size difference is remarkable: Bloch's specimen (measuring roughly 10 × 17 cm) is almost half the size of the Lincks' exhibit (which is 25 × 27 cm). The body shape of the former resembles a ball while the latter is almost triangular. The former is bulgingly blown, the latter has a skin sack resembling a rooster's comb facing downwards. When looking

34 Even if the pufferfish were still alive, there would be no external characteristics in shape and colour which could help in determining the animal's sex. I would like to thank Wolf-Eberhard Engelmann for this information.

35 For biologists, further similarities might be obvious. For laypersons, however, all objects' appearances are substantially different.

closely, the Lincks' specimen has a straight dorsal line, the Bloch's dry specimen has a curved back; the area of the head is different as well. Lastly, the Lincks' exhibit stares out with two artificial yellow glass eyes while Bloch's is "blind", having only empty eye sockets.[36]

In summary, all specimens are brownish but vary in shade. However, the wet specimen's skin shows the most detailed patterns with little white spots that cannot be found in today's appearance of the other two. The morphological features of Bloch's specimens are more or less consistent but contrarily to the shape and size of Linck's exhibit. Consequently, the specimens do not present a clear and cohesive image of what is supposed to be the same species, resulting in an ambiguous idea of what a *Tetrodon hispidus* looks like. What they do have in common, however, is their display in European collections and, hence, their function as physical proof of the species' existence. Moreover, these exhibits once were individual pufferfish swimming in distant oceans. So, what we observe are products of appropriation processes entangled in anthropocentric as well as Eurocentric hegemonies, which merit further investigation in future studies.

3 Pufferfish in Print

Marcus Elieser Bloch's work *Naturgeschichte der ausländischen Fische* (in English: 'Natural history of foreign fish')[37] serves the purpose of portraying fish in a scientific manner by presenting the common properties of individuals in a species. This book was published in nine parts and three plate volumes between 1785 and 1795, combining descriptions and depictions of 'foreign' fish. Later on, Bloch merged this with his earlier work on fish of the Prussian states (*Oeconomische Naturgeschichte der Fische*, published between 1782 and 1785) into the *Allgemeine Naturgeschichte der Fische*.[38] This encyclopaedia became known as 'the most important ichthyological work of the century'.[39] In this

36 Considering the rest of Bloch's collection, it is unlikely that his specimen ever had glass eyes. In the case of the Lincks' exhibit, it would be necessary to analyse the glass eyes of the specimen thoroughly to determine whether they were perhaps added later and if so, when.

37 Bloch Marcus Elieser, *Naturgeschichte der ausländischen Fische. Mit sechs und dreissig ausgemalten Kupfern nach Originalen. Ersther Theil* (Berlin, Marcus Elieser Bloch: 1785).

38 Paepke, *Bloch's Fish Collection* 157. Translations of the titles: *Economic Natural History of Fish* and *General Natural History of Fishes* (by D.F.).

39 Nissen, *Die zoologische Buchillustration* 153.

opus magnum based on Linnaeus's principles,[40] he described about 500 fish species of which 267 were previously unknown to the scholarly world.[41] It alone includes 432 plates with depictions of fish.[42]

Bloch's *Naturgeschichte der ausländischen Fische* offers a detailed written description of *Tetrodon hispidus* which he legitimises by making references to earlier "ichthyological" works, namely by Pierre Belon (1517–1564), Guillaume Rondelet (1507–1566), Conrad Gessner (1516–1565), Ippolito Salviani (1514–1572) and Ulisse Aldrovandi (1522–1605).[43] Next to the fish's physiognomy, Bloch describes their colours:

> Der Körper ist kurz, und wenn der Bauch aufgeblasen ist, ausserordentlich dick; man könnte daher züglicher sagen, der Fisch sei ganz Bauch, als mit Plinius, daß er ganz Kopf sei a). Der Kopf ist klein, die Lippen am Munde stark, und die Nasenlöcher ohnweit [sic!] den Augen; letztere sind klein, haben einen schwarzen Stern, der von einem goldenen Ringe umgeben wird. Der Rücken ist rund, der Schwanz kurz, und auf den Seiten ein wenig zusammengedrückt. Der ganze Körper ist bis auf den Schwanz dicht mit kleinen Stacheln besetzt. Die Grundfarbe des Fisches ist weißgrau, der Rükken [sic!] bräunlich, und auf beiden Seiten nimmt man verschiedene Streiffen [sic!] von ähnlicher Farbe wahr. Sämmtliche [sic!] Flossen sind klein, von grauer Farbe, und mit vielzweigigten [sic!] Strahlen versehen. Wir treffen diesen Fisch im mittelländischen und ostindischen Meere, auch im Nilstrom an. Er wird einen bis zwey [sic!] Fuss lang [...].

> The body is short, and when the belly is inflated, [becomes] extraordinarily big; one could therefore say more briskly that the fish is all belly, [contra] with Pliny that it is all head a). The head is small, the lips at the mouth strong, and the nostrils not far from the eyes; the last [i.e. the eyes] are small, consisting of a black star surrounded by a golden ring. The back is round, the tail short and a little compressed on the sides. The whole body, except for the tail, is densely covered with small spines. The basic colour of the fish is whitish-grey, the back is brownish, and on both

40 Paepke, *Bloch's Fish Collection* 19.

41 Ibidem 157.

42 Nissen C., *Die zoologische Buchillustration* 153. More information about the artists can be found here as well.

43 Bloch, *Naturgeschichte der ausländischen Fische* 130, 131, 132. He is also criticising previous authors by discussing the shortcomings of the preceding iconography of *Tetrodon hispidus* in their works. Ibidem 131, 132.

> sides, one perceives various stripes of a similar colour. All the fins are small, of grey colour, and with many-branched rays. We find this fish in the Mediterranean and East Indian seas, also in the Nile River. It grows one to two feet long […].[44]

This description of a one to two feet long fish with a white-greyish overall colour, brownish back, grey fins and stripes does not entirely match today's appearances of any of the three specimens. However, the 'brownish back' is a feature that all specimens exhibit. The described overall colouration and pattern are not shown on either one of the dry specimens, but do resemble the wet specimen. Its white spots, however, are not mentioned in this paragraph.

A remarkable similarity of the description to the Bloch specimens is the portrayal of the fish's body as 'short' compared to the Lincks' exhibit. The Lincks' *Tetrodon hispidus*, even though fitting a size of 'one to two feet', is arguably not 'short' in length, nor in a relative sense compared with its height. It also has no 'round back' and is overall significantly different in its body shape. It shows almost no resemblance except for the 'golden ring' around the eye (with its yellow glass eyes). In this case, the relation between the written description and the individual specimens on display does not indicate a direct correspondence between the two modes of representing the species. Although the text describes the sum of Bloch's specimens, it is impossible to map his description onto the characteristics of any of the three exhibits into a coherent image of the *Tetrodon hispidus*.

In addition to the written descriptions, detailed hand-coloured copperplate prints can be found in the same book. These images mirror the whole species, showing its main features in a stylised manner, synthesising all the characteristics of a species and thus also serving as a practical tool for classifying animals visually. Unlike *particular* specimens, these images do not constitute proof that what is being depicted actually exists; rather, they provide a *general* and schematic representation devoid of individuality. Moreover, unlike specimens, they can be reproduced and disseminated widely.

As one of these illustrations, plate 142 [Fig. 19.7] shows a creature from a side profile, protruding clearly from the colourless blank ground. With dark letters contrasting the light paper background, the lettering in the upper right corner reveals not only the plate number but also a scientific categorisation reading 'TETRODON HISPIDUS'. Below this Latin reference to Linnaeus's systematisation, three more lines with the animal's naming in German ('Der Seekröpfer'),

44 Bloch, *Naturgeschichte der ausländischen Fische* 131. Translation by D.F.

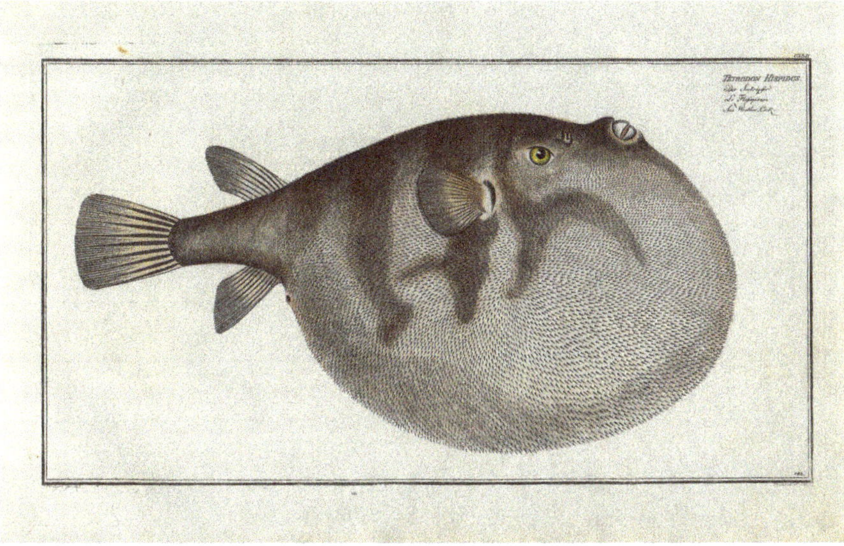

FIGURE 19.7 Johann Friedrich August Krüger jun. (inventor) and Ferdinand Schmidt (engraver), *TETRODON HISPIDUS*, copper engraving, Plate 142. From: Bloch Marcus Elieser, *Naturgeschichte der ausländischen Fische. Mit sechs und dreissig ausgemalten Kupfern nach Originalen. Ersther Theil* (Berlin, Marcus Elieser Bloch: 1785)
IMAGE SOURCE: ZENTRALBIBLIOTHEK ZÜRICH, URL: HTTPS://WWW.E-RARA.CH/ZUZ/DOI/10.3931/E-RARA-54281 (22/03/2022)

French ('flascopsaro') and English ('Sea Weather Cock') in a different font testify to the internationality of the scientific community (or at least that the author intended a broad group of recipients). These indicate that this image is a representation of the same species, *Tetrodon hispidus*, as the specimens of Bloch's collection [Figs. 19.1 and 19.6] and the dry specimen of the Lincks' [Figs. 19.2–19.4] above. To the viewer's left, a rigid, fanned tail fin is attached to a narrow, smooth, dark-coloured tail. To the right-hand side is a short, round body that makes up most of the animal, which is covered in small spines, depicted as short strokes. Its lateral fin, an almost round yellow eye and slightly opened, beak-like lips are located in the upper right quarter of the animal. As if swollen, the chest and abdomen of this fish stretch out to be disproportionately large underneath. Its colour shows different shades of brown, except for a yellow eye, a pinkish coloured cloaca, as well as a mouth that stands out from the monochrome colouring. The white reflection in the eye and its pupil transports a vivid impression. From the back to the middle of the body, dark brown patterns extend to the underside of the animal, reaching its abdomen in asymmetric, curved stripes.

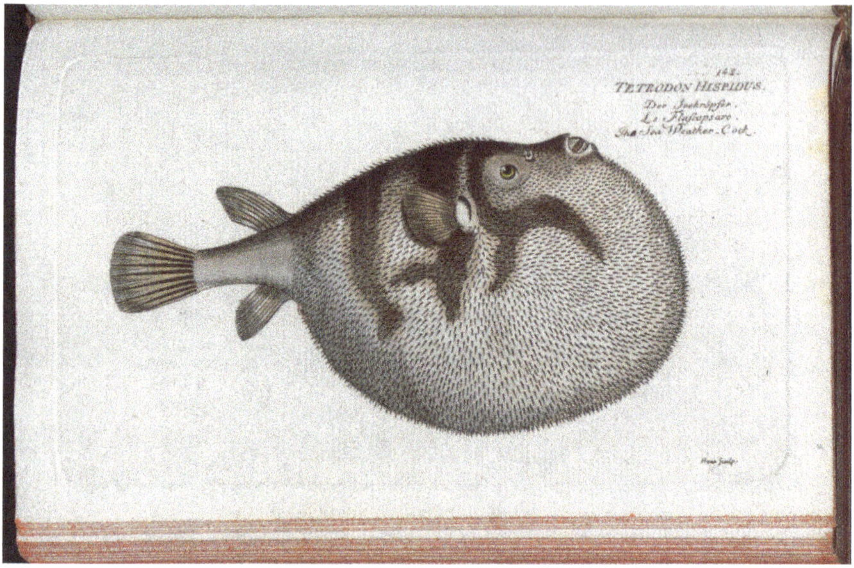

FIGURE 19.8 Peter Haas (engraver), *TETRODON HISPIDUS*, copper engraving, Plate 142. From: Bloch Marcus Elieser, *Naturgeschichte der ausländischen Fische. Mit sechs und dreissig ausgemalten Kupfern nach Originalen. Ersther Theil* (Berlin, Marcus Elieser Bloch: 1785)
IMAGE SOURCE: UNIVERSITÄTSBIBLIOTHEK FREIBERG – SLUB DRESDEN, URL: HTTP://DIGITAL.SLUB-DRESDEN.DE/ID480651450/73 (10/10/2023)

Comparing this with another edition of the print at hand, signed by the engraver Peter Haas [Fig. 19.8], further discoveries can be made. Its composition is the same as in Fig. 19.7, the spines of the pufferfish, however, are evoking a more haptic illusion as they are depicted less in number but thicker and slightly longer. This indicates another artist's hand, supposedly transferring the protruding spines of their model more dominantly into this illustration. Although there are slight differences in the depiction, like the almost hedgehog-like spines, a lighter coloured tail and the fish's eye, it is clearly referring to the same model image. This image could partly be derived by the written description of the *Tetrodon hispidus*. However, merely from the description these pictures could not have been constructed since the text lacks details with respect to body proportions, shapes of the fins and patterns.

Bloch himself left Germany only occasionally and travelled little. As a non-travelling naturalist, his immense collection emerged through his significant exchanges of goods, letters and knowledge.[45] Regularly, he relied on previous

45 For further information on Bloch's Biography: Paepke, *Bloch's Fish Collection*, especially 14–16. European naturalists like Bloch mostly came into possession of information about

books and drawings from authors like Charles Plumier (1646–1704) to synthesise existing knowledge as well as new self-made discoveries on fish into coherent representations. Bloch was meticulous in citing these external sources, with particular respect and praise for Plumier's descriptions.[46] However, the majority of Bloch's illustrations was based on eyewitnesses' descriptions and/or wet and dry preparations and *not*, as with previous fish books, on copies by other authors only.[47] He was confident in the completeness of his own collection and its function as the material basis of his scientific inquiries, and, ultimately, as the foundation of his written descriptions and illustrations.[48]

As Bloch's interest was not merely in collecting specimens but in generating ichthyological knowledge, he also relied on specimens from other collections on the occasion that his possessions were insufficient. One of them was the Lincks' in Leipzig; as Bloch notes in his publications more than once, he sent his painters to their collection to use specific specimens as models for his illustrations.[49] For the species of *Chaetodon Kleinii* (sunburst butterflyfish), as an exemplar case, the depiction was made of a bigger specimen from the collection of the Lincks since Bloch's specimen did not meet his own quality requirements.[50]

animals as well as specimens from a distance, acquired by way of the missionary colonies in these regions, by post or by buying them off merchants. Paepke, "M.E. Bloch's frühe aquatische Versuche" 34.

46 Bloch, *Naturgeschichte der ausländischen Fische* and Pietsch T.W., "Charles Plumier (1646–1704) and his drawings of French and American Fishes", *Archives of Natural History* 28.1 (2001) 1–57, here 8.

47 Bloch, *Naturgeschichte der ausländischen Fische*, preface. As Florike Egmond and Sachiko Kusukawa demonstrated in 2019, already Gessner commissioned fish drawings that were based on preserved specimens. They specifically mention the drawing of a pufferfish 'depicted after dried exemplars'. Egmond F. – Kusukawa S., "Gessner's Fish: Images as Objects", in Leu U. – Opitz P. (eds.), *Conrad Gessner (1516–1565). The Renaissance of Learning* (Berlin – Boston: 2019) 581–606, here 584. Thus, at least since the 16th-century specimens functioned as models and given that there was no major innovation in keeping exotic fish alive at the time and with the examples at hand this was still the practice circa two hundred years later.

48 Paepke, *Bloch's Fish Collection* 24, 157.

49 See exemplarily: Bloch, *Naturgeschichte der ausländischen Fische* vol. 4, 10, 12; vol. 5, 140.

50 Bloch, *Naturgeschichte der ausländischen Fische* vol. 4, 8. 'Dieser Fisch gehört in Ostindien zu Hause: Seine eigentliche Größe kann ich nicht angeben, denn ich besitze nur ein kleines Exemplar; meine Zeichnung aber ist von einem größeren, welches in der vortreflichen Sammlung des Herrn Commerzienrath Lincke zu Leipzig befindlich ist, genommen worden.' / 'This fish is at home in the East Indies: I cannot state its actual size, for I possess only a small specimen; my drawing, however, was taken from a larger one, which is in the excellent collection of Lincke from Leipzig' (Translation by D.F.). On this specimen also: Engelmann – Sterba, "Über einige interessante Objekte".

In addition to the description of the *Tetrodon hispidus*'s external features and its occurrences, Bloch's text includes different vernacular names, among them 'Seeflasche,'[51] the same name as used by Johann Heinrich Linck the Younger.[52] While this alone is not sufficient evidence for Bloch's incorporation of the Lincks' specimen, it is possible that he was consulting their collection. Another testimony of their relation is Linck the Younger's working copy of the *Index musaei linckiani*, which today is in the collections of *Leipzig University Library*. Next to the entry on the *Tetrodon hispidus*, there is a brief handwritten annotation: 'Bl. T. 142'.[53] It is reasonable to assume that 'Bl.' is an abbreviation of the name 'Bloch' and 'T.' stands for the German word 'Tafel' (engl. 'plate'), referring to Bloch's copperplate number 142 as represented in Figs. 19.7 and 19.8. Details like these point to a broader practice in the production and circulation of knowledge amongst collectors fixating and sharing knowledge in different media.

Although Linck the Younger refers to Bloch's print, recalling the Lincks' *Tetrodon hispidus* exhibit [Figs. 19.2, 19.3 and 19.4], no direct connection can be established. The similarities between this specimen and the illustration are rather generic, as only the eye colour, and the contrast in the overall brown colouration (top dark, bottom bright) equate, while they are different in brightness and pattern.[54] Both, being displayed in side profile, emphasise a huge abdomen, even though remarkably different in shape. The dried specimen's overall form resembles a triangle much more than the ball shape of the illustration. The head is directed to the right, following the straight dorsal line, whereas the printed fish conveys a curved dorsal line leading to an upward shaped mouth, evoking the impression of a slightly upward tilted head.

Unsurprisingly, this comparison yields the same results as the comparison between the written description and the Lincks' exhibit. Consequently, it is not possible to bridge the gap between Bloch's description and the illustration using the Lincks' *Tetrodon hispidus*. In contrast, Bloch's specimens not only resemble the written description but also complement it with regard to the illustration. Even though one cannot see distinguished stripes as in the print, Bloch's dry specimen [Fig. 19.6] is especially a visual match in the overall appearance. The printed fish's shape and proportion, its round back, the small head and the short tail give the impression that the illustration synthesises selected written knowledge and *selected* specimens which altogether reflect a

51 Bloch, *Naturgeschichte der ausländischen Fische* 131.

52 Linck, *Index musaei linckiani* 59.

53 Ibidem.

54 However, the dark colouration of the dry specimens could also be due to age-related darkening.

coherent image of the *Tetrodon hispidus*. The emphasis lies on *selected* as also the white spots of the wet specimen [Fig. 19.5] did not find their way into the printed fish depiction. One possible explanation for this is that Bloch prefers to follow written knowledge[55] rather than trusting the wet specimen in front of him, supposedly – in contrast to the other dry specimens – considering it an anomaly. Hence, given that previous naturalists had not mentioned them and that, as far as can be judged today, the dry specimens, likewise, did not display any spots, his white-spotted specimen seemed to be an exception and, thus, this characteristic did not seem noteworthy to Bloch.[56] This would prove that anomalies in an individual were systematically excluded in the interest of generating a universal image of the species.

That pufferfish always show a "typical" round shape is a misconception. Even though they are 'capable of inflating their abdomens with water', this only happens 'when frightened or disturbed'.[57] While both dry specimens, as well as the prints, display the blown abdomen, in his book Bloch mentions repeatedly that this is a temporary condition.[58] Thus, scientifically, the text has an advantage over the specimens as well as the illustrations, which are only presenting one moment in time. If the pufferfish is not always blown, what justification is there for an almost ubiquitous iconography that represents the species as inflated? One possible reason is that the pufferfish's ability to transform its body into a ball shape is the most distinguishing characteristic known in Europe from its century-long presentation in *Kunst- und Wunderkammern*.[59] Another possibility is its categorisation as a swimming amphibian according to Linnaeus. The inflated depiction would then be in alignment with the idea that they breathe with lung-like organs.[60] Certainly, as something special about the

55 See footnote 44.

56 This is remarkable though as one would expect him to pass this information on to his readership. Thus, another possibility is, of course, that he was not aware of this feature as the spots are not visible in his dry specimen and the spotted specimen might have only reached Bloch's collection after his books' publications. This, however, seems unlikely as other features of the wet specimen did indeed find their way into the printed illustration (e.g., its stripes).

57 Hardy G. – Jing L. – Leis J.L. – Liu M. – Matsuura K. – Shao K., "White-spotted puffer *Arothron hispidus*", *The IUCN Red List of Threatened Species* (2014), online, URL https://dx.doi.org/10.2305/IUCN.UK.2014-3.RLTS.T193699A2262231.en (22/03/2022).

58 Bloch, *Naturgeschichte der ausländischen Fische* 120, 130.

59 Further on in his description, Bloch himself names the *Tetrodon hispidus* as 'Stachelkugel' ('spiky ball'). Bloch, *Naturgeschichte der ausländischen Fische* 132.

60 This body feature leaves room for further research as there are exceptions to this rule when it comes to other species of the pufferfish, which are indeed depicted in Bloch's book, as well as other pufferfish specimens in both collections in an uninflated state.

species, it ought to be included if it is to be a universal image of the species, showing what makes the entire species unique.

4 Establishing an Iconography of an Un-spotted "*Tetrodon hispidus*"

In order to create illustrations that did justice to the species as a whole and did not merely depict individual specimens, Bloch needed an understanding of what were species-specific and what were individual characteristics of an animal in question. Thus, it was not unusual that the sketches for the later illustrations were based on several sources and exhibits. This seems not to be the case with the particular species of pufferfish under investigation. Although Bloch mentions previous authors, and we know of a connection to the Lincks' collection, there is no direct resemblance between Bloch's written and illustrative work and the Lincks' specimen strong enough to suggest a relation. Most likely, Bloch relied on his own two specimens since they were close to each other in their external features, align with written sources,[61] and appeal to the traditional shape of pufferfish.[62] It remains to be answered whether this depiction of *Tetrodon hispidus* persists as its most credible "iconography", in this case meaning a coherent image of the species,[63] especially because the Lincks' specimen proposes a different image of the fish species.

As demonstrated above, the Lincks' specimen falls out of line compared to Bloch's specimens as well as his written and pictorial description of *Tetrodon hispidus*. From today's perspective, this is easily explained: The physiognomic features strongly indicate that this specimen was wrongly classified. This "animal" more likely was once a *Triodon macropterus* rather than a *Tetrodon hispidus*.[64] This species does not have the capability to fully extend

61 His two specimens are, however, significantly shorter than the described length of the fish species ('one to two feet') in his written descriptions. Furthermore, as elaborated on above, white spots as on the wet specimen are not mentioned.

62 This observation allows the speculation of whether he possibly wanted to legitimise the truthfulness of his own preparations through their visual proximity.

63 Further on the discourse of iconography see for example: Noll T., "Ikonographie/Ikonologie", in Pfisterer U. (ed.), *Metzler Lexikon Kunstwissenschaft. Ideen, Methoden, Begriffe* (Stuttgart – Weimar: 2011) 194–198 and especially on animal iconography: Kalof L., "History of Animal Iconography", in Roscher M. – Krebber A. –Mizelle B (ed.), *Handbook of Historical Animal Studies* (Berlin: 2021) 471–492.

64 For this enlightening information, I would like to thank Edda Aßel and Peter Bartsch from the ichthyological collection in the *Museum für Naturkunde Berlin*. Already in 1999, Beyrich mentioned in passing that this specimen might be a *Triodon bursarius*. Beyrich,

their abdomen but has a bone structure shaping an abdominal crease which explains the shape [Fig. 19.4] and its difference compared to the round abdomen of Bloch's fish. However, as mentioned above, Linck the Younger seemed quite certain that his specimen was indeed a *Tetrodon hispidus* as he not only named it that way, but also added the copperplate number of Bloch's print in his *Index museai linckiani*. This is consistent insofar that the correct species was introduced formally as late as around 1830, long after Linneaus's publication on which the Lincks' rested upon.[65] As the classification system did not allow for another interpretation, to the collector standing in front of this specimen, it had to be a *Tetrodon hispidus*. This emphasises not only the difficulties in classifying exotic species in the collectors' rooms far from the fish's natural habitat. Furthermore, it underlines the importance of an encyclopaedic book like Bloch's *Naturgeschichte der ausländischen Fische* for ichthyology as a discipline evolving in the Enlightenment with universalist ambitions.

Bloch implicitly argued in his description that he is introducing a new image of the *Tetrodon hispidus* as the previous ones did not suffice.[66] As is apparent by looking at earlier depictions of the animal (e.g., Salviani's)[67] compared to Figs. 19.7 and 19.8, Bloch indeed introduced an innovative image of this fish: Even though the blown abdomen resembles earlier traditions of depicting pufferfish in European collections, his illustration diverges from previous pictures and might gain more credibility by its similarities to his own exhibits. Whereas the print is richer in detail and easy to distribute, the individual specimens are the material witnesses and tangible evidence to both the description and image. Still there is the open question of why the print does not show the white spots, visible on his own wet specimen. It is possible that Bloch deemed this exhibit an anomaly, as the spots are not present on other available specimens or written sources. So, not depicting the spots is an argument for Bloch's approach of providing a universal image of the species as a whole, portraying the *Tetrodon hispidus* as abstract as possible without losing the illustration's purpose as a

"Das Linck'sche Naturalien- und Kunstkabinett" 597. Today this term is valid as *Triodon macropterus*. Froese R. – Pauly D., "*Triodon macropterus* Lesson, 1831", *World Register of Marine Species* (2022), online, URL https://www.marinespecies.org/aphia.php?p=taxdetails&id=219917 (07/04/2022).

65 Froese – Pauly, *Triodon macropterus* (online).

66 Bloch, *Naturgeschichte der ausländischen Fische* 131–132.

67 In his description of *Tetrodon hispidus*, Bloch praises Salviani's depiction as excellent ('vorzüglich […]'). Bloch, *Naturgeschichte der ausländischen Fische* 132. Most likely referring to Salviani Ippolito, *Aquatilium animalium historiae, liber primus: cum eorumdem formis, aere excusis* (Rome, Ippolito Salviani: 1558) plate 77.

classifying tool. With his work, he incorporates the previous knowledge and thereby inscribes this specific iconography into the scholarly discourse.

What were the long-term consequences of the interplay of different media of representation (specimens and print) in the 18th-century central European mainland, far from the sea? Introducing another *Tetrodon hispidus* might provide an answer. This yet undated dry specimen from the *Naturkundemuseum im Ottoneum* in Kassel (Germany) is about 34 centimetres long and 14 centimetres wide and high [Fig. 19.9].[68] While its size is closer to the Lincks' specimen it does not share any distinguishing characteristics with it. Its blown abdomen is greyish-beige, its back is brown-yellowish and shows dark brown patterns. Its amber coloured pupils are located next to a white, beak-like dental plate. Its body is strewn with little spines. The exhibit seems, purely from its external features, to have been made after Bloch's illustration and description. The shape is emphasising the blown abdomen while the colouration with the distinguished stripes resembles the Bloch illustration's patterns and description. This would suggest a dating of the preparation, or at least its colouring, to the end of the 18th century at the earliest, or what is more likely, even later than the start of the 19th century. At this point, the preparation's dating remains speculative – unfortunately a common problem with these specimens.[69]

Today, *Tetrodon hispidus* is (re)categorised as *Arothron hispidus*. Directing the gaze away from the collection displays, study rooms and fish books raises the question of the appearance of a living animal and the closeness of this "original" to the representations discussed. The species naturally inhabits reefs, lagoons or estuaries 'to depths of at least 50 m'[70] in water temperatures of 25°C and is 'is widely distributed in the Indian and Pacific Oceans.'[71] It can reach 50 cm in length and its diet consists of algae, corals, crabs and molluscs among other things.[72] The fish consumes these using the beak-like dental plate. Fig. 19.10 shows a contemporary photograph of *Arothron hispidus* in an

68 The museum's inventory list, which was established in the second half of the 19th century, contains the following entry: '"Tetrodon hispidus" (heute: Arothron hispidus (Linnaeus, 1758), deutsch: Weißfleck-Kugelfisch)'. Unfortunately, it is not proven if this record corresponds to the specimen in Fig. 19.9. I thank Peter Mansfeld for this information (mail correspondence, 5th August, 2021).

69 In recent years, it is gratifying to observe that (early) modern animal preparations are increasingly being examined, for example by observing their insides with X-rays to gain more information around their origin and manufacturing.

70 Froese R. – Pauly D., "*Tetrodon hispidus* Linnaeus, 1758", *World Register of Marine Species* (2022), online, URL https://www.marinespecies.org/aphia.php?p=taxdetails&id=298330 (07/04/2022).

71 Hardy – Jing – Leis – Liu – Matsuura – Shao, *Arothron hispidus* (online).

72 Ibidem.

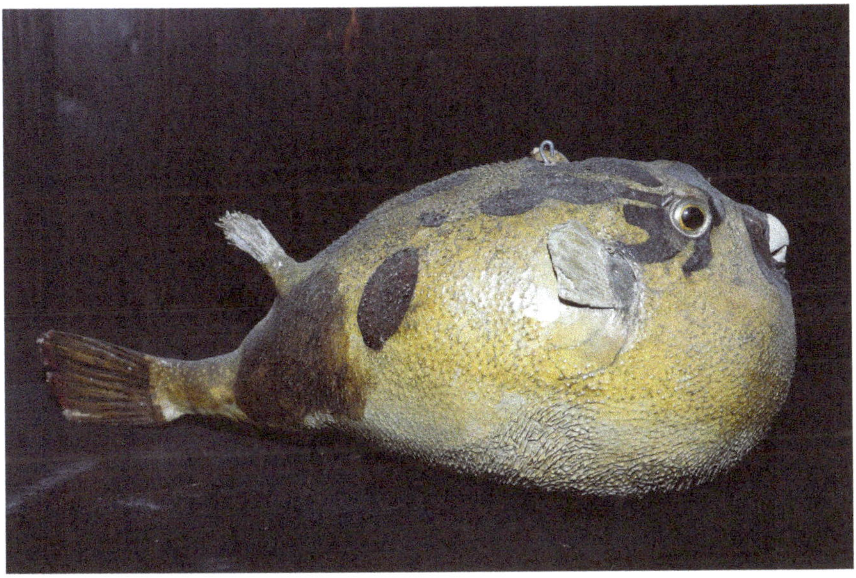

FIGURE 19.9 *Tetrodon hispidus*, dry specimen, undated, circa 34 cm × 14 cm, Naturkundemuseum im Ottoneum in Kassel (Germany)
IMAGE © PETER MANSFELD 2021

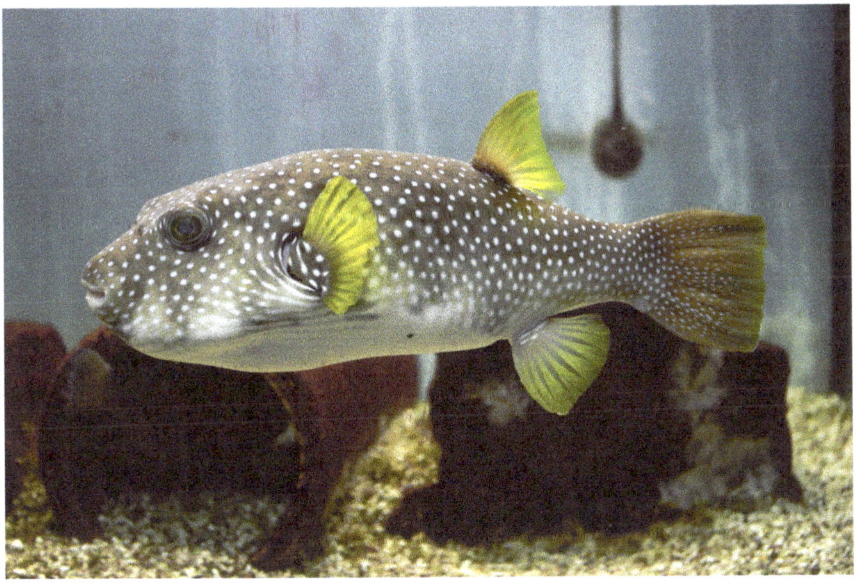

FIGURE 19.10 *Arothron hispidus*, photograph, 21st century
SOURCE: HTTPS://DE.WIKIPEDIA.ORG/WIKI/DATEI:AROTHRON_HISPIDUS_6.JPG#FILELINKS (22/03/2022)
IMAGE © WIKIMEDIA (FACTUMQUINTUS 2012 | TOGABI 2017)

aquarium. Its tough skin is greyish or brownish on top, depending on the individual's habitat, and becomes, with stripes in the transition zone, white on the ventral part. Every part except for the abdomen is covered with clearly distinguished white spots. That is why *Arothron hispidus* is vernacularly known as the white-spotted pufferfish. These eye-catching white spots confuse when recalling Bloch's description and the previous figures; all but the wet specimen [Figs. 19.1 and 19.5] representing this species with no spots at all.[73] It seems that a significant loss of information took place in the process in which the animal 'had to be selected, captured, dried, transported, sold and bought, and then finally put on display […]'[74] and so this (mis)representation was distributed in a printed illustration.

To remind ourselves, the previous representations of *Tetrodon hispidus* merged into a print illustration showing a brown-shaded pufferfish without spots. As Bloch's research output was highly influential, it is reasonable to assume that this specific print became the main reference and representation of the species far off its habitat. The Kassel specimen [Fig. 19.9] reflects this iconography shaped by taxidermy and printing. When modelling and colouring this exhibit, the preparator was – being, presumably, far away from the sea – *not* guided solely by the real fish (whose remains they would have in front of them), but by images and descriptions that circulated around collections of European naturalists.

Perhaps the fish lost its individual characteristics, namely the white spots, in transit from ocean to workshop (for example, due to conservation issues), or it was deliberately deprived of them in the picture so as to function as an exemplar organism that needed to reflect the commonly accepted image of the species (shaped by earlier written sources). In either case, whether the spots were lost during the preparation process or were intentionally omitted from the illustrations, it seems evident that Bloch's publication established an iconography of an "un-spotted" white-spotted pufferfish.

---

73 Looking at other photographs of the species underline this feature even more: Bariche M. – Constantinou C. – Sayar N., "First confirmed record of the white-spotted puffer *Arothron hispidus* (Linnaeus, 1758) in the Mediterranean Sea", *BioInvasions Records* 7.4 (2018): 433–436, here 434.

74 Rijks, "Fish out of Water" 51.

5 Summary and Conclusion

Keeping in mind the photograph of the living white-spotted pufferfish, *Arothron hispidus*, a conclusion can be drawn about the 18th-century images of "*Tetrodon hispidus*". The present chapter provided a case study of the species in the context of collecting fish in 18th-century Germany and posed the question of whether and how its (re)presentation, far from the sea, coalesced into a persistent iconography. As illustrated, the process of generating knowledge consists of abstracting and synthesising the defining characteristics that are determined by different media representing the same species and networks of collectors. The specimens attest to the existence of the species as these are (for central Europe) far from home and "exotic" enough that people often did not even know that these animals nor their different subspecies existed. The illustrations, on the other hand, represent a synthesis of the knowledge gained about that species. So, whereas the specimens (sometimes wrongly) attest to the authenticity of the species' existence, Bloch's *Naturgeschichte der ausländischen Fische* combines the ichthyological knowledge of previous authors and specimens for other collectors to classify their fishes. Its prints entail the highest degree of information and can be identified as the visual synthesis of contemporary knowledge, but cannot serve the purpose of credible representation alone as they do not depict white spots. It has been shown that the network between collectors allowed for cross-referencing and legitimising knowledge, but that it does not automatically lead to the synthesis of all information. Bloch relied on specimens of the Linck family's collection for some of his illustrations, but did not consider any of the distinguishing characteristics of their "*Tetrodon hispidus*" exhibit except for the yellow eyes.[75] This is not surprising since the two specimens in his own collection, the wet and the dry one, resembled each other and also the overall appearance described in previous sources much more closely. Specimens, so it appears, function as models for the printed image only if their appearance matches either other specimens or other sources of knowledge. They, however, can always function as exemplary organisms to legitimise knowledge. In this sense, Johann Heinrich Linck the Younger's *Index musaei linckiani* listing the exhibit of "*Tetrodon hispidus*" refers

75 The origin of the printed fish' yellow eyes would need further investigation. They might originate either from a specimen with glass eyes like the Lincks', or from a pictorial tradition, as other fish species in Bloch's as well as previous books also have this feature. It might also be an authentic feature of the fish. Bloch's written description mentions 'a golden ring' around the eyes as well. Bloch, *Naturgeschichte der ausländischen Fische* 131.

to the function of this specimen as an exemplar organism, independently of the "accuracy" of its appearance. Interestingly, on the other hand, the print's function as a reliable identification tool does not work properly; in his working copy of the *Index museai linckiani*, Linck notes that Bloch's print represents his dry specimen, although, according to today's knowledge, it shows a different species.

The media of representation – written sources as well as specimens and illustrations – gave a heterogenous image of "the" *Tetrodon hispidus*. These different ideas of the same fish coexisted and, through selection by collectors like Bloch, the different representations merged into popular illustrations which, subsequently, influenced the practice of taxidermy, manifested by the exhibit in Kassel displaying a strong resemblance to the depiction in print. Hence, all media in reciprocity led to a long-lasting depiction of the *Tetrodon hispidus* as an inflated ball-shaped fish with a brownish colour – and without spots. This, however, cannot do justice to the fish as they appear in their natural habitat and, hence, stresses the limitations of conserved specimens as legitimate sources of (the circulation of) knowledge as well as the persistence of falsely legitimised knowledge. Likewise, it also demonstrates the significance of the iconography derived from the print.

To conclude, this paper showcases the processes behind depictions of pufferfish far from the sea and how the abstract knowledge about this species circulated in written and visual sources in two exemplary German collections, consequently merging into an iconography. In a century where a quest for completeness, classification and generalisability prevailed, these pufferfish seem to have proved resistant to these universalising tendencies. The forceful transfer from their natural habitat into human collection systems was accompanied by the permanent loss of crucial information about the "real" fish. Both the knowledge about the inflated abdomen as a reaction to a state of emergency or the fish's exact appearance including the white spots was not transported with it into its afterlife. As the figures above reveal, the representations were no close match to the living animal's appearance or indeed its nature.[76] Despite the complex process of selecting and synthesising existing knowledge, an incomplete iconography of what could be called an "un-spotted" pufferfish manifested itself. On the one hand, due to the distance to the *Tetrodon hispidus*'s place of origin, the fish could not be spotted alive by most European naturalists.

---

76 As the close examination of the specimens demonstrated, shedding light on these conserved objects previously excluded from the art historical canon offers a base for rich comparisons and inquiries. Thus, this article is also a plea for more interdisciplinary approaches to these nearly forgotten objects.

The long journey to obtain the specimens resulted in significant changes to the animals' appearance. On the other hand, even though white-spotted pufferfish evidently did arrive in European collections, the selection of traits among conflicting sources of knowledge led to the image of an un-spotted brown-shaded *Tetrodon hispidus*. Thus, the circumstances in which the species was received far from the sea resulted in a long-persisting iconography of a 'white-spotted pufferfish without spots'.

Acknowledgements

This research was only possible with the support of the *Museum für Naturkunde Berlin*, namely Edda Aßel and Peter Bartsch, the *Naturkundemuseum im Ottoneum in Kassel*, namely Peter Mansfeld, the *Museum Naturalienkabinett Waldenburg*, namely Sandy Nagy and Fanny Stoye, as well as SLUB *Dresden* and *Zentralbibliothek Zürich*. Furthermore, I wish to thank my friends and colleagues for their various contributions to this article; Stephan Brakensiek, Andrea Diederichs, Nike Dreyer, Lisa Effertz, Jannik Eikmeier, Pia Emmrich, Ulrike Gehring, Annika Leuschner, Carole Martin, Colin Murtha, Marie Muschalek, Aylin Seboth and Lena Spindler. I owe special thanks to Timo Kuklau and cannot express enough gratitude for his crucial support in this project. Last but not least, thanks to all contributors of the enlightening conference *Towards a Cultural History of Early Modern Ichthyology (1500–1800)* in 2021 and especially to the editors of this volume, Florike Egmond and Paul Smith.

Bibliography

Bariche M. – Constantinou C. – Sayar N., "First confirmed record of the white-spotted puffer *Arothron hispidus* (Linnaeus, 1758) in the Mediterranean Sea", *BioInvasions Records* 7.4 (2018) 433–436.

Bauernfeind R., "Jona und der Hai. Zu einem frühneuzeitlichen Hai-Präparat zwischen Exegese und Naturgeschichte", *Zeitschrift für Kunstgeschichte* 82 (2019) 166–178.

Beyrich H., "Das Linck'sche Naturalien- und Kunstkabinett aus Leipzig, jetzt in Waldenburg (Sachsen)", in Grote A. (ed.), *Macrocosmos in Microcosmo. Die Welt in der Stube. Zur Geschichte des Sammelns 1450–1800* (Wiesbaden: 1994) 581–601.

Bloch Marcus Elieser, *Naturgeschichte der ausländischen Fische. Mit sechs und dreissig ausgemalten Kupfern nach Originalen. Ersther Theil* (Berlin, Marcus Elieser Bloch: 1785), holding institution: Zentralbibliothek Zürich.

Bredekamp H., "Bildwissenschaft", in Pfisterer U. (ed.), *Metzler Lexikon Kunstwissenschaft. Ideen, Methoden, Begriffe* (Stuttgart – Weimar: 2011) 72–75.

Dreyer N. – Fischer D., "Migration vom Ozean in Wissensordnungen des 18. Jahrhunderts. Ein Kugelfisch-Präparat des Linck'schen Naturalienkabinetts", in Ullrich J. – Middelhoff F. (eds.), *Tierstudien (Tiere und Migration)* 19 (2021) 43–54.

Egmond F. – Kusukawa S., "Gessner's Fish: Images as Objects", in Leu U. – Opitz P. (eds.), *Conrad Gessner (1516–1565). The Renaissance of Learning* (Berlin – Boston: 2019) 581–606.

Engelmann W.E. – Sterba G.W.H., "Über einige interessante Objekte in der Fischsammlung des Linck'schen Naturalienkabinetts", *Bulletin of Fish Biology* 16.1/2 (2016) 15–32.

Froese R. – Pauly D., "*Tetrodon hispidus* Linnaeus, 1758", *World Register of Marine Species* (2022), online, URL https://www.marinespecies.org/aphia.php?p=taxdetails&id=298330 (07/04/2022).

Froese R. – Pauly D., "*Triodon macropterus* Lesson, 1831", *World Register of Marine Species* (2022), online, URL https://www.marinespecies.org/aphia.php?p=taxdetails&id=219917 (07/04/2022).

Hardy G. – Jing L. – Leis J.L. – Liu M. – Matsuura K. – Shao K., "White-spotted puffer *Arothron hispidus*", *The IUCN Red List of Threatened Species* (2014), online, URL https://dx.doi.org/10.2305/IUCN.UK.2014-3.RLTS.T193699A2262231.en (22/03/2022).

Heesen A. te – Spary E.C., "Sammeln als Wissen", in Heesen A. te – Spary E.C. (eds.), *Sammeln als Wissen. Das Sammeln und seine wissenschaftsgeschichtliche Bedeutung* (Göttingen: 2001) 7–21.

Kalof L., "History of Animal Iconography", in Roscher M. – Krebber A. – Mizelle B. (ed.), *Handbook of Historical Animal Studies* (Berlin: 2021) 471–492.

Linck Heinrich, *Rerum naturalium amatoribus et admiratoribus qui huic museo praesentiam commodarunt suam officiosam memoriam spondet musei possessor 10. Heinricus Linckius* [Linck family's guest book] (Leipzig, unpublished: 1767–1809), holding institution: Museum Naturalienkabinett Waldenburg.

Linck Johann Heinrich, *Index musaei linckiani, oder kurzes schematisches Verzeichnis der vornehmsten Stücke der Linckischen Naturaliensammlung zu Leipzig. Erster Theil* (Leipzig, Buchhandlung der Gelehrten: 1783), holding institution: Leipzig University Library.

Nissen C., *Die zoologische Buchillustration. Ihre Bibliographie und Geschichte. Band II* (Stuttgart: 1978).

Noll T., "Ikonographie/Ikonologie", in Pfisterer U. (ed.), *Metzler Lexikon Kunstwissenschaft. Ideen, Methoden, Begriffe* (Stuttgart – Weimar: 2011) 194–198.

Paepke H.-J., "M.E. Bloch's frühe aquatische Versuche. Über einen Pionier der Heimtierhaltung", *T.I. Magazin* 28.129 (1996) 33–36.

Paepke H.-J., *Bloch's Fish Collection in the Museum für Naturkunde der Humboldt-Universität zu Berlin: An Illustrated Catalog and Historical Account* (Rugell: 1999).

Pietsch T.W., "Charles Plumier (1646–1704) and his drawings of French and American Fishes", *Archives of Natural History* 28.1 (2001) 1–57.

Rijks M., "Fish out of Water. Collecting Aquatic Animals in the Early Modern Period", in Rijks M. – Smith P.J. – Egmond F. (eds.), *Fish & Fiction. Aquatic Animals between Science and Imagination (1500–1900)* (Leiden: 2020) 48–61.

Ross A.S., "Recycling Embryos: Old Animal Specimens in New Museums, 1660–1840", *Journal of Social History* 52.4 (2019) 1087–1109.

Salviani Ippolito, *Aquatilium animalium historiae, liber primus: cum eorumdem formis, aere excusis* (Rome, Ippolito Salviani: 1558), holding institution: Smithsonian Libraries.

Schlegel H., "Über das Anfertigen von Amphibien- und Fischbildern", in Nissen C. (ed.), *Die zoologische Buchillustration. Ihre Bibliographie und Geschichte. Band II: Geschichte* (Stuttgart: 1978) 250–252.

Siemer S., "Naturkundliches Sammeln im 18. Jahrhundert. Ein Überblick", in Mieth K.M. – Museum Waldenburg (eds.), *Das Naturalienkabinett. Sammeln, Forschen, Zeigen* (Chemnitz: 2011) 42–54.

Trijp D. van, "The Murky Waters of Classification. Ordering Fish in Eighteenth-century Europe", in Rijks M. – Smith P.J. – Egmond F. (eds.), *Fish & Fiction. Aquatic Animals between Science and Imagination (1500–1900)* (Leiden: 2020) 76–85.

Vennen M., *Das Aquarium. Praktiken, Techniken und Medien der Wissensproduktion (1840–1910)* (Göttingen: 2018).

Wells A., "History of Animal Collections/Animal Taxonomy", in Roscher M. – Krebber A. – Mizelle B. (eds.), *Handbook of Historical Animal Studies* (Oldenbourg: 2021) 603–618.

CHAPTER 20

Louis Renard (1678/1679–1746) and His *Poissons, ecrevisses et crabes* (1719): 300 Years of One of Natural History's most Curious Colour Plate Books

Theodore W. Pietsch and Justin R. Hanisch

1 Introduction

Louis Renard's *Poissons, ecrevisses et crabes, de diverses couleurs et figures extraordinaires, que l'on trouve autour des isles Moluques et sur les côtes des terres Australes* has presented a beautiful conundrum to bibliographers and natural historians since its publication in Amsterdam in 1719. The book's 100 hand-coloured, copper-engraved plates illustrate 460 individual animals, some of which are immediately recognizable while others seem entirely fantastical. Although Renard's work went through three editions under three different publishers in the 18th century, its legacy has been predominately one of derision and dismissal, especially amongst 20th-century bibliographers and natural historians. The book's vivid yet seemingly random colouring, the inclusion of decorations such as faces and suns on some animals, and inaccurate and sometimes outlandish description of animal behaviour have all led to skepticism of the book's scientific veracity and seriousness. More recent work, however, has succeeded in identifying nearly all the book's 460 organisms at least to the level of family and has restored the importance of Renard's *Poissons* in the history and development of ichthyological literature.

Herein we provide a historical overview of Louis Renard and the three editions of his *Poissons*, also known frequently by its half title, *Histoire naturelle des plus rares curiositez de la mer des Indies*. We also discuss Renard's engravings, with a visual comparison of examples from *Poissons* with modern photographs of the species they were intended to represent. Finally, we detail the discovery of a potentially unique copy of Renard's book that somehow escaped colouring and existed essentially unrecorded until it was broken, coloured, and sold leaf-by-leaf at auction in 2020–2021. Despite its reputation as a work of ambiguous scientific merit, Renard's *Poissons* remains a significant contribution to the natural history of aquatic animals and continues to present new opportunities for research 300 years after its initial publication.

 | DOI:10.1163/9789004681187_021

2 Louis Renard and the Three Editions of His *Poissons, Ecrevisses et Crabes*

Born in the fortress town of Charlemont, northeastern France, in 1678 or 1679,[1] Louis Renard came from a Huguenot family that fled, after the revocation of the Edict of Nantes in 1685, to the Netherlands to escape religious persecution. Although no word of his parents, his early childhood, or his education seems to have survived, we do know that he settled in Amsterdam, where he became a member of the Walloon (or French Reformed) Church on 6 September 1699 and a citizen of that town on 17 September 1703.[2] By 1703, he had established himself as a bookdealer and publisher, and on 17 September of that year he became a member of the Amsterdam Guild of bookdealers, paying his dues to that organization through the year 1735.[3] Between 1704 and 1725 there appeared a number of works under the Renard name, most of which were printed in French, but some in Latin. Initially these were small publications dealing with current events, probably produced in connection with his selling of newspapers. Later, his attention turned to the production of new editions of books printed in France, the publication of maps and plates, and finally, some large plate works for which he himself supplied the text.[4]

By the mid-1720s, however, Renard's publishing efforts seem to have come to a halt. Until this time his addresses in Amsterdam were near the town hall, in the heart of the bookselling district; but after 1722 we find him far removed from there, a sure indication that books were less important in his life.[5] After 1724 his name cannot be found in the pages of the *Gazette d'Amsterdam* or the *Amsterdamsche Courant*, which at that time were two of the most important

1 In July 1703, 'Louis Renard of Charlemont' was said to be 24 years old (Gemeentelijke Archiefdienst Amsterdam (GAA), DTB 536, f. 59, 27 July 1703).

2 Schutte O., *Repertorium der buitenlandse vertegenwoordigers, residerende in Nederland 1584–1810* (The Hague: 1983) 140.

3 Universiteits-Bibliotheek Amsterdam (UBA), *Gilden Archieven*, nos. 68, 69; see also Eeghen I.H. van, *De Amsterdamse boekhandel, 1680–1725: V (1–2): De boekhandel van de Republiek 1572–1795, summary, overzichten en indices* (Amsterdam: 1978) 348.

4 Eeghen I.H. van, *De Amsterdamse boekhandel, 1680–1725: IV: Gegevens over de vervaardigers, hun internationale relaties en de uitgaven N–W, papierhandel, drukkerijen en boekverkopers in het algemeen.* (Amsterdam: 1967) 61.

5 For known addresses of Renard in Amsterdam, see Pietsch T.W., *Fishes, Crayfishes, and Crabs: Louis Renard and His Natural History of the Rarest Curiosities of the Seas of the Indies* (Baltimore, MA: 1995) 156, n. 40.

advertising journals for the Dutch book trade.[6] By 1735 he had stopped paying his dues to the Amsterdam Guild of book dealers. On 4 and 5 January 1737, books belonging to Renard were sold at auction.[7] Copperplates belonging to Renard were auctioned off on 14 and 15 July 1738,[8] and finally, more of his books were auctioned off from 16 to 21 May 1746, almost three months after his death.[9]

In addition to his occupation with books, Renard acted as a kind of spy on behalf of the British Crown. From at least 1706 until his death in February 1746, he was employed by Queen Anne,[10] and later by George I and George II,[11] to (among other things) search ships leaving Amsterdam to prevent supplies of arms and stores from reaching the Roman Catholic 'Old Pretender,' James Stuart.[12] This service, however, could hardly have been much of a secret, since as early as 1719 Renard publicly announced his title of 'agent' in his advertisements (for example in the *Gazette d'Amsterdam*) and on the title page of his publications.

In addition to bookdealing and spying, Renard had at least two other sources of income. The first of these had to do with buying and selling English bonds.[13] The second, an occupation inherited from his father-in-law, Daniel de la Feuille (1640–1709), was selling certain medicinals, the recipes for which were held in strict secrecy.[14] The names of two such products are known: one was a *tincture de tartre*, the claimed effect of which is now unknown; the other was *beaume de reunion* or *beau cene*. The latter product purportedly would

6 Eeghen I.H. van, *De Amsterdamse boekhandel, 1680–1725: I. Jean Louis de Lorme en zijn copieboek* (Amsterdam: 1960) 25–26, 47–51; Eeghen I.H. van, *De Amsterdamse boekhandel, 1680–1725: II. Uitgaven van Jean Louis de Lorme en familieleden* (Amsterdam: 1963) 261–265.

7 UBA, *Gilden Archieven*, 112, 4–5 January 1737.

8 UBA, *Gilden Archieven*, 113, 14–15 July 1738.

9 UBA, *Gilden Archieven*, 121, 16–21 May 1746.

10 *Calendar of Treasury Books*, vol. 28, part 2, appendix, p. 450 (21 August 1707).

11 Appointment as agent in Amsterdam through patent of King George I dated 8/19 April 1715; appointment as agent in Amsterdam through patent of King George II dated 18/29 July 1727 (Niedersächsische Landesbibliothek, Hanover, Cal. Br. 24, no. 3224).

12 For details of Renard's activities as a British agent, see Pietsch, *Fishes, Crayfishes, and Crabs* 9–15.

13 GAA, *Notarieel Archief* 6512, no. 291, correspondence with Gerard Bicker van Swieten, superintendent of police, dated 14 September 1720; GAA, *Desolate Boedelskamer*, no. 251, item 9, documents listing assets of Daniel Renard, son of Louis Renard, dated March 1735; GAA, *Notarieel Archief* 8041, no. 110, an act of the notary Philippe de Marolles, dated 19 October 1731, which indicates that Renard was involved in speculation in English bonds.

14 GAA, *Notarieel Archief* 6479, no. 139, pp. 1013–1015, 13 May 1704, contract between Daniel de la Feuille and Louis Renard witnessed by public notary Hendrik de Wilde, See Eeghen I.H. van., *De Amsterdamse boekhandel, 1680–1725: III. Gegevens over de vervaardigers, hun internationale relaties en de uitgaven A–M* (Amsterdam: 1965) 193.

'thoroughly heal all sorts of fractures in children within eight weeks, and in older people within a slightly longer period; a bottle sells for eight guilders, or you can agree to pay for the entire cure with a money-back guarantee: you apply a plaster, which in the evening is moistened with only ten to twenty small droplets'.[15]

Of Renard's personal life we know very little. His income as agent, combined with that acquired from his dealings in books and medicinals, no doubt allowed him to live a rather comfortable life among the upper-middle-class citizens of Dutch society. In 1742 his annual income was estimated to be 1,500 guilders, enough to provide his family with a maid-servant (*dienstbode*) and a house on the Herengracht. His social status, and thus perhaps the extent of his wealth, may also be judged by his association both in correspondence and in personal relations with some rather significant personalities of his day.[16]

However, between 1742 and his death in 1746, some unknown factor seems to have adversely affected Renard's finances. This is reflected not only by his change of address during this time (from the Herengracht to what must have been considerably more modest quarters on the Spiegelstraat) but also by comments made by a contemporary, Jacob Bicker Raye (1701–1775),[17] who noted that Renard, when he died, 'left a very mediocre capital' and that as a result 'his large family was left behind in distressed circumstances'.[18] Things were so bad that a daughter, too proud to hire herself out as a cleaning lady, committed suicide.[19] Renard's second son, Daniel (1711–?), who inherited the agent-ships of Great Britain and Hanover,[20] became so ill from stress that he had to be confined by his wife to a rest home.[21] Constant internal bickering among the siblings resulted in a long series of nasty lawsuits filed by Renard's oldest daughter, Marianne Germaine (1704–?), against her brother Daniel

15 GAA, *Amsterdamsche Courant*, 23 March 1706.

16 For details, see Pietsch, *Fishes, Crayfishes, and Crabs* 160, n. 81.

17 For Bicker Raye (who most likely would have been passed over completely by biographers had it not been that from 1732 to 1772 he kept a diary of daily happenings in his hometown of Amsterdam), see the introduction by M.G. de Boer to the published version of Bicker's diary: Bicker Raye J., *Het dagboek van Jacob Bicker Raije, 1732–1772, naar het oorspronkelijk dagboek medegedeeld door Fr. Beijerinck en M.G. de Boer.* Second Edition. (Amsterdam: 1960) v–xii.

18 Ibidem 126, 184.

19 Ibidem 184.

20 Appointed agent at Amsterdam through patent of King George I dated 25 April/6 May 1746 (Niedersächsische Hauptstaatsarchiv Hanover, Cal. Br. 24, no. 3224).

21 British Museum (BM), Add. MSS 38203, fols. 238–239, Renard, Daniel, formerly British agent at Amsterdam, papers relating to his family affairs, 1764, etc.

and his wife Johanna van Segveld (1711–?).[22] Although Bicker Raye criticized Renard for his failure to provide adequately for his family, he mentioned also that Renard was a 'very wise, friendly, and obliging man,' and 'a rather favorable judgement may be given as to his character'.[23]

3 *Poissons, Ecrevisses et Crabes*

On 9 May 1716, Louis Renard wrote to Robert Erskine (1677–1718), physician and advisor to Peter the Great and first curator of the czar's natural history collection, describing his latest publishing effort. Hoping to solicit interest from the czar and others among the nobility of St. Petersburg, he elaborated in extraordinary terms:

> I received last year [1715] a manuscript on the Moluccas, concerning the natural history of fishes of that country. There are about 500 of them, painted in their natural colors; they are more beautiful than the parrots, the butterflies, and even the flowers in our gardens. Their colors are so vivid that it is like a new miracle of nature. It is, Monsieur, the General of the Indies [Adriaen van der Stel] who has had them drawn and painted. They have worked on this for seventeen years. It is the most beautiful collection in the world for people interested in the curiosities of nature. I have already had engraved from this collection 50 large plates. The others are being done.[24]

The result of this work, which was published three years later, is Renard's (1719, 1754, 1782) great masterpiece, *Poissons, Ecrevisses et Crabes* (Fishes, Crayfishes, and Crabs) containing 100 plates divided into two parts; each part bearing the half-title "Histoire Naturelle des plus Rares Curiositez de la Mer des Indes" (Natural History of the Rarest Curiosities of the Seas of the Indies). Three editions of this book are known, all of which contain 100 color plates, bearing a total of 460 brilliantly colored copper engravings, representing 415 fishes,

22 BM, Add. MSS 38203, fols. 233–241, Renard, Daniel, formerly British agent at Amsterdam, papers relating to his family affairs, 1764, etc.: 'Narrative and State of the Case between Mr. Daniel Renard, his Spouse Jane, [Johanna] van Segveld, and Marianne Germaine Renard,' dated 13 November 1764.

23 Bicker Raye J., *Het dagboek van Jacob Bicker Raije* 126, 184.

24 Original in the Archives of the Academy of Sciences, St. Petersburg, MSS 846/44, fol. 63v. This and all other quotes throughout this chapter are taken from Pietsch, *Fishes, Crayfishes, and Crabs*.

42 crustaceans, two stick-insects, a dugong, and a mermaid. With one exception, all of the illustrations represent tropical species of the East Indies, said to have been drawn from nature on the island of Ambon in the South Moluccas by an artist, named Samuel Fallours, in the employ of the Dutch East India Company. The original drawings were obtained by Renard from various individuals who brought them to Holland in 1708 and 1715.[25]

4 The First Edition

The first edition of the "Histoire Naturelle", published in 1719[26] by Renard himself, is quite rare. Of the 100 copies originally printed,[27] only sixteen are known, and a census is currently underway to update extant copies of all editions. Six of the surviving copies are bound two-parts-in-one, but the two parts (hereafter referred to as volumes) of eight other copies (for which information is available) are bound separately. Following the half-title page, volume 1 contains a full-title page printed in red and black,[28] a two-page dedication to King George I of England bearing his coat of arms, a two-page "Avertissement de l'Editeur" that includes testimony and certification of the authenticity of the contents, and 43 color plates. The second volume is similar to the first, repeating the full-title page and "Avertissement de l'Editeur", but lacking the dedication pages, and including 57 color plates and a four-page "Table Alphabetique des Noms".

---

25 See Holthuis L.B., "Notes on Pre-Linnean Carcinology (Including the Study of Xiphosura) of the Malay Archipelago", in de Wit H.C.D. (ed.), *Rumphius Memorial Volume* (Baarn: 1959) 63–125; Pietsch T.W., "Louis Renard's fanciful fishes", *Natural History* 93 (1984) 58–67; Pietsch T.W. "Fallours, S. Fishes of the Indo-west Pacific: A Collection of Handcoloured Drawings", in Antiquariaat Junk. *Natural History & Travel*, Catalogue 241 (Amsterdam: 1986) 36–39; Pietsch T.W., "Samuel Fallours and his 'sirenne' from the province of Ambon", *Archives of Natural History* 18 (1991) 1–25; Pietsch T.W., "On the three editions of *Louis Renard's Poissons, ecrevisses et crabes: Histoire naturelle des plus rares curiositez de la Mer des Indes*", *Archives of Natural History* 20 (1993) 49–68; Pietsch, *Fishes, Crayfishes, and Crabs* 39–6.; Pietsch T.W. – Rubiano D.M., "On the Date of Publication of the First Edition of Louis Renard's *Poissons, ecrevisses et crabes: Histoire naturelle des plus rares curiositez de la Mer des Indes*", *Archives of Natural History* 15 (1988) 63–71.

26 For evidence that the first edition of Renard's book was published in 1719 rather than the often-cited date 1718, see Pietsch T.W. – Rubiano D.M., "On the Date of Publication of the First Edition of Louis Renard's *Poissons*" 63–71.

27 'I have had only one-hundred copies made in all' (Renard to J.H. von Bülow, dated Amsterdam, 19 November 1718, Niedersächsische Staats- und Universitätsbibliothek, Gottingen, qu.--4°, Cod. MS Hist. Nat. 108).

28 Pietsch – Rubiano, "On the date of publication of the first edition of Louis Renard's *Poissons*" 63–71.

One known copy of the first edition, held by the Library of the Academy of Sciences St. Petersburg, contains, in addition to the material described above, a two-page "Declaration sur cet ouvrage" (Declaration concerning this work), in which Renard certifies the authenticity of his work: 'Since this is one of the most precious works to enrich natural history since the birth of literature, I have taken great care to go to the source and to produce proofs of the facts I expose herein'.[29] Following this statement and two additional paragraphs of testimony concerning the accuracy of the drawings in volumes 1 and 2 of the "Histoire Naturelle" (in which Renard admits to some possible exaggeration, especially with respect to the mermaid), the St. Petersburg copy is signed and dated by Renard: 'Fait a Amsterdam, le 28 novembre 1719, Louis Renard, Agent de sa Majeste le Roy de la Grande-Bretagne'. Why Renard included the "Declaration" in this one copy but not in all one hundred copies, especially after having gone to the trouble and expense of having the copperplates engraved, is difficult to understand. Perhaps at the last moment he decided that the "Declaration" did little more than repeat his statements in the "Avertissement de l'Éditeur" and for that reason was best left out. On the other hand, it seems more likely that Renard became concerned that his admission of possible exaggeration and excessive altering of the truth would adversely affect the sale of his book.[30]

The title page of this first printing was undated, and considerable disagreement exists in the literature as to whether it originally appeared in 1718 or 1719, or the two volumes of the work were printed in 1718 and 1719, respectively; however, conclusive evidence that the production of the book was completed sometime during the winter or spring of 1719 was provided by Pietsch and Rubiano.[31]

29 V.A. Filov, director, Library of the Academy of Sciences St. Petersburg, personal communication, 7 April, 13 June, and 13 October 1986.

30 Renard might seem out of character in the 'Declaration sur cet ouvrage' when he confesses 'that the painter has exaggerated and excessively altered' some of the drawings of volume 2; he even goes so far as to acknowledge fear that 'the monster represented under the name of mermaid [sirenne] [...] needs to be rectified' (Pietsch, *Fishes, Crayfishes, and Crabs* 76). To be fair, it thus appears that Renard honestly tried to produce an accurate picture of the marine fauna of the East Indies (for more, see Ibidem 163–164, n. 6).

31 Pietsch – Rubiano, "On the Date of Publication of the First Edition of Louis Renard's *Poissons*" 63–71.

5 The Second Edition

The second edition of the "Histoire Naturelle", produced in 1754 by the publishing house of Reinier and Josue Ottens, Amsterdam, is only slightly more accessible than the first. Although the number originally printed is unknown, there is evidence that 100 copies were planned[32] of these, 34 copies have been located and an updated census is currently underway. This edition differs from the first in having a slightly modified title page (but, unlike the first edition, the title page appears only once), a four-page "Preface" by Aernout Vosmaer,[33] and the two-page "Declaration sur cet Ouvrage", the latter written by Renard (undated, but pre-1719) but for some unknown reason not included in the first edition (with one known exception; see above). The dedication to King George, the "Avertissement de l'Editeur" (absent in seven of 22, second edition copies examined), and the "Table Alphabetique" are identical to those found in the first edition (the "Avertissement", however, when present, appears only once), and were obviously pulled from the same original copper plates. Except for the absence of the "Avertissement" in seven of the 22 copies examined, the contents of all known copies are the same, but the order in which the material is bound is highly variable, there being at least ten different arrangements. Assuming that the original order of contents of these volumes was not confused during subsequent bindings, it would seem that the collation of the second edition was done in a highly haphazard way.[34]

32 Pietsch, "On the Three Editions of Louis Renard's *Poissons*" 52.

33 For Aernout Vosmaer, see Benthem Jutting W.S.S. van, "A Brief History of the Conchological Collections at the Zoological Museum of Amsterdam, with some Reflections on 18th-Century Shell Cabinets and their Proprietors, on the Occasion of the Centenary of the Royal Zoological Society 'Natura Artis Magistra'", *Bijdragen tot de Dierkunde* 27 (1939) 167–246, here 207; Boeseman M., "The Vicissitudes and Dispersal of Albertus Seba's Zoological Specimens", *Zoologische Mededelingen* 44 (1970) 177–206, here 180, 184–187; Pieters F.F.J.M., "Notes on the Menagerie and Zoological Cabinet of Stadholder William v of Holland, Directed by Aernout Vosmaer", *Journal of the Society for the Bibliography of Natural History* 9 (1980) 539–563; and Engel H., Hendrik Engel's Alphabetical List of Dutch Zoological Cabinets and Menageries. eds. P. Smit – A.P.M. Sanders – J.P.F. van der Veer (Amsterdam: 1986) 201–202, 293–294.

34 That the various copies of Renard's book show such a great diversity in the arrangement of their parts (except perhaps for those copies reserved for subscribers, which may have been bound early in the production process) is probably because most copies were left as sheets and loose plates, and were arranged only when an order came in; perhaps, also, the order of parts that we see today was, in each case, the individual choice of the buyer who purchased the loose parts and only later had them bound (L.B. Holthuis, Rijksmuseum van Natuurlijke Historie, Leiden, personal communication, 3 December 1991).

The circumstances surrounding the production of the second edition of Renard's book are complex and not fully understood. What little we do know is contained within six letters written by the publishers Reinier and Josue Ottens in Amsterdam to Aernout Vosmaer, the latter then living at The Hague as Director of the Menagerie and Cabinet of Prince William v. In the first of these letters, dated 16 June 1753,[35] the Ottens acknowledge that they have purchased 36 sets of uncolored plates of Renard's work, and although not specifically stated, the old coppers for the work as well. Their original intention was to have these sets colored, using a copy of the first edition as a model, and to advertise them for sale: 'Before we had the honor of talking to you [Vosmaer], it was our intention to publish them according to the enclosed advertisement':

> R. & J. Ottens, Art, Maps, & Booksellers of Amsterdam, have acquired in the last few years, by purchase from [the estate of] agent Louis Renard, not more than thirty [of the thirty-six sets mentioned elsewhere in the letter; apparently six were to be retained as reserves] copies of a beautiful and uncommon work by Renard, displaying a cabinet of fishes, crabs, and sea monsters from Ambon and other oriental regions, drawn from life, composed of one hundred plates with about 460 different pictures. All have been precisely painted in their true colors under the supervision of R. & J. Ottens, and enriched with a foreword and letters from experts that prove the authenticity of the work. Offered at a price of 70 guilders and neatly bound in an English cover.

Vosmaer evidently persuaded the Ottens to use the old coppers to print more copies, to color them as before, but to sell them with a new preface to be written by him. At first, the Ottens estimated the price for each of the 30 copies to be f 70, –, but with an expanded production of 100 copies the price could be reduced to f 50, –. It is evident from the second letter, dated Amsterdam, 18 August 1753, that Vosmaer found this price to be too high; the Ottens defended themselves, however, by explaining that the high cost of paper, and of printing and coloring the new copies, made it impossible to lower the price below f 50, –. It thus appears that the first 30 copies, with 70 additional, newly printed ones, together constituted the final plan between the Ottens and Vosmaer for the production of a second edition.

35 Six letters from Ottens in Amsterdam to Vosmaer at The Hague, dated 16 June 1753, 18 August 1753, 6 September 1753, 10 September 1753, 26 January 1754, and 6 May 1754 (Leiden University Library B.P.L. 246). See Pietsch, *Fishes, Crawfishes, and Crabs* 22–26.

The second letter goes on to ask Vosmaer if he would draft the conditions of subscription, explaining that this must 'be written in French by an able hand, and since we feel that no one is more capable, you being such a great lover of science, we take the liberty to ask in the most friendly way that you take this task upon yourself'. The third letter, dated Amsterdam, 6 September 1753, indicates that Vosmaer has fulfilled his publisher's requests: 'From your recent letter we have learned with pleasure that you have taken the trouble to make not only a description of the fish book, but also a draft of the conditions of subscription and advertisement for the newspapers'. Copies of Vosmaer's advertisement, written in both French and Dutch, are attached to this letter:

> R. & J. Ottens, booksellers in the arts of literature and painting, are presently selling, and have forwarded both at home and abroad the conditions for subscription to, the momentous and completed work having for title *Natural History of Fishes, Crayfishes and Crabs, of Diverse Coloration and Extraordinary Form, Which are to be Found About the Islands of the Moluccas and on the Coasts of Southern Lands*, etc., the whole provided with divers certificates and testimonials, divided into two parts, all the figures of which are represented and illuminated according to life, brought to light by Mr. Louis Renard, acting for His Majesty the king of Great Britain, and augmented by a preface by Mr. A. Vosmaer. Interested persons may examine the complete work at the premises of the above-mentioned R. & J. Ottens, and for the convenience of subscribers, as a sample, two copper-engraved colored plates have been sent to all parts.

The fourth letter, dated Amsterdam, 15 September 1753, contains a number of items for Vosmaer's consideration:

> We are pleased with your offer to provide a brief preface and we leave to you the reworking of the title page. Should the dedication to George I be dropped? We ask that you not say anything in the preface about the [drawings of] coloured fishes which have come recently to your attention, since we believe this would cause harm to the work.[36] Subscribers are more interested in buying something that is already completed, rather

36 Despite this plea from the Ottens to keep quiet about an additional set of colored fish drawings (besides the two sets that had gone into the making of Renard's book (see Pietsch, "Louis Renard's fanciful fishes" 62), Vosmaer could not refrain from devoting a paragraph to it in his "Preface" to the second edition: 'by chance, there fell into my hands a collection such as the present one, but which came from the Indies thirty years earlier'.

> than something that will take much more time to finish. Furthermore, we do not want to give them the impression that we can make as many books like this as we would like. We can decide to publish a third volume later if the present production goes well.

In the fifth letter, dated Amsterdam, 26 January 1754, we learn that Vosmaer's 'Preface' has been written and sent to the printer: 'Yesterday, I received a letter from Mr. Luzac,[37] informing me that your excellent preface will occupy one sheet [four pages], and the dedication, a half a sheet [two pages]. He says nothing about the title page or the certification. I am not sure whether to include the 'Declaration Concerning this Work,' since I believe your preface fills the role'. Finally, in the sixth letter, dated Amsterdam, 6 May 1754, we learn that the book has been printed and that 'there are no longer any bound copies left in stock'.[38]

To summarize the more significant information provided by this correspondence, it seems that the publishing firm of Ottens took the 30 (or 36) unbound copies purchased from Renard's estate, had the plates colored, replaced the old undated title page of Renard, and added a "Preface" provided by Vosmaer and the 'Declaration sur cet Ouvrage' of Renard. These then, together with some 70 additional copies newly printed from the original coppers (save for the new title page and "Preface"), constitute the second edition.

Still left unexplained, however, is the existence of three surviving copies of the "Histoire Naturelle" that contain Renard's original, undated title page, but at the same time contain the "Preface" of the 1754 edition.[39] On superficial

37 Mr Luzac is evidently Elias Luzac (1723–1796), a printer in Leiden who also produced some of the volumes of Albertus Seba (Eeghen I.H., *De Amterdamse … en indices* 127; L.B. Holthuis, Rijksmuseum van Natuurlijke Historie, Leiden, personal communication, 13 June 1984).

38 The fact that Ottens had no 'bound copies left in stock' does not mean that the edition was sold out. Publishers, who were also their own printers, only colored and bound copies that they could sell right away (for example, to subscribers); copies forming the larger part of the stock were colored and bound only when there was a demand for them. This also explains the differences that one sees in coloration between the various copies of a single edition (Bridson G.D.R., "From Xylography to Holography: Five Centuries of Natural History Illustration", *Archives of Natural History* 16 (1989) 121–141, here 125). The coloring was often done by whole families, even the children taking part in the work (L.B. Holthuis, Rijksmuseum van Natuurlijke Historie, Leiden, personal communication, 13 June 1984). For a brief survey of the materials and methods used for the visual presentation of natural history since the introduction of printing, see Bridson, "From Xylography to Holography".

39 Three known copies of the second edition contain the undated title page of the first edition, as well as the 'Preface' of Vosmaer dating from 1754: Hunterian Library, Glasgow University, Glasgow; Bibliothèque Nationale, Paris; Universitetsbiblioteket 2. Afd,

examination, these "hybrid" copies appear to be perfectly good, first edition copies dating from 1719 until one discovers Vosmaer's "Preface" with its reference to the *Mercure de France* of September 1749.[40] Confusion between the first and second editions has been mentioned by several authors.[41] Jacques-Charles Brunet[42] tells us that the first edition 'appeared without a date. [...] Later, a second edition was printed to which was added a preface by Vosmaer. This new printing appeared either without a date, which caused it to be confused with the first [edition], or with the date 1754, [...] [but both versions contain] the preface in which one finds a reference to the Mercure de France of September 1749'. From Brunet's remarks, and on the basis of the evidence provided by these three known hybrid copies, it must be concluded that the firm of Ottens took some of the original 30 (or 36) copies acquired from the Renard estate, bound them up with Vosmaer's newly prepared "Preface", and sold them under the old title page. But why this should have been done remains a mystery. When issuing old stocks as new it was typical at the time for publishers to immediately discard the old title page and issue the text with a new title page bearing their name; to sell a product under a previous publisher's name would seemingly have been to their disadvantage, but since the title page of 1719 carried no date, the Ottens may have thought at least initially that it could serve just as well.[43]

Copenhagen. A copy held by the Bibliothèque Centrale, Muséum national d'Histoire naturelle, Paris, is unique among all extant copies of Renard's book in having the original, undated title page of the first edition, but lacking the 'Preface' of Vosmaer. The order of contents is like that of the second edition; that is, in contrast to the first edition, the half-title pages, 'Tome I' and 'Tome II,' are used to separate only the color plates into two parts, the first containing 43 plates, the second containing 57. This collation, as well as the presence of Renard's unsigned 'Declaration sur cet Ouvrage,' indicates that this copy dates from 1754 rather than 1719.

40 This reference to the Mercure de France for September 1749 (pp. 183–184) is a 'Notice to the devotees of Natural History' of a 'Mr. Guyot, pharmacist and chemist in the employ of Mr. Pajot, the Count of Onsembray' (Louis-Leon Pajot, Count of Ons-en-Bray, a French engineer, born at Paris in 1678 and died at Bercy in 1754; *Nouvelle Biographie Generale*, 38, pp. 693–695), who holds the secret to a method by which 'he can conserve not only plants, but also all sorts of fishes and other animals with their natural colors'.

41 Haag E. – Haag E., *La France protestante, ou Vies des protestants Français qui se sont fait un nom dans l'histoire depuis les premiers temps de la réformation jusqu'à la reconnaissance du principe de la liberté des cultes par l'Assemblée nationale. Ouvrage précédé d'une notice historique sur le protestantisme en France*. Vol. 8 (Paris: 1858) 408; Brunet J.C., *Manuel du libraire et de l'amateur de livres*, vol. 4 (Paris: 1863) 1220; Graesse J.G.T., *Trésor de livres rares et précieux, ou Nouveau dictionnaire bibliographique*, vol. 6 (Berlin: 1922) 81.

42 Pietsch, *Fishes, Crayfishes, and Crabs* 25–26.

43 L.B. Holthuis, Rijksmuseum van Natuurlijke Historie, Leiden, personal communication, 13 July 1984.

6 The Third Edition

The least is known about the third edition of Renard's book, published in 1782 by the house of Abraham van Paddenburg and Willem Holtrop, Utrecht and Amsterdam.[44] Only six copies are known, and an updated census of extant copies is currently underway.[45] Its extreme rarity is probably due to the almost certain fact that its publication was never completed. A copy held by the University Library of Amsterdam tells us more about this edition than the remaining five.[46] Entirely uncut, in sheets as printed, the text of the Amsterdam copy consists of 56 pages, fourteen gatherings of four pages each, published in four fascicles, each fascicle accompanied by a printed wrapper dated 1782.[47] Despite the mention of a preface by Aernout Vosmaer on each wrapper ('Met eene Voorreden van den Wel-Edelen Heere A. Vosmaer'), these pages contain none of the front matter or "Table Alphabetique des Noms" of the first and second editions of Renard's book. Instead we find brief synonymies and descriptions in double columns, printed in Dutch and French, respectively, prepared by the Dutch physician and naturalist Pieter Boddaert.[48] Apparently a title page was never printed, but in both Dutch and French the initial lines of the first page read: 'Natural History of the Fishes, Crayfishes, and Crabs of the Indies'. This introductory text of the third edition is followed by the 43 plates (obviously pulled from the original coppers of 1719) that form volume 1 of the first two editions of Renard's book. Boddaert's descriptions refer only to the fishes and crustaceans depicted on the 43 plates of volume 1 of the first two editions of Renard's book; in correlation with this fact, the

---

44 Abraham van Paddenburg was active as a bookseller in Utrecht between 1752 and 1790, but not much more is known about him. For information about Willem Holtrop (1751–1835), who was probably the more important of the two, see *Nieuw Nederlandsch Biografisch Woordenboek*, vol. 1, 1146–1147.

45 Since a list of known, extant copies of the third edition of Renard's book was published (Pietsch – Rubiano, "On the Date of Publication of the First Edition of Louis Renard's *Poissons*" 69), one additional copy has been located (library of the Diergaarde Blijdorp, Rotterdam) bringing the total number to six.

46 The Amsterdam copy of the third edition was purchased in 1979 from the firm of Ronald Meesters, Antiquarian Bookseller, Amsterdam for Dfl. 30.000, –. For a full description from the sale catalog, see Meesters R., *Catalogue no. 22, Science and Medicine*, bookseller cat., Ronald Meesters, Antiquarian Bookseller (Amsterdam: 1979) item no. 1.

47 The printed covers or wrappers of the third edition, of which only four appear to have been published, each consist of four unnumbered pages: a sheet folded in folio, of which page 1 gives the title in Dutch, pages 2 and 3 are blank, and page 4 gives the title in French.

48 For Pieter Boddaert see Dean B., *A Bibliography of Fishes*, vol. 1 (New York: 1916) 142; Engel H. "Alphabetical list of Dutch Zoological Cabinets and Menageries", *Bijdragen tot de Dierkunde* 27 (1939) 247–356, here 259; Engel H., *Hendrik Engel's Alphabetical List* 33–34; Benthem Jutting W.S.S. van "A brief history of the conchological collections" 220.

Amsterdam copy lacks the 57 plates of Volume 2.[49] However, that a new printing of all 100 plates, as well as descriptions for the animals of volume 2, was planned, is evidenced by a note printed at the bottom of each wrapper: 'The work will contain one hundred plates and 20 or 22 sheets of printed text [each sheet having four pages]. Each fascicle will contain ten plates and will cost 12 stivers per plate'. A second known copy of the third edition was acquired in 1963 by the late L.B. Holthuis, of the Naturalis Biodiversity Center Leiden, from the English booksellers Wheldon and Wesley Limited.[50] This copy is identical to the Amsterdam copy except that it is cut and bound; the four wrappers and text pages 53 through 56 are missing. Third edition copies held by the Houghton Library of Harvard University, Cambridge, Massachusetts, the University Library of Utrecht, and the Mitchell Library, State Library of New South Wales, Sydney, are more complete than the two just described: while Boddaert's text remains the same, still describing only the figures of volume 1, all 100 plates are present. The Harvard copy, a cut and bound copy given to the university in 1915 by Daniel B. Fearing (1859–1918) of Newport, Rhode Island,[51] begins with a strange, hand-lettered title page that repeats the Dutch and French title indicated above, but includes the added phrase: 'Door een Voornaam Liefhebber by een Verzaamelt' (Collected by a Distinguished Amateur). The margins surrounding this lettering are elaborately decorated with rather crudely drawn, watercolor sketches of objects related to fishing: dip-nets and seines, a canoe paddle, a trident, and a pair of large ceramic vessels pouring water. The textual portion that remains is identical to that of the Amsterdam copy except that only the wrapper for fascicle II is present. The third edition copy held by the University Library of Utrecht remained unknown until 1985 when it was discovered incorrectly catalogued (but since corrected) as a copy of Renard's second edition of 1754. Like the Harvard copy, it too begins with a hand-lettered title page, this one apparently copied directly from that of the second edition: 'L. Renard, Histoire naturelle des plus rares curiositez de la mer des Indes ... Augm. d'une preface par A. Vosmaer. Amsterdam 1754 Avec 100 pl. col'. Boddaert's 56 pages of synonymy and description are present, as are all 100 plates. Unfortunately, nothing of the history of the Utrecht copy is known. Another third edition copy, this one in the library of the Rotterdam zoological

49 The Amsterdam copy of the third edition also lacks plates 19 and 23, and 41 through 43.

50 On Holthuis's copy, see Pieters F.F.J.M., "Histoire Naturelle des Poissons, Ecrivisses et Crabes des Indes. De uiterst zeldzame derde editie van Louis Renards Poissons, Ecrivisses et Crabes met tekst van Pieter Boddaert, gepubliceerd in 1782", in Alsemgeest A. – Fransen C. (eds.), *In krabbengang door kreeftenboeken. De* Bibliotheca Carcinologica *L.B. Holthuis* (Leiden: 2016) 81–86.

51 For Daniel Butler Fearing, see *Who Was Who in America*, 1897–1942, vol. 1, p. 389.

gardens (Diergaarde Blijdorp), discovered in 1989 and, like the Utrecht copy, incorrectly catalogued (also since corrected) as a copy of the second edition, is less complete than the other five. It consists only of plates 41 through 43 of Volume 1, and plates 1 through 41 and 43 through 56 of Volume 2, bound together with a hand-lettered title page that was undoubtedly produced by the same hand that fashioned the title page of the Utrecht copy described above: in all respects the style and content are identical, even as far as the double rendering of the lettering of 'A. Vosmaer'.[52] The third edition copy held by the Mitchell Library of the State Library of New South Wales, Sydney, Australia, differs yet again from all other versions of this printing: it begins with the first color plate of volume 1 ('Fol. 1'), facing the first page of the text that bears the title in Dutch and French; Boddaert's 56 pages of description follow, interspersed among the 43 plates of volume 1; this is then followed by the 57 plates of volume 2.

Prior to August 1963, when Professor L.B. Holthuis acquired his copy from Wheldon and Wesley Limited, the existence of a third edition was doubtful. Mulder Bosgoed, in his *Bibliotheca Ichthyologica et Piscatoria*,[53] only briefly mentioned an edition of 1782, based on a description of such a book in the catalogue of the library of Professor Jan van der Hoeven.[54] Aside from a single reference to Mulder Bosgoed's note made by L.B. Holthuis,[55] no mention is again made of a third edition until 1963 when Wheldon and Wesley advertised the book for sale in their 1963 catalogue.[56] This first published verification of the existence of a third edition was followed some years later by a mention in Claus Nissen's *Die zoologische Buchillustration*.[57] His description, however, is so imperfect it seems obvious that he could not have examined a copy himself. Nissen's bibliographic record was later followed by a brief, but somewhat more accurate, description provided by John Landwehr.[58]

52 This description, as well as a photocopy of the hand-lettered title page, of the Rotterdam copy of the third edition of Renard's book was kindly provided by A.E. Hylkema, Librarian, Diergaarde Blijdorp (personal communications, 21 September and 15 November 1989).

53 Mulder Bosgoed D., *Bibliotheca ichthyologica et piscatoria: Catalogue de livres et d'écrits sur l'histoire naturelle des poissons et des cétacés, la pisciculture, les pêches, la législation des pêches, etc* (Haarlem: 1873) 122, item no. 1913.

54 For Jan van der Hoeven see Engel H., *Hendrik Engel's Alphabetical List* 124.

55 Holthuis, "Notes on pre-Linnean carcinology" 76.

56 See Wheldon & Wesley Ltd, 1963, cat. 101, item 801.

57 Nissen C. 1969, *Die zoologische Buchillustration, ihre Bibliographie und Geschichte*, vol. 1. (Stuttgart: 1969) 336, item no. 3362.

58 Landwehr's mention of a copy of the 1782 edition in the library of the Rijksmuseum van Natuurlijke Histoire at Leiden is actually a reference to the copy in the private library of the late L.B. Holthuis, Curator Emeritus of that institution. See Landwehr J., *Studies*

7 Renard's *Poissons, Ecrevisses et Crabes:* Fact or Fiction?

All three editions of Renard's work depict 460 organisms in dazzling colour across 100 engraved plates, including 415 fishes, 41 crustaceans, 2 terrestrial insects, a dougong and a mermaid. The book represents one of the earliest contributions to the natural history of fishes of the Indonesian archipelago and was published more than 140 years before the first volume of Pieter Bleeker seminal work on the region, *Atlas Ichthyologique des Indes Orientales Néêrlandaises*.[59] The original drawings that serve as models for Renard's engravings are known and their artist[60] and contemporary owners[61] have provided testimonials to their accuracy.[62] Why, then, have the images in Renard's book elicited scorn from both natural historians and bibliographers? Indeed, the book has been described as 'crudely drawn and barbarously coloured',[63] 'inadmissible, insufficient, and often fantastic',[64] and a 'phantasmagoria ... weirdly formed and riotously coloured'.[65]

Certainly, some of this ridicule is a direct consequence of obvious embellishments on some of the fishes and crustaceans depicted in the book. Fishes and crabs are shown with suns, faces, and even plants on their bodies [Fig. 20.1]; these adornments strain the belief of even credulous readers – both contemporary and current. The descriptions accompanying some of the fishes are no less outlandish, with *Sambia*[66] described thusly by the artist Samuel Fallours:

---

in Dutch books with coloured plates published 1662–1875: Natural history, topography and travel, costumes and uniforms (The Hague: 1976) 166, item no. 160.

59 Bleeker's *Atlas Ichthyologique* was published in 9 volumes over 16 years (1862–1868) and is considered 'One of the most important treatises on [East Asian] fishes,' see Wood C.A., *An Introduction to the Literature of Vertebrate Zoology* (London: 1931) 244.

60 Samuel Fallours (dates unknown), the artist responsible for many of the original drawings copied by the engraver, provided the following testimonial (Pietsch, *Fishes, Crayfishes, and Crabs* 32): 'I [....] declare that the fishes included in this collection were drawn and painted by me [...] from nature. This was done to my best ability, not believing that human arts can express the beauty of the colours of these fishes [...]'.

61 Frederick Julius Coyett (1680–1736), a contemporary owner of original drawings, provided the following testimonial (Ibidem 32): '[...] I can assure you on my honour that they were drawn and colored as truthfully as the painter and the strength of the colours could permit, although it is impossible to obtain with a brush the brightness and admirable variety of colors that these fishes have when still alive'.

62 Ibidem 32, 43–48.

63 Dean, *Bibliography of Fishes*, vol. 3 (New York: 1923) 307.

64 Nissen C., *Schöne Fischbücher, kurze Geschichte der ichthyologischen Illustration und Bibliographie fischkundlicher Abbildungswerke* (Stuttgart: 1951) 20.

65 Dance P.S., *The Art of Natural History* (Woodstock, NY: 1978) 47–48.

66 *Sambia* is also described as the 'Walking Fish' or the 'Running Fish of Ambon,' an ambulatory characteristic that provides a clue to its identification.

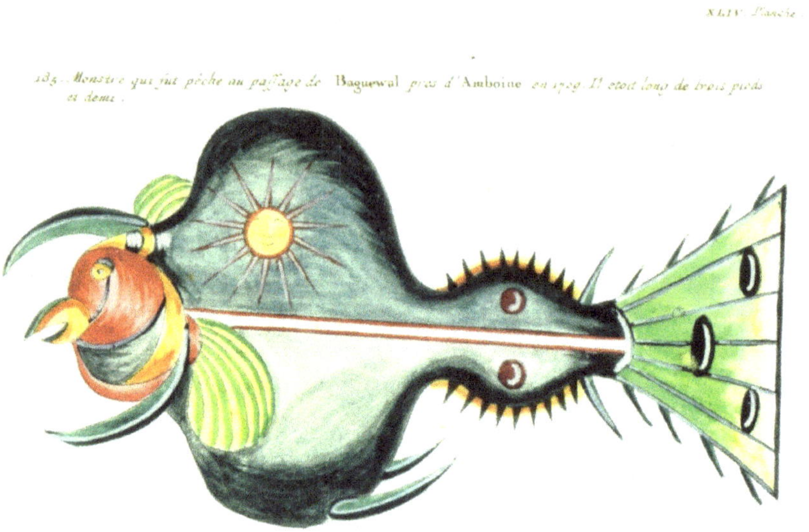

FIGURE 20.1 Plate 44, Figure 185 from the second volume of Renard's *Poissons, ecrevisses et crabes, de diverses couleurs et figures extraordinaires, que l'on trouve autour des isles Moluques et sur les côtes des terres Australes* [...] (Amsterdam, Reinier and Josue Ottens: 1754) showing a smiling sun on the back of the fish
IMAGE TAKEN FROM THE SECOND EDITION IN THE COLLECTION OF THE BAYERISCHE STAATSBIBLIOTHEK MUNICH

'I caught it on the sand and kept it alive for three days in my house, where it followed me everywhere with great familiarity, much like a little dog'.[67]

It is not certain why Fallours included these decorations in the drawings, and Renard himself was troubled that they might harm the credibility of his work.[68] However, a few 18th- and 19th-century ichthyologists were able to see beyond the embellishments and recognize the scientific merit within Renard's contribution to a little-known fauna of the era. Peter Simon Pallas (1741–1811) wrote of Renard's book:[69]

> We are taught almost daily by [the arrival of] new specimens that there are innumerable fishes, in the ocean of both East and West Indies (especially the East) that are surprising in color, form, and habits [...] It seems

67 Pietsch, *Fishes, Crayfishes, and Crabs* 114.
68 Ibidem 19.
69 Ibidem 19.

> to me less and less doubtful that most of them were [drawn] from life [...] and, for certain, I should dare affirm that as the fishes of India [...] become known, it will be an easy task to gradually classify, and refer to species all of these histrionic and monstrous representations.

Later, Georges Cuvier, writing in his monumental *Histoire naturelle des poissons* (1828), also saw merit in Renard's book, describing the images therein as 'still indispensable, either for giving an idea of the natural colours of known species, or for helping to recognize new forms that travellers bring to us daily from those so productive waters [...] Valentyn's and Renard's drawings, although they may be outlandish, nevertheless all represent real objects'.[70]

Whether the drawings represented 'real objects' was unsettled until the 1980s. While it is true that all the organisms depicted are somewhat crudely drawn, sometimes arbitrarily coloured, and not *entirely* accurate depictions of known species, there *is* truth in the images. Indeed, fully 91% of the 460 organisms depicted in the book have been identified at least to the taxonomic level of family. Of the 415 fishes, 385 (93%) can be identified, including 251 to species, 97 to genus, and 37 to family; of the 40 crustaceans, 34 (85%) can be identified, including 24 to species, 6 to genus, and 4 to family.[71] As of September 2023, Fishbase recognized approximately 4860 species of fishes in the Indonesian archipelago,[72] the result of nearly 200 years of systematic exploration in the region. The fishes depicted in Renard's *Poissons* were not necessarily collected systematically to inform natural history investigations. Instead, they were collected at the behest of colonial governors and painted – sometimes from copies of copies – and distributed by a soldier-artist with a potential eye towards profit.[73] Thus, the fact that the book depicts nearly 10% of the extant Indonesian fish fauna in enough accuracy and detail for at least partial identification supports the optimism of Pallas and Cuvier over the cynicism of cynicism of others.

How were these drawings matched with their living species? Pietsch describes 'ignor[ing] coloration and the numerous errors in number and placement of certain anatomical features, and concentrate[ing] [...] on color pattern and certain key generic and familiar characteristics' to match the

70 Ibidem 78.

71 Ibidem 78–79.

72 Fishbase is a regularly updated, international web portal devoted to the diversity of fishes. The count of Indonesian fish species given here was generated on 22 September 2023 from <https://www.fishbase.de/Country/CountryChecklist.php?c_code=360&vhabitat=all2&csub_code=&cpresence=present>.

73 Pietsch, *Fishes, Crayfishes, and Crabs* 38.

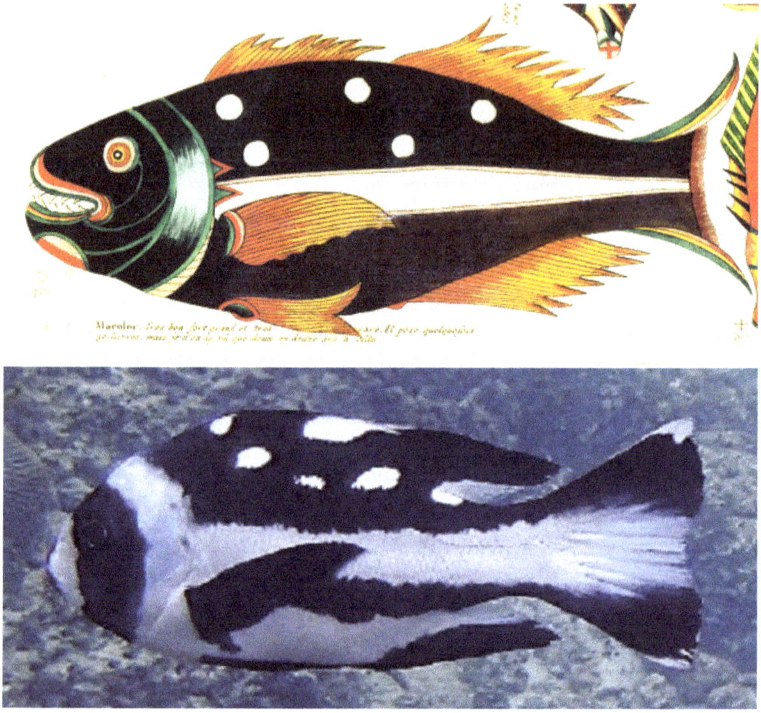

FIGURE 20.2 Plate 7, Fig. 30 *Macolor* from the second volume of Renard's *Poissons, ecrevisses et crabes, de diverses couleurs et figures extraordinaires, que l'on trouve autour des isles Moluques et sur les côtes des terres Australes* [...] (Amsterdam, Reinier and Josue Ottens: 1754) (top) and a photograph of the Black and White Snapper, *Macolor niger* (bottom) showing similarities and differences between the engraving and photograph of the species it represents
TOP IMAGE TAKEN FROM THE SECOND EDITION IN THE COLLECTION OF THE BAYERISCHE STAATSBIBLIOTHEK MUNICH. BOTTOM IMAGE © DAWN GOEBBELS OBTAINED VIA HTTPS://WWW.FISHBASE.DE/PHOTOS/UPLOADEDBY.PHP?AUTOCTR=23443&WIN=UPLOADED

engravings with existing animals.[74] The fidelity between images from Renard and living species is certainly evident when one is compared next to the other. For example, *Macolor* of Renard is a species that can be identified as the Black and White Snapper, *Macolor niger*.[75] Similarities between Renard's coloured engraving and the living specimen are clear [Fig. 20.2]. One can see the white

74 Pietsch, "Louis Renard's fanciful fishes" 67.

75 Pietsch, *Fishes, Crayfishes, and Crabs* 114–115.

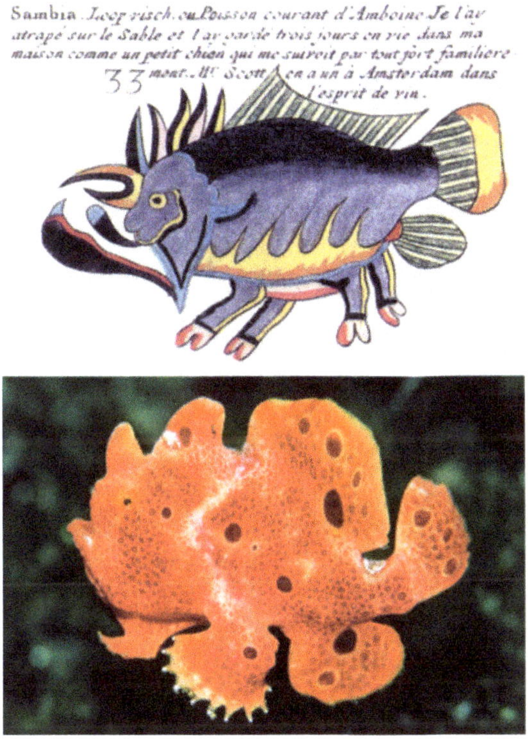

FIGURE 20.3
Plate 7, Fig. 33 *Sambia* from the second volume of Renard's *Poissons, ecrevisses et crabes, de diverses couleurs et figures extraordinaires, que l'on trouve autour des isles Moluques et sur les côtes des terres Australes* [...] (Amsterdam, Reinier and Josue Ottens: 1754) (top) and a photograph of a frogfish of the family Antennariidae (bottom) showing similarities and differences between the engraving and photograph of a member of the family it represents
TOP IMAGE TAKEN FROM THE SECOND EDITION IN THE COLLECTION OF THE BAYERISCHE STAATSBIBLIOTHEK MUNICH. BOTTOM IMAGE COURTESY OF ROGER STEENE

stripe along the body, black bands of colour, white spotting on the dorsal side, and bands of colour near the eye of both the engraving and the photograph. The inaccuracies are just as evident as the similarities, however, with oranges, reds, and yellows present in the engraving that are not characteristics of the living fish. Nevertheless, the similarities in this example allowed for a confident identification to species. Interestingly, this is a juvenile stage of *Macolor niger*; an adult stage is depicted in Renard's Plate 20, Figure 95 of volume 2, where it is called *Kakatoe*.[76] *Kakatoe* is evidently not associated with *Macolor* by Renard or Fallours.

At the other end of the taxonomic spectrum, it was only possible to identify some of Renard's engraved fishes to family. *Sambia*[77] or 'Walking Fish' [Fig. 20.3] is clearly a frogfish of the family Antennariidae,[78] but additional identification below this level is not possible. When comparing the two

76 Ibidem 140–141.
77 Ibidem 114–115.
78 Pietsch T.W. – Arnold R.J., *Frogfishes: Biodiversity, Zoogeography, and Behavioral Ecology* (Baltimore, MA: 2020).

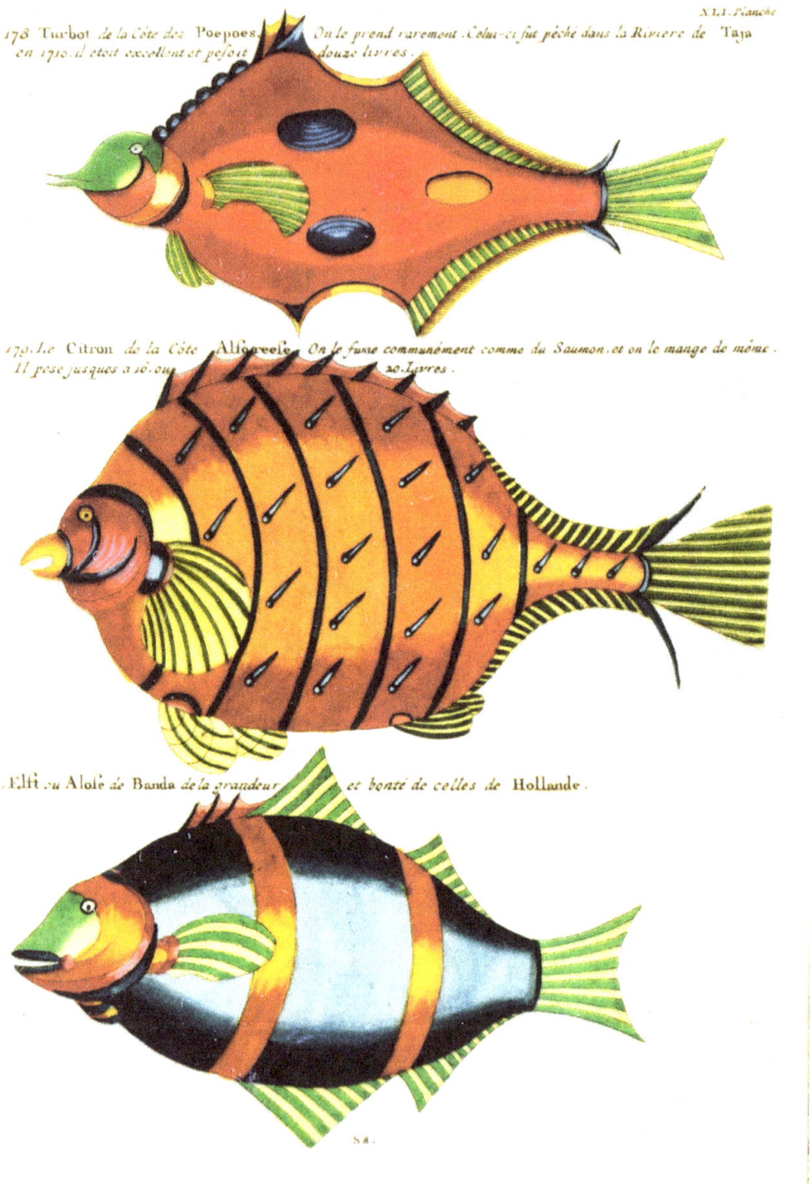

FIGURE 20.4 Plate 41, Fig. 178 (fish at the top) *Turbot de la Côte des Poepoes* from the second volume of Renard's *Poissons, ecrevisses et crabes, de diverses couleurs et figures extraordinaires, que l'on trouve autour des isles Moluques et sur les côtes des terres Australes* [...] (Amsterdam, Reinier and Josue Ottens: 1754). This creature is an example of one of the 9% of organisms in Renard's *Poissons* that could not be identified

IMAGE TAKEN FROM THE SECOND EDITION IN THE COLLECTION OF THE BAYERISCHE STAATSBIBLIOTHEK MUNICH

images, some similarities are evident including the general body shape and fin placement. Although the depictions of the fins on the ventral side of the fish in the engraving are inaccurate, their resemblance to 'legs' combined with the description of the fish as the 'walking fish' alludes to a common behaviour of the frogfishes. This group is known to 'walk' along the substrate using fins modified for this purpose. Thus, although Renard's engraving is inaccurate and the description outlandish, taken together they do reveal some understanding of the fish and its biology.

Finally, although they represent only approximately 9% of the organisms depicted in the book's 100 plates, there is a subset of engravings for which no plausible identification could be made. One example is *Turbot de la Côte des Poepoes* [Fig. 20.4]. Renard describes this fish as 'rarely caught' and the specimen depicted weighed 12 pounds.[79] However, the combination of characteristics given to this fish do not align it with any known taxonomic family.

8 Escaped from the Colourist: a Potentially Unique Copy?

Poissons, ecrevisses et crabes was issued by its publishers coloured, and there is no evidence that uncoloured copies were offered for sale. Indeed, the book is commonly recognized as the first book on fishes to be illustrated with coloured plates, and contemporary correspondence provides evidence that the book was coloured by the publishers of the first[80] and second[81] editions prior to their sale. Similar contemporary correspondence is not known for the third edition; however, all known extant copies of the third edition are also coloured. Despite no records of uncoloured copies of the book ever leaving the publishers, there is intriguing evidence that six of the 36 uncoloured copies of the first edition purchased by the Ottens were kept 'in reserve' and not coloured.[82]

79 Pietsch, *Fishes, Crayfishes, and Crabs* 182–183.

80 In letters to Sir Hans Sloane (1660–1753), Renard describes the great costs of having the books coloured prior to their sale. Indeed, Renard describes how it is difficult for him to send Sloane copies of the book for sale 'on commission' and that Renard must receive payment in advance of sending books to Sloane in England for sale. Renard did eventually send 30 coloured copies of the book to a bookseller in England (Ibidem 21).

81 Correspondence between Aernout Vosmaer and the publishers of the second edition indicate that uncoloured copies of the plates purchased from Renard's estate were coloured and that new plates pulled from the original copperplates were also coloured prior to sale. In particular, the Ottens justify the price of the book by explaining the expenses incurred, in part, by the colouring (Ibidem 23–24).

82 The Ottens corresponded with Aernout Vosmaer about purchasing 36 uncoloured copies of the first edition from Renard's estate but only discuss having 30 of these coloured (Ibidem 23).

Thirty of these 36 copies were apparently coloured and offered as part of the print run of the second edition,[83] but there is no further record of the fate of the 'reserve' copies. Thus, until recently, all extant copies known of all three editions of *Poissons* were thought to be coloured by their respective publishers and only offered for sale once coloured.

It was therefore an unexpected and significant discovery when single plates of the second edition of *Poissons* began to appear on the online auction platform, Invaluable, in 2020. These plates were offered with 'recent professional hand colouring,' implying that they had also recently existed in an uncoloured state. When we contacted the firm auctioning these plates, the firm confirmed that they were in possession of an uncoloured copy of Renard's *Poissons* that was purchased from an antiquarian bookdealer. Unfortunately, this dealer had no additional provenance on the copy, but it had been bound in recent, undecorated full vellum that preserved an old leather spine label from the book's previous binding. The firm intended to colour and auction individually each of the 100 plates from the now dismantled copy, but when we contacted the seller in November 2020, not all plates had yet been coloured. To secure as many plates as possible in their uncoloured state, a single owner purchased 23 uncoloured plates from the firm to preserve them together as a fragment of the original copy for future research. The remaining 77 plates have been coloured and dispersed via auction in 2020 and 2021. Four of these coloured examples were purchased by the owner of 23 uncoloured plates, creating a fragmentary copy of the original book with 23 uncoloured plates, 4 coloured plates, and a single text leaf (the half title to volume 2).

The provenance of this uncoloured copy is currently unknown. Although it included the title page for the second edition, it is not known when the plates in the copy were pulled from the copperplates or why it may have escaped colouring. It is possible that this copy was one of the six 'held in reserve' by the publishers of the second edition and not coloured with the other 30 copies purchased from Renard's estate. Because all plates from all three editions were pulled from the same copperplates engraved for the book's first edition, it has not yet been possible to determine which copies of the second edition include plates pulled before 1719 and purchased from Renard's estate and which include plates newly pulled for the book's 1754 second edition.

However, while studying the uncoloured plates, we noticed a detail that has apparently not previously been examined. The upper margins of Plates 29 and 31 in volume 2 contain two errant ink markings near the plate mark, likely the result of damage to the copperplates [Fig. 20.5]. These identical ink marks have been confirmed in extant colored copies of the second and third editions of

83 Ibidem 23.

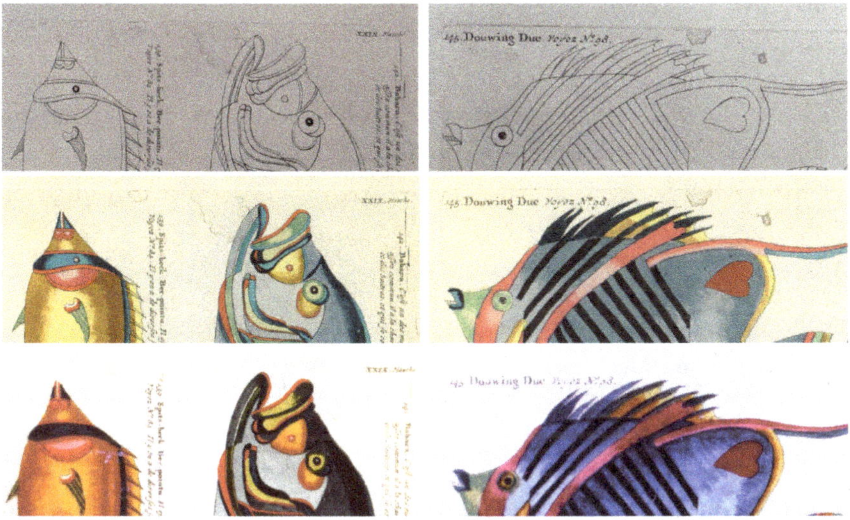

FIGURE 20.5 Details of Plate 29 (left) and 31 (right) from the second volume of the second edition of Renard's *Poissons, ecrevisses et crabes, de diverses couleurs et figures extraordinaires, que l'on trouve autour des isles Moluques et sur les côtes des terres Australes* (1754) showing errant ink markings in the uncoloured copy (top), the same markings in the copy in Harvard's Ernst Mayr Library (middle), and the absence of these markings in the copy in the Bayerische Staatsbibliothek Munich

TOP IMAGES © JUSTIN R. HANISCH. MIDDLE IMAGES FROM HARVARD UNIVERSITY IN PUBLIC DOMAIN OBTAINED VIA BIODIVERSITY HERITAGE LIBRARY. BOTTOM IMAGES TAKEN FROM THE SECOND EDITION IN THE COLLECTION OF THE BAYERISCHE STAATSBIBLIOTHEK MUNICH

the work, but in at least two copies of the first edition examined digitally, the errant ink markings are absent. Thus, if the ink markings are absent from all copies of the first edition, it is potential evidence that the damage occurred sometime between the completion of the first edition and before or during the printing of the second. If even a single copy of the first edition is found with the errant ink, then the damage to the plates likely occurred during the printing of the first edition. However, if all extant copies of the first edition do not contain the errant ink marks and if some copies of the second edition are free of the errant ink,[84] it is strong evidence that the damage to the plate occurred sometime between the printing of the first and second editions and that any second editions extant without the evidence of the damage may indeed contain plates from the first pulling in 1718–19. This would also mean the uncoloured copy is likely not one of the "reserve copies" of the first edition.

84 A copy of the second edition has been confirmed without the damage.

Of course, with only a relatively small number of copies from all editions in existence today, this investigation could only provide partial and inconclusive evidence of when this damage may have occurred. A census of extant copies is underway to determine which copies bear these errant ink markings and whether it might be a clue to the printing priority of plates in the first and second editions.

9 Conclusions

Often dismissed largely as a curiosity or a monstrosity, Louis Renard's *Poissons, ecrevisses et crabes, de diverses couleurs et figures extraordinaires, que l'on trouve autour des isles Moluques et sur les côtes des terres Australes* is instead a serious scientific effort unfortunately marred by embellishments and inaccuracies that obscure the book's rightful contribution to the ichthyological literature of 18th century Indo-Pacific Oceania. Rather than comprising a flipbook of fantasy, Renard's book accurately presents 385 fishes and 34 crustaceans at least to the level of family and many to genus or species. The book found a small audience in the 18th century, appearing in three different editions by three separate publishers but never exceeding 100 copies in an edition. Indeed, the book was expensive to produce and expensive to purchase in the 18th century,[85] and is today one of the rarest and most expensive 18th-century natural history works on fishes.

Although Renard's *Poissons* has received considerable study from its publication through to today, it still has secrets to reveal. Ongoing research into errant ink markings visible on some plates may provide new insights into the complicated printing history of this enigmatic work, and fragments of the potentially unique uncoloured copy have been preserved for future study and reflection.

Bibliography

Benthem Jutting W.S.S. van, "A Brief History of the Conchological Collections at the Zoological Museum of Amsterdam, with Some Reflections on 18th-Century Shell Cabinets and Their Proprietors, on the Occasion of the Centenary of the Royal Zoological Society 'Natura Artis Magistra'", *Bijdragen tot de Dierkunde* 27 (1939) 167–246.

85 The second edition was apparently offered at 50 guilders (Ibidem 23) or several hundred Euro in 2022 equivalent.

Bicker Raye J., *Het dagboek van Jacob Bicker Raije, 1732–1772, naar het oorspronkelijk dagboek medegedeeld door Fr. Beijerinck en M.G. de Boer*. Second Edition (Amsterdam: 1960).

Boer M.G. de, "Introduction", in Bicker Raye J., *Het dagboek van Jacob Bicker Raije, 1732–1772, naar het oorspronkelijk dagboek medegedeeld door Fr. Beijerinck en M.G. de Boer*. Second Edition (Amsterdam: 1960) v–xii.

Boeseman M., "The Vicissitudes and Dispersal of Albertus Seba's Zoological Specimens", *Zoologische Mededelingen* 44 (1970) 177–206.

Bridson G.D.R., "From Xylography to Holography: Five Centuries of Natural History Illustration", *Archives of Natural History* 16 (1989) 121–141.

Brunet J.C., *Manuel du libraire et de l'amateur de livres*, vol. 4 (Paris: 1863).

Dance P.S., *The Art of Natural History* (Woodstock, NY: 1978).

Dean B., *A Bibliography of Fishes*, vol. 1 (New York: 1916).

Dean B. *A Bibliography of Fishes*, vol. 3 (New York: 1923).

Eeghen I.H. van, *De Amsterdamse boekhandel, 1680–1725: I. Jean Louis de Lorme en zijn copieboek* (Amsterdam: 1960).

Eeghen I.H. van, *De Amsterdamse boekhandel, 1680–1725: II. Uitgaven van Jean Louis de Lorme en familieleden* (Amsterdam: 1963).

Eeghen I.H. van, *De Amsterdamse boekhandel, 1680–1725: III. Gegevens over de vervaardigers, hun internationale relaties en de uitgaven A–M* (Amsterdam: 1965).

Eeghen I.H. van, *De Amsterdamse boekhandel, 1680–1725: IV: Gegevens over de vervaardigers, hun internationale relaties en de uitgaven N–W, papierhandel, drukkerijen en boekverkopers in het algemeen.* (Amsterdam: 1967).

Eeghen I.H. van, *De Amsterdamse boekhandel, 1680–1725: V (1–2): De boekhandel van de Republiek 1572–1795, summary, overzichten en indices* (Amsterdam: 1978).

Engel H. "Alphabetical list of Dutch Zoological Cabinets and Menageries", *Bydragen tot de Dierkunde* 27 (1939) 247–356.

Engel H., *Hendrik Engel's Alphabetical List of Dutch Zoological Cabinets and Menageries*, eds., P. Smit – A.P.M. Sanders – J.P.F. van der Veer (Amsterdam: 1986).

Graesse J.G.T., *Trésor de livres rares et précieux, ou Nouveau dictionnaire bibliographique*, vol. 6 (Berlin: 1922).

Haag E. – Haag E., *La France protestante, ou Vies des protestants Français qui se sont fait un nom dans l'histoire depuis les premiers temps de la réformation jusqu'à la reconnaissance du principe de la liberté des cultes par l'Assemblée nationale. Ouvrage précédé d'une notice historique sur le protestantisme en France*, vol. 8 (Paris: 1858).

Holthuis L.B., "Notes on Pre-Linnean Carcinology (Including the Study of Xiphosura) of the Malay Archipelago", in Wit H.C.D. de (ed.), *Rumphius Memorial Volume* (Baarn: 1959) 63–125.

Landwehr J., *Studies in Dutch Books with Coloured Plates Published 1662–1875: Natural History, Topography and Travel, Costumes and Uniforms* (The Hague: 1976).

Meesters R., *Catalogue no. 22, Science and Medicine*, bookseller cat., Ronald Meesters, Antiquarian Bookseller (Amsterdam: 1979).

Mulder Bosgoed D., *Bibliotheca ichthyologica et piscatoria: Catalogue de livres et d'écrits sur l'histoire naturelle des poissons et des cétacés, la pisciculture, les pêches, la législation des pêches, etc* (Haarlem: 1873).

Nissen C., *Schöne Fischbücher, kurze Geschichte der ichthyologischen Illustration und Bibliographie fischkundlicher Abbildungswerke* (Stuttgart: 1951).

Nissen C. 1969, *Die zoologische Buchillustration, ihre Bibliographie und Geschichte*, vol. 1 (Stuttgart: 1969).

Pieters F.F.J.M., "Notes on the Menagerie and Zoological Cabinet of Stadholder William v of Holland, Directed by Aernout Vosmaer", *Journal of the Society for the Bibliography of Natural History* 9 (1980) 539–563.

Pieters F.F.J.M., "Histoire Naturelle des Poissons, Ecrivisses et Crabes des Indes. De uiterst zeldzame derde editie van Louis Renards Poissons, Ecrivisses et Crabes met tekst van Pieter Boddaert, gepubliceerd in 1782", in Alsemgeest A. – Fransen C. (eds.), *In krabbengang door kreeftenboeken. De* Bibliotheca Carcinologica *L.B. Holthuis* (Leiden: 2016) 81–86.

Pietsch T.W., "Louis Renard's Fanciful Fishes", *Natural History* 93 (1984) 58–67.

Pietsch T.W., "Fallours, S. Fishes of the Indo-west Pacific: A Collection of Handcoloured Drawings", in Antiquariaat Junk. *Natural History & Travel*, Catalogue 241 (Amsterdam: 1986) 36–39.

Pietsch T.W., "Samuel Fallours and His 'sirenne' from the Province of Ambon", *Archives of Natural History* 18 (1991) 1–25.

Pietsch T.W., "On the Three Editions of Louis Renard's *Poissons, ecrevisses et crabes: Histoire naturelle des plus rares curiositez de la Mer des Indes*", *Archives of Natural History* 20 (1993) 49–68.

Pietsch T.W., *Fishes, Crayfishes, and Crabs: Louis Renard and His Natural History of the Rarest Curiosities of the Seas of the Indies* (Baltimore, MA: 1995).

Pietsch T.W. – Arnold R.J., *Frogfishes: Biodiversity, Zoogeography, and Behavioral Ecology* (Baltimore, MA: 2020).

Pietsch T.W. – Rubiano D.M., "On the Date of Publication of the First Edition of Louis Renard's *Poissons, ecrevisses et crabes: Histoire naturelle des plus rares curiositez de la Mer des Indes*", *Archives of Natural History* 15 (1988) 63–71.

Renard Louis, *Poissons, écrevisses et crabes de diverses couleurs et figures extraordinaire, que l'on trouve autour des Isles Moluques, et sur les côtes des Terres Australes* [...] (Amsterdam, Louis Renard: [1719]).

Renard Louis, *Poissons, écrevisses et crabes de diverses couleurs et figures extraordinaire, que l'on trouve autour des Isles Moluques, et sur les côtes des Terres Australes* [...] (Amsterdam, Reinier and Josue Ottens: 1754).

Renard Louis, *Natuurlyke historie Indische Zeeën; behelsende de visschen, kreeften en krabben van verschillende kleuren en buitengewoone gedaanten, van de Moluksche Eilanden en op de kusten der Zuidlyke Landen* [...] (Utrecht – Amsterdam, Abraham van Paddenburg – Willem Holtrop: 1782).

Schutte O., *Repertorium der buitenlandse vertegenwoordigers, residerende in Nederland 1584–1810* (The Hague: 1983).

Wood C.A., *An Introduction to the Literature of Vertebrate Zoology* (London: 1931).

CHAPTER 21

Distance, Geography, and Anecdote in M.E. Bloch's *Natural History of Fishes*

Johannes Müller

Inferring biological information from a dead animal specimen is always a delicate task and both modern scientists and early modern naturalists were aware of this problem. Yet, collections of conserved specimens were the main resources for European naturalists who studied non-European water animals that could not be transported alive from other continents. The results are sometimes still visible in the European names and even in scientific nomenclature: some egg-laying species were erroneously thought to be livebearers and, as a result, still carry the species epithet *viviparus* (live-bearing), such as the African bowstripe barb (*Enteromius viviparus*).[1] Other fishes received their names from erroneous geographical attributions, such as the South American wolf fish, named *Esox malabaricus* by Marcus Elieser Bloch, who assumed its South Indian origin and named it after the Malabar Coast.[2]

To avoid such errors, European naturalists were crucially dependent on the information that came with the specimen they received and then described. Which accounts could be trusted, whose observations counted and under which circumstances were they documented and brought to Europe? Questions like these were critical in all fields of zoology but precise descriptions of fishes were particularly difficult, as not only their behavior but even their body shape and color were often impossible to reconstruct from a conserved specimen. Regardless if a water animal was stored in alcohol or as a dried specimen – which was a rarer technique of preservation – it lost most of its live features and could be used for little more than a count of fin rays and scales and, especially after the first half of the 19th century, for comparative

1 Barnard K.H., "Note on Alleged Viviparity in *Barbus viviparus*, and Description of a New Species of *Beirabarbus*", *Annals and Magazine of Natural History* 8.47 (1941) 469–471. I thank Chris Scharpf for pointing me to this reference.

2 Abdala Dergam Dos Santos J., *Phylogeography and Character Congruence Within the Hoplias Malabaricus Bloch, 1794 (Erythrinidae, Characiformes, Ostariophysil) Species Complex* (Ph.D. Dissertation: Colorado State University 1996) 4. The valid name of Bloch's *Esox malabaricus* is now *Hoplias malabaricus*. In the following, I will first refer to Bloch's original name and then add the currently valid name in brackets.

© JOHANNES MÜLLER, 2024 | DOI:10.1163/9789004681187_022

anatomical research. This chapter focuses on the aforementioned ichthyologist Marcus Elieser Bloch (1723–1799) and his attempts to make sense of the fish specimens he received from his correspondents abroad or from dealers in exotic natural objects, and the accompanying information that came with these conserved animal remains.

Bloch's contribution to ichthyology can hardly be overestimated: his *Oeconomische Naturgeschichte der Fische Deutschlands* (1782–1784) and *Naturgeschichte der ausländischen Fische* (1785–1795) (hereafter *Natural History of the German Fishes* and *Natural History of Foreign Fishes*), published in twelve volumes, was the first major attempt of an overview of known fish species according to the still recent Linnean taxonomic system, and the most extensive one before Bernard Germain de Lacépède's *Histoire naturelle des poissons* (1798–1803).[3] Bloch, who never left Europe, has traditionally been regarded a typical 'compiler', and his project to describe as much fish species as possible crucially depended on secondary information which he could never directly check or confirm.[4] His *Natural History* offers important insights into the ways in which information moved around the globe, how it was exchanged between travelers, colonial agents, missionaries and European naturalists, and how it was interpreted, reviewed and incorporated into larger systems of knowledge and learning. Addressing Bloch as a *reader* and interpreter of circulating and often unverifiable information, this chapter aims to shed light on the mechanisms of evaluation and source criticism in the natural sciences in late 18th-century Europe.

Strategies of managing, verifying and processing information have been addressed in a large body of literature on travel, geography and knowledge production, especially on the early modern period. In the wake of seminal works like Steven Shapin's *Social History of Truth* and Katherine Park's and Lorraine Daston's *Wonders and the Order of Nature*, questions of credibility and the evaluation of truth claims became central topics in the history of knowledge and science.[5] In more recent years, strategies of information management have received more attention and the ways in which scholars and naturalists organized their materials have been explored by Ann Blair, Staffan Müller-Wille,

3 Bloch Marcus Elieser, *Oeconomische Naturgeschichte der Fische Deutschlands* (Berlin, Hesse: 1782–84); *Naturgeschichte der ausländischen Fische*, 3 vols. (Berlin, Morino: 1785–1795).

4 Starr Jordan D., "The History of Ichthyology", *Science* 16.398 (1902) 241–258.

5 Shapin S., *A Social History of Truth. Civility and Science in Seventeenth-Century England* (Chicago: 1994); Daston L. – Park K., *Wonders and the Order of Nature 1150–1750* (New York: 1998), especially 215–225; 246–255; 343–350.

Isabelle Charmantier and others.[6] This important body of literature has seldom been connected to the period after 1750 – as if questions of credibility and the management of information were no longer acute and pressing. This chapter seeks fill this gap by exploring how new observational and taxonomic practices went hand in hand with the critical evaluation of sources, and how systematic knowledge production relied on all these three components. Marcus Elieser Bloch's work provides an insightful case study as it laid the base for a wide number of taxonomic and nomenclatural decisions that shape the field of ichthyology until today. In the first part, I will offer a brief sketch of his background and the social and intellectual milieu in which he operated. After a discussion of his speculations on the geographical origin of his materials, I will then discuss his use of anecdotes from which he tried to infer information about behavior of 'his' fishes and their interactions with their environment. As Bloch's use of second-hand information shows, anecdotal information continued to play an important role in ichthyological knowledge even at a point when anatomical and physiological knowledge, professionalizing research methodologies and taxonomic classification systems had thoroughly transformed ichthyology and established it as a field in its own right.[7]

1 Bloch, the Berlin *Haskalah* and Fieldwork in Prussia

Like many ichthyologists of the 17th and 18th centuries, Bloch's interest in fishes started as a personal "hobby", rather than a result of academic training in this specific field. Born into a Jewish family of modest financial means in the Franconian town of Ansbach, he only had access to academic learning at a relatively late age in his life. Even though his father was a Thora scribe, his access to non-religious literature must have been rather limited. It is often stated that Bloch only learned German and Latin at the age of twenty, a claim that is hard to believe and should probably be understood in the sense that he had not yet learned reading in the Latin alphabet, let alone managed German *Fraktur*

6 Blair A., *Too Much to Know. Managing Scholarly Information before the Modern Age* (New Haven, Connecticut: 2010); Blair A., "Note Taking as an Art of Transmission", *Critical Inquiry* 31 (2004) 85–107; Müller-Wille S. – Charmantier I., "Lists as Research Technologies", *Isis* 103.4 (2012), 743–52; Müller-Wille S. – Scharf S., "Indexing Nature: Carl Linnaeus (1707–1778) and His Fact-Gathering Strategies", *Working Papers on The Nature of Evidence: How Well Do 'Facts' Travel?* 36.8 (2009) 1–46.

7 Trijp D.R. van, *Captured on Paper. Fish Books, Natural History and Questions of Demarcation in Eighteenth-Century Europe (ca. 1680–1820)*, (Ph.D. Dissertation: Leiden University 2021) 227–232.

script.[8] In 1743, he left Ansbach for Hamburg and became the private tutor to the children of a Jewish surgeon. Little is known about Bloch's life in his twenties and thirties but his time in Hamburg brought him in touch with the medical profession and the world of learning. Following anatomy lessons in Berlin, he pursued a career as a medical doctor but could not receive a doctorate there because of his Jewish background. To obtain an official degree, he had to relocate to Frankfurt an der Oder in the early 1760s, where no regulations withheld Jewish students from graduating.

After his graduation, Bloch practiced medicine in Berlin. One of his patients, the famous philosopher Moses Mendelsohn, became a close friend and through such acquaintances, Bloch entered Jewish learned circles that promoted a reform of Jewish traditions according to Enlightenment ideas, later known as the *Haskalah*.[9] In some respects, Bloch was an exceptional figure in these circles. The interests of his Berlin circle were of a more philosophical nature and concerned questions such as the reconciliation of contemporary German philosophy with Jewish tradition and faith. Despite their interest in the natural sciences, Bloch was the only one who actually dedicated himself to natural historical research.[10] After some shorter publications on various medical topics, he focused more and more on study of fishes. As he remarked in the preface to the first volume of the *Fishes of Germany*, his surprise about the great diversity in fishes and the incongruence between ichthyological literature and his observations in Prussian lakes had inspired his work on this topic.[11]

Compared to his Berlin network of Jewish intellectuals, Bloch's scholarly work shows little traces of typical Haskalah themes or ideas.[12] Aside from a dedication of one of the *Natural History*'s parts to the Danish crown prince Frederick, 'who has made my oppressed Brethren equal to the other inhabitants',

8 Karrer C., "Marcus Elieser Bloch (1723–1799), Sein Leben und die Geschichte seiner Fischsammlung", *Sitzungsberichte der Gesellschaft Naturforschender Freunde zu Berlin* 18 (1978) 129–149, there 130–131; Trijp, *Captured on Paper* 168. The difficulties of reading German *Fraktur* for non-native speakers and readers are explicitly mentioned in Bloch's preface to the *Fishes of Germany*. See Bloch, *Naturgeschichte der Fische Deutschlands*, part 1, 5–6.

9 Schulte C., "Zur Debatte um die Anfänge der jüdischen Aufklärung", *Zeitschrift für Religions- Und Geistesgeschichte* 54.2 (2002) 122–137, there 122; Lesser R., "Dr. Marcus Elieser Bloch. Ein Jude begründet die moderne Ichthyologie", *Das achtzehnte Jahrhundert. Zeitschrift der Deutschen Gesellschaft für die Erforschung des achtzehnten Jahrhunderts* 23.2 (1999 – Special issue *Haskala. Die jüdische Aufklärung in Deutschland 1769–1812*) 238–246.

10 Keller A.G., "Science In The Early 'Haskalah'", *European Judaism* 24.2 (1991) 8–13.

11 Bloch, *Naturgeschichte der Fische Deutschlands*, vol. 1, preface, 3.

12 Schulte, "Anfänge der jüdischen Aufklärung" 135.

there are no specific references to Jewish life or traditions.[13] In fact, Frederick had not fully emancipated Denmark's Jewish inhabitants but only allowed them to learn skilled trades as apprentices, a step that had not yet been taken in Prussia and Berlin. Like Mendelssohn, Bloch's admission to the Prussian Academy of Sciences was refused because of his Jewish background – allegedly also due to the role Frederick II of Prussia who had a final say in the acceptance of new members.[14] However, the great success of the first three parts of his ichthyological work, the *Economical Natural History of the Fishes of Germany*, had founded his reputation as a renowned scholar in Prussia and the wider German-speaking world and he was invited into the Imperial natural history academy *Leopoldina* and he had already been a corresponding member of the learned societies of Berlin, Göttingen, Leipzig, Halle, and many other German cities. As his fame grew, he was also accepted into the societies of Utrecht, Haarlem, Flushing, Zürich and Saint Petersburg.[15]

Even though the *Natural History of the Fishes of Germany* and the *Foreign Fishes* show clear structural parallels, the methodology and the entire approach were fundamentally different. As Bloch recounts, he dedicated his spare time to the study of fishes in small fishing communities in Brandenburg.[16] Talking to local fishers and studying fresh-caught fishes was a method that enabled Bloch to go beyond the scholarly world of books and the descriptions of others, and his surprise about the true diversity of Prussia's and Germany's fish fauna is expressed repeatedly. Comparing his own observations to those of Linnaeus and Artedi allowed for a vast expansion of the rich European fish diversity.[17] Bloch was also able to study live specimens of native fishes in metal tubs and noted important aspects of their behavior.[18] In this respect, the *Natural*

---

13 Bloch, cited in Paepke H.-J., "Blochs Schlangenkopf- und Labyrinthfische. Ein Beitrag zum 200. Todestag von Marcus Elieser Bloch (1723–1799)", *Der Makropode* 21.1–2 (1999) 2–13, there 3; *Naturgeschichte der ausländischen Fische*, vol. 4, part 7, preface, v–vi.

14 Paepke H.-J., *Bloch's Fish Collection in the Museum für Naturkunde der Humboldt-Universität zu Berlin. An Illustrated Catalog and Historical Account* (Rugell, Liechtenstein: 1999) 15. The role of Frederick II in the Academy's admission policy with regards to Jewish candidates is disputed. In the famous case of Mendelsohn, Frederick did not receive the list in which Mendelsohn's name was mentioned and was therefore probably the decisive factor in his refusal. See Berkemann J., "Die Emanzipation der deutschen Juden und der Begriff der Toleranz", in: Enders C. – Kahlo M. (eds.), *Toleranz als Ordnungsprinzip. Die moderne Bürgergesellschaft zwischen Offenheit und Selbstaufgabe* (Leiden: Brill, 2007) 71–107, there 71.

15 Bloch, *Naturgeschichte der Fische Deutschlands*, part 1, title page; *Naturgeschichte der ausländischen Fische*, part 7, title page.

16 Bloch, *Naturgeschichte der Fische Deutschlands*, part 1, preface, 3.

17 Bloch, *Naturgeschichte der Fische Deutschlands*, part 1, preface, 2–3; 5.

18 Paepke, *Bloch's Fish Collection*, 22.

History of the Fishes of Germany is fundamentally different from the *Foreign Fishes*: in the latter, Bloch was fully dependent on the descriptions of others and he did not have the opportunity to study the animals in a freshly killed or even live state. The need to infer all the information from dead specimens and the accounts from his correspondents abroad or secondary literature made the *Foreign Fishes* a project that required hermeneutical and textual-critical skills in order to make sense of the existing information. While Bloch had little choice to use accounts he could not always fully trust, his work reflects strategies of critical and comparative reading that sometimes allowed for comprehensive theorizations of new and hitherto poorly understood biological phenomena.

2 Origin Unknown – Bloch's Speculative Biogeographies

Even before he published the *Fishes of Germany*, Bloch was already familiar with and interested in the fish faunas of other continents. To provide taxonomic context for his description of European species, he extensively referred to the zoological works of the French friar and naturalist Charles Plumier (1646–1704). Plumier, who had been appointed the French royal botanist by Louis XIV in 1693, had undertaken three research expeditions to Central America and besides his study of plants also described and drawn numerous fish species from the Americas. Bloch was well aware of the differences between American and European species, even though he did not clearly separate similar-looking fishes from different continents into different genera. As his reputation as a renowned naturalist grew – he had received a gold medal for his *Fishes of Germany* from Emperor Joseph II in 1782 – acquaintances and fellow naturalists and collectors sent him more and more conserved fish specimens, also from other parts of the world. By the end of his life, his collection had grown to more than 1.400 specimens of fish, in addition to 400 birds and many other natural objects.[19]

Once his collection had grown, Bloch actively sought to complete it and acquire as many fishes as he could get from dealers in natural object or correspondents in Asia, such as the Protestant missionary Christoph Samuel John and the physician Johann Gerhard König in South India or the botanist Paul Erdmann Isert who had travelled through West Africa and the Caribbean.[20] While Bloch was able to shed light on a number of complex ichthyological

19 Paepke, *Bloch's Fish Collection*, 16.

20 See John Christoph Samuel, "Einige Nachrichten von Trankenbar auf der Küste Koromandel. Aus einem Briefe von dem Missionarius Hrn John an Herrn Doktor Bloch in

problems by comparing different sources, one of the most basic questions proved difficult to answer: what was the fish's origin and natural habitat? The number of Bloch's errors in identifying the place or region of origin are numerous, even though he was aware that he could not always trust his sources, especially dealers in exotic natural objects. The indicated species distribution in his collection and his book project reveals much about how he imagined the geography of the world outside Europe. Besides the aforementioned misidentification of the South American wolf fish (*Hoplias malabaricus*) as a 'Malabarian' species and the confusion between several South American and African cichlids, most of the confusion on the animals' whereabouts concerned Asian fishes.[21]

One significant sequence of mistakes is Bloch's reference to Japan as the origin of his specimens. In 1786 or early 1787 he received a collection of "East Indian" fishes, which were listed and discussed in detail in the third and fourth parts of the *Natural History of Foreign Fishes*.[22] The vagueness of the origin of this collection is reflected in several references throughout the volumes and Bloch sometimes switches between "East Indian" and "Japanese".[23] A closer examination of these "Japanese" fishes reveals that only a part can in fact be found in Japan and the majority is distributed along the coasts of South and South East Asia. At the same time, Bloch described some species as Caribbean, when they were in fact native to East Asia and Japan.[24] Even the fishes that can be found in Japan have a much wider distribution and there is no clear evidence that any of his specimens actually originated from Japan.[25]

Berlin", *Berlinische Monatschrift* 20 (1792) 585–596. On Bloch correspondents in Asia, see also Trijp, *Captured on Paper* 189–191 and Paepke, *Bloch's Fish Collection* 24–25.

21 For more examples of confusions between African and American fishes, see Paepke, *Bloch's Fish Collection* 27.

22 Bloch, *Naturgeschichte der ausländischen Fische*, vol. 1, part 3, preface, fol. A2; 115.

23 Bloch, *Naturgeschichte der ausländischen Fische*, vol. 1, part 3, 115.

24 E.g. *Sparus fasciatus* (*Cheilinus fasciatus*) and *Sparus chlorourus* (*Cheilinus chlorourus*), discussed in the *Natural of Foreign Fishes*, part 5.

25 Bloch, *Naturgeschichte der ausländischen Fische*, vol. 1, part 1: *Squalus varius* (*Stegostoma fasciatum*): Indian Ocean and Pacific (also Japan). Part 2: *Takifugu ocellatus* (*Tetraodon fasciata*): Southeast Asia and Pacific (also Japan); *Muraena pinna* (probably *Muraena conger*): Europe, Atlantic Ocean. Part 3: *Chaetodon imperator* (*Pomacanthus imperator*): Indian Ocean and Pacific (also Japan); *Chaetodon guttatus* (*Siganus guttatus*): Indo-Pacific and Japan: *Chaetodon collare*: Indo-Pacific and Pacific (also Japan); *Chaetodon mesoleucos*: Red Sea, Arab Sea. Part 4: *Cephalopholis boenak* (*Bodianus boenak*): Indian Ocean and Pacific (also Japan); *Bodianus gutatus* (*Cephalopholis argus*): Indian Ocean and Pacific (also Japan); *Holocentrus ongus* (*Epinephelus ongus*): Indian Ocean and Pacific; *Scarus viridis* (*Sparisoma viride*): Caribbean/Atlantic; *Holocentrus quadrilineatus* (*Pelates quadrilineatus*): Indian Ocean and Pacific (also Japan); *Holocentrus calcarifer* (*Lates

Hans-Joachim Paepke has suggested that Bloch might have confused Japan with Java, a conclusion for which some support can be drawn from a comparison to his posthumous *Systema Ichthyologiae*, where some species, for example *Sparus fasciatus* (*Cheilinus fasciatus*), are identified as a Southeast Asian species ('Habitat in Java'), whereas the *Foreign Fishes* lists them as 'Japanese'.[26] This specific case suggests that Bloch had indeed confused Java and Japan, but it is important to pay attention to the sources that informed his geographical ideas about East and Southeast Asia.

One of Bloch's main sources on Asia was François Valentyn's *Oud en Nieuw Oost-Indiën* (*Old and New East India*), published between 1724 and 1726 in Amsterdam and Dordrecht. Valentyn was active as a Protestant minister in the Dutch East Indies, mostly in Ambon on the Moluccas, and besides his clerical duties, he dedicated his stay in Asia to a large-scale geographical description of the Malay Archipelago and the wider world surrounding it. In the title, he already made clear that his five-volume work had a wider scope than only the colonized islands under the rule of the Dutch East India Company:

> Old and new East India, containing a precise and detailed discussion of the Dutch Government in these regions, besides an extensive description of the Moluccas, Ambon, Banda, Timor, Solor, Java, and all the islands under the rule of the same administration, the Dutch directorate at Suratte as well as a description of the lives of the Great Mughals. Furthermore an

calcarifer): Indian Ocean and Pacific (also Japan): *Lutjanus Lutjanus*: Indian Ocean and Indo-Pacific; *Lutjanus hasta*: Indian Ocean and Indo-Pacific; *Lutjanus erythropterus*: Indian Ocean and Indo-Pacific, also Japan. Part 5: *Lutjanus verres*: Indian Ocean and Indo-Pacific; *Sparus fasciatus* (*Cheilinus fasciatus*): Indian Ocean and Pacific (also Japan); *Sparus chlorourus* (*Cheilinus chlorourus*): Indian Ocean and Pacific (also Japan. Bloch also assumes that it is native St. Domingo and the Caribbean); *Sparus erythrourus* (*Gerres erythrourus*): Indo-Pacific and Pacific (also Japan); *Labrus trivittatus* (*Pentapodus trivittatus*): Indo-Pacific and Pacific (also Japan); *Lutjanus bohar* (*Sparus cynodon*): Indian Ocean and Pacific; *Labrus viridis*: Atlantic Ocean/Mediterranean. Part 6: *Labrus fasciatus* (*Hemigymnus fasciatus*): Indian Ocean and Pacific (also Japan); *Trichopodus trichopterus* (*Labrus trichopterus*): Southeast Asia (only freshwater); *Labrus melapterus* (*Hemigymnus melapterus*): Indian Ocean and Pacific (also Japan); *Perca argentata* (*Lutjanus argentimaculatus*): Indian Ocean and Pacific (also Japan); *Anthias macropthalmus* (*Priacanthus hamrur*): Indian Ocean and Pacific (also Japan); *Anthias testudineus* (*Anabas testudineus*): South and Southeast Asia (only freshwater). Part 7: *Anthias orientalis* (*Plectorhinchus orientalis*): Indian Ocean/Indo-Pacific. Part 8: *Clupea sinensis* (*Tenualosa reevesii*): South Chinese Sea. The indicated distribution ranges of each species are based on data from FishBase: Froese R. – Pauly D., *FishBase. World Wide Web Electronic Publication*. www.fishbase.org (02/2022).

26 Paepke, *Bloch's Fish Collection* 27.

> informative discussion of the most important facts on the Coromandel Coast, Pegu, Arracan, Bengal, Mocha, Persia, Malacca, Sumatra, Ceylon, Malar, Celebes or Macassar, China, Japan, Taiwan or Formosa, Tonkin, Cambodia, Siam, Bali, the Cape of Good Hope and Mauritius.[27]

Valentyn's work, which significantly shaped European ideas on Asia throughout the 18th century, covered what modern historians might call the wider "Indian Ocean World" and even stretched its scope to the Northern Pacific. Besides detailed descriptions of nature, the work offered historical descriptions of Moghul India and geographical accounts of South Asian cultures and religions. However, as Siegfried Huigen has recently argued, *Old and New East India* is characterized by a telling discrepancy between specific description and generalization. While the history and culture of the various Asian regions are discussed in terms of geographical diversity, descriptions of nature are largely limited to the Moluccas, and more specifically to Ambon itself. In Valentyn's account, the island serves as a model of Southeast Asian nature in general and he expected 'nature in the East Indies to be more or less the same everywhere.'[28]

This discrepancy between cultural diversity and assumed natural uniformity informed the European image of Asian nature.[29] Reading Bloch's misidentification of Asian fishes in the light of *Old and New East India* explains the ambiguity between his references to "Japan" and "East India."[30] In at least

27 Valentyn François, *Oud en Nieuw Oost-Indiën, vervattende Een Naaukeurige en Uitvoerige Verhandelinge van Nederlands Mogentheyd in die Gewesten, benevens Eene wydluftige Beschryvinge der Moluccos, Amboina, Banda, Timor, en Solor, Java, en alle de Eylanden onder dezelve Landbestieringen behoorende; het Nederlands Comptoir op Suratte, en de Levens der Groote Mogols; als ook Een Keuryke Verhandeling van 't wezentlykste dat men behoort te weten van Choromandel, Pegu, Arracan, Bengale, Mocha, Persien, Malacca, Sumatra, Ceylon, Malabar, Celebes of Macassar, China, Japan, Tayouan of Formosa, Tonkin, Cambodia, Siam, Borneo, Bali, Kaap der Goede Hoop en van Mauritius* [...] (Dordrecht, Joannes van Braam – Amsterdam: Gerard onder de Linden, 1724–1726). The entire book title is even significantly longer.

28 On the Dutch and European reception of Valentyn's work, see Huigen S., "Repackaging East Indies Natural History in François Valentyn's Oud en Nieuw Oost-Indiën", *Early Modern Low Countries* 3/2 (2019) 234–264, there 259.

29 Huigen, "Repackaging East Indies Natural History" 258–259.

30 Another Dutch geographical work that informed Bloch's *Natural History of Foreign Fishes* was Johan Nieuhof's travel account to the Dutch East Indies and other parts of Asia. In this account, both "Japan" and "China" are sometimes used interchangeably with "East India." See Nieuhof Johan, *Joan Nieuhofs Zee en lant-reize, door verscheide gewesten van Oostindien: behelzende veele zeltzaame en wonderlijke voorvallen en geschiedenissen. Beneffens een beschrijving van lantschappen, steden, dieren, gewassen, draghten, zeden en*

four instances, Bloch refers to his "Japanese" fishes by names that are clearly of Malay origin, for example in the case of *Lutjanus Lutjanus* or *Labrus trichtopterus* (*Trichopodus trichtopterus*), whose 'Japanese' names he notes as 'Ikan lutjang' and 'Ikan Marate Djantan'.[31] As an avid reader of Valentyn – he cites the Dutch clergyman more than thirty times – Bloch must have been aware of the difference between Java and Japan but what he did not fully comprehend were the immense biogeographical differences between the Malay Archipelago and the Northern Pacific. To European readers who relied on Valentyn, Asian nature appeared as one geographical continuum in which Java, Japan, China and Taiwan were more or less interchangeable. Modern studies have often noted with surprise that Bloch virtually never travelled – even within Europe.[32] His attempts to make sense of biogeographical differences depended on literature such as Valentyn's *Old and New East India* and Johan Nieuhof's Asian travel accounts and his acquisition of a "Japanese" fish collection from a dealer in exotic *naturalia* left with him with little more clue than what he found in these Dutch writers.

When Bloch did not trust his sources, he sometimes used comparative methods to make sense of a fish's region of origin. On one fish he had bought from a dealer in exotic natural objects, *Chaetodon ciliaris*, he noted:

> The origin of this fish is East India, according to the merchant in *naturalia* from whom I bought it. I tend to believe that this information is correct since the specimen shows long dorsal and anal fins: all the fishes I find in Marcgraf, Piso and in the drawings of Father Plumier [who had all described South American fishes – JM] show long anal and dorsal fins. In the ones that are depicted in Valentyn, these long fins are rounder.[33]

---

godsdienst der inwoonders: en inzonderheit een wijtloopig verhael der stad Batavia, verciert doorgaens met verscheide koopere platen (Amsterdam, Jacob van Meurs: 1682), vol 1. The second volume offers a description of Dutch Brazil, which did not exist anymore at the time of publication. For less informed readers, the difference between "India" and "East India" might not always have been entirely clear.

31 Bloch, *Naturgeschichte der ausländischen Fische*, vol. 2, part 4, 108; part 6, 24. Other examples are *Holocentrus ongus* (*Epinephelus ongus*), called 'Ikan ongo' and *Lutjanus bohar* (*Sparus cynodon*), called 'Ikan Caccatoea Iju'. *Ikan* is the word for fish in Malay and a number of other Austronesian languages.

32 Paepke, *Bloch's Fish Collection* 15–16.

33 Bloch, *Naturgeschichte der ausländischen Fische*, vol. 1, part 3, 112: 'Dieser Fisch soll, wie mir der Naturalienhändler, von dem ich ihn gekauft habe, aus Ostindien gekommen seyn. Mir ist es wahrscheinlicher, dass er seinen Aufenthalt richtig angegeben habe, weil er mit einer langen After- und Rückenflosse versehen ist: denn fast alle Fische, die ich im *Marcgraf*, *Piso*, und in den Handzeichnungen des Pater *Plümier* finde, sind mit einer

His judgement was not correct – *Chaetodon ciliaris* is in fact a Caribbean and South American species – but such comparative methods to infer the geographical origin of an animal from its body shape is remarkable in a period when no coherent theory of biogeography was yet available. The idea that differences or similarities in fin shapes between different genera or families corresponded with their respective distribution range probably relied more on intuition but through comparisons like these, Bloch attempted to structure his collection in terms of both taxonomy and biogeography.

3 The Man-Eater and the Slave Ship – Anecdotes as a Source of Knowledge

Unable to study his "foreign" fishes live or even *in situ*, Bloch had to infer as much of information as he could from travel writing and other geographical literature. Such accounts often had an inevitable narrative and anecdotal character, which required clear strategies to use them as sources of natural-historical knowledge. Bloch was not alone in his attempt to make sense of such anecdotes. Post-Linnean zoology and botany is often seen as a break with the early modern encyclopedic, and inherently anecdotal, approach to natural history, and yet it produced its own historical anecdotes if nothing else was available.[34] Especially bigger species that left an impression with seafarers of fishers inspired historical anecdotes that repeated again and again in ichthyological literature. Body size was also a factor that could make it difficult for collectors to acquire or store entire specimens in their homes or museum. It is therefore no coincidence that the illustrations of large animals such as sharks or tuna are far less accurate than those of species were preserved specimens were available [Fig. 21.1].

The lack of complete specimens was therefore an important reason to rely more on accounts from travel writing, geographical literature or even mere hearsay to gain information about the size or the feeding behavior of a fish. Stories and anecdotes are strongly featured in Bloch's *Natural History of Foreign Fishes* and many other ichthyological accounts. Estimates of the power of their

langen Rücken- und Afterflosse abgebildet; dahingegen, die ich aus Ostindien erhalten habe, und die im *Valentyn* stehen, beinahe durchgängig dieselben Flossen abgerundet haben'.

34 One of the most notorious examples is the Welsh naturalist Thomas Pennant (1726–1798) who even tried to infer natural-historical knowledge from folksongs and poetry. See Pennant Thomas, *British Zoology* (London: Benjamin White, 1776–1777), vol. 3, for example 49; 82–83; 128; 335; 339.

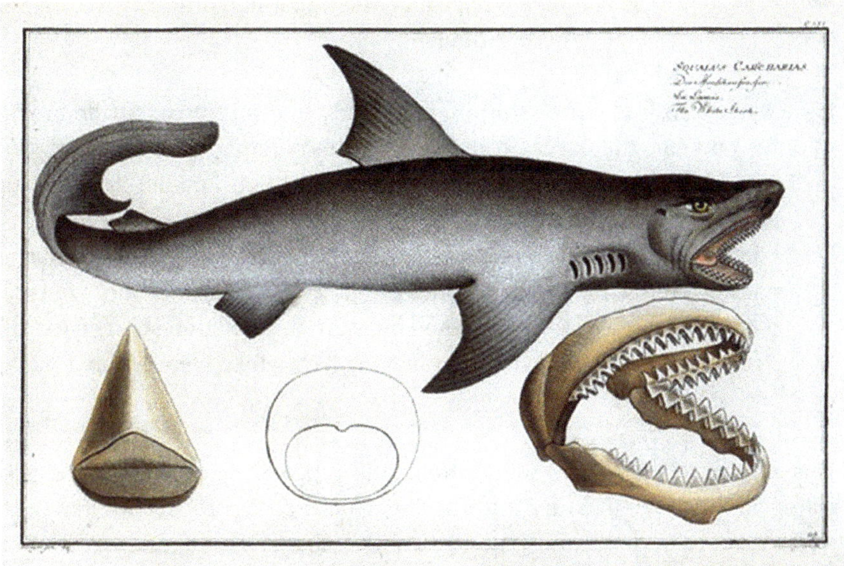

FIGURE 21.1 Great white shark (*Squalus carcharias*). In Bloch Marcus Elieser, *Ichtyologie, ou, Histoire naturelle, générale et particulière des poissons : avec des figures enluminées, dessinées d'après nature* (Berlin, Bloch – De la Garde: 1785–1797 [1787]), vol. 4, p. 127. The New York Public Library, Rare Book Division. https://digitalcollections.nypl.org/items/510d47da-695f-a3d9-e040-e00a18064a99

jaws were inferred from spectacular stories of bitten-off limbs or even deadly attacks. The section on the Great white shark (*Carcharodon* [Bloch: *Squalus*] *carcharias*), for example, is filled with accounts of that illustrate the rapacious nature of this animal: Bloch recounts how a sailor was wading in shallow water and had his leg bitten off or cites a story from Georg Forster's *Voyage Round the World* in which a caught shark tries to bite off a sailor's hand but only catches his sleeves. He continues his discussion of the shark's jaw apparatus with reports of entire seals or even fully-clothed humans that were found inside the fish's belly.[35] As Bloch concludes from such observations, the teeth of the Great white shark were only 'made to hold and bite' its prey, which was then not chewed but swallowed as a whole.[36]

These considerations bring him to one of the most-cited anecdotes in 18th- and 19th-century natural history: the story of a Guinea slave ship and the Great white shark – consistently named 'Menschenfresser' (man-eater) by Bloch. The anecdote was first mentioned in the third volume of Thomas Pennant's

35 Bloch, *Naturgeschichte der ausländischen Fische*, vol. 1, part 1, 35–37.
36 Bloch, *Naturgeschichte der ausländischen Fische*, vol. 1, part 1, 36.

British Zoology, who claimed to have received it first-hand from a slaver. Bloch quotes Pennant in the exact wording:

> A master of a Guinea-ship informed me, that a rage of suicide among his new bought slaves, from a notion the unhappy creatures had, that after death they should be restored again to their families, friends, and country. To convince them at lest that they should no re-animate their bodies, he ordered one of their corpses to be tied by the heels to a rope, and lowered it into the sea; and, tho' it was drawn up again as fast as the united force of the crew could be exerted, yet in that short space, the sharks had devoured every part but the feet, which were secured by the end of the cord.[37]

This gruesome story was widely spread and repeated in zoological and geographical literature until far into the nineteenth century.[38] Stories like these were also used in Abolitionist movements and sharks that followed slave ships became a trope in poems and anti-slavery literature. As maritime and slavery historian Marcus Rediker has argued, the idea that enslaved Africans committed suicide in order to be united with their ancestors in their home country was indeed based on some truth, and anecdotes like Pennant's should not be dismissed as mere sensationalism.[39] In Bloch, such anecdotes were not uncritically presented as clear evidence but by presenting them from a synoptic perspective and in relation to other accounts, they could be used as source of knowledge on phenomena that could not be studied by direct observation.

---

37 Pennant, *British Zoology*, vol. 3, 82–83. In Bloch's translation: 'Ein Capitain der aus Guinea Sklaven auf seinem Schiffe hatte, und wahrnahm, dass die Schwarzen deswegen den Selbstmord ausübten, weil sie glaubten, sie stünden bey den Ihrigen wieder auf, wollte sie vom Gegenteil überzeugen: er liess einen Selbstmörder, nachdem er ihm die Beine hatte festbinden lassen, in die See werfen, und ohngeachtet er mit aller möglicher Geschwindigkeit wieder herausgezogen werden sollte, so hatte ihn ein Menschenfresser verschluckt, und an den Beinen glatt abgebissen' (Bloch, *Naturgeschichte der ausländischen Fische*, vol. 1, part 1, 37).

38 Goldsmith O., *A History of the Earth, and Animated Nature* (Glasgow, Fullarton: 1837 [1774]), vol. 3, 494; Smith T., *The Naturalist's Cabinet: Containing Interesting Sketches of Natural History*, 6 vols. (London, Cundee: 1806–1807), vol. 5, 65; Gregory G., *A New and Complete Dictionary of Arts and Sciences: Including the Latest Improvement and Discovery and the Present States of Every Branch of Human Knowledge*, 2 vols. (London, Oddy: 1815), vol. 2, 697.

39 Rediker M., "History from Below the Water Line. Sharks and the Atlantic Slave Trade", *Atlantic Studies* 5.2 (2008) 285–297. Rediker makes this arguments based on reports from ship surgeons.

It is interesting that the only sources that are explicitly criticized by Bloch with regard to the Great white shark are those of other naturalists, not the non-scholarly accounts discussed above. Especially Peter Artedi and Guillaume Rondelet, who assumed a relationship between sharks and whales (Rondelet even believed that they had breasts like mammals) are relentlessly criticized. Bloch was thus not an uncritical reader but he apparently believed that stories told by 'practical men' such as sailors or fishers might contain some informative value.[40] As he notes in the preface to his *German Fishes*, he did not only write for scholars but also for readers with a more practical or economic interest in fish and it was their judgment that was taken as serious as (and sometimes perhaps more serious than) scholarly speculations.

4 The Leyden Jar and the Electric Eel – Bloch's Hermeneutics of Empirical Observation

Bloch's mode of description went far beyond noting basic anatomic features and putting them into a taxonomic framework. Fishes that were known for their extraordinary behavior were discussed in close detail and Bloch dedicated several pages to observational accounts of these species. One of the most enigmatic creatures to 18th-century European science were the African and South American electric eels, whose physiological features inspired new theories of electricity.[41] Electric fishes had long been known in Europe – some Mediterranean species were already described by Aristotle and Galen, and some Roman court physicians advised the use of live electric rays for therapeutical purposes. Electric eels from South America, however, received a renewed and greater attention and were used for different kinds of experimental research as their discharges were up to three to ten times stronger (*Electrophorus electricus* can produce shocks of more than 600 V).[42]

The existence of such enigmatic and dangerous creatures inspired a wide number of sensational accounts and sparked curiosity among European audiences. The South American electric eel also featured in Aphra Behn's novel

40 Bloch, *Naturgeschichte der ausländischen Fische*, vol. 1, part 1, 41.

41 See e.g. Koehler P. – Finger S. – Piccolino M., "The 'Eels' of South America: Mid-18th-Century Dutch Contributions to the Theory of Animal Electricity", *Journal of the History of Biology* 42.4 (2009) 715–763; Wu C.H., "Electric Fish and the Discovery of Animal Electricity: The Mystery of the Electric Fish motivated Research into Electricity and was Instrumental in the Emergence of Electrophysiology", *American Scientist* 72.6 (1984) 598–607.

42 Koehler – Finger – Piccolino, "The 'Eels' of South America" 723.

Oroonoko: or the Royal Slave (1688).[43] Experiments with these animals were first conducted in the Dutch Essequibo Colony between Surinam and British Guyana in the 18th century. Dutch colonists who had observed that electric eels could cause heavy pains and cramps, put the fish into tubs and found out that they had the capacity to kill chickens.[44] Reports of such observations were sent to the Netherlands and reached the Leiden professor Pieter van Musschenbroek, whose groundbreaking research on electricity had just resulted in his invention of the Leyden jar when he heard about the news from the Essequibo. Van Musschenbroek and others concluded that the described phenomena were indeed caused by 'animal electricity', similar to those of the new invention.

European naturalists soon learned that the electric eel offered significant research opportunities as it was one the few fishes that could survive long voyages by ship. While water animals from Asia or the Americas typically did not survive the journey as their water containers could not be sufficiently oxygenated, electric eels were tolerant to hypoxic environments. It was only later discovered that they could use the vascularized tissue of their mouth as an air-breathing organ which allowed them to survive at very low oxygen levels.[45] The first electric eel was brought to colonial North America and examined by the Scottish physician Alexander Garden in Charleston in 1774. Another eel even survived the journey to England and was there studied by John Walsh and John Hunter, who examined the fish live and then dissected it, which allowed for further theorization of electric capacity in animals.[46]

Bloch, who dedicated more than 15 pages to the electric eel (described as *Gymnotus cauda obtusa*), paid close attention to the history of these discoveries.[47] Studying all available reports on these fishes, he argued that it was not Walsh or Hunter who should be celebrated for discovering the phenomenon of animal electricity, but the Dutch colonists of Essequibo.[48] His

43 Behn Aphra, *Oroonoko: or the Royal Slave. A True History* (London: Canning, 1688) 153–154; 162–163.

44 Koehler – Finger – Piccolino, "The 'Eels' of South America" 741.

45 Graham J.B., *Air-Breathing Fishes. Evolution, Diversity, and Adaptation* (San Diego: 1997) 40.

46 Finger S., "Dr. Alexander Garden, a Linnaean in Colonial America, and the Saga of Five 'Electric Eels'", *Perspectives in Biology and Medicine* 53.3 (201) 388–406.

47 Bloch confused several African and South American electric eels in his discussion and assumed that they all belonged to the same species. See Bloch, *Naturgeschichte der ausländischen Fische*, vol. 1, part 2, 44: 'Wir treffen diesen Fisch in Guinea, Surinam, Cayenne, Peru, an den afrikanischen Küsten im Fluss Senegal, und überhaupt unter dem heißen Himmelsstrich, an'.

48 Bloch, *Naturgeschichte der ausländischen Fische*, vol. 1, part 2, 57: 'Hunter hält zwar den Walsh für den Entdecker der thierischen Elektricität, allein da dieser erst 1773 die

account quotes more than twenty sources and discusses eight observational reports in detail. In an attempt to theorize electricity in animals, he relied on a comparative and synoptic analysis of all these reports, ranging from 17th-century travel accounts to the Guyanas and the first experiments of Dutch colonists in Essequibo to the latest research on the anatomy of these animals. His analysis is divided into several steps: after describing each experiment or observational account, he summarized the most important theoretical conclusions that could be drawn from each respective report. He then tried to harmonize the findings by comparing the specific conditions under which the experiments were conducted.[49]

Comparing the different accounts of electricity in fish, Bloch noticed a number of contradictions: 1. some reports described electrical discharges even above the water surface, 2. others reported that sticks or other objects could transmit the shocks to bodies outside the water, and 3. some writers asserted that electric eels could be handled or even taken out of the water without releasing any electric discharges.[50] These problems were then be solved by a number of hypotheses, that allowed for a wider theorization of animal electricity. In order to make sense of the different experiments and observations, Bloch assumed that the fish could control its electric behavior or that it was at least dependent on its mood:

1. That the fish does not cause any adverse reactions when it is in a calm state.
2. That, on the contrary, when it is aggressive ('böse'), it will cause a shock, and that this shock will be more intense after physical irritation.
3. That a fresh fish will display this effect much stronger than one that has been stored in a container for a longer period.[51]

Furthermore, the intensity of the shocks depended on the fish's health and was caused by a strong contraction of its muscles behind its head. Its function was to catch prey and to defend the fish against enemies. As Bloch concluded,

---

Versuche mit dem Zitterrochen zu Rochelle angestellt hat, und Gravesand u. a. m. verschiedene Jahre vorher durch Versuche diefe Eigenschaft bey unserm Fisch hinlänglich erliefen hatten; so kann auch Walsh nicht für den Entdecker gehalten werden'. Laurens Storm van 's Gravesande ("Gravesand") was the governor of the Essequibo Colony and the author of the report that was sent to the Netherlands.

49 Bloch, *Naturgeschichte der ausländischen Fische*, vol. 1, part 2, 47–53.

50 Bloch, *Naturgeschichte der ausländischen Fische*, vol. 1, part 2, 52–53.

51 Bloch, *Naturgeschichte der ausländischen Fische*, vol. 1, part 2, 52–53: '1.) Dass der Fisch, wenn er ruhig ist, keine widrige Empfindungen verursache. 2.) Dass er im Gegentheil, wenn er böse ist, einen Stoß hervorbringe, und dass dieser desto heftiger sey, je mehr er vor der Berührung gereizt wird. 3) Dass ein frischer Fisch diese Würkung ungleich stärker äußere, als einer der schon lange in einem Gefäße gestanden hat'.

all these features could also be applied to the electric ray (*Torpedo sp.*) that was already known in Europe since antiquity. A comparison of the existing accounts on torpedoes and electric eels could only lead to the conclusion that both fishes used the same physiological mechanism. As he closed his argument triumphally, 'hereby all hypotheses that were offered in the last 2.000 years have become obsolete.'[52]

5 Ichthyology as *Second-Order Observation*

It is telling that most of Bloch's more theoretical conclusions were drawn from observations made by others. Theorizing such accounts was only possible by approaching them from a comparative and synoptic perspective and taking into account the specific circumstances under which they were produced. In this sense, projects such as Bloch's were dependent on what sociological systems theorists have called *second-order observation*, or the observation of the observations of others.[53] A systematic description of the fishes of the world by Linnean principles did not only involve the anatomical study of conserved specimens but an elaborate system of information management, in which knowledge was both produced and structured. Such natural-historical knowledge brought its objects and materials in relation to the often anecdotal sources of information that circulated in- and outside the world of learning. In this respect, the reform of natural history in the eighteenth century did not fully erase older encyclopedic forms of writing and documenting: stories and anecdotes remained a crucial part of ichthyological knowledge and their comparative evaluation was one of the central methods of the *Natural History of Fishes*. Bloch's work laid the base for a wide number of strictly empirical studies, such as Francis Day's survey on the fishes of India.[54] His influence on such 19th-century projects reflects how empirical observation and the interpretation of second- and sometimes third-hand accounts belonged and that textual

52 Bloch, *Naturgeschichte der ausländischen Fische*, vol. 1, part 2, 54: ' […] und sind dadurch alle Hypothesen, die man seit zweitausend Jahren erdacht hat, selbige zu erklären, unnütz geworden'.

53 See e.g. Foerster H. von, *Observing Systems* (Seaside, CA: 1981); Luhmann N., *Social Systems* (Redwood, CA: 1996).

54 Wells, Ellen B., "M.E. Bloch's Allgemeine Naturgeschichte der Fische: A Study", *Archives of Natural History* 10.1 (1981), 7–13, there 7; Day, Francis, *The Fishes of India: Being a Natural History of the Fishes Known to Inhabit the Seas and Fresh Waters of India, Burma, and Ceylon*, 2 vols. (London: Quaritch, 1875–1878).

criticism remained a key feature of zoological knowledge production in the 19th century.

Bibliography

Abdala Dergam Dos Santos J., *Phylogeography and Character Congruence Within the Hoplias Malabaricus Bloch, 1794 (Erythrinidae, Characiformes, Ostariophysil) Species Complex* (Ph.D. Dissertation: Colorado State University 1996).

Barnard K.H., "Note on Alleged Viviparity in *Barbus viviparus*, and Description of a New Species of *Beirabarbus*", *Annals and Magazine of Natural History* 8.7 (1941) 69–471.

Behn Aphra, *Oroonoko: or the Royal Slave. A True History* (London, Canning: 1688).

Berkemann J., "Die Emanzipation der deutschen Juden und der Begriff der Toleranz" in Enders C. – Kahlo M., *Toleranz als Ordnungsprinzip. Die moderne Bürgergesellschaft zwischen Offenheit und Selbstaufgabe* (Leiden: 2007) 71–107.

Blair A., "Note Taking as an Art of Transmission", *Critical Inquiry* 31 (2004) 85–107.

Blair A., *Too Much to Know. Managing Scholarly Information before the Modern Age* (New Haven, Connecticut: 2010).

Bloch Marcus Elieser, *Oeconomische Naturgeschichte der Fische Deutschlands* (Berlin, Hesse: 1782–1784).

Bloch Marcus Elieser, *Naturgeschichte der ausländischen Fische* (Berlin, Morino: 1785–1795).

Daston L. – Park K., *Wonders and the Order of Nature 1150–1750* (New York: 1998).

Day Francis, *The Fishes of India: Being a Natural History of the Fishes Known to Inhabit the Seas and Fresh Waters of India, Burma, and Ceylon*, 2 vols. (London, Quaritch: 1875–1878).

Finger S., "Dr. Alexander Garden, a Linnaean in Colonial America, and the Saga of Five 'Electric Eels'", *Perspectives in Biology and Medicine* 53.3 (2010) 388–406.

Foerster H. von, *Observing Systems* (Seaside/CA: 1981).

Froese R. – Pauly D., *FishBase. World Wide Web Electronic Publication.* www.fishbase.org (02/2022)

Goldsmith Oliver, *A History of the Earth, and Animated Nature*, 6 vols. (Glasgow: Fullarton, 1837 [1774]).

Graham J.B., *Air-Breathing Fishes. Evolution, Diversity, and Adaptation* (San Diego: 1997).

Gregory G., *A New and Complete Dictionary of Arts and Sciences: Including the Latest Improvement and Discovery and the Present States of Every Branch of Human Knowledge* (London: 1815).

Huigen S., "Repackaging East Indies Natural History in François Valentyn's Oud en Nieuw Oost-Indiën", *Early Modern Low Countries* 3.2 (2019) 234–264.

John Christoph Samuel, "Einige Nachrichten von Trankenbar auf der Küste Koromandel. Aus einem Briefe von dem Missionarius Hrn John an Herrn Doktor Bloch in Berlin", *Berlinische Monatschrift* 20 (1792).

Karrer C., "Marcus Elieser Bloch (1723 – 1799), Sein Leben und die Geschichte seiner Fischsammlung", *Sitzungsberichte der Gesellschaft Naturforschender Freunde zu Berlin* 18 (1978) 129–149.

Keller A.G., "Science In The Early 'Haskalah'", *European Judaism* 24.2 (1991) 8–13.

Koehler P. – Finger S. – Piccolino M., "The 'Eels' of South America: Mid-18th-Century Dutch Contributions to the Theory of Animal Electricity", *Journal of the History of Biology* 42.4 (2009) 715–763.

Lesser R., "Dr. Marcus Elieser Bloch. Ein Jude begründet die moderne Ichthyologie", *Das achtzehnte Jahrhundert. Zeitschrift der Deutschen Gesellschaft für die Erforschung des achtzehnten Jahrhunderts* 23.2 (1999 – Special issue Haskala. Die jüdische Aufklärung in Deutschland 1769–1812) 238–246.

Luhmann N., *Social Systems* (Redwood, CA: 1996).

Müller-Wille S. – Scharf S., "Indexing Nature: Carl Linnaeus (1707–1778) and His Fact-Gathering Strategies", *Working Papers on The Nature of Evidence: How Well Do 'Facts' Travel?* 36.8 (2009) 1–46.

Müller-Wille S. – Charmantier I., "Lists as Research Technologies", *Isis* 103.4 (2012) 743–52.

Nieuhof, Johan, *Joan Nieuhofs Zee en lant-reize, door verscheide gewesten van Oostindien: behelzende veele zeltzaame en wonderlijke voorvallen en geschiedenissen. Beneffens een beschrijving van lantschappen, steden, dieren, gewassen, draghten, zeden en godsdienst der inwoonders: en inzonderheit een wijtloopig verhael der stad Batavia, verciert doorgaens met verscheide koopere platen* (Amsterdam, Jacob van Meurs: 1682).

Paepke H.-J., *Bloch's Fish Collection in the Museum für Naturkunde der Humboldt-Universität zu Berlin. An Illustrated Catalog and Historical Account* (Rugell, Liechtenstein: 1999).

Paepke H.-J., "Blochs Schlangenkopf- und Labyrinthfische. Ein Beitrag zum 200. Todestag von Marcus Elieser Bloch (1723–1799)", *Der Makropode* 21.1–2 (1999) 2–13.

Pennant Thomas, *British Zoology* (London, Benjamin White: 1776–1777).

Rediker M., "History from Below the Water Line. Sharks and the Atlantic Slave Trade", *Atlantic Studies* 5.2 (2008) 285–297.

Schulte C., "Zur Debatte um die Anfänge der jüdischen Aufklärung", *Zeitschrift für Religions- und Geistesgeschichte* 54.2 (2002) 122–137.

Shapin S., *A Social History of Truth. Civility and Science in Seventeenth-Century England* (Chicago: 1994).

Smith T., *The Naturalist's Cabinet: Containing Interesting Sketches of Natural History* (London: 1806–1807).

Starr Jordan D., "The History of Ichthyology", *Science* 16.398 (1902) 241–258.

Trijp D.R. van, *Captured on Paper. Fish Books, Natural History and Questions of Demarcation in Eighteenth-Century Europe (ca. 1680–1820)* (Ph.D. Dissertation: Leiden University 2021).

Wells E.B., "M.E. Bloch's Allgemeine Naturgeschichte der Fische: A Study", *Archives of Natural History* 10.1 (1981) 7–13.

Valentyn, François, *Oud en Nieuw Oost-Indiën, vervattende Een Naaukeurige en Uitvoerige Verhandelinge van Nederlands Mogentheyd in die Gewesten, benevens Eene wydluftige Beschryvinge der Moluccos, Amboina, Banda, Timor, en Solor, Java, en alle de Eylanden onder dezelve Landbestieringen behoorende; het Nederlands Comptoir op Suratte, en de Levens der Groote Mogols; als ook Een Keuryke Verhandeling van 't wezentlykste dat men behoort te weten van Choromandel, Pegu, Arracan, Bengale, Mocha, Persien, Malacca, Sumatra, Ceylon, Malabar, Celebes of Macassar, China, Japan, Tayouan of Formosa, Tonkin, Cambodia, Siam, Borneo, Bali, Kaap der Goede Hoop en van Mauritius* [...] (Dordrecht, Joannes van Braam – Amsterdam, Gerard onder de Linden: 1724–1726).

Wu C.H., "Electric Fish and the Discovery of Animal Electricity: The Mystery of the Electric Fish motivated Research into Electricity and was Instrumental in the Emergence of Electrophysiology", *American Scientist* 72.6 (1984) 598–607.

CHAPTER 22

Between Science and Art: On Painted Natural Illustrations of Fish in China

Ching-Ling Wang

1 A Short History of Depicting Fish in Chinese Art

Fish have been depicted in China as early as the Neolithic period. Many archeological excavations have brought to light images of fish or fish pattern decoration on pottery made in this period. It reflects the lifestyle of a fishing and hunting society.[1] In archaeological findings of later periods and dynasties, such as Zhou (510–314 BC), Qin (221–207 BC) and Han (202 BC–220 AD), fish appear as a common motif in the decoration of objects, such as bronzes, lacquer ware, stone reliefs, textiles and so on. But although the depiction of fish in China can be traced back to Neolithic pottery, it did not become a major subject for painting until the 10th century.

The earliest textual record relating to fish as a subject of representation in painting history appears in *A Record of the Famous Painters of All the Dynasties* (*Lidai minghua ji* 歷代名畫記) written by Zhang Yanyuan 張彥遠 (815–907). Here Zhang documents an ancient cartographic painting titled *Yellow River Map with Dragon and Fish* (*Longyu hetu* 龍魚河圖), and mentions a painter by the name of Xü Miao 徐邈 (171–249) from Wei, one of the Three Kingdoms (220–280), who was proficient at painting fish.[2] According to *A Record of the Famous Paintings in Tang Dynasty* (*Tangchao minghua lu* 唐朝名畫錄) by Zhu Jingxuan 朱景玄 (fl. 8th century), the painter Zheng Qian 鄭虔 (691–759) was also a capable painter of fish whose works were praised by his contemporaries.[3]

By the period of the Five Dynasties (907–960) and the Northern Song Dynasty (960–1127), fish painting had been established as a specific genre, as is evident from the *Xuanhe Catalogue of Paintings* (*Xuanhe huapu* 宣和畫譜, 1120);

1 See Zhongguo kexueyuan kaogu yanjiuyuan 中國科學院考古研究院 ed., *Xinzhongguo de kaogu shouhuo* 新中國的考古收穫 (Beijing, Wenxu: 1961) 10; Xiaonan Yang (ed.), *The Golden Age of Chinese Archaeology: Celebrated Discoveries from the People's Republic of China* (Washington DC: 1999) 60, 64–67.

2 Zhang Yanyuan 張彥遠, *Lidai minghua ji* 歷代名畫記, reprint (Beijing: 1963) 73, 82, 104.

3 Zhu Jingxuan 朱景玄, *Tangchao minghua lu* 唐朝名畫錄, reprint in Pan Yungao 潘運告 (ed.), *Tang Wudai hualun* 唐五代畫論 (Changsha: 1997) 113.

DOI:10.1163/9789004681187_023

a catalogue of the painting collection of Emperor Huizhong (r. 1100–1126), that contains twenty chapters. The recorded paintings were divided into ten categories or genres, with 'dragons and fish' listed as one of them.[4] Famous painters such as Teng Changyou 滕昌祐 (d. 881), Xü Xi 徐熙 (886–975), Xü Chongsi 徐崇嗣 (fl. 10th century), Dong Yu 董羽 (fl. 10th century) as well as many others, enjoyed the reputation of being skilled in painting fish. That dragons and fish were seen as belonging to the same category or genre is probably due to a legend describing the transformation of a carp into a dragon or to the Buddhist and Daoist use of images of dragons and fish to pray for rain.

Not only did it become one of the major painting genres, the representation of fish also reached its peak of realistic and naturalistic depiction in the Song dynasty (960–1279). The scrolls *Fish Swimming amid Falling Flowers* (*Luohua youyu tu* 落花游魚圖, Saint Louis Art Museum), attributed to Liu Cai 劉寀 (fl. 1080–1120),[5] and *Fish and Waterweeds* (*Yuzao tu* 魚藻圖) [Fig. 22.1], attributed to Fan Anren 范安仁 (fl. mid-13th century), are considered masterpieces of fish painting.[6] Both are a symphony of rhythm and movement and depict the impression of swimming, darting, drifting fish and clusters of fish. According to zoologist Dietrich Neumann's detailed observation, in the scroll *Fish and Waterweeds*, Fan Anren depicts a total of 47 sharpbelly (*Hemiculter*), each caught in a different fleeting posture, allowing the viewer to follow and understand the typical sequence of the reproduction process of fish: from the tracking of sexually mature females (spawners) by males that are ready to mate (milters) up to the males' whirling in circles while releasing their seminal fluid over the eggs laid among aquatic plants by the females. No other painting condenses the reproduction process of fish in such a logic and vivid and at the same time animated and charmingly viewable manner.[7]

After the Song dynasty, images of fish, both in painting and in the applied arts, often had auspicious symbolic meanings, for example goldfish (*jinyu* 金魚) symbolizes "gold and jade filled in the hall (*jin yu man tang* 金玉滿堂)";

4 *Huanhe huapu* 宣和畫譜, 1120, reprint (Taipei: 1967). For painting activities during the Huizhong court and the establishment of different painting genres, see Yun-Ru Chen 陳韻如, "Hua yi yi ye: Chonggu Song Huizhongchao de huaihua huodong 畫亦藝也：重估宋徽宗朝的繪畫活動" (Ph.D. dissertation, Graduate Institute of Art History, National Taiwan University: 2008).

5 For the image of Liu Cai's *Fish Swimming amid Falling Flowers*, see: https://www.slam.org/collection/objects/32476/ (accessed on 31.01.2022).

6 Li Lincan 李霖燦, "Yucao hua de houpo sheenyi 魚藻畫的活潑生意, " in *Zhongguo meishushi gao* 中國美術史稿 (Taipei: 1987) 201–206.

7 Neumann D., "Experiencing and Depicting Nature", in Dietrich Neumann D. – Ogando J. (eds.), *Fascinated by Nature: Landscapes, Plants and Animals in the Tradition of Chinese and Japanese Painting from the Neumann-Ogando Collection* (Berlin: 2012) 22.

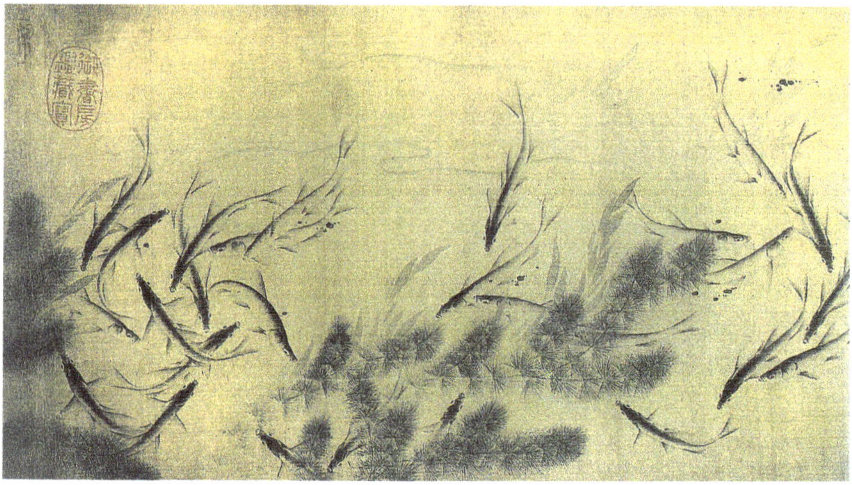

FIGURE 22.1 Fan Anren, *Fish and Waterweeds*, detail, mid-13th century, National Palace Museum, Taipei

a combination of lotus and fish symbolizes "every year ends with ample surplus (*lian nian you yu* 連年有餘)", and etc.[8] In some rare cases, the fish became a vehicle for the artist to express his emotions and philosophy of life, for instance the fish depicted in Bada Shanren's 八大山人 (Zhu Da 朱耷, 1626–1705) paintings, where the expression of the fish is portrayed in a comic and exaggerated manner, almost like a caricature, especially in the eyes through which we see the contempt and discontent of the painter towards the circumstances of his time [Fig. 22.2].[9] Alternatively, the fish were depicted in the context of evidential research and learning (*kaojü* 考據), for example Zhao Zhiqian's 趙之謙

8 For the development of fish painting in China, see Liu Zhigui 劉治貴, *Zhongguo huihua yuanliu* 中國繪畫源流 (Changsha: 2003) 418–423. For the symbolism of fish, see Noriko Miyazaki 宮崎法子, *Kacho sansuiga o yomitoku: Chugoku kaiga no imi* 花鳥・山水画を読み解く–中国絵画の意味 (Tokyo: 2003) 157–171; Hou-mei Song, *Decoded Messages: The Symbolic Language of Chinese Animal Painting* (Ohio: 2010) 207–244; Tokyo National Museum (ed.), *Kisshō: Chūgoku bijutsu ni komerareta imi* 吉祥–中国美術にこめられた意味 (Tokyo: 1998) 26–75.

9 Wang Fangyu – Barnhart R.M. – Smith J.C (eds)., *Master of the Lotus Garden : The Life and Art of Bada Shanren* (New Haven: 1990) 102–104, 128–129, 148–151; Hui-Shu Lee, "The Fish Leaves of the *Anwan Album*: Bada Shanren's Journey to a Landscape of the Past", *Ars Orientalis* 20 (1990) 69–85; Hui-Shu Lee, "Bada Shanren's Bird-and-Fish and the Art of Transformation", *Archives of Asian Art* 40 (1991) 6–26.

FIGURE 22.2 Bada Shanren (Zhu Da), *Fish and Ducks*, detail, 1689, Shanghai Museum

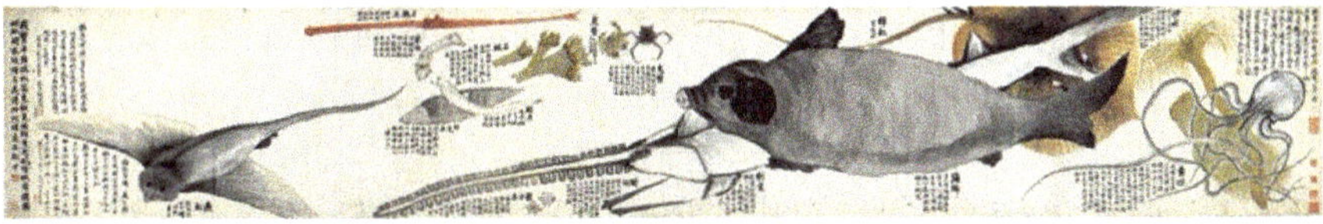

FIGURE 22.3 Zhao Zhiqian, *Extraordinary Fish*, 1861, private collection

(1829–1884) work *Extraordinary Fishes* (*Yiyu tu* 異魚圖) [Fig. 22.3], which will be discussed in the following text.[10]

From the 17th-century natural illustrations of fish emerged in China, produced by individual scholars and anonymous workshops painters in Canton. This article is a survey of painted natural illustrations of fish in China from the 17th to the 19th century and examines their development in different contexts.

2 Nie Huang's *Pictures of the Various Marine Creatures* (*Haicuo tu*)

Evidential research and learning (*kaojü* 考據 or *kaozheng* 考證) was a school and approach of study in the late Ming and Qing dynasties of China from about 1600 to 1850. The approach corresponds to the methods of modern textual studies and, on occasion, associated with empirical studies and philology. One of several examples that is of significance towards the natural illustrations of fish is Nie Huang's 聶璜 (fl. 1662–1722) work *Pictures of the Various Marine Creatures* (*Haicuo tu* 海錯圖) [Fig. 22.4] from 1698, which consists of four albums. The first three albums are now preserved in the collection of

10 Wu Chaoran 吳超然, "Zhao Zhiqian yibaliuyi nian de sanjian zhuopin *Yiyu tu, Ouzhong wuchan tujuan, Ouzhong caomu siping*: Jinshi huapai yu haipai guishu zhi shangque 趙之謙一八六一年的三件作品《異魚圖》、《甌中物產圖卷》、《甌中草木四屏》：金石畫派與海派歸屬之商榷," in Yang Dunyao 楊敦堯 (ed.), *Shibian, Xingxiang, liufeng: Zhongguo jindai huihua 1796–1949 xueshu yentaohui lunwenji* 世變・形象・流風：中國近代繪畫 1796–1949 學術研討會論文集 (Taipei: 2008) 451–469.

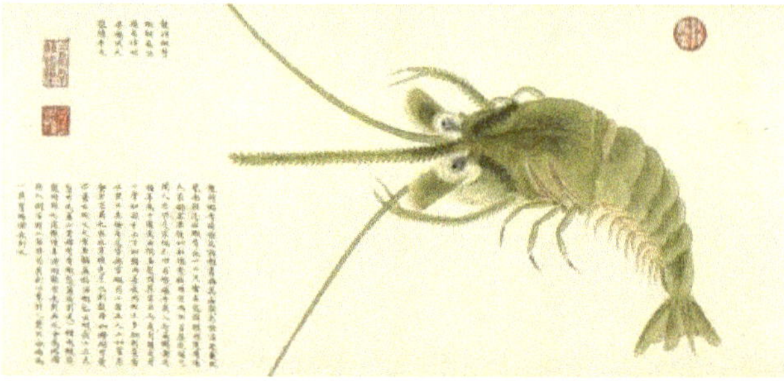

FIGURE 22.4
Nie Huang, *Pictures of Various Marine Creatures*, 1698, National Palace Museum, Taipei

the Palace Museum, Beijing and the fourth album is in the collection of the National Palace Museum, Taipei. Besides eight leaves of Nie's prefaces, poems, and postscripts placed at the beginning of the first album, the four albums contain a total of 199 leaves and depict 371 species of sea creatures and creatures around the coast, including fish, crabs, shrimps, turtles, shells, corals, insects, plants, birds, animals and so on.[11]

According to Nie Huang's own two prefaces, the reason he started to illustrate sea creatures was because it had never been done. Although there are historical books documenting fish, none of them contain illustrations. The illustrations published in the section of fish included in the books of *materia medica* (*bencao* 本草) – the tradition of biological inquiry into animals, fungi and plants, and inorganic material, such as minerals, that are used in traditional Chinese medicine – lack fidelity.[12] The illustrations appear as afterthoughts, or interpretations, rather than observations and in most cases, are inserted later. The added illustrations and text sometimes do not match. Also, during the process of transforming from painted illustration to woodblock print and reprint, certain distortions would occur, so that in Nie's opinion, the illustrations lack the fidelity of the originals.[13]

The first book on fish in China, *Illustrated Eulogies of Remarkable Fish* (*Yiyu tuzan* 異魚圖贊, preface dated 1544) by Yang Shen 楊慎 (1488–1599) appeared during the mid-16th century. Yang Shen intended his work to be a kind of

---

11 For all the images of the first three albums, see Palace Museum (ed.), *Qinggong haicuo tu* 清宮海錯圖 (Beijing: 2014); and for the fourth album, see: https://digitalarchive.npm.gov.tw/ITWaterFall (accessed on 11.12.2021).

12 Palace Museum (2014) 34–37; 40–47.

13 Sterckx R., "The Limits of Illustration: Animallia and Pharmacopeia from Guo Pu to *Bencaogangmu*", in Lo V. – Barrett P. (eds.), *Imaging Chinese Medicine* (Leiden: 2018) 135–150. For illustrations in the books of *materia medica*, see Zheng Jingsheng, "Observational Drawing and Fine Art in Chinese Materia Medica Illustration," in Lo V. – Barrett P. (eds.), *Imaging Chinese Medicine* (Leiden: 2018) 152–160.

response to the *Illustrations of Exceptional Fish* (*Yiyu tu* 異魚圖) of the time of the Southern Dynasties (420–589). The book describes 87 species of fish and 35 species of river snails, shells, conches, clams and other marine life; however, the illustrations have been lost. We might also mention Tu Benjun's 屠本畯 (1542–1622) *Notes on the Sea Creatures of Fujian* (*Minzhong haicuo shu* 閩中海錯疏) published in 1596. The book describes 167 kinds of fish and 90 different types of shells, clams, and turtles. It also includes some freshwater fish, such as carp, and various species of frogs and toads.[14] Moreover, the geographical range of the book not only covers the province of Fujian, but extends to the shores of Guangdong and Zhejiang.[15] Tu quotes from ancient texts and makes use of fisherfolk's knowledge, to which he adds his own findings. Supplements to both books were published in later periods, but none of them contain any images.[16]

It was not until Nie Huang lived at Wenzhou for almost twenty years, that he encountered and started to illustrate sea creatures. Wenzhou is located at the extreme southeast of Zhejiang province that is surrounded by mountains, the East China Sea and 436 islands. In 1687, he completed the *Illustrated Manual of Thirty Species of Crabs* (*Xiepu sanshi zhong* 蟹譜三十種) based on what he observed in Wenzhou. Later he lived in Huian and Yangzhou in the Jiangsu province, during which he often went to the coastline of Hebei province and Tianjing to see and observe sea creatures; and travelled to Yunan, Guizhou, Hubei and Hunan provinces. Eventually around 1693 he settled at Fujian province, that has a large variety of sea creatures. The *Pictures of the Various Marine Creatures* was made during Nie's stay in Fujian, in which he combined the content of his previous work, *Illustrated Manual of Thirty Species of Crabs* and what he had subsequently seen and heard, as well as what he read in books to complete this work.[17]

14 Liu Changzhi 劉昌芝, "Woguo xian cun zui zao de shuichan dongwu zhi: *Minzhong haicuo shu* 我國現存最早的水產動物志–《閩中海錯疏》", *Ziran kexueshi yanjiu* 自然科學史研究 12 (1982) 333–338; Wang Yonghou 王永厚, "Tu Benjun jiqi *Minzhong haicao shu* 屠本畯及其《閩中海錯疏》", *Zhongguo shuzhang* 中國水產 2 (1984) 29.

15 For the identification of the Latin names of the species in Tu's book, see Liu Changzhi (1998) 336–338.

16 For Yang's book, there are Hu Shi'an's 胡世安 commentary with the title *Notes on Illustrated Eulogies of Remarkable Fish* (*Yiyu tuzan jian* 異魚圖贊箋) and his supplement titled as *Supplement of Illustrated Eulogies of Remarkable Fish* (*Yiyu tuzan bu* 異魚圖贊補) of the Qing dynasty; as for Tu's book, Xü Bo 徐𤊹 (1563–1639) wrote a supplement with the title *Supplement of Notes on the Sea Creatures of Fujian* (*Minzhong haicuo buzhi* 閩中海錯補志).

17 Palace Museum (2014) 34–37; 40–47.

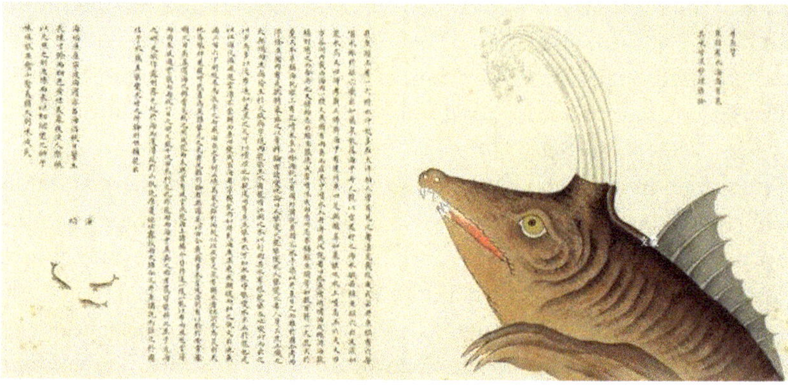

FIGURE 22.5 Nie Huang, *Pictures of Various Marine Creatures*, 1698, Palace Museum, Beijing

For each species depicted in the *Pictures of the Various Marine Creatures*, there is an illustration, its name, a poem, and a text describing its appearance, habitual behavior, and place of origin. In some cases, Nie Huang even related legends or its application in daily life, such as how to cook it and what it tastes like. Moreover, the text is based on Chinese and European sources. Some of the illustrations were clearly influenced by images in European works. For example, the image of the whale in Nie Huang's work [Fig. 22.5] was copied directly from the European missionary Ferdinandus Verbiest's (also known as Nan Huiren 南懷仁, 1623–1688) book *Illustrated Explanation of the World*, (*Kunyu tushuo* 坤輿圖說, 1674) or the world map (*Kunyu quantu*).[18] It is interesting to point out that Nie Huang's *Pictures of the Various Marine Creatures* includes both actual and imaginary creatures such as the dragon and the mermaid.[19] He also included the Chinese legends of, for example, "shrimps transforming to dragon flies", "sharks transforming to tigers", and so on; and the legend surrounding the phenomenon of mirages.

Despite the legends and some of the exaggeratedly depicted species, most of the illustrations are naturalistic depictions of the actual creatures and have

18 Palace Museum (2014) 82–83; Lai Yu-chih 賴毓芝, "Zhishi, xiangxiang yu jiaoliu: Nan Huiren *Kunyu quantu* zhi shengwu chahui yanjiu 知識、想像與交流：南懷仁《坤輿全圖》之生物插繪研究", in Dong Shaoxin 董少新 (ed.), Gantong shenshou: Zhong Xi wenhua jiaoliu Beijing xia de ganguan yu ganjue 感同身受—中西文化交流背景下的感官與感覺 (Shanghai: 2018) 141–182.

19 Wu Songfeng 吳誦芬, "*Haicuo tu* 海錯圖", *Gugong wenwu yuekan* 故宮文物月刊 363 (2016) 66–73; Zou Zhenghuan 鄒振環, "Haicuo tu yu Zhong Xi zhishi zhi jiaoliu 《海錯圖》與中西知識之交流", *Zijincheng* 紫禁城266 (2017) 124–131.

been rendered with the attempt to be true to life.[20] According to Nie himself, his method when composing the *Pictures of the Various Marine Creatures* was "first painting the illustration of the species, then identifying its name, then composing a poetry for the species, then conducting textual research of the species, and eventually making the final judgement".[21] Worth noting is that Nie also paid attention to the relative size of the species and the whole visual composition and arranged the text and image in an organic and vivid way.

In the fourth year of Emperor Yongzheng's reign (1726), Nie's *Pictures of the Various Marine Creatures* entered the Qing imperial collection and later documented in the *Second Volume of the Shiqü Catalogue of Imperial Collection of Painting and Calligraphy* (*Shiqü baoji xübian* 石渠寶笈續編) of Qing dynasty (1644–1911).[22] According to the archives of the imperial workshops of the Qing dynasty, in the third year of Emperor Qianlong's reign (1738), the Emperor Qianlong viewed the four albums of Nie's *Pictures of the Various Marine Creatures* and gave the order to remount the albums.[23] Worth noticing is that in the imperial archives Nie's *Pictures of the Various Marine Creatures* is mentioned as *Albums of Fish* (*Yupu* 魚譜), while during 1750 to 1761 in the Qing court there was a painting project of the *Albums of Beasts* (*Shoupu* 獸譜), *Albums of Birds* (*Niaopu* 鳥譜) and *Illustrations of Official Tributes* (*Zhigong tu* 職貢圖) to visualize all beings under his majesty's rule.[24] Could it be that the reason that fish are not included in this pictorial illustrating project was because of the existence of Nie's work? Nevertheless, due to entering the imperial collection Nie's work was no longer accessible, and hence Nie's work had very limited impact to the public as his study was unknown.

20 For the identification of the species in Nie's *Pictures of the Various Marine Creatures* in Chinese, English and Latin, see Palace Museum (2014) 298–305.

21 Palace Museum (2014) 36; Wu Songfeng (2016) 68.

22 *Qingding shiqü baoji xübian* 欽定石渠寶笈續編 (Qing imperial edition: 1793) 101–130; Palace Museum (2014) 13.

23 Palace Museum (2014) 13.

24 For the pictorial illustrating projects in the Qing court, see: Lai Yu-chih 賴毓芝, "Qinggong dui Ouzhou ziranshi tuxiang de zaizhi: yi Qianlongchao *Shoupu* weili 清宮對歐洲自然史圖像的再製：以乾隆朝《獸譜》為例", *Zhongyang yanjiuyuan jingdaishi yanjiusuo jikan* 中央研究院近代史研究所集刊 80 (2013) 1–75; Lai Yu-chih, "Domesticating the Global and Materializing the Unknown: A Study on Album of Beasts at the Qianlong Court", in Grasskamp A. – Juneja M. (eds.), *EuroAsian Objects: Art and Material Culture in Global Exchange, 1600–1800* (Berlin: 2018) 125–174; Lai Yu-chih, "Costuming the Empire: A Study on the Production of Tributary Paintings at the Qianlong Court in 18th Century China", in Klich L. – Zanardi T. (eds.), *Visual Typologies from the Early Modern to the Contemporary: Local Contexts and Global Practices* (New York: 2018) 90–103.

3 Han Liangqing's Commission of the *Pictures of Sea Fish* (*Haiyu tu*)

The collection of Museum für Asiatische Kunst, Staatliche Museen zu Berlin includes a newly discovered Chinese painted album titled *Pictures of Sea Fish* (*Haiyu tu* 海魚圖), dated to 1739. The album has 18 leaves in total. The first leaf is the title page with its original Chinese title *Haiyu tu* (*Pictures of Sea fish*), written in the running-script (*xingshu* 行書). The second and third pages contain a preface signed by Miechi Zhuren 蔑癡主人 (Hang Liangqing 韓良卿, d. 1746). The following 14 leaves, the album's main content, present pictorial representation of more than 130 species of fish. The final page has a colophon written by Wang Jian 王建 (fl. 18th century).

According to the preface, Han Liangqing had worked in the desert regions of China for many years when he was put in charge of coastline surveillance at Jieshi in Guangdong province. He had never seen so many different aquatic creatures, so he asked the fishermen to bring in what they caught and asked a painter to produce detailed images of them. To these depictions, he added comments based on his own investigations. The comments, which accompany the images of the fish, have a fixed format: first, he documented the name by which the fish was known locally; second, he tried to find a more common name in order to identify the fish; third, he documented the physical attributes of the fish and how it tasted.[25] For example, in the case of the red cornet fish, *mabianyu* 馬鞭魚 (*Fistularia petimba*), he wrote:

> The local name is horsewhip-fish [*mabianyu*], because its shape is like a horsewhip, the actual name unknown. Its length can be 4 to 5 *chi* 尺 [one *chi* = 33.33 cm], the part from its mouth to its eyes looks like a bamboo joint, the tip of its tail is like a line, just like the tip of a whip. Its appearance is also odd. Its flesh is tender, and the flavour is delicate. [Fig. 22.6]

Han Liangqing tasted most, but not all of the fish himself. He mentions, for example, a fish he refers to as the *bi*-fish (*biyu* 璧魚):

> The local name is *bi*-fish [*biyu*], the actual name unknown. Its shape is similar to a purse [*hebao* 荷包]. Its mouth is on the side of its stomach. Its body is thin and soft, the tip of its tail is blackish. The whole body is boneless. It can grow to around 2 to 3 *jin* 斤 [one *jin* = 604.8 g]. When cooking it, one must wait until the water is boiling, then place it into the

25 For complete translation of the preface, see Ching-Ling Wang, "On the *Picture of Sea Fish* (*Haiyu tu*)" (forthcoming).

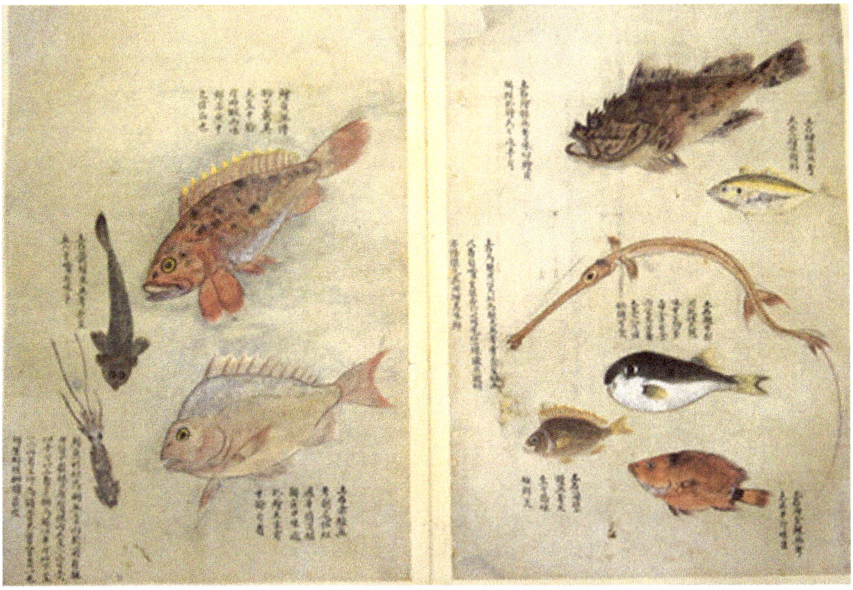

FIGURE 22.6 *Pictures of Sea Fish*, 1739, Museum für Asiatische Kunst, Staatliche Museen zu Berlin
PHOTO: CHING-LING WANG

> water, otherwise the flesh will dissolve in the water. The flavour is somewhat sweet and refreshing. I find it odd, so I don't dare to eat it. [Fig. 22.7]

He also documented some poisonous fish, for example a type of globefish that local people named *mianguai* 面乖:

> The local name is *mianguai*. Similar to globefish its character [as food] is extremely hot in nature; its flavour is sweet and refreshing, but it is poisonous. People are often poisoned because of eating it. It can grow to 7 or 8 *jin*. It can only be eaten when dried. (see Fig. 22.6)

Not all of the fish in the album are accompanied by text. Before Han Liangqing had investigated the fish, there simply was no text. According to him, this was supposed to be a long-term project; he intended to first collect all the fish, then the shells, and so on, but his project was interrupted when he was posted back to the west – *Pictures of Sea Fish* is only the first part of his ambitious project.

The depiction of the fish in the album is unique in the manner of pictorial representation in China. They have been drawn in a descriptive and naturalistic way without scheme or pattern of conventional representation. The painter

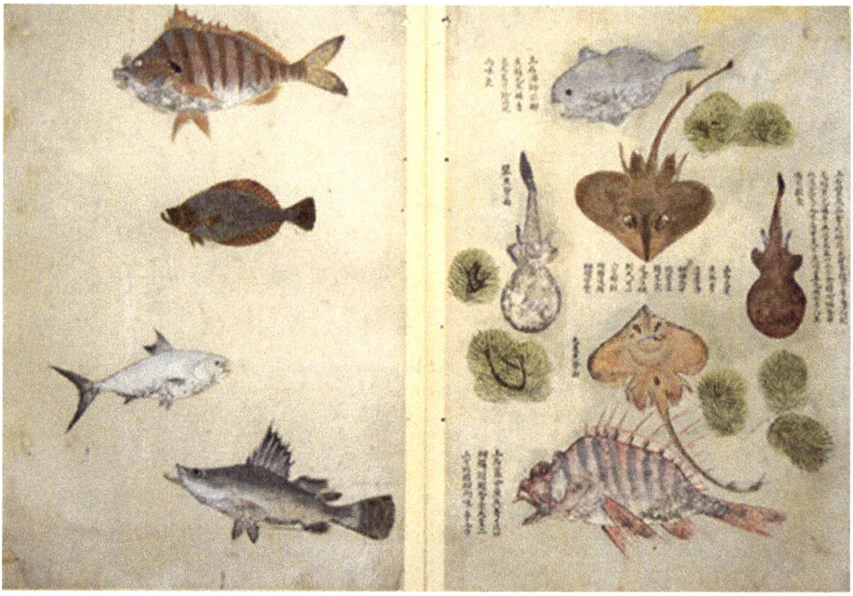

FIGURE 22.7 *Pictures of Sea Fish*, 1739, Museum für Asiatische Kunst, Staatliche Museen zu Berlin
PHOTO: CHING-LING WANG

focused on capturing the colours of the fish and their physical characteristics (such as the shape of the fins, the texture of the scales etc.) and reproduced these in a lively sketch-like manner. Most of the fish are symmetrical, that is, one side looks much the same as the other. Therefore, in the album, only one side is depicted; in cases of asymmetry, both sides are shown, as in the case of the *bi*-fish. One might say the depictions in *Pictures of Sea Fish* show the intention to be faithful to the natural appearance of the fish. The painter portrays the specimens in a realistic manner, with accuracy and objectivity, instead of pursuing a certain quality of the line or painterly effect.

Although the initiative for making *Pictures of Sea Fish* emerged from Han Liangqing's own curiosity about sea creatures that he had never seen, the painter he commissioned faithfully documented the fish and gave them their actual biological attributes. The comments made by Han Liangqing were based on his own objective observation of the appearance and subjective experiences of the taste. Both the text and the illustrations are realistic. In addition, Han set out to make a complete record, instead of selecting a few for their aesthetic qualities – he did not have a work of art in mind. Most importantly, the purpose of the album is to inform viewers that these creatures are not strange but actually quite common. This is all a world away from the traditional artistic representation of fish and has more in common with impartial scientific investigation.

The album *Pictures of Sea Fish* is an extraordinary example of the pursuit of scientific illustration and empirical science in 18th-century China. One may surmise that the album was made to form part of a book of ichthyological studies. As he himself mentioned in the preface of the album, however, the original grand project was never completed.

4 Zao Zhiqian's painting of the *Extraordinary Fish*

In 1861, the painter Zhao Zhiqian produced the painting *Extraordinary Fish* (see Fig. 22.3), a visual record of the marine creatures in the regions of Wenzhou and Ruian along the coast of Zhejiang province. Next to each creature he added his commentary. According to his own inscription on the painting:

> In the *xinyou*-year [1861] of the Xianfeng reign, Huishu [I] traveled to Wenzhou, [I] saw there are sea creatures with strange appearances, hence I depicted them on this paper and also conducted textual research into their names. This is how master painter Gu Kaizhi 顧愷之 (345–406) could depict lively things!

It is worth noting that the octopus (*zhangjü* 章矩), red cornet fish (*mabianyu* 馬鞭魚), butterfly ray (*yanhong* 燕魟), and other sea creatures are depicted realistically in this painting; on the other hand, the dolphin (*haixi* 海豨) is depicted with a pig's head and a fish's body and the box fish (*shaiziyu* 骰子魚) oddly looks like small gaming dice with the head, fins, and tail of a fish. This is probably because '*haixi*' literally means 'sea pig', and '*shaiziyu*' literally means 'dice fish', which may have to do with its boxy shape and spots, hence they were depicted this way. All of these suggest that Zhao Zhiqian did not see all the sea creatures he depicted with his own eyes as he claimed.

Zhao's painting *Extraordinary Fish*, although it can be understood in the context of evidential study, actually aims to depict fish as a local product. Besides the *Extraordinary Fish*, during his stay in Wenzhou, he created several paintings to depict local plants and fish in the region, for example, the *Plants and Trees in Wenzhou* (*Ouzhong caomu tu* 甌中草木圖, Tokyo National Museum) in four hanging scrolls and the *Local Products in Wenzhou* (*Ouzhong wuzhan tu*

甌中物產圖, Studio Rongbaozhai).[26] In these paintings, Zhao implied the natural illustrations and transformed them into an art form; hence the painting *Extraordinary Fish* is not the study of local fish, rather an accidental production of Zhao's art works.

5 The Album *Manual of Sea Oddities* (*Haiguai tu*) in the Qing Imperial Collection

There is also an anonymous and undated album *Manual of Sea Oddities* (*Haiguai tu* 海怪圖) [Fig. 22.8] now in the collection of the National Palace Museum, Taipei. It was Daniel Greenberg who first noticed this album and pointed out that its illustrations were based on those in European books on natural science, such as Conrad Gessner's (1516–1565) *Historiae animalium* (1558), John Jonston's (1603–1675) *Historia naturalis* (1649–1650) and Johann Zahn's (1641–1707) *Specula physico-mathematico-historica* (1696), all introduced into China by European missionaries. According to the inscription '*wuchen* 戊辰' on the cover of the album, he also proposed to date this album to 1688 (*wuchen*-year) and linked it to the Kangxi Emperor who took a great interest in European science.[27] There is, however, no strong evidence to support this date: the album could have been painted in 1748 or even later, in 1808 or 1868. Also, this album was not documented in the three volumes of the *Shiqü Catalogue of Imperial Collection of Painting and Calligraphy* (*Shiqü baoji* 石渠寶笈). The last volume of these catalogues was not completed until 1816, which may indicate the album only entered the imperial collection after 1816. If we follow this logic, it is most likely that the album was made in 1868.

It is also worth mentioning that the album *Manual of Sea Oddities* contains no text, only images, hence its function and purpose are unclear. It may have been made simply out of curiosity and be seen as an exotic visual wonder for its viewer's (in this case, the emperor's) amusement. Furthermore, the style of depiction in the *Manual of Sea Oddities* is close to the export painting produced in the late 18th and 19th centuries in Canton. Since the context of producing this album is unclear, the possibility exists that this album was not produced by the imperial workshop but was produced in Canton and presented to the court as a local tribute. As Yu-chih Lai pointed out, there are some paintings

26 Wan Qingli 萬青力, *Bingfei shuairuo de bainian: shijiu shiji Zhongguo huihua shi* 並非衰弱的百年：十九世紀中國繪畫史 (Taipei: 2005) 205–207; Wu Chaoran (2008).

27 See Greenberg D., "Weird Science: European Origins of the Fantastic Creatures in the Qing Court Painting, the *Manual of Sea Oddities*", in Silbergeld J. – Wang E.Y. (eds), *The Zoomorphic Imagination in Chinese Art and Culture* (Honolulu: 2016) 379–400.

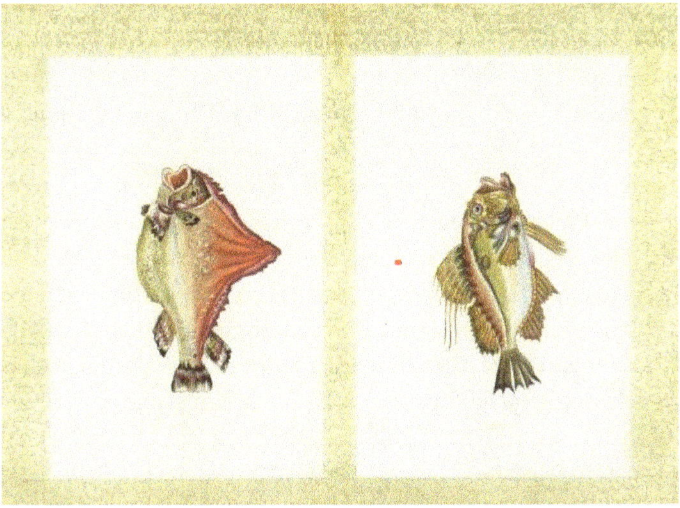

FIGURE 22.8 *Manual of See Oddities*, 1868(?), National Palace Museum, Taipei

in the imperial collection that share a style similar to the export paintings produced in Canton, but the exact context of the artistic and stylistic exchanges between Qing imperial court paining and Canton export painting requires further research.[28]

6 Jean Theodore Royer's *Twelve Albums of 288 Images of Fish from Canton*

Botanical and zoological illustrations are a part of the vast scope of export paintings produced in Canton (Guangzhou) during the second half of the eighteenth and nineteenth centuries. The collection of Museum Volkenkunde in Leiden contains twelve albums of fish, dated 1773–1776, in identical format; each album consists of 24 fish, a total of 288 illustrations, all elaborately painted in water colours, and the majority highlighted with silver, produced by

28 Lai Yu-chih in her insightful article reminds us of the interaction between the court painting and the export painting produced in Canton, see Lai Yu-Chih 賴毓芝, "Qinggong yu Guangdong waixiao huafeng de jiaohui: wumingkuan *Haidong cejing tuce* chutan 清宮與廣東外銷畫風的交會–無名款海東測景圖測初探", *Gugong wenwu yukan* 363 (2013) 74–86.

the workshops in Canton.[29] They belonged to the collection established during 1765 to 1780 by the Dutch lawyer and amateur sinologist Jean Theodore Royer (1737–1807) through his connections with the Dutch East Indian Company (*Vereenigde Oostindische Compagnie*) officials and their Chinese relations in Canton.[30]

In Royer's fish albums the fish are depicted individually in a natural setting with background on each leaf and each fish has its Chinese name written next to the illustration. The depiction is executed in a Western manner with shading and colourwash to depict the spots and patterns on the creatures in detail and to create a realistic three-dimensional representation of the fish. Although each album consists of 24 leaves, there does not seem to be a system according to which the fish were placed together. Take the first album for example: it includes crab, lobster, various carps, gold fish, several kinds of perches, different squids, loaches, marble goby (*sunkeyu* 筍殼魚, *Oxyeleotris marmorata*), rhino-fish (*xiniuyu* 犀牛魚) [Fig. 22.9], and others.[31] Therefore, content-wise each album is a random assembly of a mixture of different species, both freshwater and sea fish. Apart from this, most of the names of the fish documented in the albums are not scientific names but local names or made-up names, which makes them difficult to identify.

Although the fish in the albums are portrayed in a naturalistic and realistic manner, some of the species do not exist: for example the *pi*-fish (*piyu* 魮魚) on leaf 9 is a fish with a bird's head documented in the *Classic of Mountains and Seas* (*Shanhaijing* 山海經, *c.*475 BCE), an ancient Chinese classic text compiling mythic geography and beasts.[32] The so-called rhino-fish on leaf 23, is depicted

29 Inv. no.: RV-360–379a to RV-360–379l. For the complete images of twelve albums, see https://collectie.wereldculturen.nl/?query=search=packages=OnViewWM#/query/da27f61a-ecff-410e-9207-65a743d9c28a (acc essed on 20.12.2021). https://collectie.wereldculturen.nl/?query=search=packages=OnViewWM#/query/da27f61a-ecff-410e-9207-65a743d9c28a.

30 For Royers collection of Chinese objects, see: Jan van Campen, *De Haagse jurist Jean Theodore Royer (1737–1807) en zijn verzameling Chinese voorwerpen* (Hilversum: 2000); English version, see Jan van Campen, *Collecting China: Jean Thedore Royer (1737–1807), Collections and Chinese Studies* (Hilversum: 2021).

31 The names of the species documented in the first album (inv-no.: RV-360–379a) of the Royer's fish albums documented in sequence: *hongyu* 紅魚, *longxia* 龍蝦, *pengxie* 蟛蟹, *baihualu* 白花鱸, *jianban* 繭班, *feilu* 飛鱸, *zuanchitong* 鑽匙筒, *autouli* 鰲頭鯉, *piyu* 魮魚, *qingjiaolu* 青鮫鱸, *xioluyu* 綉盧魚, *youyu* 鱿魚, *huoli* 火鯉, *jingyu* 金魚, *ban'ao* 班鰲, *nizhui* 坭錐, *caoyu* 鰽魚, *yaoyu* 姚魚, *qishayu* 蜞殺魚, *huamuyu* 花木魚, *shaqiu* 沙鰍, *hankouli* 鉗口鯉, *xiniuyu* 犀牛魚, and *sunkeyu* 笋殼魚.

32 Inv-no.: RV-360–379a9. For the animals, plants and fish documented in the *Shanhaijing*, see Zhang Yan 張岩, *Shanhaijing yu gudai shehui* 山海經與古代社會 (Beijing: 1999) 36–98.

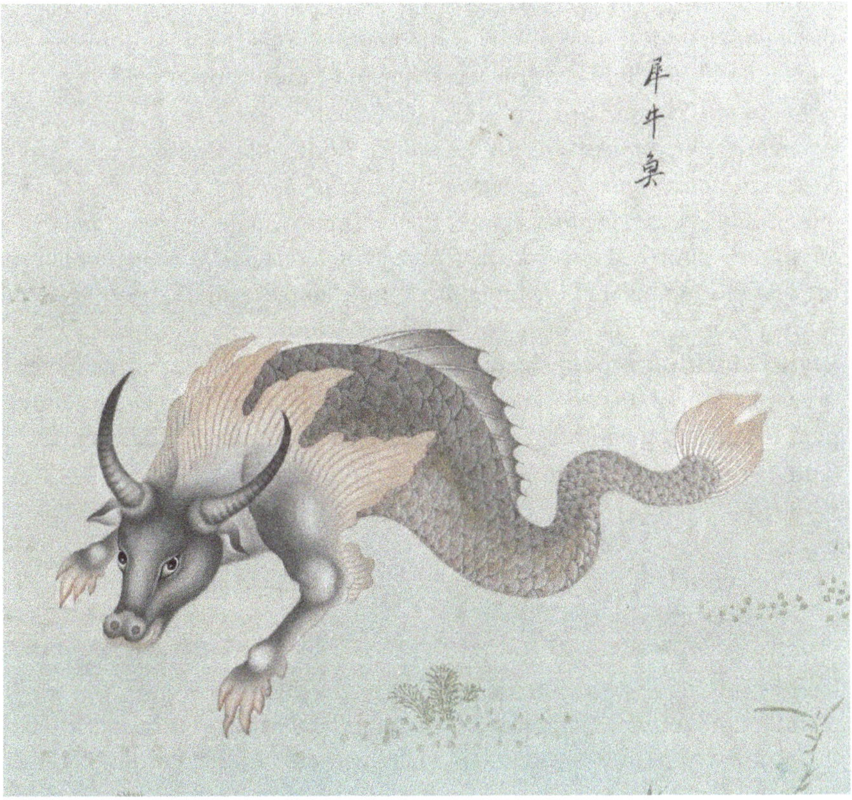

FIGURE 22.9 *Twelve Albums of 288 Images of Fish*, 1:23, 1773–1776, Museum Volkenkunde, Leiden

as a fish with a buffalo's head (and two forelegs), an image like the buffalo-fish (*qianniuyu* 潛牛魚) depicted in Nie's album, that has its source in a legend told by local people.[33] Hence the sources for the depictions of the Royer's fish albums are varied. Some of the species were probably painted from life, but many of them also were created based on the descriptions in the ancient texts or legends and some of them were fantasy or even made up, such as horse-fish (*mayu* 馬魚) and monkey-fish (houyu 猴魚) with heads of horse and monkey.[34]

This kind of export painting produced by the anonymous workshop painters in Canton, often served as souvenirs or exotic visual sources of wonder or amusement to illustrate fish in China for its Western viewers. When examining the content of all the twelve albums, one would find none of them seem

33 Palace Museum (2014) 178.
34 Inv-no.: RV-360–379d22, RV-360–379e22, Museum Volkenkunde, Leiden.

to have a subject or system. Some of the fish also repeat. One could imagine, these illustrations were probably made in a single sheet with multiple copies and randomly assembled by the workshop or by the customer's choice.

In Royer's possession there were also *Album with Mythical Animals, Ten Albums with 320 Images of Chinese Citizens, Twelve Albums with 280 Images of Historical Figures and Chinese Citizens, Twelve Albums with 288 Images of Birds, Twelve Albums with 288 Images of Insects, Twelve Albums with 288 Images of Plants*, and *Twelve Albums with 1200 Images of Plants and Minerals*, which testify that the argument of the function of these kind of paintings served as souvenirs or exotic visual sources of wonder to illustrate the image of China.[35] Royer himself used these illustrations and the texts on them as his material to learn Chinese and the knowledge about China.[36] Due to the enormous production of the various workshops, however, the quality and accuracy of Canton export paintings also vary.

7 Commission of Natural Illustrations of Fish by John Reeves in Canton

Besides serving as souvenirs or exotic visual sources of wonder, the export paintings produced in Canton also led to engagement from European scholars. It was known that foreign scholars, such as the botanist John Bradby Blake (1745–1773), the naturalist John Reeves (1774–1856) and others, were engaged in the natural illustrations production in Canton.[37] Especially Reeves appears to have been in Canton where he served as tea inspector for the British East India Company and was personally involved in the making of the botanical and zoological illustrations (including illustrations of fish) with the local workshop painters.

The fish illustrations made by Canton painters in his collection were especially commissioned by him, as in his notes he recorded the process of commission and also the four painters' names: Akut, Akam, Akew and Asung.[38]

35 Van Campen, *Collecting China* 218–220.

36 Ibid. 37–64.

37 Chen Yin 陳滢, *Lingnan huaniaohua liubian 1368–1949* 嶺南花鳥畫流變 1368–1949 (Shanghai, Shanghai guji: 2004) 265–306; Jiang Yinghe 江滢河, *Qingdai yanghua yu Guangzhou tongshang kuoan* 清代洋畫與廣州口岸 (Beijing, Zhonghua shujü: 2007) 222–226; Magee J., *Images of Nature: Chinese Art and the Reeves Collection* (London: 2011) 4–13.

38 Magee, *Images of Nature* 11; Kate Bailey, *John Reeves: Pioneering Collector of Chinese Plants and Botanical Art* (London: 2019) 108–109.

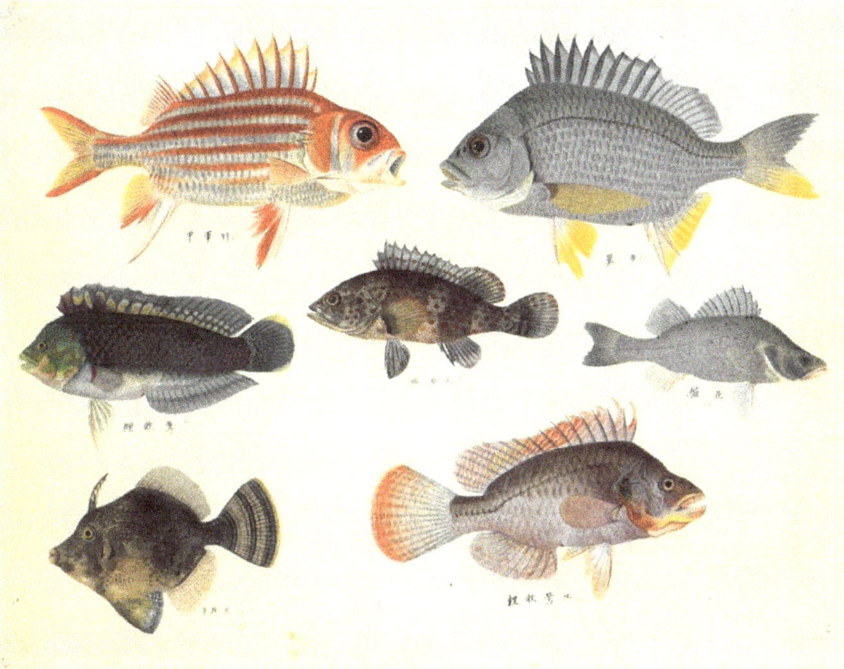

FIGURE 22.10 *Collection of Fish*, c.1826–1831, Natural History Museum, London

The fish illustrations commissioned by Reeves was for Major-General Thomas Hardwicke (1756–1835), a soldier and amateur naturalist based in India for much of his career. Reeves had four copies made.[39] The products can be seen as a combination of Chinese art (executed in the Western manner) and Western science. The artists who worked with Reeves paid special attention to depict the texture of the fish, for example, they used gold and silver powder to reproduce the iridescence of fish scales (Fig. 22.10).

The fish illustrations commissioned by Reeves were sent to England and were recognized by John Richardson (1787–1865) in his "Report on the Ichthyology of the Seas of China and Japan" in 1846:

> John Reeves, […] who was long resident at Macao, filling an important office in the employ of the India Company, with an enlightened munificence, caused beautiful coloured drawings, mostly of the natural size, to be made of no fewer than 310 species of fish which are brought to the market at Canton. These drawings are executed with correctness and

39 Bailey, *John Reeves* 115.

> finish which will be sought after in vain in the older works on ichthyology, and which are not surpassed in the plates of any large European work of the present day. The unrivalled brilliancy and effect of the colouring, and correctness of profile, render them excellent portraits of the fish[40]

Interestingly enough, the natural illustrations produced in Canton with the engagements of foreign scholars were sent back to Europe and made an impact on scholarship there. In contrast, there appears not to have been an impact in China of these paintings.

8 Conclusion

The origins of the European study of nature can be traced back to Greek and Roman antiquity, but illustration for science first flourished during the Renaissance, and was seen by contemporary scholars as a 'combination of art and science'.[41] This survey of the development of natural illustration of fish in China, however, shows a different path.

The cases of Nie Huang's *Pictures of the Various Marine Creatures* and Han Liangqi's commission of *Album of Sea Fish* are in the context of evidential study, but Nie's work entered the Qing imperial collection in 1726, since then it was not accessible to the public. Han's *Album of Sea Fish* was an unfinished project, hence both of them made little impact to the study of the ichthyology in China. Zhao Zhiqian's painting *Extraordinary Fish* aims to portray the local products in an artistic way rather than to provide material for natural study. The album *Manual of Sea Oddities* preserved in the Qing imperial collection has an unknown context in the making and requires further research. Although workshops in Canton produced natural illustrations, the quality and accuracy varies, the fish albums in Royer's collection served as his gateway to learn Chinese as a language and knowledge about China. The engagement of foreigner scholars in the production of exported natural illustrations in Canton, such as Reeves should be considered as a special commission and

40 Richardson J.M., "Report on the Ichthyology of the Seas of China and Japan", in *Report of the British Association for the Advancement of Science* (1845), cited from Bailey, *John Reeves* 113–115.

41 De Luca M.E. – Wolf G., "Ligozzis Naturstudien zwischen Kunst und Wissenschaft", in Bundeskunsthalle (ed.), *Florenz!* (Munich: 2013) 292–294; For natural history and its imagery in Europe, see O'Malley T. – Meyers A.R.W. (eds.), *The Art of Natural History: Illustrated Treatises and Botanical Paintings, 1400–1850* (New Haven: 2008).

those illustrations were custom-made to fit the foreigner customers' requirements and gave impact to ichthyology in England instead of in China.

This overview of the depiction of fish and other marine creatures in Chinese art offers a view on the varied way in which these paintings came about and the purposes for which they were made. Ranging from an attempt at scientific accuracy, to societal commentaries and entertainment purposes, the illustrations and descriptions of the various species highlight that science and art at times work in parallel but often also may proceed in different degree.

Bibliography

Bailey K., *John Reeves: Pioneering Collector of Chinese Plants and Botanical Art* (London: 2019).

Chen Yin 陳瀅, *Lingnan huaniaohua liubian 1368–1949* 嶺南花鳥畫流變 1368–1949 (Shanghai: 2004).

Chen Yun-Ru 陳韻如, "Hua yi yi ye: Chonggu Song Huizhongchao de huaihua huodong 畫亦藝也:重估宋徽宗朝的繪畫活動," Ph.D. dissertation, Graduate Institute of Art History, National Taiwan University, 2008.

De Luca M.E. – Wolf G., "Ligozzis Naturstudien zwischen Kunst und Wissenschaft", in Bundeskunsthalle (ed.), *Florenz!* (Munich: 2013) 292–294.

Greenberg D., "Weird Science: European Origins of the Fantastic Creatures in the Qing Court Painting, the *Manual of Sea Oddities*", in Silbergeld J. – Wang E.Y. (eds.), *The Zoomorphic Imagination in Chinese Art and Culture* (Honolulu: 2016) 379–400.

Huanhe huapu 宣和畫譜, 1120, reprint (Taipei: 1967).

Jiang Yinghe 江瀅河, *Qingdai yanghua yu Guangzhou tongshang kuoan* 清代洋畫與廣州口岸 (Beijing: Zhonghua shujü, 2007).

Lai Yu-Chih 賴毓芝, "Qinggong yu Guangdong waixiao huafeng de jiaohui: wumingkuan *Haidong cejing tuce* chutan 清宮與廣東外銷畫風的交會－無名款海東測景圖測初探", *Gugong wenwu yukan* 363 (2013) 74–86.

Lai Yu-chih 賴毓芝, "Qinggong dui Ouzhou ziranshi tuxiang de zaizhi: yi Qianlongchao *Shoupu* weili 清宮對歐洲自然史圖像的再製:以乾隆朝《獸譜》為例", in *Zhongyang yanjiuyuan jingdaishi yanjiusuo jikan* 中央研究院近代史研究所集刊 80 (2013) 1–75.

Lai Yu-chih, "Domesticating the Global and Materializing the Unknown: A Study on Album of Beasts at the Qianlong Court," in Anna Grasskamp A. – Juneja M. (eds.), *EuroAsian Objects: Art and Material Culture in Global Exchange, 1600–1800* (Berlin: 2018) 125–174.

Lai Yu-chih, "Costuming the Empire: A Study on the Production of Tributary Paintings at the Qianlong Court in 18th Century China", in Klich L. – Zanardi T. (eds.), *Visual*

Typologies from the Early Modern to the Contemporary: Local Contexts and Global Practices (New York: 2018) 90–103.

Lai Yu-chih 賴毓芝, "Zhishi, xiangxiang yu jiaoliu: Nan Huiren Kunyu quantu zhi shengwu chahui yanjiu 知識、想像與交流:南懷仁《坤輿全圖》之生物插繪研究," in Dong Shaoxin 董少新 (ed.), *Gantong shenshou: Zhong Xi wenhua jiaoliu Beijing xia de ganguan yu ganjue* 感同身受—中西文化交流背景下的感官與感覺 (Shanghai: 2018) 141–182.

Lee Hui-Shu, "The Fish Leaves of the *Anwan Album*: Bada Shanren's Journey to a Landscape of the Past", *Ars Orientalis* 20 (1990) 69–85.

Lee Hui-Shu, "Bada Shanren's Bird-and-Fish and the Art of Transformation", *Archives of Asian Art* 40 (1991) 6–26.

Li Lincan 李霖燦, "Yucao hua de houpo sheenyi 魚藻畫的活潑生意", in *Zhongguo meishushi gao* 中國美術史稿 (Taipei: 1987).

Liu Changzhi 劉昌芝, "Woguo xian cun zui zao de shuichan dongwu zhi: *Minzhong haicuo shu* 我國現存最早的水產動物志－《閩中海錯疏》", *Ziran kexueshi yanjiu* 自然科學史研究 12 (1982) 333–338.

Liu Zhigui 劉治貴, *Zhongguo huihua yuanliu* 中國繪畫源流 (Changsha: 2003).

Magee J., *Images of Nature: Chinese Art and the Reeves Collection* (London: 2011).

Miyazaki Noriko 宮崎法子, *Kacho sansuiga o yomitoku: Chugoku kaiga no imi* 花鳥・山水画を読み解く－中国絵画の意味 (Tokyo: 2003).

Neumann D., "Experiencing and Depicting Nature", in Neumann D. – Ogando J. (eds.), *Fascinated by Nature: Landscapes, Plants and Animals in the Tradition of Chinese and Japanese Painting from the Neumann-Ogando Collection* (Berlin: 2012) 14–30.

O'Malley T. – Meyers A.R.W. (eds.), *The Art of Natural History: Illustrated Treatises and Botanical Paintings, 1400–1850* (New Haven: 2008).

Palace Museum (ed.), *Qinggong haicuo tu* 清宮海錯圖 (Beijing: 2014).

Qingding shiqü baoji xübian 欽定石渠寶笈續編 (Qing imperial edition, 1793).

Song Hou-mei, *Decoded Messages: The Symbolic Language of Chinese Animal Painting* (Ohio: 2010).

Sterckx R., "The Limits of Illustration: Animalia and Pharmacopeia from Guo Pu to *Bencaogangmu*", in Lo V. – Barrett P. (eds.), *Imaging Chinese Medicine* (Leiden: 2018) 135–150.

Tokyo National Museum (ed.), *Kisshō: Chūgoku bijutsu ni komerareta imi* 吉祥－中国美術にこめられた意味 (Tokyo: 1998).

Van Campen J., *De Haagse jurist Jean Theodore Royer (1737–1807) en zijn verzameling Chinese voorwerpen* (Hilversum: 2000).

Van Campen J., *Collecting China: Jean Thedore Royer (1737–1807), Collections and Chinese Studies* (Hilversum: 2021).

Wan Qingli 萬青力, *Bingfei shuairuo de bainian: shijiu shiji Zhongguo huihua shi* 並非衰弱的百年:十九世紀中國繪畫史 (Taipei: 2005).

Wang Ching-Ling, "On the *Picture of Sea Fish* (*Haiyu tu*)" (forthcoming).

Wang Fangyu – Barnhart R.M. – Smith J.G. (eds.), *Master of the Lotus Garden: The Life and Art of Bada Shanren* (New Haven: 1990).

Wang Yonghou 王永厚, "Tu Benjun jiqi *Minzhong haicao shu* 屠本畯及其《閩中海錯疏》", in *Zhongguo shuzhang* 中國水產 2 (1984) 29.

Wu, Songfeng 吳誦芬, "*Haicuo tu* 海錯圖," in *Gugong wenwu yuekan* 故宮文物月刊 363 (2016) 66–73.

Wu Chaoran 吳超然, "Zhao Zhiqian yibaliuyi nian de sanjian zhuopin *Yiyu tu, Ouzhong wuchan tujuan, Ouzhong caomu siping*: Jinshi huapai yu haipai guishu zhi shangque 趙之謙一八六一年的三件作品《異魚圖》、《甌中物產圖卷》、《甌中草木四屏》: 金石畫派與海派歸屬之商榷", in Yang Dunyao 楊敦堯 (ed.), *Shibian, Xingxiang, liufeng: Zhongguo jindai huihua 1796–1949 xueshu yentaohui lunwenji* 世變・形象・流風:中國近代繪畫 1796–1949 學術研討會論文集 (Taipei: 2008) 451–469.

Yang Xiaonan (ed.), *The Golden Age of Chinese Archaeology: Celebrated Discoveries from the People's Republic of China* (Washington DC: 1999).

Zhang Yan 張岩, *Shanhaijing yu gudai shehui* 山海經與古代社會 (Beijing: 1999).

Zhang Yanyuan 張彥遠, *Lidai minghua ji* 歷代名畫記, reprint (Beijing: 1963).

Zheng Jingsheng, "Observational Drawing and Fine Art in Chinese Materia Medica Illustration", in Lo V. – Barrett P. (eds.), *Imaging Chinese Medicine* (Leiden: 2018) 152–160.

Zhongguo kexueyuan kaogu yanjiuyuan 中國科學院考古研究院 (ed.), *Xinzhongguo de kaogu shouhuo* 新中國的考古收穫 (Beijing: 1961).

Zhu Jingxuan 朱景玄, *Tangchao minghua lu* 唐朝名畫錄, reprint in Pan Yungao 潘運告 ed., *Tang Wudai hualun* 唐五代畫論 (Changsha: 1997).

Zou, Zhenghuan 鄒振環, "Haicuo tu yu Zhong Xi zhishi zhi jiaoliu 《海錯圖》與中西知識之交流", *Zijincheng* 紫禁城 266 (2017) 124–131.

https://collectie.wereldculturen.nl/?query=search=packages=OnViewWM#/query/da27f61a-ecff-410e-9207-65a743d9c28a (accessed on 20.12.2021).

https://digitalarchive.npm.gov.tw/ITWaterFall (accessed on 11.12.2021).

https://www.slam.org/collection/objects/32476/ (accessed on 31.01.2022).

CHAPTER 23

Early "Dutch" Contributions to Japanese Ichthyology

Martien J.P. van Oijen

From 1641 to 1854 the Dutch were the only European people who were allowed to enter the harbour of Nagasaki and trade with the Japanese. Via contacts with Dutch inhabitants of Deshima and imported books (Maclean, 1973, 1974) the Japanese were informed about scientific developments in the western world.[1] Especially the physicians (e.g. Kaempfer, Thunberg, von Siebold) and the "Opperhoofden" or chiefs of the Dutch factory were responsible for this transfer of knowledge. Although there were severe restrictions on contacts with the Japanese and to what Japanese persons were allowed to tell the Dutch, the Opperhoofden and especially the physicians managed to obtain knowledge about Japan.

Officially only Dutch persons, i.e. inhabitants of respectively the Republic of the Seven Provinces, the Batavian Republic, and the Kingdom of the Netherlands, were allowed to enter Japan. However, the United East-India Company (v.o.c.) (till 1797), the Dutch State, and the Dutch Trade Company (from 1827) employed many foreigners. For the Japanese, the Dutch language was the only way to recognize a Dutchman as such. Japanese interpreters at Deshima, who mastered the Dutch language, on this basis could identify a Dutchman. The physicians Kaempfer and Thunberg who had been in Dutch service for some years before they sailed to Deshima, easily passed the test. However, von Siebold, who had had little time to learn the Dutch language, was only allowed to enter the island because his fellow passengers explained his 'dialect' by stating he came from the Dutch mountains (!).

As they had been trained in the use of plants for medicines, all physicians were more interested in botany than in zoology. Plants were also much easier to preserve than animals. It was easy to obtain fish in the harbour of Nagasaki, but for the preservation of fishes, one needed arak (rice wine) that had to be shipped from Batavia. It was only when von Siebold was sent to Japan with

1 MacLean J., "Natural Science in Japan. I. Before 1830", *Annals of Science* 30. 3 (1973) 257–298; idem, "The Introduction of Books and Scientific Instruments in Japan", *Japanese Studies in the History of Science* 13 (1974) 9–86.

DOI:10.1163/9789004681187_024

a special instruction to gather information on natural products of Japan that fishes were collected in large quantities.

The present chapter provides an overview of collectors, researchers, authors, and artists, either Dutch or employed by the Dutch administration, who contributed to Japanese ichthyology between 1690 (Engelbert Kaempfer) and 1879 (Pieter Bleeker).

1 Pre-Linnean Authors

In the earliest Dutch books on Japan, fishes are hardly mentioned. In the *Beschryvinge van het Machtig Koningrijck Japan* (Description of the Mighty Kingdom of Japan) François Caron (many editions from 1645–1663) pays no attention at all to fishes. This is the more surprising when one realizes that Caron (1600–1673) lived for twenty years in Japan, a country where fish and fishery always have been essential.

Arnoldus Montanus (1625–1683) spent his whole life in the Netherlands. His large work on Japan (1669) was solely based on travel accounts. The only information on fishes in Montanus is 'The inhabitants have plenty of fish, especially roach and shad, and these they like the most.'[2] In a summary of differences between the Dutch and the Japanese, Montanus states: 'we feed on chicks and fattened fowl, they feed on fish and scalloped sea creatures.'[3] The book contains two plates with fishermen performing various fishing techniques, but these are not explained.

The German Engelbert Kaempfer (1651–1716) was the first *scientist* to write about fishes occurring in Japan. Kaempfer had wide-ranging interests and studied at various universities in Germany and Poland, but he was foremost a physician with a great interest in botany.[4]

Before he arrived as a physician at Deshima in September 1690, Kaempfer had travelled to Sweden, Russia, Persia, India, Ceylon, and Java, and had been

2 Montanus Arnoldus, *Gedenkwaerdige gesantschappen der Oost-Indische maatschappy in't Vereenigde Nederland, aan de kaisaren van Japan: vervatende wonderlyke voorvallen op de togt der Nederlandtsche gesanten: beschryving van de dorpen, sterkten, steden, landtschappen, tempels, gods-diensten, drugten, gebouwen, dieren, gewasschen, bergen, fonteinen, vereeuwde en nieuwe oorlogs-daaden der Japanders: verçiert met een groot getal afbeeldsels in Japan geteekent: getrokken uit de geschriften en reis-aanteekeningen der zelve gesanten* (Amsterdam, Jacob Meurs: 1669) 47.

3 Montanus, *Gedenkwaerdige gesantschappen* 49.

4 Holthuis L.B. – Sakai T., *Ph. F. von Siebold and Fauna Japonica. A History of Early Japanese Zoology* (Tokyo: 1970) 1–323.

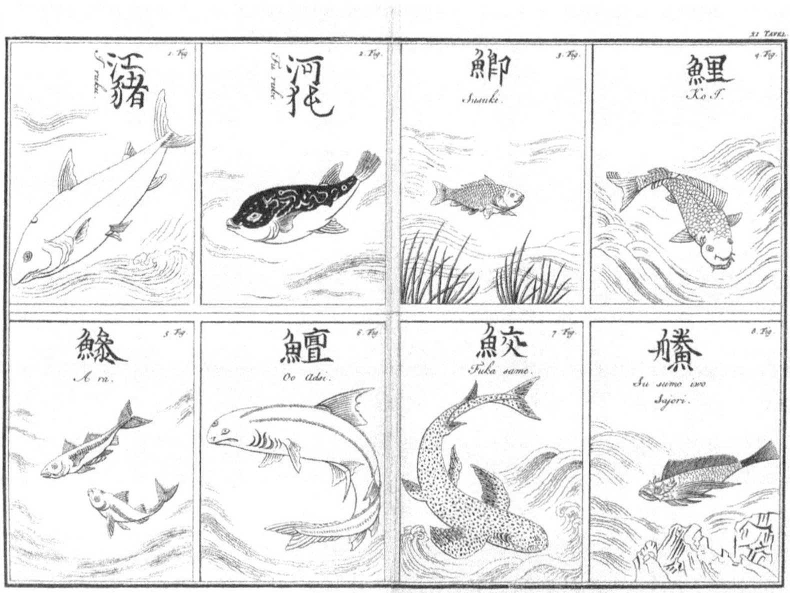

FIGURE 23.1 Kaempfer Engelbert, *De beschryving van Japan* […] (The Hague – Amsterdam, P. Gosse and J. Neaulme, Balthasar Lakeman: 1729), Plate 9. Copper engravings of Japanese woodblock prints. The first illustrations of Japanese fishes published in a European book.

in the service of the Dutch East India Company for six years. Kaempfer stayed in Japan for two years, and two times joined the journey to the court of the Shogun in Edo (Tokyo).

Most of the information on Japan that Kaempfer collected during his stay, was only published posthumously in *The History of Japan*.[5] In this book, Kaempfer listed 45 fish species, for which both the Japanese and the Dutch

5 Kaempfer Engelbert, *The History of Japan, giving an Account of the ancient and present State and Government of that Empire; of Its Temples, Palaces, Castles and other Buildings; of Its Metals, Minerals, Trees, Plants, Animals, Birds and Fishes; of The Chronology and Succession of the Emperors, Ecclesiastical and Secular; of The Original Descent, Religions, Customs, and Manufactures of the Natives, and of their Trade and Commerce with the Dutch and Chinese. Together with a Description of the Kingdom of Siam* (London, J. MacLehose and sons: 1727) and idem, *De beschryving van Japan, behelsende een verhaal van den ouden en tegenwoordigen staat en regeering van dat Ryk; van deszelfs tempels, paleysen, kasteelen en andere gebouwen: van deszelfs metalen, mineralen, boomen, planten, dieren, vogelen en visschen: van de tydrekening, en opvolging van de geestelyke en wereldlyke keyzers: van de oorsprondelyke afstamming, godsdiensten, gewoonten en handwerkselen der inboorllingen, en van hunnen koophandel met de Nederlanders en de Chineesen: Benevens eene beschryving van het koningryk Siam* (The Hague – Amsterdam, P. Gosse – J. Neaulme – Balthasar Lakeman: 1729).

names were given. Morphological characters of the fishes were not included. For a number of species, some information was provided about their taste, toxicity, or medical use. For 12 species of fish, illustrations were added [Fig. 23.1] which were copied from a Japanese book that Kaempfer had brought from Japan. Some of these illustrations and names do not match. Kaempfer cannot be blamed for these mistakes: The *History of Japan* was published long after his death, and the translator of his original German text, J.G. Scheuchtzer, was responsible for the selection and the copying of the figures.

2 The Introduction of Linnean Nomenclature in Japan

Carolus Linnaeus (1707–1778) hardly needs to be mentioned with regard to Japanese ichthyology, as the only Japanese species mentioned in the tenth edition of his *Systema naturae* was the goldfish, *Cyprinus auratus*.[6] The Dutch physician Martinus Houttuyn (1720–1798) was also a naturalist (especially a plant and bird lover) and a great admirer of Linnaeus. To make Linnaeus's work available to a larger Dutch audience, he produced between 1761 and 1773 a 37-volume work *Natuurlyke Historie of uitvoerige Beschryving der Dieren, Planten en Mineraalen, volgens het Samenstel van den Heer Linnaeus* (Natural History or Detailed Description of Animals, Plants and Minerals, According to the Compilation of Mr. Linnaeus). The first 18 volumes were dealing with zoology, the next fourteen with botany, and the last five with mineralogy.[7] As Houttuyn's name was lacking on the title page, many people thought it was only a translation of Linnaeus's *Systema Natura*. However, the content of Houttuyn's work was ten times that of the *Systema Natura*. For example, whereas Linnaeus named and described the goldfish in a few lines, Houttuyn devoted 15 pages to this species. Moreover, Houttuyn included figures of many species [Fig. 23.2]. Whereas in 2017 Akihito, the then emperor of Japan and respected ichthyologist, in an address presented at the tricentennial of the Linnean Society of London stated: 'We do not know exactly when the scientific names under the binomial nomenclature, originated by Linné, were

6 Linnaeus Carolus, *Systema naturae per regna tria natura, secundum classes, ordines, genera, species, cum characteribus, differentibus, synonymis, locis. Tomus I. Editio Decima, Reformata* (Stockholm, Lars Salvius: 1758) 322.

7 For an extensive study on the zoological part of this publication, see Boeseman M. – Ligny W. de, "Martinus Houttuyn (1720–1798) and His Contributions to the Natural Sciences, with Emphasis on Zoology", *Zoologische Verhandelingen Leiden* 349 (2004) 1–222.

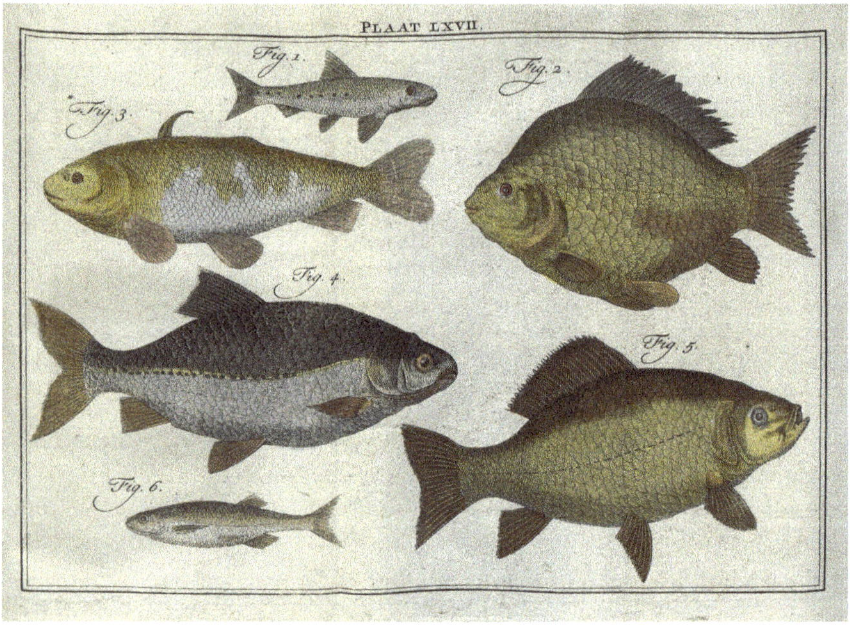

FIGURE 23.2 Houttuyn Martinus, *Natuurlyke historie* [...] (Amsterdam, Frans Houttuyn: 1765), Plate LXVII. *Fig. 1.* Rivier-Govie of Grondel; *Fig. 2.* Steenkarper of Karausch; *Fig. 3.* Chineesche Goudvisch; *Fig. 4* Blanke Voorn; *Fig. 5.* Riet Voorn of Ruisch; *Fig. 6.* Asterling of Nesteling.

introduced to Japan',[8] I believe the systematics and nomenclature of Linnaeus were introduced in Japan with Houttuyn's work.

Von Siebold in the account of his journey to the shogun of Japan in 1826 came close to the answer to this question. On March 29, near the town of Miya, he met some of his former students, including Mizutani Sukeroku, who showed him drawings of plants and animals he had made. Siebold notes:

> Two volumes of sketches, however, particularly caught my attention; it was a collection of Japanese crops, all accurately provided with Linnean names. Under each plant, the name of the genus was indicated, and of the 102 captions, I could only label four as wrong. Many of the genera indicated have been reported under the local Flora neither by Kaempfer nor by Thunberg, and some of them had not even been found by me. I very much wanted to know from him what literature he had used for

8 Akihito, "Linné and Taxonomy in Japan – On the 300th Anniversary of his Birth", *Proceedings of the Japan Academy, Series B.*, 86.3 (2010) 143–146.

> this, and heard that he only had a Dutch edition of Linné available for his research.[9]

Probably von Siebold did not know the work of Houttuyn, but evidently, this was a copy of Houttuyn's *Natuurlijke Historie*. Boeseman and de Ligny noted that the Japanese botanist Yokusai Iinuma (1783–1865) was reported to have mainly consulted Houttuyn's *Natuurlyke Historie* for his twenty volume iconography of Japanese plants.[10] In Yokusais' work a picture of the title-page of Part I, vol. 1 of Houttuyn's work suggests that he had not only the botanical volumes at his disposal but probably the whole series.

MacLean, who studied the introduction of books and scientific instruments in Japan, noted that the Japanese since 1800 possessed a rather good knowledge of botany and zoology at the time of circa 1778.[11] In that year the Dutch printer and bookseller J.A. de Chalmot (1734–1801) published an encyclopaedic work in seven volumes in which the productions of nature played an important role. For fishes he mainly relied on Buffon's *Histoire Naturelle* and Houttuyn's work, often copying complete descriptions. Both Houttuyn's volumes and De Chalmot's work were imported into Japan onwards from 1800. MacLean referring to Eikoh Shimano, concluded that the Japanese translated de Chalmot's work from 1811–1839, and: 'It was the greatest translation enterprise in the whole Edo period – the officially sponsored Kosei Shinpen'.[12] This means that basic knowledge of Linnean systematics and nomenclature from this date was even present in the Japanese language.

3 Post-Linnean Authors

The Swede Carl Peter Thunberg (1743–1828) studied medicine at the University of Upsala where he became a pupil of Linnaeus. Like his teacher, he became especially interested in botany. In 1771, by the mediation of Dr. J. Burman, professor of Botany at Amsterdam, he was offered a position with the Dutch East India Company. After having spent three years in South Africa he was appointed

9 Siebold, Philipp Franz Balthasar von, *Archiv zur Beschreibung von Japan und dessen Neben- und Schutzländern Jezo mit der südlichen Kurilen, Sachalin, Korea und den Liukiu-Inseln. Herausgegeben von seinen Sohnen: 2. Auflage*. (Würzburg und Leipzig: 1897).

10 Mizuno M. et al., *Iinuma Yokusai* (Gifu-City : 1984), vi, 1–513, 21 col. pls. [in Japanese], incl. A Bibliographical Sketch of Yokusai Iinuma, iv–vi (in English).

11 MacLean J., "The Introduction of Books and Scientific Instruments in Japan", *Japanese Studies in the History of Science* 13 (1974) 9–86, here 22.

12 MacLean, "The Introduction of Books and Scientific Instruments in Japan" 25.

physician at Deshima, where he arrived in August 1775. Thunberg taught medicine, pharmaceutical science, and natural history to Japanese doctors and interpreters, joined the court journey of 1776, and collected many plants and animals. He left Japan in December 1766.[13] Back in Batavia, he handed a collection of fishes to J.C.M. Radermacher, a high officer of the Dutch East Indian Company, amateur botanist, and one of the founders of the Bataviaasch Genootschap voor Kunsten en Wetenschappen (Batavian Society of Arts and Sciences) Radermacher sent this collection to Houttuyn in Amsterdam. In 1782, on the basis of this collection, Houttuyn published, again in Dutch, the first paper solely dealing with Japanese fishes, entitled: "Beschrijving van eenige japansche visschen en andere zee-schepselen" (Description of some Japanese fishes and other sea creatures).[14] In this paper 36 fish species were described, 21 of which were new for science. No figures of the species were added. The specimens of this collection, including the types of the new species, were dispersed and must be considered lost.[15]

After six months in Batavia during which he made a collection of Javanese plants, Thunberg sailed to Amsterdam via Ceylon, where he stayed five months to make collections, and Cape of Good Hope. From Amsterdam, he made a short trip to England where he examined collections of Kaempfer. After an absence of 8½ years, laden with collections, Thunberg finally came back to Sweden in March 1779. Thunberg, who some years after his return to Sweden succeeded Linnaeus as professor of medicine and botany, published many papers on his botanical collections; however, in his account of his voyages, he also included a list of 49 Japanese fishes.[16] Moreover, he published six short papers in which he described and figured twelve new Japanese species [Fig. 23.3].[17] Two of Thunberg's students wrote a dissertation on Japanese fishes.[18] Both are merely lists of species with a few theses.

13 Holthuis – Sakai, *Ph. F. von Siebold and Fauna Japonica*; Thunberg Carl Peter, *Travels in Europe, Asia and Africa made during the Years 1770 & 1779* (London, F. and C. Rivington: 1795–1796).

14 Houttuyn Martinus, "Beschryving van eenige Japanse visschen en andere Zeeschepselen", *Verhandelingen der Hollandsche Maatschappij van Wetenschappen Haarlem* 20 (1782) 311–350.

15 Boeseman M., "Martinus Houttuyn (1720–1798) and his Japanese fishes", *UO* 43 (1995) 1–9; Boeseman – Ligny, "Martinus Houttuyn".

16 Thunberg Carl Peter, *Resa uti Europa, Africa, Asia, förrättad åren 1770–1779*, 3 vols. (Uppsala, Joh. Edman: 1788–1794).

17 These articles, dated 1790, 1792, and 1793, are to be found in the Bibliography at the end of the present article.

18 Wernberg Olaus, *Fauna Japonica* (Ph.D. dissertation, University of Upsala: 1822); Ahlstrom Alexander Magnus, *Fauna Japonica continuata* (Ph.D. dissertation, University of Upsala: 1823).

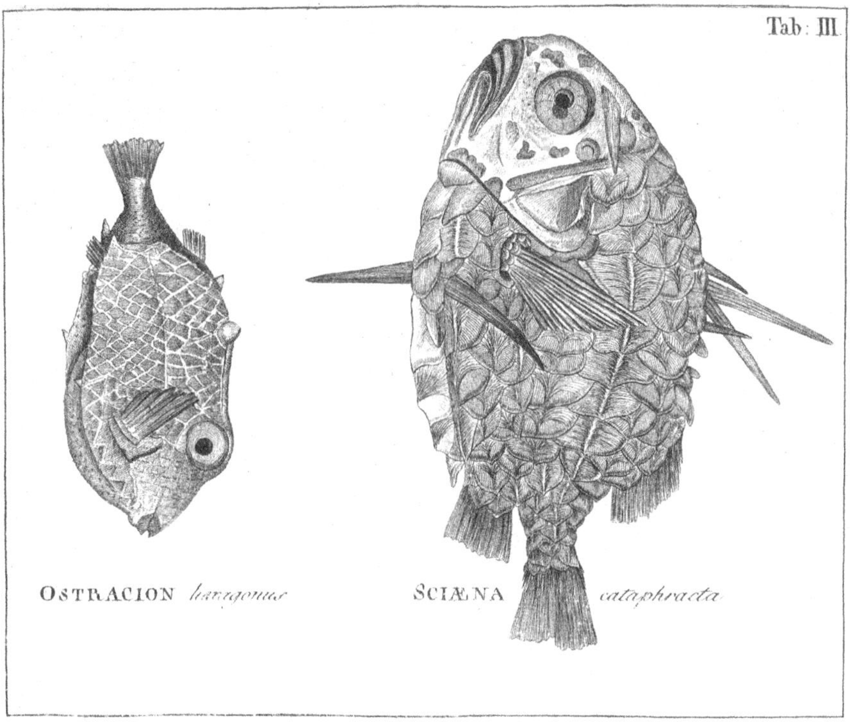

FIGURE 23.3 Illustrations of two Japanese fishes in Thunberg Carl Peter, "Beskrifning pa tvanne Fiskar ifran Japan", *Vetenskaps Academiens Nya Handlingar* 11 (1790) 106 /Der Konigl. Schwedischen Akademie Der Wissenschaften Neue Abhandlungen Aus Der Naturlehre, Haushaltungskunst Und Mechanik 11 (1790) 100–103. Both had already been named (but not figured) by Houttuyn (1782). *Ostracion hexagonis* = *Kentrocapros aculeatus* (Houttuyn) and *Sciaena cataphracta* = *Monocentrus japonicus* (Houttuyn).

In the intermezzo between Thunberg's departure and the arrival of von Siebold, Nagasaki was visited by a group of Russian and German scientists who hardly came in contact with Dutch officials, but who would indirectly have a great influence on the Dutch interfering with Japanese ichthyology: From October 1804 till April 1805 the ship Nadesha of the first Russian circumnavigation was forced to anchor in the bay of Nagasaki. However, the crew was not allowed to leave the ship. For six months, all the two naturalists, G.H. von Langsdorff and W.G. Tilesius von Tillenau, both Germans, could do, was to investigate and draw the fishes that were brought on board as food.[19] The

19 Tilesius von Tillenau Wilhelm Gottlieb von, "Description de quelques poissons observés pendant son voyage autour du monde", *Mémoires de la Société impériale des naturalistes de Moscou* 2.20 (1809) 212–249; idem, "Abbildungen und Beschreibungen einiger Fische aus

types of the new species described in this way were eaten afterward! However, only a few of Tilesius's drawings were used for actual species descriptions [Fig. 23.4]. The paintings published in Adam Johann von Krusenstern's *Atlas zur Reise um die Welt* could not be accepted for taxonomy because the species names were written in the Cyrillic alphabet. Von Langsdorff donated the few fishes he managed to save from the kitchen to the Berlin museum. Its director, Dr M.H.C. Lichtenstein, presented them to the French zoologist Georges Cuvier,[20] who subsequently described 20 new species from this collection.

The German physician Philipp Franz Balthasar von Siebold (1797–1866) stayed six years and five months in Japan, of which the last year, 1829, was spent under house arrest. Probably inspired by von Langdorff and Tilesius, during that year, he made a list of fishes brought him as provision, and selected plants from the fodder brought for his goat. Von Siebold was much more interested in botany than in zoology, yet immediately after his arrival on Deshima in August 1823, he started to investigate the zoological collection of the opperhoofd J. Cock Blomhoff, who was to leave Deshima before the end of that year. When leaving, Blomhoff not only took his collection with him, but also von Siebold's manuscript. Von Siebold's first zoological paper was published (with the wrong initials) in Batavia in 1824.[21] Regrettably, only one species of fish was present in Blomhoff's collection: the nurse shark *Squalus cirratus* Bosc. (*Ginglymostoma cirratum* (Bonnaterre 1788)).

In September 1828 when von Siebold was dismissed from research activities he wrote in a summary of his research,[22] that he had neglected the collecting of fishes, as they were hard to preserve, and because he believed that their live colours should be painted. However, his painter Toyosuke (Kawahara Keiga) was fully occupied with painting plants. Nevertheless, a total of about 700 fishes (almost all in spirits) were shipped to Leiden by von Siebold. Before

---

Japan und einiger Mollusken aus Brasilien, welche bey Gelegenheit der ersten Russischen Kaiserliche Erdumseglung lebendig beobachtet wurden", *Denkschriften der Königlichen Akademie der Wissenschaften zu München* 3 (1811–1812) 71–88; idem, *Atlas zur Reise um die Welt, unternommen auf Befehl Seiner Kaiserlichen Majestät Alexander des Ersten auf den Schiffen Nadeshda und Neva. Unter dem Commando des Captains von Krusenstern* (St. Petersburg: 1814); Langsdorf G.H. von, *Bemerkungen auf einer Reise um die Welt in den Jahren 1803–1807* (Frankfurt am Main: 1812).

20 Cuvier Georges – Valenciennes Achille, *Histoire naturelle des poissons*, vol. 4 (Paris: 1829) 261.

21 Siebold G.T. [sic] de,. *De historiae naturalis in Japonica statu, nec non de augmento emolumentisque in decursu perscrutationum expectandis dissertatio, cui accedunt Spicilegia faunae Japonicae* (Batavia: 1824).

22 Siebold, Philipp Franz Balthasar von, *Kurze Uebersicht des Gegenwärtigen Zustandes meiner wissenschaftligen Untersuchungen auf Japan* MS Japaninstituts Berlin (1828).

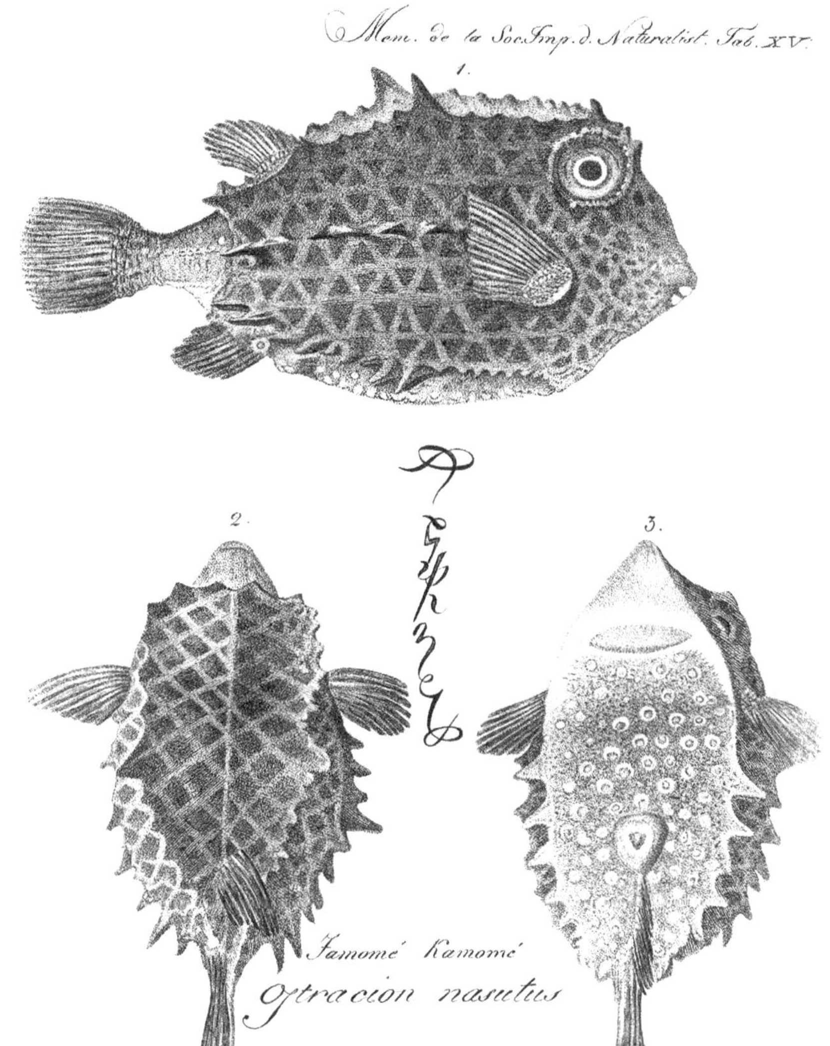

FIGURE 23.4 Tilesius von Tillenau Wilhelm Gottlieb von, "Description de quelques poissons observés pendant son voyage autour du monde", *Mémoires de la Société impériale des naturalistes de Moscou* 2.20 (1809) 212–249, plate 15. Based on drawings by W.G. von Tilesius.

he left Japan, von Siebold greatly stimulated his successor Heinrich Bürger to concentrate on making a good collection of fishes. Moreover, von Siebold was initiator, developer, and editor of the *Fauna Japonica*, of which the Fish volume was the most species-rich.

The German Heinrich Bürger (1896–1858) worked in Batavia as an apothecary before he was sent to Japan in 1825 to assist von Siebold. Although von Siebold charged him with geological and chemical research, he also trained him as a naturalist. On October 1, 1828, Bürger became responsible for all research on natural products of Japan, but only after von Siebold's departure, at the end of 1829, Bürger began his research on fish. He made detailed descriptions of 500–600 species which he numbered and had most of them painted with the same numbers by Keiga.[23] Following von Siebold's advice, he gave his species a Latin generic name and a specific name derived from the Japanese name. However, in contrast to Bürger's statement, only 200 of these descriptions reached Leiden. Bürger also made shipping lists. In four shipments, 1382 specimens (most of them stuffed) [Fig. 23.5] and 259 fish paintings by Keiga were sent to Leiden.[24] The stuffed fishes were prepared with the technique described by Temminck.[25]

According to Maclean,[26] after Bürger's departure for Java in December 1834 the Japanese Magoeits and Foské continued the research in natural science. Maclean also mentions that the inquiries on natural science were finished in 1842.

Kawahara Keiga (1786–1860?) was the only Japanese artist who was allowed to work at Deshima during his employment for the Dutch factory. Keiga had already made small paintings of plants and animals for J.G.F. van Overmeer Fischer and J. Cock Blomhoff, before he started to work for von Siebold for whom he mainly made scientific paintings of plants.[27] During the court journey, Keiga also painted landscapes for von Siebold. From 1830 onwards Bürger commissioned him to make life-size paintings of fresh fishes and crustaceans. Keiga succeeded very well in capturing the fresh colours. He probably had been shown the figures of Tilesius from von Krusenstern's Atlas [Fig. 23.6] and painted the fishes in a very European style. However, he painted them without applying shadows and a light spot in the eye.

In Leiden at the Rijksmuseum van Natuurlijke Historie, the German Hermann Schlegel (1804–1884), then curator of Vertebrates and very good draughtsman,

---

23 Bürger Heinrich, MS, without title, contains 200 descriptions of Japanese fishes (1830–1831) (Collection Naturalis Biodiversity Center).

24 Yamaguchi T. – Machida Y., "Fish specimens collected in Japan by Ph. F. von Siebold and H. Bürger and now held by the National Natuurhistorisch Museum in Leiden and other two Museums", *Calanus*, Special Number 4 (2003) 87–321.

25 Temminck Coenraad Jacob, *Voorschrift, hoedanig te handelen met voorwerpen van Natuurlijke Historie, ten einde dezelve behoorlijk te verzenden en voor bederf te bewaren; ten gebruike van het 's Rijks Museum van Natuurlijke Historie te Leyden* (Leiden: 1825).

26 MacLean, "The Introduction of Books and Scientific Instruments in Japan" 41.

27 Yamaguchi T., "Kawahara Keiga and natural history of Japan 1. Fish volume of Fauna Japonica", *Calanus* 12 (1997) 1–35.

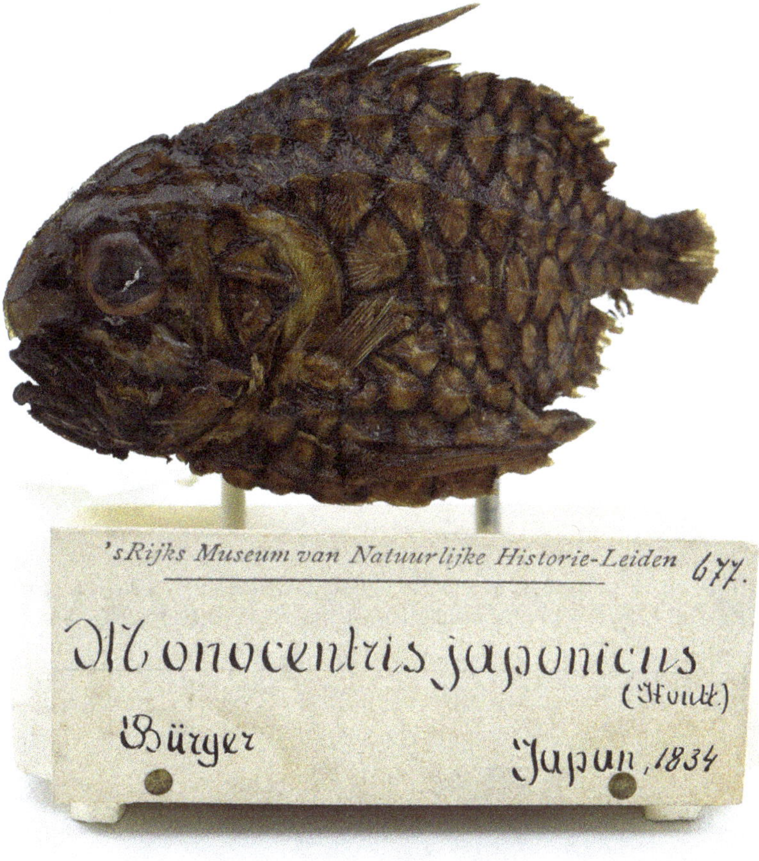

FIGURE 23.5 A stuffed Naturalis specimen of *Monocentrus japonicus* (Houttuyn 1782) RMNH.PISC.D 677, collected by Heinrich Bürger. The specimen was stuffed in Japan and shipped to Leiden without a pedestal.

investigated the fishes and described them for the *Fauna Japonica*. Although the name of C.J. Temminck, the director of the museum, is also on the title page, he did not contribute to the descriptions of the fishes.[28] The descriptions Bürger made in Japan, must at least have been helpful and parts of it were almost literally used, but Bürger's role was hardly acknowledged by Schlegel. On the basis of the collections of von Siebold and Bürger, and the *Pisces* volume of the *Fauna Japonica*, 348 species were described, 165 of which were new to science. For 20 of these, Japanese names were used in the specific name.

28 Suzanna J.A., "Levensschets van Coenraad Jacob Temminck", *Jaarboek van de Maatschappij der Nederlandsche letterkunde* (1858) 47–87, here 65.

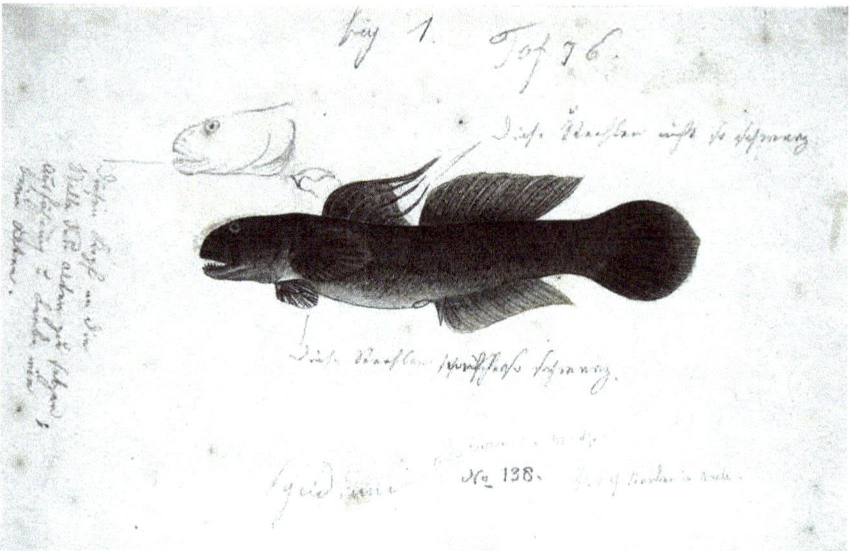

FIGURE 23.6 Painting of a Japanese goby made by Kawahara Keiga in 1831. RMNH.ART.231. Pencil remarks and outline of the head were made by Schlegel to instruct the lithographer. The number 138 refers to the description in Bürger's manuscript.

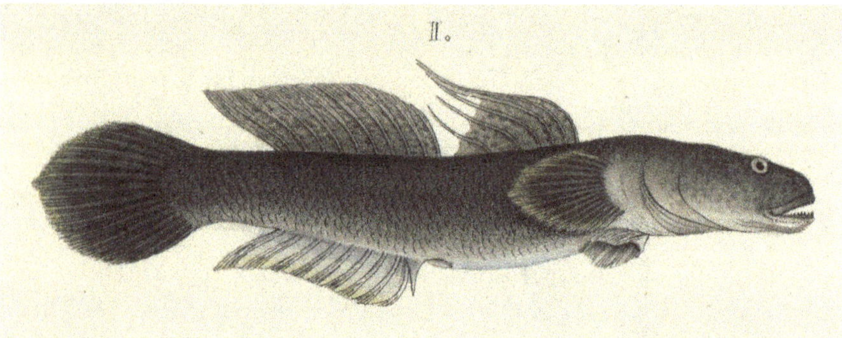

FIGURE 23.7 Temminck Coenraad Jacob – Schlegel Hermann, *Fauna Japonica, Pisces. 4* (Leiden, Arntz: 1845), Plate 76, Fig. 1. Hand coloured lithograph of Burger's species No. 138, illustrating the description of *Sicydium obscurum* = *Tridentiger obscurus*.

Most plates of fishes in the *Fauna Japonica* are based on paintings of Keiga. Schlegel added pencil notes to many of Keiga's paintings to instruct the lithographers who transferred the paintings to stone [Figs. 23.7–8]. All plates in the *Fauna Japonica* are hand-coloured lithographs, which makes all plates slightly different and unique.

Before the first instalment of the *Pisces* volume of the *Fauna Japonica* was published in 1842, the Leiden Museum was visited, in 1837 and 1838, by two

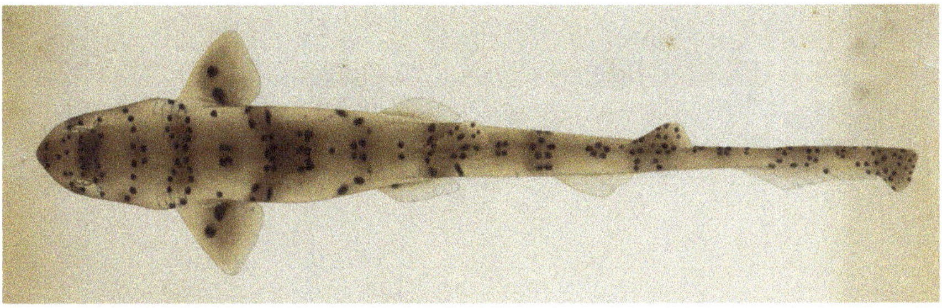

FIGURE 23.8 Painting of a Japanese shark made by Kawahara Keiga in 1831. This species had been described in 1830 by Bürger in his manuscript as *Scijllium tora.*

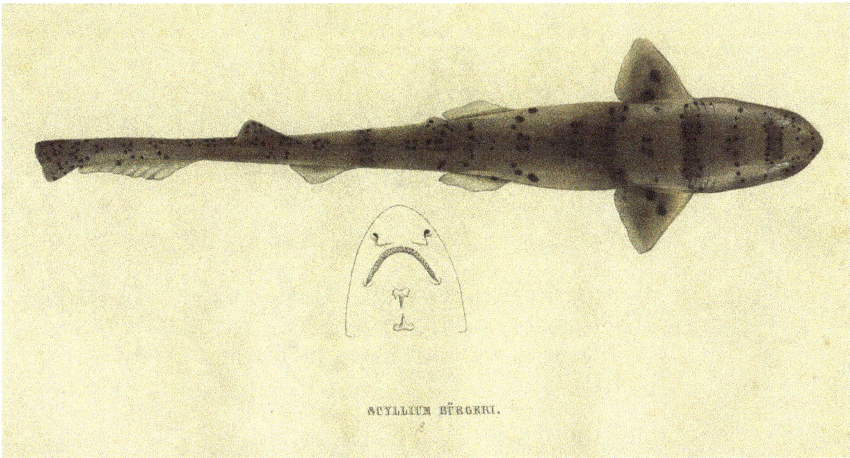

FIGURE 23.9 Müller Johannes – Henle Jacob, *Systematische Beschreibung der Plagiostomen* (Berlin: 1838–1840), Plate 2. The illustration of *Halaelurus buergeri* (Müller & Henle 1838), based on Bürger's specimen, Bürger's description and the painting by Kawahara Keiga.

German researchers, Johann Müller and Jacob Henle, who worked on a monograph of the sharks and rays. They were impressed by the specimens from Japan and at their request, Schlegel sent specimens, Keiga's paintings, and Bürger's descriptions on loan to them in Berlin. In Müller and Henle's *Plagiostomen,* published from 1839–1841, eleven species from Japan were included of which nine were new to science. All have Bürger's name as the author. Moreover, Müller and Henle honoured Bürger by naming one of the new shark species after him [Fig. 23.9].

Dr. Pieter Bleeker (1819–1878) was trained as a physician in Haarlem. During this study, he regularly visited Teylers museum where he developed a great interest in zoology. As money was lacking to obtain a degree at Leiden University, he enlisted as a health officer in the Army in the Dutch East Indies.

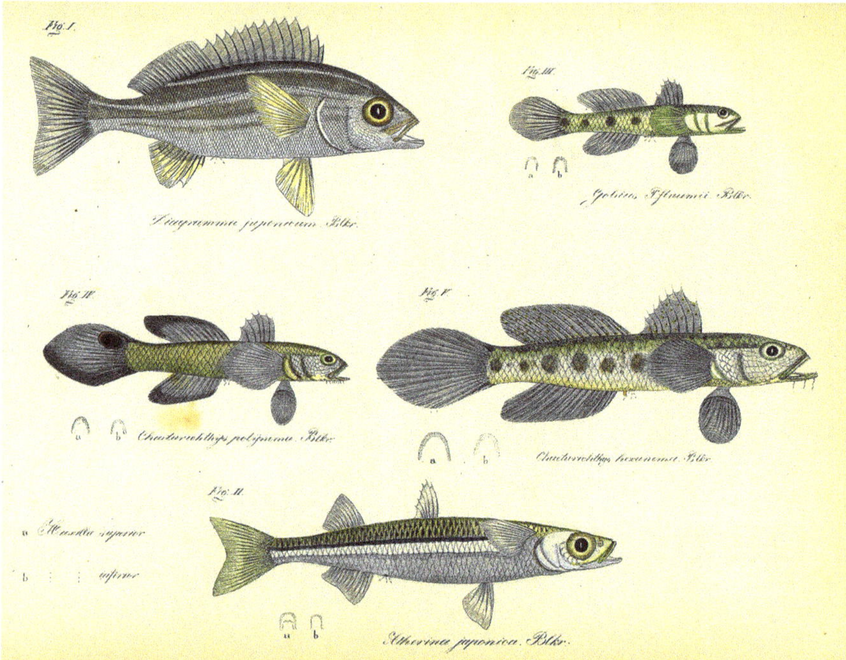

FIGURE 23.10 Bleeker Pieter, "Nalezingen op de ichthyologie van Japan", *Verhandelingen Bataviaasch Genootschap* 25 (1853) 1–56. Hand coloured lithograph with details of the dentition. Original drawings by J. Courtin.

From 1840 to 1861 he was mainly stationed in Batavia. After discovering that the fish market in Batavia harboured many specimens he could not identify, he became interested, made collections, obtained the necessary literature from Europe, and became a self-made ichthyologist. As Bleeker had no access to preserved museum specimens for comparison, he had to rely on species descriptions and figures of other ichthyologists. This made him not only very critical of the efforts of his colleagues but also of his own descriptions and the products of his artists.

Already in 1845, he writes about his plans to publish an Atlas in which all species from the Indian archipelago would be depicted. A dream he could only (partly) fulfil after he had returned to the Netherlands. By the end of his life, he had published more than 500 papers on fishes.

Bleeker did not restrict himself to the fishes of the Indian Archipelago. He also published 17 papers dealing with Japanese fishes (from 1851 to 1879, the last one posthumously).[29] Collections of fishes from Nagasaki, Edo (Tokyo), and

29 The titles of these papers are listed in the Bibliography of the present article.

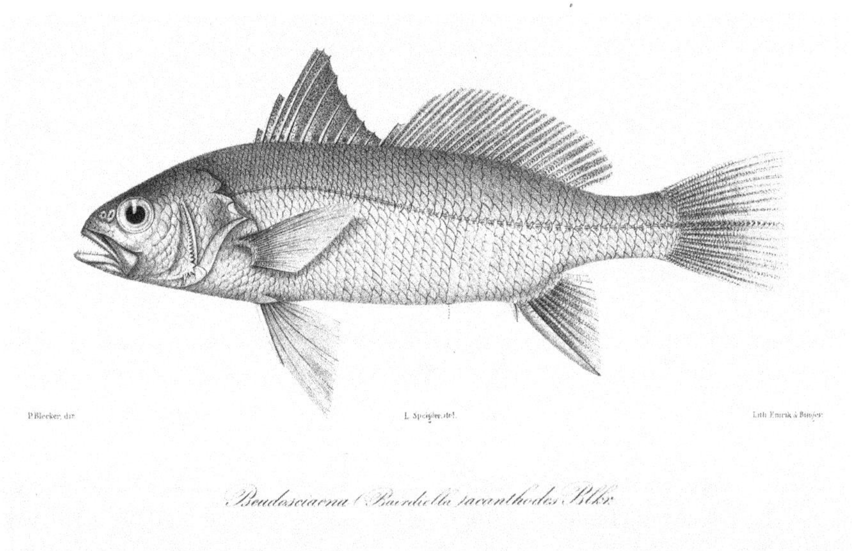

FIGURE 23.11 Detailed lithograph of *Pseudosciaena* (*Bairdella*) *acanthodes* Bleeker 1879 in Bleeker Pieter, "Énumeration des espèces de poissons actuellement connues du Japon, et description de trois espèces inédites", *Verslagen en Mededeelingen der Koninklijke Akademie van Wetenschappen Amsterdam* 18 (1879) 1–33. Original drawings by L. Speigler. This is a new species based on a specimen from the Hamburg museum. Note the shading behind the various elements of the head and the pectoral fin.

the island of Kaminoseki were sent to him by health officers and other Dutch officials he had met in Batavia. These collections were made both before and after the opening of Japan in 1854. Bleeker discovered about a hundred species that were new for the fauna of Japan and some 45 that were new for science. His descriptions are increasingly detailed, often including internal characters. Remarkably, in some papers, Bleeker writes that he has copies of figures sent from Japan by Bürger, and in one case he states he himself has the original painting, and that the copy must be in Leiden.

In contrast to the scientists discussed earlier, Bleeker presented his species in a classification system. He compared the Japanese species with related/similar species and made keys for identification. He did more than just identification, making lists of catch localities, and comparing the faunas of different areas.

The figures illustrating his papers range from very basic etchings and hand-coloured lithographs to beautiful colour lithographs and were made by his three illustrators J. Courtin, [Fig. 23.10], Ch. Engel, and L. Speigler [Fig. 23.11]. Having good illustrations was very important to Bleeker; he returned to the

Netherlands because facilities for colour lithography were lacking in Batavia. It is important to note that Bleeker based his colour descriptions on specimens that had been preserved in a solution of 30 percent ethanol for a period of several months. The illustrations of his artists were based on the same specimens and Bleeker's interpretations of the preserved colours.

4 Early Japanese Books on Fishes

The Japanese appreciation for fish as a beautiful part of nature and as the main source of food has led to fish being often depicted in books. Several examples of these books (with hand-painted or plain woodblock prints) were brought from Japan by residents of Deshima, but none of them were used for systematic works on fishes. About these books Boeseman states: 'These papers, however, consisted chiefly of numerous plates in water colours, generally very interesting and artistically perfect, but unimportant from a scientific point of view, as they are lacking in the accuracy of details'.[30] As an example, Boeseman published some plates from an anonymous Japanese work on fishes (Artiste inconnu, 1835–1840),[31] which appeared in about the same time as the *Pisces* volume of the *Fauna Japonica*. In their *Pisces* volume Temminck and Schlegel refer neither to this work nor to the beautifully illustrated Japanese books on aquatic animals brought by von Siebold.

Bleeker in two of his publications,[32] refers to Kurimoto (1838), a book with 78 figures of freshwater and marine fishes, two figures of salamanders and a rather large accompanying text.[33] Bleeker had this book translated by J.J. Hoffmann the first professor in the Japanese language in the world. The provenance of Bleeker's copy of the book is unknown, as is the current location of the book and Hoffman's translation. Bleeker was able to identify a number of species and used the catch localities.

30 Boeseman M., "Revision of the Fishes Collected by Burger and Von Siebold in Japan", *Zoologische Mededelingen* 28.1 (1947) 1–242, here 4.

31 Anonymous, *Dessins de poissons japonais. MS*, about 1835–1840 134. Boeseman, "Revision of the Fishes", pl. III–V.

32 Bleeker Pieter, "Neuvième notice sur la faune ichthyologique du Japon", *Verslagen en Mededeelingen der Koninklijke Akademie van Wetenschappen 2de Reeks* 3 (1869) 237–252; idem, "Énumeration des espèces de poissons actuellement connues du Japon, et description de trois espèces inédites", *Verslagen en Mededeelingen der Koninklijke Akademie van Wetenschappen Amsterdam* 18 (1879) 1–33.

33 Kurimoto Zuisen'in – Obuchi Tsunenori, *Kōwa gyofu* (Edo, Shijudō: 1838). Ōbuchi T., 1838. Kurimoto Zuiken's Album of Japanese Fishes. (Sijudo Publishers, Place unknown) 1–106.

Although the largest collection of Japanese fishes was housed in the Leiden Museum, Dutch researchers played no active role in educating Japanese ichthyologists after the opening of Japan. According to Abe[34] the distinction of being the first Japanese to name a native fish using Linnean nomenclature probably belongs to Kakichi Mitsukuri (1857–1909), the third Japanese professor of Zoology in Japan, who published his new species *Harriota pacifica* (now *Rhinochimarea pacifica*) in 1895.[35]

5 Synoptic Conclusion

Because of their importance as a major source of protein, much information about fishes accumulated in Japan. However, a classification system had not been developed before the arrival of Europeans. Japanese ichthyology, like ichthyology in Europe, was initiated by medical doctors and gradually developed as a separate field of Zoology. Although primarily interested in botany, physicians stationed at Deshima were the first scientists who collected Japanese fish specimens to be studied in Europe. For these physicians fishes were not a special goal, but merely one part of the till then hardly known Japanese fauna. Kaempfer and Thunberg made private collections. The elaboration and publishing of Thunberg's fishes by Houttuyn was also a private affair. The collections of Bürger and von Siebold and the paintings of Keiga were commissioned by, and thus the property of the Dutch government. Therefore, they were stored in the State Museum in Leiden and investigated and described by the then curator of Vertebrates H. Schlegel. Remarkably, the publication of the Fauna Japonica was largely a private enterprise of von Siebold. The Pisces volume of Siebold's *Fauna Japonica* and the collection that form its basis will always remain the foundation of Japanese ichthyology even though the number of Japanese fish species increased more than ten times. After the Dutch government stopped funding natural history research in Japan in 1842, collecting Japanese fishes again became a private affair. Bleeker, who was never employed as an ichthyologist, encouraged people to collect and send him fishes, which he studied and described in his spare time. Some of his Japanese fishes were

34 Abe T., "A brief history of Japanese ichthyology", in Uyeno T. – Arai R. – Taniuchi T. – Matsuura K. (eds), *Indo-Pacific Fish Biology. Proceedings of the Second International Conference* on *Indo-Pacific Fishes* (Tokyo: 1986) 1–6.

35 Mitsukuri K., "On a new genus of the chimaeroid group *Harriotta*", *Zoological Magazine* (*Tokyo*), 7.80 (1895) 97–98.

sold to the British Museum, but the majority of his collections came into possession of what is now the Naturalis Biodiversity Center.

Acknowledgements

I am indebted to the following Naturalis colleagues: Martine Hermsen for adapting the figures; Menno Hooft, Marianne van der Wal, Henk Caspers, and Karin Geenen for scanning figures from books; Gijs Baldee, Marianne, Karin, Annet Westbroek-Scheffer and Wanne de Bie for library assistance; Frank Loggen for photographs and Esther Dondorp for general support. The assistance of Matthi Forrer and Dick Raatgever is gratefully acknowledged.

Bibliography

Abe T., "A brief history of Japanese ichthyology", in Uyeno T. – Arai R. – Taniuchi T. – Matsuura K. (eds), *Indo-Pacific Fish Biology. Proceedings of the Second International Conference* on *Indo-Pacific Fishes* (Tokyo: 1986) 1–6.

Ahlstrom Alexander Magnus,. *Fauna Japonica continuata* (Ph.D. dissertation, University of Upsala: 1823).

Akihito, "Linné and Taxonomy in Japan – On the 300th Anniversary of his Birth", *Proceedings of the Japan Academy, Series B.*, 86.3 (2010) 143–146.

Anonymous, *Dessins de poissons japonais*. MS, about 1835–1840 134.

Bleeker Pieter, "Over eenige nieuwe geslachten en soorten van Makreelachtige visschen van den Indischen Archipel", *Natuurkundig Tijdschrift voor Nederlandsch Indië* 1 (1851) 341–372.

Bleeker Pieter, "Bijdrage tot de kennis der Muraenoiden en Symbranchoiden van den Indischen Archipel", *Verhandelingen Bataviaasch Genootschap* 25 (1853) 1–76.

Bleeker Pieter, "Nalezingen op de ichthyologie van Japan", *Verhandelingen Bataviaasch Genootschap* 25 (1853) 1–56.

Bleeker Pieter, "Bijdragen tot de kennis der ichthyologische fauna van Japan", *Verhandelingen van de Koninklijke Akademie van Wetenschappen* 1 (1854) 1–16.

Bleeker Pieter, "Faunae Ichthyologicae Japonicae, Species novae", *Natuurkundig Tijdschrift voor Nederlandsch-Indië* 6 (1854) 395–426.

Bleeker Pieter, "Nieuwe nalezingen op de ichthyologie van Japan", *Verhandelingen Bataviaasch Genootschap* 26 (1857) 1–132.

Bleeker Pieter, "Vierde bijdrage tot de kennis der ichthyologische fauna van Japan", *Acta Societatis Scientiarum Indo-Neerlandicae* 3 (1858) 1–46.

Bleeker Pieter, "Vijfde bijdrage tot de kennis der ichthyologische fauna van Japan", *Acta Societatis Scientiarum Indo-Neerlandicae* 5 (1858) 1–12.

Bleeker Pieter, "Vischsoorten van Japan, verzameld te Desima, door Jkhr. J.L.C. Pompe van Meerdervoort", *Natuurkundig Tijdschrift voor Nederlandsch Indië* 20 (1859) 234–236.

Bleeker Pieter, "Zesde bijdrage tot de kennis der vischfauna van Japan", *Acta Societatis Scientiarum Indo-Neerlandicae* 8 (1860) 1–104.

Bleeker Pieter, "Sur une nouvelle espèce de poisson du Japon appartenant à un nouveau Genre", *Nederlandsch Tijdschrift voor de Dierkunde* 1 (1863) 382–383.

Bleeker Pieter, "Description de quelques espèces de poissons du Japon, du Cap de Bonne Espérance et de Suriname, conservées au Musée de Leyde", *Nederlandsch Tijdschrift voor de Dierkunde* 2 (1865) 250–269.

Bleeker Pieter, "Neuvième notice sur la faune ichthyologique du Japon", *Verslagen en Mededeelingen der Koninklijke Akademie van Wetenschappen 2de Reeks* 3 (1869) 237–252.

Bleeker Pieter, "Sur le genre Parapristapoma et sur l'identité spécifique des Perca trilineata Thunb., Pristopoma japonicum CV. et Diagramma japonicum Blkr", *Archives Néerlandaises des Sciences Exactes et Naturelles* 8 (1872) 15–24.

Bleeker Pieter, "Description et figure du Lethrinus Guntheri Blkr", *Archives Néerlandaises des Sciences Exactes et Naturelles* 8 (1872) 153–154.

Bleeker Pieter, "Révision des espèces de Dentex, Synagris, Gymnocranius, Gnathodentex et Pentapus de l'Inde Achipélagique et du Japon", *Verhandelingen van de Koninklijke Akademie van Werenschappen Amsterdam* 8 (1873) 1–64.

Bleeker Pieter, "Systema Percarum revisum", *Archives Néerlandaises des Sciences Exactes et Naturelles* 9 (1876) 247–340.

Bleeker Pieter, "Révision des espèces insulindiennes de la sousfamille des Eleotriformes", *Verslagen en Mededeelingen der Koninklijke Akademie van Wetenschappen Amsterdam* 2, 11 (1877) 13–110.

Bleeker P. "Levensbericht van Pieter Bleeker door hemzelven", *Jaarboek van de Koninklijke Akademie van Wetenschappen gevestigd te Amsterdam* (1877) 1–55.

Bleeker Pieter, "Énumeration des espèces de poissons actuellement connues du Japon, et description de trois espèces inédites", *Verslagen en Mededeelingen der Koninklijke Akademie van Wetenschappen Amsterdam* 18 (1879) 1–33.

Boeseman M., "Revision of the Fishes Collected by Burger and Von Siebold in Japan", *Zoologische Mededelingen* 28.1 (1947) 1–242.

Boeseman M., "Martinus Houttuyn (1720–1798) and his Japanese fishes", *UO* 43 (1995) 1–9.

Boeseman M. – Ligny W. de, "Martinus Houttuyn (1720–1798) and His Contributions to the Natural Sciences, with emphasis on zoology", *Zoologische Verhandelingen Leiden* 349 (2004) 1–222, figs. 1–15.

Bürger Heinrich, MS, without title, contains 200 descriptions of Japanese fishes (1830–1831). Collection Naturalis Biodiversity Center.

Caron François, *Beschrijvinghe Van het Machtigh Coninckrijcke Japan, vervattende Den aert en eygenschappen van 't Landt, manieren der Volckeren, als mede hare grouwelijcke wreedtheyde teghen de Roomsche Christenen* (Amsterdam, Ioost Hartgers: 1648).

Caron François, *Beschrijvinghe van het Machtigh Koninkrijcke Japan, vervattende den aerdt en eyghenschappen van 't Landt, manieren der Volckeren, als mede hare grouwelijcke wreedtheyde teghen de Roomsche Christenen*, (Amsterdam, Ioost Hartgers: 1649)

Caron François, *Rechte beschryvinge van het Machtigh Koninghrijck van Iappan: bestaende in verscheyde Vragen, betreffende des selfs Regiering, Coophandel, maniere van Leven, strenge Justitie &c. voorgestelt door den Heer Philips Lucas, Directeur Generaal wegens de Nederlandsche Staat in India, ende de Heer Francois Caron,*[...] ('s Gravenhage, Iohannes Tongerloo: 1661).

Cuvier Georges – Valenciennes Achille, *Histoire naturelle des poissons*, vol. 4 (Paris: 1829).

Houttuyn Martinus, "Beschryving van eenige Japanse visschen en andere Zeeschepselen", *Verhandelingen der Hollandsche Maatschappij van Wetenschappen Haarlem* 20 (1782) 311–350.

Houttuyn Martinus, *Natuurlyke historie of uitvoerige beschryving der dieren, planten en mineraalen, volgens het Samenstel van den Heer Linnaeus. Met naauwkeurige Afbeeldingen.* 37 vols. (Amsterdam, Frans Houttuyn: 1761–1785).

Holthuis L.B. – Sakai T., *Ph. F. von Siebold and Fauna Japonica. A History of Early Japanese Zoology*. (Tokyo: 1970) 1–8, 1–323.

Kaempfer Engelbert, *The History of Japan, giving an Account of the ancient and present State and Government of that Empire; of Its Temples, Palaces, Castles and other Buildings; of Its Metals, Minerals, Trees, Plants, Animals, Birds and Fishes; of The Chronology and Succession of the Emperors, Ecclesiastical and Secular; of The Original Descent, Religions, Customs, and Manufactures of the Natives, and of their Trade and Commerce with the Dutch and Chinese. Together with a Description of the Kingdom of Siam* (London: J. MacLehose and sons: 1727).

Kaempfer Engelbert, *De beschryving van Japan, behelsende een verhaal van den ouden en tegenwoordigen staat en regeering van dat Ryk; van deszelfs tempels, paleysen, kasteelen en andere gebouwen: van deszelfs metalen, mineralen, boomen, planten, dieren, vogelen en visschen: van de tydrekening, en opvolging van de geestelyke en wereldlyke keyzers: van de oorsprondelyke afstamming, godsdiensten, gewoonten en handwerkselen der inboorllingen, en van hunnen koophandel met de Nederlanders en de Chineesen: Benevens eene beschryving van het koningryk Siam* ('s Gravenhage – Amsterdam, P. Gosse and J. Neaulme, Balthasar Lakeman: 1729).

Krusenstern Adam Johan von, *Reise um die Welt, in den Jahren 1803, 1804, 1805 und 1806, auf Befehl Seiner Kaiserlichen Majestat Alexanders des Ersten auf den Schiffen Nadesha und Newa.* (*Atlas*) (Berlin, Haude und Spener: 1811).

Langsdorf G.H. von, *Bemerkungen auf einer Reise um die Welt in den Jahren 1803–1807* (Frankfurt am Main: 1812).

Linnaeus Carolus, *Systema naturae per regna tria natura, secundum classes, ordines, genera, species, cum characteribus, differentibus, synonymis, locis. Tomus I. Editio Decima, Reformata* (Stockholm, Lars Salvius: 1758).

MacLean J., "Natural Science in Japan. I. Before 1830", *Annals of Science* 30. 3 (1973) 257–298.

MacLean J., "The Introduction of Books and Scientific Instruments in Japan", *Japanese Studies in the History of Science* 13 (1974) 9–86.

MacLean J., "Natural science in Japan from 1828 to 1849", *Janus* 62 (1975) 51–78.

Mitsukuri K., "On a new genus of the chimaeroid group *Harriotta*", *Zoological Magazine (Tokyo)* 7. 80 (1895) 97–98.

Mizuno M. et al., *Iinuma Yokusai* (Gifu-City: 1984).

Montanus Arnoldus, *Gedenkwaerdige gesantschappen der Oost-Indische maatschappy in't Vereenigde Nederland, aan de kaisaren van Japan: vervatende wonderlyke voorvallen op de togt der Nederlandtsche gesanten: beschryving van de dorpen, sterkten, steden, landtschappen, tempels, gods-diensten, dragten, gebouwen, dieren, gewasschen, bergen, fonteinen, vereeuwde en nieuwe oorlogs-daaden der Japanders: verçiert met een groot getal afbeeldsels in Japan geteekent: getrokken uit de geschriften en reis-aanteekeningen der zelve gesanten* (Amsterdam, Jacob Meurs: 1669).

Müller J. – Henle F.C.J., *Systematische Beschreibung der Plagiostomen* (Berlin: 1838–1840).

Ōbuchi T., *Kurimoto Zuiken's Album of Japanese Fishes*. (Sijudo Publishers, Place unknown) (1838).

Siebold G.T. [sic] de,. *De historiae naturalis in Japonica statu, nec non de augmento emolumentisque in decursu perscrutationum expectandis dissertatio, cui accedunt Spicilegia faunae Japonicae* (Batavia: 1824).

Siebold Philipp Franz Balthasar von, *Kurze Uebersicht des Gegenwärtigen Zustandes meiner wissenschaftligen Untersuchungen auf Japan* MS Japaninstituts Berlin (1828).

Siebold Philipp Franz Balthasar von, *Archiv zur Beschreibung von Japan und dessen Neben- und Schutzländern Jezo mit der südlichen Kurilen, Sachalin, Korea und den Liukiu-Inseln. Herausgegeben von seinen Sohnen: 2. Auflage*. (Würzburg und Leipzig: 1897).

Shimano, Eikoh, "The reception of Lavoisier's Chemistry in Japan", *Isis* 63. 3 (1972) 309–320.

Suzanna J.A., "Levensschets van Coenraad Jacob Temminck", *Jaarboek van de Maatschapij der Nederlandsche letterkunde* (1858) 47–87.

Temminck Coenraad Jacob, *Voorschrift, hoedanig te handelen met voorwerpen van Natuurlijke Historie, ten einde dezelve behoorlijk te verzenden en voor bederf te*

bewaren; ten gebruike van het 's Rijks Museum van Natuurlijke Historie te Leyden (Leiden: 1825).
Temminck Coenraad Jacob – Schlegel Hermann, *Fauna Japonica, Pisces. 1* (Leiden, Arntz: 1842).
Temminck Coenraad Jacob – Schlegel Hermann, *Fauna Japonica, Pisces. 2* (Leiden, Arntz: 1843).
Temminck Coenraad Jacob – Schlegel Hermann, *Fauna Japonica, Pisces. 3* (Leiden, Arntz: 1844).
Temminck Coenraad Jacob – Schlegel Hermann, *Fauna Japonica, Pisces. 4* (Leiden, Arntz: 1845).
Temminck Coenraad Jacob – Schlegel Hermann, *Fauna Japonica, Pisces. 5* (Leiden, Arntz: 1846).
Temminck Coenraad Jacob – Schlegel Hermann, *Fauna Japonica, Pisces. 6* (Leiden, Arntz: 1850).
Thunberg Carl Peter, *Resa uti Europa, Africa, Asia, förrättad åren 1770–1779*, 3 vols. (Uppsala, Joh. Edman: 1788–1794).
Thunberg Carl Peter, "Beskrifning pa tvanne Fiskar ifran Japan", *Vetenskaps Academiens Nya Handlingar* 11 (1790) 106 / *Der Konigl. Schwedischen Akademie Der Wissenschaften Neue Abhandlungen Aus Der Naturlehre, Haushaltungskunst Und Mechanik* 11 (1790) 100–103.
Thunberg Carl Peter, "Tvänne Japanske Fiskar", *Vetenskaps Academiens Nya Handlingar* 13 (1792) 29.
Thunberg Carl Peter, "Atskillige forut okande Fiskar af Abbor-Slagtet", *Vetenskaps Academiens Nya Handlingar* 13 (1792) 141–143.
Thunberg Carl Peter, "Beskrifning pa 2: ne nya Fiskar af Abbor-Slagtet ifran Japan". *Vetenskaps Academiens Nya Handlingar* 14 (1793) 55–56.
Thunberg Carl Peter, "Beskrifning pa nya Fisk-arter utaf Abbor-slagtet ifran Japan", *Vetenskaps Academiens Nya Handlingar* 14 (1793) 198–200.
Thunberg Carl Peter, "Sidsta Fortsattningen af Beskrifningen pa nya Fiskarter utaf Abbor—Slagtet ifran Japan", *Vetenskaps Academiens Nya Handlingar* 14 (1793) 296–208.
Thunberg Carl Peter, *Travels in Europe, Asia and Africa made during the Years 1770 & 1779* (London, F. and C. Rivington: 1795–1796).
Tilesius von Tillenau Wilhelm Gottlieb von, "Description de quelques poissons observés pendant son voyage autour du monde", *Mémoires de la Société impériale des naturalistes de Moscou* 2.20 (1809) 212–249.
Tilesius von Tillenau Wilhelm Gottlieb von, "Abbildungen und Beschreibungen einiger Fische aus Japan und einiger Mollusken aus Brasilien, welche bey Gelegenheit der ersten Russischen Kaiserliche Erdumseglung lebendig beobachtet wurden",

Denkschriften der Königlichen Akademie der Wissenschaften zu München 3 (1811–1812) 71–88.

Tilesius von Tillenau Wilhelm Gottlieb von, *Atlas zur Reise um die Welt, unternommen auf Befehl Seiner Kaiserlichen Majestät Alexander des Ersten auf den Schiffen Nadeshda und Neva. Unter dem Commando des Captains von Krusenstern* (St. Petersburg: 1814).

Wernberg Olaus, *Fauna Japonica* (Ph.D. dissertation, University of Upsala: 1822).

Yamaguchi T., "Kawahara Keiga and natural history of Japan 1. Fish volume of Fauna Japonica", *Calanus* 12 (1997) 1–35.

Yamaguchi T. – Machida Y., "Fish specimens collected in Japan by Ph. F. von Siebold and H. Bürger and now held by the National Natuurhistorisch Museum in Leiden and other two Museums", *Calanus*, Special Number 4 (2003) 87–321.

CHAPTER 24

Packaging Knowledge about Whales in Early Modern Japan

Doreen Mueller

For most people in early modern Japan, whales were remote but also curious creatures. One had to be at the shore to spot a living whale in the sea or to encounter the stranded body of a whale. People who had direct knowledge of whales were mostly whalers in Kii and Hizen Provinces. Seasonal migration patterns took whales close to the shore in these regions in western and southwestern Japan. Since whaling was labour-intensive and time-consuming, it was feasible largely only in these regions.[1] In addition, whalers stayed within viewing distance from the shore since the open sea was a forbidden place due to the policy of the closure of the ports. Starting in the second half of the 18th century, scholars from urban centres like Osaka in western Japan and the capital of Edo in eastern Japan travelled to whaling regions to gather empirical knowledge about whales. Being versed in several disciplines including natural history (*honzōgaku*), medicine, and Western learning (*rangaku*, literally, "Dutch Studies"), they compiled taxonomies, histories, and produced anatomical drawings of whales in illustrated horizontal scrolls, handwritten and printed books.

It might be assumed that the epistemic practices of scholars put equal weight on direct observation and on empirical drawings of whales. In fact, the link between observation and representation was not straightforward. This chapter highlights the epistemic value of copying images of whales and of referencing existing sources in the production of knowledge about whales in early modern Japan. It also argues that scholars, whalers, domain lords, and the wider population took an increasing interest in linking representations of whales to notions of prosperity. Whaling groups and domain lords in Kii and Hizen Provinces commissioned illustrated scrolls documenting the enormous efforts that went into whaling and the multifaceted benefits that local

1 Holm, F.J., *The Gods of the Sea: Whales and Coastal Communities in Northeast Japan, c.1600–1912* (Cambridge, Massachusetts: 2023).

communities derived from these.[2] The scrolls depict whalers as warrior-like figures displaying courage, physical strength, and specialised skills in the whale hunt. The scrolls also include detailed drawings of whales and their body parts. Meanwhile, urban dwellers picked up knowledge about whales through printed illustrated guides that laid out the prosperity of the realm by elaborating on famous local products (*meibutsu*).[3] Being distanced from the ontological realities of whales, they envisioned whales as curious spectacles and as sources of prosperity.

1 Negotiating Visual Accuracy and Visual Interest

Existing research on scholars' engagement with whales has highlighted *Whale Chronicle* (*Geishi*) by Kandoriya Jiemon (also Yamase Harumasa, Nanki Josuiken), first printed in 1760 in Osaka, and *Whale History* (*Geishikō*), authored by Ōtsuki Heisen (also called Ōtsuki Kiyonori) in 1808 and circulating in manuscript form. *Whale Chronicle* was the first scholarly work to focus solely on whales from the perspective of natural history studies (*Honzōgaku*). By profession, Jiemon was a merchant selling medicine and a scholar of natural history. Hailing from Kii Province, he had access to whalers' and local people's observations of whales in that province. By contrast, Ōtsuki Heisen was a samurai-bureaucrat and a scholar of Western learning specialising in Dutch Japanese translation. He hailed from Sendai in northeast Japan but was educated in the capital city of Edo which was a centre of Western learning. Heisen referenced a wide range of sources in *Whale History*, and he also went to Hizen Province to observe whaling and whales. Although both scholars used similar epistemic practices – referencing existing sources and travel to coastal regions to observe whales – their different social, geographical, and scholarly backgrounds influenced how they packaged knowledge about whales.

For Jiemon, travelling to the coast in Kii Province was as much about studying whales as it was about determining how whales might benefit people as sources of medicine, food, oil for lamps, and other products.[4] The field of natural history studies provided a useful framework for this as it registered plants

2 Mori H. – Miyazaki K., "*Nankai Hogei Emaki no Tokuchō: Kishū Chihō no Hogei Emaki to no hikaku kara* (The Characteristics of Illustrated Scrolls of Whaling in Nankai: Based on a Comparison with Illustrated Scrolls of Whaling from the Kii Region)", *Seinan Journal of Cultures* 26.2 (2012) 117–155.

3 For a detailed discussion see Goree R., *Printing Landmarks: Popular Geography and Meisho Zue in Late Tokugawa Japan* (Cambridge, Massachusetts: 2020).

4 Kandoriya Jiemon, *Geishi* (Osaka, Kashimoto Kanbee: 1794) 6.

and animals in terms of their nourishing and medicinal properties. Describing these properties took precedence over depicting plants or animals in detail. Nevertheless, as scholars became more interested in distinguishing different species in the 18th and 19th centuries, they also required greater visual accuracy and visual detail.[5] However, an increasing interest in visual detail did not apply to all things in nature equally. Scholars recorded aesthetic objects like chrysanthemums or goldfish which they kept in their house or garden to satisfy personal interest rather than respond to a need to produce practical knowledge. Unlike goldfish or chrysanthemums, whales were not appreciated for their aesthetic merits, and they were remote sea creatures that were difficult to observe closely. Whales were regarded as resources for manufacturing useful products such as whale oil. Whale oil was initially used for lamps but by the 19th century, its efficacy as a pesticide for rice crops had become more widely well-known.[6]

In this sense, Jiemon's key objective was to register whales as useful resources, noting in the preface to *Whale Chronicle*:

> all things between heaven and earth have a use. Looking at several *materia medica*, I realised that rain, frost, and snow from above as well as dirt and dung from below all have the potential to bring forth life and to cure disease. Consequently, one must record the properties [of plants and animals].[7]

This required some visual accuracy but not excessive detail. The illustrations of different whale species – fourteen in total – in *Whale Chronicle* only highlight their most conspicuous features such as overall shape and markings on the skin. Jiemon was more impressed when he saw and heard about the abundance of different whale products in Kii Province. He interviewed local people about their experiences with using these products. He was less empirical in procuring illustrations of whales. In the preface he noted that he repurposed existing images of whales. He also claimed that he ensured that the images were as accurate as possible by cross-referencing them with his own observations:

> There was a painter who had moved to the southern shore [in Kii Province], spending several years there. During that time, he closely

5 Marcon F., *The Knowledge of Nature and the Nature of Knowledge in Early Modern Japan* (Chicago: 2015) 228–250.

6 Arch J., "Whale Oil Pesticide: Natural History, Animal Resources, and Agriculture in Early Modern Japan", in Phillips D. –Kingsland S.E. (eds.) *New Perspectives on the History of Life Sciences and Agriculture*. (Switzerland: 2015) 93–111.

7 Kandoriya, *Geishi* 5.

> observed [whales] there and recorded their appearance. When he came back, he showed me the images. I looked at the images and checked them based on what I had seen and heard. I used the images that tallied with my own observations. In doing so, I kept an open mind since being narrow-minded leads to the wrong conclusions.[8]

In the preface, Jiemon did not reveal the identity of the painter, probably because the painter had been in the service of the local administration. Professional painters in official service repeatedly copied illustrated scrolls recording items that fishermen caught in the sea. Such scrolls had been produced since at least the 1660s in Kii Province.[9] Although titled *Illustrated Scroll of Whales* (*Kujira emaki*) or *Pictures of Whale* (*Kujira zu*), these scrolls also often feature other sea creatures such as dolphins, sharks, rays, and ocean sunfish. An inscription in two extant copies of a scroll that was first produced in 1723, thus predating *Whale Chronicle* by almost forty years, gives a glimpse into the possible functions of such scrolls: Probably the lord of the domain himself had requested an investigation whereupon 'an illustrated scroll of "fishes" was submitted to the customs office (Nibukuchi) in Kumano in Kii Province'.[10] The scroll thus documented items that were caught in the sea in Kumano, helping the local administration register taxable marine goods.

In leaving the painter unnamed, Jiemon obscured the original administrative purpose of the illustrations. This allowed him to reframe the illustrations as epistemological tools for promoting knowledge about whales within the field of natural history studies. He used the illustrations as anchors for providing extensive textual information on whales. For example, his description of a right whale (*semikujira*) pays attention to the history of its name:

> "*Semi*" comes from the Chinese word "*Beigan* [literally, "dry back"]", meaning that this whale swims close to the water's surface. It is so named because its back is dry from always being exposed. [...] There are two kinds. Their shape and colour are largely the same with few differences. Their colour is black with a white pattern. Their shape is smooth and long.[11]

---

8 Kandoriya, *Geishi* 6.

9 Mori – Miyazaki, "*Nankai Hogei Emaki*" 118.

10 The inscription appears in extant copies of the scroll titled *Kishū Kumanoura Shogeinozu* (*Pictures of Whales at Kumano Bay in Kii Province*) in the collections of Tokyo University Library [URL: https://da.dl.itc.u-tokyo.ac.jp/portal/assets/1de1d067-53a5-4981-a122-daa047367056] and the National Institute of Japanese Literature [URL: http://dbrec.nijl.ac.jp/DHADNac1948026.01_03400].

11 Kandoriya, *Geishi* 16.

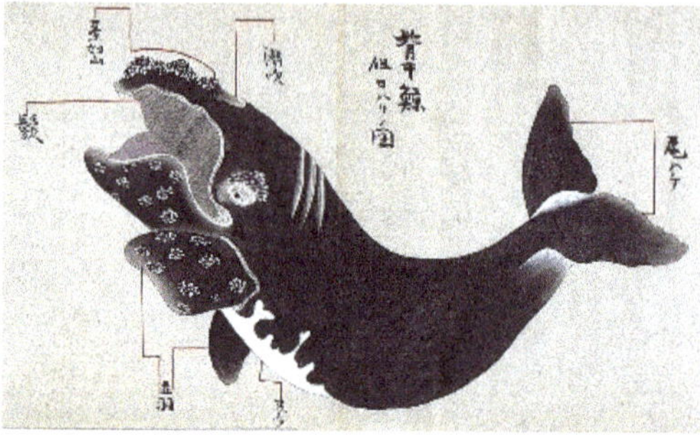

FIGURE 24.1 *Kishū Kumanoura Shogeinozu* (*Pictures of Whales at Kumano Bay in Kii Province*) Illustrated Scroll (detail), ink and colours on paper, 19th Century, Tokyo University Library (URL: https://da.dl.itc.u-tokyo.ac.jp/portal/assets/1de1d067-53a5-4981-a122-daa047367056)

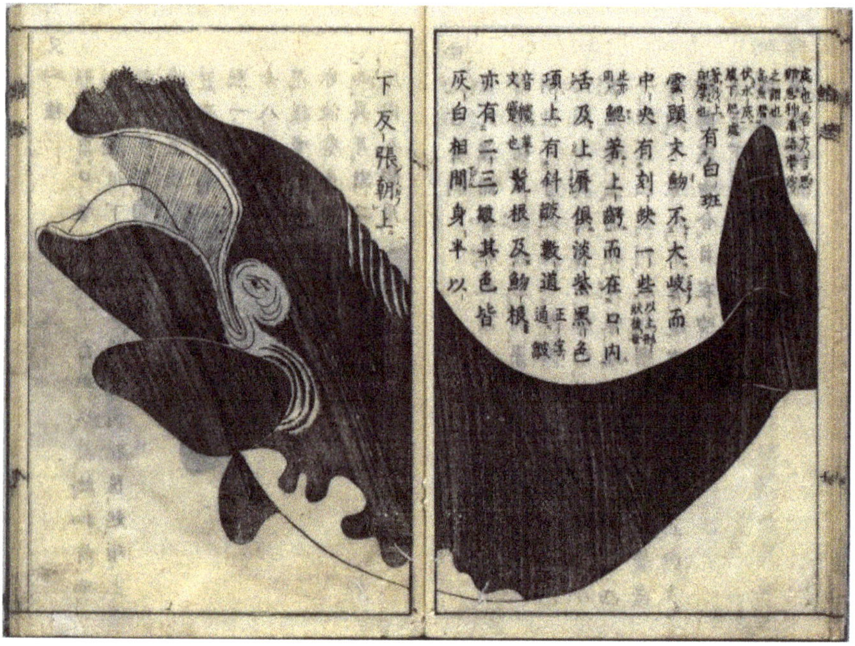

FIGURE 24.2 Double-page illustration of a right whale, Kandoriya Jiemon, *Geishi* (*Whale Chronicle*) (Osaka, Kashimoto Kanbee: 1794). Printed book. National Archives of Japan (URL: https://www.digital.archives.go.jp/file/1222480.html)

This description went beyond the brief captions that can be found in contemporary illustrated scrolls. These usually only point out conspicuous features such as blow holes or fins. Although the exact source of the illustrations in *Whale Chronicle* is not known, it is likely to have been a similar scroll to the two extant versions in Tokyo University Library and the National Institute of Japanese Literature. These feature two illustrations of a right whale resembling those in *Whale Chronicle* – one with its mouth closed and one with its mouth open displaying its baleen bristles [Fig. 24.1]. The illustration in *Whale Chronicle* [Fig. 24.2] is almost identical although the decorative pattern although the

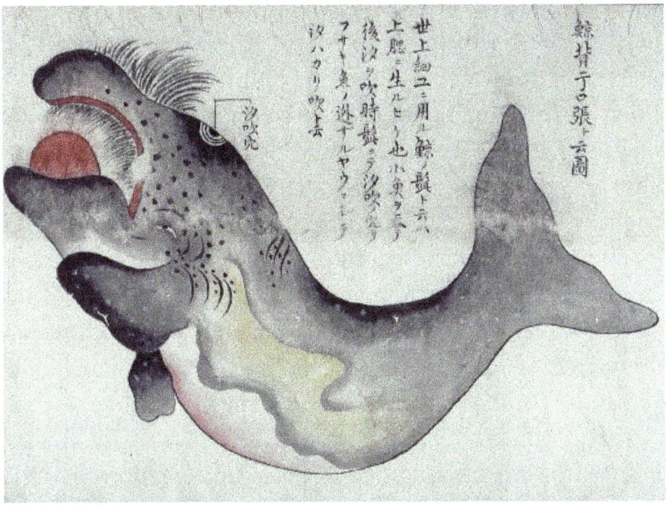

FIGURE 24.3 Sanshi Tokinari, *Kujira Emaki* (*Illustrated Scroll of Whales*), 1778, Illustrated scroll (detail), ink and colours on paper, National Institute of Japanese Literature (URL: http://dbrec.nijl.ac.jp/DHADNac1948026.01_03403)

decorative pattern evoking barnacles on the fins and the head on the fins and the head of the whale has been removed. At the same time, the illustration that shows the whale with its mouth closed in *Whale Chronicle* depicts this decorative pattern on its skin.

How can these embellishments be reconciled with Jiemon's claim to visual accuracy? Although illustrated scrolls of whales are believed to have been made under the guidance of natural history scholars,[12] they were also copied repeatedly by professional painters. In this sense, all extant whale scrolls are copies of existing scrolls.[13] Professional painters copied scrolls to preserve and to circulate them. In the process of copying, the illustrations acquired embellishments that added variation and visual interest. This image of a right whale [Fig. 24.3] closely resembles the illustration in *Whale Chronicle* apart from the fanciful colouring which is imaginary and not based on direct observation. Such embellishments, however, did not detract from the documentary value of these scrolls; they were intrinsic to documentary practices in illustrated scrolls. The level of visual accuracy was adjusted to the purpose of a particular image, and there was room for taking liberties both in illustrated scrolls and in natural history studies like *Whale Chronicle*.

12 Arch J., *Bringing Whales Ashore: Oceans and the Environment of Early Modern Japan* (Seattle: 2018) 120.

13 Mori – Miyazaki, "*Nankai Hogei Emaki*" 153–5.

In contrast to scholars' practice of empirical methods such as direct observation, painting practices were largely not so. Most professional painters were trained in copying existing paintings rather than painting objects in nature.[14] In this sense, the decorative elements of the illustrations in *Whale Chronicle* reference contemporary painting practices. They also make use of narrative techniques that added visual interest in illustrated scrolls. For example, part of the fin of the right whale [Fig. 24.2] appears on the preceding page. It is customary in illustrated scrolls to show glimpses of the action to come as the viewer unrolls a scroll, increasing a sense of anticipation.[15] Since *Whale Chronicle* was issued in print, it addressed a wider audience than scholars of natural history. Although the text of *Whale Chronicle* is written in the scholarly language of Sino-Japanese (*kanbun*), it wraps around the illustrations as in an illustrated novel. The text would not have been accessible to the wider public, but they could still enjoy the illustrations, deriving basic information about the appearance of different kinds of whales from them.

The illustrations in *Whale Chronicle* demonstrate that some attempts were made to make knowledge about remote creatures such as whales more accessible by framing them within familiar registers of reading and viewing. To achieve this aim, visual accuracy had to be balanced with the need to create visual interest through embellishments and narrative techniques. This was also necessary because these media did not function exclusively as epistemic tools. They also served documentary and narrative purposes. Images being so malleable suggests that they were not taken as seriously as text in conveying knowledge about nature. At the same time, Marcon has shown that images of plants and animals became increasingly important carriers of knowledge in early modern Japan.[16] Fukuoka has demonstrated how natural history scholars in 19th-century Nagoya practiced ink rubbing to produce faithful representations of objects in nature.[17]

Nevertheless, visual accuracy did not always require empirical painting methods. Copying and adjusting existing images could produce the same result. If an existing image accorded with a scholar's intentions, there was no

14 Kameda-Madar K., "Copying and Theory in Edo-Period Japan (1615–1868)", *Art History* 37.4 (2014) 708–727.

15 For example, flying sparks gradually turn into a raging firestorm as the viewer unrolls the illustrated scroll of the *Ban Dainagon Ekotoba* (*The Tale of Great Minister Ban*) from right to left. The set of three scrolls is in the collection of Idemitsu Museum in Tokyo. It dates to the late 12th century and is attributed to court painter Tokiwa Mitsunaga.

16 Marcon, *The Knowledge of Nature* 228–50.

17 Fukuoka M, *The Premise of Fidelity: Science, Visuality, and Representing the Real in Nineteenth-Century Japan* (Palo Alto: 2012).

need to produce a new image. This method was particularly prescient when trying to record creatures that were hard to come by and observe at close range. Adjusting the level of visual accuracy was a matter of choice and of practical necessity. Illustrations of whales in the 18th and 19th centuries demonstrate that images of things in nature did not follow a simple teleological trajectory from imaginary representations towards absolute visually accuracy. This was a complex process of negotiating empirical, documentary, and narrative interests.

At the same time, scholars increasingly valued empirical methods of gathering knowledge. The claim that a scholar had directly observed an object in nature, indicated by the frequent use of terms such as *kenmon* and *shinshi* in the prefaces and titles of manuscripts and printed books, did not necessarily mean that the image of the object was based on observation. Both *kenmon* ('seen and heard') and *shinshi* ('seen first-hand') appear in the preface of *Whale Chronicle*. Jiemon also mentions in the preface that he cross-referenced existing sources with his own observations. Contemporary epistemic practices required both referencing existing sources and empirical observation of objects in nature. Although Marcon and Fukuoka have emphasised how the latter shaped knowledge of nature in 18th- and 19th-century Japan, it is vital to acknowledge the continued importance of the former. Judging the epistemic value of an image based on its merits as an empirical drawing or in terms of its visual accuracy alone imposes a one-sided view on how knowledge was produced in early modern Japan. This also required referencing familiar registers for ordering knowledge, which were not exclusively empirical.

2 The Malleable Connotations of Anatomical Drawings of Whales

Understanding images of whales as balancing empirical observation and imaginary elements allows for reconsidering how knowledge about whales was constructed in scholarly works. In her seminal study on knowledge about whales in early modern Japan, Jakobina Arch understands Ōtsuki Heisen's *Whale History* as a less scientific work because his anatomical drawing of the inner organs of a right whale lacks visual accuracy and detail [Fig. 24.4].[18] Later anatomical drawings by Beiga in *Pictures of Six Kinds of Whales* (*Rokugei no zu*, 1847) and in the illustrated scroll titled *Isanatori ekotoba* (*Illustrated Story of Whaling*, dated 1829) which was also printed were more detailed.[19] In a similar

18 Arch, *Bringing Whales Ashore* 114.

19 Ibid. 125.

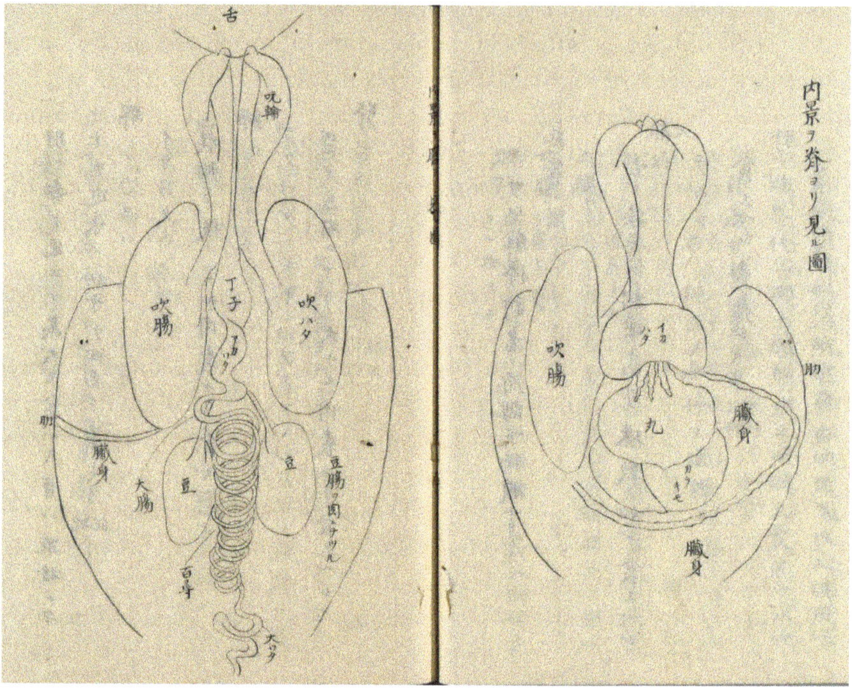

FIGURE 24.4 Double-page illustration, Ōtsuki Heisen, *Geishikō* (*Whale History*) 1808. Manuscript, National Diet Library Tokyo (URL: https://dl.ndl.go.jp/pid/2610476)

vein, Mori and Miyazaki have discussed the deficiencies of Heisen's anatomical drawing, arguing that he did not possess the skills to produce empirical drawings because he was not trained in natural history or medicine.[20] If one understands Heisen's anatomical drawing in *Whale History* as an exercise in referencing traditions of anatomical drawing in early modern Japan, one reaches a different assessment.

In *Whale History* Heisen brought together a wide range of historical, local, and scholarly sources: excerpts from Chinese and Japanese encyclopaedias, Dutch dictionaries, oral and written records from coastal whaling communities, and his own observations of whaling. Being a scholar-bureaucrat, Heisen was sent on two government missions to investigate whaling in Hizen Province

20 Mori H. – Miyazaki K., "Ōtsuki Kiyonori '*Geishikō*' to Sugita Genpaku 'Kaitai Shinsho' no Kankeisei (On the connections between Ōtsuki Kiyonori's "*Geishikō*" and Sugita Genpaku's "Kaitai Shinsho")", *Bulletin of the Kyushu University Museum* 11 (2013) 66.

in southwestern Japan.[21] By authoring *Whale History*, he sought to prove his worth as a scholar of the capital of Edo. His relation Ōtsuki Gentaku was a leading physician and scholar of Western learning who had become interested in whales when he treated a whaler. Gentaku had gained an insight into Western medicine, having revised Sugita Genpaku's translation of the *Anatomical Tables* (1725) by Johannes Kulmus.[22] He was asked to investigate whaling but sent Heisen instead as he was too busy. True to the historical intentions of its author, a large section of *Whale History* is dedicated to taking account of existing whale knowledge.

As a translator, Heisen took a keen interest in etymology. He referenced Nakano Tadao's translation of Pierre Marin's *Groot Nederduitsch and Fransch Woordenboek* (*Big Dutch French Dictionary*, published 1730 and 1752) to determine the meanings of the term *walvis*: 'Nakano Tadao says that *wal* denotes something 'striving ashore from the sea', and it also means 'building a stone wall'. Thus, *wal* is equivalent to the word for *floating reef* (*fu jiao*) in China.'[23] The fact that he did this in the first volume shows his prioritising information that was close to his expertise. In the same volume, he also included illustrations showing the shape, size, and outward features of different whale species together with information from various sources about these whales. In the entry on sperm whales, called *Makkō Kujira* (literally, "*Incense Whale*") as their colour resembles the brownish hue of incense, he quotes *Geiki* (*Record of Whales*). He notes that sperm whales can be found in the southern sea off the coast in Kii Province in southwestern Japan and that they do not migrate north. From the start, the entry emphasises that these whales are used to produce scent in the Kingdom of Ryūkyū (present Okinawa). '*They have a stone inside their stomach that is shaped like a temple bell*.'.[24] He also provides an illustration of ambergris and explains that it was used as medicine in China.

Only the fourth volume delivers new insights. Especially his anatomical drawing of whale viscera in this volume signalled that his interest in whales went beyond the usual focus of registering whales as useful resources within the framework of natural history studies. It conveyed a medical interest that identified *Whale History* as a work in the scholarly orbit of Edo which was a centre of Western medicine. Heisen tried to map out the inner organs of a right whale regardless of whether they were useful as sources of nourishment or

21 Mori – Miyazaki, "Ōtsuki Kiyonori '*Geishikō*' to Sugita Genpaku 'Kaitai Shinsho' no Kankeisei" 54–56.

22 Ogata R., *Rangaku kotohajime* (Tokyo: 1983).

23 Ōtsuki Heisen, *Geishikō* (Whale Chronicle) 1808. 20–1.

24 Ibid. 37.

medicine or not. Such an anatomical drawing had no practical use for whalers or in natural history studies. Previously, the only anatomical drawings that had been attempted by scholars in Edo represented the human body. Mapping the inner organs of whales was not a straightforward matter of faithfully recording things from direct observation. It required imagining the body of a living whale without the distorting effects of the forces of gravity on land. Translating messy ontological realities into coherent anatomical drawings was a challenge that required observational skills and imagination. Considering Heisen's lack of training in such matters, why did he attempt such a drawing?

Mori and Miyazaki have pointed out that Heisen was intrigued by the uncanny similarities between whales and the human body. He noted that whales nurse their young, and that they have eyelids.[25] However, such observations were commonly made in illustrated scrolls and in prints. The first people who recorded their observation that whales had eyelids and that they breathed air through a blow hole were whalers in western and southwestern Japan. Mayumi Itoh has demonstrated how whalers' observations of these similarities prompted them to feel remorse at hunting whales.[26] Whalers noted that whales were different from other "fish" because they had a soul. Female whales demonstrated this by fiercely protecting their calf when hunted. An illustrated manuscript titled *Ogawajima Geigei Gassen* (*Battling Whales around Ogawa Island*) recounts the story of a successful hunt of a female whale and calf in Hizen Province.[27] Although hunting parent and calf made the hunt more successful, it also caused emotional distress to whalers. In response to this, the manuscript has an addendum that explains how whalers ensured the welfare of the souls of the deceased whales.[28] At the point of death, whalers would recite the *nenbutsu*, a chant to help send the soul of the deceased whale to the Western Paradise of Amida Buddha.

Although whale bones were processed into useful products such as whale oil, some bones were held back and buried in graves where Buddhist memorial services were held for whales. The manuscript depicts a memorial service by the waterside conducted by a Buddhist priest in the presence of whalers.[29]

---

25 Mori H. – Miyazaki K., "Ōtsuki Kiyonori '*Geishikō*' to Sugita Genpaku 'Kaitai Shinsho' no Kankeisei" 67.

26 Itoh M., *The Japanese Culture of Mourning Whales: Whale Graves and Memorial Monuments in Japan*. (Singapore – Palgrave Macmillan: 2018). 179.

27 Hōshūtei Riyū, *Ogawajima Geigei Gassen* (*Battling Whales in Ogawa Island*), 1840, Kyushu University Library (URL: http://hdl.handle.net/2324/1929732).

28 Ibid. 47–57.

29 Ibid. 52. The negotiation of the religious and the commercial implications of whaling in this manuscript is discussed in detail in Ambros B.R., *Bones of Contention: Animals and Religion in Contemporary Japan* (Honolulu: 2012) 59–63.

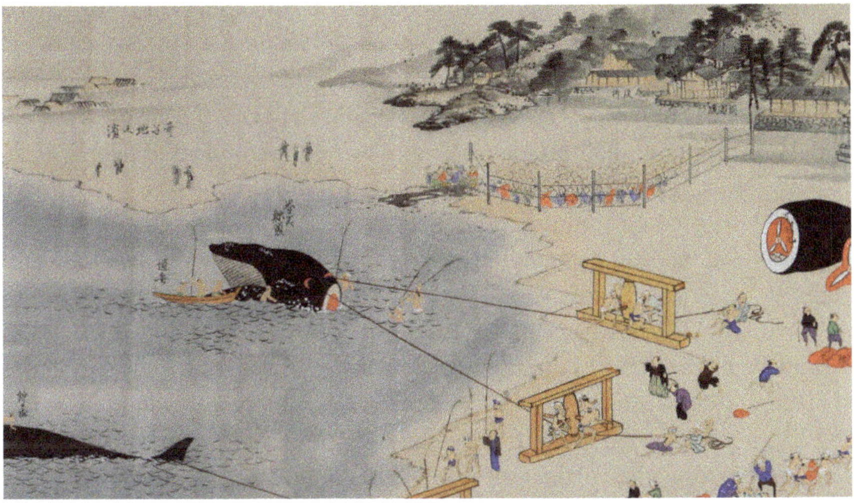

FIGURE 24.5 *Kishū Taijiura Kujira Tairyō no zu* (*Pictures of Whaling and Whales in Taiji in Kii Province*) 1861, Illustrated scroll (detail), ink and colours on paper, 30 × 1309 cm, Collection of Taiji-Chōritsu Kujira no Hakubutsukan (URL: https://kujira-digital-museum.com/en/categories/13/articles/24)

Depictions of memorial services are rare as they were usually only described in text. As this inscription in an illustrated scroll depicting whaling in Taiji in Hizen Province states:

> When the whale dies, [whalers] in whaling boats recite the six syllables [of the *nenbutsu* chant]. Whaling boats work together to pull the whale ashore with pulley-ropes. By the shore, the whale is cut up with knifes resembling naginata poles. As the water turns crimson red, a memorial service is conducted for the whale at a temple. At the shore, there are big celebrations that will surely spread to the seven shores.[30]

The accompanying illustration [Fig. 24.5] shows the body of a right whale being cut up in the water by the shore and its parts being quickly transported to nearby workstations for further processing. Guards are shown hitting local children who have come to help themselves to some of the spoils. Heisen

30 *Kishū Taijiura Kujira Tairyō no zu* (*Pictures of Whaling and Whales in Taiji in Kii Province*) 1861, Illustrated scroll, ink and colours on paper, 30 × 1309 cm, Taiji Whaling Museum, URL: https://kujira-digital-museum.com/en/categories/13/articles/24. Preface. The notion of the prosperity of the seven shores is discussed in Frank S.M., "Kuniyoshi and the Prosperity of the Seven Shores: A Garland of Japanese Woodblock Prints of Whales and Whaling, with a Short History of Whaling in Japan", *Deutsches Schiffahrtsarchiv* 25 (2002) 145–65.

copied a scroll depicting the hunt of a female whale and her calf in Ikitsuki Island in Hizen Province. He mentioned the *nenbutsu* chant in passing but he was more concerned with the uncanny resemblances between whales and the human body. Through his connection to Ōtsuki Gentaku, Heisen was aware of the Western and the Sino-Japanese traditions of anatomical drawings of the human body. It is thus understandable that his drawing of whale viscera referenced the medical tradition of depicting the human body in Edo. Including this drawing was also a statement of his identity as a man in the scholarly orbit of Edo.

Mori and Miyazaki have argued that Heisen intended his drawing of whale viscera to be as detailed as anatomical drawings of the human body and that he failed to do so because of his lack of scientific acumen. At this point, it is necessary to look beyond Heisen's shortcomings as a scholar and consider other factors that might have accounted for the lack of detail of his anatomical drawing. In early modern Japan, the human body posed less of a knowledge frontier than whale bodies. When Sugita Genpaku translated the *Anatomical Tables* by Kulmus in 1774, he grappled with the task of identifying what Latin names denoted before translating them into Sino-Japanese (*kanbun*).[31] Genpaku was able to verify the accuracy of Western anatomical drawings by observing a dissection of an actual body as he recounted in *Rangaku Kotohajime* (*The Dawn of Western Learning*) in 1814.[32] Producing knowledge about the human body required revising existing medical knowledge from the Sino-Japanese tradition. Knowing the whale body, by contrast, was a matter of closing serious knowledge gaps. In addition, as Figure 24.5 illustrates, whale bodies started being cut up in the sea. It was a challenge for anyone – scholar or not – to envision the internal structure of a whale based on these conditions.

It is possible that the anatomical drawing in *Whale History* carried symbolic connotations in conveying Heisen's aspiration to picture the inner workings of the whale body outside existing frames for envisioning whale bodies in the field of natural history and in illustrated scrolls of whaling. Considering the possibility that Heisen's drawing symbolised the expanding possibilities of knowing whales makes it necessary to compare it to representations of the human body. Heisen's anatomical drawing pictures the whale vertically with the head at the top (see Fig. 24.4). It appears that he took his cue from human anatomy as he depicted a windpipe, a pair of lungs, a pair of kidneys, and the digestive tract. These are labelled and rendered in a sketchy manner, giving no

31 Ogata R., *Rangaku kotohajime* (Tokyo: 1983).

32 Ibid.

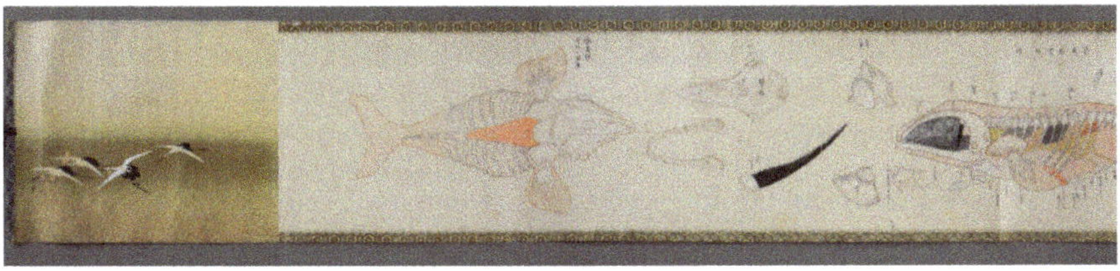

FIGURE 24.6 *Taiji Ura Kujira Ezu* (*Pictures of Whales in Taiji*) 19th Century, illustrated scroll (detail), Collection of Taiji-Chōritsu Kujira no Hakubutsukan (URL: https://kujira-digital-museum.com/ja/categories/13/articles/26)

clues as to their actual appearance.[33] There is also no frame around the organs which is suggestive of the body of the whale. The top-down orientation contrasts with the horizontal profile views of whales in illustrated scrolls and in *Whale Chronicle*. It evokes anatomical drawings of the human body in Kulmus' *Anatomical Tables*.[34] Kulmus' drawing positions the inner organs inside a naturalistically drawn human body, suggesting that the image was based on the dissection of an actual body.

Heisen's anatomical drawing makes no such suggestion, indicating that it was not meant to be a naturalistic representation. Rather, it appears to be a hypothetical exercise in considering the inner workings of the whale body as a likeness of human anatomy. Its sketchiness evoked the Sino-Japanese tradition of depicting the human body in a cartographic mode that projected the idea of the human body as a living system composed of complementary flows of energy.[35] A symbolic reading of anatomical drawings of whales is also supported by extant illustrated scrolls. An illustrated scroll in Taiji Whaling Museum opens with two anatomical drawings of a right whale [Fig. 24.6]. Unusually, the scroll reads from left to right. Its frontispiece depicts a flock of cranes against a gold plane, which suggests that it was produced to commemorate a special occasion since cranes were regarded as symbols of longevity.[36] Frontispieces conditioned how viewers interpreted the content of a scroll. The framing of anatomical drawings of whales in Heisen's *Whale History* and in

33 Arch, *Bringing Whales Ashore* 127.

34 Screech T., *The Western Scientific Gaze and Popular Imagery in Later Edo Japan: The Lens Within the Heart*. (Cambridge: 1996) 88.

35 Kuriyama S., "The Imagination of the Body and the History of Embodied Experience: The Case of Chinese Views of the Viscera", in ibid. (ed.), *The Imagination of the Body and the History of Embodied Experience* (Kyoto: 2001) 18, 26.

36 Screech T., *Obtaining Images: Art, Production and Display in Edo Japan* (London: 2012) 33–51.

illustrated scrolls suggests that their meanings were malleable and not limited to producing empirical knowledge.

3 Whales in the Urban Scopophilic Gaze

As knowledge about whales circulated more widely in the late 18th and 19th centuries, it acquired diverse connotations. In Edo, far from centres of whaling, people took a scopophilic interest in whales. Starting around the 1770s, full-colour prints came to be celebrated as famous products of Edo, so-called "brocade pictures" (*nishiki-e*) that rivalled the brilliant colours of the silks worn by the court and wealthy merchants in the old capital of Kyoto, then called Miyako. It was a fortunate coincidence that the inception of Edo's brocade pictures coincided with several memorable events. In 1798, a whale, probably a right whale (*semikujira*) or fin whale (*nagasukujira*), was caught at Edo Bay near Shinagawa. The sight of the whale was so extraordinary and so memorable that it came to be hailed as one of Edo's "Three Curious Animals" (*Daisan Chinjū*). It shared this honour with an Asian elephant that had been presented to Shogun Tokugawa Yoshimune in Edo in 1729, and a pair of camels taken by the Dutch East India Company for an audience with the Shogun in Edo in 1821. The whale was taken to the shogun's coastal residence in Edo, Hama Detached Palace, to be viewed by Shogun Tokguawa Ienari.[37]

In the popular imagination, these curious creatures acquired auspicious connotations – the Japanese name for camel (*rakuda*) rhymed with the expression "*raku da*", literally, "This is fun". Whales, on the other hand, were viewed as manifestations of the sea god Ebisu who brought riches.[38] In the visual cultures of East Asia, auspicious motifs were often depicted as complementary pairs of male and female, connoting continued prosperity. The most widely known is probably a pair of mandarin ducks symbolising conjugal harmony. As there were no visual sources for pairing whales in this way, print designers instead paired a female whale and calf.[39] A print design by Katsukawa Shuntei depicts the pair close by the shore in Shinagawa.[40] Shinagawa was known for its vibrant pleasure quarters, and the print compares the curious sight of the

37 Itoh, *The Japanese Culture of Mourning Whales* 64.

38 Holm, *The Gods of the Sea* 2.

39 The female connotations of whales are discussed in Arch J., "Heroic Whalers Hunting Whale-Mothers: Gender in the Early-Modern Japanese Whaling Industry", *Coriolis* 10.1 (2020) 48–69.

40 Katsukawa Shuntei, "Whale in the Bay at Shinagawa" 1798, woodblock print, 38.5 × 78.7 cm, Museum of Fine Arts Boston, URL: https://collections.mfa.org/objects/500404.

large whales with fashionable merrymakers in pleasure boats. The whales' uncannily large eyes ogle the goings on at the shore. A person in a pleasure boat returns the gaze by looking back through a telescope. The addition of the telescope allowed the print design to collapse the distance between the whales and the pleasure boats, showing the whales as large as possible. The telescope symbolised the scopophilic interest of city-dwellers in whales.

Existing research has focused on the indexical value of images for producing knowledge about whales, but these images also possessed symbolic meanings, feeding into notions of nature as a source of prosperity. In the 19th century, representations of whales became slightly more detailed, but this paled against the growing interest in showing the links between whaling and a region's prosperity in illustrated scrolls and in printed media. A set of scrolls commissioned by the lord of Hirado Domain illustrating the famous products of Hirado Domain in Hizen Province in southwestern Japan starts demonstrably with whaling to emphasise it as a source of the region's prosperity.[41] The scrolls were illustrated by Kizaki Morisue, a scholar of natural history. He depicted whales in detail but was less careful when illustrating other scenes such as net fishing of sardines. Mori and Miyazaki have shown that he simply copied some illustrations from existing printed illustrated guides such as Hirase Tessai's *Nihon Sankai Meibutsu zue* (*Illustrated Guide to the Famous Products of Land and Sea of Japan*).[42]

The urban population, too, combined an interest in objects in nature with a growing curiosity about methods of processing them into useful products. Since the second half of the eighteenth century, illustrated guides to local products and local places flourished. The popularity of these guides was driven by increased travel by commoners. Although the open sea was forbidden, travel over land was supported by a well-developed road network along the coast which connected major cities such as Edo, Osaka, and Kyoto. Real travel was also paralleled by imaginary travel as readers ravaged printed illustrated guides to local places and local products. As Goree has pointed out, these were intended for armchair travellers rather than being practical travel guides, but they nevertheless inspired their readers to imagine themselves sampling local products and experiencing unfamiliar local paces.[43]

A major selling point of these illustrated guides was their emphasis on authenticity based on directly experiencing local places. Authors claimed to

41 Kizaki Morisue (also Kizaki Yūken), *Hizenshū Sanbutsu zukō* (*Pictures and Explanations of Products of Hizen Province*), 1784 (URL: http://dbrec.nijl.ac.jp/DHADNac1948026.01_03713).

42 Mori – Miyazaki, *Nankai Hogei Emaki no Tokuchō* 129.

43 Goree, *Printing Landmarks* 10.

have seen local places at first-hand, which was expressed in the saying *kenmon*, literally, "seen and heard" with their own eyes and ears. Against this background, the Osaka scholar and merchant Hirase Tessai compiled a comprehensive five-volume *Illustrated Guide to the Famous Products of Land and Sea of Japan* (*Nihon Sankai Meibutsu zue*), first printed in Osaka in 1754. With this work, Tessai pioneered the genre of illustrated guides to local famous products. His illustrated guide was designed to appeal to a wider audience than just a select community of scholars who were steeped in natural history. The book has double-page illustrations describing the production of important regional products, accompanied by brief glosses providing contextual information on pictures. Both the text and the illustrations emphasise the embeddedness of local products in local practices and local environments. In emphasising the benefits that manufacturing products brought to local people, Tessai's illustrated guide humanises production processes, and it adds narrative interest. The fifth volume of the book dedicates five double-page illustrations to whaling in Kumano in Kii Province.

Like illustrated accounts of whaling, it visually narrates the trajectory of whaling from spotting the whale in the sea to bringing it ashore. In the table of contents, Tessai claims to have closely observed whales and whaling practices in Kumano and to have tasked painter Hasegawa Mitsunobu with their faithful representation:

> Whaling is a serious undertaking. Only people living near the coast know it thoroughly. For some time, I observed whaling in Kumano and enquired in detail about the processes of whaling. I have recorded (my findings) here, and Mr Hasegawa provided the images.[44]

In the gloss, Tessai informs the reader that whalers in Kumano knew five notable kinds of whales: humpback whales (*zatōkujira*), gray whales (*kokujira*), sperm whales (*makkōkujira*), right whales (*semikujira*), and fin whales (*nagasukujira*). Fin whales were the largest, and whalers in Kumano did not catch them.[45] Tessai mentions that local people took great pride in a successful hunt.[46] The illustrations make the reader understand why as they demonstrate the strenuous nature of hunting whales.

Although Mitsunobu's images were more narrative than descriptive, they drew readers' attention to the most conspicuous features of whale bodies.

44 Tessai Hirase – Hasegawa Mitsunobu, *Nihon Sankai Meibutsu* zue (Illustrated Guide to the Famous Products of Land and Sea in Japan) (Osaka, Shioya Uhei: 1797). Vol. 1, p. 4.

45 Ibid. Vol. 5, p. 13.

46 Ibid. Vol. 5, p. 17.

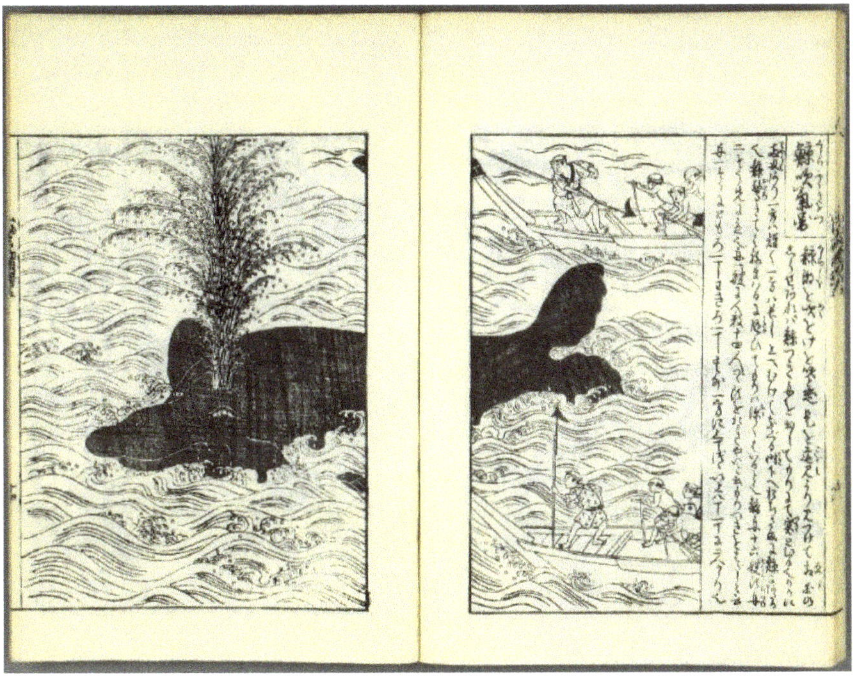

FIGURE 24.7 Double-page illustration, Tessai Hirase – Hasegawa Mitsunobu, *Nihon Sankai Meibutsu zue* (*Illustrated Guide to the Famous Products of Land and Sea in Japan*) (Osaka, Shioya Uhei: 1797). Printed book, National Diet Library Tokyo (URL: https://dl.ndl.go.jp/pid/2555436)

Plumes of water emerging from the whale's breathing hole might be spotted from the shore, but they were more impressive when viewed at close range [Fig. 24.7]. By the time a whale had been brought close to the shore, it had died, which meant that only whalers out at sea or people in possession of a telescope were able to closely observe the spectacle of its breathing. In the gloss, Tessai prepares the reader for this spectacle as he notes that lookouts on hilltops (*yamami*) looked for whales by spotting plumes of water out at sea. The illustration shows a telescope being used for this [Fig. 24.8]. Since telescopes were expensive, the addition of one in the context of spotting whales was a hint at the wealth of whalers in Kumano. The scroll (see Fig. 24.5) depicting whaling in Taiji depicts as many as three telescopes being used in the lookout.[47] The idea of prosperity was not just conveyed by depicting whales but also by showing the tools and the techniques that were being used to procure them.

47 *Kishū Taijiura Kujira Tairyō no zu* (*Pictures of Whaling and Whales in Taiji in Kii Province*) 1861.

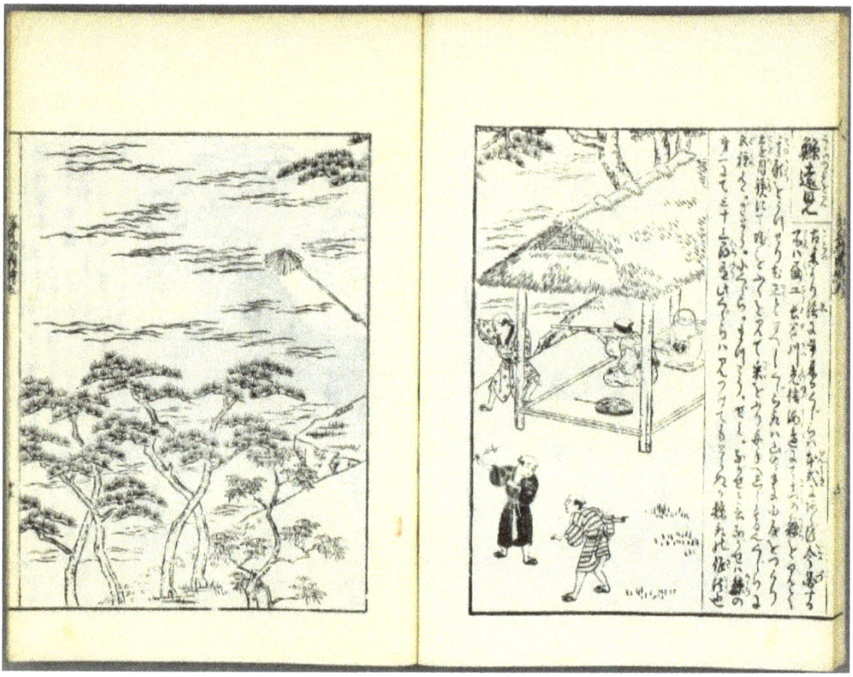

FIGURE 24.8 Double-page illustration, Tessai Hirase – Hasegawa Mitsunobu, *Nihon Sankai Meibutsu zue* (*Illustrated Guide to the Famous Products of Land and Sea in Japan*) (Osaka, Shioya Uhei: 1797). Printed book, National Diet Library Tokyo (URL: https://dl.ndl.go.jp/pid/2555436)

In addition to connoting prosperity, in printed media, especially in illustrated guides, Western telescopes symbolised a delight in looking at curious things. The presence of telescopes signified urban dwellers' scopophilic interest in the world around them. An illustrated guide to the famous places of Settsu Province depicts a party of stylish picknickers on a hill spying on other people through a telescope.[48] The illustration compares their excitement about the rich pickings offered by the telescope with nearby mushroom hunters delighting in a bumper crop of pine mushrooms (*matsutake*). Tessai's illustrated guide satisfied readers' desire to see curious things at one glance. In this sense, Mitsunobu's illustrations did not just inform about whales and whaling, but they also provided readers with otherwise unobtainable close-up views of a living whale in the sea. As the whalers encircle the whale with a net, it opens

48 Goree R., "Meisho Zue and the Mapping of Prosperity in Late Tokugawa Japan", *Cross-Currents* (2017) 90.

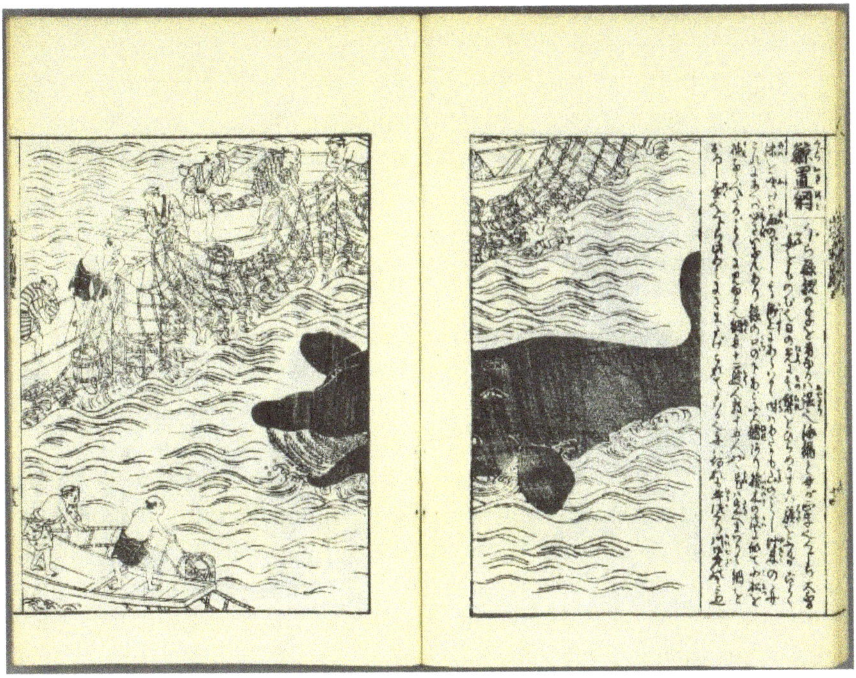

FIGURE 24.9 Double-page illustration, Tessai Hirase – Hasegawa Mitsunobu, *Nihon Sankai Meibutsu zue* (*Illustrated Guide to the Famous Products of Land and Sea in Japan*) (Osaka, Shioya Uhei: 1797). Printed book, National Diet Library Tokyo (URL: https://dl.ndl.go.jp/pid/2555436)

its mouth displaying its baleen bristles while its large eyes uncannily resemble human eyes [Fig. 24.9].

Mitsunobu's illustrations zoom in on the action, contrasting the strenuous gestures of whalers with the comparatively placid facial features of the whale. The sea thus becomes a stage for performing an engaging battle between industrious whalers and a monstrous whale. Defying gravity, the whale's gigantic body appears above the water's surface, almost fully exposing it to the curious gaze of urban readers. Goree has demonstrated that these zooming-in and zooming-out techniques were a major attraction of illustrated guides.[49] Zooming in connoted a scopophilic pleasure in satisfying one's curiosity to see and to know. In lifting the whale's body out of the water, Mitsunobu's illustrations did not just cater to the scopophilic desires of urban readers. To some degree, they also evoked the descriptive focus of books produced by natural

49 Ibid. 98.

history scholars such as *Whale Chronicle* (*Geishi*), albeit with less focus on visual accuracy. Mitsunobu balanced scientific, narrative, and scopophilic concerns in designing the illustrations. This was a response to the functions of printed illustrated guides that excited readers by offering them new ways of looking – expanding their vision with panoramic views and zooming in on details.

4 Conclusion

In 18th- and 19th-century Japan, whale knowledge was a vibrant field of intersecting local, cross-regional, Sino-Japanese, and Western epistemic practices. Illustrations of whales in printed media, manuscripts, and illustrated scrolls demonstrate that some epistemic practices – especially the practice of copying and adjusting existing images and sources – crossed social and geographic divides. Investigating illustrations of whales complicates the perception that images of animals and plants became increasingly detailed and accurate in 18th- and 19th-century Japan. It is probably more useful to argue that images of whales became more diverse, allowing scholars, painters, and the public to adjust them to their own interests.

Visual accuracy and visual detail were thus matters of negotiation rather than the ultimate goal of a teleological development towards modernity. Scholars of natural history produced documentary scrolls of whales for administrative purposes, and they compiled manuscripts that reached a wider audience in print. There was also cross-fertilisation between information that was circulating in print and in manuscript as scholars took shortcuts by copying existing images or written sources. At the same time, they also emphasised the need to observe objects directly and to stay truthful to the object in nature when representing it in image and in text. The negotiation of visual accuracy thus emerges as a subtext that needs to be acknowledged in order to get a better understanding of how knowledge about whales was produced in early modern Japan.

In particular, printed media had the power to shape whale knowledge across regional and social divides, picking up whale knowledge that derived from direct encounters between whalers and whales in western and southwestern Japan and repackaging this for a cosmopolitan urban gaze that hinged on scopophilia and curiosity about local products. Meanwhile, local practices continued to shape whale knowledge. Although the urban population at large did not take a scientific interest in whales, their engagement with whales and the coastal environment suggests a wish to push imagined and real boundaries with a view toward the world beyond the borders of the Japanese archipelago.

Bibliography

Ambros B.R., *Bones of Contention: Animals and Religion in Contemporary Japan* (Honolulu: 2012).

Arch J., "Heroic Whalers Hunting Whale-Mothers: Gender in the Early-Modern Japanese Whaling Industry", *Coriolis* 10.1 (2020) 48–69.

Arch J., "Whale Oil Pesticide: Natural History, Animal Resources, and Agriculture in Early Modern Japan", in Phillips D. – Kingsland S.E. (eds.), *New Perspectives on the History of Life Sciences and Agriculture* (Switzerland: 2015) 93–111.

Arch J., *Bringing Whales Ashore: Oceans and the Environment of Early Modern Japan* (Seattle: 2018).

Frank S.M. "Kuniyoshi and the Prosperity of the Seven Shores: A Garland of Japanese Woodblock Prints of Whales and Whaling, with a short history of whaling in Japan", *Deutsches Schiffahrtsarchiv* 25 (2002) 145–65.

Fukuoka M. *The Premise of Fidelity: Science, Visuality, and Representing the Real in Nineteenth-Century Japan* (Palo: 2012).

Goree R., "Meisho Zue and the Mapping of Prosperity in Late Tokugawa Japan", *Cross-Currents* 6.2 (2017) 404–439]

Goree R., *Printing Landmarks: Popular Geography and Meisho Zue in Late Tokugawa Japan* (Cambridge, Massachusetts: 2020).

Holm, F.J., *The Gods of the Sea: Whales and Coastal Communities in Northeast Japan, c.1600–1912* (Cambridge, Massachusetts: 2023).

Hōshūtei Riyū, *Ogawajima Geigei Gassen* (*Battling Whales in Ogawa Island*) 1840. Kyushu University Library (URL: http://hdl.handle.net/2324/1929732)

Itoh M., *The Japanese Culture of Mourning Whales: Whale Graves and Memorial Monuments in Japan* (Singapore: 2018).

Kameda-Madar K., "Copying and Theory in Edo-Period Japan (1615–1868)", *Art History* 37.4 (2014) 708–727.

Kandoriya Jiemon, *Geishi* (*Whale Chronicle*) (Osaka, Kashimoto Kanbee: 1794). National Archives of Japan (URL: https://www.digital.archives.go.jp/file/1222480.html).

Kuriyama S., "The Imagination of the Body and the History of Embodied Experience: The Case of Chinese Views of the Viscera", in ibid (ed.), *The Imagination of the Body and the History of Embodied Experience* (Kyoto: 2001) 17–29.

Marcon F., *The Knowledge of Nature and the Nature of Knowledge in Early Modern Japan* (Chicago: 2015).

Mori H. – Miyazaki K., "*Nankai Hogei Emaki no Tokuchō: Kishū Chihō no Hogei Emaki to no hikaku kara* (The Characteristics of Illustrated Scrolls of Whaling in Nankai: Based on a Comparison with Illustrated Scrolls of Whaling from the Kii Region)", *Seinan Journal of Cultures* 26.2 (2012) 117–55.

Mori H. – Miyazaki K., "*Ōtsuki Kiyonori 'Geishikō' to Sugita Genpaku 'Kaitai Shinsho' no Kankeisei* (On the connections between Ōtsuki Kiyonori's "*Geishikō*" and Sugita Genpaku's "Kaitai Shinsho")", *Bulletin of the Kyushu University Museum* 11 (2013) 53–68.

Ogata R. *Rangaku kotohajime* (Tokyo: 1983).

Ōtsuki Heisen, *Geishikō* (*Whale History*) 1808. National Diet Library Tokyo, Japan (URL: https://dl.ndl.go.jp/info:ndljp/pid/2575476).

Screech T., *The Western Scientific Gaze and Popular Imagery in Later Edo Japan: The Lens Within the Heart.* (Cambridge: 1996).

Tessai Hirase – Hasegawa Mitsunobu, *Nihon Sankai Meibutsu zue* (*Illustrated Guide to the Famous Products of Land and Sea in Japan*) (Osaka: 1797). Waseda University Library (URL: https://www.wul.waseda.ac.jp/kotenseki/html/bunk006/bunk006-02154/index.html).

CHAPTER 25

Images, Specimens, and Species: Hermann Schlegel on the Various Ways of Depicting a Fish

Robbert Striekwold

1 Introduction

In 1849, the German naturalist Hermann Schlegel (1804–1884), curator of vertebrates at the *Rijksmuseum van de Natuurlijke Historie* (National Museum for Natural History, or RMNH) in Leiden, published a short book in which he laid out the rules he believed should govern the production of good-quality natural history images. This work, titled *Verhandeling over de Vereischten van Natuurkundige Afbeeldingen* (Essay on the Requirements of Scientific Images) was Schlegel's answer to an 1845 prize question by Teylers' Second Society (a society founded in 1778 with the goal of stimulating progress in the arts and sciences), on the requirements a scientific image should fulfil to satisfy both the naturalist and the artist.[1]

Because natural history deals with a wide variety of different types of objects, the *Verhandeling* contains chapters pertaining to requirements that specifically belong to several of the branches of natural history, including one on birds, another on fishes and amphibians, and so on. But in the first chapter Schlegel begins by describing thc *goal* of natural history images in general:

> De plaats te vervangen der voorwerpen, die men zelf geene gelegenheid heeft in de natuur te zien of te onderzoeken, hen in die afbeelding te erkennen, en, zoo naauwkeurig mogelijk uit haar te kunnen afleiden, hunne gedaante of ook hunne kleuren, de verhouding hunner deelen en hunne verdere eigenschappen.

1 Schlegel H., *Verhandeling over de Vereischten van Natuurkundige Afbeeldingen* (Haarlem: 1849); Van der Velden F., "De Prijsvragen en Verhandelingen van Teylers Tweede Genootschap 1781–1866", *Teylers Museum Magazijn* 13.1 (1995) 11–15.

 | DOI:10.1163/9789004681187_026

> To replace objects one does not have the opportunity to investigate or observe in nature, recognize them in the image – their shape and colours, the proportions of their parts and other characteristics.[2]

The precise way in which this works varies between different groups, however, and even within groups. For example, Schlegel identifies four different types of botanical images: (1) an overview of plants that live in a particular region; (2) an image of the whole plant; (3) an image of plant parts, often cut in half; and (4) microscopic images. It is the second group, which he treats as botanical images proper, to which he gives most of his attention.[3] One of Schlegel's most important general rules (both for botanical and zoological images) is:

> De natuurkundige teekenaar moet zoo veel mogelijk vermijden, toevallige onregelmatigheden der voorwerpen te doen uitkomen, en hij dient in de meeste gevallen gehavende of geschondene voorwerpen zoo te teekenen als of zij gaaf en volledig waren; dikwijls moet hij zelfs de individuele eigenschappen van een voorwerp weglaten, omdat de afbeelding van een individu in de wetenschap dikwijls als het ware de geheele soort moet vertegenwoordigen.
>
> The draughtsman must avoid as much as possible to depict accidental irregularities of the objects, and should in most cases draw damaged specimens as if they were whole; often he should even leave out the individual characteristics of a specimen, because in science the image of an individual should often, as it were, represent the entire species.[4]

This rule appears to touch on an important development in 19th-century natural history, where naturalists moved from a more or less essentialist, typological understanding of species to one in which natural variation played a much larger role, reflected by the growing importance of geographical series of specimens by the mid-19th century.[5] Hansjorg Ahrens, for example, in his paper on

2 Schlegel, *Verhandeling* 3. All translations from Dutch are mine.

3 Schlegel, *Verhandeling* 56, 58.

4 Schlegel, *Verhandeling* 6–7. Translated by the author.

5 See, for instance, Daston L. – Galison P., *Objectivity* (Cambridge MA – London: 2007), ch. 2; Farber P.L., *The Emergence of Ornithology as a Scientific Discipline: 1760–1850* (Dordrecht: 1982), ch. 7; Stamos D.N., "Pre-Darwinian Taxonomy and Essentialism – A Reply to Mary Winsor", *Biology & Philosophy* 20 (2005) 79–96. Indeed, the first two directors of the RMNH, C.J. Temminck (1778–1858) and Schlegel, are generally held to represent both sides of this development, with Temminck viewing species primarily in terms of their typical, essential characteristics, and Schlegel being much more interested in variability. See Gassó Miracle M.E., "The Significance of Temminck's Work on Biogeography: Early Nineteenth

the production of natural history images at the RMNH, takes issue with this rule, singling it out as noteworthy for its apparent contradiction to Schlegel's insistence on the importance of collecting geographical series of specimens in order to fully understand species. Ahrens concludes that 'in 1846 he apparently still used the "old" species concept'.[6] However, as I will show below, this contradiction is only apparent.

2 Overview

In this chapter I will restrict myself to a discussion of the role of images in systematic natural history, the branch of natural history that concerns itself primarily with the description and naming of species, and their classification.[7] I distinguish three different types of images that were used in this field, which may be called 'species images', 'specimen images' and 'preservation images'. I believe that insufficient attention to these different categories of images has led some authors to perceive large theoretical changes where there were none, or to misidentify the cause behind a change in the use of images.

A compounding problem in the literature is the lack of balance between studies focusing on different branches of systematic natural history. If you shake a tree, historical studies on the features of botanical images will fall out by the dozens.[8] By contrast, the various branches of zoology fare much worse (with the possible exception of ornithology), and until recently palaeontology

Century Natural History in Leiden, the Netherlands", *Journal of the History of Biology* 41 (2008) 695–700; Gijzen A., *'s Rijks Museum van Natuurlijke Historie: 1820–1915* (Rotterdam: 1938) 43–63.

6 'In 1846 leunde hij kennelijk nog op het "oude" soortbegrip'. Ahrens H., "De Natuur Stilzetten op Papier. Wetenschappelijk Illustratoren van Naturalis Leggen de Dierenwereld Vast", *Tijdschrift voor Mediageschiedenis* 12 (2009) 233–274, here 240–241.

7 See Farber P.L., *Finding Order in Nature: The Naturalist Tradition from Linnaeus to E.O. Wilson* (Baltimore, MD: 2000); Gassó Miracle M.E., "On Whose Authority? Temminck's Debates on Zoological Classification and Nomenclature: 1820–1850", *Journal of the History of Biology* 44 (2011) 445–481.

8 For a small sample, see Bleichmar D., *Visible Empire: Botanical Expeditions and Visual Culture in the Hispanic Enlightenment* (Chicago: 2012); Fischer H. – Remmert V.R. – Wolschke-Bulmahn J. (eds.), *Gardens, Knowledge and the Sciences in the Early Modern Period* (Cham: 2016); Lack H.W., "The Botanical Illustrations of Franz Scheidl (*fl.* 1770–1795)", *Archives of Natural History* 47.1 (2020) 51–62; Nickelsen K., "Draughtsmen, Botanists and Nature: Constructing Eighteenth-Century Botanical Illustrations", *Studies in History and Philosophy of Biological and Biomedical Sciences* 37 (2006) 1–25; Secord A., "Pressed into Service: Specimens, Space, and Seeing in Botanical Practice", in Livingstone D.N. – Withers C.W.J. (eds.), *Geographies of Nineteenth-Century Science* (Chicago – London: 2011) 283–310.

and ichthyology were barely mentioned at all.[9] The reasons for these discrepancies are unclear, as far as I can tell, though the popularity of botanical images with art historians probably plays a role.[10] An important implication of this imbalance, however, is that generalizations about the roles of images in systematic natural history are often drawn based primarily on studies of botanical images. As I will show below, images often played different roles in different branches of natural history, so it is important to draw on a larger variety of examples in order to draw reliable general conclusions.

To this end, I will focus on two of the most neglected branches of natural history when it comes to the study of images: ichthyology and palaeontology. First I will describe and illustrate how species images, specimen images and preservation images were used in these (and other) fields, using Schlegel's essay as a point of departure. Generally speaking, species images accompanied species descriptions in published species books and mainly served to depict those characteristics of an organism that were relevant for classifying it in the system used by the author. Specimen images, by contrast, depict individual objects with all their 'accidental' characteristics, and were used primarily to communicate information about important objects, like type specimens. They rarely figured in published species descriptions, except in palaeontology. Finally, preservation images were almost never published, but played a role in the naturalists' research process by helping to record those features of an organism that were deemed important for classification, but didn't survive the process of specimen preservation.

Distinguishing between these three types of images, and keeping in mind what roles they play in the practice of natural history, can throw light on certain historical developments that have been identified in the literature. In particular, I will look at the notion of epistemic virtues, and the claim that an important shift in such virtues took place during the 19th century; and the influence of changing views about species in that same period.

---

9 For examples from ornithology, see Cooper J.A., "Edward Neale (1833–1904): Bird Illustrator", *Archives of Natural History* 46.2 (2019) 283–297; Lederer R.J., *The Art of the Bird: The History of Ornithological Art through Forty Artists* (Chicago IL: 2019). For palaeontology: Davidson J.P., *A History of Paleontology Illustration* (Bloomington IN: 2008); Dawson G., *Show me the Bone: Reconstructing Prehistoric Monsters in Nineteenth-Century Britain and America* (Chicago – London 2016). For ichthyology: Aronowksy L., "On Drawing Dead Fish", *Environmental History* 21 (2016) 542–551; Holthuis L.B. – Pietsch T.W. *Les planches inédites de poissons et autres animaux marins de l'Indo-Ouest Pacifique d'Isaac Johannes Lamotius* (Paris: 2006); Rijks M. – Smith P.J. – Egmond F. (eds.), *Fish & Fiction: Aquatic Animals between Science and Imagination (1500–1900)* (Leiden: 2018).

10 As discussed in Prince S.A. (ed.), *Of Elephants and Roses: French Natural History 1790–1830* (Philadelphia, PA: 2013).

3 Species Images

Species images seem to depict a single specimen, often in a more or less stylized manner, while some features are emphasized, and others ignored. Fishes, for instance, are almost always shown from the side (except in dorsoventrally flattened forms, like rays, which are shown from the top), with fins stretched out and features like fin rays, scales, and operculum drawn very clearly.[11] Schlegel compares two images of fishes, one drawn 'artistically' [Fig. 25.1a], the other 'scientifically' [Fig. 25.1b], and comments:

> Daar nu het bijwerk geheel en al vervalt, en de teekenaar dikwijls, om den zamengestelden vorm en ligging hunner, naar evenredigheid, kleine schubben aan te toonen, meestal zeer uitvoerig moet werken, zoo vinden zich de kunstenaars door die soort van voorstelling meestal even weinig aangetrokken, als door die van hagedissen en slangen.
>
> Since the background is left out altogether, and the artist often has to do very detailed work in order to show the composite shape and position of their relatively tiny scales, artists generally feel as little attraction to these kinds of images as to those of lizards and snakes.[12]

The function of this emphasis on a number of particular details is simple: these images accompany species descriptions and are intended to show those features, shared by the individuals of a species, that are relevant for classification. In this way, naturalists can study an organism's taxonomy, based on the image.[13]

This link between images and classification can be illustrated, perhaps somewhat surprisingly, by looking at the work of an ichthyologist who did not use illustrations at all. Swedish naturalist Peter Artedi (1705–1735), like

11 Indeed, these stylistic norms are so pervasive in 19th-century systematic natural history that even relatively subtle departures are quite noticeable. For instance, William Anderson notes of a number of ichthyological plates belonging to the American herpetologist John Edwards Holbrook, that 'the artist, whoever he/she may have been, may have had little or no previous experience in illustrating fishes. This supposition is supported by the fact that on each of the twelve fishes illustrated both pelvic fins are fully or almost fully erect, each being clearly visible, whereas the usual way of illustrating these is to display one fin only, frequently only partially erect or lying against the body.' Anderson W.D., "John Edwards Holbrook's Senckenberg Plates and the Fishes they Portray", *Archives of Natural History* 30.1 (2003) 1–12, quotation on page 10.

12 Schlegel, *Verhandeling* 43.

13 Schlegel, *Verhandeling* 5–7; Ahrens, "De Natuur Stilzetten" 235, 240.

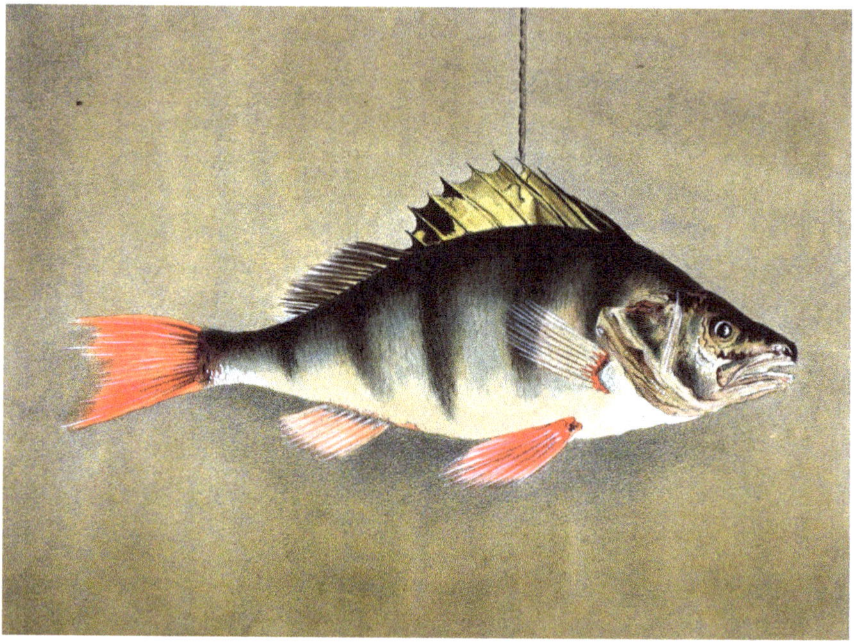

FIGURE 25.1A Hermann Schlegel, example of a perch (*Perca fluviatilis*) drawn too artfully. Lithograph, plate 12 from *Verhandeling over de Vereischten van Natuurkundige Afbeeldingen* (Haarlem: 1849)

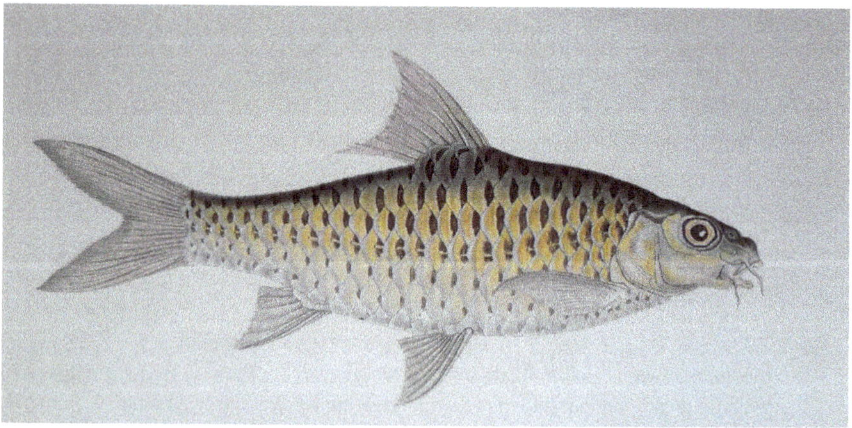

FIGURE 25.1B Hermann Schlegel, example of a carp (*Tor tambra*) drawn scientifically. Lithograph, plate 11 from *Verhandeling over de Vereischten van Natuurkundige Afbeeldingen* (Haarlem: 1849)

his friend Carl Linnaeus (1707–1778), found images distracting, even harmful, because of their unreliability – after all, two artists drawing the same specimen can produce markedly different images, and published images were often rife with errors.[14] Instead, Artedi used only those characteristics of fishes that could be described unambiguously (features like the number and relative position of fins, fin rays, and so on), and easily preserved (and observed) in collected specimens.[15] While this approach worked very well for classifying the limited number of fishes known in Artedi's time, the rapid rise in newly discovered species led to a growing need for additional characteristics that could be used in classification, including those (like colouration) that did not preserve well at all, so Artedi's image-less approach to ichthyology never caught on. Indeed, the major ichthyological publications of the rest of the 18th and 19th centuries are richly illustrated with species images.[16]

For the naturalist, species had both "essential" and "accidental" characteristics. In the context of classification, essential characteristics are those that all members of the species possess, and which can therefore be used to define it (and compare it to similar species). Accidental characteristics are possessed only by some members of a species, and are thus usually less valuable for classification purposes.[17] Deciding which features of an individual organism counted as essential for the species and which did not, was one of the primary tasks of the systematic naturalist.[18] Compare Figs. 25.2a and 25.2b, both from Bleeker's *Atlas Ichthyologique*, which represent two very similar species of

14 On examples of errors in natural history images, see Allmon W.D., "The Evolution of Accuracy in Natural History Illustration: Reversal of Printed Illustrations of Snails and Crabs in Pre-Linnaean Works Suggests Indifference to Morphological Detail", *Archives of Natural History* 34.1 (2007) 174–191; Anderson, "John Edwards Holbrook's Senckenberg Plates". Indeed, Schlegel devotes a section of most chapters of his *Verhandeling* to criticizing inaccurate images in other works.

15 Artedi P., *Ichthyologia, sive opera omnia de piscibus* (Leiden, Conradus Wishoff: 1738). See Van Trijp D., *Captured on Paper. Fish Books, Natural History and Questions of Demarcation in Eighteenth-Century Europa (ca. 1680–1820)* (Ph.D. Dissertation, Leiden University: 2021) 149–159.

16 Some of the most important are Bloch M.E., *Allgemeine Naturgeschichte der Fische*, 12 vols. (Berlin, Realschule, J. Morino: 1782–1795); Lacépède B.G., *Histoire naturelle des poissons*, 12 vols (Paris, Plassan: 1798–1803); Cuvier G. – Valenciennes A., *Histoire naturelle des poissons*, 22 vols (Paris – Strasbourg: 1828–1849); Bleeker P., *Atlas ichthyologique des Indes Orientales Néêrlandaises, publié sous les auspices du Gouvernement Colonial Néêrlandais*, 9 vols. (Amsterdam: 1862–1878).

17 Exceptions include sexually dimorphic characters, which were commonly used in bird classification. See Lederer, *Art of the Bird*.

18 See Van Neste A., "Practising Taxonomy: Joel Asaph Allen and Species-Making (W.T. Stearn Prize 2017)", *Archives of Natural History* 45.2 (2018) 197–212; Scudder S.H., "Look at your fish", *Every Saturday: A Journal of Choice Reading* (1874).

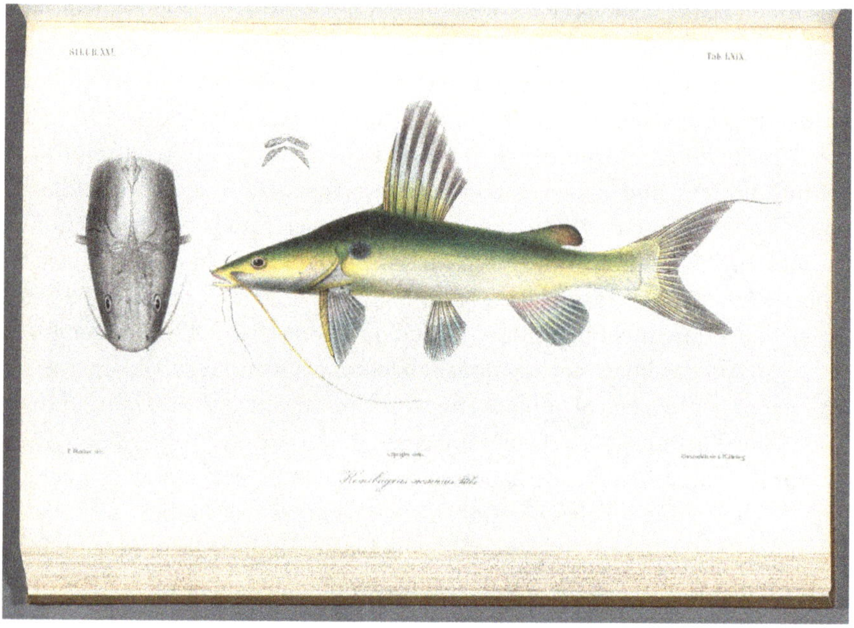

FIGURE 25.2A L. Speigler – C.W. Mieling, *Hemibagrus nemurus Blkr*. Chromolithograph, plate LXIX from Bleeker Pieter, *Atlas ichthyologique des Indes Orientales Néêrlandaises, publié sous les auspices du Gouvernement Colonial Néêrlandais* vol. 2 (Amsterdam: 1862)

catfish from the Indonesian archipelago, *Hemibagrus nemurus* and *Hemibagrus wyckii* [Figs. 25.2a–2b]. In order to distinguish species that are so alike, the naturalist has to focus on fairly subtle differences. So, for instance, Bleeker writes of *H. nemurus* that 'the tail fin is deeply incised with sharp lobes, the upper lobe is longer than the lower lobe, and commonly extends into a thread.' In *H. wyckii*, 'the tail fin is deeply incised, with sharp lobes, the outer rays of which extend into threads, with the upper thread longer than the lower thread.'[19] Features like these are very clearly depicted in the image, so the description and image complement and mutually reinforce one another.

Because of this special feature of species images, where the clear depiction of some details (but not others) is more important than realism or aesthetics, the production of species images demanded a rather intimate

19 'caudali profunde incisa lobis acutis, lobo superiore quam lobo inferiore longiore vulgo in filum producta'; 'caudali profunde incisa lobis acutis radio externo in filum producto, filo superior filo inferior longiore'. Translated by the author from Bleeker, *Atlas Ichthyologique* vol. 2, 55, 57.

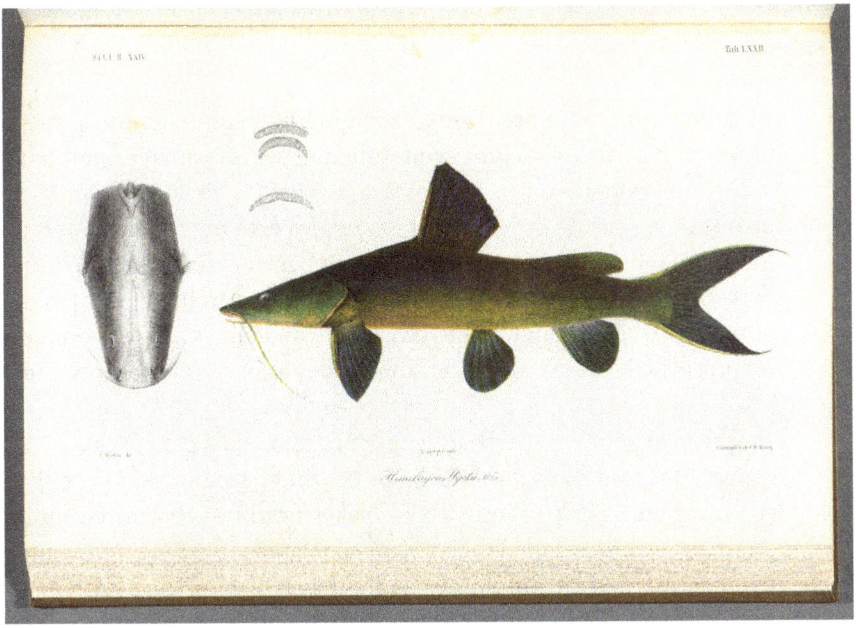

FIGURE 25.2B Speigler (delin) Mieling (litho), *Hemibagrus Wijckii Blkr*. Chromolithograph, plate LXXII from Bleeker Pieter, *Atlas ichthyologique des Indes Orientales Néêrlandaises, publié sous les auspices du Gouvernement Colonial Néêrlandais* vol. 2 (Amsterdam: 1862)

knowledge of natural history and the practical requirements of classification. So, Schlegel writes:

> Natuurkundige teekeningen moeten derhalve altijd, zoo niet door den geleerde zelven, dan toch onder diens toezigt en leiding gemaakt worden, en hij moet, in de meeste gevallen, den kunstenaar eerst vormen, want er zijn weinigen, die het aangeboren talent hebben, zelven den regten weg te vinden

> Scientific drawings should therefore always either be made by the naturalist himself, or under his supervision and direction, and in most cases he must first educate the artist, for few of them are born with the talent to find the right way by themselves.[20]

20 Schlegel, *Verhandeling* 11.

Bleeker concurs when he writes (with some frustration) of the production of the images for his *Atlas*:

> Aussi plusieurs centaines d'espèces ont été figurées de nouveau, même jusqu'à huit et dix fois, à fur et à mesure que mes dessinateurs apprenaient à représenter les objets avec plus de fidélité [...] je dois à la vérité et à ma responsabilité de ne pas laisser ignorer les extrêmes soucis causés par une coopération qui nécessitait une surveillance et une direction minutieuses et continuelles [...] je puis dire que les dessins finissaient par être mon propre ouvrage plutôt que celui du dessinateur, et que leur retouche m'a pris plus de temps que la description des espèces qu'ils représentent.
>
> Hundreds of species have been refigured again and again, even as much as eight to ten times, as my draughtsmen learned to represent the objects with more precision [...] I owe it to the sake of truth and my responsibility, not to leave unmentioned the extreme worries caused by a co-operation that necessitated a continuous, scrupulously careful supervision and instruction [...] I can say that these drawings in the end were more my own work than that of the artist, and that the retouching has taken more of my time than the description of the species they represent."[21]

4 Specimen Images

In most branches of natural history, the great majority of images accompanying species descriptions are species images. An important exception is palaeontology, where most published images represent specimens, not species. This can be illustrated by looking at the *Recherches sur les poissons fossiles* by the Swiss naturalist Louis Agassiz (1807–1873).[22] In this work, Agassiz aims to synthesize all available knowledge of fossil fishes by bringing them together in one system of classification. Looking at Fig. 25.3, however, which is the main image of *Lepidotus gigas* in the *Recherches*, it is obvious that Agassiz chose to depict fossil specimens, not species. Indeed, he points this out in his preface where he writes that they 'represent the fossil fish as I observed them, and as

21 Bleeker P., "Notice sur l'Atlas Ichthyologique des Indes Orientales Néerlandaises", *Jaarboek van de Koninklijke Nederlandse Akademie van Wetenschappen* 1877 (1878) 133–144, quotation on pages 136–137. Translation taken from Van Oijen M.J.P., "Data on the Genesis of the *Atlas Ichthyologique* from a Little Known French Paper by P. Bleeker", *The Raffles Bulletin of Zoology*, Supplement no. 13 (2005) 3–8, quotation on page 5.

22 Agassiz L., *Recherches sur les poissons fossiles*, 5 vols. (Neuchatel: 1833–1843).

they exist in the collections'.[23] This changes the interaction between text and image in a rather striking way. For instance, Agassiz writes:

> La caudale, la plus grande de toutes les nageoires, est légèrement échancrée au milieu; son lobe supérieur est un peu plus long, mais aussi plus étroit que son lobe inférieur. Elle est très-distincte dans l'exemplaire de la pl. 29, où seulement l'extrémité de son lobe supérieur est en partie enlevée. Dans plusieurs autres fragmens de la collection de M. Hartmann, on en voit différentes parties très-bien conservées.
>
> The caudal fin, the largest of all the fins, is slightly indented in the middle; its upper lobe is a little longer, but also narrower than the lower lobe. It is very distinct in the specimen in pl. 29 [Fig. 25.3], where only part of the end of its upper lobe is missing. In several other specimens in Mr. Hartmann's collection, we see other parts very well preserved.[24]

So, by using a specimen rather than a species image, Agassiz has to explain both why he chose this particular one (because, among other things, the tail was very well preserved), and how he knows about the species-defining characteristics that are missing in this specimen (there are other specimens that contain the missing parts).

Schlegel, in his short chapter on fossils (and other stony objects), confirms the importance of specimen images in palaeontology. For rather than with most zoological images, where it is important to ignore any accidental characteristics of particular specimens:

> Het komt hier vooral op de hoogst naauwkeurige uitvoering van détails aan [...] Het teekenen van fossile voorwerpen wordt intusschen in vele opzigten gemakkelijk door de natuur der voorwerpen; want zij zijn hard, zij hebben een vasten onveranderlijken vorm, en de teekenaar behoeft

23 '... représentent les poissons fossiles tels que je les ai observés et tels qu'ils existent dans les collections', in Agassiz, *Recherches* vol. 1, xiv. Agassiz distinguishes these specimen images from two other categories of images in his work: idealized anatomical studies of living fishes (for comparison), and line drawings representing *genera* of extinct fishes, extracted from the general characteristics of the species he has described. This last category deserves an article onto itself, but for the purposes of the current one it can be seen as closely related to species images, for these genus images behave the same and have a similar function to species images, just one taxonomic level higher. See Ibidem, xiv–xv.

24 Agassiz, *Recherches sur les poissons fossiles*, vol. 2, p. 238.

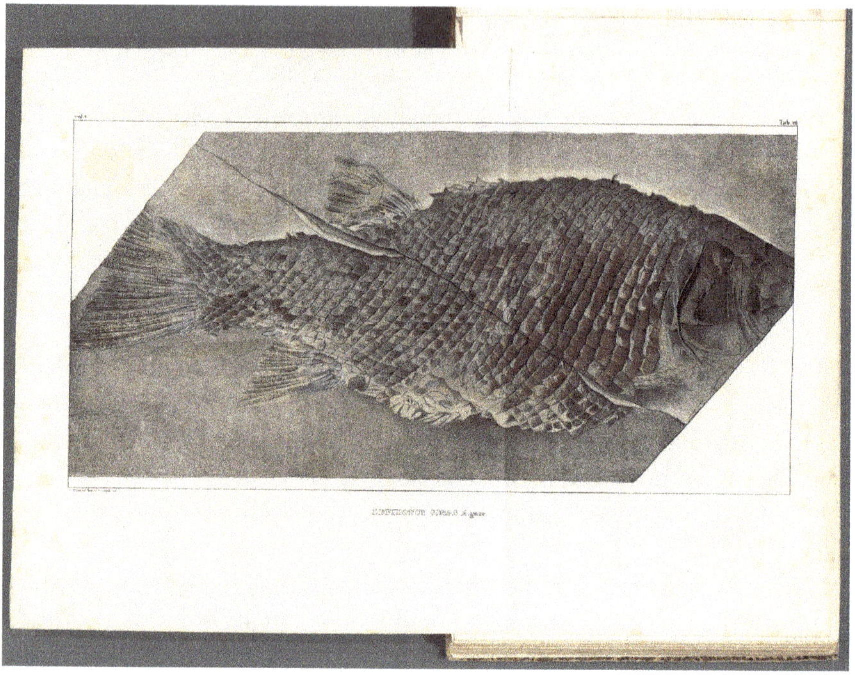

FIGURE 25.3 H. Nicolet, *Lepidotus gigas Agass*. Engraving, plate 29 from Agassiz Louis, *Recherches sur les poissons fossiles* (Neuchatel: 1833–1843)

> slechts naauwkeurig weêr te geven, hetgeen hij ziet, en eene gepaste manier van uitvoering te kiezen, ten einde volmaakt te slagen.

> here it is primarily about the very precise depiction of details [...] The drawing of fossils is made easier by the nature of the objects; for they are hard, have a fixed shape, and the artist merely needs to represent accurately that which he sees, and choose a fitting means of execution, to succeed perfectly.[25]

In other words, the interpretive steps necessary for making the species images that often strained the relationships between naturalists and the artists they employed, were unnecessary in palaeontology, which dealt primarily in the much more straightforward specimen images that depicted individual fossils.[26]

25 Schlegel, *Verhandeling* 62–63.

26 For other examples of 19th-century fossil species books with specimen images, see Cuvier G., *Recherches sur les ossemens fossiles des quadrupèdes, où l'on rétablit les*

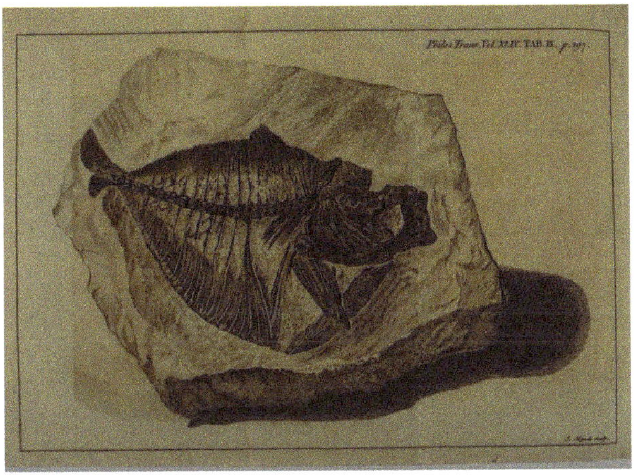

FIGURE 25.4 Arthur Pond, Fossil fish from Antigua. Engraving, plate IX from *Philosophical Transactions of the Royal Society of London* 49.1 (London: 1756)

The use of specimen images in palaeontology has a long tradition,[27] as shown for instance by Fig. 25.4, which is a mid-18th-century engraving by the artist and collector Arthur Pond [Fig. 25.4]. This image was made with the express goal of showing the object as it actually was. Pond writes: 'I have been particularly careful not to exaggerate or add the smallest trifle, by way of making it seem more complete or perfect than it is'.[28] However, in ichthyology (and other branches of zoology) published specimen images are very rare. Exceptions usually occur in reports of specimens (of known species) that are peculiar for some or other reason, for instance if they are exceptionally large, or have an unusual anatomical or physiological feature [Fig. 25.5].[29] In addition, the introduction of nomenclatural type specimens halfway through the 19th century produced a new category of objects in natural history that were

caractères de plusieurs espèces d'animaux que les révolutions du globe paroissent avoir détruites, 4 vols. (Paris: 1812); Owen R., *A History of British Fossil Reptiles*, 4 vols. (London: 1849–1884).

27 Going back to the 16th century. See Rudwick M., *The Meaning of Fossils: Episodes in the History of Palaeontology* (Chicago – London: 1976), chapters 1 and 2.

28 Pond A., "A Letter to the Right Honourable George Earl of Macclesfield, President of the Royal Society, Concerning the Stones Mentioned in the Preceding Article", *Philosophical Transactions of the Royal Society of London* 49.1 (1756) 297–298, quotation on page 298.

29 For other examples of specimen images in ichthyology, see Van Lidth de Jeude T.W., "On a Large Specimen of Orthtragoriscus on the Dutch Coast", *Notes from the Leyden Museum* XII (1890) 189–195 (an exceptionally large sunfish specimen); Hartley F., "Notes on a Specimen of Alepisaurus aesculapius Bean, From the Coast of San Luis Obispo County, California", *Proceedings of the California Academy of Sciences*, second series, 5 (1895) 49–50 (a new specimen of a very rare lancetfish).

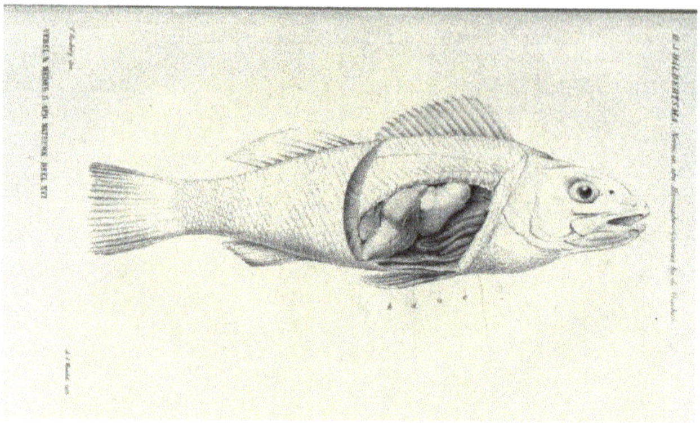

FIGURE 25.5 T. Hooiberg & A.J. Wendel, *Afbeelding van een hermaphrodieten baars uit de Brugmansche verzameling van het Anatomische kabinet te Leiden.* Lithograph, unnumbered plate from *Verslagen en Mededeelingen der Koninklijke Akademie van Wetenschappen* 16 (1863) facing p. 178

worth depicting separately *as objects*, rather than as abstracted representations of entire species.[30]

5 Preservation Images

There is, however, a third category of images in systematic natural history that is rarely published at all, but instead serves a function in the process of specimen preservation. These images tend to lack detail and instead focus on one or a small number of features of a specimen that are otherwise difficult to preserve, but are nevertheless useful for classificatory purposes. In the case of fishes, this usually came down to one feature in particular: colour. No matter how brightly coloured a fish is in life, once dead its various hues will start to fade, and when it is stuffed or stored in alcohol, it will gradually but surely take on the bland browns and greys so characteristic of preserved fish collections.[31]

30 See Secord, "Pressed into Service"; Witteveen J., "Objectivity, Historicity, Taxonomy", *Erkenntnis* 83.3 (2018) 445–463.

31 See Aronowky, "Drawing Dead Fish". Indeed, Schlegel (*Verhandeling* 44) points out that images of fishes (and amphibians) are often published on uncoloured plates, if the animals had not been observed alive but had to be drawn based only on preserved specimens.

The importance of colour is emphasized in the many guides to the proper collection and preservation of natural history specimens that were published during the 19th century. For instance, RMNH director C.J. Temminck (1778–1858) writes that '[one] should specify on a label which colours the fish had when fresh.'[32] Ideally, however, as Agassiz writes in his guide to fish collecting, 'should any collector be sufficiently familiar with painting to draw colored figures of any of these fishes, or so situated as to have some of them drawn by an artist, it would be an invaluable contribution to Natural History.'[33] Fig. 25.6a shows an example of a field sketch by the German naturalist Salomon Müller (1804–1864) [Fig. 25.6a], which is little more than an outline of the animal with its most distinctive coloured patches emphasized in black. A more elaborate, coloured example is shown in Fig. 25.6b, which is by an unknown artist but clearly shows the emphasis on colour – other taxonomically relevant characteristics, like fin rays and scales, are simply shaded in [Fig. 25.6b].[34]

Fig. 25.7 shows a more peculiar type of preservation image [Fig. 25.7]. In 1824, the great French ichthyologist Achille Valenciennes (1794–1865) visited the RMNH to study a collection of Javanese fishes brought together by two young naturalists who had tragically died before they could return to the Netherlands to publish their results.[35] Valenciennes spent a number of days at the museum to study the specimens, field notes and drawings, on the basis of which he took his own notes and made his own drawings. However, Valenciennes' (and/or possibly Sophie Duvaucel's) drawings are preservation drawings, meant to capture above all the colours of the fishes, whereas the ones he used as examples

32 "[men] duidde op de *Etiquette* daarenboven de kleuren aan, welke den visch, versch zijnde, eigen was," in Temminck C.J., *Voorschrift Hoedanig te Handelen met Voorwerpen van Natuurlijke Historie, ten Einde Dezelve Behoorlijk te Verzenden en voor Bederf te Bewaren* (Leiden: 1825) 16.

33 Agassiz L., *Directions for Collecting Fishes and Other Objects of Natural History* (Cambridge, MA: 1853) 2.

34 Both images are from the archive of the *Natuurkundige Commissie*, a commission tasked with exploring the Dutch East Indies during the period 1820–1850. The call number of Fig. 25.6A is NNM001001132_006, that of Fig. 25.6B is NNM001000530_001. For more on the *Natuurkundige Commissie*, see Veth H.J., *Overzicht van Hetgeen, in het Bijzonder door Nederland, Gedaan is voor de Kennis der Fauna van Nederlandsch Indië* (Leiden: 1879) 20–123; Weber A., "Collecting Colonial Nature. European Naturalists and the Netherlands Indies in the Early Nineteenth Century", *BMGN – Low Countries Historical Review* 134.3 (2019) 72–95.

35 The naturalists are Heinrich Kuhl (1797–1821) and Johan Conrad van Hasselt (1797–1823). See Klaver C., *Inseparable Friends in Life and Death: The Life and Work of Heinrich Kuhl (1797–1821) and Johan Conrad van Hasselt (1797–1823), students of prof. Theodorus van Swinderen* (Groningen: 2007).

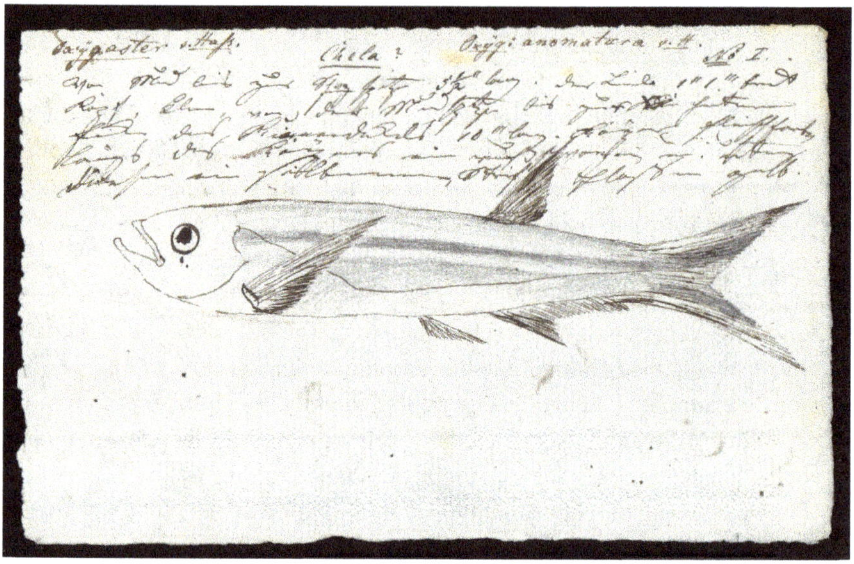

FIGURE 25.6A Salomon Müller, sketch of *Oxygaster anomatura*. Ink and pencil. Page from field notebook, *c*.1830. Available at Gassó, E. *et al.*, *Natuurkundige Commissie Archives Online* (Leiden, Brill: 2020)

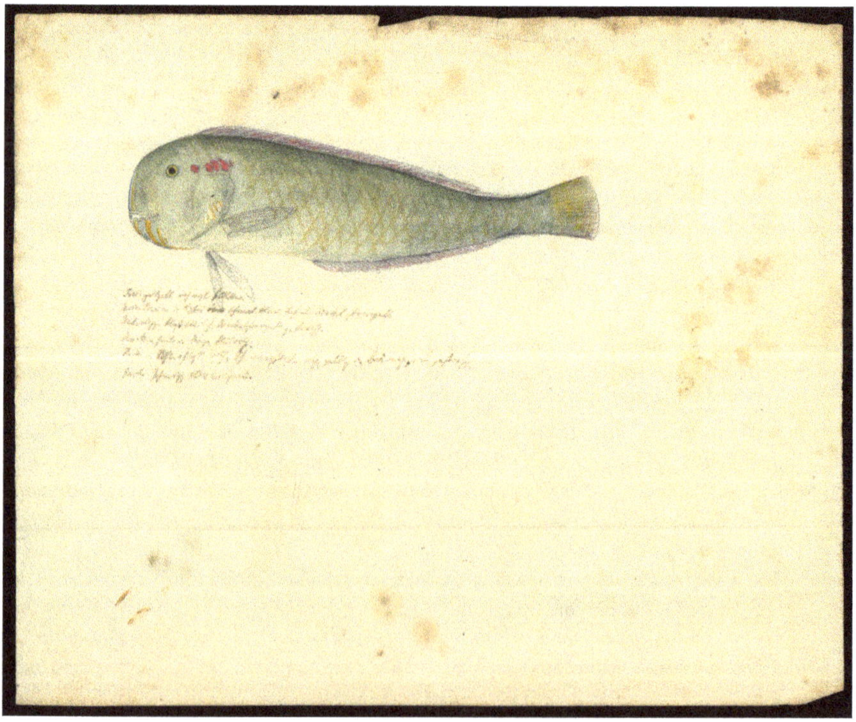

FIGURE 25.6B Unknown artist, sketch of a type of surgeonfish. Watercolour. Available at Gassó, E. *et al.*, *Natuurkundige Commissie Archives Online* (Leiden: 2020)

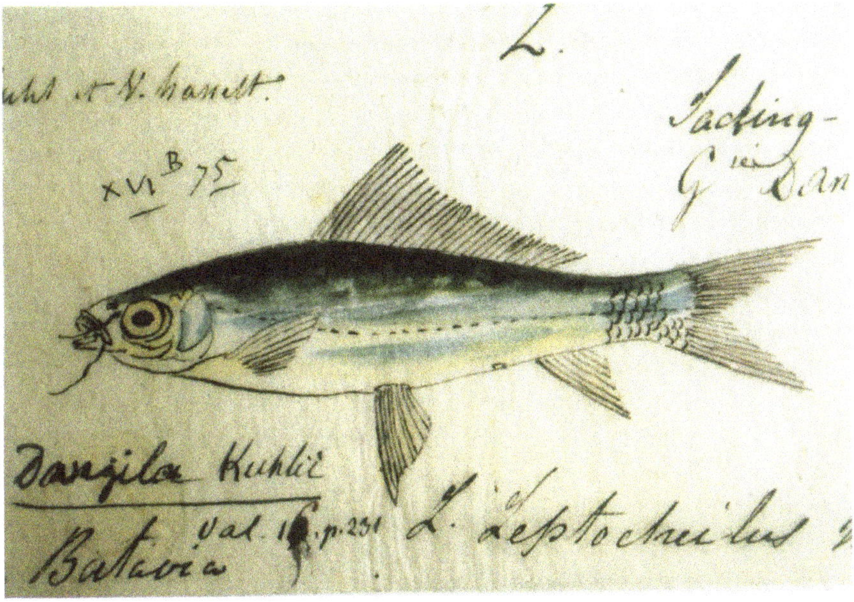

FIGURE 25.7 Achille Valenciennes and/or Sophie Duvaucel, sketch of *Labiobarbus leptocheilus*. Watercolour. Fig. 25.7 from Roberts T.R., "The Freshwater Fishes of Java, as Observed by Kuhl and van Hasselt in 1820–23", *Zoologische Verhandelingen* 285.29 (1993)

are species images, quite ready for publication. Valenciennes' preservation drawings would later aid in the production of the species descriptions and images of his and Cuvier's multi-volume *Histoire naturelle des poissons*.[36]

6 Epistemological and Metaphysical Transitions

Species images, specimen images, and preservation images. The differences between them seem quite straightforward, as do the reasons for those differences. Species images depict idealized individuals that are supposed to represent species, and are made to accompany species descriptions in works of systematic natural history. Specimen images, by contrast, aim to accurately depict individual specimens, with all their quirks and oddities, and appear whenever an author wishes to discuss a particularly important object. Finally, preservation images are usually sketches of specimens, with a strong emphasis

36 Cuvier – Valenciennes, *Histoire naturelle*; Roberts T.R., "The Freshwater Fishes of Java, as Observed by Kuhl and van Hasselt in 1820–23", *Zoologische Verhandelingen* 285.29 (1993) 1–94.

on a small number of characteristics, namely those that are important for the purpose of classification, but do not preserve well. However, while these categories spring from the practical requirements of systematic natural history, it has become popular in the historiographical literature on such images to interpret these differences in a more theoretical way, which I believe has led to some potentially confusing lines of thought. Preservation images are the least problematic in this respect (though this may be mainly due to the fact that not much has been written about them),[37] but several authors have attempted to read major theoretical shifts into the differences between species and specimen images. I will briefly discuss two such cases.

First, Lorraine Daston and Peter Galison's monumental work *Objectivity* presents a historical account of styles of scientific image-making grounded in particular epistemic virtues. As the rules by which science is done change over time, so do the images that scientists produce, is the idea. Daston & Galison identify three epistemological regimes that succeed one another, but only two are relevant here: truth-to-nature and objectivity, which in the context of natural history (in Daston & Galison's case, botany) correspond roughly to species images and specimen images, respectively.[38] Truth-to-nature has as its goal the discovery of the true nature of things regardless of their particular manifestations in the world. In natural history, this resulted in images representing ideal types: representations of species, not individuals. Objectivity, which was introduced in the 19th century and achieved dominance by the middle of that century, was concerned precisely with individual objects in all their particularity. In natural history, this resulted in images representing individual specimens.[39] To be sure, Daston & Galison do not claim that objectivity replaced truth-to-nature in an all-or-nothing fashion: 'As long as botanists insisted on figures that represented the characteristic form of a species or even genus, photographs and other mechanical images of individual plants in all their particularity would have little appeal.' However, they continue: 'Objectivity did make inroads into other areas of botanical practice such as the introduction of the "type method" in the late nineteenth and early twentieth

37 The topic of the function and history of preservation images deserves a treatment of its own. The main problem with the small literature on these images is that they are not always explicitly recognized as tools for preservation. So, when Aronowsky asks, in her paper on a collection of preservation images by the artist Joseph Drayton, 'Why did Drayton choose to portray these animals in a state he knew to be transitory?' (Aronowksy, "Drawing Dead Fish" 548–549). The answer is, presumably, because the colours he thus preserved were deemed relevant for classifying the fishes.

38 Daston – Galison, *Objectivity* 42–50.

39 Ibidem 63–68; 105–113.

centuries in order to stabilize nomenclature.' And so: 'It is no surprise that the one place where photography gained a firm foothold in botanical illustration was the representation of type specimens, in all their individuality and militant objectivity'.[40]

It is indeed no surprise, for the introduction of nomenclatural type specimens in botany meant that there was now a category of specimens that were so significant that it made sense to publish images of them. The problem is that nobody in the 18th century wished to picture individual herbarium specimens, because this field dealt in species, not specimens.[41] In the 19th century, the individual herbarium specimen became an object of interest, so it got depicted as an object. There is no evidence of a change in underlying epistemic virtues here, just a change in the meaning of a certain class of objects.[42] Compare this with palaeontology, which had a much stronger tradition of specimen images than botany. As discussed above, naturalists were reluctant to make species (truth-to-nature) depictions of fossils, instead depicting them with all their particular variations, cracks, missing parts, and so on.[43] Intriguingly, however, and quite contrary to Daston & Galison's case, truth-to-nature images became *more* prevalent in 19th century palaeontology, because the increasingly popular methods of comparative anatomy allowed for much more confident reconstructions of extinct species.[44] Thus, generally speaking, during the 19th century and before, whenever an individual specimen had to be

40 Ibidem 109; 111. See Secord, "Pressed into Service", for a more detailed treatment of illustrations and types in botanical practice. The type method involved the identification of a single specimen (or a few) as the undisputed namebearer(s) of a species. That is, any disputes about what a particular plant or animal should be called could be solved unambiguously by reference to this particular object, the type specimen, stored safely in a collection. See Farber P.L., "The Type-Concept in Zoology During the First Half of the Nineteenth Century", *Journal of the History of Biology* 9.1 (1976) 93–119; Daston L., "Type Specimens and Scientific Memory", *Critical Inquiry* 31 (2004) 153–182; Witteveen J., "Suppressing Synonymy With a Homonym: The Emergence of the Nomenclatural Type Concept in Nineteenth Century Natural History" *Journal of the History of Biology* 49 (2016) 135–189.

41 Müller-Wille S., "Collection and Collation: Theory and Practice of Linnaean Botany", *Studies in History and Philosophy of Biological and Biomedical Sciences* 38 (2007) 541–562; Secord, "Pressed into Service".

42 See Witteveen, "Objectivity, Historicity, Taxonomy" for a related critique of Daston & Galison's treatment of type specimens.

43 Because of their uniqueness, fossils were often quite valuable (especially those of large vertebrates), and specimen images thus played an important role in making these singular objects available by proxy to other naturalists. See Rudwick M., "Georges Cuvier's Paper Museum of Fossil Bones", *Archives of Natural History* 27.1 (2000) 51–68.

44 Dawson, *Show me the Bone* 198–207.

depicted, it was depicted objectively. Whenever a species had to be depicted, it was depicted abstractly, in a fashion resembling what Daston & Galison call truth-to-nature. Rather than a broad development from truth-to-nature to objectivity during the 19th century, we observe a correspondence between what needs to be depicted (specimens or species) and the manner of depicting it (objectively or not). This in turn relates to the particular problems confronting particular branches of natural history, not to more or less universal epistemic virtues.

Another 19th-century transition, one that has undeniable relevance to natural history, is the introduction of evolutionary thinking about species. Most importantly, the malleability of species, and the focus on variability within species in evolution generally moved metaphysical views about species into a less essentialist direction during the 19th century.[45] However, the interactions between a naturalist's theoretical views on the nature of species and their practice of describing them for the purpose of classification are far from straightforward. In this light it is unfortunate that much of the literature on the 18th- and 19th-century natural history of species has a strong focus on the metaphysical side of the equation. For instance, even though there is an extensive literature devoted to every aspect of the scientific works of Charles Darwin (1809–1882), including his thinking about the metaphysics of species, very little has been written about the practical aspects of his approach to delimiting and describing species.[46] As I hope to have shown above, there is no clear transition during the 19th century in the essentialism inherent in the abstract species images that grace ichthyological publications.[47] Indeed, the primary goals of systematic natural history – the description and classification of species – in a sense requires stable species definitions. By finding out which features of

45 Richards R.A., *The Species Problem: A Philosophical Analysis* (Cambridge: 2010); Wilkins J.S., *Species: A History of the Idea* (Berkeley – Los Angeles – London: 2009).

46 David Stamos is one of the few authors writing on Darwin's species concepts who pays an appreciable amount of attention to the barnacle monograph, Darwin's primary taxonomic work, in Stamos D.N., *Darwin and the Nature of Species* (Albany NY: 2007); Stamos D.N., "Darwin's Species Concept Revisited", in Pavlinov I.Y. (ed.), *The Species Problem: Ongoing Issues* (Rijeka: 2013) 251–280. He writes: '[I] go beyond Darwin's *definitions* of "species" in his writings and pay careful attention to his *use*. In the famous words of Wittgenstein which I here paraphrase, when it comes to what a word means in a language community *don't ask for the meaning, ask for the use*'. Stamos, "Darwin's Species Concept" 253.

47 Indeed, by the turn of the century ichthyological species books still use species images. See, for instance, Starr Jordan D. – Warren Evermann B., *The Fishes of North and Middle America: A Descriptive Catalogue*, part 4 (Washington: 1900); Weber M., *Siboga Expeditie* vol. 57: *Die Fische der Siboga-Expedition* (Leiden: 1913); Wyville Thomson C., *Report on the Scientific Results of the Challenger Expedition. Zoology*, vol. 1 (London: 1880).

a species *don't vary*, it is possible to create an abstract character cluster that lets naturalists recognize a species, both in collections and in nature, and tell it apart from members of other, similar species.[48] As Thomas Henry Huxley (1825–1895) puts it:

> Animals and plants are divided into groups, which become gradually smaller [...] until at length we come to the smallest groups of animals which can be defined one from the other by constant characters, which are not sexual; and these are what naturalists call species in practice, *whatever they may do in theory* [emphasis added].[49]

7 Conclusion

Schlegel pulls no punches in his judgements about the images produced by his colleagues. When discussing the famous images from the *British Salmonidae* by Sir William Jardine,[50] he tries to be respectful in his dismissal:

> De visschen zijn voorgesteld te liggen op den voorgrond van een landschap. Deze platen zijn zeer schilderachtig behandeld, en getuigen van het talent des schrijvers als kunstenaar. De voorwerpen zijn echter niet zelden in verkorting voorgesteld, de vinnen niet uitgespannen, en de détails meestal opgeofferd aan de schilderachtige uitwerking van het geheel; en het is om deze redenen, dat zijne platen meer den naam van fraaije prenten, dan van wetenschappelijke teekeningen verdienen.

48 There is a growing literature on the practical aspects of the production of species accounts in systematic natural history. See, for instance, Charmantier I. – Müller-Wille S., "Carl Linnaeus's Botanical Paper Slips", *Intellectual History Review* 24.2 (2014) 215–238; Van Neste, "Practising Taxonomy". As Kendig and Witteveen write: 'Attending to taxonomic practices allows one to discover implicit norms in taxonomic information processing activities that remain hidden from the more abstract theoretical or metaphysical treatments.' Kendig C. – Witteveen J. 2020. "The History and Philosophy of Taxonomy as an Information Science", *History & Philosophy of Life Sciences* 42.3 (2020) 40, quotation on page 3.

49 Huxley T.H., *Man's Place in Nature and Other Essays* (London: 1906) 226. He continues: 'If in a state of nature you find any two groups of living beings, which are separated one from the other by some constantly-recurring characteristic, I don't care how slight or trivial, so long as it is defined and constant, and does not depend on sexual peculiarities, then all naturalists agree in calling them two species; that is what is meant by the word species – that is to say, it is, for the practical naturalist, a mere question of structural differences.'

50 Jardine W., *British Salmonidae* (Edinburgh: 1839–1842).

> The fishes are depicted lying in the foreground of a landscape. These plates are very artistic, and showcase the author's talent as an artist. However, the objects often show foreshortening, the fins are not spread, and the details are usually sacrificed for the sake of the artistic effect of the whole; it is for these reasons that his plates are beautiful more than they are scientific.[51]

In other words, Jardine's plates may be fine works of art, but they are useless as species images.

The practical requirements of images in systematic natural history depend on what the image is meant to convey, the taxonomic group being treated, and a host of other factors I have no room to go into here. I hope to have shown that it is useful to draw a fairly rigorous distinction between species images, specimen images and preservation images (though they can and do sometimes overlap), as each plays a different role in natural history, and some confusion about these roles exists in the scholarly literature. Moreover, different taxonomic groups also yield different rules for proper conduct in the production of images. For instance, making species images from preserved ichthyological specimens consists for an important part in leaving out 'accidental' characteristics and emphasizing 'essential' ones, whereas in botany it often consists of combining features from several specimens into one, in order to show several growth stages or season-specific features in a single figure. I have emphasized these subfield-specific rules by focusing my discussion on ichthyology and palaeontology, which have traditionally been understudied. General statements about developments in the rules of image-making in natural history thus need to be supported by examples across the discipline.

References

Archive Natuurkundige Commissie, MMNAT, in Naturalis Biodiversity Center.

Agassiz Louis, *Recherches sur les poissons fossiles*, 5 vols. (Neuchatel: 1833–1843).

Agassiz Louis, *Directions for Collecting Fishes and Other Objects of Natural History* (Cambridge MA: 1853).

Ahrens H., "De Natuur Stilzetten op Papier. Wetenschappelijk Illustratoren van Naturalis Leggen de Dierenwereld Vast", *Tijdschrift voor Mediageschiedenis* 12 (2009) 233–274.

51 Schlegel, *Verhandeling* 44.

Allmon W.D., "The Evolution of Accuracy in Natural History Illustration: Reversal of Printed Illustrations of Snails and Crabs in Pre-Linnaean Works Suggests Indifference to Morphological Detail", *Archives of Natural History* 34.1 (2007) 174–191.

Anderson W.D., "John Edwards Holbrook's Senckenberg Plates and the Fishes they Portray", *Archives of Natural History* 30.1 (2003) 1–12.

Aronowksy L., "On Drawing Dead Fish", *Environmental History* 21 (2016) 542–551.

Artedi Peter, *Ichthyologia, sive opera omnia de piscibus* (Leiden, Conradus Wishoff: 1738).

Bleeker Pieter, *Atlas ichthyologique des Indes Orientales Néêrlandaises, publié sous les auspices du Gouvernement Colonial Néêrlandais*, 9 vols. (Amsterdam: 1862–1878).

Bleeker Pieter, "Notice sur l'Atlas Ichthyologique des Indes Orientales Néerlandaises", *Jaarboek van de Koninklijke Nederlandse Akademie van Wetenschappen* 1877 (1878) 133–144.

Bleichmar D., *Visible Empire: Botanical Expeditions and Visual Culture in the Hispanic Enlightenment* (Chicago: 2012).

Bloch Marcus Elieser, *Allgemeine Naturgeschichte der Fische*, 12 vols (Berlin, Realschule, J. Morino: 1782–1795).

Charmantier I. – Müller-Wille S., "Carl Linnaeus's Botanical Paper Slips", *Intellectual History Review* 24.2 (2014) 215–238.

Cooper J.A., "Edward Neale (1833–1904): Bird Illustrator", *Archives of Natural History* 46.2 (2019) 283–297.

Cuvier Georges, *Recherches sur les ossemens fossiles des quadrupèdes, où l'on rétablit les caractères de plusieurs espèces d'animaux que les révolutions du globe paroissent avoir détruites*, 4 vols (Paris: 1812).

Cuvier Georges, – Valenciennes Achille, *Histoire naturelle des poissons*, 22 vols (Paris – Strasbourg: 1828–1849).

Daston L., "Type Specimens and Scientific Memory", *Critical Inquiry* 31 (2004) 153–182.

Daston L. – Galison P., *Objectivity* (Cambridge MA – London: 2007).

Davidson J.P., *A History of Paleontology Illustration* (Bloomington IN: 2008).

Dawson G., *Show me the Bone: Reconstructing Prehistoric Monsters in Nineteenth-Century Britain and America* (Chicago – London 2016).

Farber P.L., "The Type-Concept in Zoology During the First Half of the Nineteenth Century", *Journal of the History of Biology* 9.1 (1976) 93–119.

Farber P.L., *The Emergence of Ornithology as a Scientific Discipline: 1760–1850* (Dordrecht: 1982).

Farber P.L., *Finding Order in Nature: The Naturalist Tradition from Linnaeus to E.O. Wilson* (Baltimore MD: 2000).

Fischer H. – Remmert V.R. – Wolschke-Bulmahn J. (eds.), *Gardens, Knowledge and the Sciences in the Early Modern Period* (Switzerland: 2016).

Gassó Miracle M.E., "The Significance of Temminck's Work on Biogeography: Early Nineteenth Century Natural History in Leiden, the Netherlands", *Journal of the History of Biology* 41 (2008) 677–716.

Gassó Miracle M.E., "On whose authority? Temminck's debates on Zoological Classification and Nomenclature: 1820–1850", *Journal of the History of Biology* 44 (2011) 445–481.

Gijzen A., *'s Rijks Museum van Natuurlijke Historie: 1820–1915* (Rotterdam: 1938).

Halbertsma H.J., "Normaal en Abnormal: Hermaphroditismus bij de Visschen", *Verslagen en Mededeelingen der Koninklijke Akademie van Wetenschappen* 16 (1863) 165–178.

Hartley F., "Notes on a Specimen of Alepisaurus aesculapius Bean, From the Coast of San Luis Obispo County, California", *Proceedings of the California Academy of Sciences*, second series, 5 (1895) 49–50.

Holthuis L.B. – Pietsch T.W., *Les planches inédites de poissons et autres animaux marins de l'Indo-Ouest Pacifique d'Isaac Johannes Lamotius* (Paris: 2006).

Huxley T.H., *Man's Place in Nature and Other Essays* (London: 1906).

Jardine William, *British Salmonidae* (Edinburgh: 1839–1842).

Kendig C. – Witteveen J. 2020. "The history and philosophy of taxonomy as an information science", *History & Philosophy of Life Sciences* 42.3 (2020) 40.

Klaver C., *Inseparable Friends in Life and Death: The Life and Work of Heinrich Kuhl (1797–1821) and Johan Conrad van Hasselt (1797–1823), students of prof. Theodorus van Swinderen* (Groningen: 2007).

Lacépède Bernard Germain de, *Histoire naturelle des poissons*, 12 vols (Paris, Plassant: 1798–1803).

Lack H.W., "The Botanical Illustrations of Franz Scheidl (*fl.* 1770–1795)", *Archives of Natural History* 47.1 (2020) 51–62.

Lederer R.J., *The Art of the Bird: The History of Ornithological Art through Forty Artists* (Chicago IL: 2019).

Lidth de Jeude T.W. van, "On a Large Specimen of Orthtragoriscus on the Dutch Coast", *Notes from the Leyden Museum* XII (1890) 189–195.

Müller-Wille S., "Collection and Collation: Theory and Practice of Linnaean Botany", *Studies in History and Philosophy of Biological and Biomedical Sciences* 38 (2007) 541–562.

Neste A. van, "Practising Taxonomy: Joel Asaph Allen and Species-Making (W.T. Stearn Prize 2017)", *Archives of Natural History* 45.2 (2018) 197–212.

Nickelsen K., "Draughtsmen, Botanists and Nature: Constructing Eighteenth-Century Botanical Illustrations", *Studies in History and Philosophy of Biological and Biomedical Sciences* 37 (2006) 1–25.

Oijen M.J.P. van, "Data on the Genesis of the Atlas Ichthyologique From a Little Known French Paper by P. Bleeker", *The Raffles Bulletin of Zoology*, Supplement no. 13 (2005) 3–8.

Owen R., *A History of British Fossil Reptiles*, 4 vols (London: 1849–1884).

Pond Arthur., "A Letter to the Right Honourable George Earl of Macclesfield, President of the Royal Society, Concerning the Stones Mentioned in the Preceding Article", *Philosophical Transactions of the Royal Society of London* 49.1 (1756) 297–298.

Prince S.A. (ed.), *Of Elephants and Roses: French Natural History 1790–1830* (Philadelphia PA: 2013).

Richards R.A., *The Species Problem: A Philosophical Analysis* (Cambridge: 2010).

Rijks M. – Smith P.J. – Egmond F., (eds.), *Fish & Fiction: Aquatic Animals between Science and Imagination (1500–1900)* (Leiden: 2018).

Roberts T.R., "The Freshwater Fishes of Java, as Observed by Kuhl and van Hasselt in 1820–23", *Zoologische Verhandelingen* 285.29 (1993) 1–94.

Rudwick M., *The Meaning of Fossils: Episodes in the History of Palaeontology* (Chicago – London: 1976).

Rudwick M., "Georges Cuvier's Paper Museum of Fossil Bones", *Archives of Natural History* 27.1 (2000) 51–68.

Schlegel Hermann, *Verhandeling over de Vereischten van Natuurkundige Afbeeldingen* (Haarlem: 1849).

Scudder S.H., "Look at your fish", *Every Saturday: A Journal of Choice Reading* (1874).

Secord A., "Pressed into Service: Specimens, Space, and Seeing in Botanical Practice", in Livingstone D.N. – Withers C.W.J. (eds.), *Geographies of Nineteenth-Century Science.* (Chicago – London: 2011) 283–310.

Stamos D.N., "Pre-Darwinian Taxonomy and Essentialism – A Reply to Mary Winsor", *Biology & Philosophy* 20 (2005) 79–96.

Stamos D.N., *Darwin and the Nature of Species* (Albany NY: 2007).

Stamos D.N., "Darwin's Species Concept Revisited", in Pavlinov I.Y. (ed.), *The Species Problem: Ongoing Issues* (Rijeka: 2013) 251–280.

Starr Jordan D. – Warren Evermann B., *The Fishes of North and Middle America: A Descriptive Catalogue*, part 4 (Washington: 1900).

Temminck Coenraad Jacob, *Voorschrift Hoedanig te Handelen met Voorwerpen van Natuurlijke Historie, ten Einde Dezelve Behoorlijk te Verzenden en voor Bederf te Bewaren* (Leiden: 1825).

Trijp D. van, *Captured on Paper. Fish Books, Natural History and Questions of Demarcation in Eighteenth-Century Europa (ca. 1680–1820)* (Ph.D. Dissertation, Leiden University: 2021).

Velden F. van der, "De Prijsvragen en Verhandelingen van Teylers Tweede Genootschap 1781–1866 (VI): Tekenkunde." *Teylers Museum Magazijn* 13.1 (1995) 11–15.

Veth H.J., *Overzicht van Hetgeen, in het Bijzonder door Nederland, Gedaan is voor de Kennis der Fauna van Nederlandsch Indië* (Leiden: 1879).

Weber A., "Collecting Colonial Nature. European Naturalists and the Netherlands Indies in the Early Nineteenth Century", *BMGN – Low Countries Historical Review* 134.3 (2019) 72–95.

Weber M., *Siboga Expeditie* vol. 57: *Die Fische der Siboga-Expedition* (Leiden: 1913).
Wilkins J.S., *Species: A History of the Idea* (Berkeley – Los Angeles – London: 2009).
Witteveen J., "Suppressing Synonymy With a Homonym: The Emergence of the Nomenclatural Type Concept in Nineteenth Century Natural History", *Journal of the History of Biology* 49 (2016) 135–189.
Witteveen J., "Objectivity, Historicity, Taxonomy", *Erkenntnis* 83.3 (2018) 445–463.
Wyville Thomson C., *Report on the Scientific Results of the Challenger Expedition. Zoology*, vol. 1 (London: 1880).

Index Nominum

Index of Aquatic Animals

Note: this *Index of Aquatic Animals* contains scientific and non-scientific English and Latin names. Examples: Common sole (scientific English), Sole (non-scientific English), *Sola* (non-scientific Latin), *Sola sola* (scientific Latin).

www.ingramcontent.com/pod-product-compliance
Lightning Source LLC
LaVergne TN
LVHW012052011026
849162LV00018B/185
* 9 7 8 9 0 0 4 6 8 1 1 7 0 *